S0-ACL-773

Subscriber Loop Signaling and Transmission Handbook

IEEE Telecommunications Handbook Series

Whitham D. Reeve, *Series Editor*

The *IEEE Telecommunications Handbook Series* is designed to provide the engineer and technical practitioner with working information in the three basic fields of telecommunications: inside plant, outside plant, and administration and regulatory. This integrated series of handbooks provides practical information on the link between field experience and formal telecommunications industry standards and practices. These books are essential tools for engineers and technical practitioners who require day-to-day engineering and technical information on telecommunication systems.

Published Books in the Series

Boucher *Traffic System Design Handbook*, 1993
Reeve *Subscriber Loop Signaling and Transmission Handbook: Analog*, 1992
Reeve *Subscriber Loop Signaling and Transmission Handbook: Digital*, 1995

Future Books in the Series

Bhatnagar *Network Synchronization*
Graham *Telecommunications Power Systems*
Nellist *Telecommunications Wiring for Commercial Buildings*
Nellist *Understanding Telecommunications and Lightwave Systems*

If you are interested in becoming an author, contributor, or reviewer of a book in this series, or if you would like additional information about forthcoming titles, please contact:

Whitham D. Reeve
Series Editor, IEEE Telecommunications Handbook Series
Reeve Engineers
P.O. Box 190225
Anchorage, Alaska 99519-0225
907/243-2262
E-mail—*by CompuServe:* 71011.3642@compuserve.com
by Internet: w.reeve@ieee.org

Subscriber Loop Signaling and Transmission Handbook
Digital

Whitham D. Reeve
Reeve Engineers

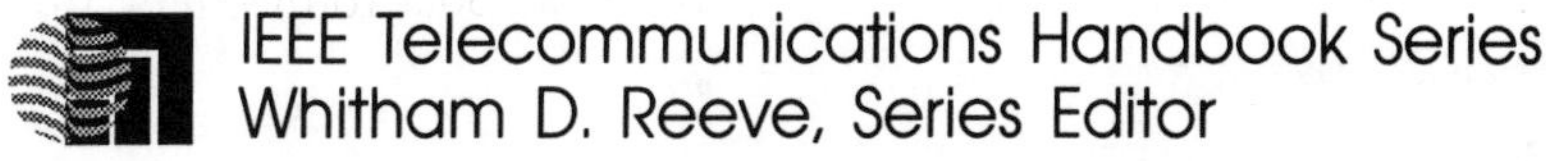

The Institute of Electrical and Electronics Engineers, Inc., New York

IEEE Press
445 Hoes Lane, PO Box 1331
Piscataway, NJ 08855-1331

IEEE PRESS Editorial Board
John B. Anderson, *Editor in Chief*

R. S. Blicq
M. Eden
D. M. Etter
G. F. Hoffnagle
R. F. Hoyt
J. D. Irwin
S. Kartalopoulos
P. Laplante
A. J. Laub
M. Lightner
J. M. F. Moura
I. Peden
E. Sánchez-Sinencio
L. Shaw
D. J. Wells

Dudley R. Kay, *Director of Book Publishing*
Carrie Briggs, *Administrative Assistant*
Lisa S. Mizrahi, *Review and Publicity Coordinator*

Technical Reviewers

Steven Cowles, *GCI Network Systems*
Thomas Croda, *Sprint Communications*
Chester Damby, *Entergy*
C. David Dow, *PennTech*
Sam Fowler, *Questar Corporation*
Michael Newman, *CSI Telecommunications Engineers*
Roy Thompson, *TDS Telecom*

This book may be purchased at a discount from the publisher when ordered in bulk quantities. For more information, contact:

IEEE PRESS Marketing
Attn: Special Sales
P.O. Box 1331/445 Hoes Lane
Piscataway, NJ 08855-1331
1-800-678-IEEE

© 1995 by the Institute of Electrical and Electronics Engineers, Inc.
345 East 47th Street, New York, NY 10017-2394

All rights reserved. No part of this book may be reproduced in any form, nor may it be stored in a retrieval system or transmitted in any form, without written permission from the publisher.

Printed in the United States of America

10 9 8 7 6 5 4 3 2 1

ISBN 0-7803-0440-3
IEEE Order Number: PC3376

Library of Congress Cataloging-in-Publication Data

Reeve, Whitham D.
Subscriber loop signaling and transmission handbook : digital / Whitham D. Reeve.
p. cm. — (Telecommunications handbook series)
Includes bibliographical references and index.
ISBN 0-7803-0440-3 (case)
1. Telephone switching systems, Electronic. 2. Telecommunication--Standards. 3. Digital communications. I. IEEE Communications Society. II. Title. III. Series.
TK6397.R443 1995
621.387′83--dc20

94-30809
CIP

Contents

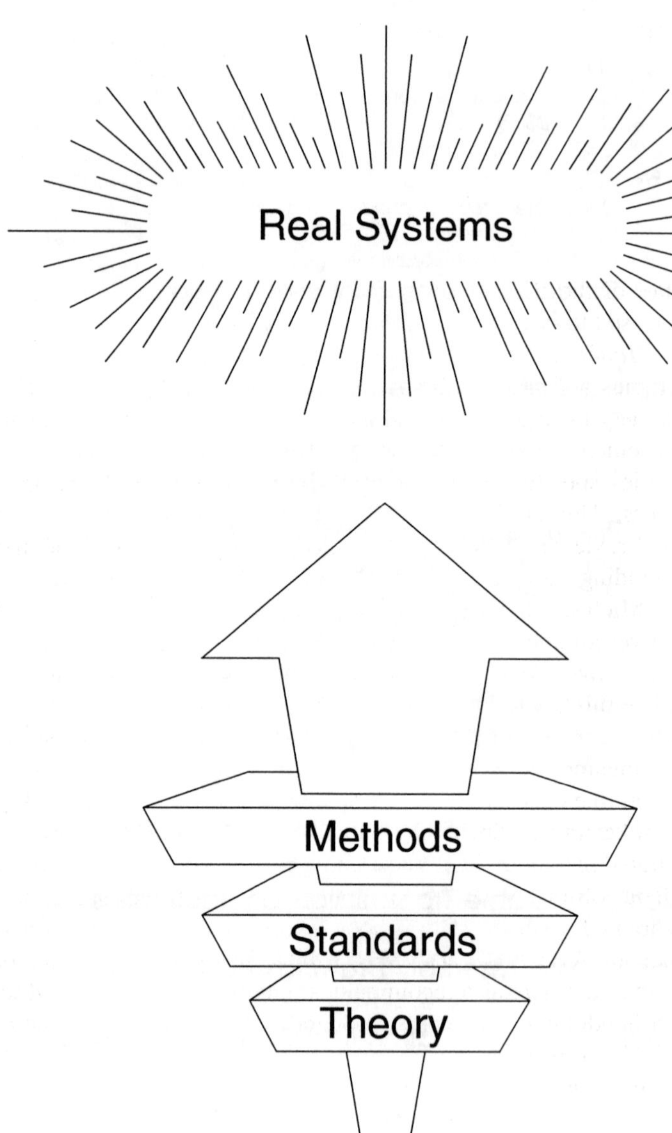
Real Systems
Methods
Standards
Theory

Preface

As editor of the IEEE Telecommunications Handbook Series and as author, I am doubly pleased to have finally finished this book. *Subscriber Loop Signaling and Transmission Handbook: Digital* is a book about the practical aspects of digital loops and related topics and is *not* a book about communications theory. This book will appeal to engineering and technical practitioners in all fields of telecommunications.

A requirement, which I wrote into the original IEEE Telecommunications Handbook Series specifications, is that all series books shall be reviewed by practicing engineers. This book meets that requirement nicely. Several people, the official technical reviewers listed on the copyright page, reviewed the manuscript for this book, including Steve Cowles, Thomas Croda, Chester Damby, C. David Dow, Sam Fowler, Michael Newman and Roy Thompson. I am grateful for their useful and constructive comments and suggestions. Steve Cowles and Mike Newman, in particular, maintained personal contact with me throughout the almost three years it took to produce this book. In addition, Dave Lester (with Bell Canada in Toronto, Ontario) provided many useful suggestions for Chapter 11, and Len Thomas (with Sprint in Burlingame, California) happily gave me permission to use some of his own material on line codes and frame structures in Chapter 3. Finally, I think readers and authors alike underestimate the amount of effort that a publisher puts into a heavily illustrated handbook like this. Debbie Graffox and many others at IEEE Press did an excellent job of converting my manuscript into a very useful handbook.

Why should I wish to write a book like this and have others published as a coherent handbook series? The answer is, I think, that practicing engineers are largely neglected in the commercial telecommunications press. I have talked to and corresponded with hundreds of engineers throughout the United States and Canada, and have yet to find one who did not agree with me. This book, the two others already published in the series, and future books to be published will help fill the gap.

I will be glad to hear from readers about this book or others in the series. This invitation is also a call for potential authors and reviewers who would like to publish a handbook in the series. My address and telephone number are provided on the Series page at the front of this book.

Whitham D. Reeve

Introduction

Digital transmission systems were initially designed to make more efficient use of imbedded metallic (copper) trunk cable plant for transmission of analog signals between switching systems. Later, the digital systems evolved into Digital Loop Carrier (DLC), which served the same purpose but in the exchange cable plant, or subscriber loop. About the same time, the demand for transmission of digital data between telecommunications system users began to rise steeply.

Engineers found that digital transmission systems had more robust and stable error performance than analog systems of equal channel capacity. Furthermore, the error rate of digital loops designed to specific criteria could be made arbitrarily small and easily controlled. This made digital loops a natural medium for transmitting digital data.

Digital loop transmission *systems* are easy to design and to implement, provided that the characteristics and limitations of the loop plant are understood. A digital loop frequently is the end-link or tail circuit of a longhaul telecommunications channel. Therefore, the understanding of digital loops sometimes goes beyond the loop itself. This is particularly true of timing and synchronization of network elements.

Telecommunications is a popular subject among both authors and readers. Of all the telecommunications books and papers written, however, there are very few that provide practical guidelines for people who directly engineer, implement, operate, and maintain systems, particularly digital loop systems, in the field on a day-to-day basis. This book is written to bridge the gap between the theoretical aspects of digital loop transmission and the actual systems, while taking into account current methods and standards. It is written for the practitioner who needs to know how to design, install, use, and maintain digital loops. This is not a book on digital communications theory.

Any person interested in the technical and qualitative aspects of the digital loop and its application—engineers, managers, construction experts, telecommunications system users, and students—will find this book informative and up-to-date. A great number of technical references are cited to complete the discussions about practical applications by providing the standards and practices upon which the applications are based. Considerable effort has been made to restrict the theoretical aspects of digital transmission while covering in detail those areas over which the designer, builder, and user have some control in a practical sense. This book is suitable for continuing education and the development of short courses and seminars.

This book complements *Subscriber Loop Signaling and Transmission Handbook: Analog*, published by IEEE PRESS in 1992. Whereas the *Analog* book emphasizes the characteristics of loops used in analog environments, the *Digital* book emphasizes the digital environment.

There are eleven chapters and four appendices in this book. Chapter 1 deals with introductory subjects, including transmission concepts and signal conversion, and the differences between analog and digital transmission. The analog-to-digital and digital-to-analog conversion processes also are presented.

In Chapter 2, the emphasis shifts exclusively to digital loop applications and interfaces. Included are discussions on the types of services that would be carried on a digital loop and the different types of digital loop carrier and their operational modes. This chapter introduces specific technologies, such as the Integrated Services Digital Network Digital Subscriber Line (ISDN DSL), the High Bit-Rate Digital Subscriber Line (HDSL), and subrate digital loops (which comprise the end-links of what is commonly known as the Digital Data System, or DDS). The ubiquitous repeatered T1-carrier is described, as are digital loop applications of optical fibers.

Chapter 3 describes in detail the different frame structures and line codes used with digital loops. Frame structures are important because they determine the arrangement of individual channel time slots. They also identify or carry signaling information in a digital transmission system. Line codes determine the spectral characteristics of the baseband digital bit streams on the transmission line.

Timing and synchronization is perhaps the most fundamental requirement of any digital transmission system or network. Chapter 4 describes the requirements and characteristics of synchronization as it applies to the digital loop and to the network in general. Jitter and wander and their effect on digital loops are discussed also. Chapter 4 includes practical implementation information lacking in most books on the subject.

All transmission systems are bombarded with interferring signals, and desired signals are subjected to all kinds of distortion. The digital loop is no exception. Chapter 5 discusses noise and distortion specific to the loop and the errors they create in digital loop transmission systems. Error rates and their effect on transmission quality and performance are discussed in Chapter 6, which also cites numerous industry-accepted requirements.

Loop transmission media, including metallic twisted pairs and fiber optics, are discussed in the next two chapters. Chapter 7, or Part I of the transmission media discussion, is dedicated to physical and electrical characteristics of twisted pair cables. Chapter 8, Part II of the transmission media discussion, focuses on the physical and electrical characteristics of fiber optic cables used in the loop.

With a knowledge of systems, impairments, and transmission media established, the focus on actual loop engineering can commence. Repeatered T1-carrier is the oldest digital technology still used in the loop, and the associated transmission engineering requirements have remained virtually unchanged. Repeatered T1-carrier engineering is described in detail in Chapter 9.

The transmission engineering requirements for all other digital loop technologies are given in Chapter 10. This includes the DSL and HDSL, subrate digital loops, and fiber optics. That all these other technologies and systems can be discussed in a single chapter, whereas repeatered T1-carrier requires its own chapter, indicates the difference in engineering effort associated with the technologies. Nevertheless, repeatered T1-carrier is a well-known technology and will be used in the loop for some time to come.

The final chapter, Chapter 11, discusses premises cabling systems. Premises cabling systems connect the subscriber loop to user equipment, and knowledge of these systems will be helpful to all practitioners. This chapter provides detailed technical descriptions of the types of cables recently standardized, particularly by the well-known industry standard EIA/TIA-568 and others. Of particular interest are ISDN applications on user premises. ISDN, as a collection of services, has different engineering requirements in different parts of the telecommunications network. The loop portion is described in Chapter 10, and the user premises portion is described in Chapter 11.

The four appendices are a collection of useful reference information. Appendix A describes the nomenclature used by AT&T to designate different cable types and is provided because of the wide use of AT&T cables throughout the United States and elsewhere. Appendix B gives the primary and secondary electrical characteristics of typical exchange cables at various temperatures over the frequency range of 1 Hz to 5 MHz.

Appendix C provides twisted pair cable core assembly drawings and color coding information for exchange cables. This information is useful when it is necessary to physically separate different analog and digital services that share a common cable sheath. Pair groups and supergroups are shown for all common cable sizes. Finally, Appendix D provides operational information for patching and reconnecting digital lines at the digital signal cross-connect (DSX) point. While the appendix focuses on lines operating at the 1.544 Mbps (DS-1) rate, it applies to other rates as well.

Note: This book makes numerous references to recommendations (standards) published by International Telegraph and Telephone Consultative Committee, commonly called CCITT. CCITT ceased to exist in 1993 (after the bulk of this book was completed) and was replaced by the International Telecommunication Union Telecommunication Standardization Sector, or ITU-T. At the same time, ITU-T made major changes in the way it publishes standards and recommendations. The last block release of CCITT's recommendations was in the *Blue Book*. These block releases were updated on a four-year schedule, but this is not the case with ITU-T. ITU-T publishes new recommendations and updates them on an individual basis rather than in block form.

Subscriber Loop Signaling and Transmission Handbook

1 Subscriber Loop Transmission Concepts and Signal Conversion

Chapter 1 Acronyms

A/D	analog-to-digital
ADM	adaptive delta modulation
ADPCM	adaptive differential pulse code modulation
AM	amplitude modulation
ANSI	American National Standards Institute
APC	adaptive predictive coding
APK	amplitude-phase keying
ASBC	adaptive subband coding
ATM	asynchronous transfer mode
BER	bit error rate or bit error ratio
BETRS	basic exchange telecommunications radio service
CCITT	The International Telegraph and Telephone Consultative Committee
codec	COder-DECoder
CVSDM	continuously variable slope delta modulation
D/A	digital-to-analog
dc	direct current
DCME	digital compression multiplex equipment
DLC	digital loop carrier
DM	delta modulation
DPCM	differential pulse code modulation
DSI	digital speech interpolation
DTE	data terminal equipment
DTMF	dual tone multifrequency
FCC	Federal Communications Commission
FDM	frequency division multiplexing
FEC	forward error correction
FM	frequency modulation
FSK	frequency-shift keying
FTTC	fiber-to-the-curb
FTTH	fiber-to-the-home
HDSL	high bit-rate digital subscriber line
HDTV	high-definition television
ISDN	integrated services digital network
LD-CELP	low delay code-excited linear prediction
LPC	linear predictive coding
MOS	mean opinion score
MP-LPC	multipulse linear predictive coding
PAM	pulse-amplitude modulation
PCM	pulse code modulation
PM	phase modulation
PSK	phase-shift keying
PSTN	public switched telephone network
QDU	quantizing distortion unit
RF	radio frequency
rms	root mean square

continued

SDR	signal-to-quantizing distortion ratio	TASI	time assignment speech interpolation
SE-LPC	stochastically excited linear predictive coding	TDM	time division multiplexing
SNR	signal-to-noise ratio	TLP	transmission level point
SONET	synchronous optical network	VNL	via net loss
SPE	synchronous payload envelope	VQL	variable quantizing level
		VT	virtual tributary

1.1 Subscriber Loops

A subscriber loop is that part of a telecommunication transmission system between a subscriber's premises and the serving central office. Subscribers can be individuals or businesses, and in this book the term *subscriber* is used interchangeably with end-user, or just user. Loops consist of twisted metallic cable pairs, radio links, or optical fibers and serve as end-links in the communication channel between users. The loop usually carries low-traffic volumes compared to interoffice facilities.

Loops can take on many forms, from a simple twisted cable pair carrying an analog voicegrade signal between two points to a complex network of optical fibers and associated interfaces carrying combinations of voicegrade and video signals and high-speed digital data. This book concentrates on landline digital loops. A detailed treatment of analog loops is given in a companion volume [1]. Some digital loop configurations are shown in Figure 1-1.

The twisted pair loop of Figure 1-1(a) is ubiquitous—it appears almost everywhere in the public network. In digital applications, it carries digital signals covering the range from near dc to over 3 Mbps, with bandwidths extending upwards of 6 MHz. When multiplexers are used with digital loops, a number of individual analog and digital channels can be combined into an aggregate data stream. Digital loop carrier (DLC), conceptually shown in Figure 1-1(b), is a typical system that uses this technique.

Where it is uneconomical to install twisted pair or fiber optic cable, such as in rural areas, digital or analog radio systems are used, as shown in Figure 1-1(c). With digital radio, analog voice frequency signals at remote terminals (subscriber stations) are digitally encoded and then used to modulate a radio frequency (RF) carrier for transmission to a base station (central office station) that serves one subscriber in point-to-point applications or many subscribers in point-to-multipoint applications. Examples of digital radios used in subscriber loop applications are those licensed by the Federal Communications Commission (FCC) as Basic Exchange Telecommunications Radio Service (BETRS) under the Rural Radio Service rules [2].

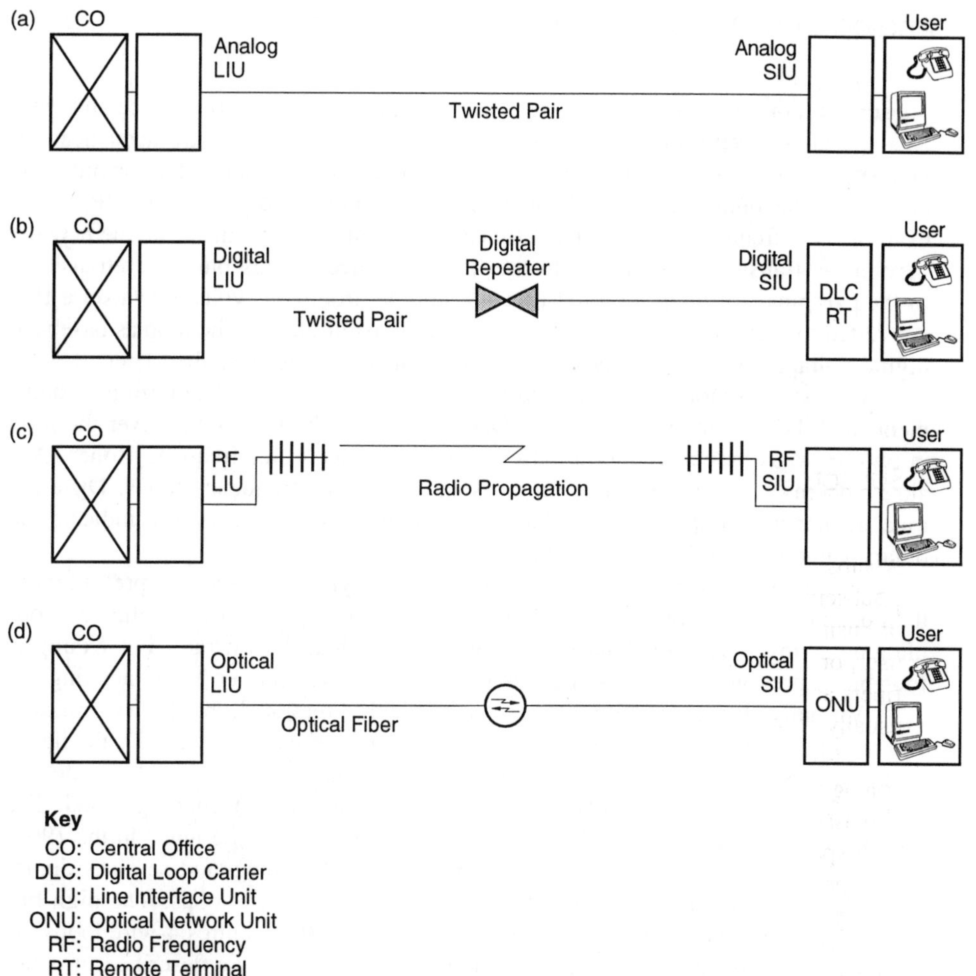

Figure 1-1 Subscriber loop configurations: (a) Analog twisted pair loop; (b) Digital loop carrier twisted pair loop; (c) Radio-derived loop; (d) Optical fiber loop

The most widely used and most mature technology used in loops is the metallic twisted pair. The newest and fastest growing segment of the loop plant uses fiber optic cables. A typical configuration is shown in Figure 1-1(d). The transmission rates presently used in optical fiber loop applications are nominally 1.5 Mbps, 6.3 Mbps, and 44.7 Mbps, although fiber-to-the-home (FTTH) and fiber-to-the-curb (FTTC) systems may use much higher rates and possibly wideband analog signals, as well (for example, television). By proper selection of high-speed line interface cards, many DLC systems may be connected directly to optical fibers. Some systems interface with the Synchronous Optical Network (SONET), which has tributary applications in the digital loop environment.

1.2 Analog and Digital Transmission

Subscriber loops carry two basic types of telecommunications traffic: voice and data. Analog voltages representing voice signals are produced by an electro-acoustical device such as a microphone in a telephone set. When such signals are transmitted on an analog telecommunications channel, only temporary frequency translations (for example, by frequency division multiplexing equipment) are made to them while they are in-transit. To be usable to the listener, voice signals must be retranslated to their original form. In a digital transmission environment, voice signals are first converted to digitally encoded signals and then transmitted over the loop as baseband digital voltage pulses. Various voice encoding methods are described later.

Data signals (for example, from a computer port or other data terminal equipment, or DTE) are inherently digital. Data signals can be transmitted over the loop in two ways: as baseband digital pulses or as modulated voice frequency tones. Analog modems convert the source digital signals to voice frequency tones. Different tone frequencies, amplitudes, or phases, or a combination, are used to indicate the binary value of each bit or group of bits.

When a single frequency, amplitude, or phase symbol is used to represent more than one bit (for example, two, three, or four bits), the modem uses multilevel modulation, giving a symbol (baud) rate that is lower than the bit rate. For example, four bits represented by one symbol is the same as four bits per baud encoding.

Virtually all present digital loop systems are bidirectional; that is, transmission and reception can be made simultaneously at each terminal. A twisted pair digital loop requires either two cable pairs (one transmit and one receive, also called 4-wire) or one cable pair (also called 2-wire), depending on the technology used. The familiar T1-carrier and subrate digital data systems, which were designed in the 1960s and 1970s, respectively, require 4-wire loops, while the integrated services digital network (ISDN) loop, designed in the 1980s, requires 2-wire loops. The high bit-rate digital subscriber line (HDSL), designed in the 1990s, requires 4-wire loops or, optionally for reduced bit rate, 2-wire loops. These systems are described in detail in later chapters.

1.3 Baseband and Passband Signals

A baseband signal is a signal that has frequency components approaching zero frequency, or *direct current* (dc). A baseband signal is not changed by frequency or phase translation; that is, it is does not consist of a modulated carrier.

A passband, or modulated, signal, on the other hand, is a carrier signal whose amplitude, frequency, or phase (or some combination of all three) is altered in response to a baseband signal. This causes the carrier or its sideband components to convey the information contained in the baseband signal. The modulation process translates the baseband signal frequency to a new frequency centered on the carrier but displaced from it. Figure 1-2 shows baseband and modulated signals.

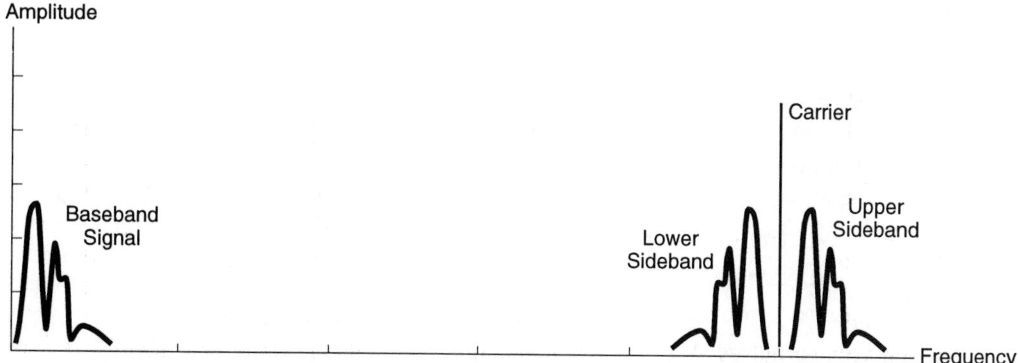

Figure 1-2 Baseband and modulated signals

1.4 Analog and Digital Modulation Methods

Analog carrier signals can be modulated by other analog signals (analog modulation) or by digital signals (digital modulation). Some analog modulation methods are AM (amplitude modulation), FM (frequency modulation), and PM (phase modulation). A commercial radio broadcast transmitter is an example of a device that uses an analog signal (voice or music) to modulate another analog signal (carrier) using either AM or FM.

A data set or modem is an example of a device that uses a digital signal (digital source data) to modulate an analog signal (carrier). When a digital signal is used in this application, the process is called keying. Modems typically use phase-shift keying (PSK), frequency-shift keying (FSK), or amplitude-phase keying (APK), or a combination, to convert the digital signals to analog. The foregoing analog and digital modulation methods are not the subject of this book; there is an abundance of literature on the subject. For example, see [3] for theoretical treatment and [4] for the application of modulation in modem design.

Modulation as used above is a specific form of signal encoding (frequency or phase translation) for the purposes of transmitting analog or digital information on an analog telecommunication channel. The modulation concept is not restricted to analog channels. Digital channels also can be used to transmit both analog and digital source information, and, for many applications, digital telecommunication channels are preferred.

The initial choice of a modulation method (analog or digital) is determined by the noise and bandwidth characteristics of the transmission channel to be used. The ultimate choice is one of economics. In some systems, particularly existing networks, this choice is a matter of convenience or policy. In many applications, digital transmission channels have significant advantages over analog channels. These are:*

---

*Adapted from [5] with additions.

- Ease of multiplexing
- Ease of signaling
- Use of modern technology
- Integration of transmission and switching
- Uniformity of transmission format
- Reduction of residual transmission loss variation of analog source signals
- Complete signal regeneration
- Operation at low signal-to-noise/interference ratios
- Better error performance
- Performance monitoring capability
- Accommodation of a number of diverse telecommunication services on a single circuit
- Ease of encryption

Although these advantages are significant, the disadvantages of digital transmission cannot be overlooked. Some disadvantages are:

- Imprecise analog-to-digital conversion of analog source signals (voice and modem) to digital format, which introduces nonlinear distortion
- Requirement for analog interfaces, particularly for voiceband services
- Increased bandwidth requirements
- Increased echo problems due to increased delay
- Need for precise synchronization of interconnected digital circuits

When analog source signals are transmitted on a digital channel, a *channel coding* technique called pulse code modulation (PCM) frequently is used. A theoretical development of PCM can be found in [5] and [6].

PCM refers to the use of a code to represent the instantaneous value of a sampled analog waveform. In other words, the analog waveform and the information contained in it are converted to a digital waveform carrying the same information. The conversion from analog to digital is called encoding; the conversion from digital to analog is called decoding. A device that performs these functions is called a codec (COder-DECoder). Modern digital transmission systems use different forms of PCM including delta modulation (DM), adaptive predictive coding (APC), differential PCM (DPCM), adaptive differential PCM (ADPCM), or linear predictive coding (LPC). These techniques differ in the specific method used to encode or decode the analog signal.

1.5 Digital Hierarchy

The digital hierarchy defines the different levels of digital multiplexing. The hierarchy used in North America, shown in Table 1-1, is based on the characteristics

Table 1-1 Digital Hierarchy

Digital Signal Level	Bit Rate	Equivalent 4 kHz Voice Channels[a]	Typical Transmission Media[b]
DS-0	64.00 kbps	1	TP
DS-1	1.544 Mbps	24	TP
DS-1C	3.152 Mbps	48	TP
DS-2	6.312 Mbps	96	FO, RD, CX
DS-3	44.736 Mbps	672	FO, RD, CX
DS-4	274.176 Mbps	4,032	FO, CX

[a]Using 64 kbps encoding.

[b]TP = twisted pair cable, FO = fiber optic cable, RD = radio, CX = coaxial cable.

and optimum information rates of the available transmission media at the time the hierarchy was conceived. The hierarchy segregates the various levels of information processing rates into well-defined blocks. The blocks at each level, shown conceptually in Figure 1-3, consist of:

- Analog-to-digital and digital-to-analog signal conversion terminals (channel banks)
- Multiplexers (source and subaggregate)
- Digital processing terminals
- Cross-connects
- Transmission facilities

At present, twisted pair digital loops carry traffic at the DS-1C, DS-1, and DS-0 rates or less (subrate). Many optical fiber transmission systems operate at the DS-2 rate, and many DS-3 rate systems are deployed in the network between central offices and user premises. The number of these higher-speed applications is expected to grow, and the DS-3 rate soon will be considered a common digital loop transmission rate. Not part of the North American digital hierarchy is the CEPT-1 (also called E1) rate of 2.048 Mbps, common in Europe. This rate has found some application in North America, particularly in private networks that cross international boundaries.

It is important to distinguish between correct terminology and colloquial usage. The terms "DS-1" and "T1" often are used interchangeably, but this is incorrect. The term DS-1 refers to a particular digital speed in the digital hierarchy, while the term T1, or T1-carrier, refers to a digital transmission system that happens to operate at the DS-1 rate. T1 is called "repeatered T1-carrier" in this book to clearly distinguish it from other digital loop transmission systems operating at the DS-1 rate.

Another hierarchy exists for SONET. Although SONET is beyond the scope of this book, a table of rates is given in Table 1-2 for reference and comparison. In SONET, the payload signals are carried in the synchronous payload envelope (SPE). Various quantities of DS-1 and other rate signals can be mapped as virtual tributaries (VTs) into the SPE. These are given in Table 1-3 for VT1 through VT6.

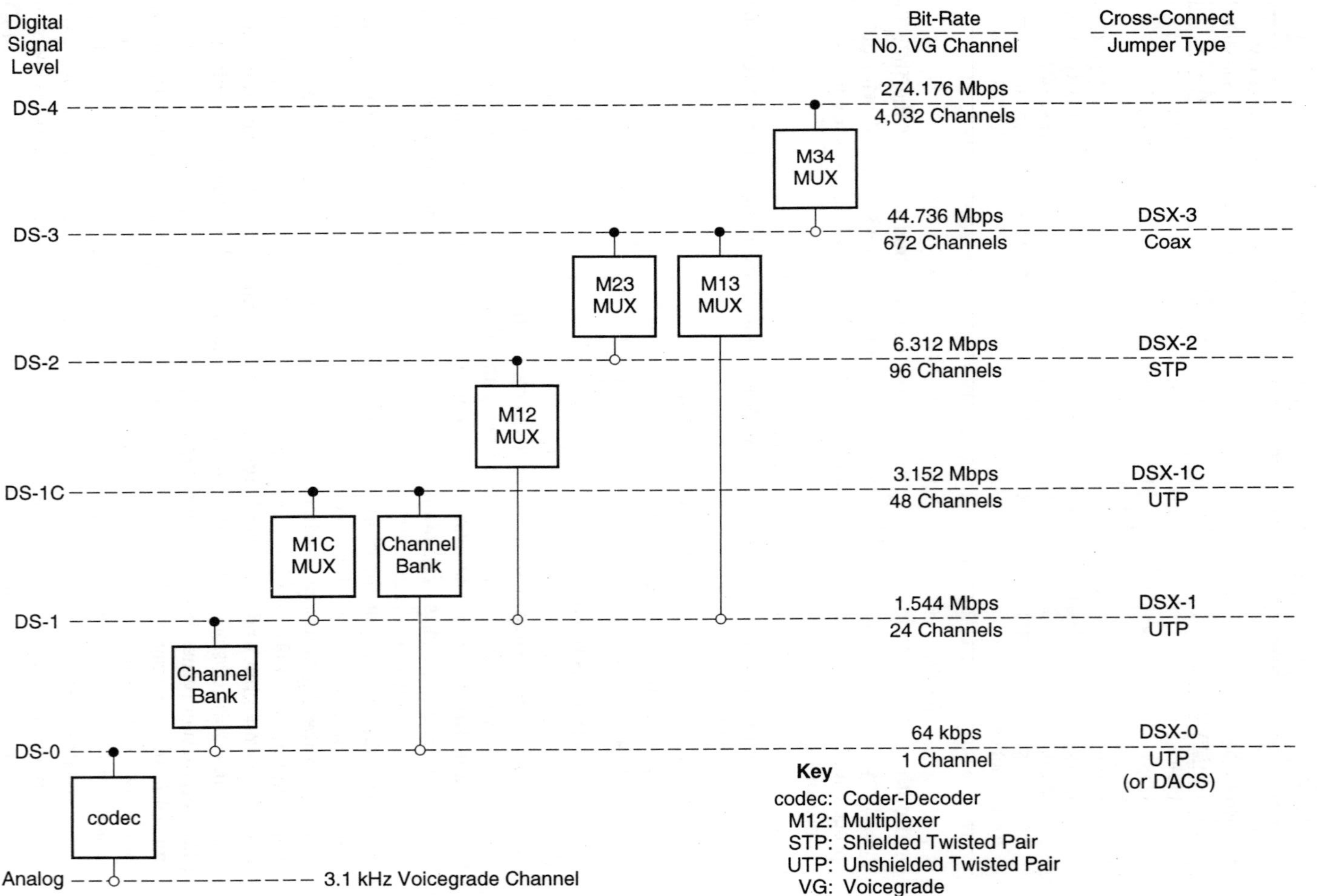

Figure 1-3 Digital hierarchy building blocks

Table 1-2 SONET Hierarchy

Optical Carrier Level	Optical Line Rate	Synchronous Transport Signal Level
OC-1	51.840 Mbps	STS-1
OC-3	155.520 Mbps	STS-3
OC-9	466.560 Mbps	STS-9
OC-12	622.080 Mbps	STS-12
OC-18	933.120 Mbps	STS-18
OC-24	1,244.160 Mbps	STS-24
OC-36	1,866.240 Mbps	STS-36
OC-48	2,488.320 Mbps	STS-48

Table 1-3 SONET Synchronous Payload Envelope

Virtual Tributary	VT Rate	Payload Rate	Designation	Number Mapped into SPE
VT1.5	1.728 Mbps	1.544 Mbps	DS-1	28
VT2	2.304 Mbps	2.048 Mbps	CEPT-1	21
VT3	3.456 Mbps	3.152 Mbps	DS-1C	14
VT6	6.912 Mbps	6.312 Mbps	DS-2	7

1.5.1 Multiplexing

Multiplexing is the process of combining multiple-user channels of information into a single, larger network channel. With analog multiplexing, several small bandwidth channels can be combined using frequency division multiplexing (FDM) into a larger bandwidth channel. Here, the bandwidth is measured in hertz (Hz). With digital multiplexing, the individual user, or source, channels are combined into a single, larger digital stream (aggregate) using time division multiplexing (TDM). TDM is shown conceptually in Figure 1-4.

In TDM the bandwidth is measured in terms of bit rate, or bits per second. The correlation between the bit rate and the actual transmission bandwidth in Hz depends on the digital coding scheme used. When a nominal 4 kHz analog channel is encoded at an 8 kHz rate into eight bits, the resulting digital bandwidth is 8 kHz $\times$ 8 bits = 64 kbps, or DS-0 rate. If transmitted as a baseband signal, the DS-0 signal has significant spectrum components beyond 64 kHz. Line coding and spectrum requirements are discussed in Chapter 3.

Multiplexers can be classified into two basic categories for the purposes of the following discussion:

- Multiplexers that have as inputs source information (for example, user data or voice). These will be called *source multiplexers* in the following discussion.
- Multiplexers that have as inputs the low aggregate rates from source multiplexers. These will be called *subaggregate multiplexers*.

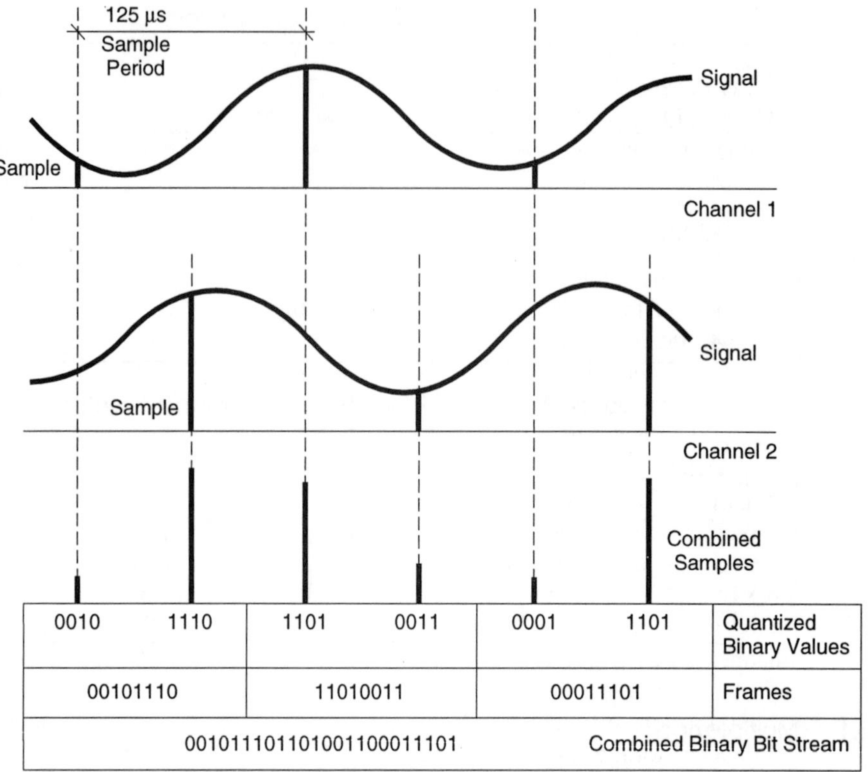

Figure 1-4 Time division multiplexing concepts

With source multiplexers, the source signals can be analog or digital. Where the source signals are analog, an analog-to-digital (A/D) converter is required to convert the signals to the required digital format for transmission. A digital-to-analog (D/A) converter is required at the other end of the circuit to restore the signal to its analog form. A/D and D/A conversion is discussed later. Early source multiplexers were called channel banks, and source signals were exclusively analog. Modern source multiplexers have the capability of accepting both analog and digital source signals with the proper choice of plug-in channel units.

Subaggregate multiplexers are classified by a simple nomenclature, M*xy*, where *x* is the low-speed multiplex input level, typically from a source multiplexer, and *y* is the high-speed output level. For example, an M13 multiplexer will accept as inputs 28 DS-1 rate signals and as output, one DS-3 rate signal. An M23 multiplexer will accept as inputs seven DS-2 rate signals and as output, one DS-3 rate signal. Finally, an MX3 multiplexer will accept as inputs a combination of DS-1 and DS-2 rate signals and as output, one DS-3 signal. Other combinations are possible, limited only by one's (or a vendor's) imagination.

A typical channel bank will encode each of 24 voice channels at a 64 kbps rate (DS-0) and multiplex these into a single aggregate digital stream at a 1.544

Mbps rate (DS-1). The DS-1 rate includes the payload at 1.536 Mbps (24 × 64 kbps) plus framing overhead at 8 kbps. Early multiplexers had a filter and sampler for each channel. The pulse-amplitude-modulated (PAM) signals from each sampler were placed on a PAM bus and combined through TDM into a quantizer (codec), which was shared among all 24 channels.

Modern channel banks and intelligent multiplexers have a codec per channel, and the combining of all channels is done on a PCM bus. This arrangement allows more flexible signaling protocols and accommodation of different types of user-side digital channel units. Additional signal processing is required of the PCM signal at the output of the codec to condition it according to the specific requirements of the transmission medium to be used.

Voice encoding does not have to be at the DS-0 rate. Encoders operating at lower bit rates are common. A typical speed is 32 kbps. An encoder operating at 32 kbps will effectively double the number of voice channels on a DS-1 rate digital loop carrier from 24 to 48. Voice encoding is discussed later.

Several independent source data channels can be combined into a single digital stream, but the aggregate rate does not necessarily have to be the sum of the individual rates; the aggregate can be higher or lower. In many multiplexers, the aggregate rate is higher than the sum of the source data rates because of signaling and framing overhead. In many situations, there is some diversity in the time any given channel is active due to the "bursty" nature of digital data traffic. A statistical multiplexer may be used to fill the gaps in one channel with data from another channel to obtain more efficient use of the aggregate channel. Statistical multiplexing uses temporary storage registers, or buffers, which introduce delay.

The opposite of multiplexing is inverse multiplexing, which is used to pool relatively inexpensive, low-speed switched and dedicated digital circuits on the network-side to support high-speed aggregate signals on the user-side. Such a multiplexer spreads the higher-speed user's data stream across multiple lower-speed network circuits. For example, a 384 kbps signal on the source (user-) side of the inverse multiplexer can be allocated to two 56 kbps dedicated circuits and five 56 kbps switched digital circuits on the network-side. This is illustrated in Figure 1-5.

As traffic demand on the user-side of the inverse multiplexer increases, the multiplexer automatically dials up additional 56 kbps switched circuits until enough network capacity is connected end-to-end. This might occur when a video conference is scheduled, as shown in Figure 1-6. During low-traffic periods, the switched 56 kbps circuits are idle. This dynamic bandwidth allocation is sometimes called "rubber bandwidth" [8].

At the sending end, the inverse multiplexer segments the user data and sends these data over the individual dedicated and switched 56 kbps channels. At the receiving end, the inverse multiplexer accepts the data from the individual channels and reorders the segments. The inverse multiplexer buffers the bit streams at the receiving end to compensate for the slight variations in transit times across the individual network channels. Depending on its sophistication, the inverse multiplexer may monitor the integrity of the connections and take action to replace failed or failing channels with new dialed channels.

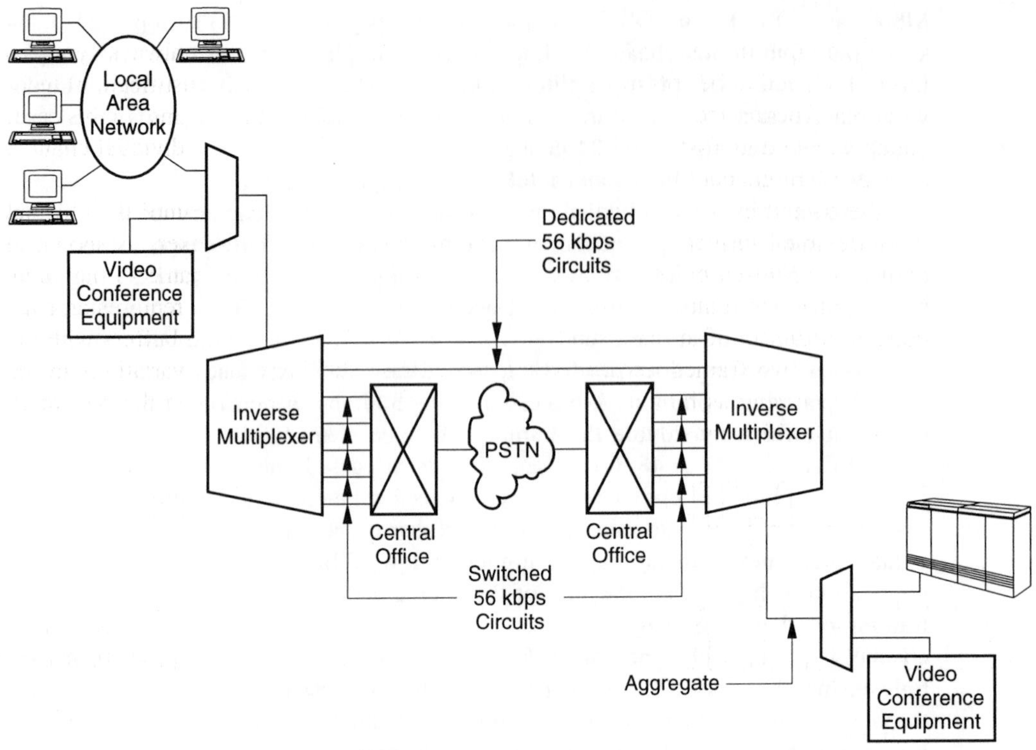

Figure 1-5 Inverse multiplexing

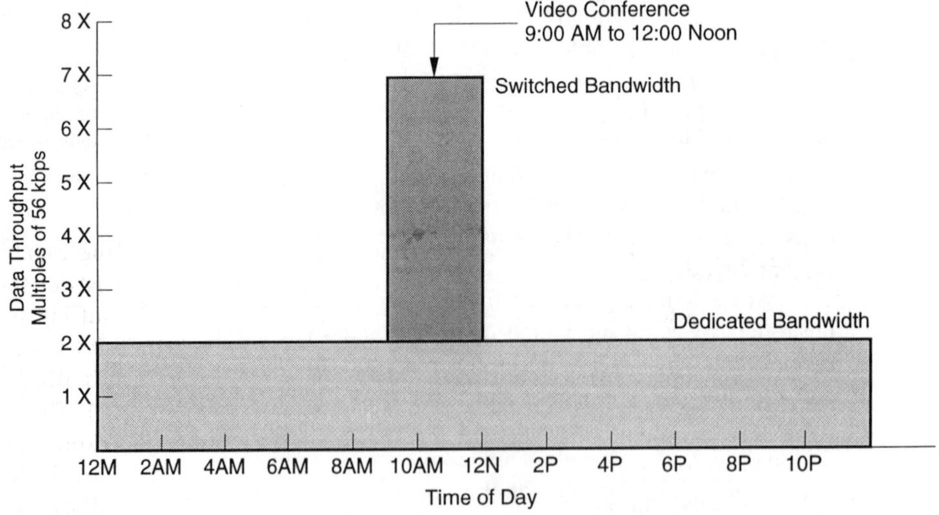

Figure 1-6 Dynamic bandwidth allocation

1.5.2 Pulse Stuffing

The input digital signals to subaggregate multiplexers are not always synchronized, nor need to be. Multiplexers use pulse stuffing (also called justification) techniques to synchronize several asynchronous or independent data streams. In this case, the aggregate digital stream has a higher rate than the sum of the individual channel rates to accommodate the extra pulses.

Extra pulses are inserted in the individual incoming streams until the channel rates are equal to that of a locally generated clock in the multiplexer, as shown in Figure 1-7. Stuffed pulses are inserted at fixed locations of each frame so they may be identified and removed at the far-end multiplexer. Pulse stuffing requires an elastic store or buffer of at least one frame in the multiplexer, although buffers with the capacity of two frames frequently are used. Pulse stuffing causes variations in the bit stream timing as bits are added. Pulse de-stuffing leaves gaps in the recovered bit stream, which must be smoothed out.

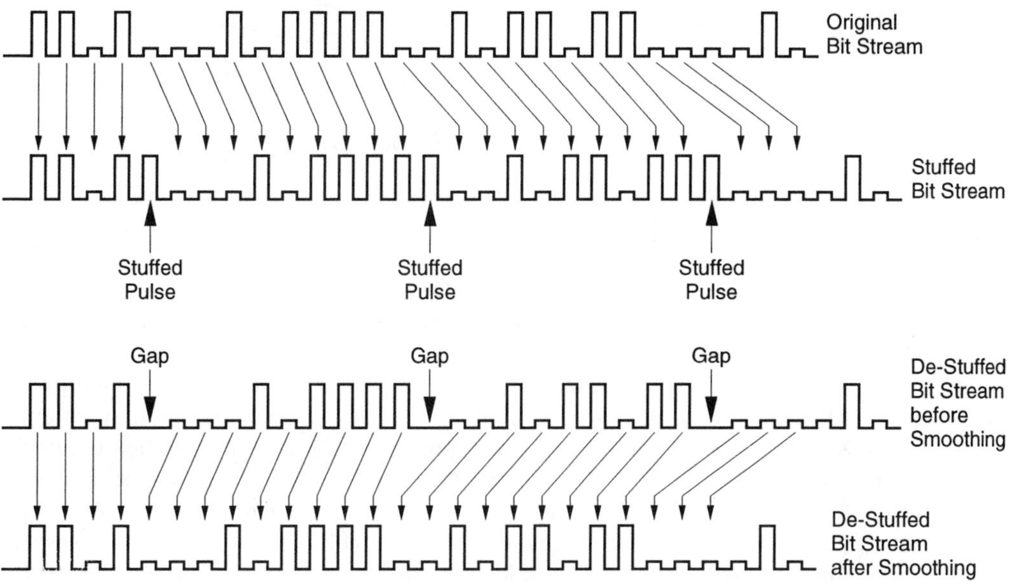

Figure 1-7 Pulse stuffing

The digital hierarchy is based on multiplexers with bit stuffing capabilities at the DS-2, DS-3, and DS-4 rates. For example, the DS-2 rate consists of four DS-1 signals with a total rate of 6.176 Mbps (4 × 1.544 Mbps). Bit stuffing is used to raise the rate of each DS-1 to 1,545,796 bps. In addition to the DS-1 signals, the DS-2 rate includes 128,816 bps of overhead for alignment and bit stuffing control. This brings the aggregate to 6.312 Mbps (4 × 1,545,796 + 128,816). The aggregate is an even multiple of the 8 kHz sampling rate. An M12 multiplexer, which multiplexes four DS-1 signals to one DS-2 signal, is illustrated in Figure 1-8(a).

Similarly, the aggregate rate for a DS-3 signal is 44.736 Mbps, which can consist of a multiplexed combination of seven DS-2 (from an M23 multiplexer) or

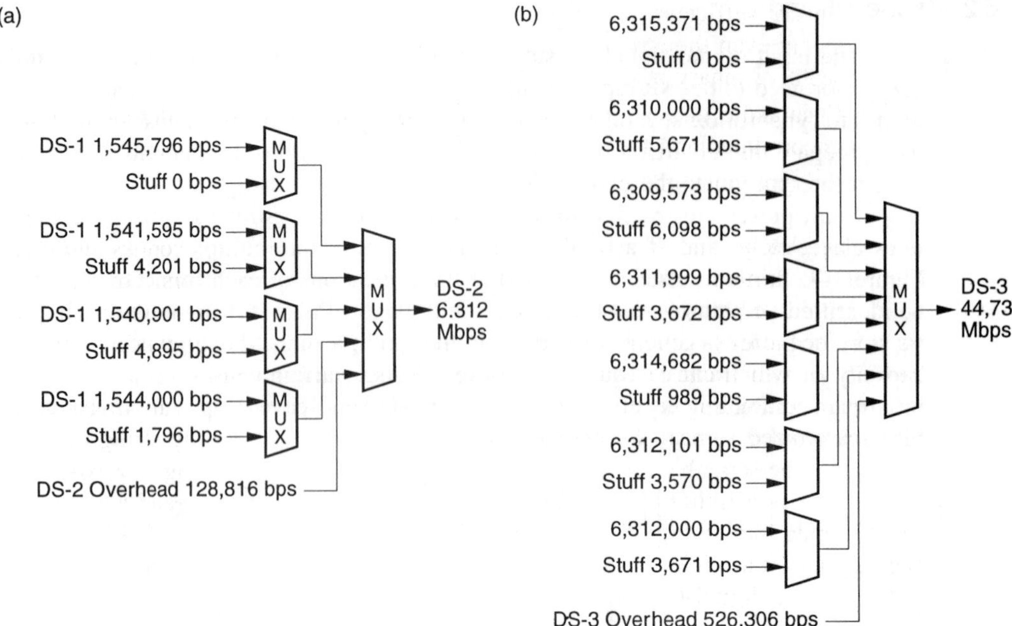

Figure 1-8 Multiplexer with bit stuffing: (a) M12 multiplexer; (b) M23 multiplexer

28 DS-1 (from an M13 multiplexer) signals. Multiplexers with a DS-3 aggregate usually are capable of mixing any appropriate combination of DS-1 and DS-2 signals (for example, five DS-2 and eight DS-1).

To arrive at the DS-3 rate, groups of four DS-1 signals are first multiplexed into a DS-2 signal as previously described. The seven DS-2 signals may be asynchronous because they are formed in different, unsynchronized multiplex terminals, so additional stuffing is required to bring them to the DS-3 rate. Bit stuffing is used to raise the rate of each DS-2 signal from nominal 6,312,000 bps to 6,315,671 bps. In addition to the stuffed DS-2 signals, the DS-3 includes 526,306 bps of overhead for alignment, error checking, multiplex control, and bit stuffing control. This brings the aggregate to 44.736 Mbps (7 × 6,315,671 + 526,306), which also is an even multiple of the 8 kHz sampling rate.* An M23 multiplexer is illustrated in Figure 1-8(b).

1.5.3 Scrambling

Scrambling is used on digital transmission systems to:

- Prevent the transmission of repetitive data patterns, which can cause discrete spectral components and crosstalk interference

---

*The totals do not add exactly due to rounding errors of fractional rates.

- Ensure sufficient timing energy is transmitted and available in the received signal even though the source bit string may contain a large number of consecutive binary zeros (no pulses)
- Ensure the data signals in the two directions of a 2-wire digital loop are uncorrelated and thereby do not bias the echo canceller adaptation
- Encrypt voice and data signals in secure applications

Completely randomized (scrambled) data provide an almost continuous spectrum and continuous spread of energy at the output of multiplexers. On the other hand, repetitive patterns, such as repeating binary-ones sequences or alternating binary 1-0 sequences, generate line spectra (or concentration of energy at a particular frequency), which can lead to crosstalk or interference problems.

Data scrambling ensures that relatively short repetition patterns are transformed into randomized signals. Scrambling does not guarantee that a long string of zeros will always be scrambled into anything other than another long string of zeros. Similarly, there is a finite probability that a random sequence of ones and zeros could be scrambled into a long string of zeros. Nevertheless, scrambling is used in systems that depend on a minimum-ones density for timing extraction. The ones density requirements of digital loop systems are discussed in Chapter 3.

1.6 Analog Signal Conversion and Channel Coding

1.6.1 A/D Conversion

Channel banks and intelligent multiplexers belong to a general class called primary multiplexers. Primary multiplexers convert analog signals (voice signals or tones from a modem) to digital and back again to analog using PCM. The conversion process is called A/D and D/A conversion.

Early channel banks performed little more than voice encoding and decoding, but later versions added digital interfaces. Some primary multiplexers have a large range of features, including direct digital interfaces, digital bandwidth allocation, and digital cross-connect capabilities, and are popularly called intelligent multiplexers.

PCM is a method of digitizing or quantizing an analog signal waveform. As in any A/D conversion, the quantization process produces an approximation of the signal sample. Such an approximation can introduce an error into the digital representation because of the finite number of bits available to represent the value. This error can be made arbitrarily small by representing the approximation by a large number of bits, thereby providing a high level of precision.

In practice, there is a tradeoff between the amount of error and the amount of data representing each signal sample. The goal of the A/D process is to quantize the signal sample in the smallest number of bits that result in a tolerable error. A linear quantization with 12 to 14 bits is the minimum required to produce an accurate digital representation of speech signals over their full range.

The number of bits required is reduced to eight by exploiting the nonlinear characteristics of human hearing. The ear is more sensitive to quantization noise in low-level signals than to noise in large signals. This nonlinearity can be matched by a logarithmic quantization function, which adjusts the data size in proportion to the input signal. Two quantization functions, or encoding laws, are used in telephony applications: the μ-law and A-law. The μ-law is used in North America and Japan, whereas the A-law is used in Europe.

A/D conversion using PCM, shown in the upper part of Figure 1-9, requires three discrete steps, in order:

1. Filtering
2. Sampling
3. Quantization

A device that provides the A/D conversion functions is called an encoder. The encoder filter is used to limit the bandwidth of the source analog signal. In voice telephony, sampling is done at 8,000 Hz to comply with the Nyquist sampling theorem, which requires sampling at twice the highest frequency of interest.* Ideal filters would allow capturing a full 4 kHz of bandwidth at the 8 kHz sampling rate.

Since ideal filters are not available, practical filters are built to start attenuating the upper frequency at approximately 3,400 Hz. This effectively provides a guardband. If proper filtering is not used, the frequency components higher than 4 kHz

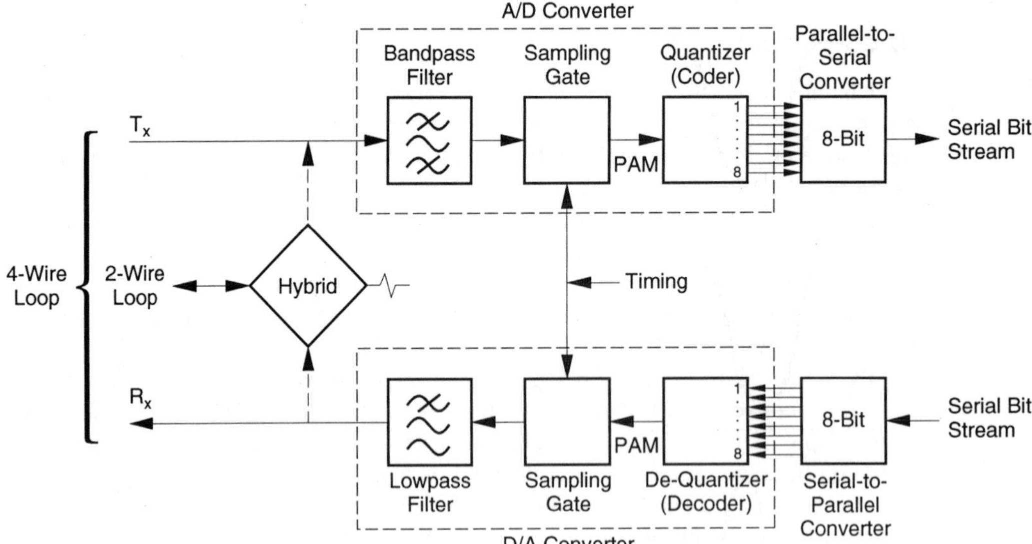

Figure 1-9 A/D converter

*The Nyquist sampling theorem is considered the basis for information theory. Further information can be found in [9].

will lead to aliasing, or foldover of the higher-frequency components into the lower components at the codec output.

Aliasing is illustrated in Figure 1-10. Assume the sampling rate is 8 kHz (8,000 samples per second). A 2 kHz input signal frequency is shown in line (a). The 8 kHz sampling rate provides four samples during each cycle of this waveform. The critical signal frequency of 4 kHz is shown in line (b). Such a waveform frequency is sampled, at most, twice during each cycle. Two samples are sufficient to recover accurately the original waveform at the decoder.

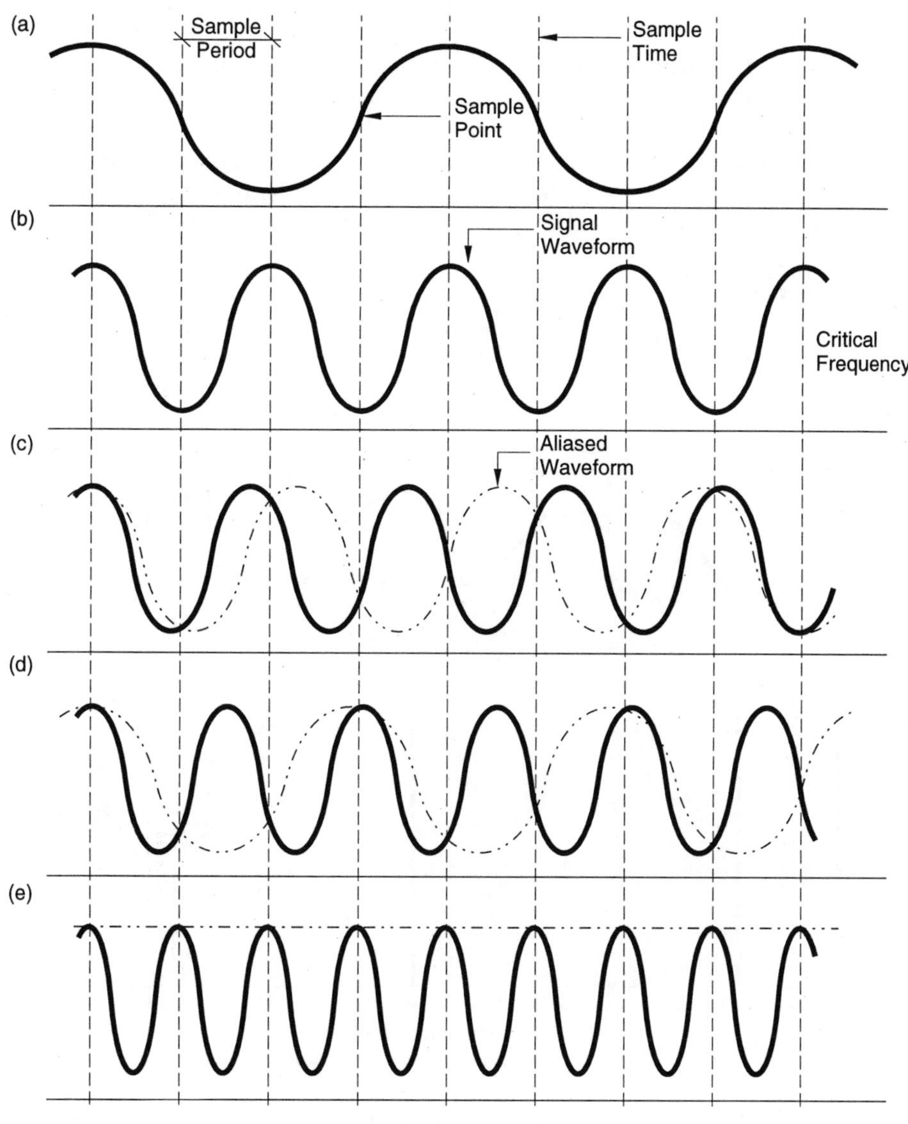

Figure 1-10 Aliasing

In line (c), the signal frequency, shown by a solid line, is raised to approximately 4.5 kHz. When decoded, the sampled points will produce not only the original 4.5 kHz signal waveform, but also the aliased waveform at (8 kHz − 4.5 kHz =) 3.5 kHz, which is shown by the dotted line. The input signal is further increased to approximately 5.2 kHz, shown by the solid line in line (d). The original waveform and (8 kHz − 5.2 kHz =) 2.8 kHz waveform will be produced in the decoder. In line (e), the input signal is 8 kHz (solid line), which results in zero frequency, or dc (dashed line) at the decoder output.

By properly filtering all frequency components greater than one-half the sampling frequency (in this case, greater than 4 kHz), the amplitude of aliased waveforms will be reduced to acceptably low levels.

In voicegrade applications, the source signal is limited to the range of 300 to 3,400 Hz or a bandwidth of 3,100 Hz (usually called a nominal 4 kHz channel). The lower frequency cutoff is used to reduce powerline hum. The performances of typical 2-wire and 4-wire voice channel filters are shown in Figure 1-11.

The periodic samplings are said to capture the time and amplitude information of the original signal. The output of the sampler is a string of pulses whose ampli-

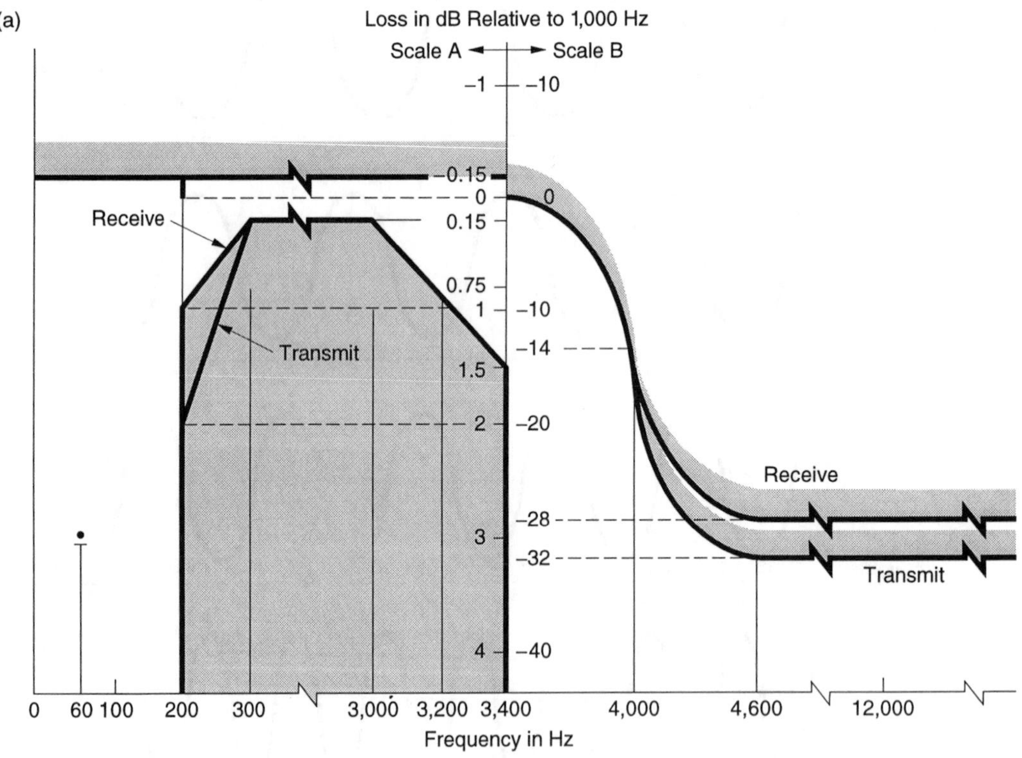

Figure 1-11 Voice channel filter performance, [10]: (a) 4-Wire; (b) 2-Wire
© 1982 AT&T. Reprinted with permission

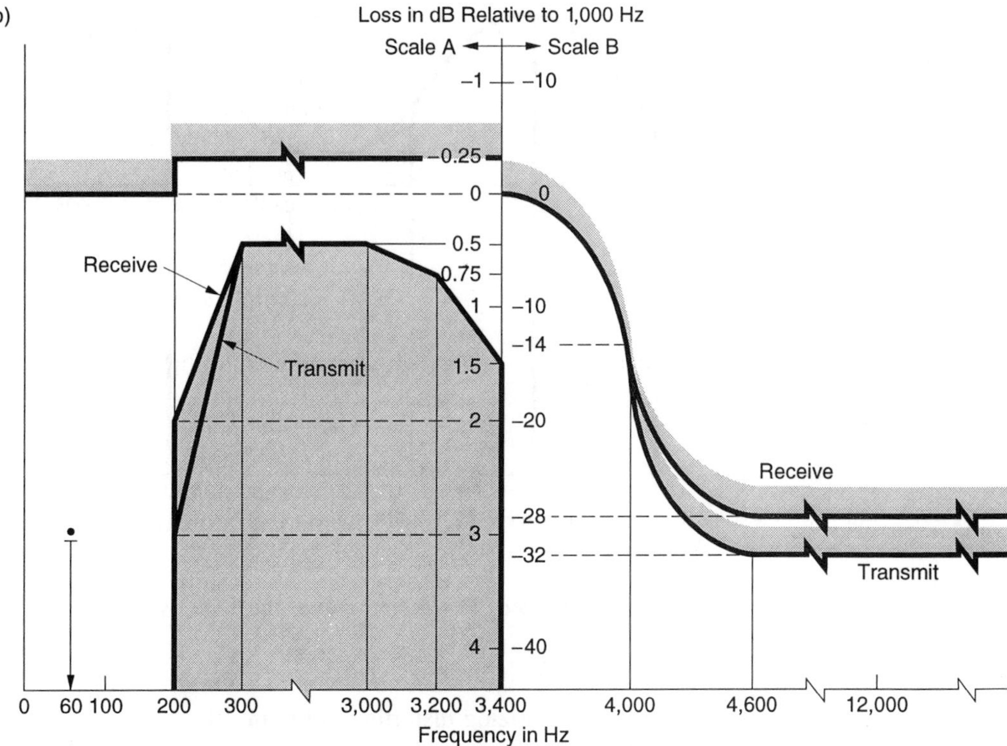

Figure 1-11 Continued

tudes correspond to the amplitude of the analog source signal at sampling time. This pulse string is called a PAM signal, as shown in Figure 1-12.

The output of the sampler is fed into a quantizer. The quantizer converts each PAM signal to an 8-bit binary code word. The code word is a digital representation of the amplitude information at sample time. Since an analog source signal is continuous, quantization naturally limits the resolution of the signal's value at any particular sample time, as shown in Figure 1-13. The decision values shown in this illustration represent the quantizing interval boundaries. Therefore, an input signal falling anywhere between two decision values will be encoded to the quantized value, which is halfway between the two corresponding decision values. The distance (or voltage) between each decision value is the same for a linear encoder but different for a nonlinear encoder, as will be seen.

The typical digital transmission system uses a code word length of eight bits, which limits the resolution to 2^8, or 256 individual binary values (± 0 through 127). This resolution defines the error of the sampled signal. Such errors are manifested in the form of distortion (noise) when the digital signal is converted back to analog.

Signal-to-quantizing distortion ratio (SDR) defines the performance of an encoder. When the analog signal peak meets but does not exceed the coding range,

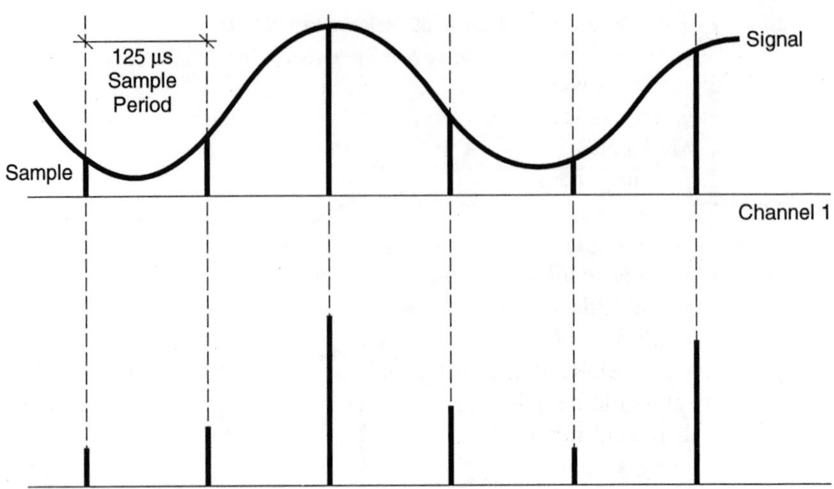

Figure 1-12 PAM signal

the encoder is said to be fully loaded. For a sine wave, the SDR is given as*

$$SDR_{dB} = 6n + 1.8 \text{ dB}$$

where SDR_{dB} = the signal-to-quantizing distortion ration in dB
n = the number of quantizing bits

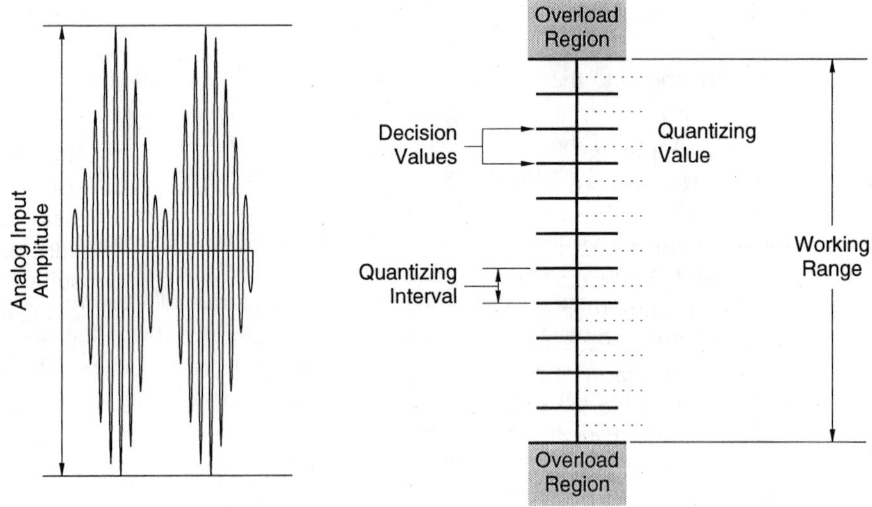

Figure 1-13 Quantizer resolution

*For a derivation of this equation, see [12].

A fully loaded 8-bit encoder will provide a best-case SDR of almost 50 dB, as seen from this expression. With a linear encoder, however, the SDR suffers considerably with lower-amplitude input signals because the noise power from linear encoding error is the same for all input amplitudes (the noise stays the same as the signal is reduced).

Different generations of channel banks have been called D1, D2, D3, D4, and (the most modern) D5. The D1 and D2 channel banks initially used a 7-bit word length for the encoded analog (voice) signal and one bit for signaling. The D3, D4, and D5 channel banks assigned all eight bits to the encoded voice signal in all but the sixth and 12th signaling frames (giving an effective 7-5/6 bits throughout the 12 frames).* The change from 7-bit word to 8-bit word provided about 5 dB improvement in SDR. If the full eight bits are used for voice encoding, a 6 dB improvement would be gained over a 7-bit encoder, as seen from the previous equation.

The typical range of voice level for any particular talker is about 25 dB. The range among a number of talkers is about 40 dB [13]. To cover the very large majority of talkers (99%), encoders are designed to have a 50 dB dynamic range. As previously mentioned, the average distortion noise power of a linear encoder is independent of input signal amplitude. Therefore, in voice applications, this encoder would give a 40 to 50 dB difference in the SDR for a range of talkers and talker volume levels (amplitudes).

EXAMPLE 1-1

Assume a fully loaded 8-bit linear encoder has a 50 dB SDR. Find the SDR for the lowest expected talker volume (level 40 dB down from fully loaded). For purposes of this problem, assume the talker power is equivalent to sine-wave power.

The full-load signal-to-quantizing distortion ratio, SDR_f, can be defined as 10 log (Psf/Pd) where Psf is the full-load signal power and Pd is the quantizing distortion power. If the signal power drops by 40 dB, it is the same as subtracting 40 dB from this expression. Quantizing distortion power remains the same. Therefore, the new SDR $= SDR_f - 40$ dB $= 50$ dB $- 40$ dB $= 10$ dB.

The amplitude distribution of voice signals is not uniform; that is, some amplitudes are more probable than others. Figure 1-14 shows the statistical distribution of instantaneous speech amplitudes relative to the nominal root mean square (rms) amplitude for a single talker. The graph shows that low amplitudes occur with a greater probability than high amplitudes. For example, the probability that the instantaneous amplitude will exceed the rms value (ratio of 1.0) is less than 15%, and more than 50% of the time the instantaneous amplitude will be less than one-fourth the rms value (ratio of 0.25).

It is possible to optimize the average SDR by using a nonlinear encoder that favors the weak signal amplitudes at the expense of the higher amplitudes. Such an encoder provides a higher SDR for more probable low amplitudes or, equivalently,

---

*In every sixth and 12th frame, only seven bits are used for source signal encoding; the eighth bit is used for supervisory signaling. See Chapter 3 for further discussion of channel bank types and framing.

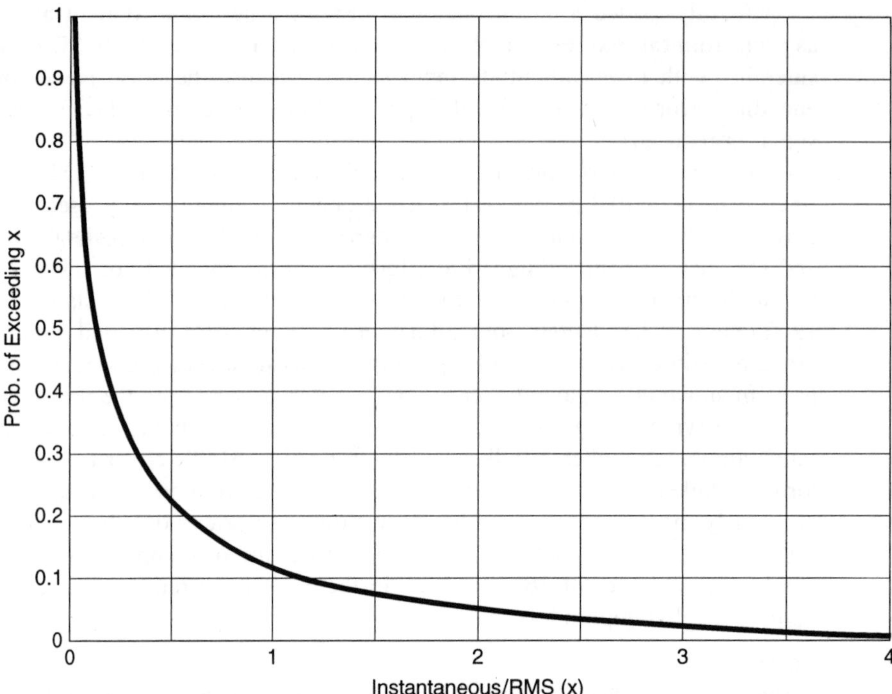

Figure 1-14 Speech amplitude distribution, adapted from [14]

lower distortion at low amplitudes. Similarly, this encoder provides a lower SDR for less probable high amplitudes or, equivalently, higher distortion at higher amplitudes. Therefore, more of the speech energy, which is at the lower amplitudes, is encoded using "fine-grained" quantizing to reduce the overall noise.

In effect, the nonlinear encoder *com*presses the signal and the associated decoder ex*pands* the signal, so the process is called *companding*. There are a number of ways to implement the compression/expansion process, including mapping, predistortion, and digital processing [13]. The transfer and error characteristics of a linear encoder are shown in Figure 1-15(a). The characteristics of a nonlinear encoder typically used in telecommunications are shown in Figure 1-15(b).

The nonlinear encoders used in North America have a modified logarithmic compression characteristic, called μ-law, such that the normalized output as a function of the input is given by

$$Output_{\mu}(\mathrm{i}) = sign(\mathrm{i})\left[\frac{\ln(1 + \mu|\mathrm{i}|)}{\ln(1 + \mu)}\right]$$

where $\mu = 255$ and i is the input signal.*

*In early 7-bit encoders, $\mu = 100$ was used.

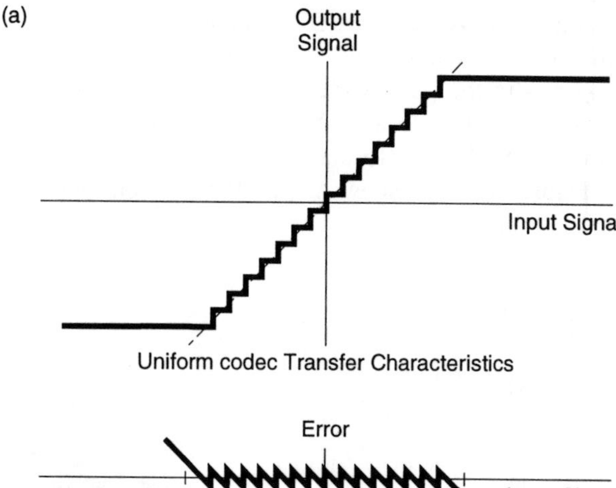

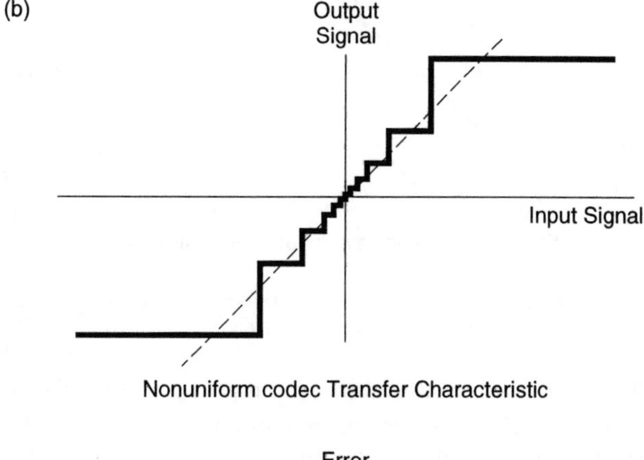

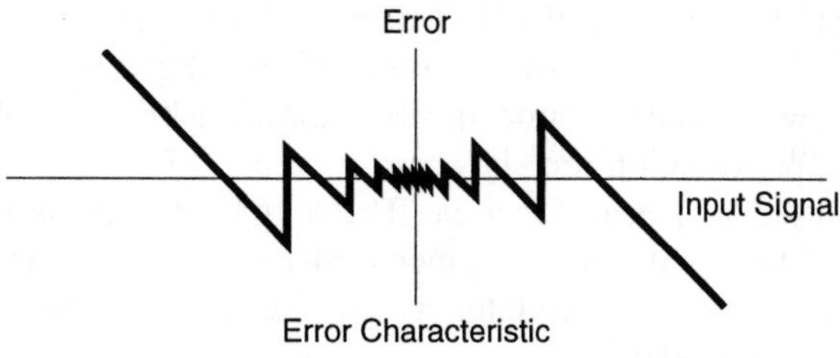

Figure 1-15 Encoder transfer characteristics, [13]: (a) Uniform encoder; (b) Nonuniform encoder
© 1982 AT&T. Reprinted with permission

With small signal levels, around −40 dBm0, this encoding function provides about 25 dB improvement in SDR over a linear encoder, as seen in Figure 1-16.* Referring back to the expression for SDR, this is equivalent to about four additional quantizing bits in the pulse code. Therefore, for an 8-bit code, the μ-law encoding provides an equivalent to a 12-bit code for low signal levels. At an input level of −10 dBm0, the SDR performance of the μ-law encoder is the same as a linear encoder. Figure 1-16 also shows the SDR for a nonsinusoidal signal that has a speech-like amplitude distribution.

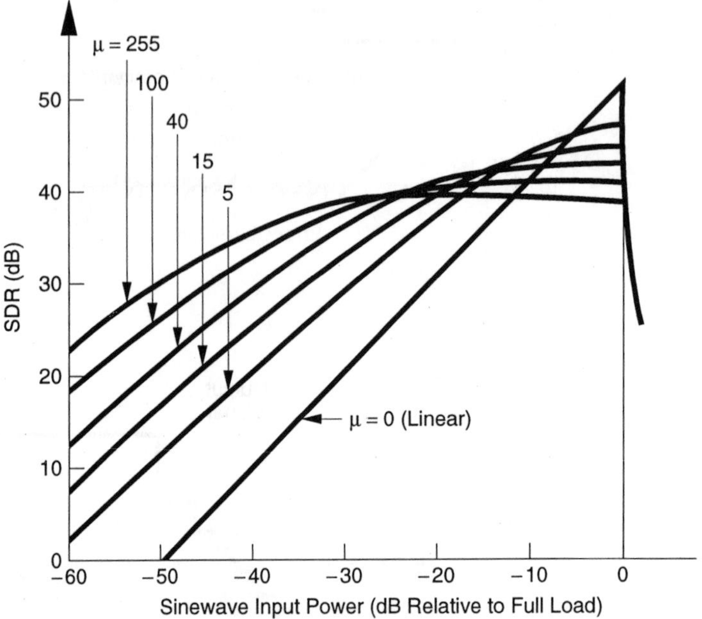

Figure 1-16 8-Bit μ-law codec SDR performance, [13]
© 1982 BELLCORE. Reprinted with permission

The maximum average signal input power to a μ-law encoder is approximately +3 dBm0 (3.17 dBm theoretical maximum). Signals with greater power are simply encoded at a level equal to +3 dBm and, in effect, are clipped at this level.

The idealized μ-law function defined above is approximated in real encoders by piecewise, linear segments. Both the ideal function and a 15-segment piecewise linear approximation are shown in Figure 1-17. The 15-segment μ-law approximation is used in North America. The length of successive segments increases by a factor of two, with eight segments of each polarity. The segments about zero are collinear, so these are counted as one. See [11] for the recommended characteristics of a μ-law encoder.

---

*The nomenclature dBm0 stands for decibels with respect to 1 milliwatt referred to the zero transmission level point (0 TLP). The 0 TLP normally is taken as the (theoretical) center of central office switching systems. See [1] for a more thorough discussion of TLP.

In early channel banks, the piecewise linear approximations were required because of technology limitations. Such approximations, however, are still used in modern equipment for compatibility and because it is possible to convert from a linear code to a compressed code or from a compressed code to a linear code without any additional impairments. This simplifies mixing ADPCM (or other types of PCM) signals on PCM transmission systems and converting PCM values derived from other encoding laws (for example, the A-law used in Europe) to the μ-law. The A-law is defined in [11]. Also, the linear code allows for precise digital pads and level control in digital loop terminal equipment.

Analog samples are encoded using an 8-bit folded (signed) binary code, as shown below. Table 1-4 gives the accompanying code set.

D1 D2 D3 D4 D5 D6 D7 D8

where D1 = sign (polarity) bit (binary zero for negative, binary one for positive)

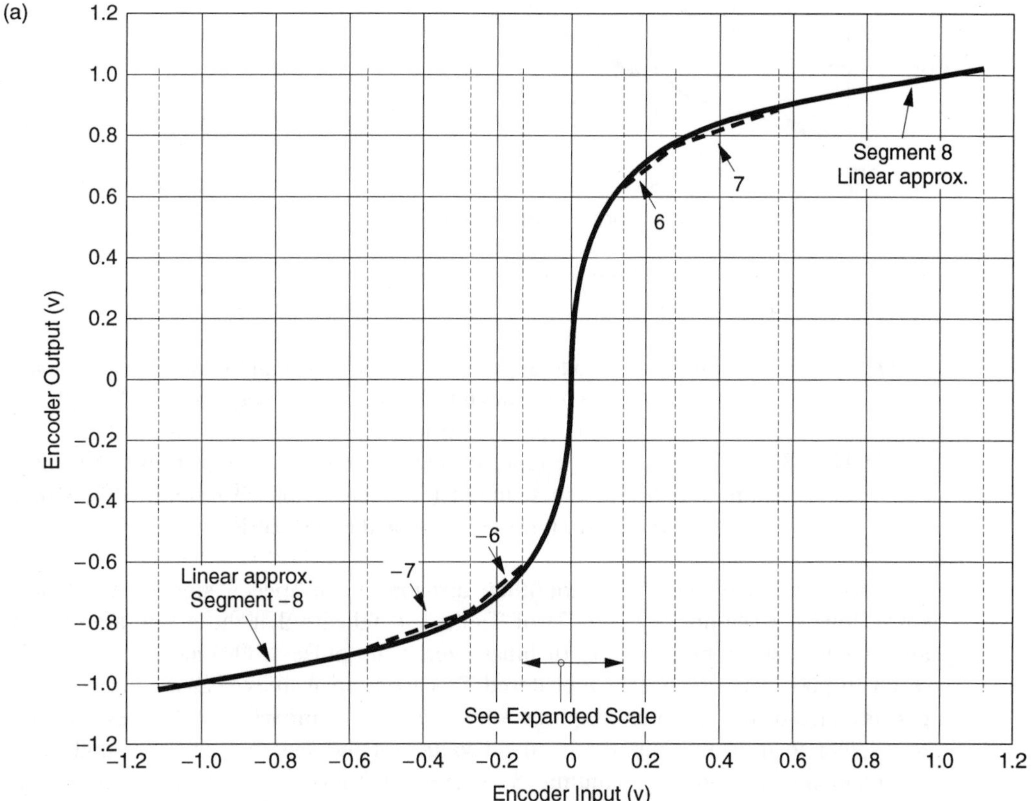

Figure 1-17 Idealized and 15-segment approximated compression curves: (a) Idealized; (b) 15-Segment

(b)

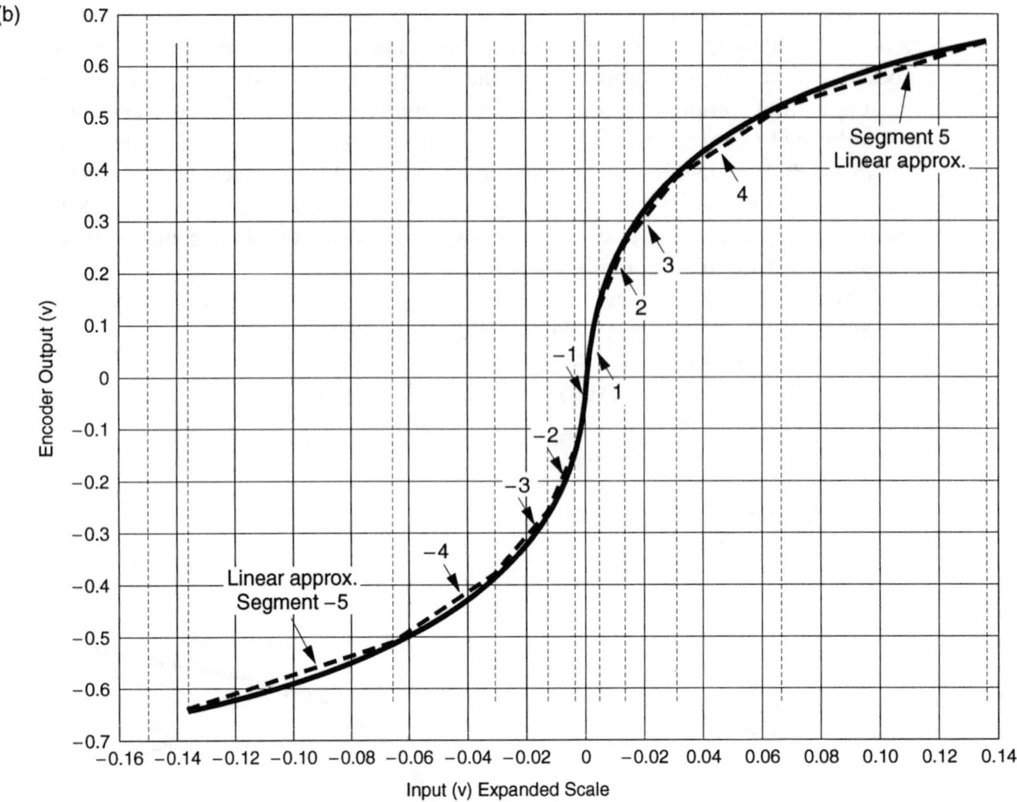

Figure 1-17 Continued

D2 D3 D4 = 3-bit binary code representing one segment of an 8-segment approximation to the magnitude (numbered segments 1 through 8), where D2 is the most significant bit of the segment code

D5 D6 D7 D8 = 4-bit binary code representing a 16-step approximation to the magnitude within the segment (numbered steps 0 through 15), where D5 is the most significant bit of the step code

As previously seen, the amplitude distribution of speech signals favors low signal levels. Columns (6) and (7) of Table 1-4 indicate that these low amplitudes are encoded with a high density of binary zeros (no pulses). Ordinarily, this would result in poor performance of repeatered T1-carrier span lines because the clocking performance of the repeaters requires a high density of binary ones (pulses). Therefore, prior to transmission, the encoded values of the segment and step codes are complemented as shown in column (8) to create a high density of binary ones. The sign bit is left unchanged. The decoder output shown in column (9) is the average of the normalized input values. This table is valid for both positive and negative values.

Table 1-4 Coding for $\mu = 255$, 15-Segment Companding

Segment (1)	Segment Endpoints (2)	Step Size (3)	Normalized Input Range (4)	Decision Value (5)	Segment Code D2 D3 D4 (6)	Step Code D5 D6 D7 D8 (7)	Transmitted Code Word (8)	Decoder Output (9)
	0		0–1	0		0 0 0 0	1 1 1 1 1 1 1	0
			1–3	1		0 0 0 1	1 1 1 1 1 1 0	2
1		2	.	.	0 0 0	.	.	.
			.	.		.	.	.
			27–29	14		1 1 1 0	1 1 1 0 0 0 1	28
	31		29–31	15		1 1 1 1	1 1 1 0 0 0 0	30
	31		31–35	16		0 0 0 0	1 1 0 1 1 1 1	33
			35–39	17		0 0 0 1	1 1 0 1 1 1 0	37
2		4	.	.	0 0 1	.	.	.
			.	.		.	.	.
			87–91	30		1 1 1 0	1 1 0 0 0 0 1	89
	95		91–95	31		1 1 1 1	1 1 0 0 0 0 0	93
	95		95–103	32		0 0 0 0	1 0 1 1 1 1 1	99
			103–111	33		0 0 0 1	1 0 1 1 1 1 0	107
3		8	.	.	0 1 0	.	.	.
			.	.		.	.	.
			207–215	46		1 1 1 0	1 0 1 0 0 0 1	211
	223		215–223	47		1 1 1 1	1 0 1 0 0 0 0	219
	223		223–239	48		0 0 0 0	1 0 0 1 1 1 1	231
			239–255	49		0 0 0 1	1 0 0 1 1 1 0	247
4		16	.	.	0 1 1	.	.	.
			.	.		.	.	.
			447–463	62		1 1 1 0	1 0 0 0 0 0 1	455
	479		463–479	63		1 1 1 1	1 0 0 0 0 0 0	471

Table 1-4 Continued

Segment (1)	Segment Endpoints (2)	Step Size (3)	Normalized Input Range (4)	Decision Value (5)	Segment Code D2 D3 D4 (6)	Step Code D5 D6 D7 D8 (7)	Transmitted Code Word (8)	Decoder Output (9)
	479		479–511	64		0 0 0 0	0 1 1 1 1 1 1	495
			511–543	65		0 0 0 1	0 1 1 1 1 1 0	527
5		32	. .	. .	1 0 0	. .	. .	. .
			927–959	78		1 1 1 0	0 1 1 0 0 0 1	943
	991		959–991	79		1 1 1 1	0 1 1 0 0 0 0	975
	991		991–1055	80		0 0 0 0	0 1 0 1 1 1 1	1023
			1055–1119	81		0 0 0 1	0 1 0 1 1 1 0	1087
6		64	. .	. .	1 0 1	. .	. .	. .
			1887–1951	94		1 1 1 0	0 1 0 0 0 0 1	1919
	2015		1951–2015	95		1 1 1 1	0 1 0 0 0 0 0	1983
	2015		2015–2143	96		0 0 0 0	0 0 1 1 1 1 1	2079
			2143–2271	97		0 0 0 1	0 0 1 1 1 1 0	2207
7		128	. .	. .	1 1 0	. .	. .	. .
			3807–3935	110		1 1 1 0	0 0 1 0 0 0 1	3871
	4063		3935–4063	111		1 1 1 1	0 0 1 0 0 0 1	3999
	4063		4063–4319	112		0 0 0 0	0 0 0 1 1 1 1	4191
			4319–4575	113		0 0 0 1	0 0 0 1 1 1 0	4447
8		256	. .	. .	1 1 1	. .	. .	. .
			7647–7903	126		1 1 1 0	0 0 0 0 0 0 1	7775
	8159		7903–8159	127		1 1 1 1	0 0 0 0 0 0 0	8031

Note: D1 = binary 1 for positive inputs and binary 0 for negative inputs

In Table 1-4, the segments are defined in column (1). By inspection of Table 1-4, segment 1 has 16 equal steps, as do the other segments; but the input voltage—represented by the normalized input range in column (4)—between each step is the smallest. The normalized range of ±8,159 is chosen so all magnitudes are represented by an integer. It is easy to visualize the purpose of the normalized units if each unit is considered to be some voltage, say 1 mV. For step 1, the step size is two normalized units and for step 8, 256 units, as shown in column (3). Therefore, an input signal falling in step 1 at sampling time is encoded with a resolution of two normalized units; similarly, an input signal falling in step 8 will be encoded with a considerably coarser resolution of 256 units.

The endpoints of each segment are shown in column (2). The decision value shown in column (5) defines the boundary between adjacent quantizing intervals. Column (6) gives the binary representation of the corresponding segment and column (7) gives the binary representation of the corresponding step (also called quantizing level). The transmitted code word in column (8) is the complement (or inverse) of the D2 through D8 bits. Finally, column (9) shows the output of the decoder at the far-end for each range of normalized inputs. For example, if the input at the encoder falls in the normalized range of 479 to 511, the decoder output always will be 495 normalized units.

A voltage level of 0 volts may be encoded as binary 1 0 0 0 0 0 0 0 or 0 0 0 0 0 0 0 0 and transmitted as 1 1 1 1 1 1 1 1 or 0 1 1 1 1 1 1 1 depending on the background noise at the sampling instant. The first bit is the polarity bit and is not shown in the table. The coarsest quantizing structure is at segment 8 of each polarity. A negative voltage peak at the maximum input power level will be encoded as binary 0 1 1 1 1 1 1 1 (negative polarity, segment 8, step 15) and transmitted as 0 0 0 0 0 0 0 0. If this occurs in an analog channel unit, zero code suppression is used, which changes the all-zeros word to binary 0 0 0 0 0 0 1 0 (negative polarity, segment 8, step 13). This modification to an all-zeros code is not done when a channel is used to transmit digital data in a clear channel format. An example will illustrate the use of Table 1-4.

EXAMPLE 1-2

A −10 dBm0 test tone level is inserted into the channel unit of a T1-carrier channel bank, which uses a $\mu = 255$, 15-segment encoder. Assume the maximum input power level of the encoder is +3.0 dBm0. Find the transmitted code word, output voltage, and voltage quantizing error at the negative peak of the waveform at the far-end decoder.

The maximum input power level of 3.0 dBm0 equates to an absolute power of 1.995 mW. Assuming a 600 Ω impedance, the maximum input voltage is 1.0941 Vrms. The peak value of a sinusoidal waveform is 1.414 times its rms value. For the maximum input voltage, the negative peak is at −1.0941 Vrms × 1.414 = −1.5471 V-peak. This corresponds to the normalized input value of 8,159 units, as shown in column (4). The absolute power of a −10 dBm0 test tone is 0.1 mW and the voltage is 0.2449 Vrms. The negative peak voltage is −0.3464 V-peak. The ratio of the two voltages is 0.3464/1.5471 = 0.2239. Therefore, the test tone is 22.39% of the maximum allowed input voltage and, in normalized units, is represented by 0.2239 × −8,159 units = −1,827 units. The sign bit is binary zero for negative polarity. In columns (2) and (4), the value falls between segment endpoints 991 and 2,015, which corresponds to segment 6 in column (1). The segment code for segment 6 is binary 1 0 1, as found in column (6). By simple extension of the values in columns (4) and (7), the step code corresponding to the input value 1,827 is found to be binary 1 1 0 1, which corresponds to an input range of 1,823 to 1,887. Therefore, the complete 8-bit code word is binary 0 1 0 1 1 1 0 1. The actual transmitted code word is the complement, or 0 0 1 0 0 0 1 0 (the sign bit is preserved). The decoder output is the center of the input range values (found by a simple arithmetic average), or $-(1{,}823 + 1{,}887)/2 = -1{,}855$. The decoder output value corresponds to an output voltage of $(1{,}855/8{,}159) \times -1.5471$ V-peak = 0.3517 V-peak. Therefore, the voltage quantizing error is $0.3517 - 0.3464 = 5.3$ mV.

EXAMPLE 1-3

Figure 1-18 shows approximately 1 ms of a 600 Hz sinusoidal input signal and a graphical representation of a 15-segment, $\mu = 255$ encoder. Three samples are taken in succession at a rate of 8,000 samples/second. Find the transmitted code word corresponding to each sample.

All three samples are negative, so bit D1 = binary zero for each. The first sample is in negative segment 8 (D2 D3 D4 = binary 1 1 1) and falls closest to step 12 (D5 D6 D7 D8 = 1 1 0 0). All bits, except the sign bit, are complemented, giving a transmitted code word of 0 0 0 0 0 0 1 1. The transmitted code word for the other samples can be found by a similar analysis. The three encoded samples are shown in Table 1-5.

Each voicegrade channel encoded at the DS-0 rate requires one of 24 available timeslots in a DS-1 rate system. Using a lower bit-rate encoding, two or more voicegrade channels can be inserted into one timeslot. Similarly, but in an opposite sense, channel units for wideband audio circuits, such as are required between a radio or television studio and a broadcast transmitter site, require more than one timeslot. The equivalent timeslots for various analog circuit types are shown in Table 1-6. The DS-0 encoding rate is the basic building block for any channelized system and not just systems operating with a DS-1 aggregate rate.

The final step in the A/D conversion process is the conversion of the parallel 8-bit words into a serial PCM bit stream by a parallel-to-serial converter. This converter can be as simple as a parallel load shift register in which the eight data bits are bulk loaded from the quantizer and then output one bit at a time. In repeatered T1-carrier systems and other channelized systems operating at the DS-1 rate, the individual bit streams from 24 separate timeslots are mixed to provide a 192-bit data stream. A framing bit is then added in each 24-channel group, as shown in Figure 1-19. An idle voice channel (that is, a channel not equipped with a channel unit) is encoded as all binary ones.

The following summarizes the steps required to encode 24 individual analog signals using PCM and standard (D4 channel bank) framing for transmission at the DS-1 rate:

1. The input signal is filtered to remove all frequency components above about 3.4 kHz to prevent aliasing and below approximately 200 Hz to remove powerline hum.
2. The input signal voltage is sampled at a rate of 8,000 samples/second, resulting in 8,000 pulse-amplitude-modulated voltage samples/second.

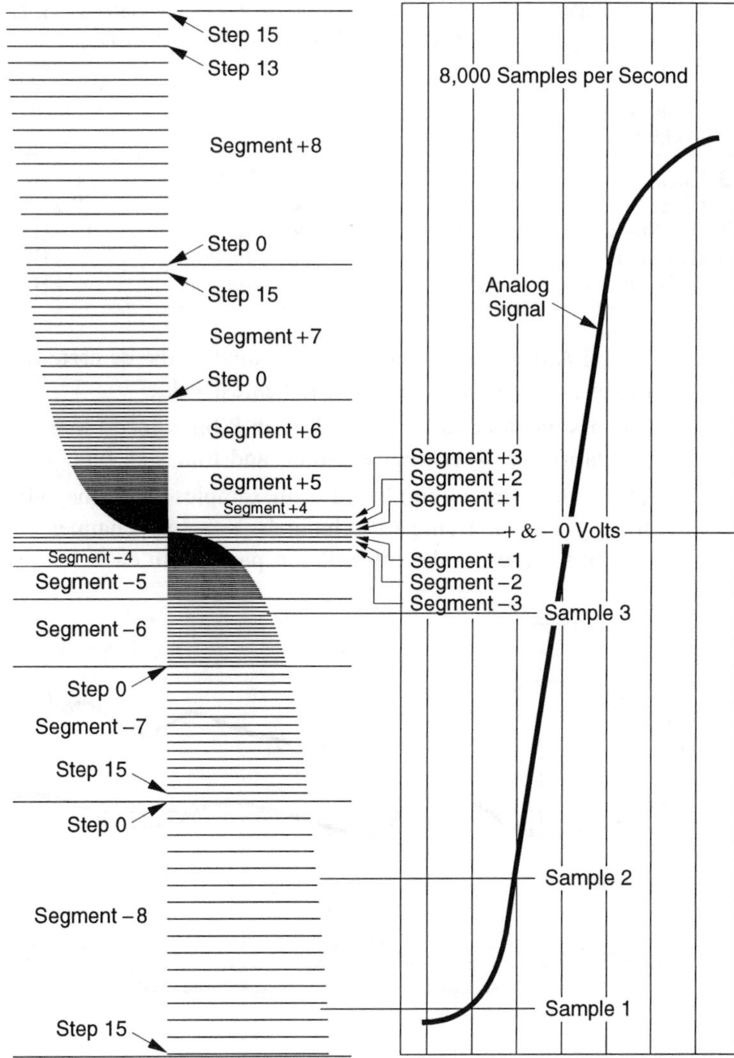

Figure 1-18 Illustration for Example 1-3

Table 1-5 Encoded Values for Example 1-3

Sample	Polarity Bit D1	Segment Bits D2 D3 D4	Step Bits D5 D6 D7 D8	Transmitted Code Word
Sample 1	0	1 1 1	1 1 0 0	0 0 0 0 0 0 1 1
Sample 2	0	1 1 1	0 1 0 1	0 0 0 0 1 0 1 0
Sample 3	0	1 0 1	0 1 0 0	0 0 1 0 1 0 1 1

Table 1-6 Voiceband and Audio Program Wideband Program Channel Characteristics

Circuit Bandwidth	Encoding	Equivalent Timeslots	Channel Sampling Rate	Bit Rate per Channel
3.1 kHz	ADPCM	$\frac{1}{2}$ per direction	8 kHz	32 kbps
3.1 kHz	PCM	1 per direction	8 kHz	64 kbps
5 kHz Studio	PCM	2 per direction	16 kHz	128 kbps
8 kHz Studio	PCM	4 per direction	32 kHz	256 kbps
15 kHz Studio	PCM	6 per direction	48 kHz	384 kbps

3. Each PAM voltage sample is compared to the discrete quantizing levels of the encoder and assigned to the nearest value.
4. The assigned value is encoded as an 8-bit signed binary word composed of a polarity bit, three segment bits, and four step bits.
5. The 8-bit word is combined with samples from the other 23 channels to form a 192-bit frame (24 channels × 8 bits/channel = 192 bits). These 192 bits represent the information payload in each frame.

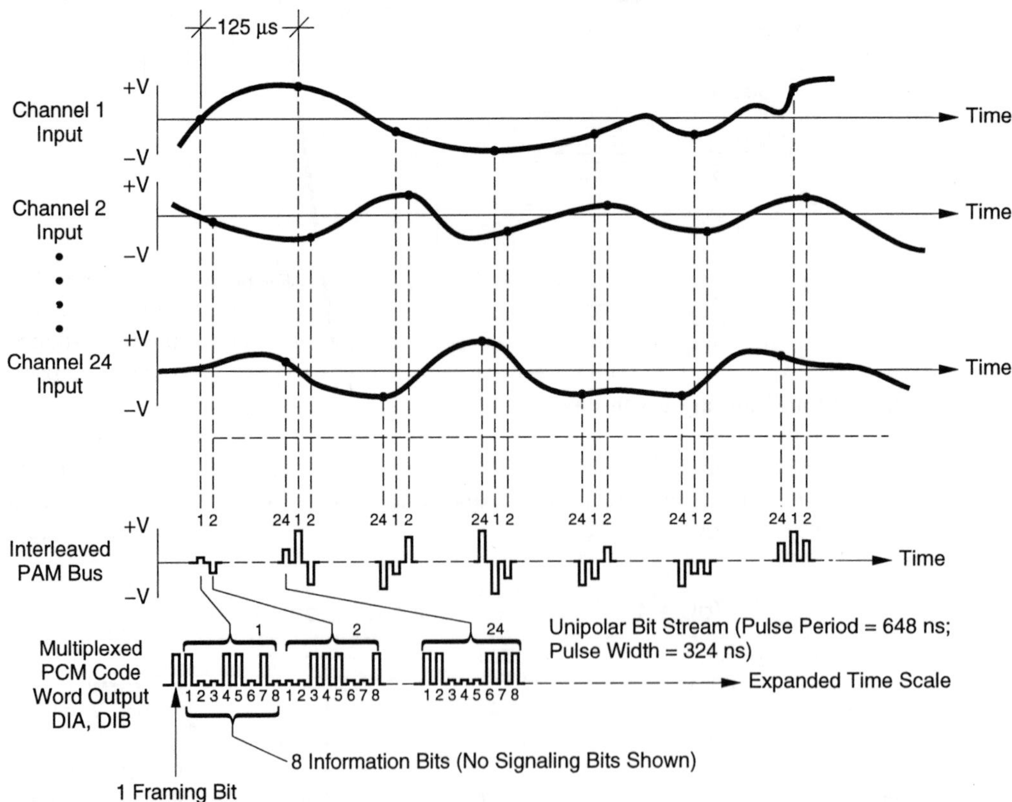

Figure 1-19 DS-1 rate system with 193-bit serial data frame

6. A single framing bit is added to the beginning of the 192 payload to complete the total frame of 193 bits. The frame width is 125 μs.
7. The 193-bit frame is transmitted at a rate of 1.544 Mbps (24 channels/frame × 8 bits/sample/channel × 8,000 samples/second + 1 framing bit/frame × 8,000 frames/second = 193 bits/frame × 8,000 frames/second = 1.544 Mbps).

1.6.2 D/A Conversion

D/A conversion, or decoding, of a signal is a mirror image of the encoding process. The serial bit stream is converted to parallel 8-bit words and then converted to a serial string of PAM signals of equivalent amplitude. The PAM signals are then filtered to extract the time and amplitude relationships of the original signal and to re-establish its continuity.

An incoming digital code word is converted to a PAM signal with an exact amplitude (within tolerance limits of the electronics) for each code word value. The limitations of the decoder are different than the encoder in that there are no additional quantizing errors introduced at the decoder. In the encoder, the signal was converted to a code word with a discrete step value, even though the signal fell anywhere within the range covered by that step.

1.7 Voice and Voiceband Encoding

1.7.1 Introduction

The largest proportion of *traffic* carried on telecommunication transmission systems is voice-type traffic.* On active circuits during the busiest hour of the transmission system, voice is present about 40% of the time; the rest of the time silence is encoded. As a result, it can be said that the largest proportion of *signals* comprise silence throughout any given day.

Digitally encoding a voice signal increases the bandwidth required for transmission. This is a clear disadvantage not to be overlooked during the design or choice of a transmission system. However, there are a number of advantages to digital encoding that usually far outweigh the bandwidth disadvantage, as previously discussed.

Voice encoding can be accomplished by three basic coder types:

- Waveform coders
- Vocoders
- Hybrid coders

---

*One inter-exchange carrier that carries traffic between the United States and the Far East reports that the majority of its traffic is facsimile [15].

With waveform coders, the actual voice waveform is encoded. A typical waveform encoder will use PCM, as previously discussed. The recovery of the signal at the output of the decoder provides an explicit approximation of the signal at the input to the encoder. With vocoders, a speech production model is parametrically summarized, and only the parameters are then digitized. Therefore, only a summary of the speech information is transmitted where it is recovered and converted to speech-like sounds at the other end. A hybrid coder uses a combination of a waveform coder and vocoder.

A typical vocoder uses LPC techniques. In the speech model, a time varying digital filter is used. The filter coefficients represent the vocal tract parameters. The filter is driven by a function, which, for voiced speech sounds, is a train of unit pulses at the fundamental, or pitch, frequency of the voice. For unvoiced sounds, the driving function is random noise with a flat spectrum. Voiced sounds are produced by vibration of the vocal chords and unvoiced sounds are produced by the lips and tongue without vocal chord vibration.

Figure 1-20 shows a block diagram of an LPC system. The speech input signal in a 15 to 20 ms window is encoded as three components: (a) a set of filter coefficients; (b) whether the signal is voiced or unvoiced; and (3) pitch period if voiced. These components, in the form of digital code words, are transmitted to the far-end through a digital channel. Several speech channels can be multiplexed into a single DS-0 rate digital channel.

The coding system predicts the value of the speech input signal based on the weighted average of the previous input samples. The number of input samples equals

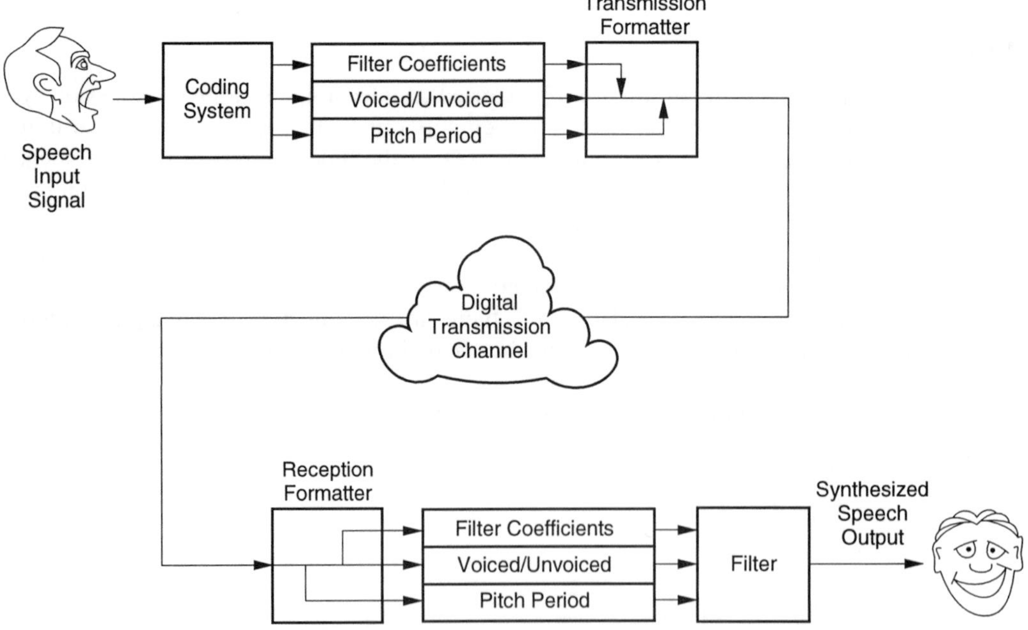

Figure 1-20 Linear predictive coding

the number of filter coefficients. The difference between actual speech input and the predicted value is the prediction error. The problem of linear prediction is to determine a set of filter coefficients in any window that minimizes the mean-squared prediction error.

At the receiving end, the LPC system synthesizes a speech signal by creating a digital filter with the received filter coefficients and driving the filter either with a train of unit impulses at the given pitch (for voiced sounds) or random noise (for unvoiced sounds).

The quality achievable by voice encoding largely depends on the type of encoding as well as the bit rate. As would be expected, the higher bit rates give higher quality for a given encoder type. Speech quality is generally related to digital transmission rate, as shown in Figure 1-21. This figure uses the terms *broadcast quality, toll quality* (or *network quality*), *communications quality,* and *synthetic quality.* Broadcast quality is what one expects from a commercial radio station. Toll quality and network quality are what the general public expects when it uses the public switched telephone network (PSTN). Toll quality can be quantified in terms of a 30 dB or better signal-to-noise ratio (SNR) and 3.1 kHz bandwidth.

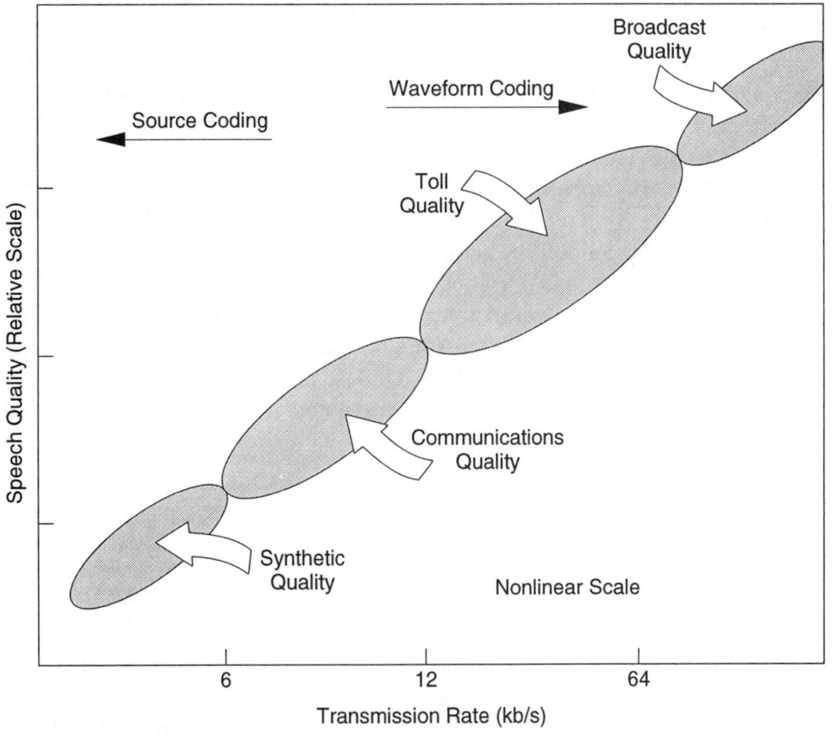

Figure 1-21 Speech quality versus digital transmission rate, [16] Kitawaki and Nagabuchi, © 1988 Institute of Electrical and Electronics Engineers. Reprinted with permission

The term "communications quality" is not as well quantified as toll quality, but it is what would be acceptable on radio communication links for pilots, and military and mobile radio users. A particular voice can be recognized on a communications quality link. Encoding rates for communication quality vary from about 5 kbps to 14 kbps. Synthetic quality, while intelligible, lacks natural voice sounds, and the person talking is not recognizable. The encoding rate for synthetic quality is below approximately 5 kbps.

The subjective aspect at the output of a voice decoder is often categorized by a mean opinion score (MOS), as shown in Table 1-7. Such a score would be given by tests using a large number of speakers, inputs, and listeners. The MOS is a perceptual value that relates the listener's expectations to the distortion in a voice signal.

PCM voice encoding at a 64 kbps rate has become somewhat of a standard against which other encoding methods are judged. A 64 kbps PCM encoder gives an MOS of 4.5 on a single encode/decode link, with an MOS of 4.0 after eight stages [17]. At present, the encoding rate for toll-quality voice encoders falls between 24 kbps and 64 kbps, although 16 kbps encoders have been developed with the required quality. Significantly lower encoding rates will provide the required quality as technologies and the economics of these technologies improve.

Toll-quality voice encoding presently uses waveform coders. Communications quality uses both waveform and hybrid encoders. Synthetic-quality voice encoding currently uses vocoders. Advances in technology and further understanding of the speech process will lead to the use of vocoders with communications, and then toll quality. The quality of the recovered signal at the output of the decoder depends on the SDR (previously discussed), which can be quantified by tests, as well as by subjective evaluations by listeners.

A quality measure used in network planning is the quantizing distortion unit (QDU). The 8-bit PCM codec pair is assigned one QDU to cover quantizing distortion. Under guidelines being developed by The International Telegraph and Telephone Consultative Committee (CCITT), a standards body,14 QDUs of impairment are allowed for an overall international connection [16]. Up to five QDUs are allowed for each national extension and four QDUs, for the international circuit. A 7-bit PCM codec is assigned three QDUs, and a 32 kbps ADPCM codec that meets CCITT recommendation G.721 is assigned 3.5 QDUs. The proposed QDU measurement method assumes a linear additive process, which is inaccurate in the region of larger QDU numbers [16].

Table 1-7 Mean Opinion Score

Mean Opinion Score (MOS)	Descriptive Value	Distortion
1	Bad	Very annoying and objectionable
2	Poor	Annoying but not objectionable
3	Good	Perceptible and slightly annoying
4	Fair	Just perceptible but not annoying
5	Excellent	Imperceptible

1.7.2 Compression

The encoding of any analog source signal is more efficient the lower the bit rate. A linear encoding scheme (for example, linear PCM at 64 kbps) gives a benchmark against which other encoding schemes can be measured. A compact representation of a source signal that results in a lower bit rate can be considered to be compressed. Compression is achieved by the removal of redundancy and reduction of irrelevancy [18].

All encoding methods strive to minimize the bit rate through some measure of compression. There are several dimensions to defining the performance of these encoders, including complexity, delay, signal quality, and bit rate. Ignoring complexity and delay, the performance of an encoder can be demonstrated by measuring the resulting signal quality for a given bit rate or by giving a specified quality at a lower bit rate.

As will be shown in the next section, there are a number of voice encoding methods. This also is true of encoding methods used on relatively wideband nonvoice signals, such as video, where compression is crucial to economical signal transmission. For example, the uncompressed bit rate for a single channel of high-definition television (HDTV) is over a gigabit per second. This single television signal is equivalent to over 20,000 uncompressed voice channels.

1.7.3 Voiceband Encoding Tradeoffs

Low bit-rate encoders suffer from disadvantages that restrict their use or require other special network design considerations. Some low bit-rate encoders perform badly (become very noisy or completely unusable) when the transmission system has relatively high error rates that otherwise would be acceptable with PCM or ADPCM. For example, a bit error rate (BER) of 1E-3 or 1E-2 will cause degraded transmission with PCM or ADPCM, but some low bit-rate encoders will not work at all.

Systems that use low bit-rate encoding are vulnerable to the rapid buildup of distortion through successive (tandem) encode/decode stages. Encoders with these characteristics are said to lack the robustness of 64 kbps PCM encoders. The low bit rates are achieved by optimizing the encoder for voice signals, which can lead to inadequate performance with nonvoice signals such as multifrequency signaling tones, modems, and facsimile signals.

Almost any encoding scheme has some delay associated with it due to the inherent characteristics of filters, samplers, quantizers, and other coding devices such as digital signal processors intermediate to and at each end of a digital path. The delay often is necessary to buffer enough speech to exploit its redundancy and allow predictive coding to take place accurately.

Delay and its effects on echo are closely related. Echo with delay greater than 5 ms can be annoying on voice calls and can cause talker and listener confusion. Delayed echo traditionally has been made less annoying by inserting loss in the receiver end of the connection. The more delay, the more inserted loss is necessary for echo at a given level to be tolerable. For example, a 10 ms delay requires approximately 10 dB of echo path loss on a typical connection for the connection to be considered reasonably good. A delay of 265 ms requires approximately 40 dB.

In the public analog network, the via net loss (VNL) method is used to determine the amount of required loss. The VNL method is described in [1]. In a lossless digital network, to which all telecommunication systems are evolving, echo is tightly controlled by echo cancellers and by minimizing the encoding delays.

Delay also can cause improper echo canceller operation, and excessively delayed signals between modems can lead to modem and DTE operation problems and reduced data throughput. Modern modem standards include echo cancellation methods (for example, CCITT V.32 and V.32bis; see [19]).

Echo can be split into two components: near echo caused by the near local loop, and far echo caused by the remote local loop. The possible echo span is in the order of tens of milliseconds for each loop, but the difference between the near and far echoes may be in the order of hundreds of milliseconds. To account for this, echo cancellers typically have two sections with a bulk delay processor between them. Low bit-rate voiceband encoders, which typically are located in the local loops, introduce additional delay and widen the near and far echo spans. Faulty operation will result if the echo canceller is not designed for the larger local echo spans (also called end-link delay).

Present low bit-rate encoders (< 16 kbps) have a one-way encode/decode delay of 20 to 40 ms, or more. CCITT has targeted a maximum delay of 5 ms (objective < 2 ms) for the next generation of 16 kbps toll-quality encoders [20].

There is a high amount of complexity associated with any low bit-rate encoding scheme. This complexity and its associated cost, in conjunction with the limitations mentioned above, lead to compromises in system design. Table 1-8 summarizes some of the tradeoffs and characteristics.

Table 1-8 Tradeoffs in Digital Voice Encoding[a]

Encoder	Type	Bit Rate	Quality	MOS	Complexity	Delay
PCM	Waveform	64 kbps	Toll	Good-excellent	0.0	0 ms
ADPCM	Waveform	32 kbps	Toll	Good	0.1	0 ms
LD-CELP	Hybrid	16 kbps	Toll	Good	0.1	2 ms
ASBC	Waveform	16 kbps	Comm.-Toll	Good	1.0	25 ms
MP-LPC	Hybrid	8 kbps	Comm.	Fair-good	10.0	35 ms
SE-LPC	Hybrid	4 kbps	Syn.-Comm.	Fair	100.0	35 ms
LPC	Vocoder	2 kbps	Syn.	Bad-poor	1.0	35 ms

[a]Source: [17]

1.7.4 ADPCM

As defined earlier, an extension of the PCM method is called ADPCM. ADPCM uses the predictable and redundant nature of speech to lower the overall encoding requirements. With this method, an algorithm is used to predict the source voice signal. The error between the actual signal and prediction is then coded. Full-amplitude, 8-bit encoding is not needed. In 32 kbps ADPCM, only four bits are used to encode the error. Detailed information on the 32 kbps ADPCM algorithms can be found in [11,21–23].

Typically, a PCM-to-ADPCM translator-encoder (transcoder) is inserted into a PCM system to increase its voice channel capacity. The ADPCM encoder accepts

64 kbps PCM values as input, and the ADPCM decoder outputs 64 kbps PCM values.

ADPCM works as follows: after an analog signal has been encoded according to 8-bit μ-law PCM requirements, it is converted to 14-bit uniform (linear) PCM by simple code word conversion. At this point the uniform PCM signal is subtracted from an estimate of the signal to give a difference signal. An adaptive 15-level quantizer assigns a 4-bit code word to the value of the difference signal. The 4-bit code word is transmitted to the far-end multiplexer at the standard 8 kHz sampling rate. An inverse quantizer at the far-end produces a quantized difference signal from the code word. A signal estimate is obtained from an adaptive predictor, which operates on both the reconstructed signal and the quantized difference signal, thus completing a feedback loop. The building blocks of an ADPCM encoder and decoder are shown in Figure 1-22.

The decoder uses a structure similar to the feedback portion of the encoder together with a uniform PCM to μ-law PCM converter. Both the encoder and decoder update their internal variables based only on the generated ADPCM value, which

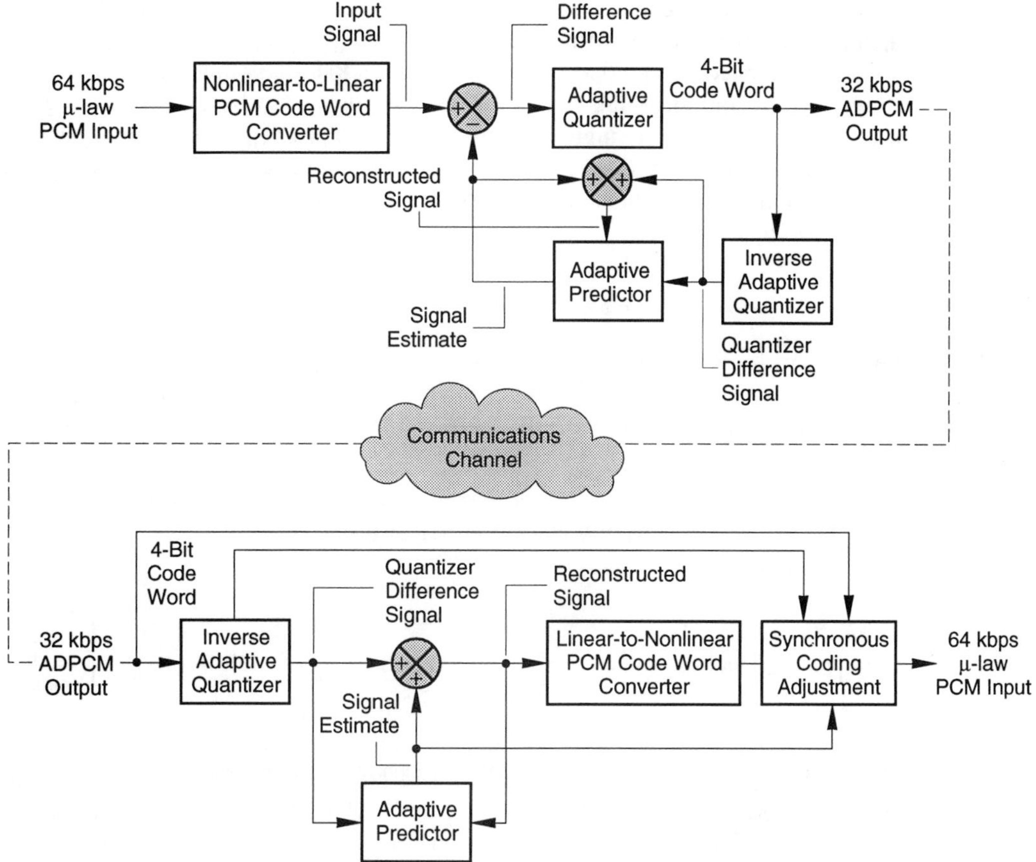

Figure 1-22 ADPCM encoder and decoder, [11]: (a) ADPCM encoder; (b) ADPCM decoder

ensures the encoder and decoder operate in synchrony without the need to send additional data. This is the function of the synchronous adjustment block. The synchronous coding adjustment reduces cumulative distortion on tandem codings (for example, ADPCM pair to PCM pair to ADPCM pair, which would be counted as three tandem connections). In order for this adjustment to work properly, there must be no intermediate digital processes such as digital pads, echo cancellers, digital speech interpolation, or μ-law to A-law converters [11].

A full ADPCM decoder is embedded within the encoder to ensure that all variables are updated based on the same data. In the receiving decoder as well as the decoder embedded in the transmitting encoder, the transmitted ADPCM value is used to update the inverse adaptive quantizer. This, in turn, produces a dequantized version of the difference signal. The dequantized value is added to the value generated by the adaptive predictor to produce the reconstructed speech sample at the decoder output.

The adaptive predictor computes a weighted average of the last six dequantized difference values and the last two predicted values. The ADPCM algorithm also includes a tone and transition detector. If a tone is detected (for example, modem or dual tone multifrequency (DTMF) tones), the predictor is changed to allow rapid adaptation and improved performance.

Since the encoder and decoder operate at an 8 kHz sampling rate and four bits are used, the overall encoding rate is 8 kHz × 4 bits = 32 kbps. Two such signals can be multiplexed into a single 64 kbps timeslot of a DS-1 rate channel bank or multiplexer, as shown in Figure 1-23. Various encoding levels can be used in the adaptive quantizers to achieve other encoding rates, such as 3-bit for 24 kbps and 5-bit for 40 kbps.

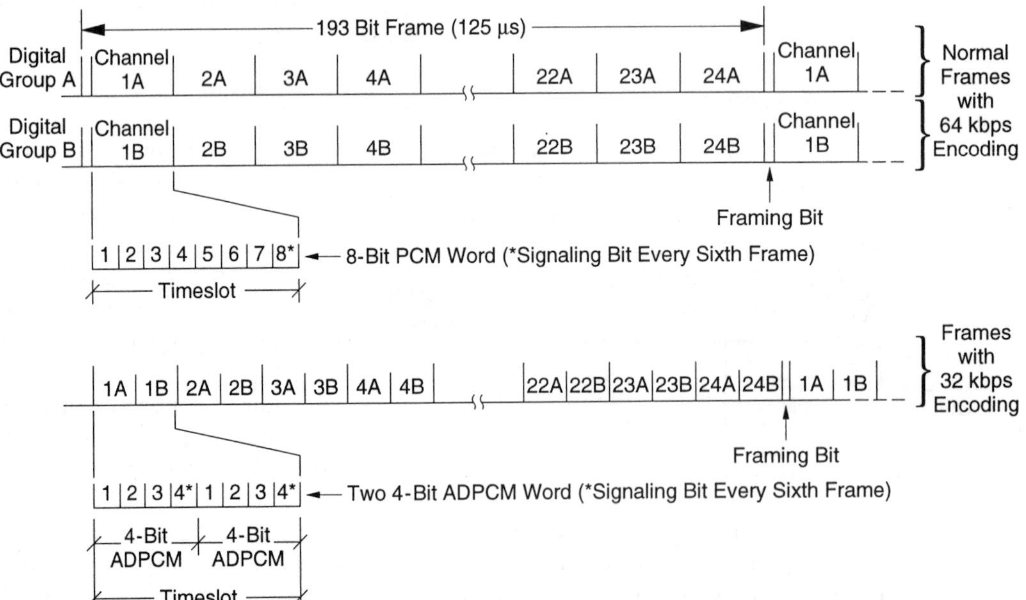

Figure 1-23 32 kbps voice channel multiplexing

The 32 kbps ADPCM was originally designed for voice applications. The voice quality of 32 kbps ADPCM is indistinguishable from 64 kbps PCM provided the system BER is better than 1E-4. The performance of 32 kbps ADPCM is better than PCM at higher BER, as shown in Figure 1-24. Both CCITT and the American National Standards Institute (ANSI) recommended that algorithms for 32 kbps ADPCM are robust enough to operate at 1E-3 BER [11,22]. However, nonvoice source signals and applications require special considerations, as will be discussed in Chapter 2.

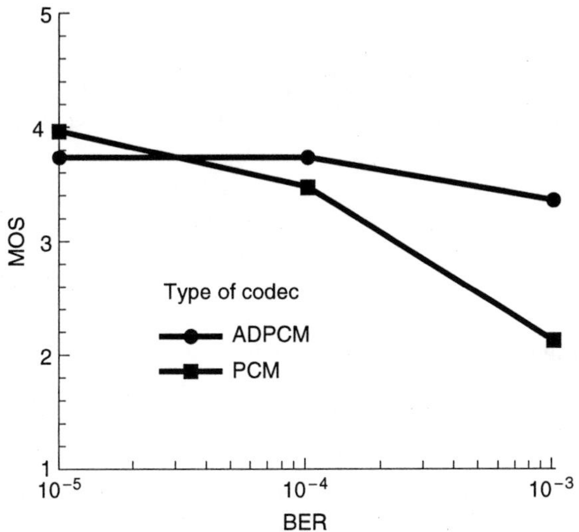

Figure 1-24 Performance of 32 kbps ADPCM and 64 kbps PCM, [24] © 1992 Artech House. Reprinted with permission

1.7.5 Delta Modulation (DM)

DM is a special case of differential PCM. DM is used to encode the difference between the analog source signal and a prediction of it, but only one bit per sample is used to indicate the error. This bit specifies the sign (positive or negative) of the error and so indicates whether the source signal has increased or decreased since the last sample. As with all modulation schemes that encode a difference signal, DM is subject to slope overload, as shown in Figure 1-25. Its main advantage is in the reduction in encoding rate and subsequent more efficient use of a digital transmission facility.

A variety of delta coding methods are used, including adaptive delta modulation (ADM) and continuously variable slope delta modulation (CVSDM). ADM uses a variable step size, whereas CVSDM encodes the slope differences rather than amplitude differences.

DM systems are found in older multiplexers used with digital loop systems. The exact protocols and encoding algorithms usually are proprietary. This leads to compatibility problems when trying to mix equipment types and manufacturers. DM was used in a number of digital loop systems in the 1970s and 1980s but has given way to ADPCM in many systems. ADPCM has been standardized by ANSI and

(a)

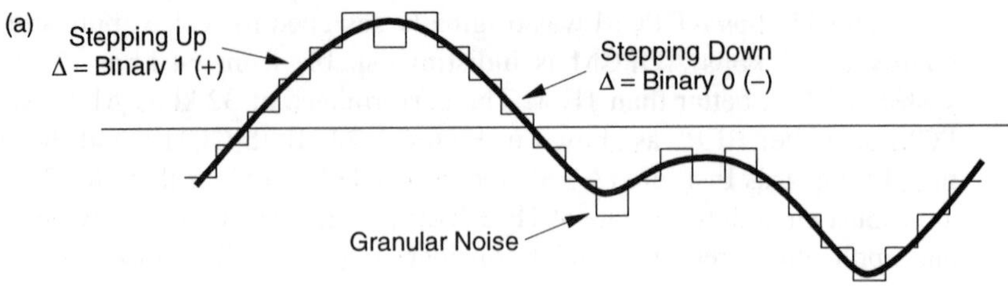

(b)

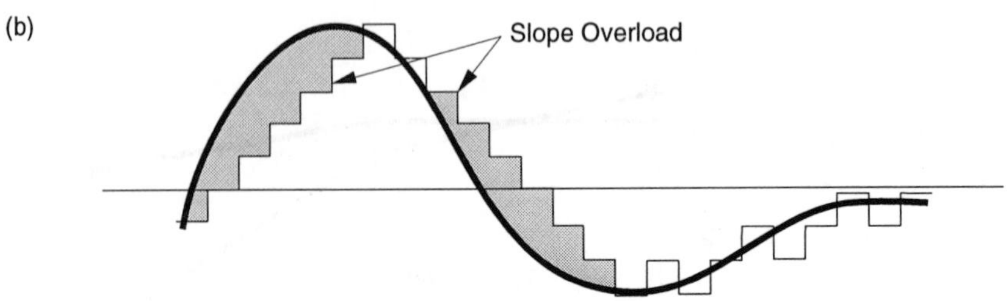

Figure 1-25 Delta Modulation, [5]: (a) DM encoding; (b) DM slope overload
© 1992 John Wiley & Sons. Reprinted with permission

CCITT. The chances of end-to-end equipment compatibility are greater with the ADPCM coding scheme, but specific implementations by different manufacturers still may lead to compatibility problems.

1.7.6 Digital Speech Interpolation (DSI)

The capacity of a digital transmission system that is used exclusively for voice can be doubled by taking advantage of the fact that either direction of transmission is used only about 40% of the time. That is, while one party is talking, the other is listening; also, normal speech contains many natural pauses [25]. A technique called time assignment speech interpolation (TASI) allows the speech power from more than one conversation to be interwoven. The technique was originally applied to intercontinental analog transmission systems using submarine cables. When it is applied to digitized voice, it is called digital speech interpolation (DSI). In this regard, DSI can be considered compression or multiplexing, and the equipment that performs this function is called digital compression multiplex equipment, or DCME.

With DSI, a channel in the transmission system is only connected when the talker is active. Early systems using DSI were not usable for analog data communications except at low speeds, if at all. Modern DCME includes circuitry to detect modem and facsimile tones, demodulate the data signals, and multiplex the data channels into one or more DS-0 rate signals. The far-end reverses the process to regenerate the original signals and deliver them unchanged. Also, the character of voice traffic after it has been compressed using DSI is no longer deterministic but, instead, is statistical and bursty. This has no direct effect on digital loop systems currently in use but can affect the delay and packet loss in asynchronous transfer mode (ATM) systems presently under development [26].

1.7.7 Variable Quantizing Level (VQL)

Variable quantizing level (VQL) is a technique used to encode voice at a 32 kbps rate [27]. It provides a measure of redundancy in the transmitted code word, which makes the technique relatively robust in the face of line errors. However, VQL introduces delay in the order of 6 ms at each encoding point. The added delay is detrimental to transmission quality when echo is present. Therefore, echo cancellers are required in combination with the VQL equipment and, in some implementations, are built-in. VQL algorithms and techniques are not standardized, which precludes mixing different manufacturers' equipment at opposite ends of a circuit. However, the general principles are similar and will be explained here.

In order to achieve a variable quantizing level, the nominal 300 to 3,400 Hz speech signal is first filtered to reduce its high-end response to 3,000 Hz. The speech signal is then sampled at 6,666.67 samples per second using regular PCM encoding techniques. Each block of 40 samples is processed every 6 ms, giving a block processing rate of 166.67 blocks per second. The amplitude of the largest signal in the 40-sample block is encoded into an 8-bit word and then divided into 11 positive and 11 negative equal levels or steps. Each of the 40 samples is then assigned a new 5-bit word, which corresponds to the closest level in the 11 steps (four bits) plus a sign bit.

Every 6 ms, corresponding to the block sampling rate, a code word is formed from a header word, which is formed from the maximum amplitude word (six bits), signaling bits (two bits) and forward error correction bits (four bits), plus the 5-bit words for each of the 40 samples. The resulting code word is 212 bits long, as follows:

$$\text{Maximum amplitude (6 bits)} + \text{signaling (2 bits)} + \text{forward error correction (4 bits)} + 40 \text{ samples} \times 5 \text{ bits/sample} = 212 \text{ bits per block}$$

The forward error correction (FEC) bits protect the maximum amplitude word and two signaling bits, allowing the receiver to detect and correct any single bit error in the header. Since the block processing rate is 166.67 blocks/second and there are 212 bits per block, the overall rate is 35,333.3 bits/second. This exceeds the standard 32 kbps rate required to put two voice channels in a 64 kbps time slot, so further processing efficiency is required.

As previously noted, each sample can assume one of 22 values (five bits). Two such samples taken together can form one of $22 \times 22 = 484$ values. This is equivalent to nine bits, or 4.5 bits per sample (versus five bits for the single sample). This results in a shortened code word, as follows:

$$\text{Maximum amplitude (6 bits)} + \text{signaling (2 bits)} + \text{FEC (4 bits)} + 40 \text{ samples} \times 4.5 \text{ bits/sample} = 192 \text{ bits per block}$$

Since the block rate is 166.67 blocks/second, the overall code word rate is 166.67 blocks/second × 192 bits/block = 32,000 kbps, which is the required rate.

The VQL encoding technique gives an almost constant signal-to-distortion ratio for varying signal levels. As a result, many voiceband modems operating at 9.6 kbps and below may work on a VQL circuit. However, certain modems that require a circuit frequency response up to 3,400 Hz can be adversely affected. In general, successful modem operation above 4.8 kbps is highly dependent on the modulation techniques used.

References

[1] Reeve, W. *Subscriber Loop Signaling and Transmission Handbook: Analog*. IEEE Press, 1992.

[2] *Public Mobile Service*. 47 CFR Part 22, subpart H. Federal Communications Commission, Oct. 1990. Available from USGPO.

[3] Taub, H., Schilling, D. *Principles of Communication Systems,* 2nd Ed. McGraw-Hill Book Co., 1986.

[4] Bingham, J. *The Theory and Practice of Modem Design*. John Wiley & Sons, 1988.

[5] Bellamy, J. *Digital Telephony,* 2nd Ed. John Wiley & Sons, 1991.

[6] Peebles, P. *Digital Communication Systems*. Prentice-Hall, 1987.

[7] Martin, J. *Future Developments in Telecommunications,* 2nd Ed. Prentice-Hall, 1977.

[8] Duncanson, J. "Inverse Multiplexing." IEEE Communications Magazine, Vol. 32, No. 4, April 1994.

[9] Bylanski, P., Ingram, D. *Digital Transmission Systems,* 2nd Ed. IEE Telecommunications Series 4, Peter Peregrinus Ltd., 1980.

[10] *Digital Channel Bank Requirements and Objectives*. AT&T Technical Reference TR43801, Nov. 1982.

[11] *General Aspects of Digital Transmission Systems: Terminal Equipments*. Recommendations G.700–G.772. CCITT Blue Book, Vol. III.4, 1989. Available from National Technical Information Service, Order No. PB89-143887.

[12] Smith, D. *Digital Transmission Systems*. Van Nostrand Reinhold, 1985.

[13] *Transmission Systems for Communications*. Bell Telephone Laboratories Technical Staff, 1982. Available from AT&T Customer Information Center.

[14] "The Transmission of PCM Over Cable." The Lenkurt Demodulator, Vol. 12, No. 1, Jan. 1963.

[15] Personal communication from Len Thomas of Sprint, May 13, 1992.

[16] Kitawaki, N., Nagabuchi, H. "Quality Assessment of Speech Coding and Speech Synthesis Systems." IEEE Communications Magazine, Oct. 1988.

[17] Bartee, T., Editor. *Digital Communications*. Howard W. Sams & Co., 1986.

[18] Jayant, N. "Signal Compression: Technology Targets and Research Direc-

tions." IEEE Journal on Selected Areas in Communications, Vol. 10, No. 5, June 1992.

[19] *Data Communications Over the Telephone Network.* Series V Recommendations, CCITT Blue Book, Vol. VIII.1, 1989. Available from National Technical Information Service.

[20] Chen, J., Atal, B., et al. "A Robust Low-Delay CELP Speech Coder at 16 KB/S." Advances in Speech Coding. Kluwer Academic Publishers, 1991.

[21] American National Standard for Telecommunications. *Digital Processing of Voice-Band Signals—Line Format for 32-kbit/s Adaptive Differential Pulse-Code Modulation (ADPCM).* ANSI T1.302-1989.

[22] American National Standard for Telecommunications. *Digital Processing of Voice-Band Signals—Algorithms for 24-, 32-, and 40-kbit/s Adaptive Differential Pulse-Code Modulation (ADPCM).* ANSI T1.303-1989.

[23] American National Standard for Telecommunications. *Network Performance—Tandem Encoding Limits for 32 kbit/s Adaptive Differential Pulse-Code Modulation (ADPCM).* ANSI T1.501-1988.

[24] Gruber, J., Williams, G. *Transmission Performance of Evolving Telecommunications Networks.* Artech House, 1992.

[25] Keiser, B., Strange, E. *Digital Telephony and Network Integration.* Van Nostrand Reinhold, 1985.

[26] Sriram, K., et al. "Voice Packetization and Compression in Broadband ATM Networks." IEEE Journal on Selected Areas in Communications, Vol. 9, No. 3, April 1991.

[27] Held, G. *Digital Networking and T-Carrier Multiplexing.* John Wiley & Sons, 1990.

2 Digital Loop Applications and Interfaces

Chapter 2 Acronyms

2B1Q	2-binary, 1-quaternary
A/D	analog-to-digital
ADSL	asymmetric digital subscriber line
AMI	alternate mark inversion
ANSI	American National Standards Institute
APS	automatic protection switching
BER	bit error rate
BETRS	basic exchange telecommunications radio service
BORSCHT	SLIC functions: battery feed; overvoltage protection; ringing; signaling; coding; hybrid; test
BRA	basic rate access
BRI	basic rate interface
CCITT	The International Telegraph and Telephone Consultative Committee
CD	compact disc
CFR	Code of Federal Regulations
CI	customer installation
codec	COder-DECoder
COR	central office repeater
COT	central office terminal
CPE	customer premises equipment
CSA	carrier serving area
CSMA/CD	carrier sense multiple access with collision detection
CSU	channel service unit
D/A	digital-to-analog
DCE	data communication equipment
DCS	digital cross-connect system
DFE	decision feedback equalization, -izer
DID	direct inward dialing
DLC	digital loop carrier
DOD	direct outward dialing
DP	data port
DSL	digital subscriber line
DSP	digital signal processing
DSS	digital central office switching system
DSU	data service unit
DTE	data terminal equipment
E	receive signaling
E/O	electrical-to-optical
ECSA	Exchange Carriers Standards Association (Now known as ATIS: Alliance for Telecommunications Industry Solutions.)
EIA	Electronic Industries Association
ESA	emergency stand-alone
fax	facsimile transmission

continued

FCC	Federal Communications Commission
FITL	fiber-in-the-loop
FT1	fractional T1-carrier
FTTC	fiber-to-the-curb
FTTH	fiber-to-the-home
HDSL	high bit-rate digital subscriber line
HDT	host digital terminal
HDTV	high-definition television
HSG	line repeater housing
HTU-C	high bit-rate terminal unit, central office
HTU-R	high bit-rate terminal unit, remote
IDLC	integrated digital loop carrier
ISDN	integrated services digital network
LAN	local area network
LED	light-emitting diode
LSI	large-scale integration
LT	line termination
LTS	line termination shelf
M	transmit signaling
MAN	metropolitan area network
MJU	multipoint junction unit
NI	network interface
NT	network termination
O/E	optical-to-electrical
OCU	office channel unit
ONU	optical network unit
OSS	operational support system
OVP	overvoltage protection
PBX	private branch exchange
PC	personal computer
PCS	personal communications services
PON	passive optical network
POTS	"plain old telephone service"
PRA	primary rate access
PRI	primary rate interface
PSTN	public switched telephone network
PTC	positive temperature coefficient
PTN	public telephone network
RJ	registered jack
RLM	remote line module
RLS	remote line switch
RST	remote switching terminal
RT	remote terminal
SJ	standard jack
SLIC	subscriber line interface circuit
SONET	synchronous optical network
SRDL	subrate digital loop
TCM	time compression multiplexing
TIA	Telecommunications Industry Association
UDLC	universal digital loop carrier
USOC	universal service ordering codes
UVG	universal voicegrade
VCR	video cassette recorder
VLSI	very large scale integration
WAN	wide-area network

2.1 Introduction

This chapter emphasizes digital loop applications that use metallic twisted pairs and optical fibers. Twisted pairs currently are used with digital systems operating at subrates (<64 kbps) up to the DS-1C rate (3.152 Mbps), although some early systems also operated at the DS-2 rate (6.312 Mbps). Special attention will be paid to the four most common twisted pair digital loops: repeatered T1-carrier; subrate digital loops (SRDL), including switched 56 kbps loops; integrated services digital network (ISDN) digital subscriber lines (DSLs); and high bit-rate digital subscriber lines (HDSLs). The asymmetric digital subscriber line (ADSL) is a forthcoming technology and is described briefly.

Optical fibers presently are used in the loop at DS-1 (1.544 Mbps), DS-2, and DS-3 (44.736 Mpbs) rates and in interoffice and long-haul applications at much higher rates. Interoffice and long-haul applications include the synchronous optical network, or SONET, which eventually will be deployed in loop applications as well. The optical fiber loop transmission design procedures are virtually identical at the DS-1, DS-2, and DS-3 loop speeds, the only difference being the optical fiber terminal equipment line interface units (cards), which determine loss budgets and loop lengths. Optical fiber loop speeds are slow compared to speeds used in interoffice, SONET, and long-haul applications and do not come close to approaching optical fiber transmission capacity. The additional design considerations at the higher speeds are beyond the scope of this book.

Digital loop radio systems licensed under the Federal Communications Commission (FCC) rules for basic exchange telecommunication radio service (BETRS) are used in rural areas. Cellular and personal communications services (PCS) are used in urban and suburban areas.* Digital loop technologies based on free space radio propagation are beyond the scope of this book, as well.

2.2 Services

Digital loops are transparent to services that fall within the bit-rate (bandwidth) capabilities of the particular digital loop technology used. Different technologies have been optimized for different transmission media. For example, the HDSL is optimized for twisted pair loops in the carrier serving area (that is, within 12 kilofeet of a central office or remote terminal). The ISDN DSL (also called basic rate interface, or BRI) is optimized for twisted pair loops within 18 kilofeet of the central office or remote terminal, but ISDN bearer channels can be carried as payload within

*BETRS is licensed under 47 CFR Part 22, Subpart H, Rural Radio Service, while cellular is licensed under 47 CFR Part 22, Subpart J, Cellular Radio Service. Regulations for the personal communications services are issued under 47 CFR Part 99, Personal Communications Services, and Part 20, Commercial Mobile Radio Services. CFR is Code of Federal Regulations.

the aggregate of higher bit-rate transmission systems. Optical fiber transmission systems can be substituted for twisted pairs in almost any situation by the proper selection of interfaces.

In a typical application, many lower-speed user data services are combined by multiplexing onto a higher-speed channel for presentation to the network. Where user bandwidth requirements vary greatly over the course of a day or week, it is sometimes advantageous to use inverse multiplexing, as was discussed in Chapter 1. Such a situation exists when only occasional video conferences are needed or data transmission requirements peak for short periods during certain times of the day.

Inverse multiplexing allows higher-speed data services, typically 384 kbps video conferences, to be spread among several lower-speed data channels, usually switched subrate digital loops at 56 kbps. Although the ideal inverse multiplexer allocates bandwidth on a demand or real-time basis, current technological restrictions require network access to be static, at least on a call-by-call basis, when the bandwidth is needed. Usually, a minimum bandwidth level is provided to the user by a dedicated subrate digital loop, and the inverse multiplexer is used to set up additional bandwidth using switched 56 kbps circuits.

Services multiplexed, or inverse multiplexed, and then transported over digital loops can be conveniently categorized by their characteristics, including *type, bit rate* and *bit-rate variability, symmetry, directional characteristics,* and *connection* characteristics. Each category can be further broken down into the attributes shown in Table 2-1.

Voice telephony includes any voicegrade service (nominal 4 kHz analog channel with usable 3.1 kHz bandwidth) such as private lines, regular switched lines ("plain old telephone service," or POTS), coin lines, private branch exchanges (PBXs), key telephone system lines, and voice applications of ISDN. In the present network, voice telephony is bidirectional, constant bit rate except in capacity-limited transmission systems such as satellite and intercontinental submarine coaxial cables. As dynamic bandwidth allocation becomes inherent to the overall network, voice telephony will be transported at variable bit rates whereby speech pauses will be encoded at bit rates lower than speech itself, or speech on one channel will be interwoven with speech pauses on another channel.

Wideband voice (bandwidth above 5 kHz) is a service that gives higher quality than the voice telephony just described. This higher quality is desirable on long voice teleconference calls. The bandwidth is 7 kHz in this case and is available through the ISDN [1]. Other audio services, such as high-fidelity (hi-fi) music on compact disc (CD), require a bandwidth of 20 kHz. Studio audio materials for AM radio stations require a 5 to 8 kHz bandwidth; FM radio and television stations require a 15 kHz bandwidth. The distribution of these services most likely will remain constant bit rate, although compression algorithms offer *CD-like* quality at significantly lower rates than uncompressed hi-fi.

Data services are primarily associated with local area networks (LANs) that are interconnected with metropolitan and wide-area networks (MANs and WANs, respectively) but also include offset networks such as Teletex. The text attribute implies interactive, or bidirectional, data file transfer, such as message transmission and interactive computer forums where at least one of the participants is a human.

Table 2-1 Service Attributes

Service Characteristic	Type	Bit-Rate Variability	Symmetry	Direction	Connection
Attributes of Characteristic	Audio	Constant	Symmetric	Bidirectional	Point-to-point
	• Voice telephony	Variable	Asymmetric	Unidirectional	Point-to-multipoint
	• Wideband				Broadcast
	Data				
	• Text				
	• File				
	Image				
	• Fax (text/line)				
	• Fax (photo/gray scale/color)				
	• Progressive image				
	Video				
	• Freeze frame				
	• Video telephony				
	• Video conference				
	• Full-motion video				
	• HDTV				

File transfer implies asymmetric bidirectional, or even unidirectional, data file transfer between computers.

Facsimile (*fax*) transmissions fall into the *image* category. This includes relatively low-resolution text and line drawings as well as higher-resolution gray scale and color photographs. Also included in the image category are *progressive* images, which involve an initial coding at a low bit rate to allow rapid image (picture) access followed, if needed, by additional stages of transmission for higher quality. This often is called *telebrowsing* [1]. All facsimile transmissions are conducive to compression to achieve lower transmission times.

The *video* category includes *freeze frame*, which is characterized by a relatively low temporal resolution of one frame per second up to approximately six frames per second. Another name for this service is video phone or image phone. *Full-motion video* includes varying qualities or resolutions. At the lower end, the quality is equal to that obtained during the playback of a video tape on a consumer video-cassette recorder (VCR). At the upper end, the quality is equal to that obtained in a commercial television studio. Between freeze frame and full-motion video lies *video telephony*, characterized by low sharpness and low levels of motion activity, such as would be obtained in the head-and-shoulders view of one person. The term video telephony is used to differentiate this type of service from *video conference*, which is characterized by almost full-motion quality needed with groups of people. Finally, at the high end of the quality spectrum is high-definition television, or HDTV, which requires extremely high bandwidths if left uncompressed.

It is possible for the bit rate used with a given service to be constant or variable. A typical voice circuit connection across a conventional digital loop carrier (DLC) system is an example of a constant bit-rate attribute. A variable bit-rate attribute is assigned to a service that uses a different bit rate for each call or changes the bit rate dynamically during a call. The dynamic bit rate is used in some radio-based personal communication services presently under development.

Symmetry defines the directional characteristics of the bit rate such that the rate of transmission in each direction is separately specified. A symmetric attribute is assigned to any bidirectional service that has the same bit rate in both directions. Some services have a higher bit rate in one direction than the other, and this service is assigned an asymmetric attribute. A unidirectional service is a limiting case of an asymmetric service.

Connection characteristics describe the multitude of users that are interconnected by a service. A point-to-multipoint attribute is assigned when the connection is from one user to a predetermined or restricted group of other users. The point-to-point connection is a limiting case in that the group is of size 1. A broadcast attribute is another limiting case. Here, the user group is very large and may not be restricted in any way.

Current service applications are largely based on a constant bit-rate, bidirectional, point-to-point, circuit switched network. However, both public and private networks are rapidly evolving to a general information transport system having variable bit-rate, on-demand capabilities using both packet and circuit switching. Not all services require symmetric operation, and point-to-multipoint and broadcast capabilities are becoming commonplace.

Table 2-2 Examples of Service Characteristics

Service	To Subscriber Bit Rate	From Subscriber Bit Rate	Symmetry	Connection	Direction
Voice telephony	2.4 kbps to 64 kbps	2.4 kbps to 64 kbps	Symmetric	Point-to-point	Bidirectional
Hi-fi distribution	16 kbps to 128 kbps	None	Asymmetric	Point-to-point	Unidirectional
Teletex	300 bps to 64 kbps	300 bps to 64 kbps	Asymmetric per call	Point-to-point	Bidirectional
Interactive data	300 bps to 64 kbps	300 bps to 64 kbps	Symmetric	Point-to-point or multipoint	Bidirectional
File transfer	300 bps to 1.5 Mbps	300 bps to 1.5 Mbps	Asymmetric per call	Point-to-point	Bidirectional
Facsimile	9.6 kbps to 64 kbps	None	Asymmetric	Point-to-point	Unidirectional
Freeze frame video	9.6 kbps	9.6 kbps	Symmetric	Point-to-point	Bidirectional
Video telephony	64 kbps to 1.5 Mbps	64 kbps to 1.5 Mbps	Symmetric	Point-to-point	Bidirectional
HDTV	20 Mbps	None	Asymmetric	Broadcast	Unidirectional
Multimedia	300 bps to 140 Mbps	300 bps to 64 kbps	Asymmetric per call	Point-to-point	Bidirectional

Hi-fi = high fidelity.

In context of these definitions, Table 2-2 compares the attributes of a number of more specific service categories. This tabulation is representative but not exhaustive. It is interesting to note that practically all the source signals are analog, which are digitally encoded at the originating point for transport and then decoded at the terminating point. Various compression methods are used to achieve desired signal qualities within the constraints of the available transport bandwidth.

Not specifically identified in Table 2-1, but listed in Table 2-2, are multimedia services. Multimedia is a combination of virtually every other service in a pick-and-choose manner. That is, the user has access to a wide variety of services and is allocated the bandwidth necessary for the particular session. Another name for multimedia is *motion videotex*. It is clear from comparing this table to the digital bandwidth provided by current digital loops that the full range of services is not currently supported by the loop plant; however, this will change with advancing technology.

Specific implementations are available for the various service types to meet individual requirements. For example, the generic "voice telephony" includes a large number of specific interfaces and implementations such as switched-message POTS, coin services, centrex, and a multitude of voicegrade private line services. The generic "interactive data" include switched and private line subrate digital data, ISDN, and DS-1 rate services. Many of these implementations are discussed later in this chapter.

It is estimated that approximately 70% of the access lines in service in the United States serve residential establishments [2]. From a telecommunication infrastructure point of view, this represents a major investment and major national resource. The telecommunications industry is experiencing the demand for advanced technology and advanced services to the home. This is one of the fundamental requirements behind the migration of the metallic twisted pair loop to the fiber optic loop as service transport mechanisms.

2.3 Subscriber Line Interface Circuit

The subscriber line interface circuit, or SLIC, provides the interface between the analog loop and a digital central office switching system (DSS), remote switching terminal (RST), or DLC remote terminal. The DLC and RST are discussed in the following sections. At the DSS, the SLIC is a plug-in printed circuit card having one to twelve individual line circuits per card (depending on the manufacturer).

While not strictly a digital loop interface, the SLIC provides the necessary conversion between the analog services discussed in the previous section and the digital loop transport systems on which this book focuses. Therefore, this section will cover the SLIC so the reader may obtain a more complete knowledge of transport of analog services on digital loop systems.

The functions of the SLIC, colloquially known as BORSCHT, are categorized as follows:

- Battery feed (B)
- Overvoltage protection (O)
- Ringing (R)
- Signaling (S)
- Coding (C)
- Hybrid (H)
- Test (T)

Each of these will be described in the following paragraphs (Figure 2-1).

The battery feed functions supply dc to the loop through balanced feed resistances or current sources. All modern central office and digital loop carrier SLICs use current limiting techniques to maintain no more than 45 to 75 mA of loop current in a short-circuited or low-resistance loop. Higher loop resistances result in lower loop currents. The relationship between loop resistance and loop current is nonlinear. When SLICs are deployed in pair gain systems, the loop current usually is no more than 20 to 25 mA. This low current severely limits the distance between the remote terminal and the user's telephone set. In fact, telephone sets with light-emitting diode (LED) line status indicators will not work with only 20 to 25 mA available powering current. In central office and DLC SLICs, the open-circuit loop powering voltage is a nominal −48 Vdc, while in pair gain systems, the open-circuit voltage is 10 to 14 Vdc.

Overvoltage protection (OVP) is required in virtually every SLIC connected to outside plant cable pairs. The OVP limits or isolates foreign voltages on the cable pair from central office equipment or remote terminal equipment. Foreign voltages may be caused by lightning (either direct or nearby strikes), power cross, powerline induction, or other sources of electromagnetic interference. The OVP implementation in SLICs is classified as secondary protection (as distinguished from primary protection devices on central office main distribution frames and subscriber premises). Usually OVP in SLICs limits the voltages to approximately 70 V and handles surge currents of a few tens of amps.

All SLICs have some means of limiting surge currents. These typically are resistor fuses, positive temperature coefficient (PTC) resistors, or combination fuses and resistors. Any series element in the loop, including current-limiting devices, must be very well balanced to limit conversion of longitudinal noise currents to metallic noise currents.

The ringing function of the SLIC applies ringing voltage to the loop through a relay, crosspoint switch or silicon-controlled rectifiers. Also included in the ringing function is answer detection to initiate ring-trip action (cessation of ringing). The ringing voltage generator normally is external to the SLIC. The ringing cadence, or ringing and silent intervals, may be controlled internally by the SLIC or externally.

Coding is the conversion of analog signals to digital (A/D) and digital signals to analog (D/A). These are the functions of the codec, or COder-DECoder. Codec operation was described in Chapter 1.

The hybrid converts the 2-wire, bidirectional loop to a 4-wire path with separate, unidirectional transmit and receive circuits. The transmit and receive paths in

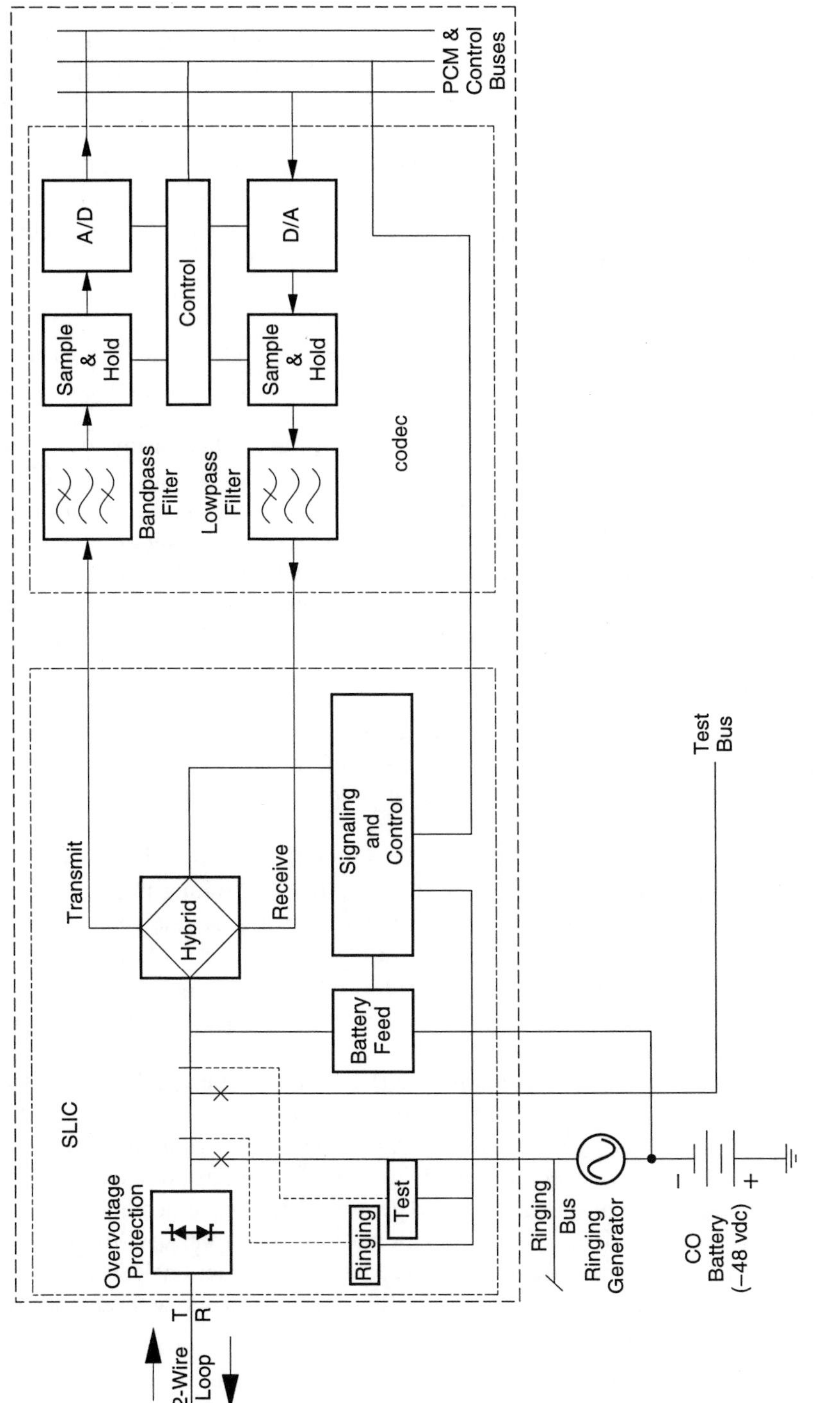

Figure 2-1 Subscriber line interface circuit

all digital switching and transmission systems are inherently separate, and this commonly is called 4-wire. The hybrid also provides impedance matching to the 2-wire loop, which acts to control echo. The hybrid can be implemented electronically using digital signal processing (DSP) techniques. Frequently, hybrids include both a hybrid balance technique and an echo canceller to control echo.

The final BORSCHT function is test. The test function provides access to the loop and SLIC from an external test bus. This access usually is through relays or crosspoint switches. The test access also can be used to provide line circuit sparing in central office switching systems.

2.4 DLC

The implementation of service transport mechanisms can take many forms, including DLC systems. The two fundamental types of DLC are the universal digital loop carrier (UDLC) and integrated digital loop carrier (IDLC). These are illustrated conceptually in Figure 2-2. The UDLC in Figure 2-2(a) uses analog or low-speed digital channel units at both the central office terminal and remote terminal and, therefore, will work with any type of central office switching system or other end-office interface depending on the application. With UDLC, switched services are connected to the switching system analog interfaces such as line and trunk circuits. Nonswitched services are connected to their respective nonswitched analog interfaces. Only one industry standard exists for the UDLC [3]. Many manufacturers build UDLC equipment both conforming and not conforming to the standard.

The IDLC in Figure 2-2(b) uses an integrated terminal in the central office switching system so A/D and D/A conversion is not required at the switch interface. The individual switched lines are converted to a digital format in the end-office line circuit, multiplexed with other lines or traffic from incoming trunks, switched and presented to the IDLC interface at a standardized digital rate (presently the DS-1 rate). Switched services use the end-office switching network while nonswitched services pass through to a separate interface. The IDLC gives significant savings in equipment and power and provides better performance from both user and operational support system perspectives.

Two industry standards exist for the IDLC. The earliest, called by its simplified abbreviation *TR8*, refers to a BELLCORE technical reference that specifies the SLC®-96 interface, which is a 96-channel subscriber loop carrier system designed by AT&T* [4]. The most current standard, abbreviated *TR303*, refers to another BELLCORE technical reference that specifies a more generic IDLC implementation [5]. TR8 does not specify extensive interfaces to operational support systems, whereas TR303 does. Many details in both standards are loosely defined or optional. This means a particular manufacturer's remote terminal (RT) may not work with another manufacturer's central office switching system.

---

*SLC®-96 is a trademark of AT&T.

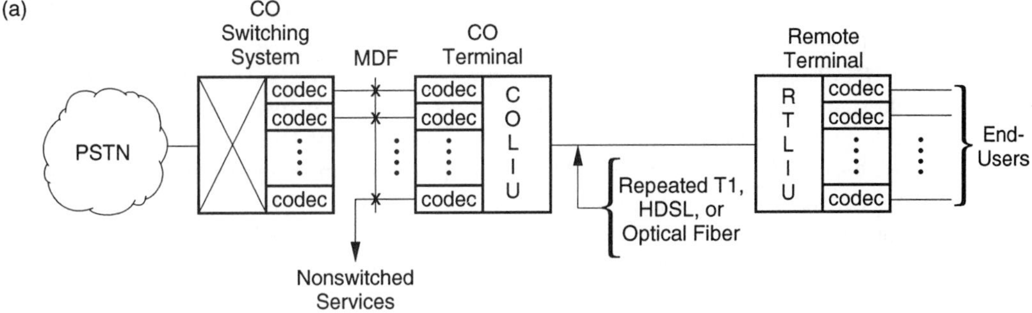

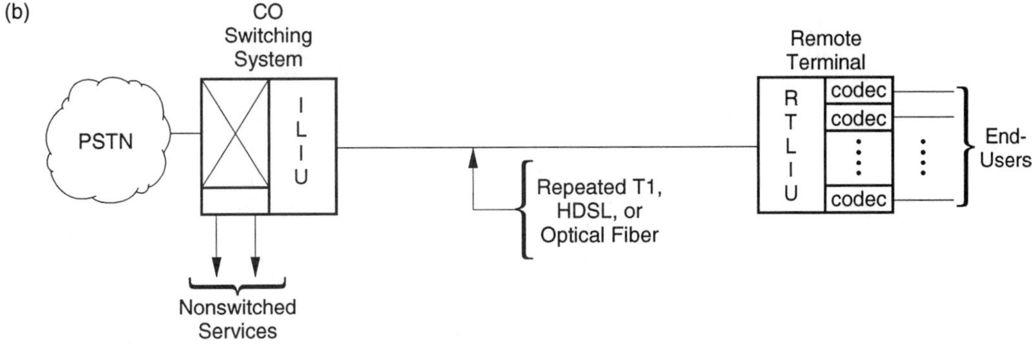

Key

PSTN: Public Switched Telephoned Network
COLIU: Central Office Line Interface Unit
RTLIU: Remote Terminal Line Interface Unit
ILIU: Integrated Line Interface Unit
codec: Coder-Decoder

Figure 2-2 Universal and integrated DLC: (a) UDLC; (b) IDLC

In many situations the IDLC equipment required to support the various services is manufacturer dependent. That is, a particular hardware or equipment set will use proprietary interfaces, protocols, or transmission methods—a practice that precludes mixing equipment from different manufacturers. All modern digital transmission equipment depends, to some degree, on software or firmware, which can lead to incompatibilities even among one manufacturer's equipment set. For example, equipment built with one firmware version may not be compatible with equipment from the same manufacturer that has another firmware version. Standards help to minimize but do not always eliminate these types of problems in complex systems.

DLC RTs can be connected to the serving central office by repeatered T1-carrier, HDSL, or optical fibers. In addition to POTS-type services, the RTs provide digital connectivity for subrate digital loops and the ISDN DSL, as well. A limited form of DLC, called a digital pair gain device, can use an ISDN DSL-like loop transmission scheme, although the pair gain devices are not actually connected to the ISDN.

DLC systems traditionally have been used to relieve twisted pair feeder cables in the outside plant. For example, early DLC systems substituted for coarse-gauge feeder cables on long, rural routes. As the economics of DLC improved, they were employed closer to the central office, typically where congested right-of-way or other factors made DLC an economical alternative to new conduit systems or new cable installations.

Improvements in mass production techniques and large-scale integrated (LSI) circuits decreased the cost of DLC over a relatively short time after their introduction. This, coupled with expanded service capabilities, made DLC more attractive in the urban and suburban environments. Initially, individual DLC systems were installed adjacent to serving area interfaces or feeder control points and provided derived cable pairs for a given distribution area. The advance of high-rate digital loop transmission facilities (for example, at the DS-3 rate) has allowed multiple DLC systems to be installed at a site to serve several distribution areas.

User requirements for digital loops at subrates and higher, and the greatly increased capabilities of modern DLC systems, require a systematic approach to DLC deployment. The deployment area of DLC commonly is called a *carrier serving area* (CSA). A CSA "is a distinct geographic area capable of being served by one theoretical DLC Remote Terminal site" [3]. Other terms, such as *digital serving area* and *electronic serving area*, are given to the area within reach of digital services. These areas are not restricted to DLC but also apply to the area within reach of the digital loop equipment installed in the central office.

At the time the CSA was conceived, the highest-rate digital loop operated at 56 kbps. Therefore, a CSA sectionalizes cable routes such that all customers along any such route can be provided with 56 kbps service without using repeaters. It is not necessary to initially install DLC systems to serve all conceivable CSAs. However, as feeder relief becomes necessary, and DLC is economical, the CSA is activated. In areas where the CSA is properly planned, activation is nothing more than administrative sectionalization of the outside plant.

The typical DLC system can be operated in either a concentrated or nonconcentrated mode. In the concentrated mode, the DLC RT is connected to more users than there are channels in the transmission link connecting the RT to the end-office switching system, as shown in Figure 2-3(a). A typical system can serve 192 users with as little as one span line of 24 channels. Such a system provides a concentration ratio of $192/24 = 8$. Another currently available system serves 48 users with 24 channels, for a ratio of $48/24 = 2$. Low-traffic areas can be served quite efficiently in the concentrated mode.

In the concentrated mode, channels are assigned to callers on a demand basis and only for the duration of the call. Concentration implies blocking; that is, if there is a large enough simultaneous demand for the limited channels between the RT and host office, some users will be denied service or will receive delayed service (for example, dialtone delay). The capacity of a system depends on the acceptable probability of blocking in the host-RT channels, typically 0.1 to 0.5%. Traffic design is covered in [6].

(a)

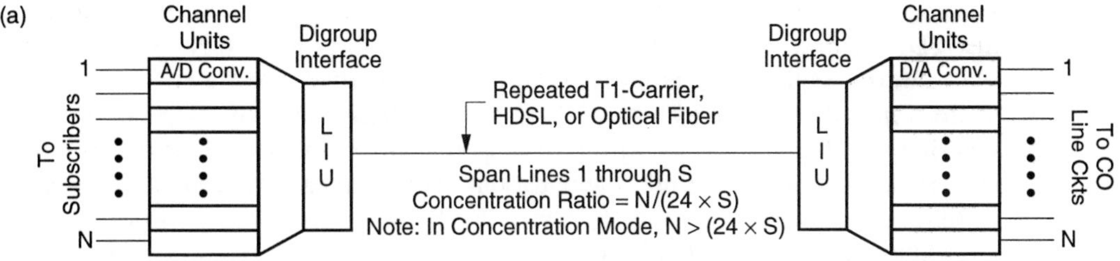

(b)

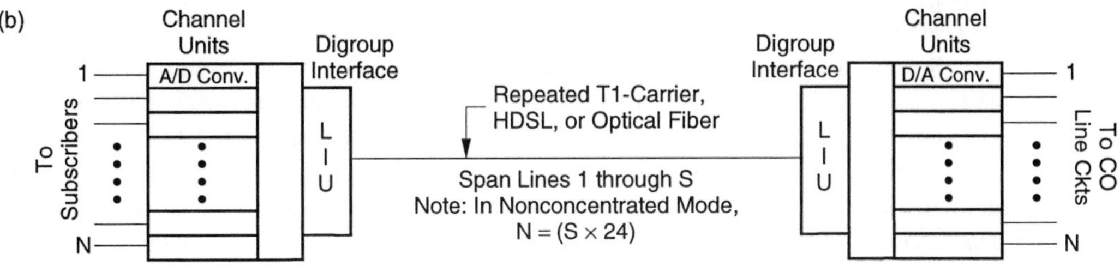

Figure 2-3 DLC operational modes: (a) Concentrated; (b) Nonconcentrated

To increase the traffic handling capacity, some concentrated systems have intra-RT calling capability. Without this feature, two channels would be tied up during a call from one user in an RT to another user in that same RT. With intra-RT calling, the host office detects that an originating call at the RT needs to terminate in the same RT. The host then signals the RT to establish a local link, thus freeing the host-RT channels that otherwise would be used.

The apparent increase in traffic handling capability between the host and RT depends on the intra-RT calling rate. This is difficult to predict unless specific traffic measurements are available, which seldom is the case. When no other information is available, a 5 to 10% intra-RT calling rate can be used. Where a high community of interest is known to exist between users connected to a concentrating DLC, a 35 to 40% intra-RT calling rate may be more appropriate. If the intra-RT calling rate is overestimated, the traffic carrying requirements of the host-RT link will be underestimated, which will result in congestion in the host-RT link.

In the nonconcentrated mode, shown in Figure 2-3(b), each user is preassigned a channel on the span line and is dedicated to it. Preassignment is necessary if the traffic level is high, if the user requires a full-time circuit, or if a special nonswitched service is to be provided. DLC systems using the nonconcentrated mode are inherently less expensive than systems with concentration capabilities, but more span lines are required. As is the case with most telecommunication decisions, the choice of modes and equipment is based on engineering analysis of current and future requirements.

To provide redundancy in the transmission link, a minimum of two span lines is usually provided between the RT and host office, even though the second span line may not be needed on a traffic basis. In this case, the second line is a spare (also called a *protection span*). It is normal practice to provide 1:N sparing, where one spare (protection) line is provided to protect N active lines. In typical systems, N varies from 1 to 14 (or as high as 28 in some installations). For example, systems requiring very high reliability will have 1:1 protection; that is, each active span will have a dedicated spare. In other applications, one spare will be provided for up to 28 active span lines.

All modern DLC systems provide a wide range of circuit types or channel units that are equivalent to circuit types available in central office switching systems. As would be expected, these circuit implementations generally are matched to specific service types. Some circuit implementations are meant to operate only within the CSA environment, whereas others can extend beyond the CSA.

Some DLC systems have a "universal voicegrade" (UVG) channel unit. The UVG channel unit places many functions of different service implementations on one physical channel unit. A highly universal and functional UVG channel unit will not require a field visit by a technician to the RT to change from one service implementation to another (for example, from a direct outward dialing (DOD) trunk to a direct inward dialing (DID) trunk or to a coin line). Instead, the change is made through software from a centralized service center. The labor savings are meant to offset the higher cost of the UVG. There is a tradeoff, however, between functionality and complexity (and, therefore, cost) of the channel unit, and a *single* UVG channel unit probably cannot be built that will serve the very large number of interface functions required by end-users and the telecommunication industry alike.

2.5 Digital Pair Gain Devices

A limited form of DLC is the digital pair gain device, or digital-added main line. These lines use a loop transmission scheme identical to the ISDN DSL U-interface (discussed later in this chapter), although these devices are not connected to the ISDN. A typical application is shown in Figure 2-4. The digital pair gain devices require a central office line circuit for each line and do not provide integrated DLC interfaces.

In a typical system, up to 12 central office terminal (COT) cards may be installed in an equipment shelf. At least two commercially available systems use cards that are mechanically interchangeable with analog subscriber carrier cards. The shelf derives its power from the central office dc power system. Each card provides two digital loops on a single 2-wire cable pair. The cable pair carries two 64 kbps B-channels and one 16 kbps D-channel. The RT derives its power through the loop from the COT. A block diagram showing the major components is given in Figure 2-5.

The integrated circuit implementation of the U-interface logic is very similar to the ISDN DSL U-interface. The analog interfaces are very similar to the SLIC previously described. The signaling and control functions in at least one commercial

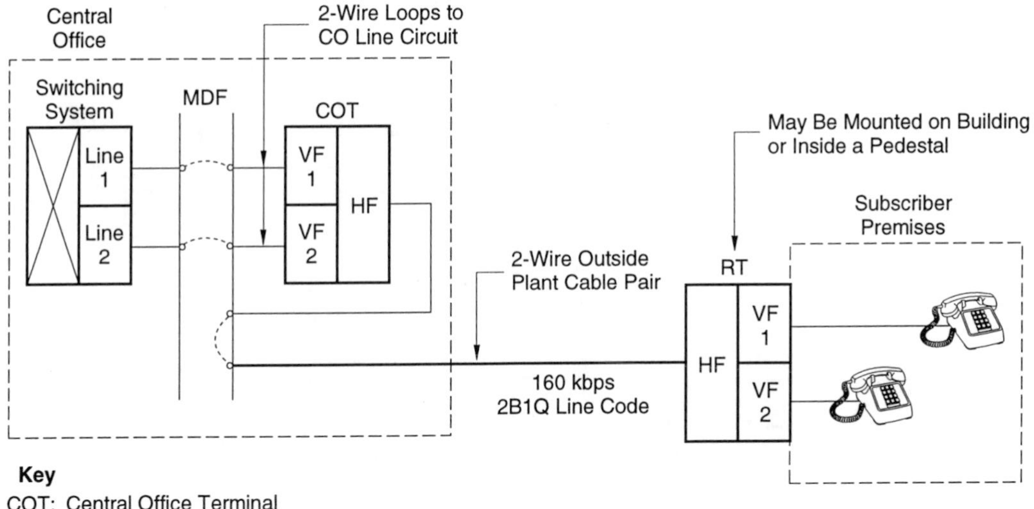

Key
COT: Central Office Terminal
RT: Remote Terminal
MDF: Main Distributing Frame
HF: High Frequency Interface
VF: Voice Frequency Interface

Figure 2-4 Digital pair gain device application

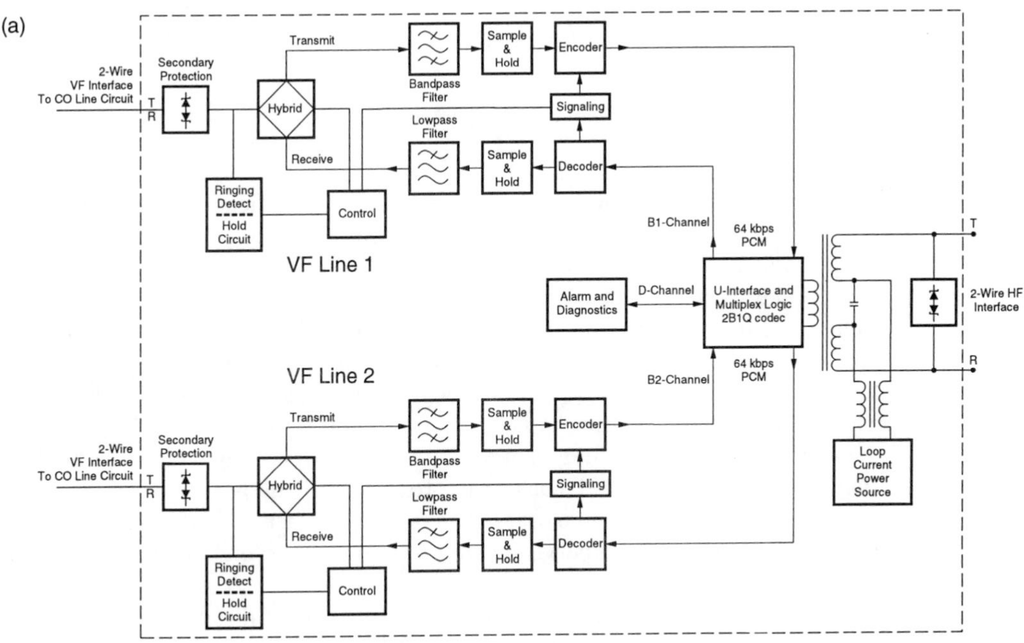

Figure 2-5 Digital pair gain block diagram: (a) COT; (b) RT

(b)

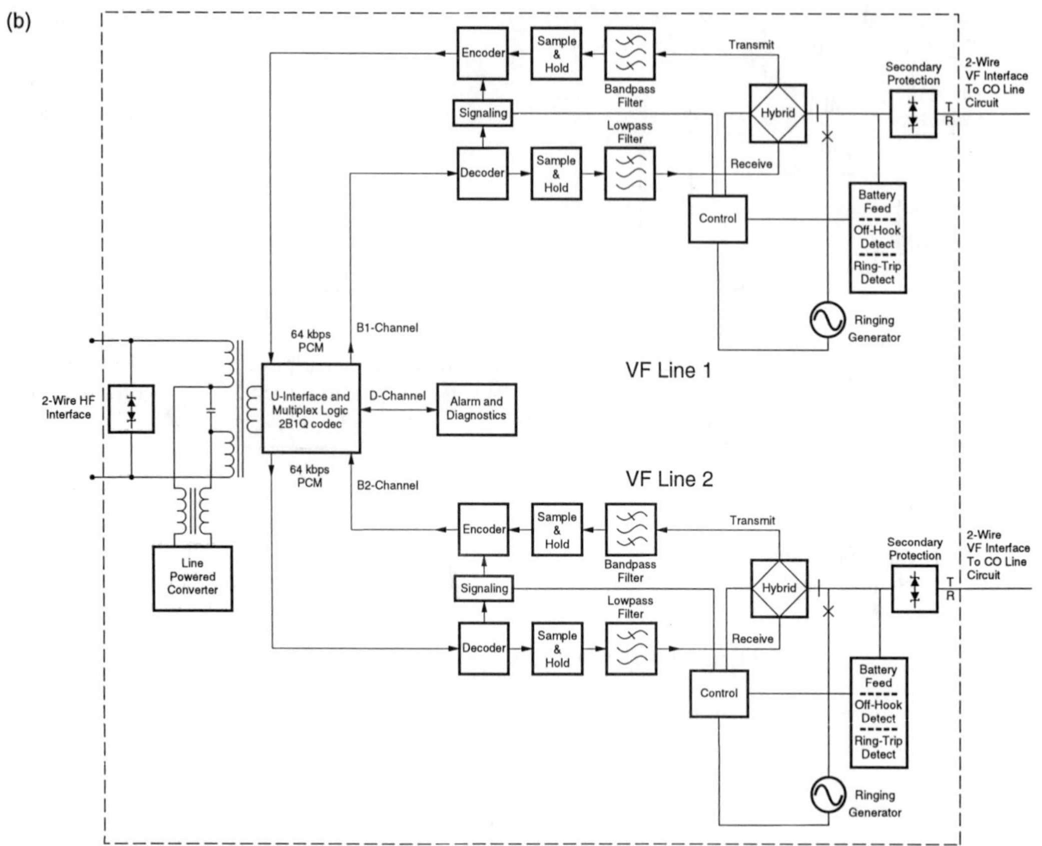

Figure 2-5 Continued

implementation of the digital pair gain device differ significantly from the ISDN DSL in that ringing detection, ring-trip, ring forward, off-hook detection, and other signaling are carried with the 64 kbps bit stream for each channel using robbed bit signaling. Other systems transport this signaling in the D-channel.

On a call from the network to a subscriber served by the digital pair gain system, the central office terminal shown in Figure 2-5(a) detects ringing voltage applied by the central office line circuit. The ringing is converted to a signaling bit in the appropriate channel and sent to the RT shown in Figure 2-5(b). The signaling bit is decoded and activates the ringing generator in the corresponding channel at the RT. Off-hook, or answer, at the RT causes ring-trip. The off-hook is encoded as a signaling bit and sent back to the central office terminal, where it is decoded and used to provide off-hook answer supervision (in the form of a hold-circuit) to the central office line circuit.

On a call from a digital pair gain RT channel unit, the off-hook seizure is detected and sent back to the appropriate channel in the central office terminal. The off-hook signaling bit is decoded and closes a hold-circuit across the central office line circuit. At this point, the digital pair gain channel is transparent and allows transmission of call progress tones (such as dialtone and ringback tones) and addressing. If dial pulse addressing is used, the central office terminal reflects the loop disconnect signals detected at the RT. Upon call completion, the on-hook at the RT is reflected back to the central office terminal to return it to the idle state.

2.6 RSTs

RSTs are extensions of a host digital switching system network and can be considered a special form of DLC. However, there are two significant differences between DLC and RSTs (also called RLM for remote line module or RLS for remote line switch). First, the RST actually extends the host switching network to a remote location, whereas the DLC is more of an appendage to the network. Second, all RSTs optionally provide some form of local switching in case the host-RST links fail. This is sometimes called emergency stand-alone (ESA) capability and is a more advanced feature than the intra-RT calling feature used in some concentrated DLC systems. With the stand-alone feature, the RST provides digit receivers and software configured translations to process calls within the area served by the RST. Special line features that require host processing are lost during the host-RST link failure. The ESA feature is an especially attractive feature where the RST serves a suburban community or isolated pockets of subscribers with a high community of interest.

All RSTs provide some measure of traffic concentration, and all RSTs connect to the host switching system via DS-1 rate transmission links. The DS-1 links may be repeatered T1-carrier span lines, HDSL, or may be embedded in a higher-speed optical fiber link, say 45 Mbps. The fiber optic terminal may be mounted in the RST enclosure or cabinet.

Concentration is illustrated in Figure 2-6, where there are fewer channels between the host switching system and the RST than there are users connected to the

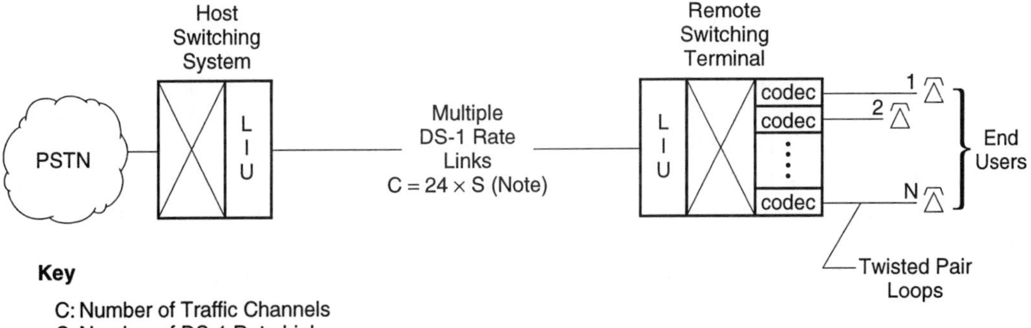

Figure 2-6 RST

RST. Concentration ratios as high as 10:1 are common, although the maximum ratio depends on the average traffic per line, as for the DLC previously described. Some RSTs use one of the 24 channels in each of the first two host-RST digital groups for control. This means that traffic calculations are based on 23 channels rather than 24 for each of the first two links. Subsequent links usually have the full 24 available for traffic. RST traffic design is discussed in [6].

2.7 SRDLs

Some digital loop systems provide a direct interface for data terminal equipment (DTE), including mainframe or minicomputers, personal computers (PCs), and automatic teller machines. This is a characteristic of the SRDLs commonly called Digital Data System (also called DATAPHONE® Digital Service by AT&T). The SRDL primarily is used as a *synchronous* telecommunications link between two computers or between computers and data terminals. SRDLs were first offered by AT&T in the mid-1970s. The system is characterized by high availability and low error rates. This high performance and resulting high cost comes through redundant backbone facilities and, in some implementations, the use of excess bandwidth to allow error detection and correction.

A typical application is shown in Figure 2-7. Only a brief description is given; the individual components are explained in greater detail later in this section. In the figure, a minicomputer is connected to the individual *asynchronous* channels of a statistical multiplexer (*stat-mux*) by EIA/TIA-232E (commonly called RS232) standard interfaces. Various standard rates of 75 bps to 19.2 kbps are used as shown. Although a stat-mux is shown, any type of multiplexer with the proper interfaces can be used. The mainframe is connected to the synchronous channels of the stat-mux at rates from 2.4 to 64 kbps by The International Telegraph and Telephone Consultative Committee (CCITT) V.35 interfaces. The composite signal from the stat-mux is connected through a V.35 interface to the SRDL channel service unit/ data service unit (CSU/DSU) on the user premises.

The CSU/DSU is connected to the SRDL at a rate that is equal to or greater than the composite input. The composite data rate is transmitted to the office channel unit (OCU) in the serving central office and through the subrate digital network. The network is transparent at this point. At the other end, similar equipment connects to individual workstations or PCs. The connection is full-duplex and completely synchronous from stat-mux to stat-mux. Certain channels of the stat-mux are shown as synchronous, which means they are synchronized with the subrate digital network.

SRDLs provide data payload transmission at the rates shown in Table 2-3. This table also shows the loop transmission rate, which includes payload data plus overhead (framing and shared control/secondary channel data). The service may be configured as one-way or two-way (half-duplex or full-duplex), with multipoint and secondary channel capabilities.

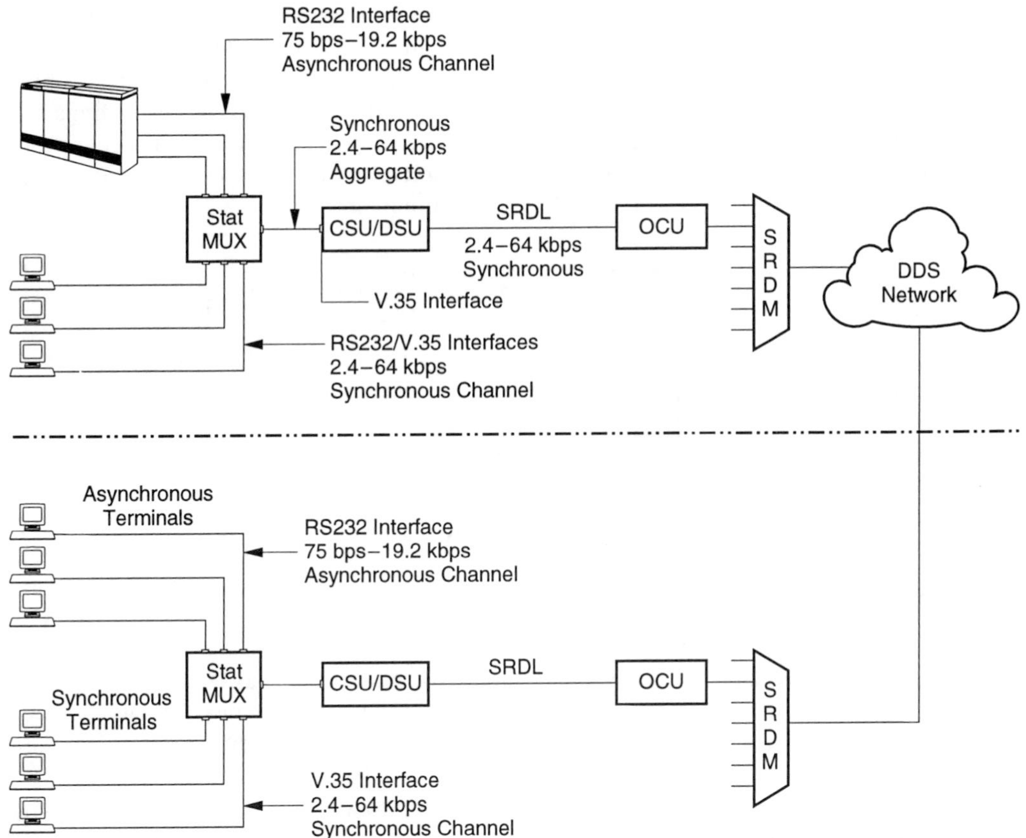

Figure 2-7 Application of the SRDL

Table 2-3 Digital Data Service Transmission Rates

Payload Rate (kbps)	Secondary Channel?	Overhead Rate (kbps)	Line Rate (kbps)
2.4	No	0	2.4
2.4	Yes	0.8	3.2
4.8	No	0	4.8
4.8	Yes	1.6	6.4
9.6	No	0	9.6
9.6	Yes	3.2	12.8
56.0	No	0	56.0
56.0	Yes	16.0	72.0
64.0	No	8.0	72.0

Half-duplex and full-duplex operation are illustrated in Figure 2-8. A half-duplex circuit, shown in Figure 2-8(a), operates in a two-way mode, but transmissions are made in only one direction at any given time. At time 1, the circuit transmits data from left to right while the other direction is idle. At time 2, the circuit transmits from right to left while the other direction is idle. This is similar to the familiar push-to-talk radio. Figure 2-8(b) shows a full-duplex circuit. This circuit arrangement allows data transmission in both directions simultaneously.

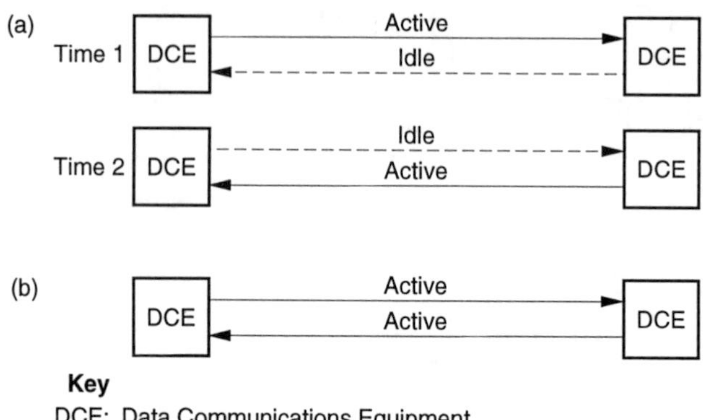

Figure 2-8 Half-duplex and full-duplex transmission: (a) Half-duplex. Transmission in only one direction at a given time; (b) Full-duplex. Transmission in both directions at given time

Multipoint service, also called point-to-multipoint service, is illustrated in Figure 2-9. Multipoint services use multipoint junction units (MJUs) at bridging locations to provide a transmission path between one point (the *control point*, or *master station*) and N other points (*remote points*, or *outlying stations*). The most rudimentary form of multipoint service is where N = 1, which is a point-to-point circuit.

Where N ≥ 2, data are broadcast from the control point to N remote points. This is straightforward and does not require special data signal protocols between circuit points. Simple address polling is used to ensure that the appropriate remote point recognizes data traffic destined for it. This polling also allows the control point to receive data from remote points, which would not otherwise be possible due to mutual interference of incoming data streams.

With polling, a particular remote point will not transmit until it receives an order from the control point. This order is provided through a unique polling address for each remote point. Since each remote point transmits in turn, there is no adverse intermingling of the serial data streams at the MJU or control point, and the data streams can be multiplexed properly and delivered. This scheme precludes direct data transmission between remote points. Therefore, any data to be delivered from one remote point to another must first be delivered to the control station, where it is rebroadcast. All points in a multipoint circuit must operate at the same data rate.

Secondary channel capabilities are another feature of the SRDL. A secondary data channel is provided that is separate from but in parallel with the primary or

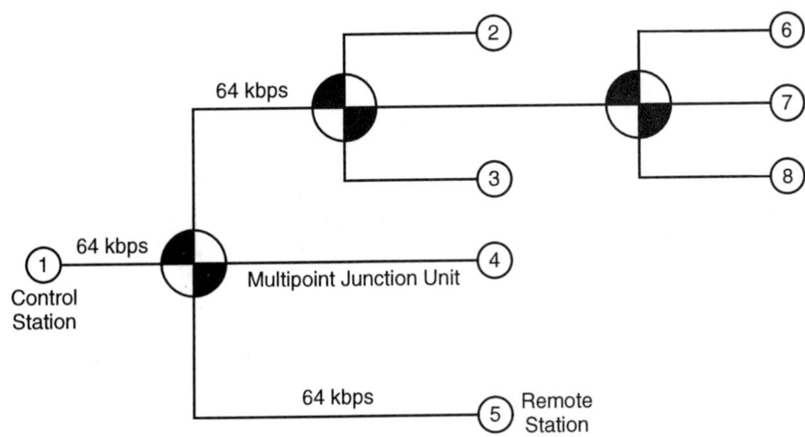

Figure 2-9 Typical SRDL multipoint circuit

payload channel. In this context, channel means separate data bits and not separate paths. In the SRDL, the secondary channel bits are shared with the network control bits. The secondary channel feature is not always required by the user. If it is used at all, the secondary channel is used for controlling the *user's* data network (as opposed to the subrate digital network) or for passing performance information between user points. The signaling format for the primary and secondary channels is covered in Chapter 3.

At the customer premises, the digital loop interface is provided by a DSU or CSU. The digital loop is terminated at the central office with an OCU. Loopback capabilities are provided at key locations at the loop interfaces for troubleshooting. The general configurations are shown in Figure 2-10. The OCUDP, or OCU data port, shown in this figure is a channel unit card used in T1-carrier channel banks. The OCUDP actually is a more typical application in central offices than stand-alone OCUs. All interface units (DSU, CSU, OCU) terminate the digital loop in a similar manner and provide loopback capability, but the specific functions differ. The SRDL interface is 4-wire with separate transmit and receive directions.

The OCU, illustrated in Figure 2-11, is the most complicated interface device. The basic function of the OCU is to convert the digital loop line code to the 64 kbps (DS-0) intra-office line code. It includes digital loop termination and intra-office loop termination as well as a means of clocking and synchronizing the Customer Premises Equipment (CPE) at the other end of the loop. The OCU is connected to the central office clock, which is highly accurate and traceable to a stratum 1 Primary Reference Source. This requires synchronous timing circuits, a rate conversion (or matching) buffer, and a retiming buffer, as shown. The incoming rate conversion buffer repeats the subrate input as necessary to obtain a 64 kbps DS-0A rate. For example, a 9.6 kbps subrate is repeated five times. The outgoing buffer strips repetitive bits to obtain the subrate output. A voting scheme provides error correction. Ancillary functions in the OCU are far-end loopback control, near-end loopback, and sealing current.

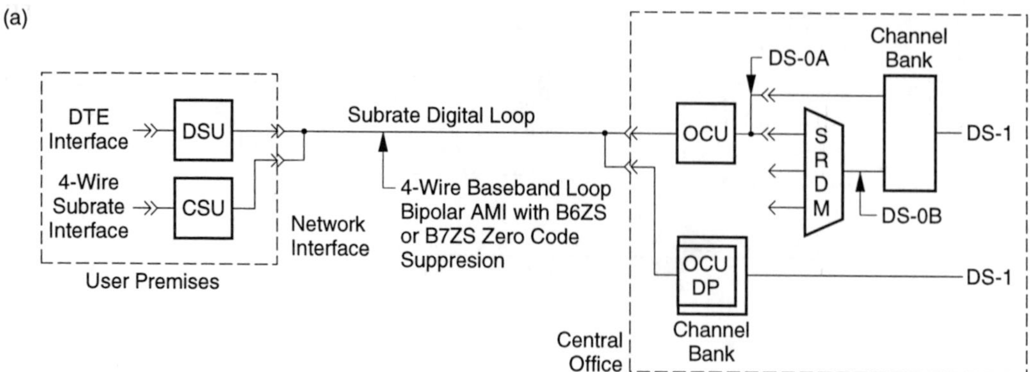

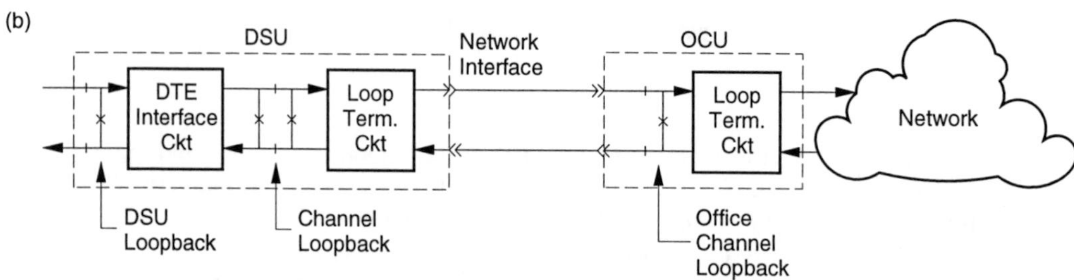

Figure 2-10 SRDL interfaces and loopbacks: (a) SRDL interfaces; (b) Digital loopbacks used with the SRDL

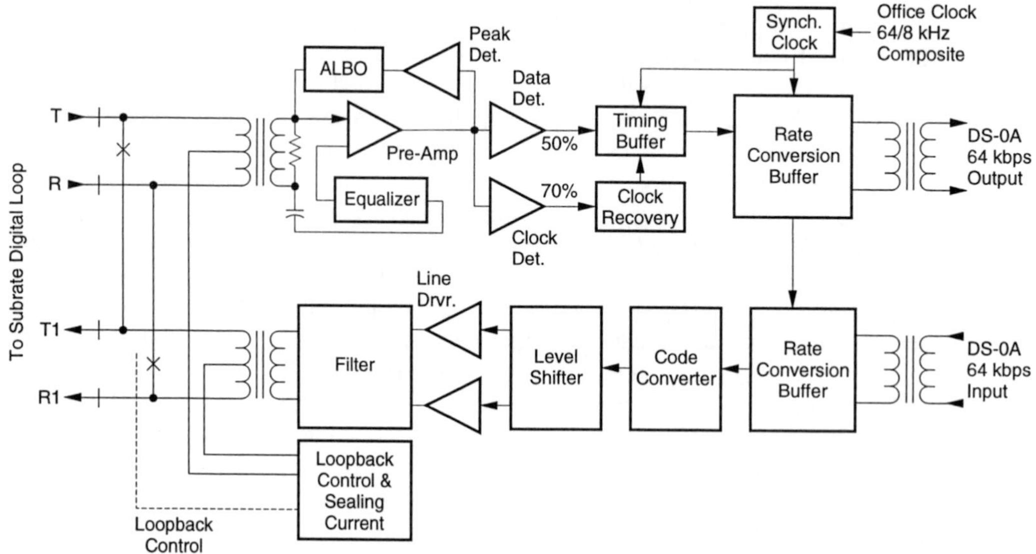

Figure 2-11 OCU block diagram

The basic function of the DSU is to convert the standard unformatted DTE interface signals to the baseband bipolar alternate mark inversion (AMI) line code, which is synchronized with the network. A block diagram of the DSU is shown in Figure 2-12. In addition to two types of loopback (channel and DSU loopback), the DSU also includes an encoder and decoder section with a data transmitter and receiver and clock recovery circuitry. On the CPE side, the DSU provides a standardized data interface such as EIA-232E or CCITT V.35. Generally, the EIA-232E interface is used for the 2.4, 4.8, 9.6, and 19.2 kbps data rates, and the V.35 interface is used for 56 kbps and 64 kbps [10,11]. Modern equipment provides other interfaces as well [12].

The CSU only provides a loop terminating section with channel loopback capability, as shown in Figure 2-13. It has no encoding or decoding capability and no clock recovery. The CPE interface is a simple 4-wire connection, and any equipment connected to the CSU must provide the necessary additional functionality of the DSU including clock recovery and line coding. The CSU is used when these capabilities are built into the data terminal equipment.

Subrate digital networks are designed to have relatively high reliability when compared to other digital networks, such as those carrying a large proportion of voice traffic; however, the subrate digital network has a finite probability of failure.

In systems or connections where even higher availability than normal is needed, dial backup lines provide an inexpensive solution. In this case, the CSU/DSU is equipped with a regular POTS line interface, as shown in Figure 2-14. When the CSU/DSU detects a network failure or degraded performance, it seizes the dial backup line, automatically detects dialtone, and calls through the public switched telephone network (PSTN) to the CSU/DSU at the far-end. The far-end CSU/DSU detects ringing and automatically answers the line. A sequence of message exchanges takes

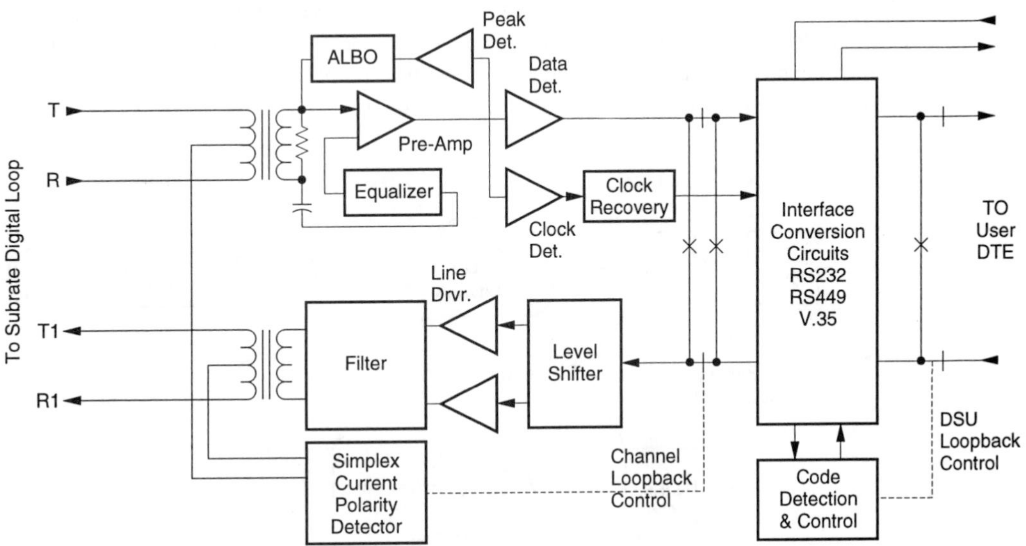

Figure 2-12 DSU block diagram

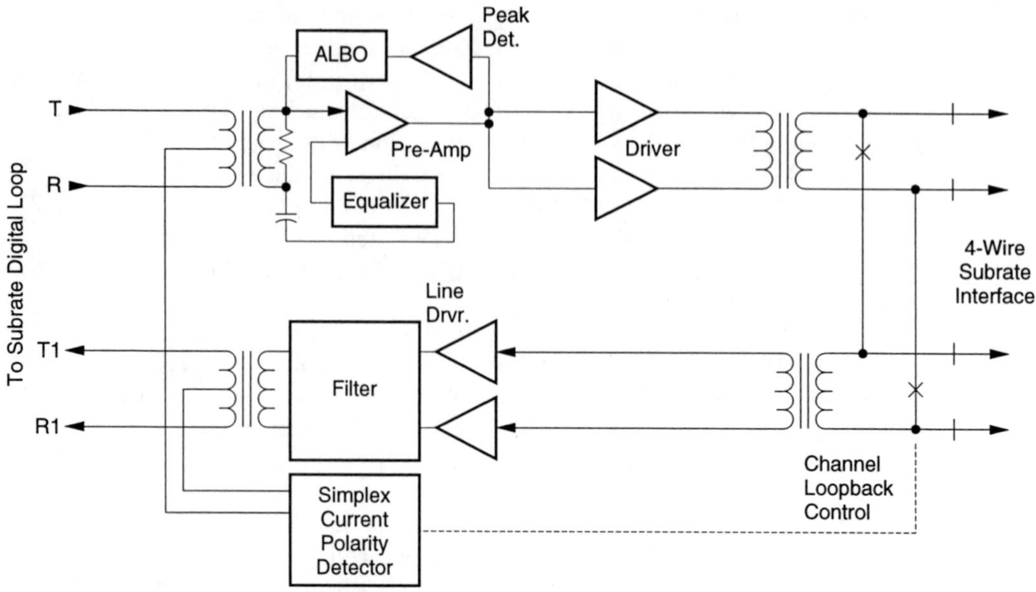

Figure 2-13 CSU block diagram

place between the two CSU/DSUs, and the data traffic is rerouted over the PSTN line.

The CSU/DSUs normally adapt to the highest reliable transmission rate over the PSTN. This can be as high as 28.8 kbps, but the CSU/DSUs automatically will drop back to lower rates if the line quality is not good enough. Meanwhile, the CSU/DSUs monitor the SRDL. When the loop returns to service, the traffic is routed over the digital loop as before, and the dial-up PSTN connections are dropped. Dial backup schemes such as this are not restricted to SRDLs; many analog private line modems have dial backup capabilities that operate similarly.

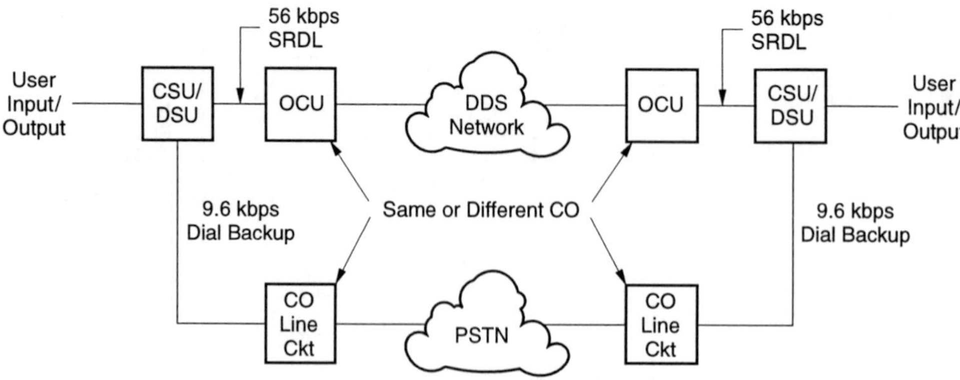

Figure 2-14 Dial backup for SRDLs

2.8 Switched 56 Loops

Switched 56 kbps loops are a subset of the SRDLs already described, but they differ significantly in their configuration with respect to the network. The SRDLs previously described are dedicated 4-wire loops that connect one user location with the serving central office and are extended, on a dedicated, full-time basis, to another user location. Switched 56 loops, on the other hand, use either a 2-wire or 4-wire loop (depending on the technology) to connect one user with the central office and are extended on a call basis to another user location.

Dedicated 4-wire SRDLs operate at varying standard line rates closely related to the user payload rate, whereas switched 56 loops operate at fixed line rates. The switched 56 loop rates depend on whether the service is 2-wire or 4-wire. Both loop types are fully synchronous but, with the proper terminal equipment, can also interface with asynchronous data terminal equipment at standard user rates up to 19.2 kbps.

The 4-wire switched 56 loops operate at a 72 kbps line rate and provide a user payload rate up to 56 kbps (some implementations provide user rates up to 64 kbps). Lower user rates are available by proper selection of the CSU/DSU at the user location. These lower rates are mapped into a 56 kbps data stream for transmission to the serving central office. The higher-rate data stream is switched through the network as if the user rate were a full 56 kbps. At the called user location, the lower user rate is stripped from the incoming 56 kbps data stream. Compatible equipment must be used at each end.

The switched 56 service provides data dialtone, and calls are dialed in the normal way. When the far-end answers, the near-end and far-end CSU/DSUs follow a rate adaptation protocol and exchange transmission parameters such as asynchronous/synchronous, bit rate, and clock type. Upon satisfactory completion of this training sequence, data transmission begins.

The error performance of switched 56 circuits is comparable to dedicated digital services. The 2-wire switched 56 loops operate at a line rate of 160 kbps using time compression multiplexing (TCM). TCM provides full-duplex transmission over a 2-wire loop by transmitting the data at the full 160 kbps rate in only one direction at a time. The data are buffered so the bursts are smoothed out and the data stream appears continuous from end to end. The most common commercial implementation of the 2-wire, TCM-switched 56 line is called DATAPATH®.*

The 2-wire switched 56 implementation provides a 64 kbps data channel (of which 56 kbps are available to the user) plus an 8 kbps signaling channel to and from the central office. The actual data stream consists of one start bit, eight signaling bits, 64 user bits, and one stop bit. The 8-bit bidirectional signaling bytes are exchanged via 1 ms frames and provide a channel for call setup and takedown and access to central office features. If necessary, the user interface may be extended beyond the normal reach of a central office switched 56 line card by using DLC. In

*DATAPATH is a trademark of Northern Telecom Inc.

this case, the user data are embedded in one of the 64 kbps channels of an interconnecting transmission system operating at the DS-1 rate. The error performance of the switched 56 circuit then depends on the performance of the transmission system (normally better than 3E-7). The loop normally is engineered to provide a bit error rate (BER) better than 4E-8.

2.9 ISDN DSL

The ISDN is based on circuit-switched, full-duplex "Bearer" B-channels and packet-switched, full-duplex "Data" D-channels. Examples of circuit-switched channels are voice and data, the latter including file downloading and transfer. Packet-switched data include interactive "bursty" transmissions, such as would occur between two on-line PC terminal users. The D-channel also carries signaling and network control messages.

An integrated digital network, such as provided by the ISDN, has many advantages. For example, the ISDN has a limited number of physical and electrical interfaces, which greatly simplifies circuit provisioning. The universal digital connectivity provided by the ISDN allows:

- Terminal equipment portability
- Nationwide calling number delivery (especially attractive for telemarketing)
- Wide-area networking
- Call-by-call service selection
- Uniform wiring plan
- Easier moves and changes
- Voice and data transmission on a single interface

ISDN digital lines presently consist of two interfaces: BRI and primary rate interface, or PRI.* The ISDN PRI operates at a line rate of 1.544 Mbps with a payload of up to 1.536 Mbps over any new or existing DS-1 rate transmission systems that provide the specified error performance (for example, repeatered T1-carrier span lines or HDSL). The ISDN PRI can provide up to 24 B-channels or combinations of B- and D-channels. The only difference between the ISDN PRI and repeatered T1-carrier, from a transport perspective, is in the terminal equipment or channel units in the multiplexer, which are specially designed to provide the ISDN channels. Table 2-4 shows some of the standardized ISDN PRI channel combinations [7].

ISDN BRI lines commonly are called DSLs, and this terminology is used throughout this book to differentiate the ISDN line from the SRDL. The DSL op-

*These interfaces also are called BRA (basic rate access) and PRA (primary rate access), respectively.

Table 2-4 ISDN PRI Channel Combinations

DESIGNATION	N	B (64 kbps)	D (64 kbps)	PAYLOAD
N × B + D	≤23	Yes	Yes	128 kbps–1.536 Mbps[a]
N × B	≤24	Yes	No	64 kbps–1.536 Mbps[a]
N × H0	4	No	No	384 kbps[b]
N × H0 + D	3	No	Yes	384 kbps[c]
H1	—	No	No	1.536 Mbps[d]
H10	—	No	Yes	1.536 Mbps[e]

[a]The rate for each B-channel and D-channel is 64 kbps. The D-channel is used for circuit switching by the ISDN.

[b]4 × H0-channels, where the rate for each H0-channel is 384 kbps. The H0-channel does not carry any signaling information for circuit switching.

[c]3 × H0-channels plus one D-channel, where the D-channel rate is 64 kbps.

[d]1 × H1-channel, where the H1-channel rate is 1,536 kbps. The H1-channel does not carry any signaling information for circuit switching.

[e]1 × H10-channel plus one D-channel, where the H10-channel rate is 1,472 kbps and the D-channel rate is 64 kbps.

erates at a line rate of 160 kbps with a payload of 144 kbps. The payload is designated 2B + D, which indicates two 64 kbps B-channels and one 16 kbps D-channel. Even though the line rate is the same as the switched 56 loop using TCM technology, the operation of the DSL is quite different.

DSLs connect at the central office switching system through a digital line termination (LT) to a 2-wire loop and at the subscriber premises through the network termination (NT) or U-interface. A typical DSL is shown in Figure 2-15. Also shown in this figure are the S-interface and T-interface, which connect the digital line to the premises wiring and terminal equipment. The S- and T-interfaces are discussed in Chapter 11.

The DSL is a 2-wire loop and therefore provides metallic access to almost any existing subscriber. Certain limitations, described in Chapter 10, are imposed on the DSL to provide this almost ubiquitous service. To separate the transmit and receive directions, and provide the equivalent of 4-wire, bidirectional service, a hybrid and echo canceller are used at each end, as shown in Figure 2-16. The design BER is 1E-7, or better.

Perhaps the three most fundamental technical advancements that allow the ISDN DSL (and other state-of-the-art digital loop systems) to be viable are:

- Very large scale integration (VLSI)
- Decision feedback equalization (DFE) techniques
- Echo cancelling

VLSI provides an economical packaging of the very complex electronic circuits, and DFE provides a means of reducing intersymbol interference caused by the far-from-perfect transmission characteristics of twisted pairs. Echo canceling allows full duplex digital transmission on a single twisted pair. Economical decision feedback equalization and echo cancelling would not be possible without VLSI.

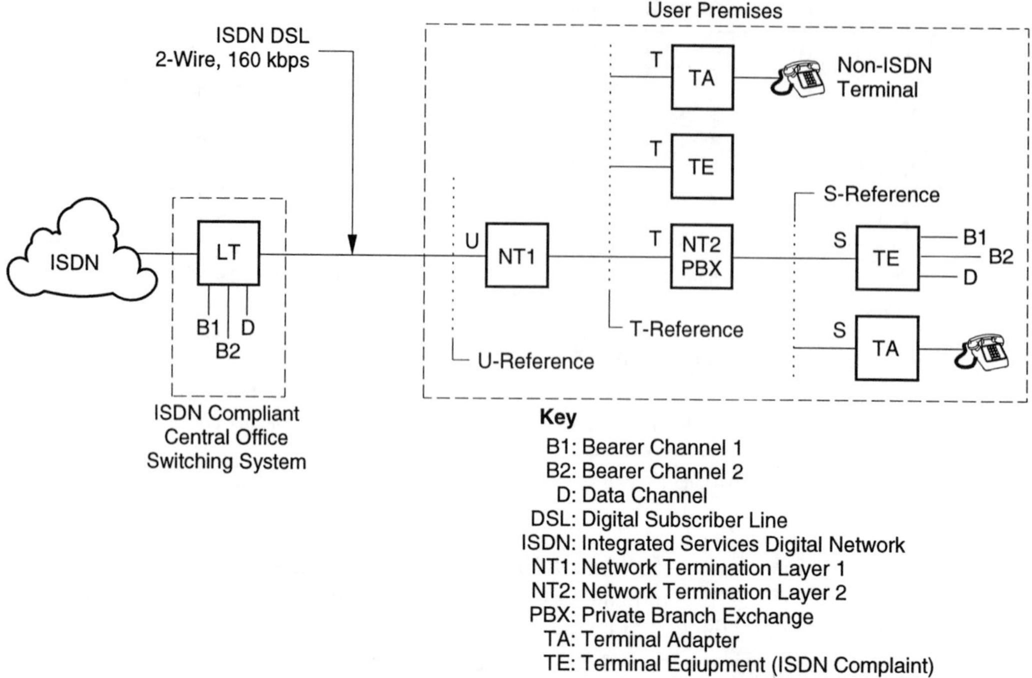

Figure 2-15 Typical DSL

In Figure 2-16 the transmitted B-channels and D-channel are first buffered and then combined with error detection and maintenance channel bits. These are then scrambled and formatted according to the BRI framing scheme. A 2-binary, 1-quaternary (2B1Q) line coding scheme is used to provide a 2 bps/baud symbol rate. The BRI framing and 2B1Q line code are described in Chapter 3. After filtering, to shape the pulses and remove unneeded high frequencies, the framed and encoded signals are transformer coupled to the 2-wire loop.

The received signal from the loop includes the signal sent from the far-end as well as part of the transmitted signal reflected back into the DSL receiver because of unavoidable impedance mismatch with the 2-wire loop. This latter unintentional signal is echo. The circuit shown in Figure 2-16 uses two echo cancelling methods. One is an analog adaptive hybrid consisting of a prefilter and hybrid balance blocks. The adaptive hybrid provides about 20 dB of echo cancellation. The other method is a fully digital echo canceller, which rejects the remaining echo from the received signal at the summing point. The echo canceller block shown is actually a very complex VLSI circuit using digital signal processing (DSP) techniques.

The DFE processes the far-end signal to remove intersymbol interference and allow correct detection of the received data. Timing also is recovered from the incoming waveform. Proper received clock recovery is needed for optimum echo cancelling, equalizing, and received data extracting. After the received data is decoded, it is descrambled. The B-channels and D-channel are extracted and buffered for presentation to the digital bus.

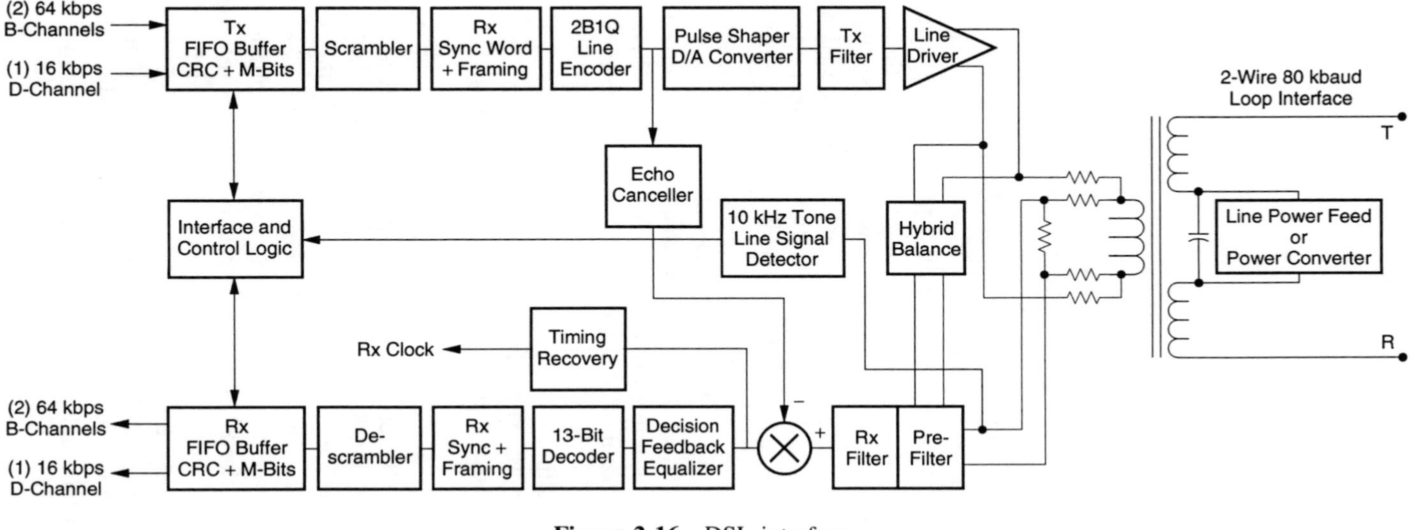

Figure 2-16 DSL interface

If the DSL circuit is implemented at the central office LT, it provides line power feed on the loop side of the coupling transformer. At the subscriber NT, the loop side of the coupling transformer is connected to a dc-dc converter for equipment powering.

2.10 HDSL

Users of the public network are demanding higher digital bandwidths and quicker installation periods than are available with present digital loop systems, particularly repeatered T1-carrier. These digital loop systems are expensive to engineer and install. The HDSL, on the other hand, provides a DS-1 rate transmission system but without the expense and time-consuming installation and engineering intervals. HDSL was conceived to meet interim customer and economic requirements between now and the time that optical fibers are fully deployed in the loop plant.

Production HDSL systems are being deployed in the network as this book is being written. A preliminary industry standard for HDSL has been developed, and commercial systems are based on this standard [8]. The HDSL is based on the use of state-of-the-art VLSI technology in the HDSL terminals. The advanced technology allows the HDSL to be installed on existing twisted pair loops meeting only certain minimum criteria, as discussed below.

The HDSL is a repeaterless digital line providing a payload of 1.536 Mbps, which is equivalent to the existing repeatered T1-carrier span line. Although the HDSL is designed as a low-cost alternative to repeatered T1-carrier, in its basic form it is only suitable for limited distances. However, at least one manufacturer builds an HDSL repeater to extend the operational distance. As will be seen in later chapters, T1-carrier requires extensive engineering effort. T1-carrier also requires extensive installation effort, particularly for repeaters and line qualifying. The HDSL requires neither extensive engineering nor line qualifying in the large majority of installations.

The HDSL technology was developed from the ISDN DSL previously described; however, whereas the DSL operates with existing loops up to 18 kilofeet, the HDSL operates with loops from nine to twelve kilofeet, depending on the cable gauge. This length restriction falls within the engineering concept of a CSA. The CSA, which was described previously, provides pre-engineered loops that meet specific length requirements and, indirectly, specific transmission requirements that are accommodated by the HDSL. The center of a CSA is the central office or remote DLC terminal. Therefore, proper network planning will allow CSAs to be developed so that all customers are within reach of the HDSL.

HDSLs connect at the central office through high bit-rate terminal units (HTU-C) to two 2-wire loops and at the subscriber premises through remote high bit-rate terminal units (HTU-R).* A typical HDSL and its interfaces are shown in Figure 2-17. Each loop carries half the payload (768 kbps) plus overhead for a total

---

*This terminology is used in reference [8].

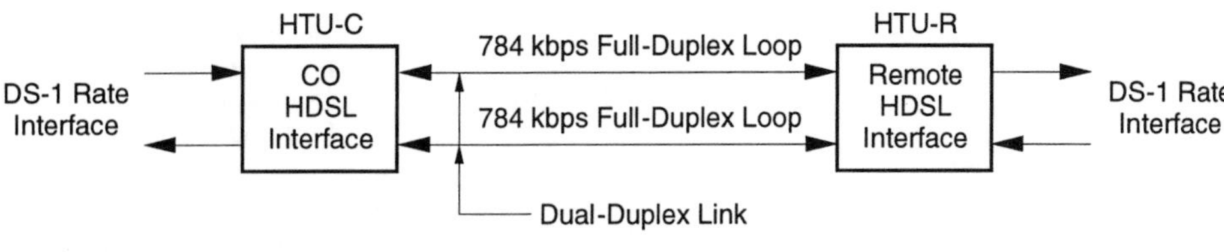

Figure 2-17 Typical HDSL

rate on each loop of 784 kbps. This arrangement is called *dual-duplex* in the literature [8].

The HDSL operates on loops meeting existing CSA engineering rules and therefore provides access to a large percentage of existing subscribers. As with ISDN DSL, a hybrid and echo canceller are used at each end to separate the transmit and receive directions on each of the 2-wire loops (Figure 2-16). In order to provide reliable and economical service, at the specified 1E-7 BER or better, the HDSL interfaces are designed to operate over a wide range of cable pair lengths (up to 12 kilofeet maximum), types, and gauges, and with a variety of bridged taps.

To meet future needs for broadband services, the telecommunication industry's goal is to use optical fiber technology instead of twisted pair technology in the loop. Deployment of optical fibers occurs especially where an existing loop plant requires rehabilitation. However, the deployment of optical fibers has been slowed because the present technology does not have cost parity with existing twisted pair technology. This is the so-called "rehab dilemma."

There is at least one proposal by a large telephone operating company to use the HDSL in loop plant rehabilitation work as an interim, cost-effective solution to this dilemma [9]. With this proposal, optical network units (ONUs) are deployed at the curb, as would be the case with fiber-to-the-curb (FTTC) topology. However, instead of installing optical fibers between the central office and the curb, an HDSL transceiver replaces the optical line interface in the ONU. This is connected to the corresponding HDSL terminal in the central office through the existing twisted pair loop plant. Thus, the life of the existing loop plant is extended until the optical fiber network topology reaches cost parity with new twisted pair loop plant; no new twisted pair loop plant need be deployed.

2.11 Repeatered T1-Carrier

Repeatered T1-carrier systems are used in point-to-point configurations only, particularly between two central offices, between a central office and an RT, or between two end-users. The system consists of LT and span line components, as shown in Figure 2-18. Operational components include the LT shelf (LTS) and central office

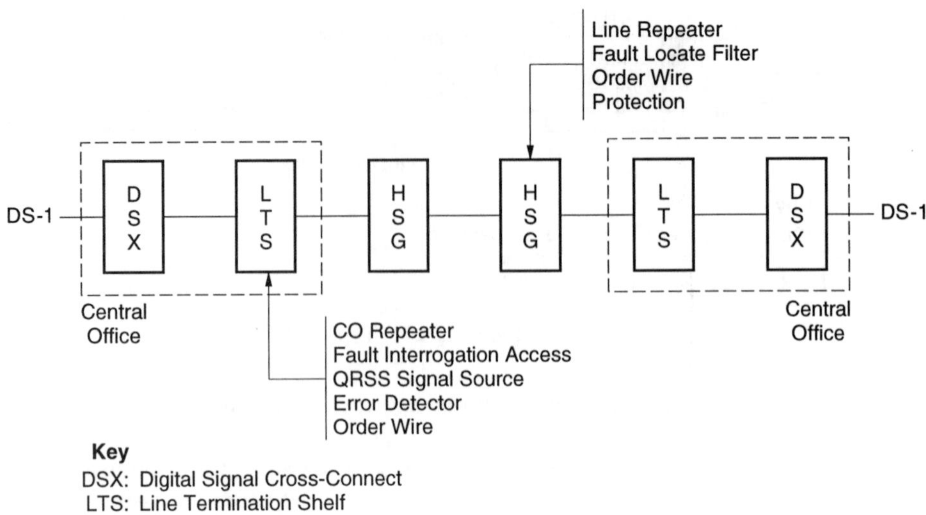

Figure 2-18 Repeatered T1-carrier components

repeaters (CORs) at each end of the span line, and line repeater housings (HSGs) spaced a regular intervals along the span.

Repeatered T1-carrier LTSs normally have provisions for line fault locating. These provisions include an interrogation access unit, test signal source and test signal detector, and order wire unit. The interrogation access unit allows a special interrogation signal to be injected from an external test set into the line. This signal is received at each repeater, filtered, and sent back to the interrogation access unit on a dedicated fault-locate cable pair. Basic fault location is described below.

Automatic protection switching (APS) can be provided as a part of LTs. The APS switches out a failed span and switches in a protection (spare) span. An example of an APS is shown in Figure 2-19.

2.12 Fractional T1-Carrier (FT1)

Fractional T1 (FT1) systems provide channels in multiples of the 64 kbps DS-0 rate. For example, an end-user may require a channel operating at 256 kbps (4 × 64 kbps) to bridge two LANs together. A typical FT1 application is shown in Figure 2-20, in which circuits are carried within the public network at 256 kbps (four DS-0). In most implementations of FT1, the local access portions, or loops, operate at the full DS-1 rate. However, some manufacturers make equipment capable of baseband transmission in the loop plant at FT1 rates. The individual DS-0 channels may or may not be in contiguous timeslots. The channel allocation scheme is determined by the channel service units and multiplexers in the circuit.

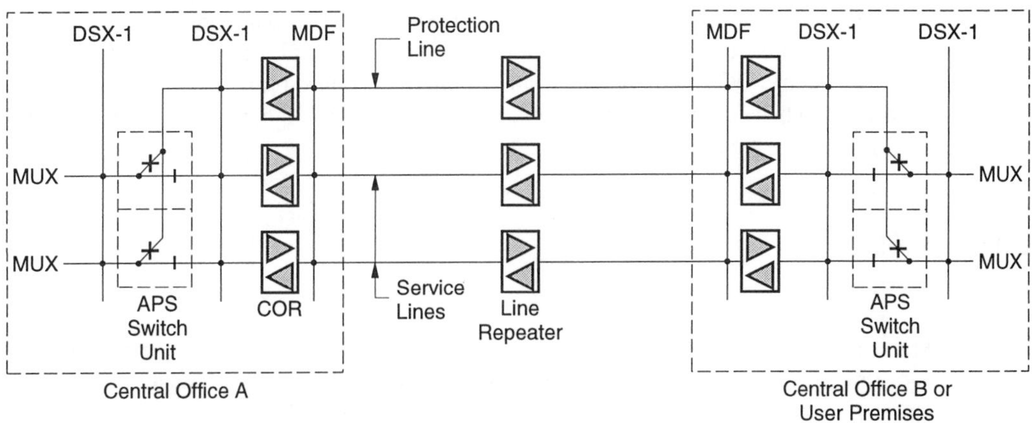

Figure 2-19 Span line with APS

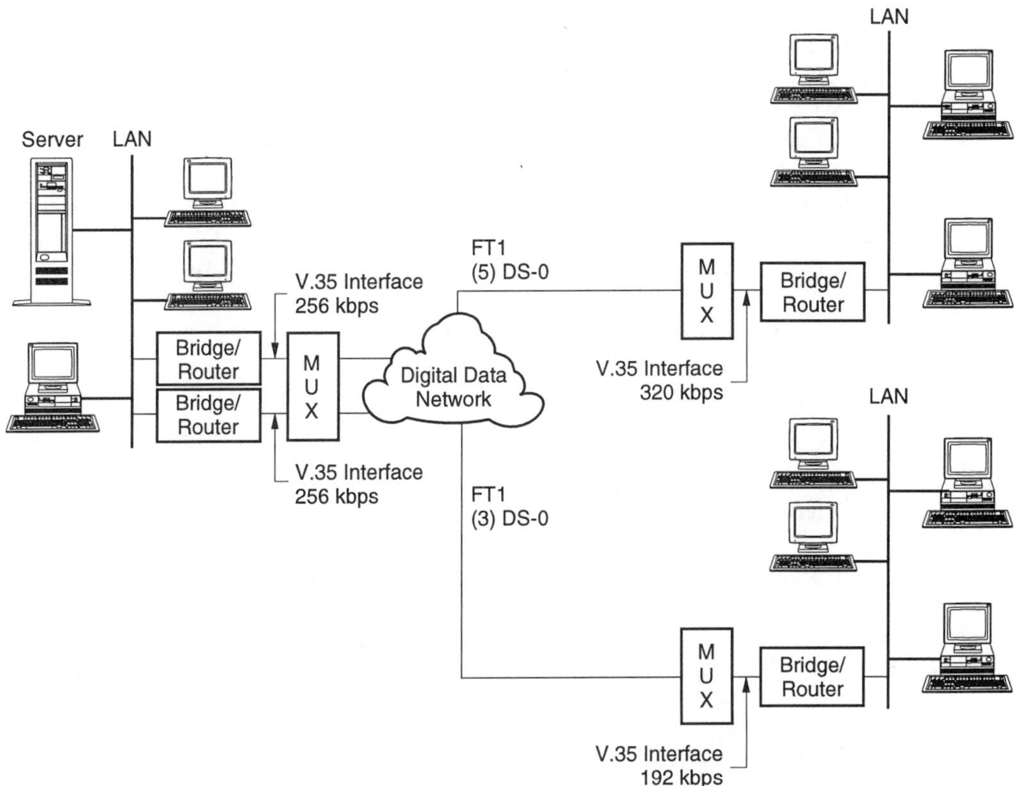

Figure 2-20 FT1

Where the loop operates at the full DS-1 rate, FT1 can provide significant savings, but only in the interoffice portions of the circuit. Generally, a digital cross-connect system (DCS) consolidates individual DS-0 rate circuits in the serving central offices. These consolidated circuits are transmitted at higher rates over backbone facilities, where economies of scale are significant.

2.13 ADSL

ADSL is a new technology. It is being field tested and is not in large-size production as of this writing; therefore, this section will be limited to a brief description. ADSL applications are meant to fall between the ISDN DSL and HDSL. The ISDN DSL uses one cable pair and provides full-duplex 144 kbps payload, and the HDSL uses two cable pairs and provides full-duplex 1.536 Mbps payload.

The ADSL provides simplex 1.536 Mbps payload toward the subscriber, simplex low-speed control and data channel toward the serving central office from the subscriber, and full-duplex POTS, all in one cable pair. The actual line rate for ADSL is expected to be closer to 1.6 Mbps to provide overhead and control signaling.

The maximum route distances for the ISDN DSL, HDSL, and ADSL are 18, 12, and 18 kilofeet, respectively. The ADSL will give the same coverage as the ISDN DSL. The ADSL is a passband system, while all other digital loop systems rely on baseband transmission techniques. The basic technical attributes for these three digital loops are summarized in Figure 2-21.

The asymmetric nature of ADSL limits its applications to a certain telecommunications market segment, namely the residential market. ADSL is designed to transport certain types of applications to the home, including [13]:

- Educational video from remote classrooms
- Multimedia
- High-speed data (up to 1.536 Mbps)
- "Movie-on-demand" and broadcast entertainment video, which has been digitized and compressed
- Interactive images and games

The ADSL is a repeaterless technology, which, like the ISDN DSL and HDSL, is supposed to be installed without extensive engineering and installation effort. However, where ADSL, ISDN DSL, HDSL, and repeatered T1-carrier are installed in the same cable, spectrum compatibility problems have been shown to exist [14]. Because the high bandwidth part of the ADSL is above 50 kHz, it is not expected to interfere with voiceband services or SRDLs. The spectrum compatibility issues probably will be resolved by pair assignment restrictions, refinement in the line and source coding of the ADSL, and advances in DSP and receiver design.

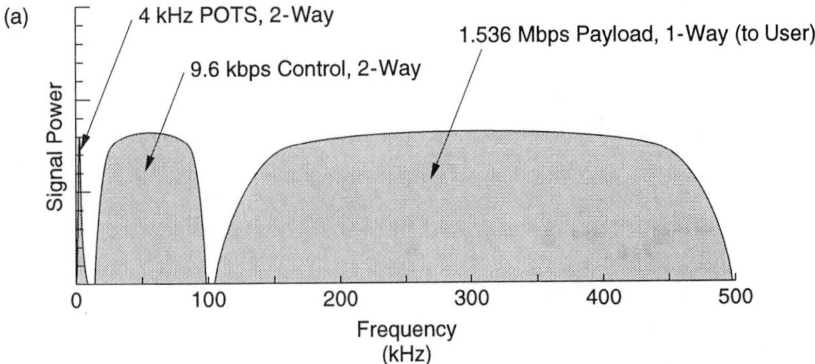

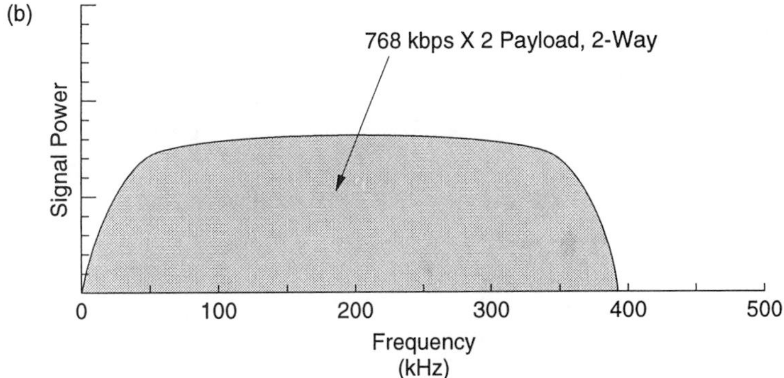

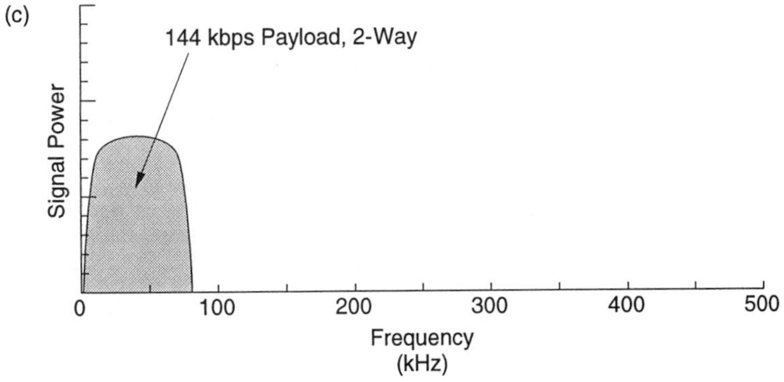

Figure 2-21 Spectral occupancy comparison: (a) ADSL; (b) HDSL; (c) ISDN DSL

2.14 Loop Applications of Optical Fibers

The application of optical fibers in subscriber service is called fiber-in-the-loop, or FITL. This acronym indicates either a generalized or application-specific architecture, as will be explained. FTTC refers to a subscriber loop application in which fiber optic cables are used exclusively in the loop except in the service drop; that is, the optical fibers are not used from the pedestal at the curb to the home or office building. In this case, the service drop uses existing technology such as twisted pair, coaxial cable, or both.

Another architecture is fiber-to-the-home, or FTTH, which describes an optical fiber path all the way from the central office to the subscriber's premises. Figure 2-22 illustrates each concept and Figure 2-23 illustrates the basic architectures envisioned for FTTC and FTTH.

A particular FITL architecture, which has become an industry standard, uses a passive optical network (PON) to connect a host digital terminal (HDT) to multiple

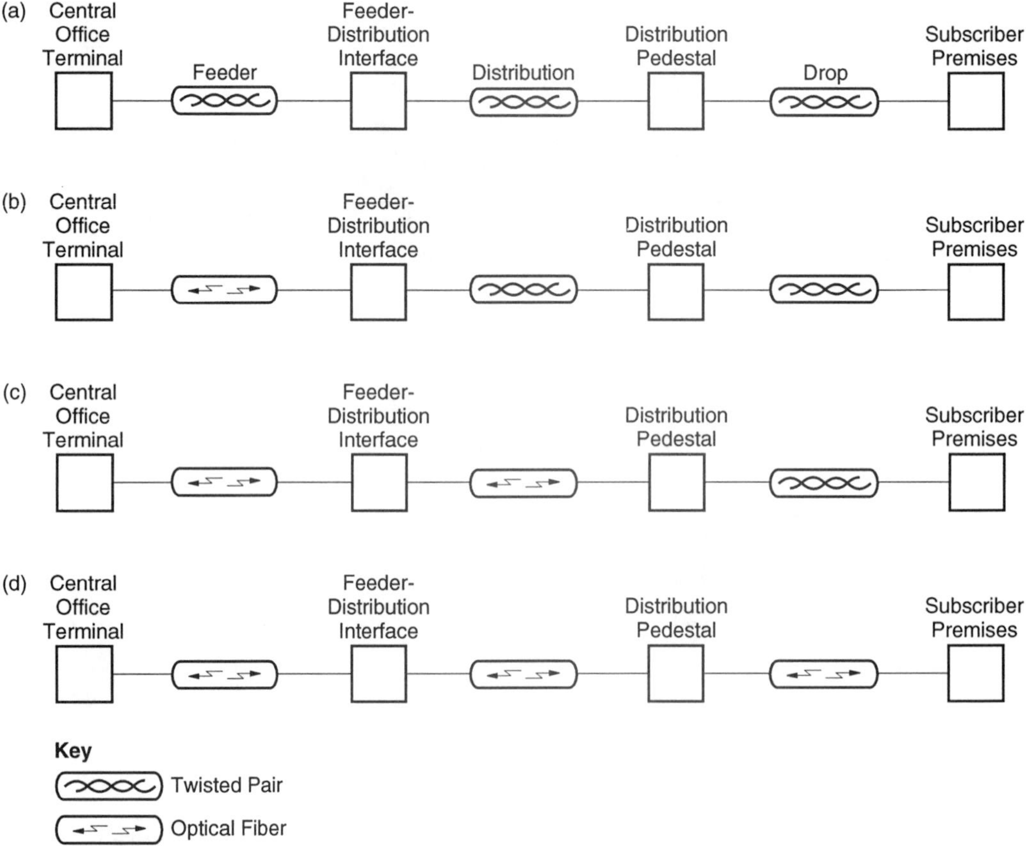

Figure 2-22 Loop application technologies: (a) Twisted pair metallic loop; (b) FITL; (c) FTTC; (d) FTTH

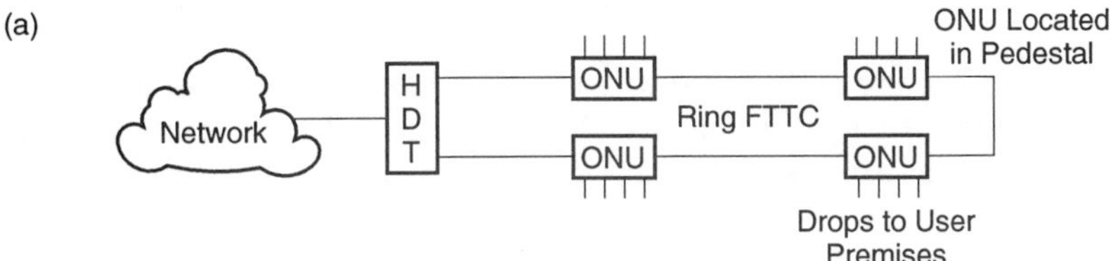

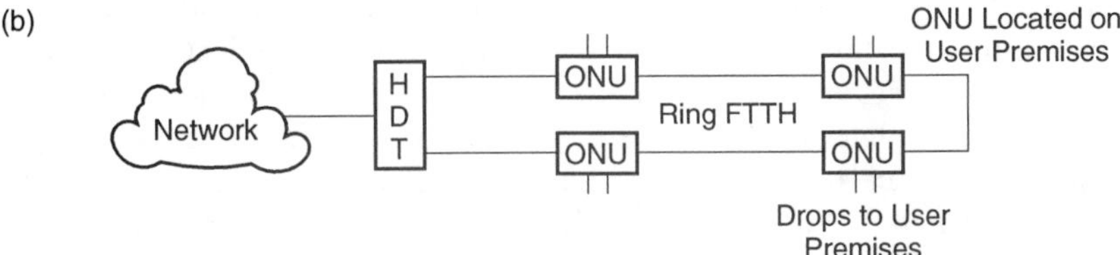

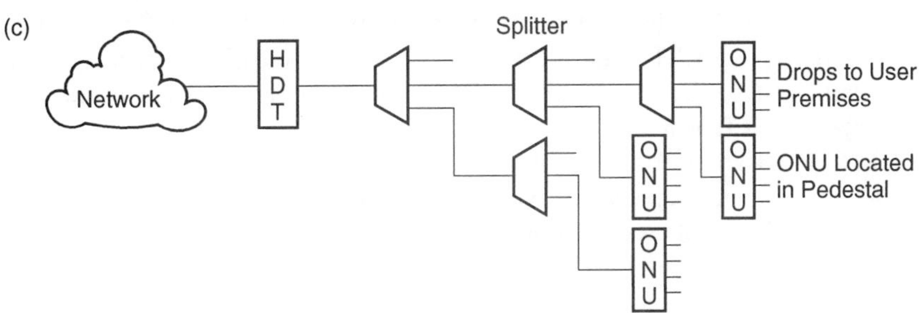

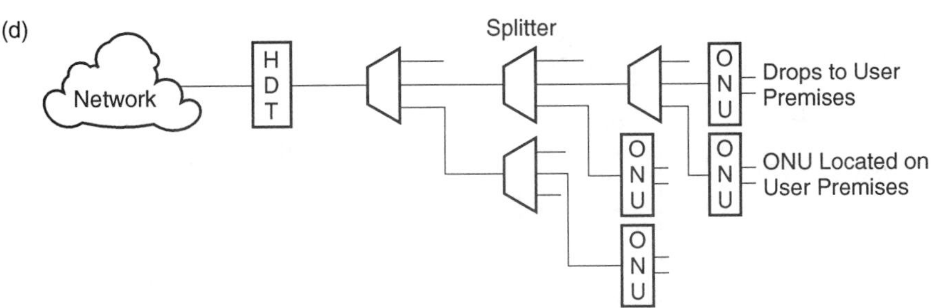

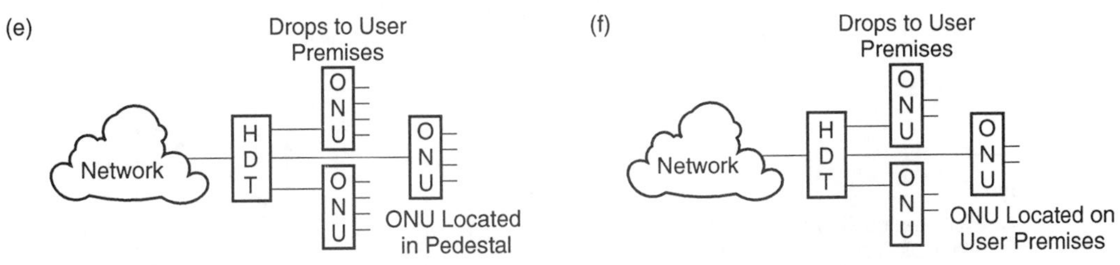

Figure 2-23 FTTC and FTTH architectures: (a) Ring FTTC; (b) Ring FTTH; (c) Tree and branch FTTC; (d) Tree and branch FTTH; (e) Star FTTC; (f) Star FTTH

ONUs [15]. This architecture is designed to serve single-family or multi-family living units and is a form of FTTC in that the ONUs are located at the curb and use a twisted pair and coaxial cable (or wireless) service drop. One implementation of a PON is illustrated in Figure 2-24. The coupler (also called splitter) allows point-to-multipoint operation without active components. The couplers have a 1:N splitting ratio, where $2 \leq N \leq 32$.

All existing telecommunication interfaces at households are analog and none are optical. Since the transmission method for all future telecommunication services is digital, at least two types of conversion are required in delivering these services. The analog signal at the interface must be converted to a digital signal (A/D and D/A conversion) and the corresponding electrical signal must be converted to optical (E/O and O/E conversion). Each ONU provides these conversion functions in addition to multiplexing the individual services onto the high-speed transmission facility, alarms, and operation support system interfaces.

The HDT is located in the central office or collocated with the RT associated with a DLC system or an RST. The HDT may or may not interface directly to the RT. The transmission facility between the HDT and the central office usually is shared with the DLC RT or RST.

The FTTC PON architecture can provide an economical upgrade path to the FTTH architecture. To upgrade, a passive splitter replaces the FTTC ONU located in the pedestal, and fiber optic cable replaces the coaxial or twisted pair drops from the pedestal to the home. The HDT also must be replaced or upgraded. One study

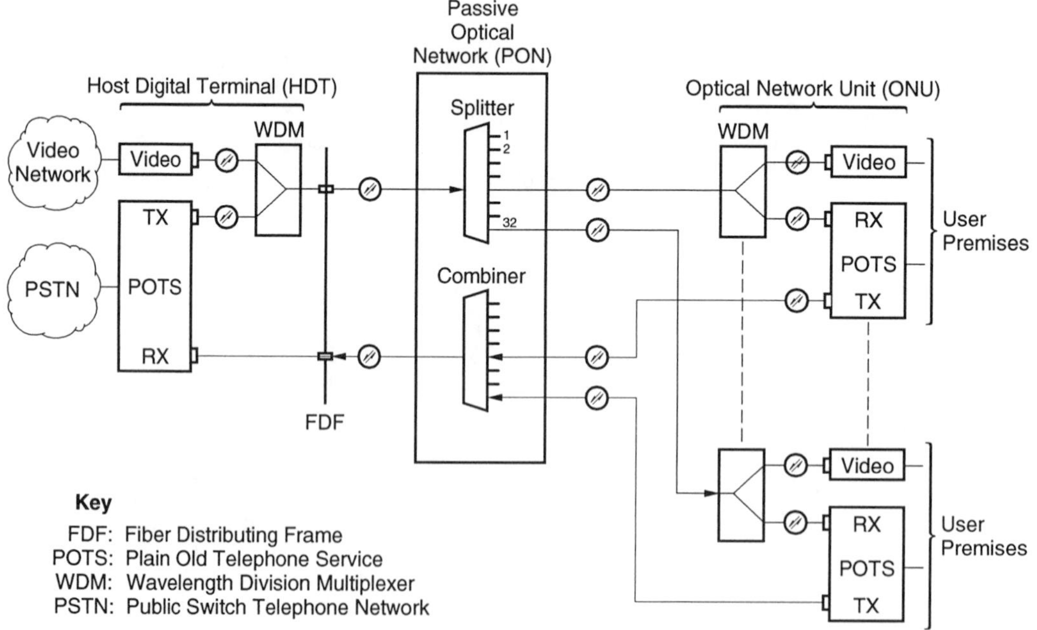

Figure 2-24 PON implementation

indicates it is economical to replace the ONU, which serves four homes in the FTTC architecture, with a 1:8 passive splitter, which serves the ONUs for eight homes [16]. Another study shows there is little difference between the cost of 1:8 and 1:16 ratios [17].

The upgrade, or migration, from FTTC to FTTH is easier to describe than implement. There are some significant planning issues that must be resolved. Some of these are:

- ONU powering method (central powering, local powering, hybrid)
- Effectiveness and life of battery backup systems used to power the ONUs
- Re-use of the HDT and ONU electronics
- Re-use of the ONU pedestals for passive splitters
- System operating margin after upgrade to FTTH (system losses are higher)

One significant difference between the FITL system and digital loop systems based on twisted pair cables is the specified error rate. The HDSL and ISDN DSL both are specified to provide BER less than 10E-7. Repeatered T1-carrier systems were originally designed for 10E-6 BER, although this can be improved to an almost arbitrarily small rate by reducing the repeater spacing. The FITL objective BER performance is 10E-10 and the specified maximum is 10E-9. Such a small error rate is possible with FITL because of the characteristics of optical fibers. Optical fibers are not sensitive to crosstalk and other external interference that degrade the performance of systems based on twisted pairs.

2.15 Problems Associated with Digital Loop Planning*

Proper planning is a very important consideration in digital loop design because the loop infrastructure has a long life (15 to 40 years), is expensive to construct, and must be compatible with future (and, many times, unknown) transmission systems and services.

The intent of loop design is to provide satisfactory and economically efficient transmission quality. The adequacy of any loop design method is directly attributable to the planning associated with the services to be provided on the loop. The planning process, as described here, is specifically oriented toward the loop outside plant facility.

Figure 2-25 shows the features identified in the loop planning process. A wire center serves some overall area. This area is broken into sections that meet network requirements and loop technology application strategies and allow convenient administration. These sections may overlap and change over time. A wire center may be considered to include DLC remote terminals and remote switching terminals, or these

*Portions of this section were taken from [18].

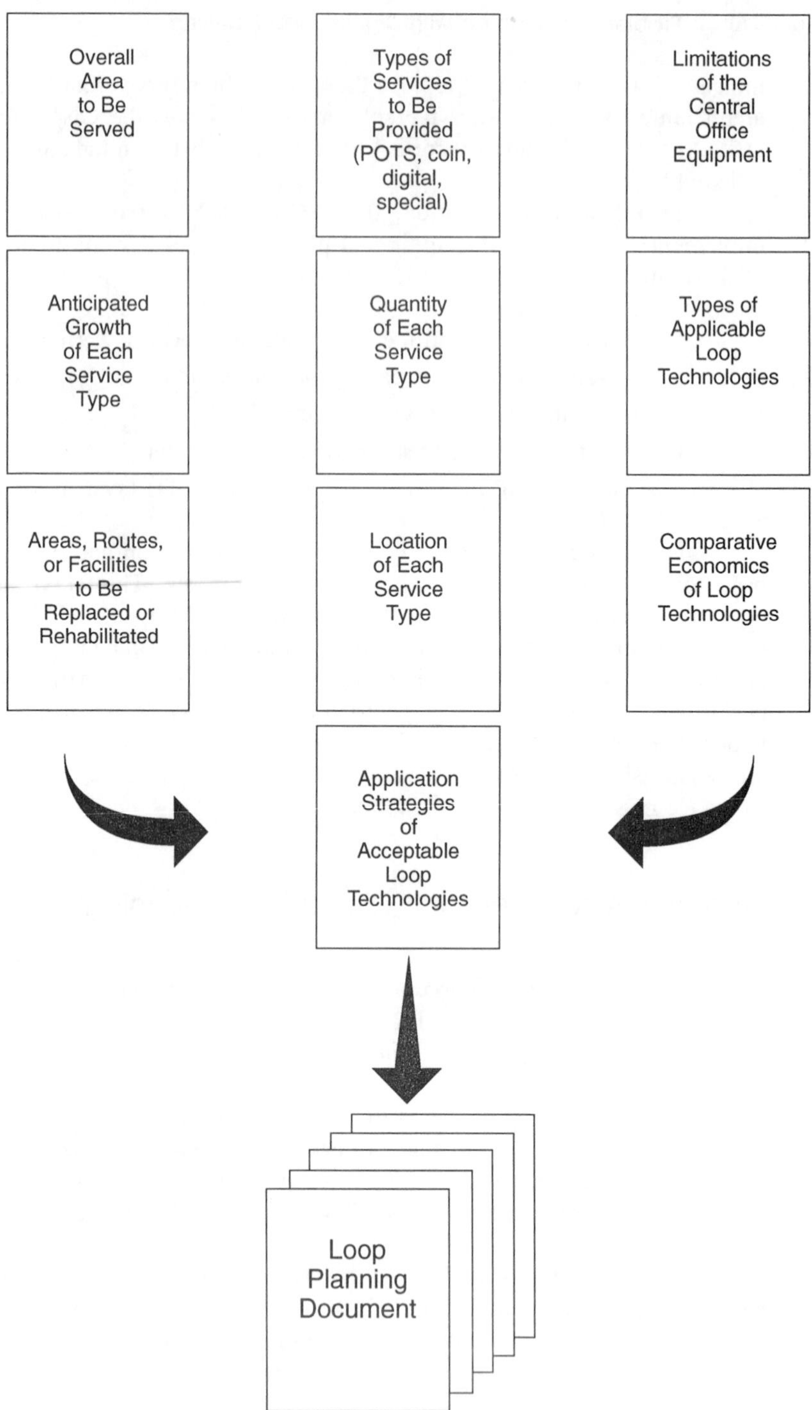

Figure 2-25 Digital loop planning, [18]
Reeve, © 1992 Institute of Electrical and Electronics Engineers. Reprinted with permission

systems may be considered as wire centers in and among themselves. A wire center area, illustrated in Figure 2-26, includes the following sections:

- Distribution area, which defines the developed or soon-to-be-developed geographical area served by the wire center.
- Rural area, which defines the thinly developed areas within the wire center boundaries. The rural area overlaps the distribution area and, over time, will decrease as population increases and land is developed into suburban or urban uses.
- CSA (also called electronic serving area), which defines the areas with presently available or planned digital connectivity. CSAs are planned ultimately to cover the entire distribution area. The area not served directly by the central office is served via DLC RTs or digital pair gain devices. RTs and digital pair gain devices are deployed and the CSA established as needs for digital services arise.
- Ultimate serving area, which defines the ultimate area to be served by the wire center.

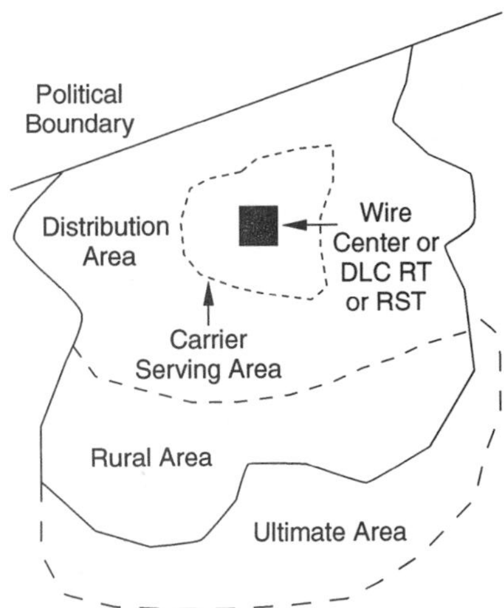

Figure 2-26 Wire center area, [18]
Reeve, © 1992 Institute of Electrical and Electronics Engineers. Reprinted with permission

The type of land development shown in Figure 2-27, both present and future, will impact the service types and therefore the loop technology used. Services can be delivered to the user by different loop types, some of which are shown in Figure 2-27. Current digital loop technologies are essentially transparent to the services carried by them. Nevertheless, it is still somewhat difficult to determine in advance the most appropriate loop technology in the wire center area over long time periods.

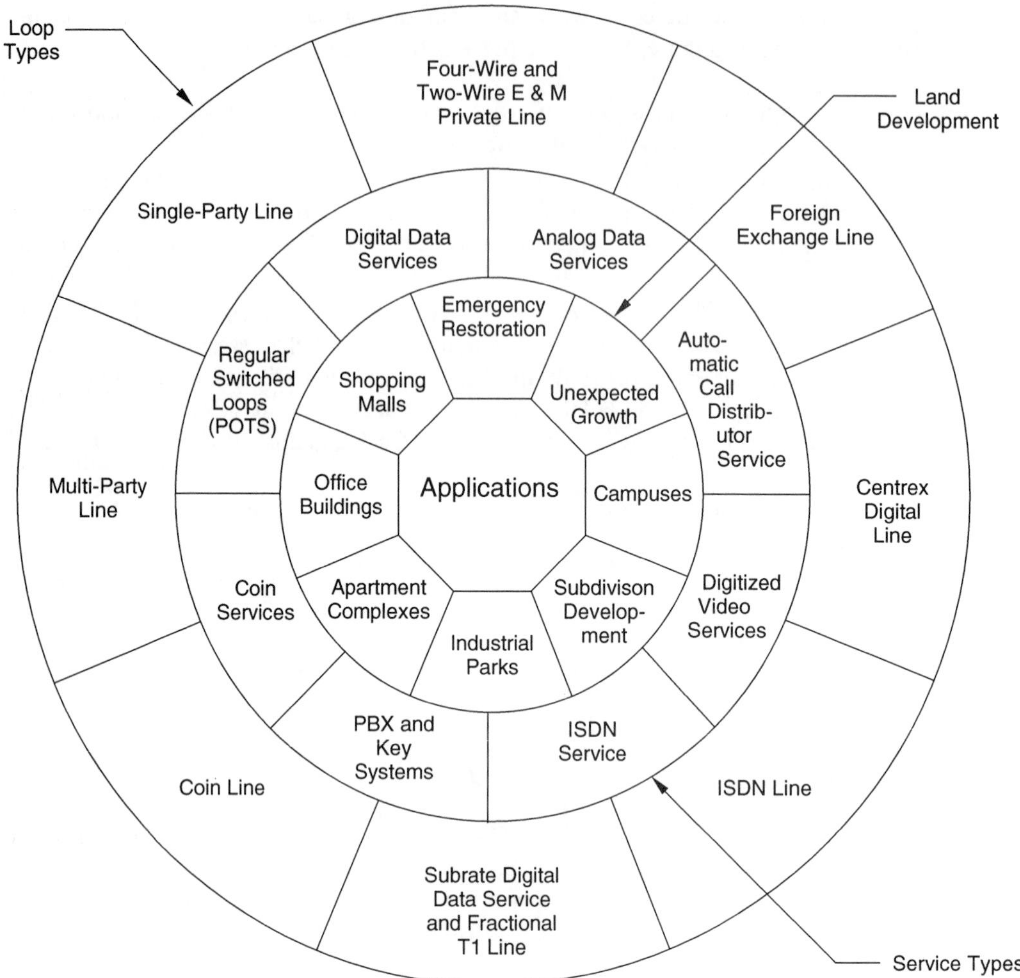

Figure 2-27 Applications within a wire center area, [18] Reeve, © 1992 Institute of Electrical and Electronics Engineers. Reprinted with permission

The already difficult planning task is further complicated by sizing considerations because:

- Growth and demand for various services over long periods are difficult to predict, and the loop technology to be used is highly dependent on these factors.
- Loop technologies are changing, and more options are becoming, and will continue to become, available. All of these technologies reduce or otherwise influence the outside plant cabling requirement.

- A highly detailed knowledge is required of land use, demographics, and area economy, and this information is rarely readily available.
- The regulatory environment can change, which can lead to deregulation and competition in previously regulated services. This increases the threat of facility bypass and stranded investment.

A well-planned service forecast accounts for these effects and becomes the basis for loop plant sizing. Large companies use internally developed software tools to mechanize the forecasts and associated economic analyses of loop plant alternatives. When these programs or detailed information are not available, the following general sizing rules for loop facilities can be used for planning purposes:

- Allot the equivalent of two cable pairs to each living unit in a developed or soon-to-be-developed area.
- Allot the equivalent of two cable pairs to each 100 sq. feet of building space not used for living purposes.

The above rules require "equivalent" cable pairs. This means the chosen loop technology should provide these pairs in distribution facilities serving end-user premises and, depending on the interface, also at the wire center. The wire center requirements will vary with the amount and type of IDLC system or RSTs used. The actual facility between the wire center and the user or groups of users may be twisted cable pairs, a radio system, optical fibers, or a combination. Using the appropriate technology, it is possible to eliminate completely the need for metallic connections between the end-user and the central office.

The loop facilities plan is written to provide as much detail as is desired, but progressively more detail is always needed as the implementation time of a particular technology approaches. Economic evaluations of the various loop technologies is vital in the planning stages. Acceptable technologies are then chosen and strategies developed for their application and deployment. Not to be forgotten in the choice of technology is the impact had on operational support systems (OSSs).

The loop design process takes into account these deployment strategies. All present strategies include the application of twisted pair metallic cables and optical fiber cables for service transport. Repeatered T1-carrier, SRDLs, ISDN DSLs, and HDSL are loop transport mechanisms.

To ensure ubiquitous digital connectivity, the application strategies of current and foreseen technologies require certain changes be made in the design, installation, and administration of new and existing loop plant. Some of these changes are:

- Eliminate all analog subscriber carrier systems from existing loop plant and from future deployment plans.
- Reduce the total length of bridged taps, or eliminate them altogether, by reducing the application of multiple plant.
- Eliminate ready-access terminals in aerial plant and open pedestals in buried plant; *use fixed-count terminals* and *good housekeeping* (sealed) *pedestals* to reduce craft-caused troubles and service interruptions.

- Remove all loop treatment, including load coils and build-out capacitors, voice frequency repeaters, and loop extenders, from loops within the CSA.

In this list, fixed-count terminals are splice points between distribution cables and subscriber drops. The fixed-count terminology means specific pairs are preassigned to that terminal or pedestal location. Good housekeeping pedestals have physical barriers to prevent craft personnel from disturbing splice points and accessing cable pairs not assigned to that pedestal.

2.16 Jack and Plug Connector Interfaces

Subpart F of Title 47, Code of Federal Regulations (CFR), part 68 of the FCC rules, defines the physical shape and size of the modular plugs and jacks used to connect premises wiring systems or terminal equipment, the so-called "customer installation," to the public telephone network (PTN) or PSTN interface [19]. Subpart F also gives wiring patterns for universal service ordering codes (USOC), which are the familiar "RJ" (registered jack) series of plugs and jacks.

There are connector wiring configurations used at the network interface that do not fall under the scope of the FCC regulations. These connectors generally have been adopted under an industry standard and are given an "SJ" (standard jack) designation by Committee T1 of the Exchange Carriers Standards Association (ECSA) [20].* Other connector types and configurations are used to connect user terminal equipment to premises wiring systems or LAN equipment. Many of these have been standardized, whereas others are specific to vendor equipment.

The FCC specifies three basic connector types: 6-position, 8-position, and 50-position, where the positions may or may not be equipped with electrical contacts. Not all positions on a connector need to be wired. The 8-position connector is specified in keyed and unkeyed versions. Figure 2-28 shows the basic connector types (a) and a few common wiring configurations for digital circuits associated with modular connectors (b) through (f). Typical applications for these connectors are described in the following paragraphs. Not all connector types or wiring patterns are described here. A more complete description can be found in [19, 20].

RJ-41 and RJ-45: Although not shown in Figure 2-28, the RJ-41S and RJ-45S connectors are used to connect data communication equipment (DCE), such as modems and LAN bridges and routers, to the PTN. The connectors are keyed and are configured to provide proper transmission levels at the network interface (NI). An 8-position connector frequently is used in twisted pair premises wiring systems to interconnect computer workstations and LANs. In applications like these, the connector is frequently, but incorrectly, called an RJ-45. This designation is incorrect because RJ-45 refers to a wiring pattern that is not consistent with LAN connectors.

RJ-48: An 8-position connector is used as the NI for repeatered T1-carrier loops and HDSL (both DS-1 rate, 1.544 Mbps), SRDLs (2.4 kbps to 64.0 kbps), and LANs

---

*ECSA is now known as Alliance for Telecommunications Industry Solutions (ATIS).

on user premises. A 50-position ribbon connector also is specified for some of these applications. The 8-position connector is designated RJ-48C or RJ-48X, depending on the requirements for a shorting bar, and the 50-position connectors are designated RJ-48H, RJ-48M, and RJ-48T (the RJ-48M is configured for 8 circuits and the RJ-48H and RJ-48T are configured for 12 circuits). The RJ-48S is a keyed version of the 8-position connector. Some connectors have contacts dedicated to shield integrity. Additional details of the DS-1 interface can be found in [20,21].

The 8-position connector is widely used in LAN applications that use twisted pair to connect workstations. It is specified in the IEEE standard for the carrier sense multiple access with collision detection (CSMA/CD) LAN, widely known as Ethernet or 10BaseT [22]. The same connector is used with the token ring LAN specified in another IEEE standard (this standard also specifies a 6-position connector) [23]. The 8-position modular connector also is used with the ISDN DSL NI defined in [24]. This particular wiring configuration is designated SJA11.

As would be expected, all of the interfaces described in the previous paragraph are wired differently. Finally, the 8-position connector is specified in the EIA/TIA structured wiring standard, which assigns each conductor of a 4-pair cable to one of the eight positions (rather than function/position assignment) [25]. Two wiring configurations are specified, called T568A ("TIA") and T568B ("AT&T"), respectively. The latter indicates the influence AT&T has on standards-making bodies.

The only 6-position digital NI connector is the SJA-48, which is used in 2-wire applications of the switched digital access line operating at 56 kbps (or lower

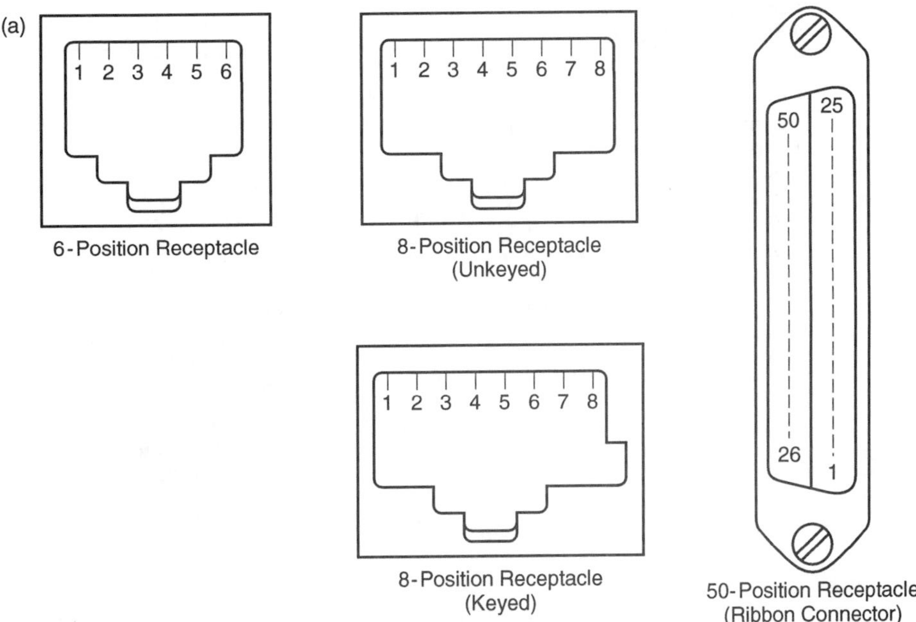

Figure 2-28 Wiring configurations for typical plugs and jacks: (a) 6-, 8-, and 50-position jacks; (b) RJ-48C; (c) RJ-48M; (d) RJ-48S; (e) RJ-48T; (f) RJ-48X

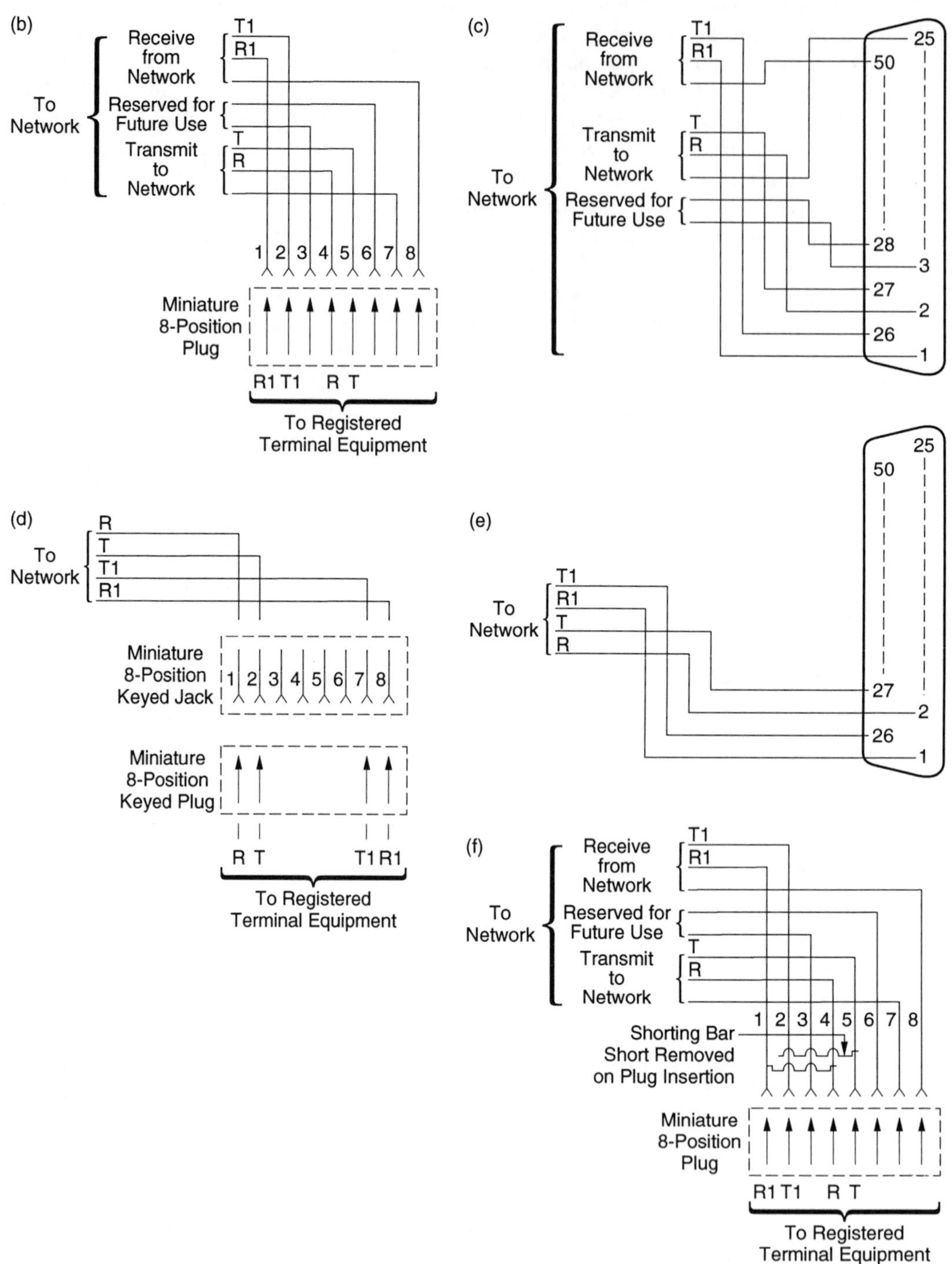

Figure 2-28 Continued

subrate speeds). A 4-wire version of the switched digital access line uses the SJA-56 or SJA-57 connector. The SJA-56 is an 8-position connector for single-line applications and the SJA-57 is a 50-position connector that allows access to 12 switched digital lines. The SJA-56 is similar to the RJ-48S and the SJA-57 is similar to the RJ-48T.

Tables 2-5, 2-6, and 2-7 can be used to distinguish the different wiring configurations. In addition to the digital circuit connectors described above, these tables show the connectors commonly used with analog interfaces as well. A blank position in the tables indicates no connection.

When circuits have separate transmit and receive circuits (4-wire), the directions normally are indicated by the nomenclature T, R and T1, R1. T, R indicates tip, ring—*transmit* and T1, R1 indicates tip, ring—*receive*. The direction (transmit or receive) usually is taken with respect to the customer installation (CI) on the user premises; that is, the T, R leads are used to transmit from the user CI to the user interface. Where circuits are interconnected in tandem, the sense of direction must be maintained. Normally, this means the T, R leads from the CI (transmit) are connected to the T1, R1 of the NI (receive), but not always. The exchange carrier usually defines the direction and designation of the leads for their equipment at the NI. Variations in nomenclature exist, especially where carrier transmission equipment is involved. Therefore, it is more important to use a common direction reference than it is to use a common nomenclature.

Table 2-5 6-Position Connector

	Contact Number					
Type	1	2	3	4	5	6
RJ-11C/W			R	T		
RJ-14C/W		T-2	R-1	T-1	R-2	
Token Ring (IEEE 802.5) [23]		TX−	RX+	RX−	TX+	
SJA-48 [20]			R	T		

2.17 Digital Loop Equipment Operating Constraints

2.17.1 Operating Voltages

There are three nominal voltages used to power telecommunications equipment. Equipment should be designed to meet specified requirements throughout the anticipated voltage ranges for each nominal voltage, as follows:

- Nominal 120 Vac with range A and range B utilization voltages of 110 to 126 and 106 to 127 volts, respectively, as specified in ANSI Standard C84.1 [26]. Range A voltages are encountered in any normally operating electrical system. Range B voltages extend beyond range A voltages and are "limited in extent, frequency and duration." Typical equipment will operate satisfactory over slightly wider limits than range B, typically 117 Vac ± 10%.

Table 2-6 8-Position Connector

Type	Contact Number								Keyed
	1	2	3	4	5	6	7	8	
RJ-31X[a]	R1			R	T			T1	No
RJ-41S	R (FL)	T (FL)		R (PL)	T (PL)		PR	PC	Yes
RJ-45S				R (PL)	T (PL)		PR	PC	Yes
RJ-48C	R1	T1		R	T		SI1	SI	No
RJ-48S	R	T					T1	R1	Yes
RJ-48X	R1	T1		R	T		SI1	SI	No
T568A (TIA) [25]	T-3	R-3	T-2	R-1	T-1	R-2	T-4	R-4	No
T568B (AT&T) [25]	T-2	R-2	T-3	R-1	T-1	R-3	T-4	R-4	No
10BaseT (IEEE 802.3) [22]	TD+	TD−	RD+			RD−			No
Token Ring (IEEE 802.5) [23]			TX−	RX+	RX−	TX+			No
SJA-11 (ANSI T1.601) [24]				T/R	R/T				No
SJA-56 [20]	R	T					T1	R1	Yes

[a]The RJ-31X is considered obsolete as a NI connector.

Table 2-7 50-Position Connector

Type	Contact Number												
	1	26	2	27	3	28	·	23	48	24	49	25	50
RJ-21X	R-1	T-1	R-2	T-2	R-3	T-3	·	R-23	T-23	R-24	T-24	R-25	T-25
RJ-23X[a]	R-1	T-1	A1-1	A-1	R-2	T-2	·	R-12	T-12	A1-12	A-12		
RJ-48H	R1-1	T1-1	R1-2	T1-2	R1-3	T1-3	·	R-10	T-10	R-11	T-11	R-12	T-12
RJ-48M	R1-1	T1-1	R-1	T-1			·	R-8	T-8				
RJ-48T	R1-1	T1-1	R-1	T-1	R1-2	T1-2	·	R1-12	T1-12	R-12	T-12		
SJA-57	R1-1	T1-1	R-1	T-1	R1-2	T1-2	·	R1-12	T1-12	R-12	T-12		

[a]The RJ-23X is considered obsolete as an NI connector.

Key for Tables 2-5, 2-6, and 2-7:

T	Tip (for a 4-wire circuit, the direction is from network to customer interface)
R	Ring (for a 4-wire circuit, the direction is from network to customer interface)
T1	Tip (for a 4-wire circuit, the direction is transmit to network from customer interface)
R1	Ring (for a 4-wire circuit, the direction is transmit to network from customer interface)
T-1	Tip, Line 1
R-1	Ring, Line 1
T-2	Tip, Line 2
R-2	Ring, Line 2
FL	Fixed loss loop connection
PL	Programmable loss loop connection
PR	Pad programming resistor connection
PC	Pad programming resistor common connection
SI	Shield integrity (from network cable)
SI1	Shield integrity (to network cable)
A1	Lamp control lead, common (applies to 1A2 electromechanical key systems)
A	Lamp control lead (applies to 1A2 electromechanical key systems)

- Nominal −48 Vdc with range from −42 to −56 V. The −42 V corresponds to 1.75 V/cell in the 24-cell battery commonly used in telecommunications systems. Low-voltage disconnects in dc power systems normally are set to −42 V to prevent damage to battery cells from overdischarge. The −56 V corresponds to 2.33 V/cell, which is considered to be the upper equalizing voltage limit for lead-acid battery cells (wet-cells). Many telecommunications power systems use maintenance-free batteries with suspended or gelled electrolyte. Because of their electrochemical makeup, maintenance-free batteries operate at slightly higher voltages than regular wet-cell batteries. To meet the upper equipment voltage requirement of 56 V, these batteries consist of 23 cells in many installations rather than 24 cells. Slightly different values for the lower and upper voltage limits are given in a recently published industry standard [27]. These are 42.75 and 56.7 V and are measured at the load equipment terminals. The lower limit is chosen to accommodate existing equipment and the higher limit is based on a cell voltage of 2.35 V, a drop of 0.0 V in the power cables, and a +0.5% voltage regulation.
- Nominal −24 Vdc with range from −21 to −28 V. These voltages correspond to the normal operating voltages of a 12-cell battery. The same per-cell limits apply as with the 24-cell battery. The industry standard cited in the previous paragraph gives the upper and lower voltage limits as 20.0 and 28.3 V. The lower limit assumes a 1.0 V drop in the power cables. The upper limit is based on the same parameters as given for the nominal 48 V system.

2.17.2 Environmental Requirements

Loop equipment can be installed in several different environments—central office, indoor, and outdoor. Central offices normally have the most stringently controlled environments of these three. The temperature and humidity ranges that can be expected in central offices are shown in Table 2-8. In this table, "short term" means no more than 72 consecutive hours and 15 days per year, in aggregate.

Indoor equipment should meet performance requirements over a temperature range of 0 to +50 °C. Outdoor equipment should perform over a temperature range of −40 to +60 °C. In arctic or subarctic environments, a lower temperature of −50 to −60 °C is sometimes used. Indoor equipment sometimes is adapted for outdoor environments by installing it in an enclosure, with temperature-control devices such as fans for high-temperature control and heaters for low-temperature control. The

Table 2-8 Central Office Environmental Ranges[a]

Parameter	Condition	Range
Temperature	Continuous	+4 to +38 °C
	Short term	+2 to +49 °C
Humidity	Continuous	20 to 55%
	Short term	20 to 80%

[a]Source: [28]

normal relative humidity range for indoor equipment is 0 to 95%. Outdoor equipment has to operate over a range of 0 to 100%.

To accommodate installations in mountainous regions, equipment should be designed to meet specified requirements at altitudes of −1,000 feet up to 15,000 feet. Low-voltage equipment has little trouble meeting this requirement.

References

[1] Jayant, N. "Signal Compression: Technology Targets and Research Directions." IEEE Journal on Selected Areas in Communications, Vol. 10, No. 5, June 1992.

[2] Harrington, G. *Say 'Goodbye' to LECs As We Know Them*. Telephone Engineer & Management, Aug. 1, 1992.

[3] *Functional Criteria for Digital Loop Carrier Systems*. BELLCORE Technical Reference TR-TSY-00057, Nov. 1988. Available from BELLCORE Customer Service.

[4] *Digital Interface between the SLC®96 Digital Loop Carrier System and Local Digital Switch*. BELLCORE Technical Reference TR-TSY-000008, Aug. 1987. Available from BELLCORE Customer Service.

[5] *Integrated Digital Loop Carrier System Generic Requirements, Objectives, and Interface*. BELLCORE Technical Reference TR-TSY-000303, Sept. 1986 (with revisions and supplements). Available from BELLCORE Customer Service.

[6] Boucher, J.R. *Traffic System Design Handbook*. IEEE Press, 1993.

[7] *High-Capacity Digital Special Access Service—Transmission Parameter Limits and Interface Combinations*. BELLCORE Technical Reference TR-INS-000342, Feb. 1991. Available from BELLCORE Customer Service.

[8] *Generic Requirements for High-Bit-Rate Digital Subscriber Lines*. BELLCORE Technical Advisory TA-NWT-001210, Oct. 1991.

[9] Kostalek, R. *Using HDSL Technology as a Transition Strategy to FITL*. Presentation No. 112 at 1992 SuperCom/ICC'92, Chicago, Illinois.

[10] *Interface between Data Terminal Equipment and Data Circuit-Terminating Equipment Employing Serial Binary Data Interchange*. EIA/TIA-232E. July 1991.

[11] *Data Communications Over the Telephone Network*. Series V Recommendations (Study Group XVII), Vol. VIII, Fascicle VIII.1. CCITT Blue Book. Available from National Technical Information Service (NTIS).

[12] *General Purpose 37-Position and 9-Position Interface for Data Terminal Equipment and Data Circuit-Terminating Equipment Employing Serial Binary Data Interchange*. EIA-449. Dec. 1984.

[13] *Asymmetric Digital Subscriber Line (ADSL): Technology and System Considerations*. BELLCORE Special Report SR-TSV-002240, June 1992.

[14] Sistanizadeh, K. *Spectral Compatibility of ADSL with Basic Rate DSLs, HDSLs, and T1 Lines*. Section 4 of BELLCORE Special Report SR-TSV-002240, June 1992.

[15] *Generic Requirements and Objectives for Fiber in the Loop Systems*. BELLCORE Technical Reference TR-NWT-000909, Dec. 1991.

[16] Horn, S. "When Fiber Comes Marching Home," *TE&M* magazine, Feb. 1, 1992.

[17] Okada, K., Shinohara, H. "Fiber Optic Subscriber Systems." IEEE LTS magazine, Vol. 3, No. 4, Nov. 1992.

[18] Reeve, W. *Subscriber Loop Signaling and Transmission: Analog*. IEEE Press, 1992.

[19] Code of Federal Regulations. *Title 47: Telecommunication, Part 68—Connection of Terminal Equipment to the Telephone Network*. US Government Printing Office.

[20] *Carrier to Customer Installation Interface Connector Wiring Configuration Catalog*. Technical Report #5, prepared by T1E1.3 Working Group on Connectors and Wiring Arrangements, Exchange Carrier Standards Association, June 1990.

[21] American National Standard for Telecommunications. *Carrier-to-Customer Installation, DS1 Metallic Interface*. ANSI T1.403-1989.

[22] Local and Metropolitan Area Networks, 8023 supplement. *System Considerations for Multisegment 10 Mb/s Baseband Networks (Section 13), Twisted-Pair Medium Attachment Unit (MAU), and Baseband Medium, Type 10BASE-T (Section 14)*. Institute of Electrical and Electronics Engineers. IEEE Standard 802.3i-1990, New York.

[23] *IEEE Recommended Practice for Use of Unshielded Twisted Pair Cable (UTP) for Token Ring Data Transmission at 4 Mb/s*. Institute of Electrical and Electronics Engineers. IEEE Standard 802.5b-1991, New York.

[24] American National Standard for Telecommunications. *Integrated Services Digital Network (ISDN)—Basic Access Interface for Use on Metallic Loops for Application on the Network Side of the NT (Layer 1 Specification)*. ANSI T1.601-1988, New York.

[25] *Commercial Building Telecommunications Wiring Standard*. Electronic Industries Association. EIA/TIA-568, 1991.

[26] American National Standard for Electric Power Systems and Equipment. *Voltage Ratings (60 Hz)*. ANSI C84.1-1982, New York. Available from ANSI.

[27] American National Standard for Telecommunications. *Voltage Levels for DC-Powered Equipment—Used in the Telecommunications Environment*. ANSI T1.315-1994.

[28] American National Standard for Telecommunications. *Telephone Central Office Equipment: Ambient Temperature and Humidity Requirements*. ANSI T1.304-1989.

3 Digital Loop Frame Structures and Line Codes

Chapter 3 Acronyms

2B1Q	2-binary, 1-quaternary line	DLC	digital loop carrier
act	activation (bit)	DP	dial pulse
aib	alarm indicator bit	DPO	dial pulse originating
AIS	alarm indication signal	DPT	dial pulse terminating
AMI	alternate mark inversion	DS-[ ]	digital signal-level [variable]
ANSI	American National Standards Institute	DSL	digital subscriber line
AWG	American wire gauge	DSX	digital signal cross-connect
B8ZS	bipolar with 8-zero substitution	E	received signaling
BNZS	bipolar with N-zero substitution	eoc	embedded operations channel
BPV	bipolar violation	ESF	extended superframe
BRA	basic rate access	FAC	frame alignment channel
BRI	basic rate interface	FAS	frame alignment signal
CCC	clear channel capability	febe	far-end block error (bit)
CCITT	The International Telegraph and Telephone Consultative Committee	FPS	framing pattern sequence
CMB	cyclic redundancy check message block	FXO	foreign exchange office
codec	COder-DECoder	FXS	foreign exchange station
CPE	customer premises equipment	HDSL	high bit-rate digital subscriber line
CRC, crc	cyclic redundancy check	hrp	HDSL repeater present (bit)
CSC	common signaling channel	HTU-C	central office high bit-rate terminal unit
cso	cold-start only (bit)	HTU-R	remote high bit-rate terminal unit
CSU	channel service unit	I-bit	identification bit
DDS	digital data service	IDLC	integrated digital loop carrier
dea	deactivation (bit)	ISDN	integrated services digital network
DL	data link	ISW	inverted synchronization word

continued

LAPD	link access procedure for the D-channel	ps	power status (bit)
LOFA	loss-of-frame alignment	QRSS	quasi-random signal source
LOF	loss-of-frame	RZ	return-to-zero
LT	line termination	SAC	superframe alignment channel
M	transmit signaling	SF	superframe format
MAS	multiframe alignment signal	SRDL	subrate digital loop
MF	multifrequency	SRDM	subrate data multiplexer
NI	network interface	SW	synchronization word
NRZ	nonreturn-to-zero	T	transparent
NT	network termination	TABS	telemetry asynchronous block serial protocol
ntm	network termination 1 in test mode (bit)	TMC	timeslot management channel
OCU	office channel unit	uib	unspecified indicator (bit)
PAM	pulse-amplitude modulated, -tion	VAZO	violating all-zero octet
PBX	private branch exchange	VF	voice frequency
pps	pulse per second	ZBTSI	zero-byte timeslot interchange
PRI	primary rate interface		

3.1 Frame Structures

Frame structures specify the distinctive, and usually cyclic, arrangement of timeslots in a digital transmission line for the purpose of identifying individual timeslots in a frame of information. Frame structures also are called *framing formats,* especially in early literature associated with T1-carrier. Framing of digital loop signals accounts for the need to synchronize and identify data bits at four levels:

- Bit level
- Byte level
- Frame level
- System or network level

The information transmitted on digital loops is composed of individual data bits. Error-free performance depends on the precise control of the bit rates throughout the network. This means that the loop must be synchronized with the network at the bit level. The number of bits transmitted are never more than the number of bits received, and the number of bits received cannot exceed the temporary storage capacity of loop interfaces, which, for T1-carrier, is usually a 386 bit frame buffer.

Information is processed at the byte level (unique strings or groups of 8-bit words) by digital cross-connect and switching systems. In order for these systems to identify unique strings of bits, the serial data stream must include frame synchronization. A frame usually includes a number of data bytes. In multichannel loops, such as T1-carrier, each byte will belong to a different channel. Once the frame is identified, the bytes belonging to each channel can be identified.

Framing exists at every level of the digital multiplex hierarchy. A group of frames from a lower multiplex level are combined using a pre-established pattern into a higher level. The higher-level frame includes information on the location of each lower-level frame so that the lower-level frames can be identified and removed at their destination. Different network types will have different framing requirements. It is beyond the scope of this book to discuss the framing patterns for levels beyond those used in digital loops.

3.1.1 Frame Structure Types

The exact framing structure of data strings depends on the type of digital loop. Of particular interest are T1-carrier, subrate digital data services, integrated services digital network (ISDN) loops, and the high bit-rate digital subscriber line (HDSL). Even though each of these types has different requirements, they share transmission facilities in many situations. The framing structure for each loop type is described in the following sections.

3.1.2 T1-Carrier Frame Structure*

The first widely used digital transmission scheme was T1-carrier. In this book, the associated loops are called repeatered T1-carrier. T1-carrier was originally designed to combine (multiplex) digitally encoded voice signals from 24 analog channels onto a data stream operating at an aggregate rate of 1.544 Mbps. The structure of this data stream spells out the relationship between each of the channels. As T1-carrier evolved, the formatting changed to accommodate the network requirements. A variety of structures have been introduced over time, and these will be discussed (the term *framing format* is used interchangeably with *frame structure* for consistency with early nomenclature):

- D1 Format
- D2 Format
- Superframe (SF, D3, or D4) Format
- Extended Superframe (ESF or Fe) Format
- CCITT G.704 Format

---

*Several references were used to develop this section: [1–4].

- SLC®96* and Integrated Digital Loop Carrier Format
- Unformatted T1

The nomenclature used to describe the T1-carrier frame structure is based on the D-type digital channel banks developed for the transmission system. The first generation of T1-carrier channel bank was called D1. The D2, D3, D4, and D5 channel banks evolved later, with each succeeding digital channel bank providing more capabilities and using more advanced technologies. The D1 channel bank was designed for exchange applications, specifically interoffice trunking, with circuit lengths less than about 25 miles. Variations of the D1 are called D1A, D1B, D1C, and D1D.

The second-generation channel bank was called D2, and it was designed for inter-exchange toll trunking applications. D2 offered a significant improvement in transmission quality over D1. The next generation, D3, was developed to use integrated circuits for cost reduction. Since D3 used a different frame structure than D1A and D1B, the D1D channel bank was developed as a retrofit to the earlier channel banks to make them end-to-end compatible with D3. D1C was developed for a special application requiring remote operator capability.

The D4 channel bank uses the same frame structure as D3, but was optimized as a 48-channel terminal for T1C applications. However, the D4 channel bank can be optioned for D3 operation. Finally, the latest generation, or D5 terminal system, was developed with microprocessor-controlled maintenance and administrative functions.

All D-type channel banks convert between analog (voice frequency, VF) channels and DS-1 rate (1.544 Mbps) aggregate serial bit streams. The D4 and D5 channel banks also have the capability of converting to DS-1C rate (3.152 Mbps) bit streams with 48 channels. Furthermore, two D4 channel banks may be connected together to provide a DS-2 rate (6.312 Mbps) bit stream with 96 channels. The later-generation channel banks also are capable of directly accepting multiple digital data signals at the channel level and multiplexing them into the aggregate bit stream.

Each of 24 VF channels connected to the channel bank is sampled at an 8 kHz rate and then converted to a 7- or 8-bit word. When 7-bit encoding is used, an 8th bit is available for signaling and supervision. Therefore, each channel has an 8-bit output, which provides a total of 192 bits for the 24 channels. A framing bit is added at the beginning of each group of 192 bits, to give a total of 193 bits in a frame. Although the frame patterns for the various channel bank types are similar, the allocation of bits and framing bit values vary, and for this reason are incompatible unless equipped with software or hardware options.

Each bit is placed into a predetermined timeslot in the data stream. The bits—or more specifically, the timeslots—associated with a particular channel are mapped (Table 3-1) for the various D-type channel banks. The different mapping introduces additional incompatibility between types unless they have mapping options.

---

*SLC®96 is a registered trademark of AT&T.

Table 3-1 T-Carrier Timeslot Assignments

TIMESLOT	D1 FORMAT (INTERLEAVED CHANNEL)	D2 FORMAT (PSEUDO-RANDOM CHANNEL)	D1D, D3, D4, AND D5 FORMAT (SEQUENTIAL CHANNEL)
1	1	12	1
2	13	13	2
3	2	1	3
4	14	17	4
5	3	5	5
6	15	21	6
7	4	9	7
8	16	15	8
9	5	3	9
10	17	19	10
11	6	7	11
12	18	23	12
13	7	11	13
14	19	14	14
15	8	2	15
16	20	18	16
17	9	6	17
18	21	22	18
19	10	10	19
20	22	16	20
21	11	4	21
22	23	20	22
23	12	8	23
24	24	24	24

3.1.2.1 D1 and D2 Frame Structure

The framing characteristics of the D1 and D2 channel banks are similar, but they are incompatible in other ways unless digital signal processing is used. From a modern network perspective, D1 and D2 are of historical interest only; however, these early-generation systems set the stage for later versions and are discussed here for that reason.

In the basic D1 system, seven bits are used for digitally encoded voice and one bit for signaling and supervision, as shown in Figure 3-1(a). The bit 1 timeslot in each word is reserved for signaling and supervisory information for the previous channel. Bits 2 through 8 are the data bits associated with the digitally encoded voice for that channel.

The receiving terminal must be able to identify properly the frame of 192 data bits, so a 193rd time slot is inserted for the framing bit. The framing bit alternates between a binary zero and a binary one in succeeding frames, as shown in Table 3-2. As a result, the aggregate data stream has a stable signal component at one-half the frame rate. Since the frames occur at an 8 kHz rate, the alternating frame pulses produce a strong 4 kHz component. The framing circuitry in the line receiver

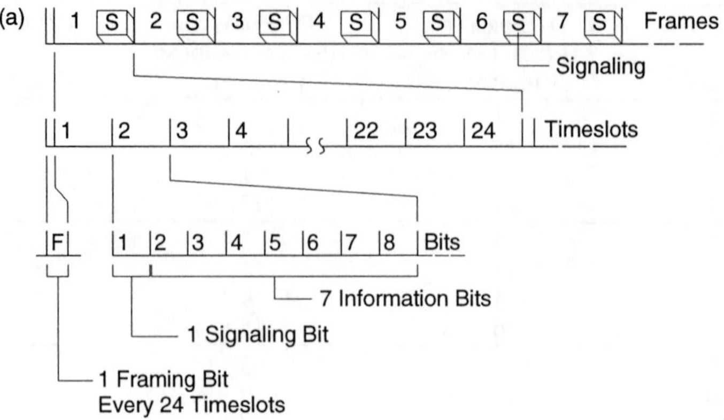

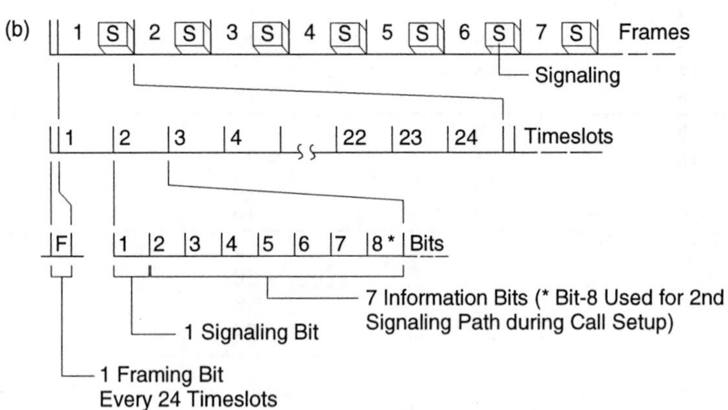

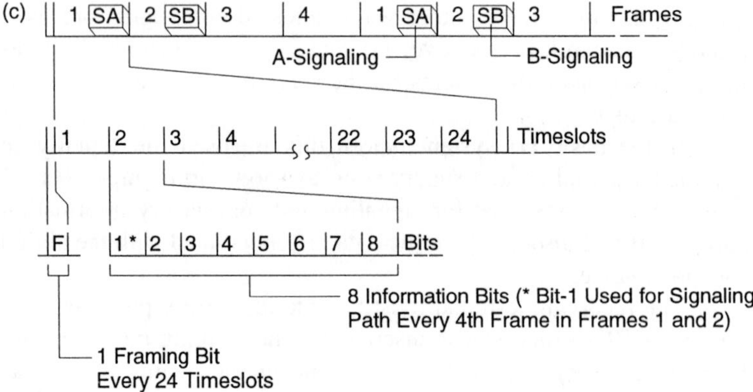

Figure 3-1 D1 framing format: (a) Basic D1 framing format; (b) D1A framing format; (c) D1B framing format

Table 3-2 D1, D1A, and D1B Frame Structure

Frame No.	F1	F2	F3	F4	F1	F2	F3	F4
D1 Framing Bits	0	1	0	1	0	1	0	1
D1A Signaling Frame	SaSb	SaSb	SaSb	SaSb	SaSb	SaSb	SaSb	SaSb
D1B Signaling Frame	Sa	Sb			Sa	Sb		

Sa = Path A signaling (bit 1 of each frame for D1A; bit 1 of frame 1 for D1B)

Sb = Path B signaling (bit 8 of each frame for D1A except after call setup; bit 1 of frame 2 for D1B)

locks onto the frame rate. If synchronization is lost, the receiver slips one bit per frame until synchronization is regained. Two complete frames, or 386 bits, are checked before an alarm is initiated.

The D1A and D1B channel banks use a diode-based compander with a μ = 100 companding characteristic followed by a linear coder-decoder (codec). In order to ensure adequate clock recovery in T1 line and office repeaters, the D1 channel bank suppresses an all-zeros code in any of the channels. Such a code has a small, but finite, probability with voice encoding. The all-zeros code was replaced by substituting the code word

$$0\,0\,0\,0\,0\,0\,1\,0$$

whenever the code word

$$0\,0\,0\,0\,0\,0\,0\,0$$

appeared in any channel position. Only the 7th bit is affected. This bit 7 zero code suppression is used in all *voice* applications, including analog channel units used with D2, D3, D4, and D5 channel banks.

A single bit, Sa, in each channel is dedicated to transmitting signaling and supervisory information from one terminal to another, as shown in Table 3-2. Certain types of signaling require two signaling paths in each direction, but only one bit is available in the basic D1 frame structure. To accommodate this requirement, two separate signaling schemes were developed, D1A and D1B.

With D1A, the 8th bit (least significant bit) in each channel is borrowed to provide the second signaling path, or Sb in Table 3-2. See Figure 3-1(b) for the D1A bit layout within the frame structure. Once the call is established and answer supervision provided, the bit is returned to exclusive use in transmitting the encoded voice. However, with certain calls that do not provide answer supervision (for example, "free" calls), the 8th bit continues to be used for signaling. This results in a significant degradation in decoded voice quality, since only six bits are being used in that channel for voice encoding.

D1B framing overcame the limitations of D1A by using the bit 1 timeslot in one frame for one signaling path and the same timeslot in the next frame for the second signaling path. Both signaling paths are inhibited during frames 3 and 4 to prevent the receiver from locking onto a false frame set up by statistically possible signaling bit patterns that mimic the alternating one-zero framing pattern. This results in the four-frame pattern shown in Table 3-2 and Figure 3-1(c). With D1B, each of the two signaling paths are sampled every fourth frame, or every 500 μs. This provided about 100 samples during a typical dial pulse interval.

The D2 format was developed to provide transmission quality improvement, so T1-carrier could be used in inter-toll applications. It provided a significant improvement in quantizing noise over the D1 format (approximately 4 dB improvement). The D2 channel versus timeslot assignment differs from D1, as previously seen. Because of these improvements and changes, D2 channel banks are not compatible with D1 on an end-to-end basis. Nevertheless, both use interchangeable T1 repeatered span lines.

As with D1 framing, D2 framing has 24 channels with eight bits per channel and a framing bit in the 193rd timeslot. However, D2 uses the full eight bits for data (voice encoding) in five out of every six frames. In the 6th frame, seven bits (bits 1 through 7) are available for data and one bit (bit 8, or least significant bit) is "robbed" for signaling. This results in an average of 7-5/6 bit encoding for a voice signal. Also, a 15-segment, $\mu = 255$ nonlinear codec is used, which provides companding and PCM conversion in one unit. As with the D1 frame structure, an all-zeros code is suppressed in any channel position by substituting binary 0 0 0 0 0 0 1 0 for 0 0 0 0 0 0 0 0 if the latter occurs. In voice transmission, this substitution does not cause a noticeable increase in quantizing distortion.

Since two signaling paths are required by many channel interfaces, a superframe of 12 frames is used with D2. This means that frame 6 will have one signaling path, Sa, and frame 12 will have the other signaling path, Sb. To allow frame identification and synchronization, the frames are divided into odd and even frames. In the odd frames, the framing bit alternates between a binary one and binary zero, which allows the framing circuitry to lock and maintain synchronization. In the even frames, the framing bit follows a binary 0 0 1 1 1 0 . . . pattern, as shown in Table 3-3. This pattern identifies the 6th frame as following a 0-to-1 transition and the 12th frame as following a 1-to-0 transition of the even framing bit. The D2 frame structure is shown in Figure 3-2.

Table 3-3 D2 Framing Bit Pattern

Frame No.	F1	F2	F3	F4	F5	F6	F7	F8	F9	F10	F11	F12
Odd-Frame Bit	1		0		1		0		1		0	
Even-Frame Bit		0		0		1		1		1		0
Composite Pattern	1	0	0	0	1	1	0	1	1	1	0	0
Signaling Frame						Sa						Sb

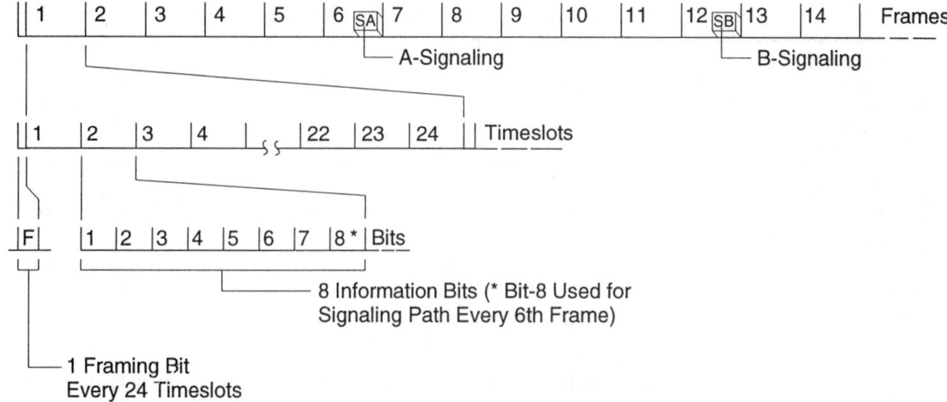

Figure 3-2 D2 and SFs

3.1.2.2 D1D, D3, D4, and D5 Superframe Frame Structures

The D1D, D3, D4, and D5 frame structures are identical in pattern and function; the only difference is the channel bank vintage. Each uses a 7-5/6 bit, 15-segment, μ-law, nonlinear codec for analog VF channels, and associated zero-code suppression, as with D2. End-to-end operation between the D1D and D2 is conceivable, but the timeslot versus channel assignments are different, which can cause administrative problems. For example, with D2, timeslot 1 is channel 12, whereas with D1D, timeslot 1 is channel 1. The D3, D4, and D5 channel banks normally can be optioned for any of the timeslot assignments and thus can be made compatible with D2 framing. The D1D, D3, D4, and D5 channel banks are incompatible with D1A, D1B, or D1C.

Because of the similarity between the D1D, D3, D4, and D5, all will be called the superframe frame structure, or SF, in the following discussion. The SF consists of 12 frames, as shown in Figure 3-1, with frames 6 and 12 carrying signaling and supervisory information in the 8th bit position of each channel. With the introduction of the SF nomenclature, the 6th and 12th frames also became known as signaling frame A (A-channel with corresponding A signaling bit) and signaling frame B (B-channel with corresponding B signaling bit), respectively. The SF was developed by the Bell System, but it has been completely standardized by industry [5–10].

The superframe is partitioned into 12 frames × 193 bits/frame = 2,316 bits. Each frame contains 192 bits of payload and one bit of framing overhead (in this case called F-bit). The 192 bits of payload can be subdivided into 24 blocks or channels of eight bits (1 byte) each, although this is not necessary for the superframe. The payload bits can be used in one or more channels, as determined by the multiplexer at each end of the circuit.

The 12 F-bits in a superframe are divided into the frame alignment channel (FAC), or terminal framing FAC-bits, F1 through F6, and the superframe alignment channel (SAC), or signaling framing SAC-bits, S1 through S6. The FAC-bits and SAC-bits are similar to the odd and even frame bits described for the D2 format. The superframe is shown in Table 3-4.

Table 3-4 Superframe Channels

FRAME No.	F1	F2	F3	F4	F5	F6	F7	F8	F9	F10	F11	F12
FAC-BITS	1		0		1		0		1		0	
SAC-BITS		0		0		1		1		1		0
COMPOSITE PATTERN	1	0	0	0	1	1	0	1	1	1	0	0
SIGNALING FRAME						Note						Note

Note: Either no signaling, Sa, or Sa and Sb; see text.

The FAC-bits are used to locate the 24 payload channels within a frame. The SAC-bits are used to locate the 12 frames within a superframe and to locate the signaling bits. The frame used for signaling is located whenever the SAC-bit changes from a binary zero to a binary one (frame 6) and from binary one to binary zero (frame 12).

The signaling frame can be used in one of three ways:

- No signaling (transparent)
- 2-state signaling (A-channel signaling)
- 4-state signaling (A-channel and B-channel signaling)

The 2-state and 4-state signaling options are called *robbed bit signaling* because the 8th bit (least significant bit) of frames 6 and 12 is used for signaling rather than payload. This is illustrated in Table 3-5. The 2-state option provides a 1,333 bps signaling channel (A-channel), and the 4-state option provides two independent 667 bps signaling channels (A-channel in frame 6 and B-channel in frame 12). These signaling rates are not to be confused with the pulsing rates of dial-pulse channel units. Dial pulses are sampled at the signaling channel rate of 667 or 1,333 samples/s and inserted directly into the signaling channel. Sampling leads to pulse edge distortion when the dial pulses are reconstructed at the far-end terminal. Dial-pulse signaling normally is limited to 20 pulses per second (pps) to keep distortion within acceptable limits. With transparent signaling, no bits are used for signaling and all bits in each channel of every frame are available for payload.

Each signaling channel can convey on-hook and off-hook supervision in both directions. This is the primary use in channel banks in which the supervisory status of individual channels is transmitted from one terminal to the other. Some circuit types require a simple 2-state, on-hook or off-hook status (for example, a dial pulse originating (DPO) channel unit, or 4-wire E&M* channel unit), whereas others require a 4-state status (for example, a foreign exchange circuit that requires ringing as well as supervision, or coin line circuits that require several supervisory functions).

---

*E&M is a type of signaling that uses receive (E) and transmit (M) leads to indicate the supervision and signaling conditions at the far- and near-ends of a circuit, respectively.

Table 3-5 Superframe Signaling Options

Frame No.	Transparent Signaling (Option T)		Robbed Bit Signaling			
	Payload Bits	Signaling Bits	Payload Bits	Signaling Bits	2-State	4-State
1	bits 1–8	None	bits 1–8	None	N/A	N/A
2	bits 1–8	None	bits 1–8	None	N/A	N/A
3	bits 1–8	None	bits 1–8	None	N/A	N/A
4	bits 1–8	None	bits 1–8	None	N/A	N/A
5	bits 1–8	None	bits 1–8	None	N/A	N/A
6	bits 1–8	None	bits 1–7	bit 7	A-channel	A-Channel
7	bits 1–8	None	bits 1–8	None	N/A	N/A
8	bits 1–8	None	bits 1–8	None	N/A	N/A
9	bits 1–8	None	bits 1–8	None	N/A	N/A
10	bits 1–8	None	bits 1–8	None	N/A	N/A
11	bits 1–8	None	bits 1–8	None	N/A	N/A
12	bits 1–8	None	bits 1–7	bit 7	A-channel	B-channel
			Signaling Channel Rate, Each Channel		1,333 bps	667 bps

Several examples of how the A and B signaling channels are used are given in the following tables. These tables use office A/office Z terminology to designate each end of the circuit. Office A is called "west" and office Z is called "east" when there are no intermediate offices. Tables 3-6 and 3-7 show 2-state (A-channel) sig-

Table 3-6 DPO and DPT Signaling

(a) Both Ends Idle

Office A	Channel Unit	Digital Channel	Channel Unit	Office Z
CO[a] trunk circuit	DPO	Signaling state	DPT	CO trunk circuit
Loop open	Transmit →	A = 0 B = 0	Loop open	→ Receive
Receive ←	Normal battery polarity	A = 0 B = 0	← Transmit	Normal battery polarity

(b) Office A Originates

Office A	Channel Unit	Digital Channel	Channel Unit	Office Z	
CO trunk circuit	DPO	Signaling state	DPT	CO trunk circuit	Remarks
Receive ←	Normal battery polarity	A = 0 B = 0	← Transmit	Normal battery polarity	Office Z idle
Loop closure	Transmit →	A = 1 B = 1	Loop closure	→ Receive	Office A seizes trunk
Loop closure	Transmit →	A = 1[b] B = 1[b]	Loop closure	→ Receive	Office A dialing
Receive ←	Reverse battery polarity	A = 1 B = 1	← Transmit	Reverse battery polarity	Office Z answer

(c) Both Ends Busy

Office A	Channel Unit	Digital Channel	Channel Unit	Office Z
CO trunk circuit	DPO	Signaling state	DPT	CO trunk circuit
Loop closure	Transmit →	A = 1 B = 1	Loop closure	→ Receive
Receive ←	Reverse battery polarity	A = 1 B = 1	← Transmit	Reverse battery polarity

[a]CO = central office

[b]If office A uses dial pulse signaling, the A-bit and B-bit follow the dial pulses (0 during on-hook interval and 1 during off-hook interval).

Table 3-7 E&M Signaling

(a) Both Ends Idle

Office A	Channel Unit	Digital Channel	Channel Unit	Office Z
CO[a] trunk circuit	E&M	Signaling state	E&M	CO trunk circuit
M-lead ground or open	Transmit →	A = 0 B = 0	E-lead open	→ Receive
Receive ←	E-lead open	A = 0 B = 0	← Transmit	M-lead ground or open

(b) Office A Originates

Office A	Channel Unit	Digital Channel	Channel Unit	Office Z	
CO trunk circuit	E&M	Signaling state	E&M	CO trunk circuit	Remarks
Receive ←	E-lead open	A = 0 B = 0	← Transmit	M-lead ground or open	Office Z idle
M-lead battery or looped	Transmit →	A = 1 B = 1	E-lead ground or looped	→ Receive	Office A seizes trunk
M-lead battery or looped	Transmit →	A = 1[b] B = 1[b]	E-lead ground or looped	→ Receive	Office A dialing
Receive ←	E-lead ground or looped	A = 1 B = 1	← Transmit	M-lead battery or looped	Office Z answer

(c) Office Z Originates

Office A	Channel Unit	Digital Channel	Channel Unit	Office Z	
CO trunk circuit	E&M	Signaling state	E&M	CO trunk circuit	Remarks
M-lead ground or open	Transmit →	A = 0 B = 0	E-lead open	→ Receive	Office A idle
Receive ←	E-lead ground or looped	A = 1 B = 1	← Transmit	M-lead battery or looped	Office Z seizes trunk
Receive ←	E-lead ground or looped	A = 1[c] B = 1[c]	← Transmit	M-lead battery or looped	Office Z dialing
M-lead battery or looped	Transmit →	A = 1 B = 1	E-lead ground or looped	→ Receive	Office A answer

Table 3-7 Continued

(d) Both Ends Busy

Office A	Channel Unit	Digital Channel	Channel Unit	Office Z
CO trunk circuit	E&M	Signaling state	E&M	CO trunk circuit
M-lead battery or looped	Transmit →	A = 1 B = 1	E-lead ground or looped	→ Receive
Receive ←	E-lead ground or looped	A = 1 B = 1	← Transmit	M-lead battery or looped

[a]CO = central office

[b]If office A uses dial pulse signaling, the A-bit and B-bit follow the dial pulses (0 during on-hook interval and 1 during off-hook interval).

[c]If office Z uses dial pulse signaling, the A-bit and B-bit follow the dial pulses (0 during on-hook interval and 1 during off-hook interval).

naling. Although these only require the A-channel, by convention the signaling is shown using both A-bit and B-bit terminology, where both bits are always the same. It should be remembered that each direction of transmission has independent A and B signaling channels.

Table 3-6(a) through (c) shows various conditions in a system using DPO and dial pulse terminating (DPT) channel units. The DPO channel unit is frequently used with 2-wire, one-way outgoing analog trunk circuits in interoffice and direct inward dial applications. The DPT is used at the other end of the circuit and is connected to a 2-wire, one-way incoming trunk circuit. The channel units will accommodate either dial pulse (DP) or multifrequency (MF) signaling.

In Table 3-6(a), both ends are idle. The A-bit (and, by convention, the B-bit) is set to binary zero in both directions. At office A, an idle trunk is indicated to the DPO by an open loop circuit on the tip and ring leads. At office Z, an idle trunk is indicated to the DPT by a normal battery polarity on the tip and ring leads.

Table 3-6(b) shows that office Z is initially idle with a subsequent seizure by office A. When office A seizes the trunk by closing the loop, the A-bit (and B-bit) changes to a binary-one condition as shown in Table 3-6(b). If office A uses dial pulse signaling, the A-bit (and B-bit) changes state in response to the dial pulses. Office Z transmits off-hook supervision by reversing the battery polarity on the tip and ring leads. This condition is detected by the DPT channel unit, which changes the A-bit (and B-bit) state to binary one. Once a call has been set up and both ends are busy, the signaling states are as shown in Table 3-6(c).

The E&M channel units shown in Table 3-7(a) through (d) use a similar 2-state protocol. In this case, however, either end of the circuit may originate a call. E&M is the most widely used trunk signaling and supervision method. In Table 3-7(a), both ends are idle, and the A-bit (and B-bit) is set to binary zero. There are several types of E&M signaling available; Type I and Type II are the most common. With Type I E&M, an idle M-lead is indicated by a ground, and with Type II E&M, an idle M-lead is indicated by a open loop between the M-lead and SB-lead. Similarly, with Type I E&M, an idle E-lead is indicated by an open circuit on the lead, and with Type II E&M, an idle E-lead is indicated by an open loop between the E-lead and SG-lead.

When office A seizes the circuit, as in Table 3-7(b), it changes the condition of the M-lead. With Type I E&M, seizure is indicated by an M-lead battery, and with Type II, seizure is indicated by a closed loop on the M-lead and SB-lead. The E&M channel unit in office A responds to the change in the trunk circuit M-lead by changing the A-bit (and B-bit) in the digital channel toward office Z to a binary one. When the E&M channel unit in office Z receives the new A-bit state, it reflects the M-lead condition in Office A by changing the condition of the E-lead toward the trunk circuit at the office Z end. With Type I E&M, a busy (off-hook) E-lead condition is indicated by a ground, and with Type II E&M, an off-hook is indicated by a closed loop between the E-lead and SG-lead. When office Z answers, it changes its trunk M-lead to a busy condition (indicated by A-bit set to binary one) that is received at office A on the E-lead.

A similar sequence takes place when office Z originates a call, as shown in Table 3-7(c). When both offices are busy, as with a call in progress, the conditions shown in Table 3-7(d) apply.

Examples of 4-state signaling (A- and B-channels) are given in Tables 3-8 and 3-9. These examples show foreign exchange lines in which dialtone in office A is made available to a station in a distant office Z. The foreign exchange office (FXO) channel is used at the end providing the dialtone (office A in this case) and the foreign exchange station (FXS) channel unit is used at the station end (office Z). The station may be a regular telephone set or a private branch exchange (PBX), and either loop-start or ground-start supervision may be used.*

In applications where the dialtone is provided to a PBX, ground-start supervision normally is used to reduce the likelihood of call collision, or glare. FX ground-start is shown in Table 3-8(a) through (d). The values shown for the A and B signaling channels during various call conditions are as given in an industry standard [10]. However, some manufacturer's implementations may differ slightly, especially with foreign exchange channel units.

Table 3-8 FX Ground Start

(a) Both Ends Idle

Office A	Channel Unit	Digital Channel	Channel Unit	Office Z
CO[a] line circuit	FXO/GS	Signaling state	FXS/GS	Station
Tip = no ground, no ringing	Transmit →	A = 1 B = 1	Tip = no ground, no ringing	→ Receive
Receive ←	Loop open Ring = no ground	A = 0 B = 1	← Transmit	Loop open Ring = no ground

*Additional details of loop-start and ground-start supervision may be found in [44].

Table 3-8 Continued

(b) Station (Office Z) Originates

Office A	Channel Unit	Digital Channel	Channel Unit	Office Z	
CO line circuit	FXO/GS	Signaling state	FXS/GS	Station	Remarks
Tip = no ground, no ringing	Transmit →	A = 1 B = 1	Tip = no ground, no ringing	→ Receive	CO idle
Receive ←	Loop open Ring = ground	A = 0 B = 0	← Transmit	Loop open Ring = ground	Station seizes loop
Tip = ground, no ringing	Transmit →	A = 0 B = 1	Tip = ground, no ringing	→ Receive	CO dialtone
Receive ←	Loop closure Ring = no ground	A = 1[b] B = 1	← Transmit	Loop closure Ring = no ground	Station dialing

(c) Central Office (Office A) Originates (Rings Station)

Office A	Channel Unit	Digital Channel	Channel Unit	Office Z	
CO line circuit	FXO/GS	Signaling state	FXS/GS	Station	Remarks
Receive ←	Loop open Ring = no ground	A = 0 B = 1	← Transmit	Loop open Ring = no ground	Station idle
Tip = ground, no ringing	Transmit →	A = 0 B = 1	Tip = ground, no ringing	→ Receive	CO seizes loop
Tip = ground, ringing	Transmit →	A = 0 B = 0[c]	Tip = ground, ringing	→ Receive	Ring station
Receive ←	Loop closure Ring = no ground	A = 1 B = 1	← Transmit	Loop closure Ring = no ground	Station answer
Tip = ground, no ringing	Transmit →	A = 0 B = 1	Tip = ground, no ringing	→ Receive	CO trip ringing

Table 3-8 Continued

(d) Both Ends Busy

OFFICE A	CHANNEL UNIT	DIGITAL CHANNEL	CHANNEL UNIT	OFFICE Z
CO line circuit	FXO/GS	Signaling state	FXS/GS	Station
Tip = ground, no ringing	Transmit →	A = 0 B = 1	Tip = ground, no ringing	→ Receive
Receive ←	Loop closure Ring = no ground	A = 1 B = 1	← Transmit	Loop closure Ring = no ground

[a]CO = central office

[b]If the station uses dial pulse signaling, the A-bit follows the dial pulses (0 during loop open interval and 1 during loop closure interval).

[c]The B-bit from office A follows the ringing cycle; that is, B = 0 when ringing is present and 1 when no ringing is present.

With ground start, two conditions must be monitored at office A: the tip lead and ringing, which are conveyed over the signaling channel via the A-bit and B-bit, respectively. The tip-lead may have a no-ground condition (A = 1) or ground condition (A = 0), and there may be ringing voltage present (B = 0) or no ringing (B = 1). The B-bit follows the ringing on-off cycle so that the ringing cadence at office A is repeated at office Z. At office Z, the conditions to be conveyed are loop open (A = 0), loop closed (A = 1), ground on ring-lead (B = 0), and no ground on ring-lead (B = 1). The tables show various call types and may be followed similarly to the previous tables for DPO/DPT and E&M signaling.

When the foreign exchange channel units are used for loop start, the number of circuit conditions to be conveyed on the signaling channel are fewer than for ground start. Although it is not required at the FXS, the A-bit always is sent as a binary zero from the FXO. Similarly, the B-bit always is sent as a binary one from the FXS, although it is not required at the FXO. With loop start, the only conditions to be sent from the office A FXO are ringing voltage present on the tip and ring leads (B = 0) or no ringing (B = 1). From the office Z FXS, the only conditions are loop open (A = 0) or loop closed (A = 1). The various circuit conditions and call states are shown in Table 3-9(a) through (d). Note that with FX loop start, the central office line circuit has no way of conveying its busy/idle status to the other end.

The A- and B-channels also can be used to send messages at a 1,333 or 667 bps rate between one terminal and another for alarm and administrative purposes. In other words, instead of simple binary (on/off) signaling, groups of A-bits and B-bits can form messages of any practical length. In the superframe, these messages are specific to each manufacturer. A local terminal (say, a digital switching system) by one manufacturer will not be compatible with a remote terminal (say, a digital loop carrier) by another manufacturer. The SF does not have any standardized performance monitoring messages as in the extended superframe (ESF) described in the next section.

Table 3-9 FX Loop Start

(a) Both Ends Idle

Office A	Channel Unit	Digital Channel	Channel Unit	Office Z
CO[a] line circuit	FXO/LS	Signaling state	FXS/LS	Station
No ringing	Transmit →	A = 0 B = 1	No ringing	→ Receive
Receive ←	Loop open	A = 0 B = 1	← Transmit	Loop open

(b) Station (Office Z) Originates

Office A	Channel Unit	Digital Channel	Channel Unit	Office Z	
CO line circuit	FXO/LS	Signaling state	FXS/LS	Station	Remarks
No ringing	Transmit →	A = 0 B = 1	No ringing	→ Receive	CO idle
Receive ←	Loop closure	A = 1 B = 1	← Transmit	Loop closure	Station seizes loop
No ringing	Transmit →	A = 0 B = 1	No ringing	→ Receive	CO dialtone
Receive ←	Loop closure	A = 1[b] B = 1	← Transmit	Loop closure	Station dialing

(c) Central Office (Office A) Originates (Rings Station)

Office A	Channel Unit	Digital Channel	Channel Unit	Office Z	
CO line circuit	FXO/LS	Signaling state	FXS/LS	Station	Remarks
Receive ←	Loop open	A = 0 B = 1	← Transmit	Loop open	Station idle
Ringing	Transmit →	A = 0 B = 0[c]	Ringing	→ Receive	Ring station
Receive ←	Loop closure	A = 1 B = 1	← Transmit	Loop closure	Station answer
No ringing	Transmit →	A = 0 B = 1	No ringing	→ Receive	CO trip ringing

Table 3-9 Continued

(d) Both Ends Busy

Office A	Channel Unit	Digital Channel	Channel Unit	Office Z
CO line circuit	FXO/LS	Signaling state	FXS/LS	Station
No ringing	Transmit →	A = 0 B = 1	No ringing	→ Receive
Receive ←	Loop closure	A = 1 B = 1	← Transmit	Loop closure

[a]CO = central office

[b]If the station uses dial pulse signaling, the A-bit follows the dial pulses (0 during loop open interval and 1 during loop closure interval).

[c]The B-bit from office A follows the ringing cycle; that is, B = 0 when ringing is present and 1 when no ringing is present.

3.1.2.3 Extended Superframe

The extended superframe, commonly known as ESF but sometimes called F_e format, is specified by industry standards [5,11–14]. Figure 3-3 shows the basic structure of the ESF. The ESF is partitioned into 24 frames of 193 bits each for a total of 24 × 193 = 4,632 bits. Of the 193 bits, 192 are payload bits and one bit is framing overhead (F-bit). The 192 payload bits normally are divided into 24 channels of eight bits each, as with the superframe already described.

The 24 F-bits in the ESF are divided into three channels: extended superframe alignment channel (also called framing pattern sequence, FPS, or terminal synchronization channel, F1 through F6), cyclic redundancy check (CRC, C1 through C6), and data link (DL, M1 through M12). The sequential relationship between these bits is shown in Table 3-10.

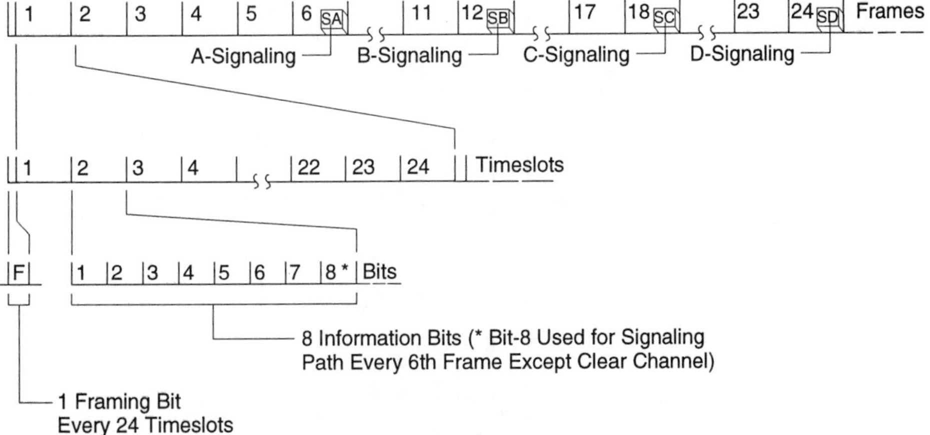

Figure 3-3 Extended SF

Table 3-10 Extended Superframe Channels

Frame No.	F1	F2	F3	F4	F5	F6	F7	F8	F9	F10	F11	F12	F13	F14	F15	F16	F17	F18	F19	F20	F21	F22	F23	F24
FPS				0				0				1				0				1				1
C-Bits		C				C				C				C				C				C		
M-Bits	M		M		M		M		M		M		M		M		M		M		M		M	
Composite F-Bits	M1	C1	M2	F1	M3	C2	M4	F2	M5	C3	M6	F3	M7	C4	M8	F4	M9	C5	M10	F5	M11	C6	M12	F6
Signaling Frame						Note						Note						Note						Note

Note: Either no signaling, Sa, Sa and Sb, or Sa, Sb, Sc, and Sd; see text.

Each frame is sent at an 8 kbps rate. Since the FPS bits appear in every fourth F-bit position, the FPS bit rate is 2 kbps. The FPS-bits are used to locate the 24 frames and to indicate the frame boundaries, position of other F-bits, and position of the signaling frames and signaling channels.

The CRC bits also are sent in every fourth position, or at a 2 kbps rate. The CRC channel carries a six-bit cyclic redundancy check code (CRC-6), which is used to monitor the performance of the ESF. Performance monitoring is discussed in Chapter 6. The value of each CRC bit is calculated from the CRC-6 message block (CMB), which is the sequence of 4,632 bits making up the *preceding* extended superframe. For calculation purposes, all F-bits are set to binary one and all payload bits are left unchanged.

A 4 kbps DL is used to send performance information and other messages. When the DL is not used (idle), the M-bits follow the repeating idle code of 0 1 1 1 1 1 1 0 . . . . The DL carries performance information and control signals across the network interface in both directions. Performance information in one direction describes the transmission quality in the other direction. Two signal formats are used:

- Bit-oriented signals (repeated bit patterns)
- Message-oriented signals using a predefined protocol

The bit-oriented messages include CRC error event, severely errored framing event, frame-synchronization bit error event, line code violation event, and controlled slip event. These are explained in Chapter 6. Performance messages conform either to the CCITT Q.921 link access procedure for the D-Channel (LAPD, also called the ANSI T1.403 protocol) or the messaging procedure described in AT&T Technical Reference TR54016, which is called the telemetry asynchronous block serial (TABS) protocol [11,12,15]. Almost all modern digital terminal equipment can be optioned for either performance message protocol. The bit-oriented signals have priority and will pre-empt performance messages.

The difference between the two message protocols is as follows: the AT&T TR54016 messaging procedure requires the customer installation to detect, organize, and store performance data in 15-minute and 24-hour intervals. This data usually is developed in the channel service unit (CSU). It is sent to the other end of the ESF link on request using the CCITT X.25 protocol [16]. The ANSI T1.403 protocol quantizes the ESF error parameters over one-second intervals and sends the performance data for the current and previous three seconds (four seconds total), every second using an LAPD message. Both AT&T TR54016 and ANSI T1.403 base performance on a one-second window. ANSI T1.403 derived parameters can be translated into the AT&T TR54016 format at the receive end, but the reverse is not true. Table 3-11 compares ANSI T1.403 and AT&T TR54016.

The extended superframe has four signaling options designated T (transparent), 2-state, 4-state, and 16-state. Table 3-12 shows the frame position and channel type for each option. The transparent, 2-state, and 4-state signaling options are identical to those used in the superframe. In addition, the ESF uses frames 18 and 24 in addition to frames 6 and 12 to obtain a 16-state signaling channel.

Table 3-11 Comparison of ANSI T1.403 and AT&T TR54016 Messaging

ITEM	ANSI T1.403	AT&T TR54016
Data link	12-bit	12-bit
Loopback	In-band or FDL	In-band or FDL
Performance statistics:		
CRC errors	Yes	Yes
Out-of-frame	Yes	Yes
Facility data link (FDL)	LAPD	TABS
Protocol	CCITT Q.921	CCITT X.25
Send interval	1 s	On request
Performance window	1 s	1 s
Data storage	4 s	15 min, 24 h
Message protocol translation	Yes	No

CRC = cyclic redundancy check

SF and ESF are not directly compatible on the same link. However, an SF link may be interconnected to an ESF link through a digital cross-connect system, line monitor unit, or channel service unit that is capable of making the necessary conversion.

3.1.2.4 CCITT G.704

The CCITT G.704 recommendations for T1 framing are basically the same as those already discussed for SF and ESF [17]; however, subtle differences exist, which are explained in this section. Table 1 of recommendation G.704 covers the structure of the 24-frame multiframe (similar to the ESF), and Tables 2 and 5 demonstrate the structure of the 12-frame multiframe (similar to the superframe). In the following discussion, references to the CCITT G.704 framing will be called the 24-frame multiframe or 12-frame multiframe for comparison with the ESF and superframe, respectively.

The CCITT recommendations use a slightly different nomenclature than already discussed for the same basic frame structures of SF and ESF. In the 12-frame multiframe, the F-bits are called the frame alignment signal (or FAS in place of the FAC bits) and multiframe alignment signal (or MAS in place of the SAC bits). The framing patterns are identical; that is, the FAS follows an alternating 1 0 1 0 1 0 pattern and the MAS follows a 0 0 1 1 1 0 pattern. The two patterns are interwoven identically to the FAC and SAC bits of the SF.

CCITT G.704 12-frame multiframe only specifies two channel-associated signaling options, A-channel or A- and B-channel, which operate at 1,333 bps and 667 bps rates, respectively, similar to the superframe. Transparent signaling is not specified.

The final difference between the SF and the G.704 12-frame multiframe concerns alarm signaling. Whereas the SF uses an all-ones signal for the alarm indication signal (AIS) and bit 2 of each channel for Yellow Alarm, a G.704-compatible terminal only has provisions for signaling alarm indications to the other end by chang-

Table 3-12 Extended Superframe Signaling Options

Frame No.	Transparent Signaling		Robbed Bit Signaling				
	Payload Bits	Signaling Bit	Payload Bits	Signaling Bit	2-State	4-State	16-State
1	bits 1–8	None	bits 1–8	None	N/A	N/A	N/A
2	bits 1–8	None	bits 1–8	None	N/A	N/A	N/A
3	bits 1–8	None	bits 1–8	None	N/A	N/A	N/A
4	bits 1–8	None	bits 1–8	None	N/A	N/A	N/A
5	bits 1–8	None	bits 1–8	None	N/A	N/A	N/A
6	bits 1–8	None	bits 1–7	bit 8	A-channel	A-channel	A-channel
7	bits 1–8	None	bits 1–8	None	N/A	N/A	N/A
8	bits 1–8	None	bits 1–8	None	N/A	N/A	N/A
9	bits 1–8	None	bits 1–8	None	N/A	N/A	N/A
10	bits 1–8	None	bits 1–8	None	N/A	N/A	N/A
11	bits 1–8	None	bits 1–8	None	N/A	N/A	N/A
12	bits 1–8	None	bits 1–7	bit 8	A-channel	B-channel	B-channel
13	bits 1–8	None	bits 1–8	None	N/A	N/A	N/A
14	bits 1–8	None	bits 1–8	None	N/A	N/A	N/A
15	bits 1–8	None	bits 1–8	None	N/A	N/A	N/A
16	bits 1–8	None	bits 1–8	None	N/A	N/A	N/A
17	bits 1–8	None	bits 1–8	None	N/A	N/A	N/A
18	bits 1–8	None	bits 1–7	bit 8	A-channel	A-channel	C-channel
19	bits 1–8	None	bits 1–8	None	N/A	N/A	N/A
20	bits 1–8	None	bits 1–8	None	N/A	N/A	N/A
21	bits 1–8	None	bits 1–8	None	N/A	N/A	N/A
22	bits 1–8	None	bits 1–8	None	N/A	N/A	N/A
23	bits 1–8	None	bits 1–8	None	N/A	N/A	N/A
24	bits 1–8	None	bits 1–7	bit 8	A-channel	B-channel	D-channel
			Signaling Channel Rate		1,333 bps	667 bps	333 bps

ing the MAS bit in frame 12 from a binary zero to a binary one. This method is not a part of the superframe standard previously described.

The 24-frame multiframe is virtually identical to the ESF except, in G.704, the idle code for the DL (M-bits) is not specified. Some minor terminology differences exist. For example, the loss-of-frame (LOF) condition in the SF and ESF is called loss-of-frame alignment (LFA) in G.704. The relevant attributes of the North American framing formats and G.704 are summarized in Tables 3-13 and 3-14.

Table 3-13 Comparison of SF with G.704 Multiframe

ITEM	SF	G.704 12-FRAME MULTIFRAME
Framing bit nomenclature	Frame alignment channel Superframe alignment channel	Frame alignment signal Multiframe alignment signal
Channel-associated signaling	Transparent, A, A-B	A, A-B
Alarm signaling		
AIS	Unframed all-ones	Frame 12 framing bit set to one
Yellow Alarm	Bit-2 set to zero in each channel	None
Frame loss nomenclature	Loss-of-frame (LOF)	Loss-of-frame alignment (LOFA)

Table 3-14 Comparison of ESF with G.704 Multiframe

ITEM	ESF	G.704 24-FRAME MULTIFRAME
Framing bit nomenclature	ESF alignment channel CRC channel Data link channel	Frame alignment signal Multiframe alignment signal Facility data link
Channel-associated signaling	Transparent, A, A-B	A, A-B
Alarm signaling		
AIS	Unframed all-ones	Frame 12 framing bit set to one
Yellow Alarm	Repeating 8-ones + 8-zeros on data link channel	None
Frame loss nomenclature	Loss-of-frame (LOF)	Loss-of-frame alignment (LOFA)
Data link idle code	Repeating 0 1 1 1 1 1 1 0 . . .	Not specified

3.1.2.5 SLC®96 DLC and TR303 Integrated DLC Frame Structure

The SLC®96 is an early version of integrated digital loop carrier (IDLC). The framing structure is an extension of the SF and is commonly called TR8 after the industry standard [18]. The TR8 superframe consists of 72 frames of 193 bits each. The FAC (terminal framing) bits are retained and have a repeating pattern of 1 0 1

0 1 0 . . . . However, the SAC bits form a DL to carry system status between the central office and remote terminal. A DL frame is nine ms long (72 frames × 125 μs/frame) and forms a structure composed of 36 bits, shown in Table 3-15.

The first 12 bits (sync bits) are used for synchronization in a 0 0 0 1 1 1 0 0 0 1 1 1 . . . pattern. The sync bits are followed by 11 concentrator field bits C1–C11 used to assign channels dynamically on a demand basis. Following the concentrator bits are a 3-bit spoiler field sp12–sp14, consisting of the repeating pattern 0 1 0 . . . , which is used to prevent receiver misframing. A 3-bit maintenance field, M15–M17, controls channel- and drop-side testing. A 2-bit alarm field, A1–A2, carries alarm information and associated control commands. A 4-bit protection line switch field, P20–P23, controls switching of the protection line. A 1-bit spoiler field, SP24, is always binary one and is used to prevent receiver misframing.

The more advanced digital loop carrier (DLC) systems follow the industry standard commonly called TR303 [19]. The frame structure is based on the ESF with synchronization, error detection, and facility DL, as previously described. The 4-kbps facility data link is used to transmit operations and far-end performance messages under normal conditions and then pre-empted during abnormal conditions to transmit alarm conditions and protection line switch commands.

The TR303 interface uses either a *hybrid signaling* or *out-of-band signaling method*. In the hybrid signaling method, robbed bit signaling via the ABCD signaling channels is used for call supervision, and one of the twenty-four 64 kbps voice channels is used as a timeslot management channel (TMC) for timeslot assignment of the lines on a per-call basis. With the out-of-band method, a 64 kbps channel handles timeslot assignments as well as call supervision. This channel is configured as a common signaling channel (CSC). Both the TMC and CSC use the LAPD protocol.

Table 3-15 TR8 Signaling Alignment Channel DL

	Sync Word	Concentrator Field	Spoiler Field	Maintenance Field	Alarm Field	Protection Switch Field	Spoiler Field
No. of Bits	12	11	3	3	2	4	1
Bit Positions	1–12	13–23	24–26	27–29	30–31	32–35	36
Designation	SW	C1–C11	SP12–SP14	M15–M17	A1–A2	P20–P23	SP24

3.1.2.6 Unframed T1

Unframed T1 is used when digital bandwidth is required that has no relationship to the normal D-type, 24-channel frame structure. Although the name implies that no framing whatsoever is used, in fact some type of frame structure has to be used to identify boundaries of bit groups. However, this frame structure does not follow industry standards and is always proprietary to a manufacturer. This means the terminals at each end of a circuit must be made by the same manufacturer, although this does not always guarantee compatibility, possibly because of different hardware or software vintages.

Even though a terminal may use an unstructured signal in the sense of normal SF or ESF, the transmitted signal must still meet the minimum ones-density requirements if it is connected to a repeatered T1-carrier span line using twisted pairs. In fact, any transmission system that uses the alternate mark inversion (AMI) line code without some inherent means of zero code suppression has a ones-density requirement. Two methods currently used to ensure ones-density on T1-carrier are B8ZS and ZBTSI, both of which are discussed in later sections. The ones-density requirement generally does not apply to fiber optic links or terrestrial microwave radio links unless these systems recover the clock from the DS-1 rate signal.

Fractional T1 itself is not a transmission system, except in some proprietary systems. In loop applications, the full bandwidth of a DS-1 rate signal (T1-carrier) with SF or ESF is used for transmission, but only a partial number of channels are assigned to the end-user. These channels are assigned in multiples of 64 kbps, or N × 64 kbps, where N = 1 to 24.

3.1.3 ISDN Primary Rate Interface

The ISDN primary rate inferface (PRI) uses the same basic frame structure as the ESF including FAC, CRC, and DL; however, the individual payload channels are allocated in a different way, as shown in Table 3-16.

Table 3-16 ISDN PRI Channel Combinations

DESIGNATION	N	B (64 kbps)	D (64 kbps)	PAYLOAD
N × B + D	≤23	Yes	Yes	128 kbps–1.536 Mbps
N × B	≤24	Yes	No	64 kbps–1.536 Mbps
N × H0	4	No	No	384 kbps
N × H0 + D	3	No	Yes	384 kbps
H1	—	No	No	1.536 Mbps
H10	—	No	Yes	1.536 Mbps

3.1.4 ISDN Basic Rate Interface

The frame structure specification for the basic access ISDN (2B + D) is described in industry standards [20–22]. This is variously called BRI (basic rate interface) or BRA (basic rate access) and is implemented on the ISDN digital subscriber line (DSL). The format uses frames and superframes (also called multiframes). The basic frame has a length of 1.5 ms and is composed of 240 bits, giving a bit rate of 160 kbps (80 kbaud). These bits are partitioned into:

- one slot for a synchronization word (also called frame word) of 18 bits
- 12 slots of 2B + D payload of 18 bits each (total 216 bits)
- one slot of frame overhead (M-channel or control link) of 6 bits

All bits except the synchronization word are scrambled before they are coded. Bit pairs are formed into quaternary symbols (quats), using the 2-binary, 1-quaternary line (2B1Q) code, as shown in Table 3-17. Conversion to quats is a function of line coding, which is discussed later in this chapter. The frame structure with the

Table 3-17 Quaternary Symbols

First Bit (Sign)	Second Bit (Magnitude)	Quat
1	0	+3
1	1	+1
0	1	−1
0	0	−3

corresponding quats is shown in Table 3-18. This table also references other tables that describe each function in greater detail.

An ISDN DSL superframe consists of eight frames, as shown in Table 3-19. The superframe length is 8×1.5 ms = 12 ms. The inverted synchronization word (ISW) is used at the beginning of the first frame of the superframe and a normal synchronization word (SW) is used at the beginning of the subsequent seven frames. The ISW and SW are composed as shown in Table 3-20.

Twelve fields of 2B + D payload follow the SW in each frame. Each field consists of 18 bits as follows: B1 = 8 bits; B2 = 8 bits; D = 2 bits. These fields are completely transparent to user data, except during startup. It is not necessary for each channel to be used. Unused channels are filled with an idle code provided by the equipment connected to the network interface (NI). The format for the payload is shown in Table 3-21. The quats are generated after the payload is scrambled.

Superframes are numbered A, B, C, . . . in the line termination (LT) to network termination 1 (NT1) direction and 1, 2, 3, . . . in the NT1 to LT direction.

Table 3-18 ISDN BRI Frame Structure

Frame	ISW/SW	12 × (2B + D)	M1 through M6
Function	Synchronization word	Payload	Overhead
No. of bits	18	216	6
Bit positions	1–18	19–234	235–240
No. of quats	9	108	3
Quat positions	1–9	10–117	118–120
Signaling rate, bps	12 kbps	144 kbps	4 kbps
Signaling rate, baud	6 kbaud	72 kbaud	2 kbaud
Reference table	Table 3-16	Table 3-17	Table 3-19

Table 3-19 ISDN BRI Superframe

Frame 1	Frame 2	3	4	5	6	7	Frame 8
ISW + 12 × (2B + D) + M	SW + 12 × (2B + D) + M	..	..	..	..	..	SW + 12 × (2B + D) + M

Table 3-20 IDSN DSL Synchronization Word Frame Structure Specification

ISW	+3	+3	−3	−3	−3	+3	−3	+3	+3	Quats
	1 0	1 0	0 0	0 0	0 0	1 0	0 0	1 0	1 0	Bits
SW	−3	−3	+3	+3	+3	−3	+3	−3	−3	Quats
	0 0	0 0	1 0	1 0	1 0	0 0	1 0	0 0	0 0	Bits

Table 3-21 2B + D Payload Frame Structure Specification

DATA	B1				B2				D
BIT PAIRS	$b_{11}b_{12}$	$b_{13}b_{14}$	$b_{15}b_{16}$	$b_{17}b_{18}$	$b_{21}b_{22}$	$b_{23}b_{24}$	$b_{25}b_{26}$	$b_{27}b_{28}$	d_1d_2
NO. OF BITS	8				8				2
QUATS	q_1	q_2	q_3	q_4	q_5	q_6	q_7	q_8	q_9
NO. OF QUATS	4				4				1

The frames and superframes are identical in both directions except for the allocation of M-bits in the control link. There are a total of 48 M-bits in each superframe. These bits are used for a variety of purposes and are allocated as follows:

- Embedded operations channel (eoc = 24 bits at 2 kbps)
- 12-bit CRC (CRC-12 = 12 bits at 1 kbs)
- Activation (act bit)
- Deactivation (dea bit)
- Far-end block error (febe bit)
- Cold-start only (cso bit)
- NT1 in test mode (ntm bit)
- Power status (ps1, ps2 bits)
- Spare bits

The act, dea, febe, cso, ntm, ps, and spare bits total 12 bits per superframe (1 kbps). Table 3-22 shows which M-bits are used in each transmission direction. The various M-bits are interwoven in a predefined manner, as shown in Table 3-23. For clarity, this table does not show the payload bits, 2B + D, which are placed between the frame word and M-bits, as previously explained. All bits set to binary one are spare bits reserved for future use. Table 3-24 gives the nomenclature and rules for the bits not already explained.

Table 3-22 M-Bit Directional Characteristics

M-BIT CHANNEL	LT TO NT1	NT1 TO LT
eoc	Yes	Yes
crc-12	Yes	Yes
act	Yes	Yes
febe	Yes	Yes
dea	Yes	No
cso	No	Yes
ntm	No	Yes
ps1, ps2	No	Yes

Table 3-23 M-Bit Frame Structure Specification

FRAME NO.	FRAME WORD	M1	M2	M3	M4	M5	M6
			LINE TERMINATION TO NT1				
1	ISW	eoc_{a1}	eoc_{a2}	eoc_{a3}	act	1	1
2	SW	eoc_{dm}	eoc_{i1}	eoc_{i2}	dea	1	febe
3	SW	eoc_{i3}	eoc_{i4}	eoc_{i5}	1	crc_1	crc_2
4	SW	eoc_{i6}	eoc_{i7}	eoc_{i8}	1	crc_3	crc_4
5	SW	eoc_{a1}	eoc_{a2}	eoc_{a3}	1	crc_5	crc_6
6	SW	eoc_{dm}	eoc_{i1}	eoc_{i2}	1	crc_7	crc_8
7	SW	eoc_{i3}	eoc_{i4}	eoc_{i5}	1	crc_9	crc_{10}
8	SW	eoc_{i6}	eoc_{i7}	eoc_{i8}	1	crc_{11}	crc_{12}
			NT1 TO LINE TERMINATION				
1	ISW	eoc_{a1}	eoc_{a2}	eoc_{a3}	act	1	1
2	SW	eoc_{dm}	eoc_{i1}	eoc_{i2}	ps1	1	febe
3	SW	eoc_{i3}	eoc_{i4}	eoc_{i5}	ps2	crc_1	crc_2
4	SW	eoc_{i6}	eoc_{i7}	eoc_{i8}	ntm	crc_3	crc_4
5	SW	eoc_{a1}	eoc_{a2}	eoc_{a3}	cso	crc_5	crc_6
6	SW	eoc_{dm}	eoc_{i1}	eoc_{i2}	1	crc_7	crc_8
7	SW	eoc_{i3}	eoc_{i4}	eoc_{i5}	1	crc_9	crc_{10}
8	SW	eoc_{i6}	eoc_{i7}	eoc_{i8}	1	crc_{11}	crc_{12}

Table 3-24 Basic M-Bit Nomenclature and Rules

M-BIT	RULE
eoc_{a1}	Embedded operations channel address bit 1; 3 bits total may be used to address up to 7 locations
eoc_{dm}	Embedded operations channel data/message indicator; if binary 1, an operational message is present in the information field; if binary 0, the information field contains numerical data
eoc_{i1}	Embedded operations channel information bit 1; up to 8 bits per address; eoc operates in a repetitive command/response mode
act	Activation bit set to binary 1 during activation
dea	Deactivation bit set to binary 0 to announce deactivation
ps1, ps2	Power status bits set to binary 1 to indicate normal power in primary (ps1) and secondary (ps2) supplies; if both set to binary 0, NT1 may shortly cease operation
ntm	NT1 in test mode bit set to binary 0 to indicate test mode
cso	Cold-start-only bit set to binary 1 to indicate NT1 has cold-start-only transceiver
febe	Far-end block error bit set to binary 0 to indicate a CRC error was received

3.1.5 Subrate Digital Loop

The subrate digital loop (SRDL) is a part of the overall digital system network, also referred to as the digital data service (DDS). This section describes the frame structure of the DDS, including the SRDL. The frame structure for the DDS is a three-level byte structure that depends on the network multiplex level, as follows:

1. Local loop (SRDL) byte (at user rate plus overhead, if any)
2. Network byte (at 64 kbps plus overhead)
3. Transmission byte (at 64 kbps plus overhead)

Although this book emphasizes the local loop portion, the network and transmission portions of the system are explained to broaden the understanding of this specific system. Industry standards give subrate digital data specifications in detail [23–33].

The discussion immediately following on subrate digital data systems applies to end-user rates of 2.4, 4.8, and 9.6 kbps. Specific systems may provide speeds of 19.2 kbps and 38.4 kbps, also. These all are called *subrate* speeds. End-user rates of 56 and 64 kbps are covered later.

Two types of transmission can take place at the local loop level: unframed subrate digital loops (also known as *standard DDS*) and subrate with secondary channel. Standard DDS, as a network, does not require any framing structure. End-user data is transmitted to and received from the network on the local loop as straight data with no framing, except for framing that may be embedded in the end-user data. Bipolar violations are used to transmit network control information and for zero-code suppression, as explained in later sections of this chapter. The line rate equals the end-user bit rate in the case of standard DDS.

When subrate digital loops are used with secondary channel signaling, a byte-level format is used to differentiate between the primary and secondary channels. With the secondary channel, the local loop byte includes a total of eight bits: six data bits (D-bits), a secondary channel framing bit (F-bit), and a shared secondary channel/control bit (C-bit), as shown in Table 3-25. The line rate, Table 3-26, is always higher than the user rate due to the extra overhead from the secondary channel. Line rate is calculated from

$$\text{Line rate (bps)} = [\text{End-user bit rate (bps)}/6 \text{ bits}] \times 8 \text{ bits}$$

The secondary channel framing bit (F-bit) sequence is a repeating 1 0 1 1 0 0 . . . structure; it is used to provide character alignment for the seconday channel. The network control bit (C-bit) is shared with the secondary channel information bit (S-bit) in a repeating S C C . . . pattern, as shown in Table 3-27. The S-bit and C-bit follow the framing (F-bits) and data bits (D-bits). Bipolar violations are not used with the secondary channel for network control and zero-code suppression, as in subrate digital data without secondary channel.

The local loop bytes are accepted by the office channel unit (OCU) in central office A at user rates (plus overhead if the secondary channel is used). The OCU

Table 3-25 Subrate Digital Loop Byte Structures for 2.4 kbps through 9.6 kpbs User Rates

Type	Byte	Secondary Channel
Local loop	D_1 D_2 D_3 D_4 D_5 D_6 D_1 D_2 D_3 D_4 D_5 D_6 F C/S	Without secondary channel With secondary channel[a]
Network DS-0A	1 D_1 D_2 D_3 D_4 D_5 D_6 C/S	With or without secondary channel[b]
Transmission DS-0B	I D_1 D_2 D_3 D_4 D_5 D_6 C/S	With or without secondary channel[c]

[a] D_i is the primary channel data bit, F is the secondary channel framing bit, and C/S is the shared control/secondary channel bit.

[b] The first bit position is set to binary 1, D_i, F, and C/S are as defined above. If secondary channel is not used, the S-bit takes on "don't care" values.

[c] I is the subrate framing bit (also called user identification bit), D_i and C/S are as defined above.

Table 3-26 Line Rate for Subrate Digital Data with and without Secondary Channel

End-User Rate	Secondary Channel?	Overhead	Line Rate
2.4 kbps	No	None	2.4 kbps
2.4 kbps	Yes	0.8 kbps	3.2 kbps
4.8 kbps	No	None	4.8 kbps
4.8 kbps	Yes	1.6 kbps	6.4 kbps
9.6 kbps	No	None	9.6 kbps
9.6 kpbs	Yes	3.2 kbps	12.8 kbps

Table 3-27 Shared Network Control (C-Bit) and Secondary Channel (S-Bit)

Loop Byte 1	Loop Byte 2	Loop Byte 3	Loop Byte 4
D D D D D D F S	D D D D D D F C	D D D D D D F C	Repeat Byte 1

converts the input from the loop to the intra-office (network) rate of 64 kbps, which is called the DS-0A rate. The DS-0A is a bipolar nonreturn-to-zero, 100% duty cycle square wave. The conversion from a loop byte structure to a network byte structure at the DS-0A rate requires two steps: first, the byte structure is rearranged by adding a binary one bit before each group of six end-user bits, deleting the framing bit (if secondary channel is used), and adding a shared control/secondary channel bit after the end-user data bits. If the secondary channel is not used, the S-bits take on "don't care" values. Second, simple byte stuffing is used whereby the resulting 8-bit byte is repeated as necessary to obtain the 64 kbps DS-0A rate. The framing bit, which was removed from the byte structure, is reinserted at the other end, when the byte is transmitted by the OCU in central office B to user B.

The byte stuffing is such that the data and overhead from a 9.6 kbps user rate is repeated five times (12.8 kbps × 5 = 64 kbps). Similarly, the 4.8 kbps user rate is repeated ten times (6.4 kbps × 10) and 2.4 kbps repeated 20 times (3.2 kbps ×

20). The entire DS-0A bandwidth is occupied by data from a single end-user. The network byte structure is shown in Table 3-25.

To get to the transmission level of the network, the DS-0A outputs from several OCUs are combined by a subrate data multiplexer (SRDM). The output of the SRDM also is 64 kbps, but the byte structure is at the transmission level and designated DS-0B. The transmission byte structure is shown in Table 3-25. The DS-0B signal is electrically identical to the DS-0A signal except the message content is different. One difference is in the prefix bit. At the DS-0A rate, this bit is always set to binary one. At the DS-0B rate, this bit is replaced by a user rate identification bit (I-bit). The I-bit patterns for various end-user subrates are shown in Table 3-28. The patterns are unique for each rate.

Table 3-28 I-Bit End-User Rate Identification Patterns

User (Payload) Rate	I-Bit Pattern
9.6 kbps	0 1 1 0 0 0 1 1 0 0 0 1 1 0 0 0 1 1 0 0 . . .
4.8 kbps	0 1 1 0 0 1 0 1 0 0 0 1 1 0 0 1 0 1 0 0 . . .
2.4 kbps	0 1 1 0 0 1 0 1 0 0 1 1 1 0 0 0 0 1 0 0 . . .

The other difference is the end-user subrate data redundancy is removed when the DS-0B byte is formed. A given DS-0B SRDM multiplexes the subrates from a number of different users (OCUs) so that the data from each end-user appear only once. The number of subrate inputs is restricted to the highest subrate used in that multiplexer. For example, if the highest user rate of any of the channels is 9.6 kbps, the maximum number of inputs to the SRDM is five, as shown in Figure 3-4.

Not all channels have to be at the 9.6 kbps rate; four of the channels could be any mixture of 9.6, 4.8, and 2.4 kbps user rates. Each byte of a 4.8 kbps channel is transmitted twice, and each byte of a 2.4 kbps channel is transmitted four times, which, in this case, retains some of the redundancy at the subrate input. If the highest user rate is 4.8 kbps, the maximum number of inputs to the SRDM is ten. Similarly, if the highest user rate is 2.4 kbps, the maximum number of inputs is 20. An example of the multiplexing formats is shown in Figure 3-5.

The loop byte structures for 56 and 64 kbps user data rates are slightly different than those previously described. For these rates, a seventh data bit (D_7) is added for 56 and 64 kbps and an eighth data bit (D_8) replaces the C-bit for 64 kbps, as shown in Table 3-29. The F-bit pattern repeats (1 0 1 1 0 0 . . .) and, when used, the C-bit performs the same functions as for the subrates. The most significant difference is the absence of a prefix bit in the network and transmission bytes at these higher speeds. Secondary channel operation is not supported by 64 kbps service. Table 3-30 shows the end-user rates and corresponding line rates for 56 and 64 kbps services.

Another stage of multiplexing takes place for interoffice transmission whereby DS-0A and DS-0B signals are multiplexed together to form a DS-1 (1.544 Mbps) signal. A variety of multiplexers can be used in this application. One type (T1DM) allows up to 23 DS-0A or DS-0B signals to be multiplexed together. The 24th channel is used for framing and synchronization. Another type (T1WB4) accepts inputs from partially filled D4 channel banks (which contain encoded voicegrade signals)

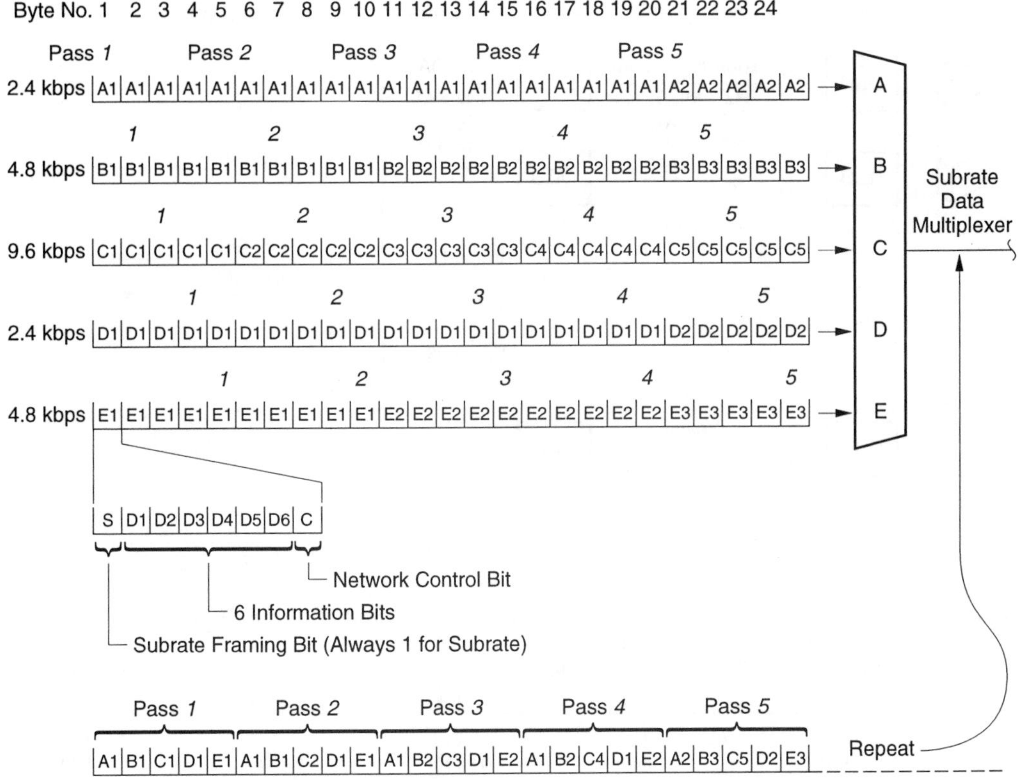

Figure 3-4 SRDM restricted to five inputs with 9.6 kbps

and up to 12 DS-0A or DS-0B signals. Other T1 multiplexers (generically called channel banks) are capable of mixing data and voice signals in less restrictive combinations.

The digital data system can be used with more user rates than previously described, but these are directly put into the network at the DS-0B rate; that is, after the initial stages of subrate multiplexing. Intermediate user rates of 19.2, 28.8, and 38.4 kbps are presently supported [24]. These rates are handled by combining two, three, or four adjacent 9.6 kbps bytes out of the 5-byte DS-0B signal structure, respectively.

3.1.6 High Bit-Rate Digital Subscriber Line

As of this writing (1994), the formatting specification for the HDSL is described in a BELLCORE technical advisory, which is a preliminary document written by BELLCORE for its client companies [34]. In spite of the standard's preliminary nature, at least four independent manufacturers are producing HDSL products, but these products are not compatible with each other on an end-to-end basis.

The formatting specification described below applies to the HDSL operation after initial startup synchronization has taken place and reliable payload transport is

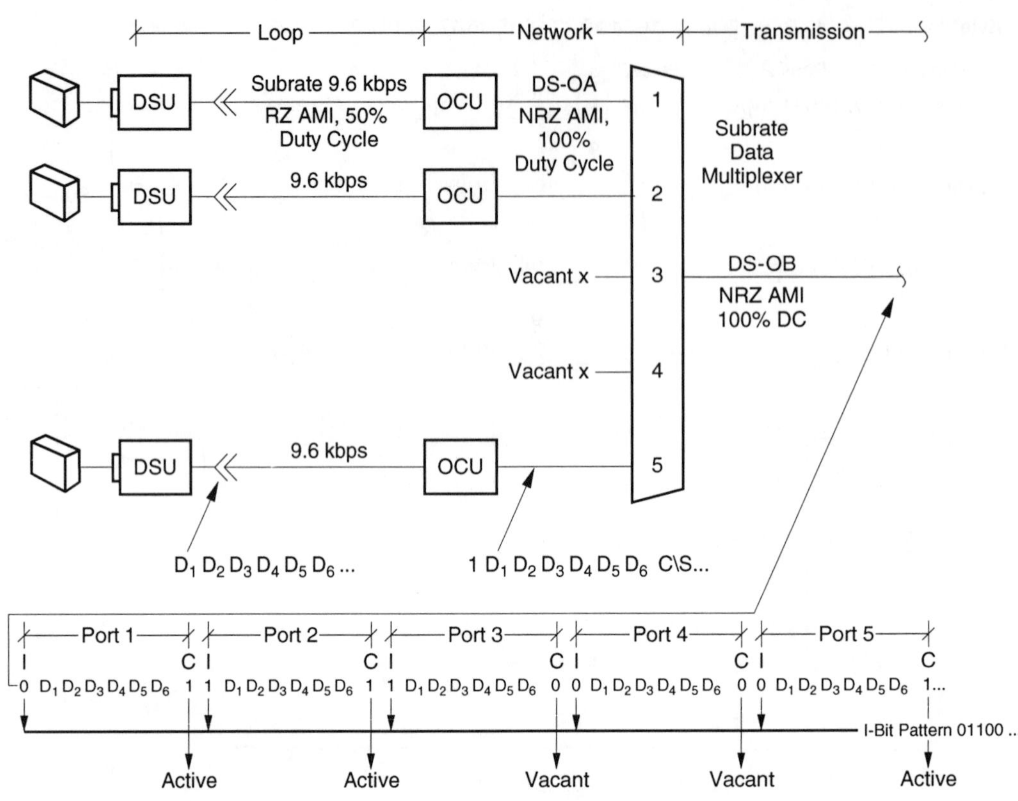

Figure 3-5 Subrate multiplexing example

Table 3-29 Byte Structures for 56 kbps and 64 kbps User Rates

TYPE	BYTE	SECONDARY CHANNEL
56 kbps local loop	D_1 D_2 D_3 D_4 D_5 D_6 D_7	Without secondary channel
	D_1 D_2 D_3 D_4 D_5 D_6 D_7 F C/S	With secondary channel
56 kbps network DS-0A	D_1 D_2 D_3 D_4 D_5 D_6 D_7 C/S	With or without secondary channel
56 kbps transmission DS-0B	D_1 D_2 D_3 D_4 D_5 D_6 D_7 C/S	With or without secondary channel
64 kbps local loop	D_1 D_2 D_3 D_4 D_5 D_6 D_7 F D_8	Secondary channel not available
64 kbps network DS-0A	D_1 D_2 D_3 D_4 D_5 D_6 D_7 D_8	See above
64 kbps transmission DS-0B	D_1 D_2 D_3 D_4 D_5 D_6 D_7 D_8	See above

The nomenclature is the same as Table 3-25.

Table 3-30 Line Rate for 56 and 64 kbps Digital Data with and without Secondary Channel

End-User Rate	Secondary Channel?	Overhead	Line Rate
56 kbps	No	0	56 kbps
56 kbps	Yes	16.0 kbps	72 kbps
64 kbps	Not available	8.0 kbps	72 kbps

possible. The basic frame has a nominal length of 6 ms and is composed of an average of 4,704 bits, giving a bit rate of 784 kbps (392 kbaud). Any individual frame will have either zero or four stuffing bits, for a total length of 6 − (1/392) ms or 6 + (1/392) ms, respectively. The stuff bits are necessary to accommodate frequency differences between the two DS-1 rate directions and between the DS-1 rates and the HDSL system clock rate. Since the HDSL uses the 2B1Q line code, bit pairs are formed into quats. Table 3-31 summarizes the frame in terms of bits, quats, and time length under average and stuffed conditions.

The bits are chronologically partitioned into:

- Fourteen bits (seven quats) of sync word
- Two bits (one quat) of HDSL overhead
- Ninety-seven bits (48-1/2 quats) of HDSL payload (12 DS-1 rate data blocks)
- Ten bits (five quats) of HDSL overhead
- Ninety-seven bits (48-1/2 quats) of HDSL payload (another 12 DS-1 rate data blocks)
- Ten bits (five quats) of HDSL overhead
- and so on, until a total of 388 bits (194 quats) of HDSL payload (48 DS-1 rate blocks) and 46 bits (23 quats) of HDSL overhead, including sync word, have been transmitted. At this point, either zero or four bits (two quats) of stuffing are inserted depending on (1) whether the incoming DS-1 rate is fast or slow relative to the internal HDSL clock, and (2) the status of the data buffers.

All bits except the sync word and stuff bits are scrambled before they are coded. One of the benefits of scrambling is that it tends to reduce repetition of symbols with the same polarity. However, since the sync word and stuff bits are not scrambled, the possibility of dc unbalance exists over time. The stuff bits do not carry intrinsic information but act as temporal spacers in the frame. Therefore, they may be used to improve the dc balance by offsetting any bias in the transmitted symbols.

Table 3-31 HDSL Frame Length

Condition	Length (Ms)	Length (Bits)	Length (Quats)
Average	6	4,704	2,352
2-quat stuffing	6.00255 . . .	4,706	2,353
No stuffing	5.99745 . . .	4,702	2,351

The bit pair to quaternary conversion previously shown for the ISDN DSL is repeated in Table 3-32 for completeness in this section. This table applies to all bit pairs except those associated with the sync word. The sync words use the "Double Barker Code" sequence shown in Table 3-33 [34].

The HDSL consists of two identical loops. The same basic frame *structure* is used in both loops for both directions. The only difference is in the payload; loop 1 carries bidirectional DS-1 rate data blocks for channels 1 through 12 (including original DS-1 framing) and loop 2 carries bidirectional DS-1 rate data blocks for channels 13 through 24 (including a duplicate of the original DS-1 framing). However, even though the HDSL frame structure is the same for each loop, the functional assignments of individual bits are not necessarily the same. The frame structure with the corresponding quats is shown in Table 3-34. This table also references other tables that describe each function in greater detail, including the synchronization word structure shown in Table 3-35.

There are a total of 32 overhead bits (not including stuff bits) in each frame. These bits are used for a variety of purposes and are allocated as follows:

- Alarm indicator bit (aib bit)
- Far-end block error (febe bit)
- Embedded operations channel (eoc—12 bits at 2 kbps)
- Six-bit cyclic redundancy check (CRC6—six bits at 1 kbs)
- Power status (ps1, ps2 bits)
- Digital signal, level 1 (or, DS-1) error (ds1e bit)
- HDSL repeater present (hrp bit)
- Unspecified indicator bits (uib—eight bits)

The febe, aib, ds1e, hrp, ps, and spare bits total 14 bits per frame (2.333 kbps). Not all of these bits are used in both directions. Table 3-36 shows which overhead

Table 3-32 Quaternary Symbols Except for Sync Word Quat

First Bit (Sign)	Second Bit (Magnitude)	Quat
1	0	+3
1	1	+1
0	1	−1
0	0	−3

Table 3-33 Double Barker Code Quaternary Symbols for Sync Word Quat

First Bit (Sign)	Second Bit (Magnitude)	Quat	Valid in Sync Word
1	1	+3	Yes
0	1	+1	No
1	0	−1	No
0	0	−3	Yes

Table 3-34 HDSL Frame Structure

Function	SW	HDSL Overhead	DS-1 Rate Data Block	HDSL Overhead	3 × DS-1 Rate Data Block and HDSL Overhead	Stuffing
	Sync Word	Overhead	Payload	Overhead	Payload + overhead	Stuffing
No. of bits Bit positions	14 1–14	2 15,16	1,164 17–1,180	10 1,181–1,190	3,512 1,191–4,702	4 4,703–4,706
No. of quats Quat positions	7 1–7	1 8	582 9–590	5 591–595	1,756 596–2,351	2 2,352–2,353
Signaling rate, bps Signaling rate, baud	2,333 bps 1,167 kbaud	333 bps 167 baud	194 kbps 97 kbaud	1.667 kbps 0.833 kbaud		333 bps 167 baud
Reference table	Table 3-35	Table 3-37	see SF and ESF format	Table 3-37	see SF and ESF format and Table 3-37	Table 3-37

SF = superframe; ESF = extended superframe

Table 3-35 HDSL Synchronization Word Structure Specification

Loop #1	Normal SW	+3	+3	+3	−3	−3	+3	−3	Quats
		1 1	1 1	1 1	0 0	0 0	1 1	0 0	Bits
Loop #2	Time Reverse SW	−3	+3	−3	−3	+3	+3	+3	Quats
		0 0	1 1	0 0	0 0	1 1	1 1	1 1	Bits

Table 3-36 HDSL Overhead Bit Directional Characteristics

Channel	HTU-C to HTU-R	HTU-R to HTU-C	Repeater to HTU-R and Repeater to HTU-C
losd	Yes	Yes	Transparent
febe	Yes	Yes	Transparent
eoc	Yes	Yes	Transparent
crc-6	Yes	Yes	Transparent
ps1, ps2	No	Yes	Transparent
ds1e	Yes	Yes	Transparent
hrp	No	No	Yes
uib	Yes	Yes	Transparent

HTU-C = central office high bit-rate terminal unit; HTU-R = remote high bit-rate terminal unit

bits are used in each direction. The various overhead bits are interwoven in a predefined manner, as shown in Table 3-37. For clarity, this table does not show the payload bits, which are placed between the overhead bits, as previously explained.

All bits marked as uib were unspecified at the time the technical advisory was issued (1991) and can be considered spare bits reserved for future use. Table 3-38 gives the nomenclature and rules used with the overhead bits.

3.2 Line Coding

A line code is a set of rules "that defines the equivalence between sets of digits presented for transmission and the corresponding sequence of signal elements transmitted over that channel" [17]. Line codes are chosen to suit the characteristics of a transmission channel. In digital systems that handle voicegrade (analog) traffic, two levels of coding take place (source coding and line coding); this section is concerned only with line coding.* The AMI line code is a well-known example and one of several line codes covered in this chapter.

There is no necessary relationship between line codes and frame structures, yet it is logical to discuss them together. Whereas line codes provide bit transparency,

*Source coding was discussed in Chapter 1.

Table 3-37 HDSL Ovehead Bit Format Specification

Bit No.	Quat No.	Pos. 1	Pos. 2	Pos. 3	Pos. 4	Pos. 5	Pos. 6	Pos. 7	Pos. 8	Pos. 9	Pos. 10
HTU-C to HTU-R											
15–16	8	losd	febe	N/A	N/A	N/A	N/A	N/A	N/A	N/A	N/A
1,181–1,190	591–595	eoc_{01}	eoc_{02}	eoc_{03}	eoc_{04}	crc_1	crc_2	uib	uib	ds1e	eoc_5
2,355–2,364	1,178–1,182	eoc_{06}	eoc_{07}	eoc_{08}	eoc_{09}	crc_3	crc_4	uib	uib	uib	uib
3,529–3,538	1,765–1,769	eoc_{10}	eoc_{11}	eoc_{12}	eoc_{13}	crc_5	crc_6	uib	uib	uib	uib
4,703–4,706	2,352–2,353	stq_{1s}	stq_{1m}	stq_{2s}	stq_{2m}	N/A	N/A	N/A	N/A	N/A	N/A
HTU-R to HTU-C											
15–16	8	losd	febe	N/A	N/A	N/A	N/A	N/A	N/A	N/A	N/A
1,181–1,190	591–595	eoc_{01}	eoc_{02}	eoc_{03}	eoc_{04}	crc_1	crc_2	ps1	ps2	ds1e	eoc_5
2,355–2,364	1,178–1,182	eoc_{06}	eoc_{07}	eoc_{08}	eoc_{09}	crc_3	crc_4	hrp	uib	uib	uib
3,529–3,538	1,765–1,769	eoc_{10}	eoc_{11}	eoc_{12}	eoc_{13}	crc_5	crc_6	uib	uib	uib	uib
4,703–4,706	2,352–2,353	stq_{1s}	stq_{1m}	stq_{2s}	stq_{2m}	N/A	N/A	N/A	N/A	N/A	N/A

N/A means not applicable; Pos. = position

Table 3-38 HDSL Overhead Bit Nomenclature and Rules

Overhead Bit	Rule
losd	Loss-of-signal DS-1 bit set to binary 1 to indicate normal status and binary 0 to indicate alarm or alert status
febe	Far-end block error bit set to binary 0 to indicate a crc error was received
eoc_{ij}	Embedded operations channel bits; functions remain to be specified
crc_i	The cyclic redundancy check bits cover the previous HDSL frame in any given direction; the crc calculation does not include the sync word, crc bits, or stuff bits
ps1, ps2	Power status bits set to binary 1 to indicate normal power in primary (ps1) and secondary (ps2) supplies; if both set to binary 0, remote may shortly cease operation ("dying gasp")
ds1e	If there are no BPV (excluding B8ZS encoding) and no DS-1 rate framing errors on the incoming DS-1 rate signals during the preceding HDSL frame, the DS-1 rate error bit, ds1e, is set to binary 1
hrp	The HTU-C and HTU-R both set the HDSL repeater present bit, hpr, to binary 1; a repeater, if present, resets the hrp bit to binary 0 in both directions
uib	Unspecified indicator bits remain to be specified
stq_{xy}	The binary values of the stuff quat bits, stq, are not specified, but the value of each bit in the pair must be the same; the stq bits may be used to provide dc balance

that is, a given bit value going into the transmission system always comes out the same value, frame structures specify the demarcation point between individual frames, words, bytes, and bits. This section covers the most common line codes used in practical, working digital loop transmission systems, which are subrate digital loop (SRDL), ISDN DSL, HDSL, and repeatered T1-carrier. Additional line codes are discussed in the literature [35–37].

3.2.1 Typical Line Codes

The line codes of particular interest in practical digital loop transmission systems are given and defined in Table 3-39; each will be covered in detail. Each of these line codes serves some special purpose (a given line code may serve more than one purpose):

Table 3-39 Digital Loop Line Codes

Line Code	Examples
Return-to-zero (RZ)	Repeatered T1-carrier, SRDL
Nonreturn-to-zero (NRZ)	Optical fiber
Alternate mark inversion (AMI)	Repeatered T1-carrier
Bipolar with N-zero substitution (BNZS)	Repeatered T1-carrier, SRDL
Zero-Byte timeslot interchange (ZBTSI)	Repeatered T1-carrier
Two binary, one quaternary (2B1Q)	ISDN DSL, HDSL

1. To provide signal redundancy to improve error performance (AMI)
2. To limit or tailor signal spectrum (RZ, AMI, 2B1Q)
3. To ensure minimum pulse density requirements are met for proper timing recovery under all bit sequence conditions (clear channel capability) (B8ZS, ZBTSI)

3.2.1.1 Return-to-Zero and Nonreturn-to-Zero

The return-to-zero (RZ) and nonreturn-to-zero (NRZ) are fundamental line codes upon which the other line codes in this section are based. An RZ line code uses pulses that have less than 100% duty cycle and therefore return to a zero voltage value before the next timeslot. An NRZ line code usually has 100% duty cycle, which means the voltage value is maintained throughout the pulse period. RZ and NRZ line codes are compared to a unipolar bit stream in Figure 3-6.

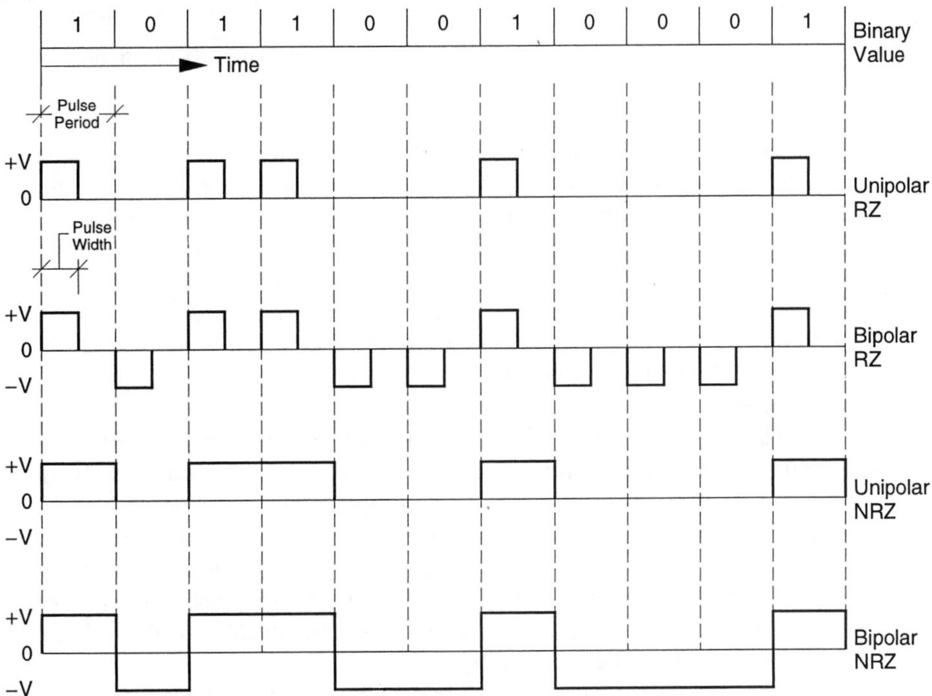

Figure 3-6 RZ and NRZ line codes

3.2.1.2 AMI

AMI is used in SRDLs and in repeatered T1-carrier span lines. Binary ones are sent as alternate voltage pulses with opposite polarity and binary zeros are sent as zero voltage (no pulse). This is compared to a binary (unipolar) digital signal in Figure 3-7. Binary streams can contain a large dc component, as shown. It is desirable in digital loop transmission systems to eliminate the dc component from the

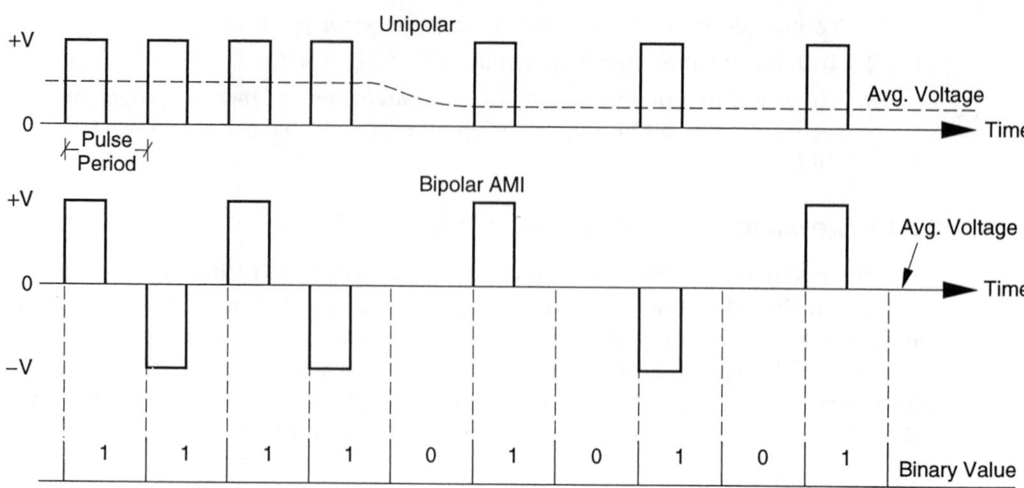

Figure 3-7 Average voltage of unipolar and bipolar AMI bit streams

information-carrying signal for several reasons. First, all cables used in these applications carry somewhat large longitudinal currents from powerline induction and other sources. Terminating digital loops in transformers at each end protects and isolates the electronic signaling circuitry from the resulting longitudinal voltages, but transformers do not couple dc. Second, the actual value of the signal's dc component will be affected by temperature and cable length, and these variations may interfere with recovery of the signal content at the receiver. Finally, dc is used for powering, control, and sealing current on digital loops, which precludes its use as a signal component.

The dc component of the signal is easily removed by converting the unipolar signals to a bipolar AMI format, whereby a pulse of one polarity is always followed by a pulse of opposite polarity. Such a bipolar line code not only eliminates any restriction on the use of coupling transformers, but also shifts the signal spectrum to lower frequencies, which can be transmitted with less loss on twisted pair cables.

Another characteristic of AMI is that the actual pulse only occupies one-half the bit interval. This gives each pulse a 50% duty cycle, and the voltage returns to a zero value before the next bit position.

AMI provides a measure of redundancy to the binary ones transmitted over the channel. Since the coding rule always requires alternating pulse polarity for successive binary ones, any two sequential pulses received that have the same polarity indicate the rule has been violated and an error has occurred. Such errors, illustrated in Figure 3-8, are called bipolar violations (BPVs). BPVs are sometimes intentionally inserted in the transmitted bit stream, as will be discussed later. A repeatered T1-carrier is a digital loop system that uses the number of BPVs to indicate a measure of system performance. The subrate digital data system uses BPVs on the local loop for network control and zero-code suppression.

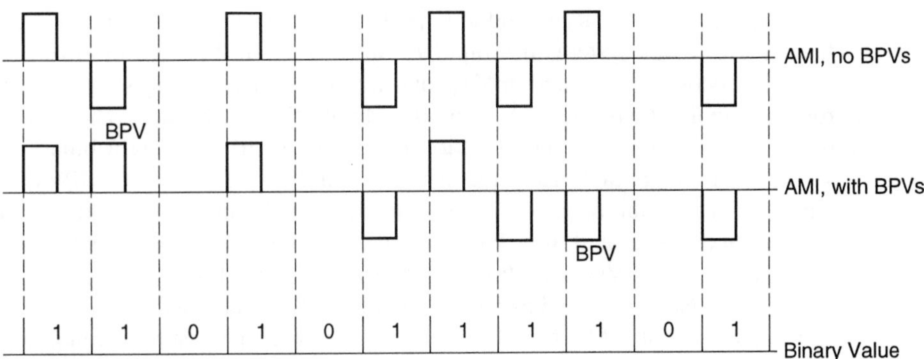

Figure 3-8 BPVs

3.2.1.3 Bipolar with N-Zero Substitution (BNZS)*

Clock timing information is transmitted by the pulses in the bipolar AMI line code, and this clock can be recovered provided the data stream has a sufficient density of binary-ones. If the data stream consists of a long series of binary zeros (no pulses), no clocking information is present for the duration of the series. If this series length exceeds the design time of the pulsed clock in the regenerator, synchronization will be lost. Long series of binary zeros could easily exist when the digital loop is used to transmit data (but are unlikely with encoded voice).

The pulse density of a repeatered T1-carrier must meet certain minimum requirements so the clock can be recovered from the data stream at regenerators. Since the first production versions of a repeatered T1-carrier, the average pulse density requirement has been at least 12.5% (at least N binary ones in each and every time window of 8(N + 1) timeslots, where N = 1 to 24), with no more than 15 consecutive binary zeros.

A violation of the pulse density requirement would result, for example, if two consecutive data bytes (16 bits) were all binary zeros. This constraint reflects the early exclusive use of T1-carrier for voice traffic, in which case a binary one was inserted in place of a binary zero in bit position 7 to provide the needed density. The recovered signal suffered little (ear-) detectable distortion. However, when a repeatered T1-carrier is used for data traffic, this constraint prevents clear channel capability (CCC) for data, where bit sequence independence is an inherent requirement. Without CCC, only seven of the eight bits in each channel are available for source data, which reduces the data rate from 64 kbps (8 bits × 8 kHz encoding rate) to 56 kbps (7 bits × 8 kHz encoding rate).

Various techniques are used to ensure that average density requirements are met while still providing clear channel capability (that is, no constraints on the value of any data bit). For example, early systems transmitted data on every other channel, and ones were inserted in all data bit positions on adjacent channels. This guaranteed

*The author is grateful to Len Thomas for his contributions to this section.

the average pulse density would always be 50%; however, this resulted in inefficient facility use. Later systems use bipolar with 8-zero substitution, or B8ZS.

To accommodate random data and the possibility of long strings of binary zeros, the bipolar AMI line code is modified with the B8ZS code. The output circuits of a terminal, such as a data multiplexer, detect a string of eight binary zeros and substitute a unique and easily identified "invalid" code pattern of BPVs, which includes at least four pulses. This is illustrated in Figure 3-9. At the input of the terminal at the other end, the clock is first recovered and the known pattern of invalid data (pattern of BPVs) is replaced by a string of eight binary zeros.

B8ZS may be viewed as an enhancement of bipolar AMI. It has all the features and advantages of AMI, but also includes a feature that assures a minimum density of binary ones, even with 100% clear channel data. Whenever a string of eight binary zeros is detected, for example—0000 0000—, the code 00 0VB 0VB is substituted. Here, B represents a normal bipolar pulse in bit positions 5 and 8 that does not belong there, and V represents a BPV (that is, it has the same polarity as the preceeding pulse) in bit positions 4 and 7.

B8ZS may cause problems peculiar to a repeatered T1-carrier. If the line receiver is improperly optioned for AMI and the line transmitter is optioned for B8ZS, each string of eight zeros will result in four bit errors and two BPV errors at the receiver. User data or framing patterns may be corrupted by the errors, depending

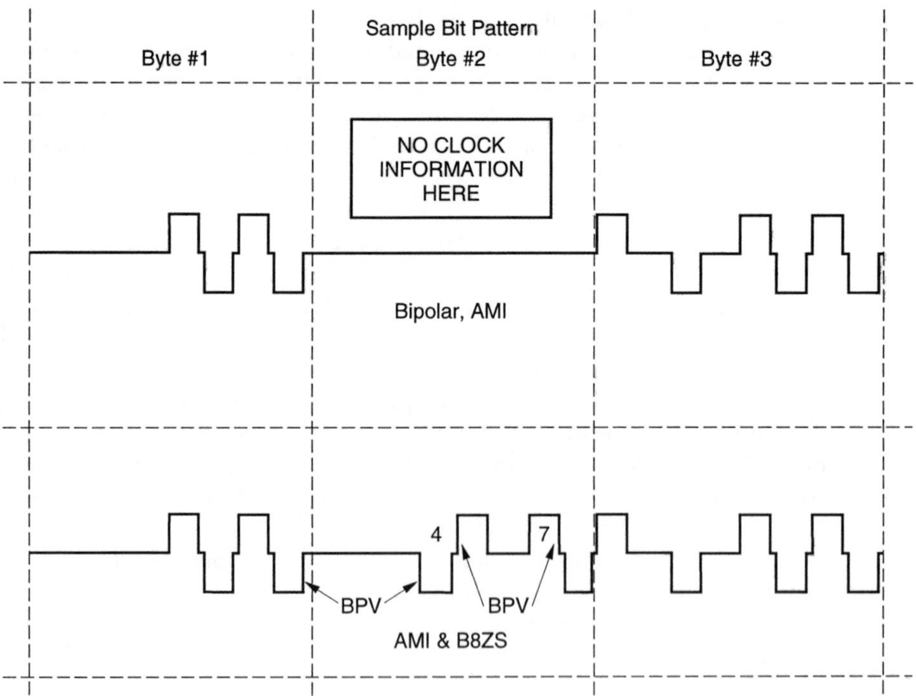

Figure 3-9 B8ZS

on their position in the data stream. If the errors occur on channels carrying encoded voice signals, the errors are manifested as impulse noise.

If the transmitter is optioned for AMI and the receiver is optioned for B8ZS, no problems occur because the B8ZS code never will be generated. However, the transmitter may transmit a data stream with insufficient pulse density.

The case of improper option selection deserves more attention because it is a real problem encountered in the field. If an option error has been made, the symptom will be the detection of BPVs. An example of this condition is shown in Figure 3-10. The actual source binary data stream is shown in Figure 3-10(a), with three groups of eight bits labeled bytes 1, 2, and 3. Byte 2 contains eight consecutive binary zeros. Figure 3-10(b) shows these data when the binary values are encoded

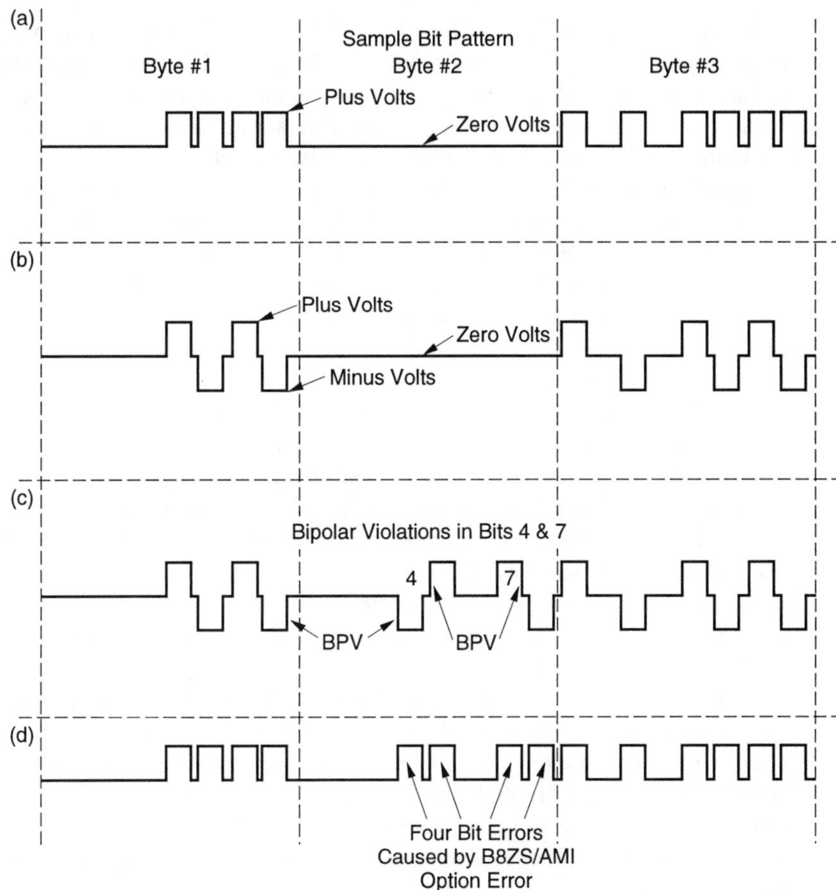

Figure 3-10 B8ZS and AMI option error: (a) Unipolar representation of sample bit pattern; (b) Bipolar AMI; (c) B8ZS; (d) Unipolar data at other end
Source data courtesy of Len Thomas

by the bipolar AMI line code. Byte 2 has insufficient binary zeros, which is a violation of the pulse density requirements. Nevertheless, byte 2 is a valid data byte.

Figure 3-10(c) shows the same source data encoded by the B8ZS line code. This signal has the required pulse density but contains two BPVs and four binary zeros recoded into pulses. Improperly optioned terminals may:

- Discard the BPVs and substitute pulses to provide the specified pulse density
- Raise an alarm due to errors caused by the BPVs
- Cause span line protection to be invoked
- Simply convert the BPVs to binary ones (thus giving four errors, as shown in Figure 3-10(d))

The terminal equipment must have B8ZS capability if clear channel data is to be transmitted. Older T1-carrier channel banks do not have this capability. In a channel bank without B8ZS capability, the source data stream must be limited to 56 kbps. In this case, only seven data bits in any channel are used; data bit 8 is always set to a binary one to provide the necessary pulse density.

B8ZS may not be compatible with many existing repeatered T1-carrier transmission systems, especially where the BPVs are used for alarm purposes or for control of span line protection systems. Also, the B8ZS line code is only applicable to the transmission of a DS1 signal via twisted pair media, where line and office repeaters recover the clock for regeneration. B8ZS is not transmitted over multiplexers at higher levels in the digital hierarchy, nor is this required with optical fiber systems operating at the DS-1 rate.

B8ZS extends from one end of a twisted pair T1-carrier span line to the other and is therefore a local loop issue, not a network issue. Where it is impractical to use B8ZS due to technology or equipment design restrictions, an alternate solution, called zero-byte timeslot interchange (ZBTSI), first mentioned in Table 3-35, may be used. However, B8ZS is considered to be the only long-term solution to providing clear channel capability; ZBTSI is an interim solution. The purchase of new DS-1 rate equipment that does not support B8ZS is not recommended. A technical report on the performance of AMI signals through B8ZS-optioned equipment can be found in [38].

The subrate digital loop uses B6ZS and B7ZS for zero-code suppression. B6ZS is used with end-user rates of 2.4, 4.8, and 9.6 kbps without secondary channel. With B6ZS, any sequence of six consecutive zeros (0 0 0 0 0 0) is replaced by 0 0 0 X 0 V. When the end-user rate is 56 kbps without secondary channel, any sequence of seven consecutive zeros (0 0 0 0 0 0 0) is replaced by 0 0 0 0 X 0 V, which is a B7ZS line code. Zero-code suppression is not required, nor used, at 64 kbps end-user rates.

An undesirable dc component could arise with the unrestricted use of the bipolar violations. Therefore, the X timeslot is used to encode a pulse or no pulse in such a way that successive violations (Vi) alternate in polarity. The X denotes a 0 or B. The value of X is chosen to keep the total number of Bs odd, which ensures that the desired alternation of V is achieved. Thus, if the number of B pulses since the last violation is odd, the X-bit is set to zero; if the number of B pulses since the last violation is even, the X-bit is set to one.

The subrate digital loop also uses BPVs at all end-user rates except 64 kbps for simple network control functions at the local loop level. For example, an idle loop transmits the sequence B B B X 0 Vi at 2.4, 4.8, and 9.6 kbps and B B B B X 0 Vi at 56 kbps. Table 3-40 shows the minimum average pulse density for the subrate digital loop and T1-carrier.

Table 3-40 Minimum Average Pulse Density

Loop Speed	Minimum Average Pulse Density	Maximum No. of Consecutive Zeros
3.2 kbps (2.4 kbps user rate)	1 pulse in 16 pulse periods	23
6.4 kbps (4.8 kbps user rate)	1 pulse in 16 pulse periods	23
12.8 kbps (9.6 kbps user rate)	1 pulse in 16 pulse periods	23
72 kbps (56 & 64 kbps user rate)	1 pulse in 18 pulse periods	26
1.544 Mbps (1.536 Mbps user rate)	N pulses in 8(N + 1) pulse periods	15

*3.2.1.4 ZBTSI**

ZBTSI is applicable to a repeatered T1-carrier only and is not used in any other existing digital loop systems. Like B8ZS, ZBTSI ensures the pulse density meets minimum requirements. With ZBTSI, at least 23 bits in the 192 data bits of each data frame will be a binary one (pulse), and there will be no more than 15 consecutive binary zeros.

An encoder optioned for ZBTSI requires 2 kbps of overhead information. It will alter the source data and will use the frame format to signal the decoder of this alteration. Unlike bipolar AMI and B8ZS, ZBTSI is an end-to-end (encoder to decoder) signal conversion protocol that is passed through the network at all multiplex levels and systems, including optical fiber transmission systems. ZBTSI does not alter the line code in any way, and therefore is compatible with both B8ZS and bipolar AMI.

The ZBTSI encoder operates as follows:†

1. SF or ESF framing is recovered by the ZBTSI encoder.
2. Frame data are stripped from the incoming data stream and delayed for a time equal to the processing time of the encoder.
3. Framing bits 1, 5, 9, 13, 17, and 21 of the ESF or 1, 5, and 9 of the SF are renamed "Z" (ZBTSI frame flag) bits and set to binary one; all other bits are routed around the encoder. In the SF, framing bits 5 and 9 are normally set to binary one except under certain conditions; if altered by the ZBTSI function, they will be handled as frame errors through the net-

---

*The author is grateful to Len Thomas for his contributions to this section.

†The complete ZBTSI algorithm is described in [5].

work and corrected by the ZBTSI decoder. With ESF, performance data are routed around the encoder provided they are communicated via framing bits 3, 7, 11, 15, 19, and 23. The ESF function must precede the ZBTSI encoder at the source-end and follow the ZBTSI decoder at the sink-end.

4. Source data from four consecutive frames are organized into 96 units of eight bits each, which is a block of 768 bits. Each overhead bit is associated with each 768-bit block.
5. Source data are scrambled by a fixed pattern scrambler. This assures that long strings of binary zeros are evenly distributed. Scrambling does not guarantee that a string of binary zeros will be scrambled in such a way that the pulse density requirements are met, but it does ensure randomized data.
6. The identity of the data is lost in the scrambling process. Therefore, a unit of eight bits will be called an octet rather than a byte in the following discussion to differentiate source data from the scrambled data. The 798 scrambled source data bits are organized into 96 octets. Octet 96 is relocated so it precedes octet 1.
7. The stored data are examined to determine if any of the following violations have occurred (called violating all-zero octet, or VAZO):
   a. Octet 1 and octet 95 contain all-binary zeros
   b. The all-zero octet is not octet 1, 95, or 96 and it combines with its adjacent "octs" to form a data string with 15 or more consecutive binary zeros
   c. The all-zero octet is not octet 1, 95, or 96 and the data in either adjacent octet contain fewer than two binary ones
   d. The all-zero octet is octet 96 and either there is at least one other VAZO in that data block (96 octet group) or the data in octet 1 contain fewer than two binary ones
8. If no violations have occurred, data are transmitted in the scrambled and shifted order just described. If one or more VAZO exists:
   a. The "Z" framing bit for that group of 96 octets is changed from a binary one to binary zero.
   b. The location (octet number) of the VAZO is determined. This number, which is between 1 and 96, will be encoded into a 7-bit binary code.
9. The data in octet 96 are stored in a temporary memory and the location of the first VAZO is stored in its place using the 7-bit code. The address of the first VAZO will always be carried in the octet immediately following the "Z" bit. The first bit in octet 96 will be used to signify if this is the only violating octet. If it is, the first bit will be set to a binary one. The next seven bits will identify the octet number of the VAZO and that location will contain the original data from octet 96.
10. If there is more than one VAZO, the first bit in octet 96 will be set to binary zero and the next seven bits will be the address of the first VAZO. The new data in the location of the first VAZO will be a binary one if

there are more VAZOs and a binary one if there are only two VAZOs. The location of the second VAZO will be stored in the location of the first VAZO, and again, the first bit in this octet is used to identify, if there are still more VAZOs. This process continues until all VAZOs are identified. The last octet will contain data from octet 96.

11. After the locations of all VAZOs have been encoded, and the data for octet 96 has been stored in the location of the last VAZO, the data frame is inserted in its proper location and transmitted.
12. ZBTSI decoding reverses the foregoing process. Decoding is initiated at the receiver if the "Z" bit indicates the group is encoded. The first octet points to the location of the first VAZO and specifies whether any additional octets were overwritten. If more than one VAZO was removed, the first addressed octet contains the address of the second VAZO and an indication of whether the decoding process should be continued in a similar manner. The last addressed location contains data bits that were removed from octet 1, and these are replaced last.

3.2.1.5 2B1Q

The 2B1Q line code is in present use in the ISDN BRI and the HDSL [22,34]. This discussion will concentrate on the ISDN BRI application, although it is equally applicable to the HDSL. The 2B1Q line code is a 4-level, pulse-amplitude modulated (PAM) code with no redundancy. With ISDN BRI, the user data stream is composed of two 64 kbps B-channels and one 16 kbps D-channel.

As user data (2B + D) are presented to the ISDN line transmitter, the binary digits are composed into bit pairs for conversion to quaternary (quat) symbols, as was discussed in the framing section above. Other bits for maintenance purposes also are composed into bit pairs and converted to quats. The first bit of a quat is called the sign bit, and the second is the magnitude bit. Table 3-41 shows the relationship between each possible pair of scrambled bits and its corresponding quaternary symbol. At this point, the quats are considered symbol names and not numeric values. At the receiver, each quat is converted to a pair of bits, descrambled, and formed into a bit stream of B- and D-channel bits and M-channel bits by reversing the process just described.

The line code appears at the transmitter output, as shown conceptually in Figure 3-11 for the bit stream shown. All bit values shown are after scrambling. The conceptual squarewave output bears no resemblence to an actual transmitted bit stream, which consists of shaped pulses.

Table 3-41 Quaternary Symbols

Sign Bit (First)	Magnitude Bit (Second)	Quat
1	0	+3
1	1	+1
0	1	−1
0	0	−3

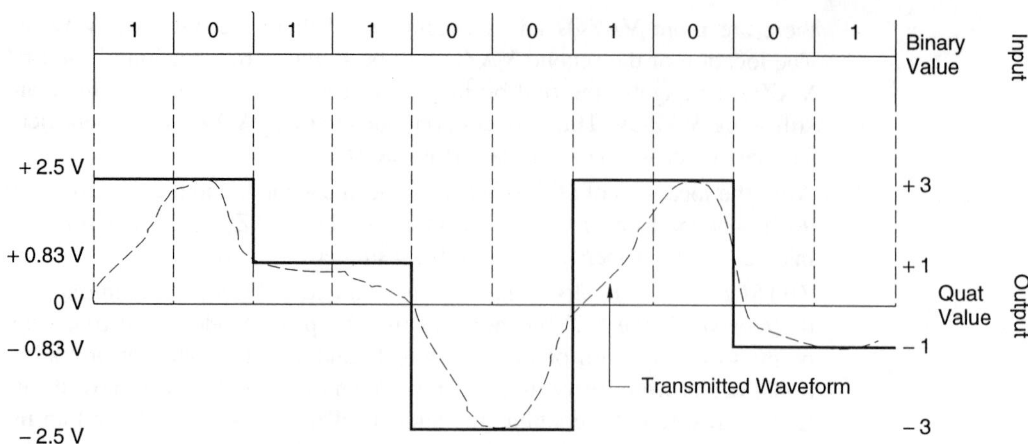

Figure 3-11 ISDN DSL quaternary symbol output

3.3 Spectrum Management and Compatibility

In order to make maximum use of telecommunications facilities, twisted pair cables frequently carry a variety of digital and analog services. The line code, pulse characteristics, and, of course, the pulse repetition rates determine the spectral characteristics of digital loop signals. The signals on one pair are coupled to other pairs by crosstalk, causing interference. This interference can be reduced by:

- Designing the spectrum of each service to limit interfering components
- Maximizing crosstalk loss in cables at all frequencies of mutual interest

The actual analysis of spectrum compatibility considers both of these requirements as a single problem. The following sections discuss the waveforms and frequency spectra of common digital loops. The loop designer has little control over the spectrum used by the various loop types because the spectrum is defined when the system characteristics are defined.

3.3.1 Waveforms and Frequency Spectra

The frequency spectra of signals used on digital loops depend on several factors:

- Encoding scheme
- Pulse density
- Presence (or absence) of repetitive patterns
- Pulse shape

- Pulse repetition rate
- Number of logical levels transmitted

Most of these characteristics are constrained by the interference requirements of digital and analog loops. The characteristics of repeatered T1-carrier, SRDL, ISDN, and HDSL loops are described in the following sections.

3.3.1.1 Repeatered T1-Carrier

Repeatered T1-carrier uses a bipolar AMI line code with 50% duty cycle pulses, as previously discussed. The maximum of its power spectrum is at about one-half the pulse repetition frequency, or 772 kHz. If the signal is a continuous string of pulses (binary ones), approximately 81% of the power is at 772 kHz, with the balance at odd harmonics. A completely random bit stream has no discrete frequency components. For a random signal, the probability of a binary one and a binary zero are both one-half. Since binary ones are encoded as an alternate positive or negative pulse, the probability of each polarity is one-quarter. The power spectral density into load R for such a pulse stream is given by [39]:

$$P(f) = \{|G(f)|^2/RT\} \times (1 - \cos(2\pi fT)W/Hz$$

where P(f) = power spectral density
G(f) = Fourier Transform of the pulse shape, V/Hz
R = 100 Ω
T = pulse repetition period, seconds ($1/1.544 \times 10^6$)
f = frequency, Hz

The Fourier Transform of a square pulse of amplitude V and pulse width T/2 is given as [40]:

$$G(f) = V \sin(\pi fT/2)/\pi f$$

By substitution, the spectral density becomes

$$P(f) = \{V^2/RT\pi^2 f^2\}\sin^2(\pi fT/2)[1 - \cos(2\pi fT)]W/Hz$$

The spectrum will change if the probability of binary ones and zeros changes. A higher ones density translates into more power concentration at and around the pulse repetition frequency. For a random signal with varying ones density, the spectral density becomes [37]:

$$P(f) = \{4p(1 - p)(V^2/RT\pi^2 f^2)\sin^2(\pi fT/2)[1 - \cos(2\pi fT)]\}/\{1 + (2p - 1)^2 + 2(2p - 1)\cos(2\pi fT)\}$$

where p = probability of binary 1

The average power spectrum is plotted in Figure 3-12 for various probabilities of binary ones. This is not the spectrum that would be observed with a spectrum analyzer at the output of a T1-carrier line interface. Such an observation would include the effects of the low-pass filter in the output of the line driver.

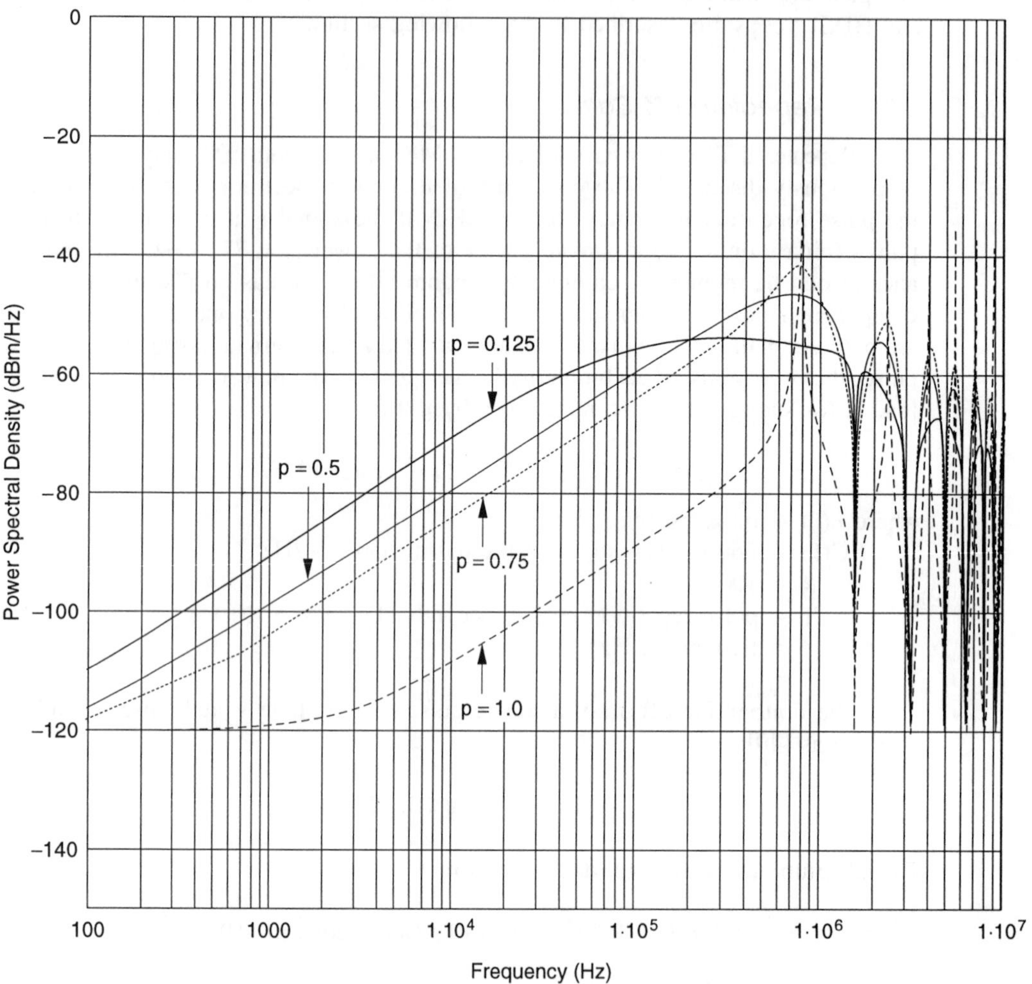

Figure 3-12 Unfiltered power spectral density for repeatered T1-carrier

Figure 3-13(a) shows a plot taken from a spectrum analyzer connected at the DSX-1 panel in a digital transmission system carrying live traffic.* Figure 3-13(b) shows a plot taken at the DSX-1 in the same system carrying the bit pattern from a quasi-random signal source (QRSS). If a signal contains repetitious patterns in either ones or zeros, the spectrum will contain discrete power frequencies or power nulls,

*The digital signal cross-connect (DSX-1)/level one panel is described in Chapter 9.

(a)

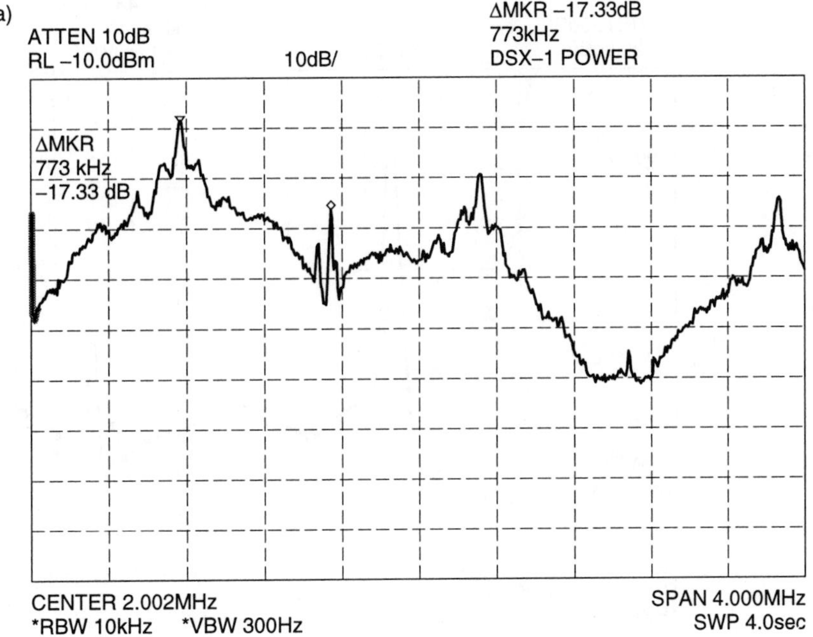

(b)

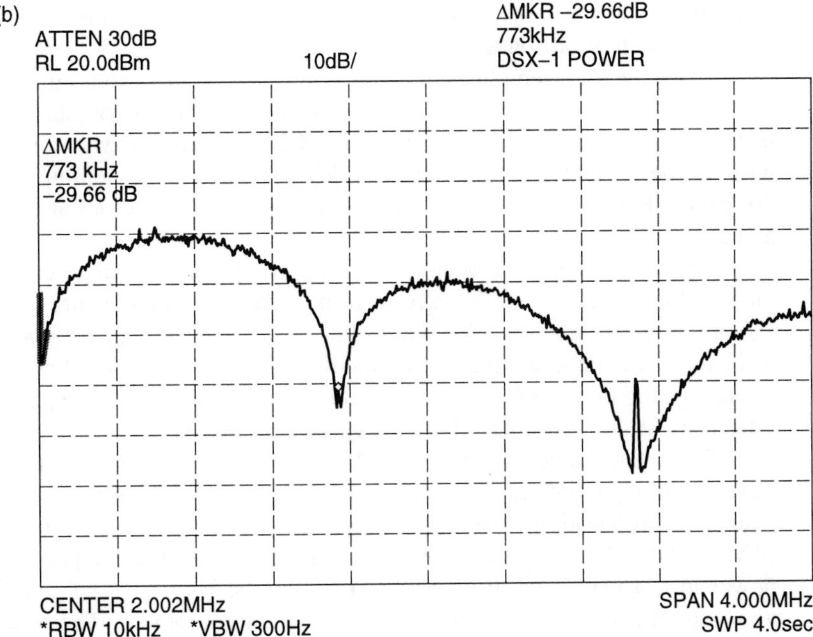

Figure 3-13 Spectrum analyzer output at DSX-1 for T1-carrier: (a) Live traffic; (b) QRSS bit pattern; (c) All-ones bit pattern

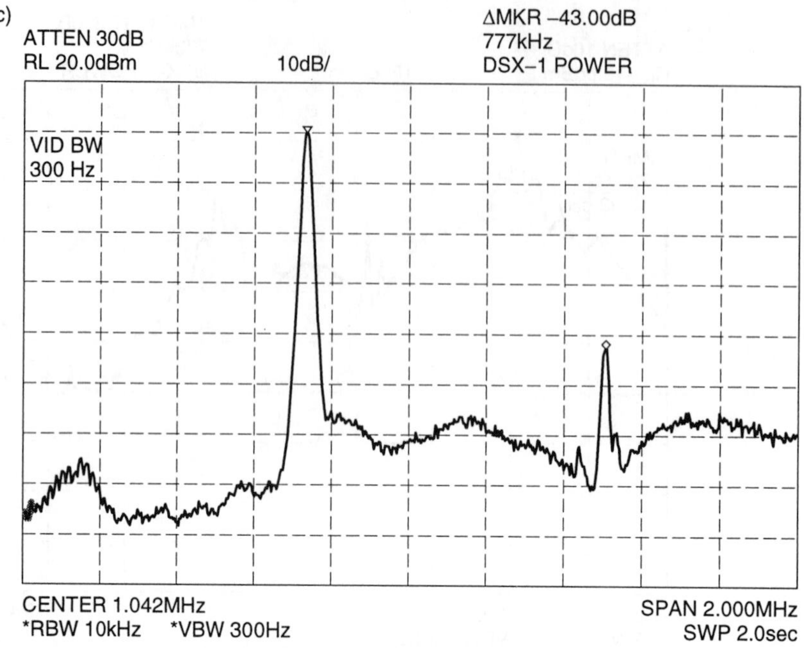

Figure 3-13 Continued

respectively, at the repetition frequencies. Figure 3-13(c) shows the spectrum analyzer plot with an all-ones test pattern.

The waveform templates of isolated pulses used in various equipment operating at the DS-1 rate are shown in Figure 3-14. This figure shows pulse shapes for both new (a) and old (b) equipment at the DSX-1 and at the network interface (c). The pulse shape for new equipment at the DSX-1 applies to any equipment connected to this point. For example, the HDSL provides a standard output at the central office and NIs.

The amplitudes have been normalized in Figure 3-14. In all cases, the nominal pulse voltage is 2.4 to 3.6 V, zero-to-peak, measured at the center of the pulse (zero UI on the templates). A pulse within this range may be scaled by a constant factor to fit the normalized template amplitudes of 1.0. There is an allowable 0.5 dB difference in the power of the positive and negative pulses. This equates to a 0.15 V difference in amplitude of otherwise symmetric pulses.

The pulse shapes of Figure 3-14 appear at a DSX-1 panel after the pulse has traveled through 655 feet of reference cable from or otherwise equalized by the equipment in question. The reference cable is taken to be 22 American wire gauge (AWG) ABAM twisted pair.* When measured at the DSX-1 point, the maximum power in a 3 kHz band around 772 kHz must be between 12.6 and 17.9 dBm for new equipment and 12.4 to 18.0 dBm for older equipment. The power in a 3 kHz band around 1.544 MHz must be at least 29 dB below the power at 772 kHz [41].

*See Appendix A for a definition of ABAM.

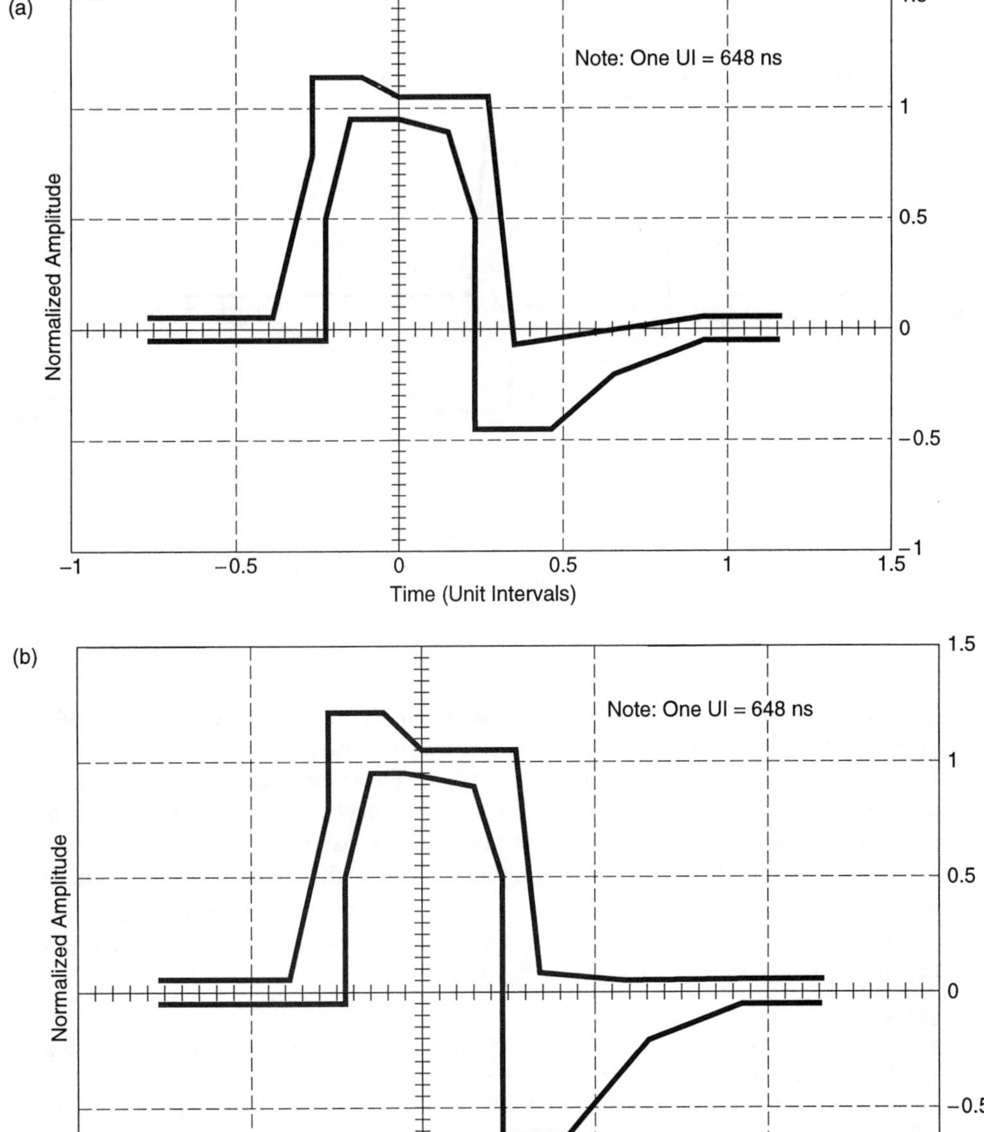

Figure 3-14 T1-carrier isolated pulse shape: (a) New equipment, [41]; (b) Older equipment, [41]; (c) Network interface, [11]

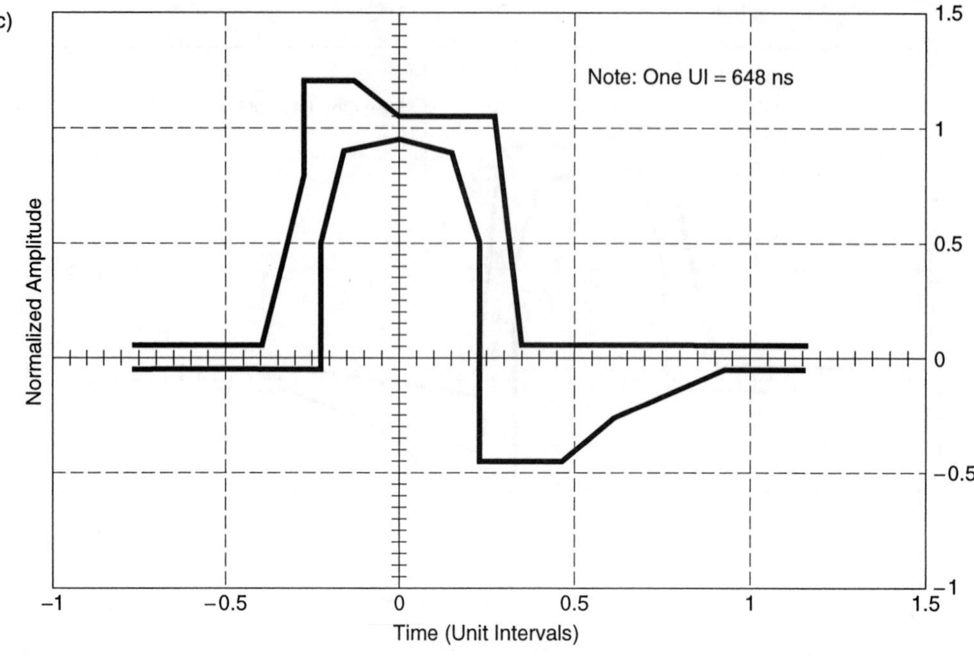

Figure 3-14 Continued

The nominal base-to-peak (or just peak) amplitude of the pulse is 3 V, which gives a peak-to-peak waveform of 6 V. For a continuous stream of pulses, the peak amplitude of the 772 kHz component is 90% of the pulse height, or 2.7 V, and the peak power is +18.63 dBm. The 772 kHz component is used for engineering purposes to determine loop loss and other performance parameters.

The pulse is attenuated as it propagates down the loop. Due to regenerator design, the minimum loss required to achieve proper equalization usually is taken as 5 to 7 dB, and the maximum loss is approximately 33 to 35 dB. The actual values depend on the repeaters used.

EXAMPLE 3-1

Find the base-to-peak amplitude of the 772 kHz component at the input to a regenerator installed at the end of a loop with 26 dB loss.

The peak power is +18.63 dBm − 26 dB = −7.37 dBm. This is equivalent to 0.18 mW. The impedance at the input of the regenerator is 100 Ω, so the peak voltage is 135 mV.

3.3.1.2 SRDL

The power spectral density of the SRDL has approximately the same shape as previously shown for the repeatered T1-carrier systems. The line code is identical (bipolar AMI), but the output filtering of the SRDL shapes the transmitted pulses, giving a slightly different spectral characteristic. This lowpass filtering reduces the output such that at 1.3 × line rate, the power is down by 3 dB and continues to decrease at 20 dB/decade.

Two specific bands require additional rejection to reduce interference to analog carrier systems with carriers at 28 ± 4 kHz and 76 ± 4 kHz operating in the same cable [42]. The band reject filter is only required for the 2.4, 4.8, and 9.6 kbps rates [25]. For practical purposes, the spectral density functions previously defined for the repeatered T1-carrier can be used for the SRDL, with adjustments for the different pulse repetition periods and termination impedance and filtering. The unfiltered spectral density function is repeated here. For a bit stream with equiprobability binary ones and zeros:

$$P(f) = \{V^2/RT\pi^2f^2\}\sin^2(\pi fT/2)[1 - \cos(2\pi fT)]W/Hz$$

and for different binary ones density probability

$$P(f) = \frac{4p(1 - p)(V^2/RT\pi^2f^2)\sin^2(\pi fT/2)[1 - \cos(2\pi fT)]}{1 + (2p - 1)^2 + 2(2p - 1)\cos(2\pi fT)}$$

where p = probability of binary 1
V = pulse amplitude from Table 3-42, V
R = 135 Ω
T = pulse repetition period from Table 3-42, s

The pulse levels and maximum average power used with the SRDL depend on the line bit rate (payload plus overhead), as shown in Table 3-42.

Table 3-42 SRDL Output Pulse Characteristics

LINE BIT RATE	OUTPUT AMPLITUDE (INTO 135 Ω LOAD)	PULSE REPETITION PERIOD, T	MAXIMUM AVERAGE POWER
2.4 kbps	1.40 volts	4.17×10^{-4} s	+6 dBm
3.2 kbps	1.40 volts	3.13×10^{-4} s	+6 dBm
4.8 kbps	1.40 volts	2.08×10^{-4} s	+6 dBm
6.4 kbps	1.40 volts	1.56×10^{-4} s	+6 dBm
9.6 kbps	0.79 volts	1.04×10^{-4} s	0 dBm
12.8 kbps	0.79 volts	7.81×10^{-5} s	0 dBm
64.0 kbps	1.40 volts	1.56×10^{-5} s	+6 dBm
72.0 kbps	1.40 volts	1.39×10^{-5} s	+6 dBm

3.3.1.3 ISDN DSL

The ISDN DSL interface uses scrambling, as previously described. This gives an almost continuous spectrum that, for the 80 kbaud 2B1Q line code, extends out to several hundred kHz. However, all DSL transmitters are equipped with a filter (shown in Figure 3-15) that reduces the spectrum above 50 kHz such that the power

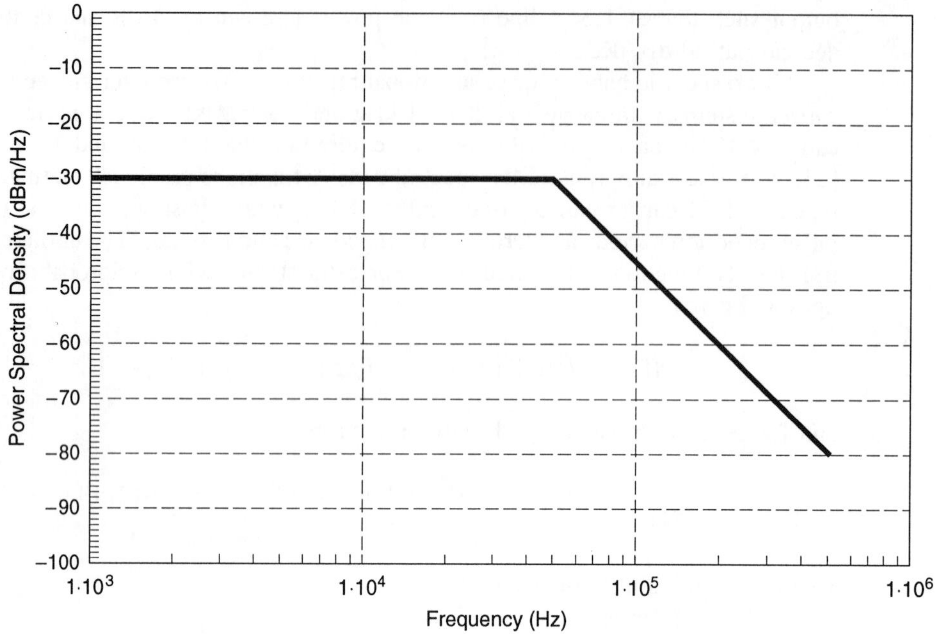

Figure 3-15 Filter characteristics for ISDN DSL, [20]

at approximately 80 kHz is 10 dB down from the 50 kHz value (−50 dB/decade), with a floor at 500 kHz [22].

The spectral density is given as [20]:

$$P(f) = \{(5/9)V^2/RT\pi^2 f^2\}\sin^2(2\pi f T)W/Hz$$

where V = 2.33 V
R = 135 Ω
T = pulse repetition period, s (1/80 × 10^3)
f = frequency, Hz

The average power spectrum is plotted in Figure 3-16.

The waveform template for an isolated ISDN DSL pulse is shown in Figure 3-17. Such a pulse shape would appear at the DSL NI (looking toward the central office). The base-to-peak (or just peak) amplitude of the pulse depends on the quaternary symbol, as shown in the table accompanying the pulse template. The average power of a random bit stream is approximately +13.5 dBm over a frequency range of 0 to 160 kHz. The nominal pulse voltage corresponding to each quaternary symbol is given in Table 3-43.

Table 3-43 ISDN DSL Pulse Voltage

Quat	Pulse Voltage
+3	2.5 V
+1	$\frac{5}{6}$ V
−1	$-\frac{5}{6}$ V
−3	−2.5 V

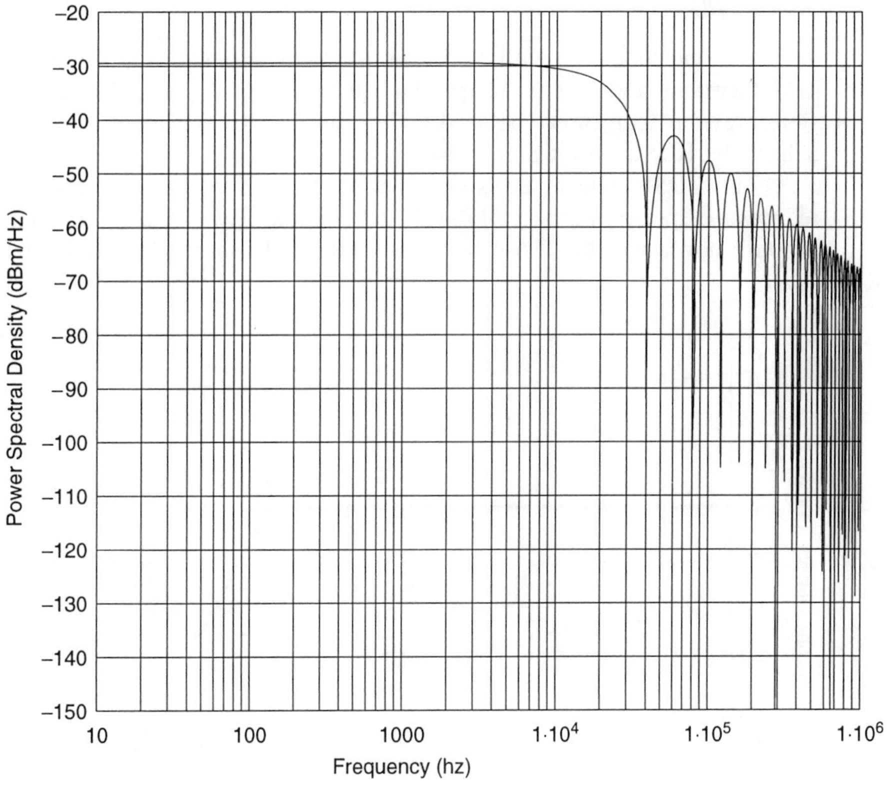

Figure 3-16 Average power spectrum for ISDN DSL

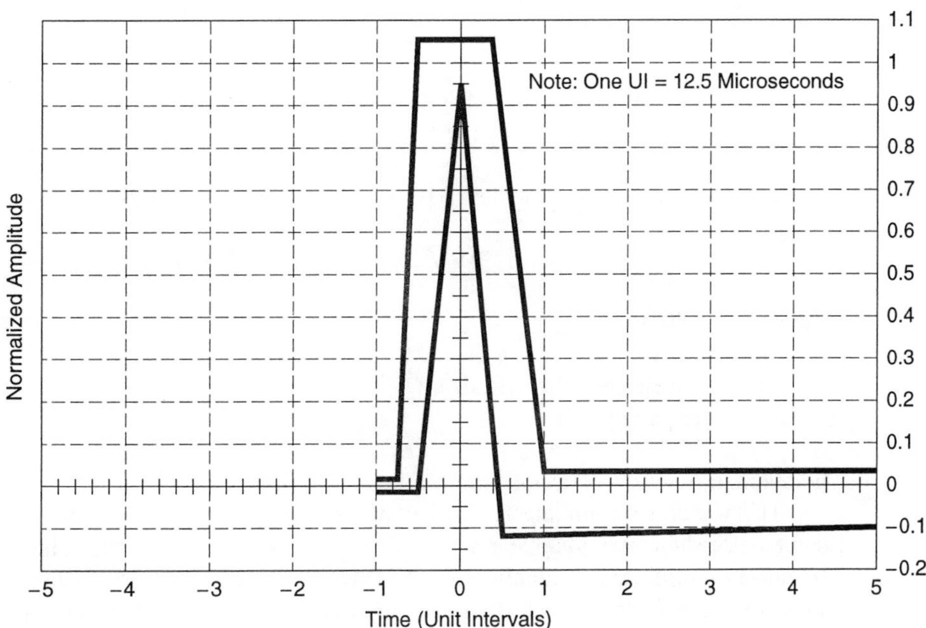

Figure 3-17 ISDN DSL isolated pulse shape, [20]

3.3.1.4 HDSL

The line code used with the HDSL is identical to that used with the ISDN DSL [34]; therefore, the spectral shape characteristics also are identical and the expressions previously given can be used with adjustment for pulse repetition period. Each half of the HDSL operates at 784 kbps or 392 kbaud.

The HDSL transmitter is equipped with a filter (shown in Figure 3-18) that reduces the spectrum above 196 kHz such that the power at approximately 260 kHz is 10 dB down from the 196 kHz value (−80 dB/decade), with a floor at 1.96 MHz.

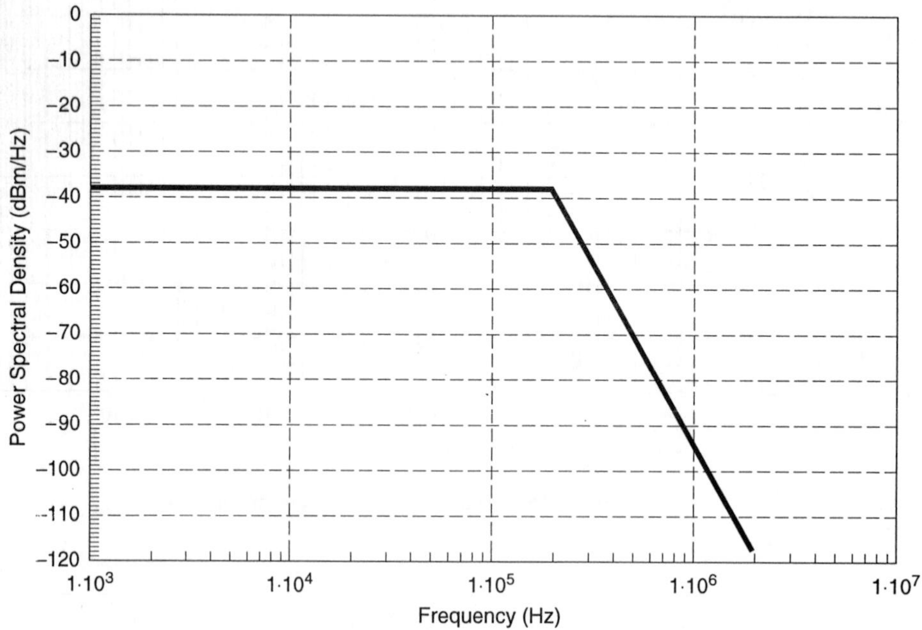

Figure 3-18 Filter characteristics for HDSL, [34]

The spectral density for the 2B1Q line code is given as [20]:

$$P(f) = \{(5/9)V^2/RT\pi^2f^2\}\sin^2(2\pi fT)\,W/Hz$$

where V = 2.40 V
R = 135 Ω
T = pulse repetition period, s $(1/392 \times 10^3)$
f = frequency, Hz

The average power spectrum is plotted in Figure 3-19.

The waveform template for an isolated HDSL pulse is shown in Figure 3-20. Such a pulse shape would appear at the HDSL NI (looking toward the central office). The pulse amplitudes are identical to the ISDN DSL previously described. The average power of a random bit stream is approximately +13.5 dBm over a range of

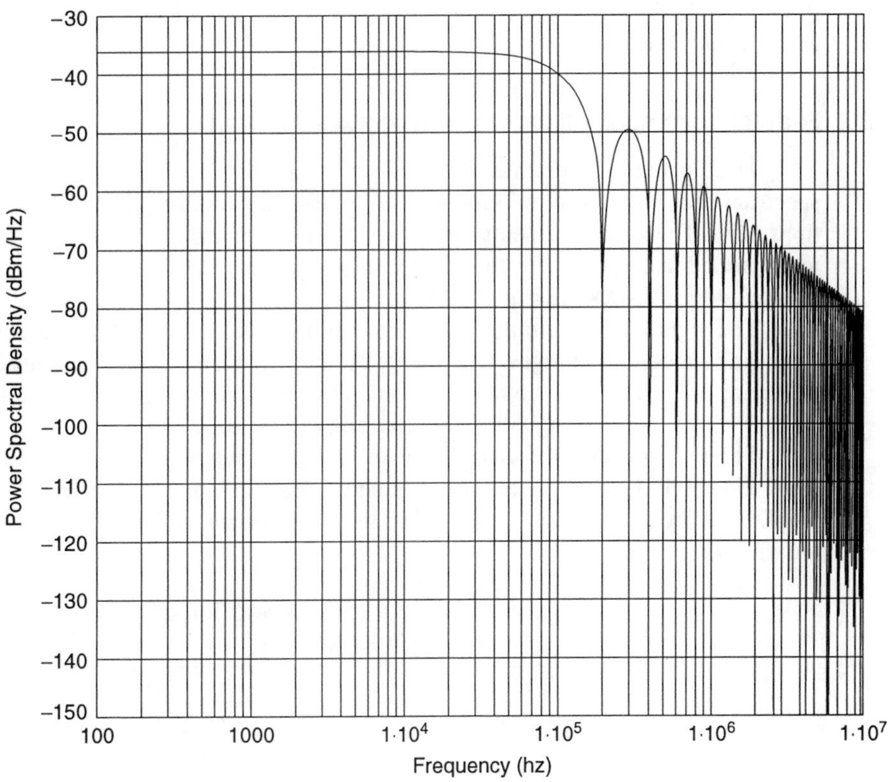

Figure 3-19 Average power spectrum for HDSL using 2B1Q line code

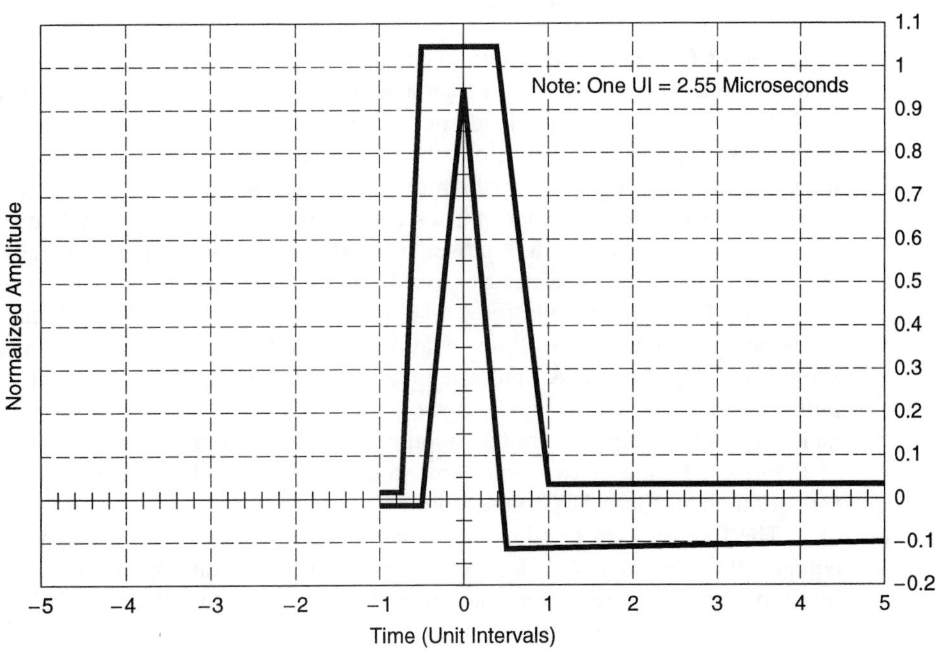

Figure 3-20 HDSL isolated pulse shape for 2B1Q line code, [34]

0 to 784 kHz. The pulse voltages for the HDSL are identical to those for the ISDN DSL (Table 3-43).

3.3.2 Federal Communications Commission Requirements

The Federal Communications Commission (FCC) has specified a number of spectrum management requirements in 47 CFR Part 68.308, Signal Power Limitiations [43]. These requirements apply to customer-owned equipment and are specifically aimed at limiting interference by subrate and 1.544 Mbps digital services and do not address ISDN BRI nor HDSL.

The signal power limitations extend from 10 Hz to 6 MHz and cover longitudinal and metallic voltages. The basic, but not exhaustive, requirements are shown in Figure 3-21. This illustration does not show the requirements for specific subrate or high bit-rate digital signals, which are described below.

The FCC defines the maximum *driving* pulse amplitudes for subrate services from 2.4 to 56.0 kbps (these are payload rates), as shown in Table 3-44. These amplitudes are not directly comparable to those shown in Table 3-42 because they are unfiltered.

Table 3-44 Subrate Driving Pulse Amplitudes[a]

Payload Rate	Amplitude
2.4 kbps	1.66 V
4.8 kbps	1.66 V
9.6 kbps	0.83 V
56.0 kbps	1.66 V

[a]Source: [43]

The FCC requires 50% duty cycle rectangular pulses of the amplitudes shown in the table to be first passed through a single real-pole, low-pass filter. The pulses for 2.4, 4.8, and 9.6 kbps payload rates require additional filtering, which is identical to that discussed in the section on subrate digital loops. The resulting pulse amplitudes are used as templates to which the actual terminal equipment pulses are compared. The terminal equipment pulses cannot exceed the pulse template peak amplitudes by 10%. The average power in a random signal with equiprobable binary ones and zeros is the same, as discussed in the section on subrate digital loops.

The FCC requirements for 1.544 Mbps services are similar to industry standards but are stated in a different way, as shown in Figure 3-22. This illustration shows the FCC "Option A" pulse template, which can be compared to the industry pulse template defined in [11] and shown in Figure 3-14(c). The FCC "Option B" and "Option C" templates are the "Option A" pulse passed through the filter (shown in Figure 3-23) one or two times, respectively. Table 3-45 provides additional information about these FCC pulse options.

The filter is used to adjust the output voltage of the customer-owned terminal equipment for varying cable losses between the terminal and the first public network line repeater, and also to limit the spectrum at higher frequencies. The attenuator

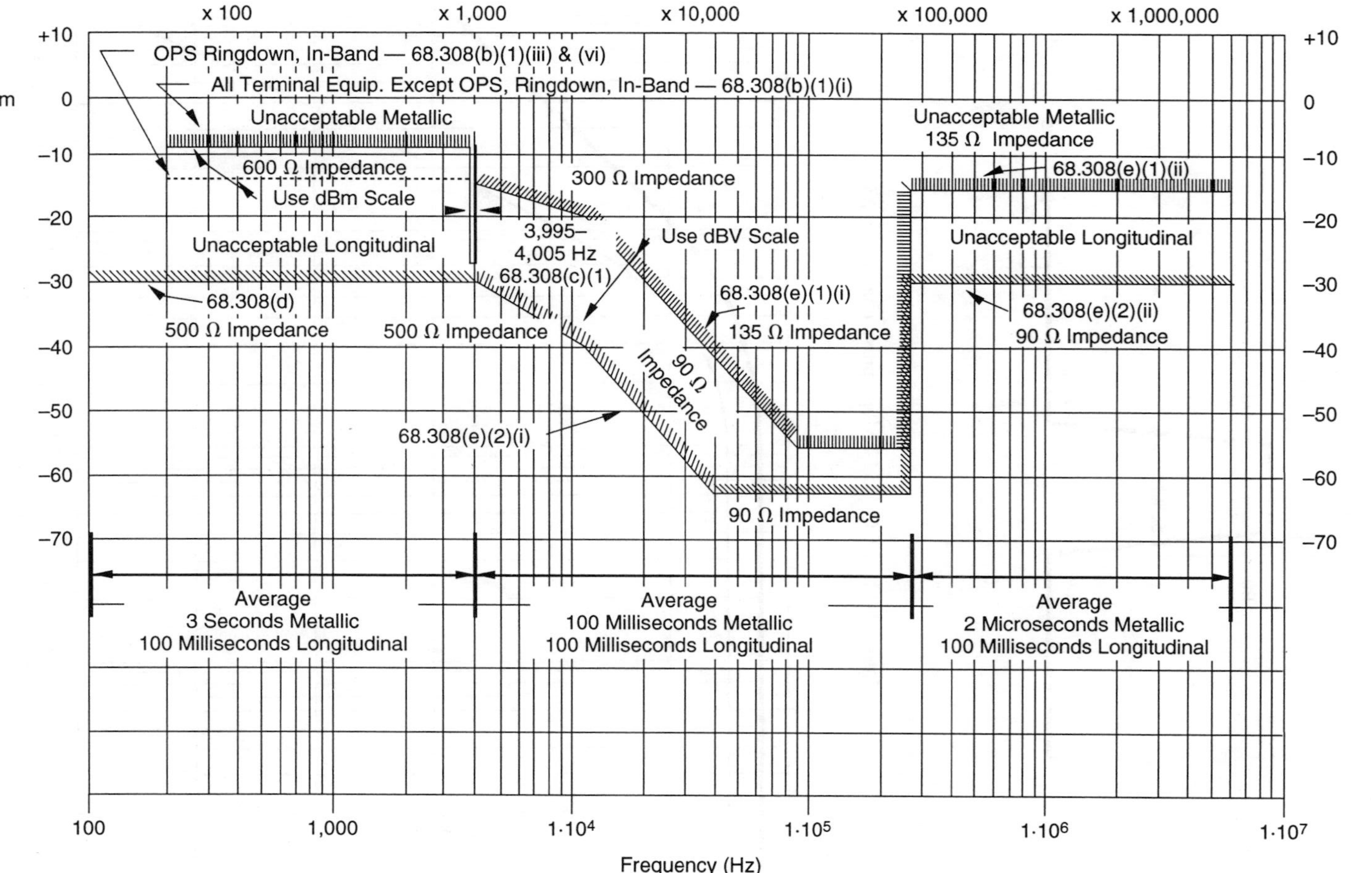

Figure 3-21 FCC signal power limitations per Par. 68.308 [44]
Reeve, © 1992 Institute of Electrical and Electronics Engineers.
Reprinted with permission

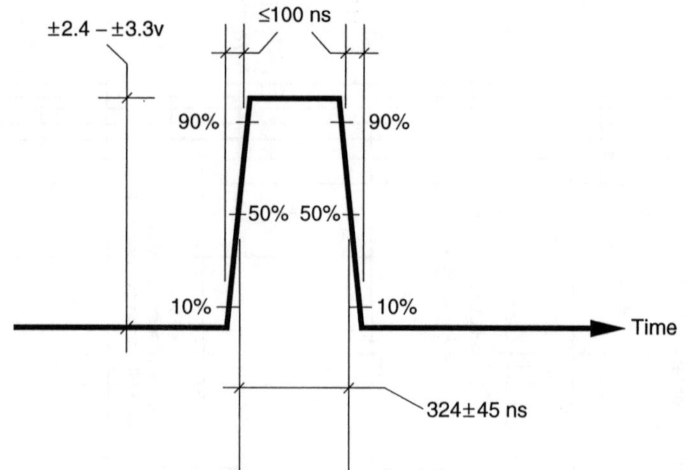

Figure 3-22 FCC 1.544 Mbps pulse characteristics

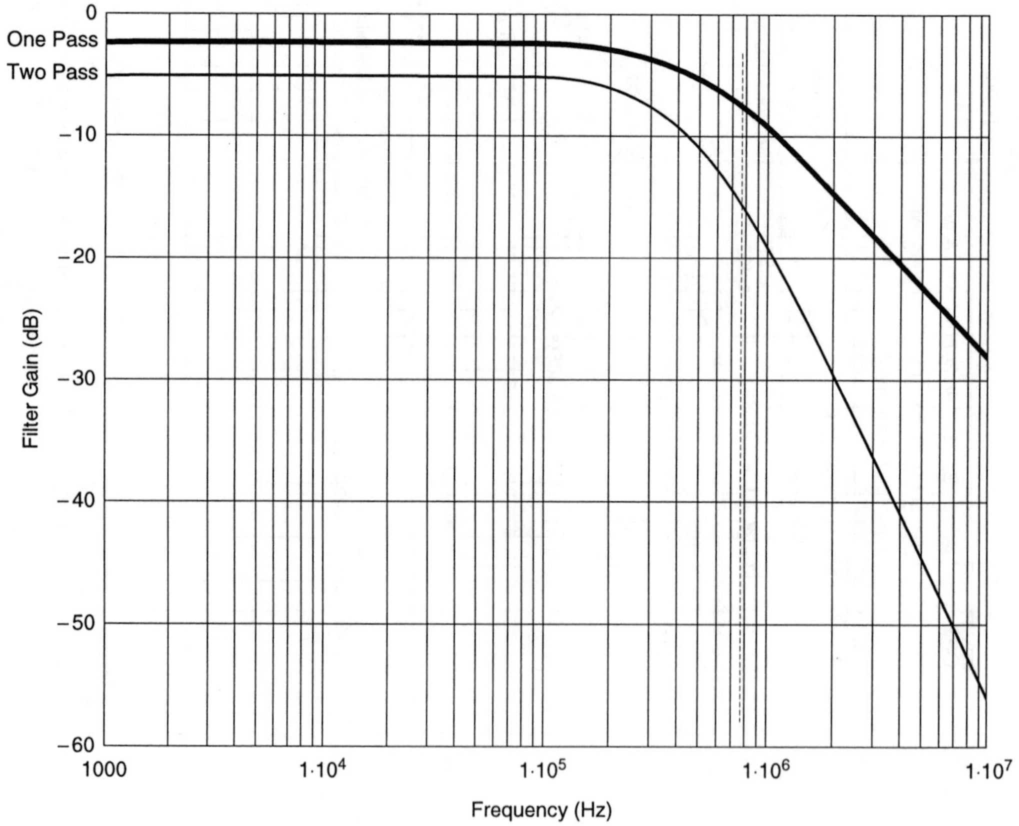

Figure 3-23 FCC 1.544 Mbps filter

Table 3-45 FCC Pulse Option Application

Cable Loss @ 772 kHz	Customer Premises Terminal Equipment		
	Pulse Option	Attenuator Loss @ 772 kHz	Power in 3 kHz Band Around 772 kHz
15 to 22 dB	A	0 dB	+12 to +19 dBm
7.5 to 15 dB	B	7.5 dB	+4.5 to +11.5 dBm
0 to 7.5 dB	C	15.0 dB	−3 to +4 dBm

normally is part of a CSU installed on the customer premises. Table 3-45 shows the application of pulse options for ranges of cable losses. This table also shows the maximum allowable output power for an all-ones signal sequence. The FCC requires the power in a 3 kHz band around 1.544 MHz be 25 dB below that shown for 772 kHz. This is 4 dB higher than allowed at the DSX-1 point, by industry standards [41], but equal to that specified at customer installations by industry standards [11].

Most services operating at the 1.544 Mbps rate carry encoded analog signals, including voice and analog modem tones. The maximum equivalent power of encoded analog signals (analog modem signals, but not live voice) placed on the digital services cannot exceed −12 dBm0 when averaged over any 3-second interval. Analog tones used for network control cannot exceed −3 dBm0 when averaged over any 3-second interval. See [44] for more information about analog signals.

References

[1] *D1, D2, D3, D4 Digital Channel Banks and D5 Digital Terminal System Application Engineering—Carrier Engineering*. AT&T Practice 855-351-103.

[2] *Lenkurt Demodulator,* Vol. 3. GTE Lenkurt, Inc., 1976.

[3] *Bell System Technical Journal,* Vol. 44, No. 7, Sept. 1965.

[4] Henning, H., Pan, J. *D2 Channel Bank: System Aspects*. Bell System Technical Journal, Vol. 51, No. 8, Oct. 1972.

[5] American National Standard for Telecommunications. *Digital Hierarchy—Formats Specifications*. ANSI T1.107-1988.

[6] *ACCUNET® T1.5 Service Description and Interface Specifications*. AT&T Technical Reference TR62411, Dec. 1988. Available from AT&T Corporate Mailings.

[7] *Access Specification for High Capacity (DS1/DS3) Dedicated Digital Services*. AT&T Technical Reference TR62415, June 1989. Available from AT&T Corporate Mailings.

[8] *High Capacity Digital Service (1.544 Mb/s) Interface Generic Reqiurements for End Users*. BELLCORE Technical Reference TR-NPL-000054, April 1989. Available from BELLCORE Customer Service.

[9] *Digroup Terminal and Digital Interface Frame Technical Reference and Compatibility Specification*. AT&T Compatibility Bulletin 123, Aug. 1981. Available from AT&T Corporate Mailings.

[10] *Digital Channel Bank Requirements and Objectives*. Bell System Transmission Engineering, BELLCORE Technical Reference TR43801, Nov. 1982. Available from BELLCORE Customer Service.

[11] American National Standard for Telecommunications. *Carrier-to-Customer Installation, DS1 Metallic Interface*. ANSI T1.403-1989.

[12] *Requirements for Interfacing Digital Terminal Equipment to Services Employing the Extended Superframe Format*. AT&T Technical Reference TR54016, Sept. 1989. Available from AT&T Corporate Mailings.

[13] *Extended Superframe Format (ESF) Interface Specification*. BELLCORE Technical Reference TR-TSY-000194, Dec. 1987. Available from BELLCORE Customer Service.

[14] *Digroup Terminal and Digital Interface Frame Technical Reference and Compatibility Specification*. AT&T CB123, Aug. 1981. Available from AT&T Corporate Mailings.

[15] *Digital Subscriber Signaling System No. 1 (DSS 1), Data Link Layer*. Recommendations Q.920–Q.921. CCITT Blue Book, Vol. VI, Fascicle VI.10, Geneva, 1989. Available from NTIS.

[16] *Data Communication Networks: Services and Facilities Interfaces*. Recommendations X.1–X.32. CCITT Blue Book, Vol. VIII, Fascicle VIII.2, Geneva, 1989. Available from NTIS.

[17] *General Aspects of Digital Transmission Systems; Terminal Equipments*. Recommendations G.700–G.772. CCITT Blue Book, Vol. III, Fascicle III.4, Geneva, 1989.

[18] *Digital Interface Between the SLC®96* Digital Loop Carrier System and Local Digital Switch. BELLCORE Technical Reference TR-TSY-000008, Aug. 1987. Available from BELLCORE Customer Service.

[19] *Integrated Digital Loop Carrier System Generic Requirements, Objectives, and Interface*. BELLCORE Technical Reference TR-TSY-000303, Sept. 1986, revision 4, Aug. 1991.

[20] American National Standard for Telecommunications. *Integrated Services Digital Network (ISDN): Basic Access Interface for Use on Metallic Loops for Application on the Network Side of the NT (Layer 1 Specification)*. ANSI T1.601-1988.

[21] *Digital Networks, Digital Sections and Digital Line Systems*. CCITT Blue Book, Vol. III, Fascicle III.5, Geneva, 1989. Available from NTIS, Order No. PB89-143895.

[22] *Generic Requirements for ISDN Basic Access Digital Subscriber Lines*. BELLCORE Technical Reference TR-NWT-000393, Jan. 1991. Available from BELLCORE Customer Service.

[23] American National Standard for Telecommunications. *Digital Hierarchy: Supplement to Formats Specifications (Synchronous Digital Data Format)*. ANSI T1.107b-1991.

[24] *Subrate Data Multiplexing, A Service Function of DATAPHONE®* Digital Service. AT&T Technical Reference TR54075, Nov. 1988.

[25] *Secondary Channel in the Digital Data System: Channel Interface Requirements.* BELLCORE Technical Reference TR-NPL-000157, BELLCORE, April 1986. Available from BELLCORE Customer Service.

[26] *Digital Data System Channel Interface Specifications.* AT&T Technical Reference PUB 41021, March 1987. Available from AT&T Corporate Mailings.

[27] *Digital Data System Channel Interface Specification.* AT&T Technical Reference PUB 62310, Nov. 1987; Addendum 1, Jan. 1988; Addendum 2, Oct. 1989; Addendum 3, Dec. 1989. Available from AT&T Corporate Mailings.

[28] *D3 and D4 Subrate Dataport Channel Unit Technical Reference and Compatibility Specification.* AT&T Compatibility Bulletin No. 126, April 1981. Available from AT&T Corporate Mailings.

[29] *D3 and D4 56 KB Dataport Channel Unit Technical Reference and Compatibility Specification.* AT&T Compatibility Bulletin No. 141, April 1981. Available from AT&T Corporate Mailings.

[30] *Digital Channel Banks: Requirements for Dataport Channel Unit Functions.* BELLCORE Technical Advisory TA-TSY-000077, 1986. Available from BELLCORE Documents Registrar.

[31] *Generic Requirements for the Digital Data System (DDS) Network Office Channel Unit.* BELLCORE Technical Advisory TA-TSY-000083, April 1986. Available from BELLCORE Documents Registrar.

[32] *Generic Requirements for the Subrate Multiplexer.* BELLCORE Technical Advisory TA-TSY-000189, April 1986. Available from BELLCORE Documents Registrar.

[33] *Digital Data System (DDS) Multipoint Junction Unit (MJU) Requirements.* BELLCORE Technical Advisory TA-TSY-000192, April 1986. Available from BELLCORE Documents Registrar.

[34] *Generic Requirements for High-Bit-Rate Digital Subscriber Lines.* BELLCORE Technical Advisory TA-NWT-001210, Oct. 1991. Available from BELLCORE Customer Service.

[35] Bellamy, J. *Digital Telephony*, 2nd Ed. John Wiley & Sons, 1991.

[36] Smith, D. *Digital Transmission Systems.* Van Nostrand Reinhold, 1985.

[37] Bylanski, P., Ingram, D. *Digital Transmission Systems*, 2nd Ed. IEE Telecommunications Series 4, Peter Peregrinus Ltd., 1980.

[38] Exchange Carrier Standards Association. *A Technical Report on the Performance of AMI Signals Through B8ZS Optioned Equipment Across Network Boundaries.* Committee T1—Telecommunications Report No. 2. Prepared by T1M1.3 Working Group on Testing and Operations Support Systems, 1989.

[39] Cravis, H., Crater, T. "Engineering of T1 Carrier System Repeatered Lines," Bell System Technical Journal Vol. XLII, March 1963, p. 431.

[40] Jordan, E., Ed. *Reference Data for Engineers: Radio, Electronics, Computer, and Communications,* 7th Ed. Howard W. Sams & Co., 1985.

[41] American National Standard for Telecommunications. *Digital Hierarchy: Electrical Interfaces.* ANSI T1.102-1987.

[42] *Digital Data System Channel Interface Specification*. AT&T Technical Reference TR62310. Nov. 1987. Available from AT&T Corporate Mailings.

[43] *Connection of Terminal Equipment to the Telephone Network*. 47 CFR Part 68. Available from USGPO.

[44] Reeve, W. *Subscriber Loop Signaling and Transmission Handbook: Analog*. IEEE Press, 1992.

4 Timing and Synchronization

Chapter 4 Acronyms

ABAM	air core 83 nF/mile, solid polyolefin-insulated, 22 AWG, aluminum-shielded cable
ADPCM	adaptive differential pulse code modulation
AMI	alternate mark inversion
ANSI	American National Standards Institute
AWG	American wire gauge
BITS	building-integrated timing supply
BPV	bipolar violation
CC	composite clock
COFA	change-of-frame alignment
COT	central office terminal
CRC, crc	cyclic redundancy check
DCS	digital cross-connect system
DDS	digital data system
DLC	digital loop carrier
DSL	digital subscriber line
DSS	digital switching system
DSX-1	digital signal cross-connect–DS-1 rate
EO	end-office
ESF	extended superframe format
F	frequency
FIFO	first in, first out
GPS	global positioning system
HDSL	high bit-rate digital subscriber line
IDLC	integrated digital loop carrier
ISDN	integrated services digital network
LORAN-C	LOng RAnge Navigation system, C-type
MTBF	mean time between failure
MTIE	maximum time interval error
PCM	pulse code modulation
ppm	parts per million
PRS	primary reference source
PTN	public telephone network
RST	remote switching terminal
RZ	return-to-zero
SF	superframe format
SONET	synchronous optical network
SRDL	subrate digital loop
TIE	time interval error
TSG	timing signal generator
UDLC	universal digital loop carrier
UI	unit interval; time duration of 1 bit; inverse of bit rate
UIp-p	unit interval, peak-to-peak
UTC	universal coordinated time

4.1 Introduction

Timing and synchronization are fundamental to any digital transmission and switching network, including digital loops. Timing can be thought of as the separation of events according to some time scale, and synchronization is the provision of a stable timing reference to multiple digital network elements that generate, switch, or retime the digital bit stream in a network.

In a synchronized network, the *significant instants* of the transmitted and received data streams are separated by the same nominal time interval, and the received waveform is sampled at the same nominal frequency as the frequency of the transmitted waveform. By using the word *nominal* instead of *exactly*, this definition recognizes that real telecommunication systems do not operate under ideal conditions, and variations in time intervals and frequency exist. Synchronized network design, as discussed in this chapter, attempts to minimize the negative effects of these time and frequency variations.

In all digital telecommunication transmission systems, timing is encoded with the transmitted data signal. At the far-end receiving location, timing is recovered from the transmitted signal and is used to decode the incoming data. If the recovered timing is sufficiently stable, it may be used to time (clock) the signal on the return path (back to the original sending location), or it may be passed on to the next (downstream) digital element in the network.

Synchronization is maintained at multiple levels in a network:

- Frequency (bit) synchronization
- Frame synchronization
- Phase (byte or word) synchronization
- Network synchronization

With *frequency synchronization,* the receiver clock at the distant end of the circuit is designed to follow the incoming signal. This is called clock recovery. If the digital loop is terminated in channel banks used for voice transmission, frequency synchronization between the transmit and receive circuits at a given end is not needed. Clock recovery is needed only for received signal decoding, and the digital signal rate in one direction does not need to be exactly the same as the signal rate in the other direction.

Once frequency synchronization is achieved, the receiver must then achieve *frame synchronization* to identify the beginning and end of frame boundaries. Framing usually is established by bit pattern recognition, where the bits to be recognized are widely and uniformly distributed in the serial bit stream. In the DS-1 rate (1.544 Mbps) signal used with T1-carrier, every 193rd bit is a framing bit, as discussed in Chapter 3. In contrast, synchronous data signals in nontelephony applications usually are framed by flags, which are groups of eight contiguous bits. The notable exception is the synchronous optical network (SONET), in which 16 contiguous bits are used to define the beginning of the STS-1 frame.* Framing information introduced at the transmitter adds overhead to the bit stream. At the DS-1 rate, this overhead is slightly more than one half of 1% (1/193).

Once the frame boundaries are found, the receiver may simply count clock pulses to locate individual code words or data bytes and achieve *phase synchronization*. These code words are most commonly called timeslots. In superframe formatting (SF) and extended superframe formatting (ESF) framed DS-1 links, these timeslots (or bytes) represent individual channels. Phase synchronization also is used in these links to recover the signaling bits in robbed bit signaling schemes.

From an overall network perspective, one additional level is needed for *network synchronization*. That is, to transmit, switch, and receive digital information in an integrated digital environment, the clock frequencies in each interconnected node—even those that do not derive synchronization from the interconnection—must be the same, or at least be able to handle the asynchrony. Asynchrony may be handled by precise frequency control, buffering, and a synchronization hierarchy that properly allocates slips. These methods are discussed later.

All digital loop systems use serial bit transmission, whereby each bit has a predefined relationship in time to the bits on each side of it. When line receivers are synchronized with line transmitters, bit extraction can be done accurately, as shown in Figure 4-1(a). In this illustration, the received data stream is a continuous series of alternating pulses, and the receive clock pulses are perfectly aligned with each received pulse. When the receivers and transmitters are not synchronized, the receiver may attempt to read a bit value at a time no bit is present, as shown during the second and third clock pulses in Figure 4-1(b). This leads to transmission impairments, which manifest themselves as frame slips and data errors.

In a given network, synchronization functions typically are provided in a master-slave relationship such that the highest network level provides master timing, and lower levels are operated in slave timing modes. All digital network elements are designed to cope with the loss of master timing signals by various fallback schemes, such as clock free-run and holdover timing, according to the network element's relative importance in the synchronization hierarchy described in the next section. Lack of attention to timing and synchronization can lead to poor transmission quality at best, or complete network collapse in extreme cases.

Digital loop systems normally are at the lowest timing level in a network. Digital loop systems may be operated independently of other network elements, as shown in Figure 4-2. As digital networks evolve, timing islands, such as independent digital loop carrier (DLC) systems, will give way to completely interconnected and synchronized network elements. For independent elements, the accuracy and stability of the clock is not important as long as the other end can adequately buffer the incoming bit stream and lock to the incoming bit rate. The latter requirement is called *pull-in range*. Pull-in range is a measure of the maximum input frequency deviation

---

*Although SONET is finding applications in the loop, these discussions are beyond the scope of this book.

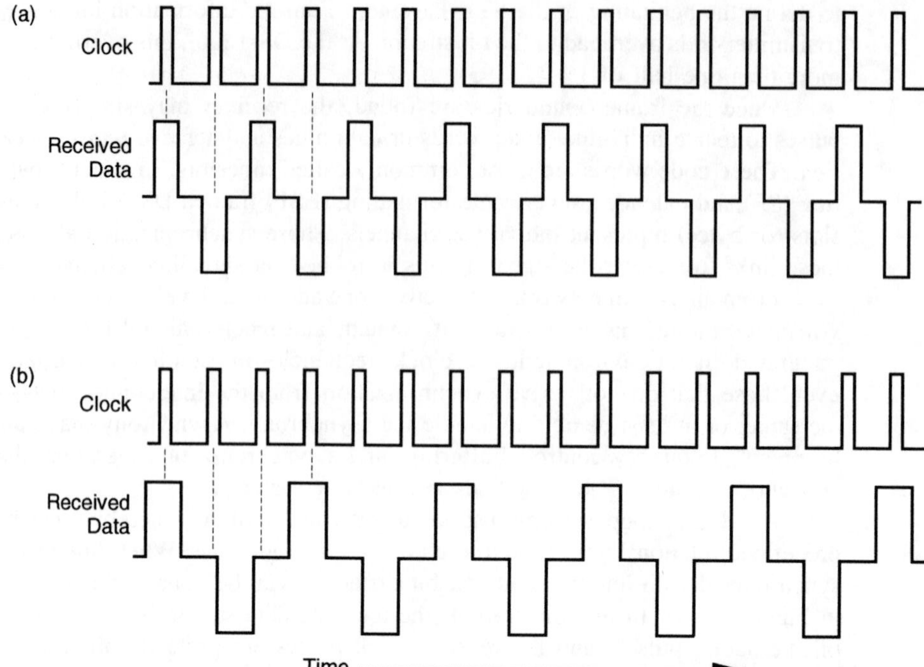

Figure 4-1 Synchronization of serial bit streams: (a) Received data synchronized with clock [clock frequency = received data rate]; (b) Received data not synchronized with clock [clock frequency > received data rate]

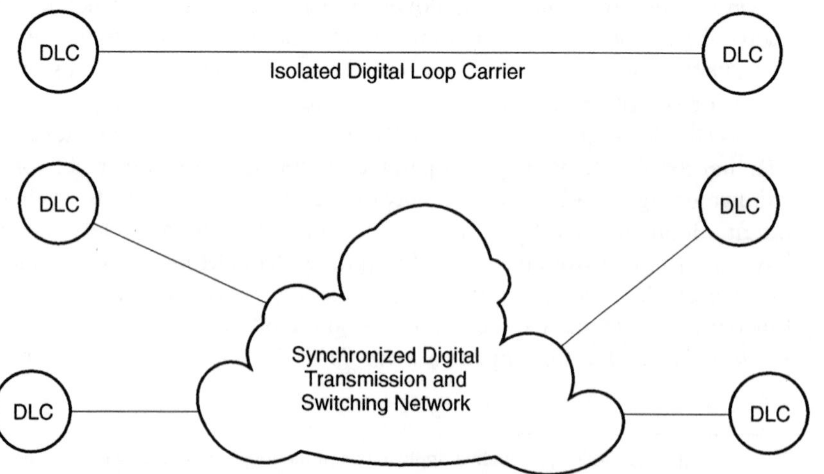

Figure 4-2 Isolated and network-synchronized DLC systems

that the receive clock can overcome so it can pull itself into synchronization with the transmit clock.

In a typical situation, one end of the loop provides the master clock and the other end derives its timing in the slave mode (called *loop timing*). This is illustrated in Figure 4-3(a). The west digital terminal (here a channel bank) provides a transmit clock at frequency F(west), which determines the line rate received by the east terminal (another channel bank). The east terminal then synchronizes its transmit clock at frequency F(east) to the incoming clock rate, such that F(east) = F(west). The digital stream never leaves the small network of two terminals, and it is not necessary to synchronize with any other network element.

Another common timing method for independent digital loops is illustrated in Figure 4-3(b). Here, each digital terminal is equipped with respective transmit clocks F(east) and F(west), which determine the line rate for each direction. These clocks do not have to be synchronized because each line receiver is clocked independently

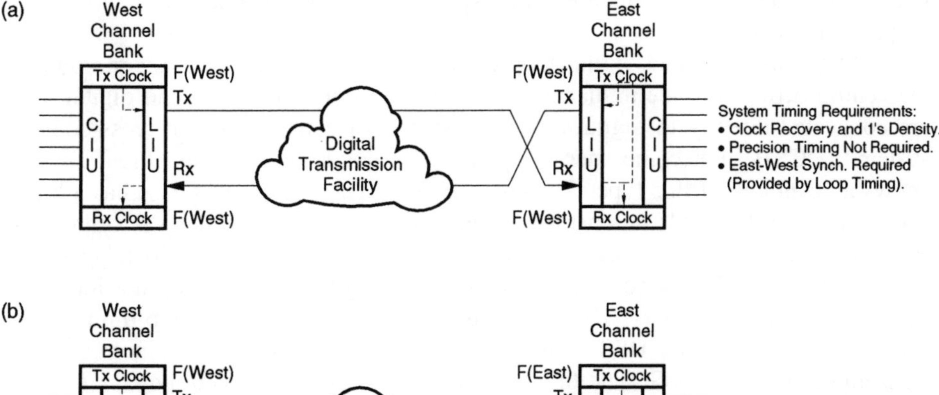

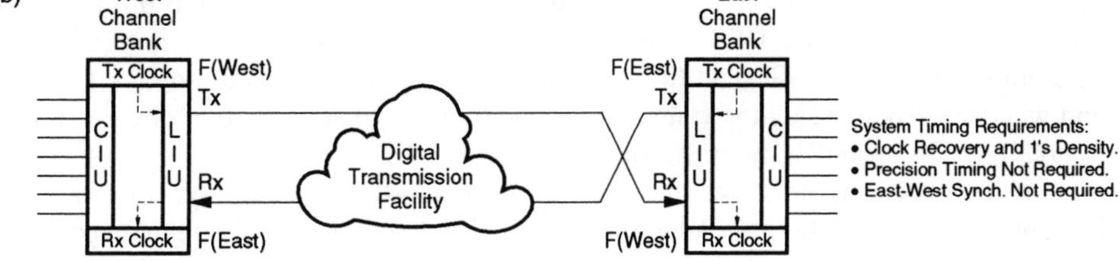

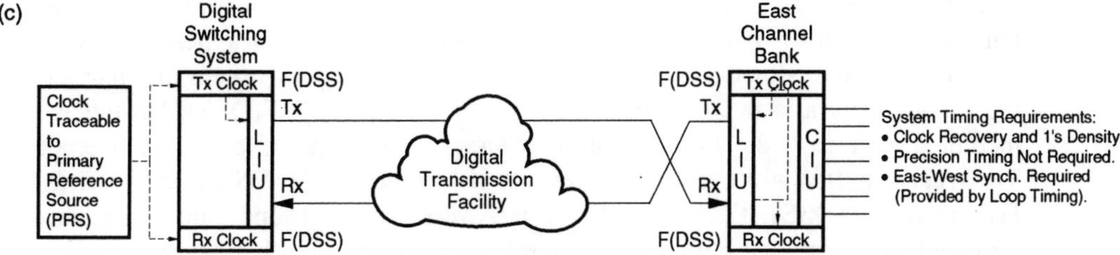

Figure 4-3 Digital loop synchronization: (a) Network isolated channel banks with loop timing; (b) Network isolated channel banks with unequal clock rates; (c) Network synchronized channel bank with loop timing

of its corresponding transmitter by the incoming bit stream. In this case, F(east) ≠ F(west). This method is adequate when voiceband signals or asynchronous digital data are transmitted over the digital loop, but it does not meet the requirements of fully synchronized digital data transmission services.

Modern universal digital loop carrier (UDLC), channel bank, and intelligent multiplexer clocks meet minimum accuracy requirements such that the transmitted bit stream is 1.544 Mbps ± 50 bps in the free-run mode. To compensate for variations in media propagation characteristics and other anomalies, the corresponding receivers function properly over a wider range, typically 1.544 Mbps ± 200 bps. The transmission rate of older equipment may be 1.544 Mbps ± 200 bps, with a correspondingly wider receiver characteristic. The rate accuracy at a digital signal cross-connect (DSX-1) point is specified in parts per million (ppm) such that the deviation from the 1.544 Mbps rate is no more than ±32 ppm, and equipment connected to the DSX-1 must be able to accept deviations of ±130 ppm. When converted to bits per second, these specifications provide the same tolerances. The foregoing apply to self-timed (independent) systems and systems with stratum 4 clocks (explained later).

Where an integrated digital loop carrier (IDLC) is used, the DLC terminal at the central office end is eliminated; and when the remote terminal and digital loop are connected directly to digital switching systems (DSSs) or digital cross-connect systems (DCSs), it becomes necessary to synchronize all elements. Figure 4-3(c) shows an example of this type of interconnection. The DSS in this case has a high-accuracy clock, traceable to a primary reference source (PRS), which determines the transmitted line rate to the DLC systems. Although a PRS traceable clock is shown, a clock of this quality is not necessary for switching systems that are not interconnected with other networks. The remote terminal is operated with loop timing. The line rate received at the DSS from the remote terminal is the same nominal rate transmitted from the DSS. The line receiver at the DSS only has to cope with jitter and wander introduced by the transmission media and clock extraction circuits at the remote end. Jitter and wander are discussed later.

If the DLC remote terminal is not loop timed—because of, say, a mistake in the option settings—there will be synchronization problems in the signal received at the DSS. Depending on the characteristics of the remote terminal clock, these synchronization problems simply may be sensitive to the time of day or otherwise intermittent. In the worse case, the DLC will be completely inoperative. Such problems are difficult to find because all other aspects of the signal appear satisfactory.

In the above situation, the DSS clocks the received data stream into its incoming frame buffer at a rate determined by its clock frequency, F(DSS), but the actual incoming bit rate is the free-running clock rate of the remote terminal, F(remote). The receive buffer will eventually overflow if F(remote) > F(DSS), or underflow if F(remote) < F(DSS). Each time the buffer overflows, a group of bits (number determined by buffer size but usually limited to one frame) will be lost; and each time the buffer underflows, a complete data group will be repeated.

An example of buffer operation (in this case, a DS-1 frame buffer) is shown in Figure 4-4. At the start of the example, the buffer is shown with 100 bits (Figure 4-4(a)). In Figure 4-4(b), the bit rate at which data are clocked into the buffer,

F(data), is less than the rate at which data are clocked out of the buffer by the clock, F(buffer), that is, F(data) < F(buffer), and the number of bits stored decreases to 50 after a time determined by the difference between F(data) and F(buffer). Conversely, the buffer has filled to 150 bits in Figure 4-4(c) when F(data) > F(buffer). Again, the time required for the buffer to increase to 150 bits is determined by the difference between the two clock rates.

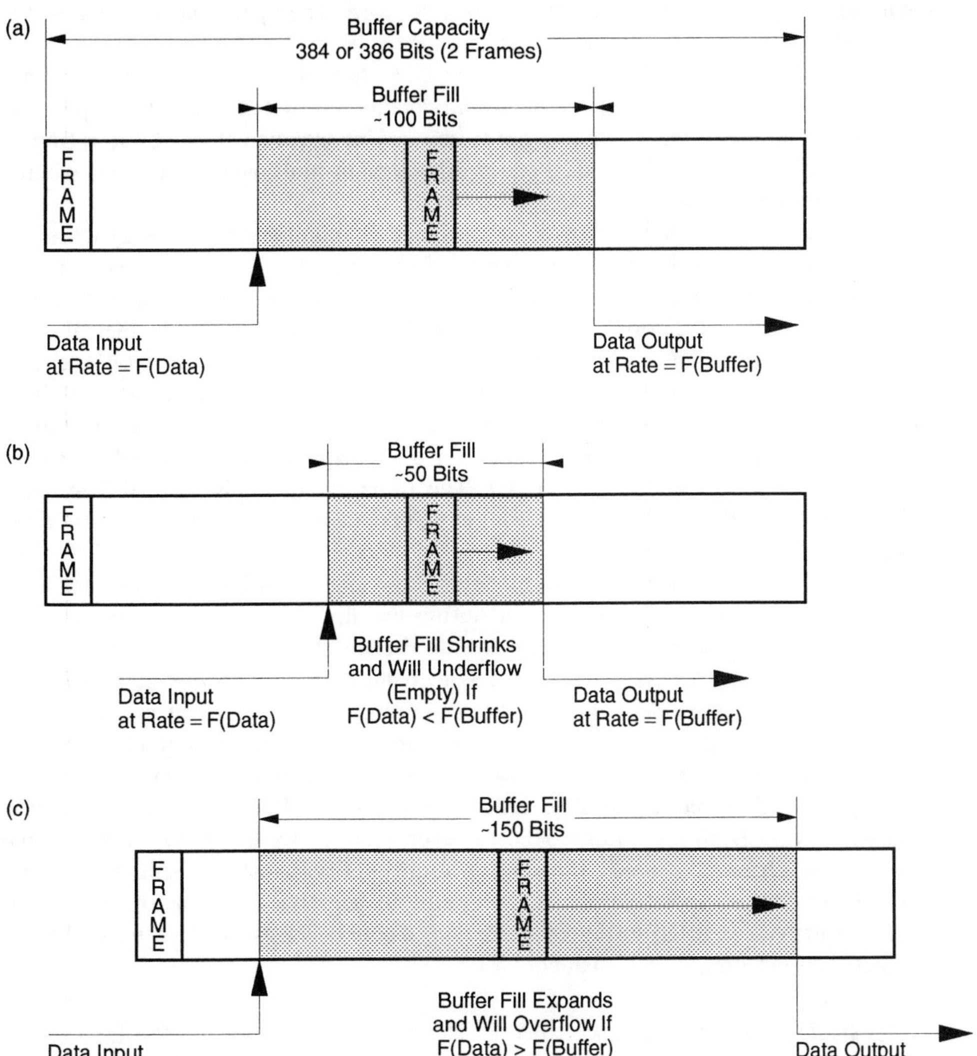

Figure 4-4 Frame buffer underflow and overflow: (a) Frame buffer stable with F(data) = F(buffer); (b) Frame buffer shrinking with F(data) < F(buffer); (c) Frame buffer expanding with F(data) > F(buffer)

In the first case discussed above, the buffer will eventually empty, and the next frame is used to provide the output as well as to fill the buffer (the data frame is duplicated). In the second case, the buffer will eventually overflow. When this happens, the data frame is deleted and the buffer fill reduced to zero. Meanwhile, frame alignment must be maintained so signaling bits and other overhead (such as cyclic redundancy check, or CRC; and facility data link) can be correctly identified in the downstream equipment. Because the amount of data in the buffer grows or shrinks with varying clock rates, it is called an *elastic store*. The operation is similar to first in, first out (FIFO) registers with variable size.

The loss or repetition of exactly one data frame is called a *controlled slip* or just *slip*. When a slip occurs, the incoming data are corrupted, but if the slip is truly controlled, it only affects data and not framing. Uncontrolled slips lead to reframing (change-of-frame alignment, COFA) of downstream multiplex equipment, whereas controlled slips do not.

All digital switching systems are designed to monitor closely the synchronization performance of digital loops connected to them. The DSS provides a maintenance alarm if the number of controlled slips exceeds a preset maintenance threshold during the measurement interval. If the controlled slip rate degrades further, the loop is automatically taken out of service. A similar action is taken for COFA events. COFA events are counted, and when the count exceeds the maintenance threshold, a maintenance alarm is given. When the count exceeds the out-of-service threshold, the loop is taken out of service. Table 4-1 gives typical values for IDLC systems connected to digital end-offices (EOs). The same values apply to UDLC systems, except that the DLC central office terminal (COT) takes the appropriate action rather than the DSS.

If the DLC is used for voice, a "click" or "pop" will be heard with each slip if there is active conversation present during the slip. If the conversation is silent, the deletion or repetition of silence is still silence. If data are being transmitted, errors will occur. A slip will cause a voiceband data modem eye pattern to shift instantly by 125 μseconds while the Q of the modem timing circuit maintains sampling times in the previous positions. Depending on how the modem adapts itself to the line, this shift either will cause transient readjustments of the adaptive equalizer settings or will force a complete retraining sequence. Recovery times have been shown to vary from six milliseconds to seven seconds for 1,200 to 9,600 bps modems. The phase of a 1,004 Hz tone on a channel will instantly shift by approximately 45 degrees (positive for frame repetition, negative for frame deletion). This phase shift can be used to detect a slip [1]. Table 4-2 lists how a single slip may affect a circuit on a digital channel.

Table 4-1 Service Thresholds for Digital Loops Operating at the DS-1 Rate

Event	Consequence	Maintenance Threshold[a]	Out-of-Service Threshold[a]
Controlled slip	No loss-of-frame alignment	4 slips	255 slips
Uncontrolled slip	Loss-of-frame alignment	17 COFA	511 COFA

[a]Threshold based on a 24-hour period.

Table 4-2 Effect of a Single Slip on a Digital Circuit

Circuit Type	Effect
Voice, PCM encoding	No effect during silent periods; otherwise, audible clicks
Voice, ADPCM encoding	No effect during silent periods; otherwise, audible clicks
Encrypted voice	Possible retransmit encryption key
Program voice	Audible clicks
Facsimile	Corruption of 4 to 8 scan lines or dropped call
Video	Picture outage or freeze frame for several seconds, loud "pop" on audio
Subrate digital data	Block retransmission, deletion, or repetition of data depending on protocol and size of data block
Voiceband data	Carrier dropout or 12 to several tens of thousands of errors, depending on the encoding level and adaptation algorithms used by the modem [1]

PCM = pulse code modulation; ADPCM = adaptive differential pulse code modulation

A single frame slip in a digital loop using SF or ESF could lead to serious problems when robbed bit signaling is used. As described in Chapter 3, these formats consist of 12 and 24 frames, respectively, into which signaling bits are placed at specified locations. The loss or replication of a frame leads to COFA. In this case, there is a 1/6 chance the least significant bit in a new frame will be robbed for signaling. This can drop the signal-to-distortion ratio by up to 1.8 dB if enough reframes are made.

Networks can experience more serious problems than simple frame slips. Although networks are designed to tolerate limited bit errors, the true loss of a single bit—for example, reception of 191 data bits rather than the required 192 in an SF or ESF DS-1 rate line—will cause a loss-of-frame alignment, and the transmission system is lost until it can reframe. During the reframing process, all transmitted bits are received in error.

Slips usually are referenced to the DS-1 rate, which processes 8,000 frames/second each with 192 data bits. Buffers used with equipment operating at the DS-1 rate are specified to store at least one frame (193 bits or 125 μseconds) plus 19 to 90 μseconds hysteresis (30 to 138 bits) [2,3]. Typical buffers store two frames (384 or 386 bits) to avoid multiple slips in one direction; however, when a DS-1 port is connected to a satellite link, larger buffering is needed to compensate for atmospheric variations and Doppler shift from satellite motion in its geosynchronous orbit. A similar situation exists for long terrestrial microwave radio links, but for reasons of multipath and atmospheric variations. In satellite transmission systems, a typical receive frame buffer will hold approximately 4,000 bits.

Some equipment does "prebuffering," as shown in Figure 4-5, to filter out, or attenuate, jitter on the incoming clock. This jitter filter typically is configured to handle 32 unit intervals (UI) peak-to-peak jitter at the remote (loop-timed) end and double that at the local (master clock) receiving end.* Such a filter uses a phase-

*A unit interval is the time duration of one bit, or inverse of the bit rate.

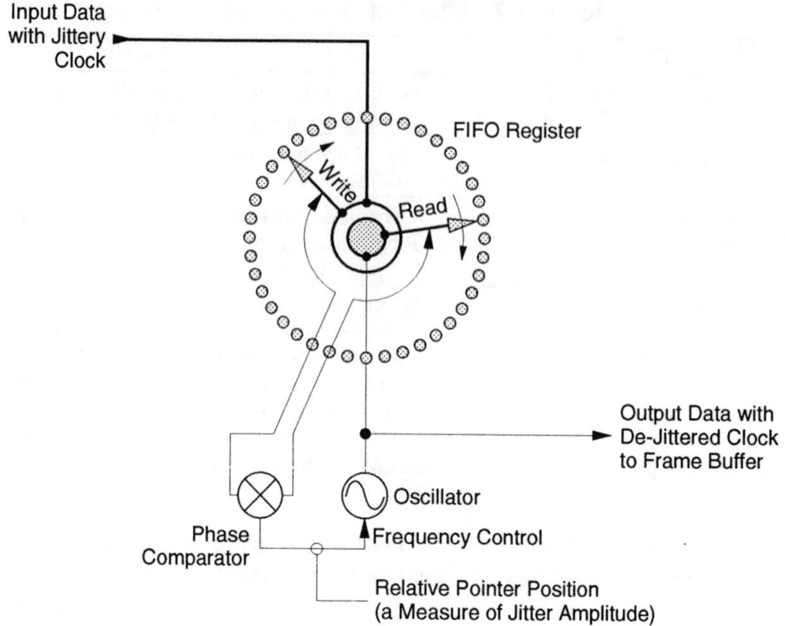

Figure 4-5 Jitter attenuator

locked loop oscillator, the frequency of which converges to the average frequency of the incoming clock. The de-jittered clock is used to clock data into the frame buffer, which is a normal elastic store of 384 bits, as previously shown. A small, 32-bit FIFO register can easily overflow or underflow under abnormal conditions. However, it is better to pass the jittered clock onto the frame buffer than it is to lose data from overflow or underflow in the jitter attenuator FIFO register. Therefore, the FIFO register clock is adjusted rapidly either down or up from its nominal rate to prevent this from happening [4].

For a given maximum variation between input and output clocks, the mean time between slips is controlled by the buffer size [5]. The delay introduced by the buffer, however, is proportional to its size. Delay causes problems with echo on voice circuits and decreases the throughput of data circuits. Therefore, it is desirable to minimize the buffer size, and thus the delay introduced by it. A tradeoff is made between tolerating large clock variations, which requires large buffers and introduces delay, or improving clock accuracy and accumulating less delay from smaller buffers. The trend is toward improving clock accuracy.

Slips are inevitable in any practical network, but their number is controlled through careful synchronization design. The synchronization hierarchy is designed for an objective limit of one slip in five hours on an end-to-end network connection [6]. This objective is met by greatly limiting slips at higher synchronization levels and allowing a greater number of slips at lower levels, where they are most likely to occur. The present public networks are better than the objective by large measure.

Many private networks maintain their own timing sources, and the public network is used to transmit data asynchronously between private network nodes or for access to the public network. However, when the private networks connect to public network DSSs, the private network must either accept synchronization from the public network or provide a timing source equal in quality to that used in the public network. In the latter case, the private and public networks are timed plesiochronously (defined more precisely later). The specific issues and requirements of private digital network synchronization are given in [3], although this reference mirrors almost all the synchronization requirements of public networks.

4.2 Clocking and Timekeeping

Timekeeping requires not only accurate frequency control but also accurate time transfer from some master source. Therefore, timekeeping involves comparing two or more clocks to ensure that their time difference is kept within predefined limits over a specified interval. Such a system either can provide relative (synchronized) or absolute timekeeping.

In a *synchronized* system, each clock is in step with every other clock in the system, but there is no requirement that these clocks be synchronized to some master clock outside the system. The *absolute* system keeps each clock in step with every other clock in the system and also with some master clock outside the system. In large telecommunications networks, the absolute system is used and synchronization is kept with an internationally accepted time scale, such as universal coordinated time (UTC). This is the function of the PRS, defined earlier and described in greater detail later. Smaller networks that have no external time reference use a simple synchronized system.

The accuracy and stability of clocks used in digital network synchronization follow a predefined hierarchy [6]. These clocks count the number of events that occur from some arbitrary starting point and allow synchronization, which ensures that all interconnected network components are operating on the same time scale.

Network nodes in the United States public telephone network (PTN) follow industry standards, which require they be synchronized to timing references that are traceable to a primary reference source [8]. A PRS:

- Provides a timing signal that is accurate to within 1E-11 over a long time
- Is verified to UTC*
- Is used to synchronize other clocks within a network

---

*UTC provides real-time verification. Compare this to a recent survey of real-time clocks in the Internet system, which is composed of a large number of PCs and host terminals. Of over 20,000 time transfers, "about half the replies had errors greater than 2 minutes, while 10% had errors greater than 4 hours. A few had errors over two weeks." [9]

Notwithstanding the desirability of overall network synchronization, timing islands, or isolated networks, exist as a rule. That is, not all digital telecommunication systems are tied to a large network or even the public network, and not all networks are tied together. This latter situation is especially apparent when international networks are considered.* Each country wants to control the timing functions of its national network, so the master-slave relationship does not exist across international boundaries. The interworking of such networks requires a standardized and tightly controlled approach to inter-network synchronization. National networks normally are internally synchronous, and when connected to other synchronous networks across international boundaries, are *plesiochronous*. Plesiochronous operation occurs whenever two independently synchronized networks are interconnected and no timing is passed between them, as shown in Figure 4-6. Of course, to meet slip rate objectives, each network must be synchronized by a sufficiently accurate clock, in this case a PRS.

Clock characteristics for the synchronization hierarchy are shown in Table 4-3. As seen from this table, there are four levels of synchronization, which are

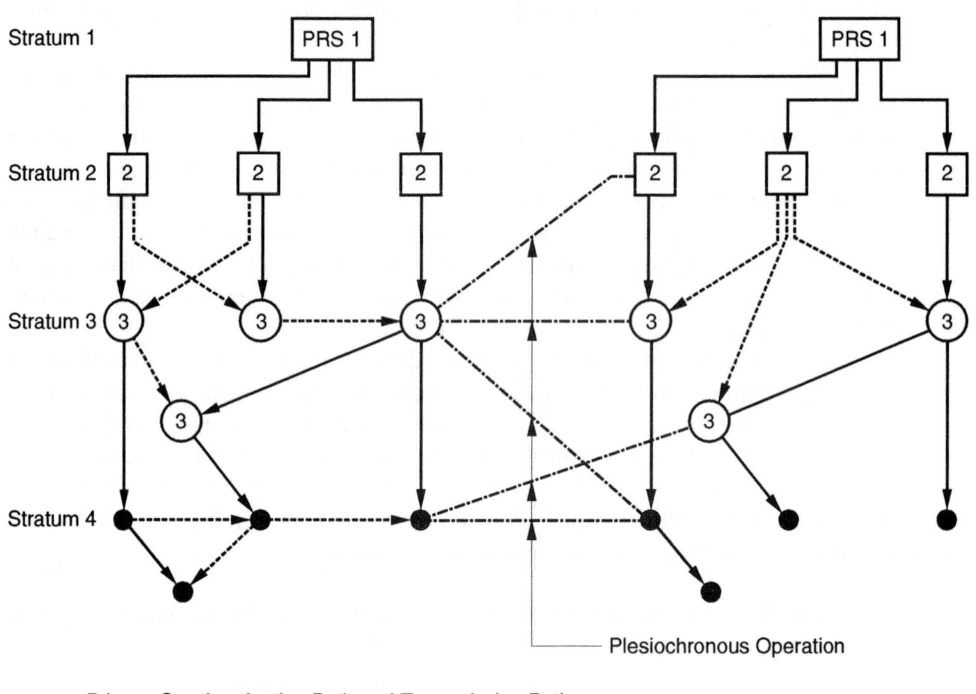

Figure 4-6 Plesiochronous networks

*Large-scale network synchronization is beyond the scope of this book. The reader is referred to [10] for a description of international synchronization issues and to [6] for a description of some national (U.S.) issues.

Table 4-3 Digital Synchronization Clock Strata[a]

Stratum	Minimum Accuracy	Equivalent Accuracy @ 1.544 MHz	Minimum Stability (24-Hour Holdover)	Pull-In Range	Slip Interval @ 8,000 Frames/Second[b]
1	$\pm 1 \times 10^{-11}$	±0.000015 Hz	Not applicable	None	72 days
2	$\pm 1.6 \times 10^{-8}$	±0.025 Hz	$\pm 1 \times 10^{-10}$/day	$\pm 1.6 \times 10^{-8}$	14.5 days
3E	$\pm 1.0 \times 10^{-6}$	±1.54 Hz	$\pm 1 \times 10^{-8}$/day	$\pm 4.6 \times 10^{-6}$	3.5 hours
3	$\pm 4.6 \times 10^{-6}$	±7 Hz	$\pm 3.7 \times 10^{-7}$/day	$\pm 4.6 \times 10^{-6}$	5.6 minutes
4	$\pm 3.2 \times 10^{-5}$	±50 Hz	Not applicable	$\pm 3.2 \times 10^{-5}$	3.9 seconds

[a]Source data adapted from [6] with additions from [7]

[b]First 24 hours if stability specified; assumes network elements operating plesiochronously (one with stratum 1 accuracy).

directly related to the quality of the reference clocks. The characteristics of typical commercially available clock sources, including strata 1, 2, and 3, are shown in Table 4-4. In this table, the slip interval is the time between slips at a frame rate of 8 kHz. Some manufacturers build so-called *enhanced* stratum 3 products. These are meant to provide better performance than clocks meeting minimum stratum 3 requirements but at a cost between that of stratum 2 and stratum 3 equipment.

The major characteristics for clocks used in digital telecommunications networks are:

- Accuracy of the frequency source
- Stability of the frequency source (also called holdover)
- Ability of the local frequency source to synchronize with other clocks (pull-in range)

Figure 4-7 shows the synchronization hierarchy of a hypothetical set of network nodes. At the top of the hierarchy is the most accurate synchronization source, stratum 1. A network synchronized top to bottom will exhibit stratum 1 accuracy throughout. That is, once a clock is synchronized to a source traceable to a stratum 1 clock, that clock will exhibit stratum 1 accuracy of ±1E-11. If this clock loses its reference, it will revert to the accuracy of its own stratum level. For example, say a stratum 3 clock uses a stratum 2 clock as its reference, and this stratum 2 clock, in turn, uses a stratum 1 clock as its reference. All clocks will exhibit stratum 1 accuracy as long as they are synchronized. If the stratum 2 clock fails and the stratum 3 clock has no backup reference, the stratum 3 clock now provides stratum 3 accuracy to the network components synchronized to it.

To improve reliability, each node can be provided with a primary and secondary frequency reference, as shown in Figure 4-7 (all nodes but one have two reference inputs in this figure). In general, primary and secondary references are provided according to the protection rules in the next section. The illustration shows both primary and secondary references for stratum 4 clocks. In practical systems, the primary and secondary references, if provided for stratum 4 clocks, are obtained from the same source but over different paths. For example, a DLC system normally operates at the stratum 4 level and has two DS-1 rate interfaces, either for traffic reasons or redundancy. Either interface can provide the clocking, but these interfaces normally do not obtain the reference frequency from different sources.

The clock stratum number for each node is chosen based on an economic evaluation and consideration of the slip requirements at that node. Obviously, a stratum 1 clock is the most expensive, but it also provides the lowest slip rate. It is neither feasible nor necessary to provide such a clock at each and every node.

The stratum 1 clock normally provides a timing reference for stratum 2 clocks. Stratum 2 clocks are located at inter-exchange carrier digital switching centers, important network hubs, or high-capacity transmission centers. The stratum 2 clocks transmit synchronization to stratum 3 clocks or directly to stratum 4 clocks. As a rule, a lower stratum number must never derive its timing from a higher stratum number; only from a lower or equal stratum number. Also, under normal conditions,

Table 4-4 Characteristics of Typical Commercially Available Clock Sources

Parameter	Stratum 1	Stratum 2	Stratum 3
Type	LORAN-C or GPS receiver or cesium beam oscillator	Rubidium atomic oscillator	Temperature-compensated or oven-controlled crystal oscillator
Pull-In	Not applicable	$\pm 4.6 \times 10^{-8}$	$\pm 1.5 \times 10^{-5}$
Holdover, 24 Hours	Not applicable	$\pm 7.5 \times 10^{-11}$	$\pm 1 \times 10^{-7}$
Free-Run, 24 Hours	1×10^{-12}	$\pm 4 \times 10^{-11}$	$\pm 5 \times 10^{-7}$
Input Reference	DS-1 signal, SF or ESF, bridged or terminated 100 Ω 64/8 kbps composite clock, 135 Ω terminated E1 signal, HDB3 framed or unframed, bridged or terminated, 75 or 120 Ω 1 to 10 MHz sinewave, 50 Ω terminated		
Timing Outputs	DS-1 signal, all-ones, SF or ESF, 100 Ω 64/8 kbps composite clock, 135 Ω E1 signal, all-ones, framed or unframed HDB3, 75 or 120 Ω 8 kHz to 2.048 MHz sinewave, 50 or 75 Ω 4 kbps to 2.048 Mbps, RS422 or TTL		

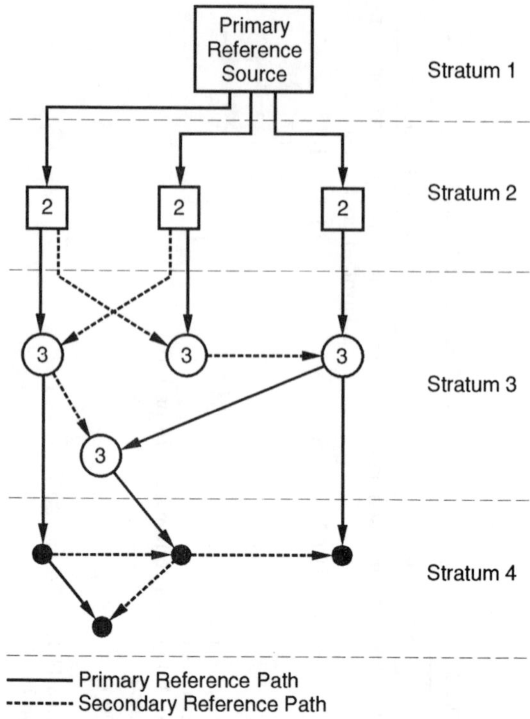

Figure 4-7 Synchronization hierarchy

all clock signals in a network must be *traceable* to a stratum 1 primary reference frequency source.

Synchronization schemes that allow *timing loops*, either during normal or failure conditions, must be avoided because this allowance will lead to network instability.* An undesirable timing loop is shown in Figure 4-8. This illustration actually shows two problems: First, a timing loop exists; the lower stratum 3 clock receives its reference from the stratum 2 clock, which receives its reference from the upper stratum 3 clock, which in turn receives its reference from the lower stratum 3 clock, resulting in a timing loop condition. Second, the stratum 2 clock receives its reference from a stratum of lower quality, in this case the upper stratum 3 clock. The synchronization rules described later prevent timing loops and stratum violations, at least in principle.

Figure 4-9 shows a typical stratum assignment in a network involving end-offices, DSSs, remote switching terminals (RSTs), DCSs, DLC, and a variety of digital loops. The clock accuracy normally provided at the customer interface, or user, under all conditions is stratum 4—the lowest in the hierarchy. For interfaces synchronized to a DSS having a stratum 3 clock or better, the clock accuracy is

*Timing loops are not to be confused with loop timing, which is an acceptable timing method discussed earlier.

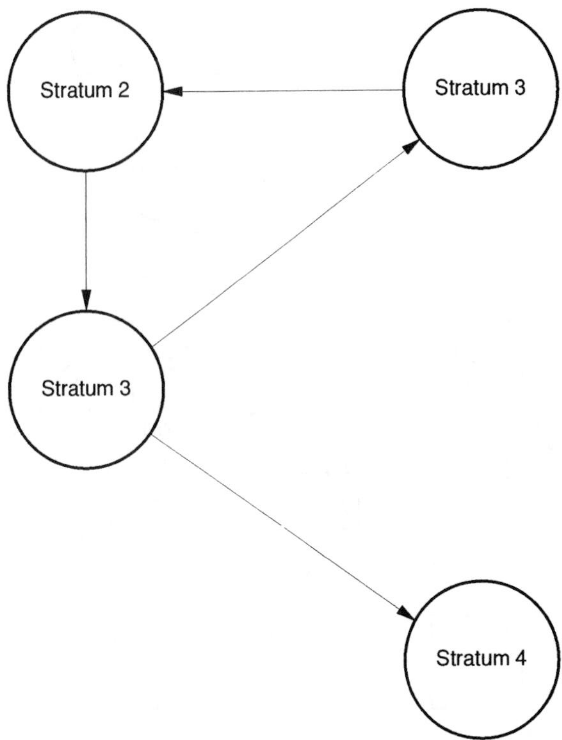

Figure 4-8 Undesirable timing loop condition

equal to that stratum level. Higher accuracies are available at customer interfaces, but for economic reasons stratum 3 or better are not provisioned as a matter of course on digital loops. Nevertheless, many large public network customers that operate their own large private networks have their own PRSs.

The facilities carrying synchronization signals do not necessarily have to be the same facilities carrying traffic. In Figure 4-9, only the synchronization paths are shown. Traffic paths can be, and usually are, much more diverse.

Existing digital transmission facilities normally are used to distribute network timing and synchronization. Typically, these facilities operate at the DS-1 line rate of 1.544 Mbps. This is convenient and economical because timing information is imbedded in the bit stream along with the message bits. At the lower stratum levels (higher quality), the DS-1 rate links used to carry synchronization signals normally do not carry traffic.

There presently exist digital end-offices with clocks that do not meet the basic requirements for any stratum level other than stratum 4. For example, there are many DSSs that use a clock with stratum 3 stability but do not meet stratum 3 pull-in range requirements. Systems with these characteristics are identified during the network engineering process and treated as special cases from a synchronization point of view.

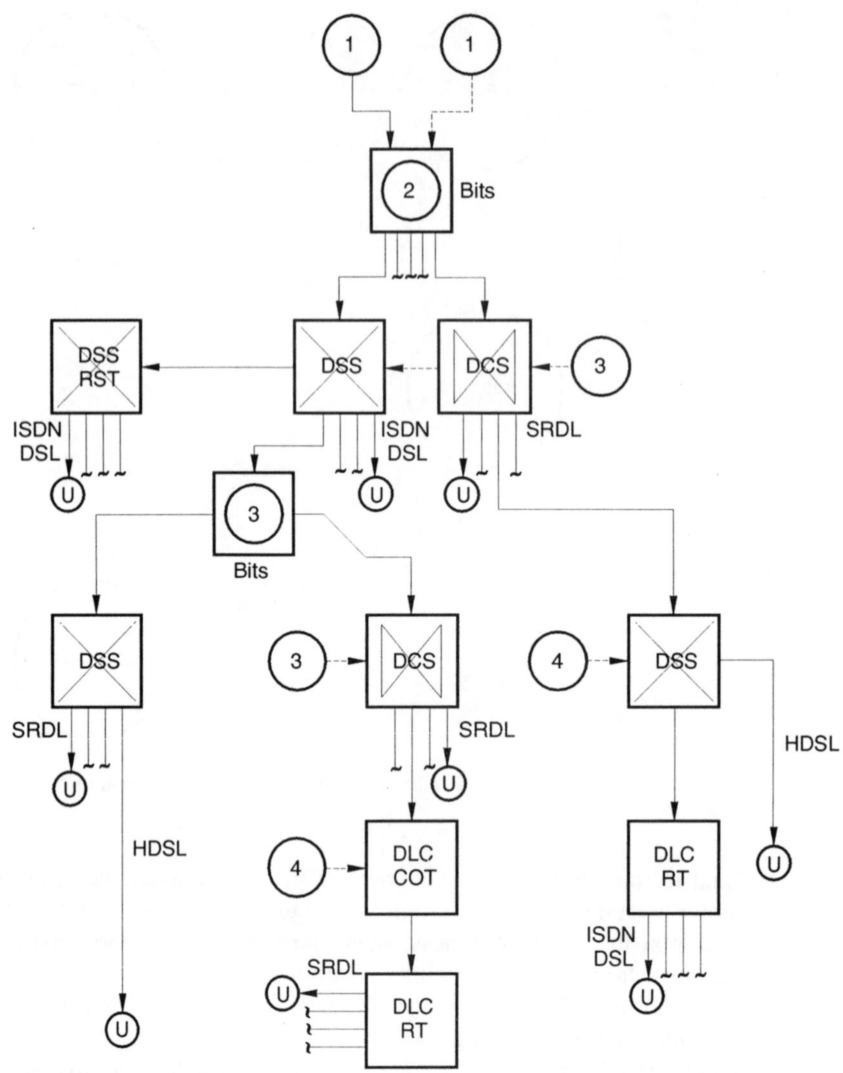

——— Primary Reference Path
- - - - - Secondary Reference Path

Key

DSS: Digital Switching System
RST: Remote Switching Terminal
DCS: Digital Cross-Connect System
DLC: Digital Loop Carrier
COT: Central Office Terminal
RT: Remote Terminal
U: User
DSL: Digital Subscriber Line
SRDL: Subrate Digital Loop
HDSL: High Bit-Rate Digital Subscriber Line

Figure 4-9 Typical clock stratum assignment

Any practical network will experience failures either in local clock sources or external synchronizing sources. To minimize the effects of these failures, strata 2, 3, and 4 nodes can be equipped with primary and secondary synchronization sources. If both the primary and secondary sources fail, the holdover accuracy of the strata 2 and 3 clocks is high enough to allow continued network operation without undue performance degradation for a limited time, based on slip performance requirements. Stratum 4 clocks do not have a specified holdover accuracy and cannot be depended on to operate without a large number of slips after synchronization source failure.

When a reference fails or is removed for some reason, the local clocks drift at a rate determined by their holdover accuracy. Slips will occur at an increasing rate determined by the drift rate. Therefore, to minimize the network degradation that results from synchronizing reference signal failures and resulting slips, it is necessary to consider the length and diversity of synchronization transport facilities. Sometimes it is necessary to elevate the clock stratum at a node to compensate for relatively unreliable facilities. Provisioning rules for synchronization facilities are discussed later.

Digital transmission facilities are subject to a variety of environmental and operational influences that tend to degrade synchronization performance. For example, when inter-building or inter-node cables are heated and cooled throughout the day and night, the propagation velocity changes (*diurnal wander*). This is especially noticeable in very long fiber optic links where diurnal and seasonal wander can lead to phase variations of 5E-10, or more [11]. The effects also are noticeable on twisted pair, coaxial cables, and digital radio systems [5]. The effect is to cause phase variations between the clocks at each end, which can lead to slips. Also, jitter introduced by line repeaters and multiplexers can affect timing extraction. Buffering in line receivers normally limits jitter and wander effects and, in some designs, filters these out, as previously discussed [12].

Digital loop systems are synchronized to the serving DSS end-office (class 5 switching systems) or DCS clocks, which operate at the stratum 3 level.* During loss of one of the references for the stratum 3 clock, the clock will automatically switch to the backup reference and its output will not drift more than 3.5E-9 during the first 24-hour period. With loss of both references, the stratum 3 clock will drift no more than 3.7E-7 during any 24-hour period [6]. Digital loops (for example, DLC or channel banks) free-run at the stratum 4 level; if, however, they are synchronized to the EO, no slips will occur.

EXAMPLE 4-1

Calculate the maximum number of slips that will occur in a DSS operating at the DS-1 rate and normally synchronized to a stratum 3 clock when (1) the primary reference frequency is lost and (2) when both frequency references are lost.

1. At the DS-1 rate, there are 8,000 frames/second × 24 hours/day × 3,600 seconds/hour = 6.9E8 frames/day. If a stratum 3 clock loses its primary reference, it should

*Assuming Stratum 3 operation of an EO or DCS is not always valid. Stratum 3 clocks in some systems are not normally provided unless specified by the system customer.

drift no more than 3.5E-9 during the first 24-hour period. In this case, the number of slips = 3.5E-9 × 6.9E8 = 2.4 slips/day or one slip in 10 hours.

2. A stratum 3 clock should drift no more than 3.7E-7 after both sources fail. In this case, the number of slips = 3.7E-7 × 6.9E8 = 255 slips/day (one slip every 5.6 minutes, average).

4.3 Engineering Rules for Synchronization

Table 4-5 shows the general engineering rules that apply to network synchronization [6]. The application of redundancy in the reference sources for each clock stratum is shown in Table 4-6 [6].

Table 4-5 Synchronization Engineering Rules

RULE	DESCRIPTION
1	Synchronize a given node or network to a clock with equal to or better quality; conversely, never synchronize to a lower-quality clock.
2	When a choice between more than one clock is available, choose the clock with the highest quality.
3	Choose the facilities with the highest overall reliability, in terms of outages and quality, for the primary and secondary reference frequencies.
4	Avoid facilities with high provisioning turnover, long length, and aerial cable.
5	Choose diverse routes and facilities for the primary and secondary references; avoid putting both primary and secondary references on facilities that share the same media (for example, same cable or same radio system).
6	Under no circumstances allow timing loops in any combination of primary and secondary reference facilities.
7	Operate two interconnected networks plesiochronously if timing is not to be passed between the two and the both networks have a PRS-traceable timing source.

Table 4-6 Reference Source Protection Rules

CLOCK STRATUM	RULE
1	No protection requirements are specified in published industry standards. However, some form of redundancy is always provided.
2	Both primary and secondary frequency references normally are provided. The holdover performance of stratum 2 clocks is such that an automatic switch to a secondary reference is not needed but may be desirable. Means for manual switching between sources is required.
3	Both primary and secondary frequency references are required. An automatic switch to a secondary source is required.
4	Both primary and secondary frequency references normally are not provided but redundant sources and automatic switching may be desirable. Degraded performance under source failure can be expected when no redundancy is provided.

4.4 Timing Impairments

Impairments introduced by the transmission medium and the received clock recovery circuits and introduced by clock generation circuits themselves limit not only whether a clock signal can be recovered but also the number of times a recovered clock may be used as a timing signal for transmission to the next network element. The results of all impairments are jitter and wander in the recovered clock signal. The principal sources of these timing impairments are:

- Clock circuit instability
- Distortion of the transmitted signal waveform caused by attenuation and distortion in the transmission line
- Noise in the electronic circuits
- Crosstalk and impulse noise in the transmission line
- Wander and jitter caused by higher-order multiplexers (when used)
- Changes in propagation delay caused by temperature change of the transmission medium (metallic twisted pairs or optical fibers)

Of particular interest is clock instability, which is measured in terms of time interval error (TIE) and maximum time interval error (MTIE). TIE is a short-term phase variation in clock frequency at any given instant in a measurement interval. MTIE is the peak-to-peak error for all measurements in the measurement interval. Both TIE and MTIE are illustrated in Figure 4-10. Figure 4-11 shows the present requirements for the permissible MTIE versus measurement period for a PRS. Intervals of 10 or 100 seconds are used in TIE and MTIE measurements [12].

Jitter and wander are variations in the phase position of a pulse with respect to its ideal position in time. The difference between jitter and wander is the rate at which they are specified; wander is considered a timing phase change at a less than 10 Hz rate, and jitter is considered a phase change at a greater than 10 Hz rate. In general, jitter affects timing recovery and wander affects buffer fill. Jitter, especially in clock recovery circuits, is related to the bit patterns transmitted. For example, clock recovery circuits can tolerate more jitter when receiving a signal with a higher pulse density than a lower pulse density. Jitter and wander are the most troublesome of transmission impairments because they have a cumulative effect as they propagate through a network and, as a result, are difficult to troubleshoot and resolve.

There are several types of jitter and wander [12]:

- Systematic jitter due to pulse detection and regeneration processes in repeaters and line receivers
- Waiting time jitter due to deletion of stuffed overhead bits in multiplexers
- Phase noise jitter and clock instability due to imperfect oscillators
- Diurnal wander due to temperature variations of transmission media

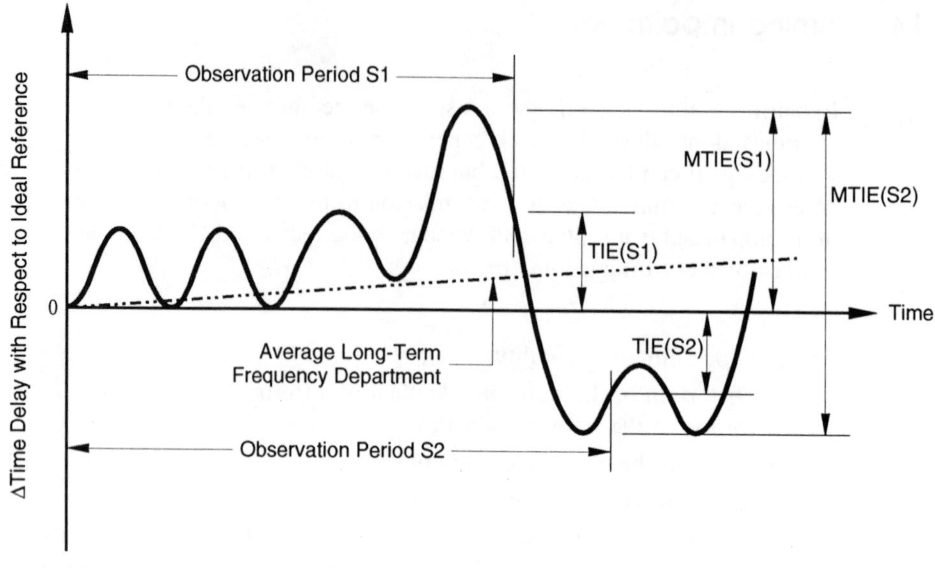

Key

MTIE(S1): Maximum Time Interval Error during Measurement Interval S1
MTIE(S2): Maximum Time Interval Error during Measurement Interval S2
TIE(S1): Time Interval Error at the End of Measurement Interval S1
TIE(S2): Time Interval Error at the End of Measurement Interval S2

Figure 4-10 Time interval error measurements

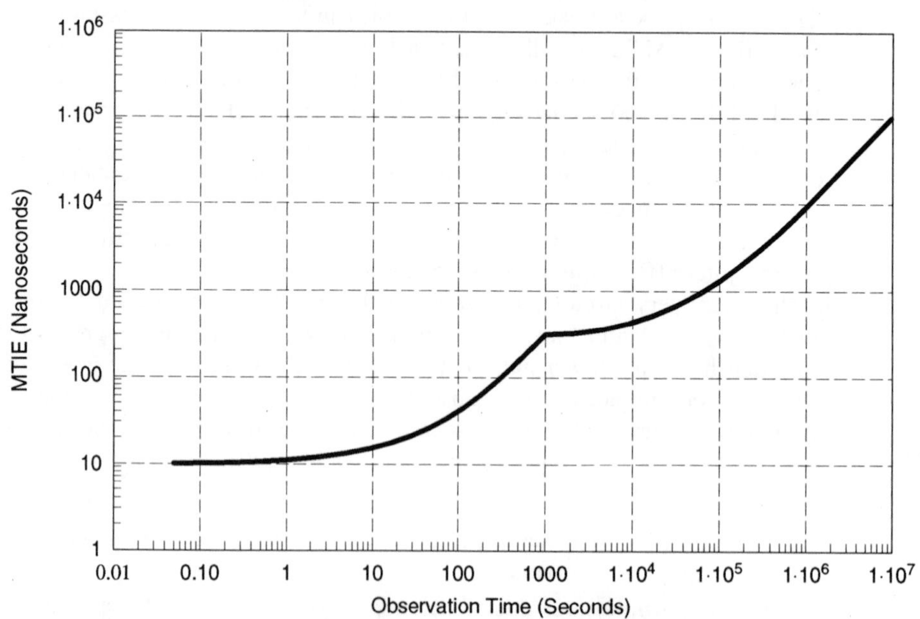

Figure 4-11 American National Standards Institute (ANSI) requirements for maximum time interval error, [8]

Figure 4-12 illustrates jitter (or wander) effects. Figure 4-12(a) shows a transmitted bipolar alternate mark inversion (AMI) bit stream, where each bit is perfectly aligned at the center of the bit interval. Figure 4-12(b) shows the received bit stream, which includes the effects of sinusoidal jitter having the waveform of Figure 4-12(c). The propagation delay is not shown. Finally, Figure 4-12(d) shows how a repetitive bit stream would appear on an oscilloscope; the shaded areas indicate some of the possible pulse positions due to positive and negative jitter amplitude.

Jitter amplitude is the magnitude of phase variation, and jitter frequency is the rate at which the phase is changing. Amplitude is measured in UIs, defined earler as the time duration of one bit, or inverse of the bit rate. One UI is equivalent

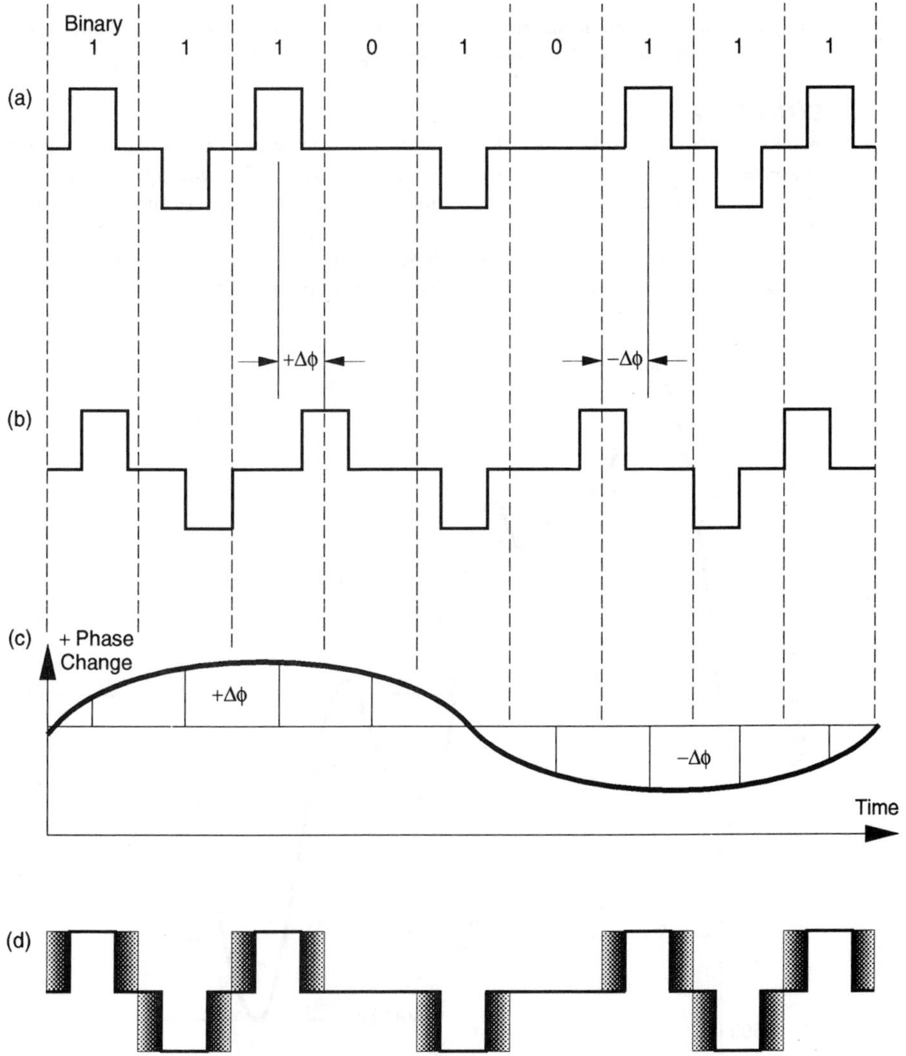

Figure 4-12 Jitter effects on a serial bit stream

to 360 degrees of phase change. For DS-1 rate transmission, one UI is equal to 1/1.544E-6 bps = 648 nseconds/bit. UIs for typical digital loop systems are shown in Table 4-7.

Table 4-7 Unit Intervals for Typical Digital Loop Systems

DIGITAL LOOP TYPE	LINE RATE	UNIT INTERVAL
Repeatered T1-Carrier	1.544 Mbps	648 nseconds
HDSL	784 kbps	5.2 μseconds
ISDN DSL	160 kbps	12.5 μseconds
SRDL	72 kbps	28 μseconds
SRDL	3.2 kbps	625 μseconds

HDSL = high bit-rate digital subscriber line; ISDN = integrated services digital network; DSL = digital subscriber line; SRDL = subrate digital loop

EXAMPLE 4-2

A DLC system operating at the DS-1 rate is synchronized to a clock with the observed wander characteristics shown in Figure 4-13. Assuming the DLC is equipped with a 220-bit buffer and a perfect clock, determine the number of slips that may occur during the observation interval.

The buffer will store the equivalent of 220 UIs. From the figure, as the clock phase changes to its first positive peak of 480 UI, two controlled slips will occur. As the clock

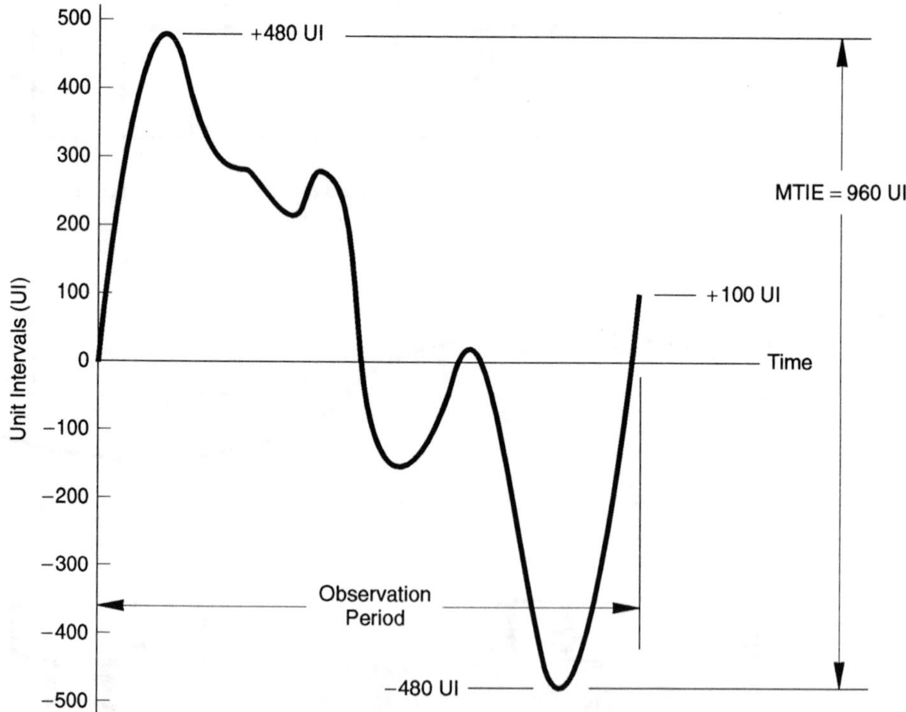

Figure 4-13 Clock wander for Example 4-2

moves to its opposite peak of −480 UI, another four slips will occur. Finally, as the clock returns to its nominal frequency, there will be two more slips. The total number of slips will be eight. Note that the TIE measured at the end of the observation (100 UI) would indicate no slips, and the MTIE of 960 UI would indicate four slips.

Frequently, several DS-1 signals from physically separated sources are combined in a multiplexer. These signals are asynchronous with respect to each other. Jitter is introduced by multiplexers with DS-1 inputs and DS-2 and DS-3 aggregate outputs. The framing formats used at these rates were explained in Chapter 3. When several such asynchronous DS-1 signals are received by an M-12 multiplexer and combined into a DS-2 output, stuff bits are introduced to compensate for the slightly different input signal rates. A similar situation exists with M-13 and M-23 multiplexers to compensate for the varying input rates. The aggregate rates of these multiplexers are not normally referenced to any precision source (other than the internal multiplexer clock) [14].

The actual transmission rate of the DS-1 signal is not critical as far as the multiplexer is concerned because stuff bits can be used as required to combine the signals properly. However, if the DS-1 signals terminate in a DSS or DCS, problems can arise. In the process of demultiplexing the digital signals from the higher aggregate rate to the lower DS-1 rate, all stuff bits and other aggregate overhead bits are discarded. This results in timing discontinuities (waiting time jitter) in the lower rate signal. In digital fiber optic networks, the principal sources of jitter and wander are the asynchronous multiplexers used in the network [14].

Some digital switching systems use an internal clock and transmission rate of 2.048 Mbps, which corresponds to the E1, 30-voice channel, 2-signaling channel system used in most of Europe. When these switches are interfaced to a line operating at the DS-1 rate, the clock conversion introduces transmit jitter—typically 30 nseconds peak-to-peak. This is similar to the waiting time jitter discussed above; the extra channels essentially contain stuff bits, which are discarded before transmission on the line.

A receiving clock must be able to absorb or tolerate some amount of jitter on the incoming signal. This tolerance is variously specified in industry standards but usually expresses the relation of a particular jitter level to bits in error, loss of synchronization, or ability to synchronize or resynchronize. One frequently cited standard specifies the jitter tolerance at the network interface, as shown in Figure 4-14(a) [15]. This applies to signals from the public network to a user's installation at the DS-1 rate. Figure 4-14(b) shows the provisional requirements for DLC cited in another widely accepted industry standard [16].

The clock recovery circuits in the line receiver must be able to synchronize to a signal with jitter at or below the lines shown on the figures, and then deliver error-free performance. The associated tests for this jitter tolerance are made in a "noise-free environment," and therefore are not field measurements. Modern commercial clock recovery circuits meet these requirements with a wide margin.

The jitter transfer function of a piece of equipment expresses the amount of jitter that is delivered from the input to the output. The output can be another downstream network element or back to the source, as in loop timing. It is desirable that

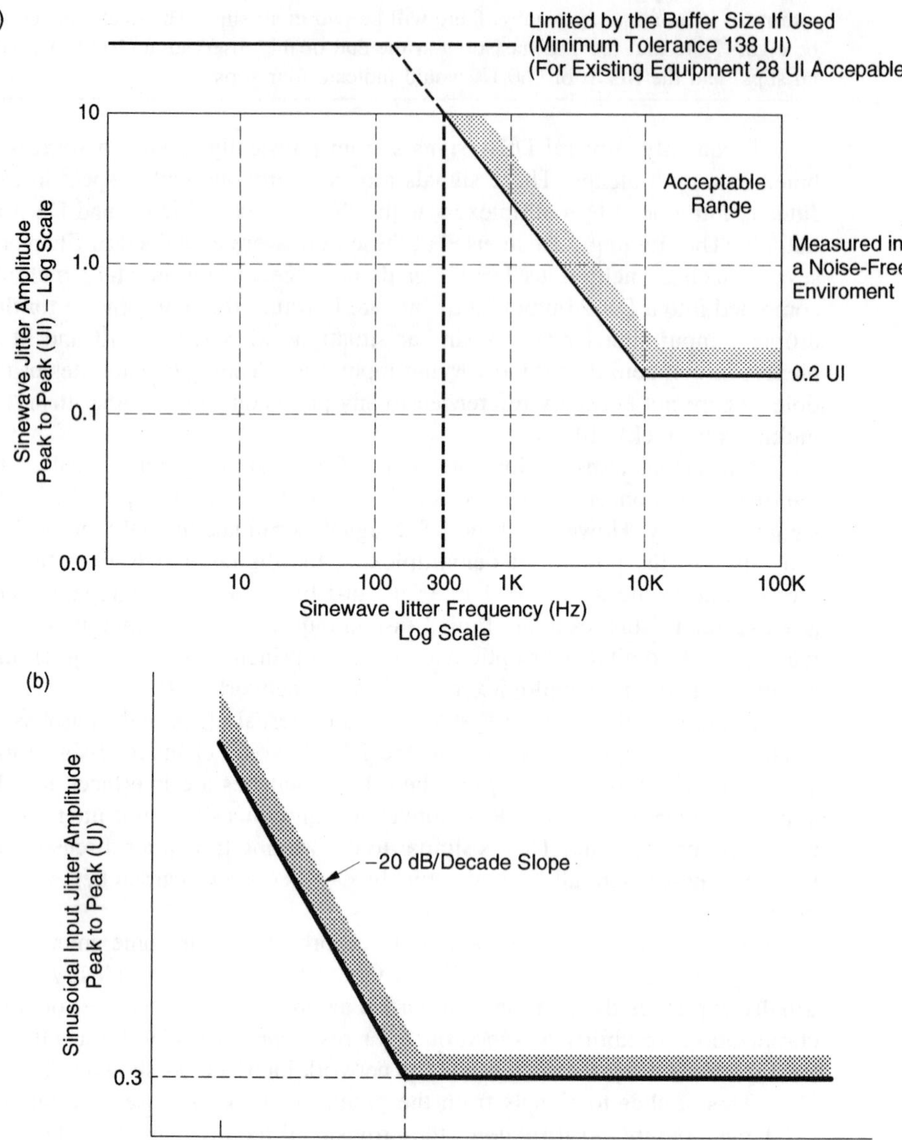

Figure 4-14 Sinusoidal input jitter tolerance digital loop systems operating at the DS-1 rate: (a) Sinewave jitter amplitude vs. jitter frequency, [15]; (b) Sinewave jitter amplitude vs. jitter frequency, [16]
© 1988 AT&T. Reprinted with permission;
© 1988 BELLCORE. Reprinted with permission

jitter be attenuated when passed through the timing circuits of digital loop systems. Industry requirements for jitter transfer are still being studied; however, a widely accepted industry standard requires that jitter at a frequency of 200 Hz in the incoming signal be attenuated by at least 0 dB (that is, the jitter must not be increased) when that signal clocks the outgoing signal and the jitter response falls off at 20 dB/decade for higher jitter frequencies [15]. The jitter transfer function of digital loop equipment operating at the DS-1 rate is seldom specified in manufacturer data sheets, but most equipment provides a general statement that the equipment "meets TR62411 requirements."

Wander for a DS-1 rate signal is specified over two intervals: 24 hours and 15 minutes. The industry standard requires that wander in either direction (from network to user and from user to network) be less than 28 UI, peak-to-peak (UIp-p) over any 24-hour period and less than 5 UIp-p in any 15-minute interval [17]. The wander requirements for other than DS-1 rate signals have not been specified in industry standards.

4.5 System Clocks

In all but the simplest installations, a centralized local clock is used. This clock typically receives its timing from a composite clock source or DS-1 synchronizing signal source and provides a means of distributing the clock to equipment in the building requiring synchronization. Most equipment can accept either a composite clock (CC) or DS-1 clock by setting an option switch or jumper. The basic concept of intra-building clock distribution is illustrated in Figure 4-15.

The CC puts out a 64 kbps (DS-0 rate) bipolar AMI signal with 62.5% (5/8) duty cycle (return-to-zero, RZ) and an intentional bipolar violation (BPV) every eighth pulse, or clock bit, as shown in Figure 4-16(a). The intentional BPV provides a strong 8 kHz frequency component, so in effect, the CC provides two clocks in one signal. The BPV is used to mark the DS-0 rate network control bit used with SRDLs (also called digital data system, DDS). The DS-1 synchronizing signal is a normal bipolar AMI, 50% duty cycle (RZ), D4 formatted (framed) signal with all ones placed in each channelized 8-bit byte. The DS-1 synchronizing signal is shown in Figure 4-16(b).

Clocks with these characteristics are part of a timing signal generator (TSG) unit, which is shown in Figure 4-17 [13]. The basic timing standard in this case is an oven-controlled crystal oscillator. The typical TSG can accept either a CC source or DS-1 synchronizing signal source as reference frequency input to the timing interface. A primary reference and secondary reference are provided, as previously explained. When a DS-1 synchronizing signal is used as a timing reference, the clock must be compatible with the ones-density restrictions associated with any DS-1 signal.

Normally, both reference sources must be the same type (either composite clock or DS-1 signal). The timing generator portion of the TSG synchronizes with the

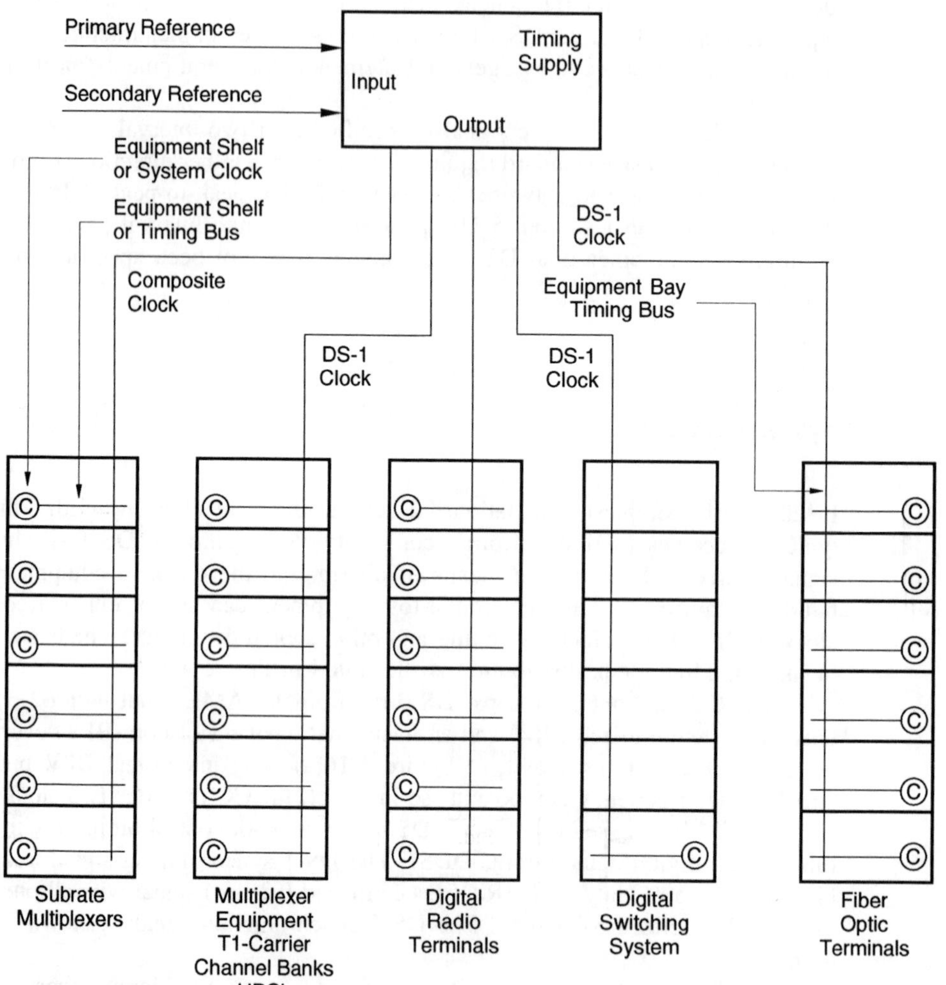

Note: Depending on Equipment Requirements, Clock Distribution within a Bay may be by Daisy-Chain to Adjacent Shelves or by Dedicated Clock Cable from Timing Supply.

Figure 4-15 Intra-building clock distribution

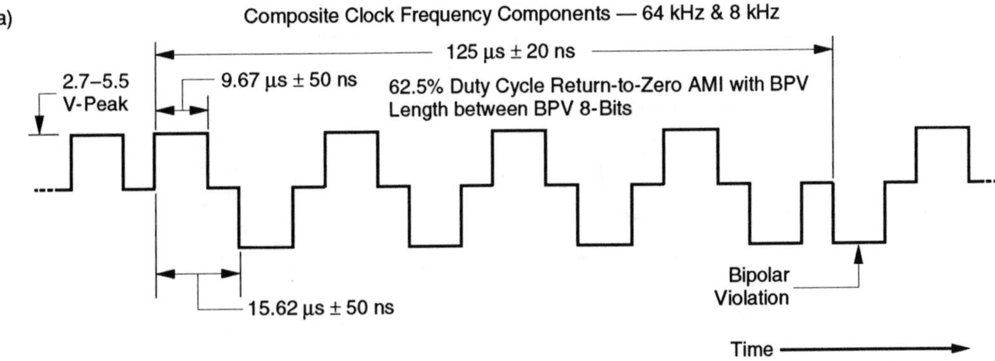

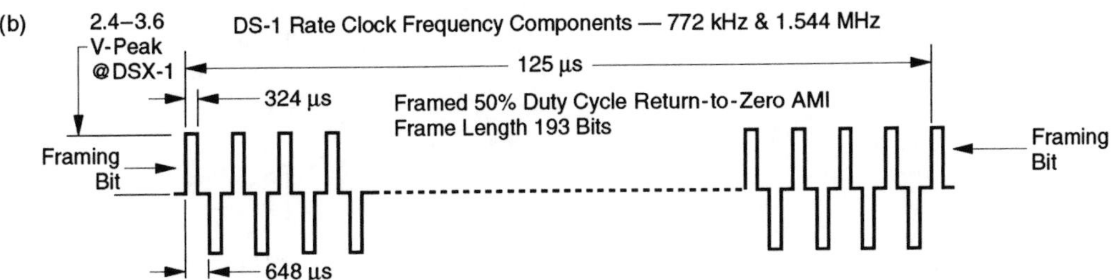

Figure 4-16 (a) Composite clock waveform; (b) DS-1 rate clock waveform

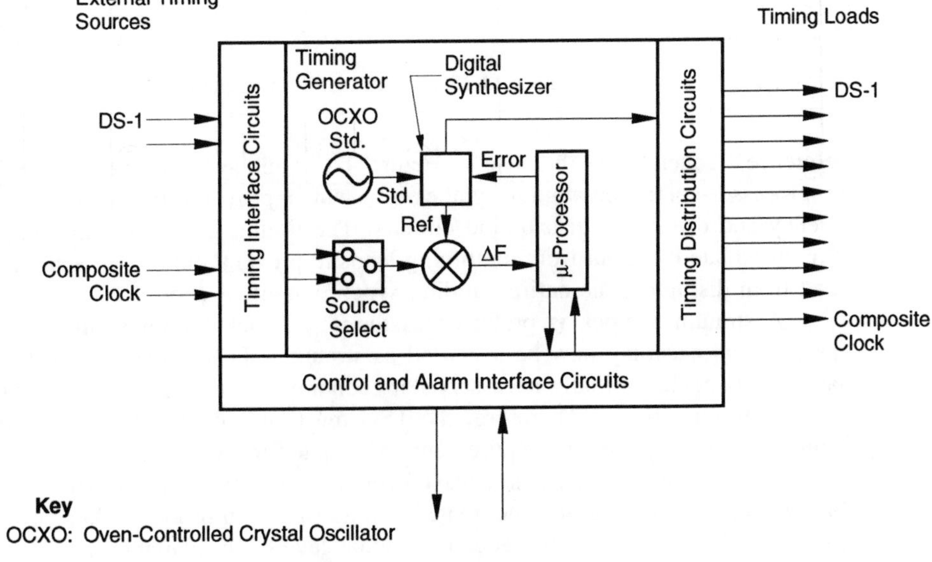

Figure 4-17 Timing signal generator

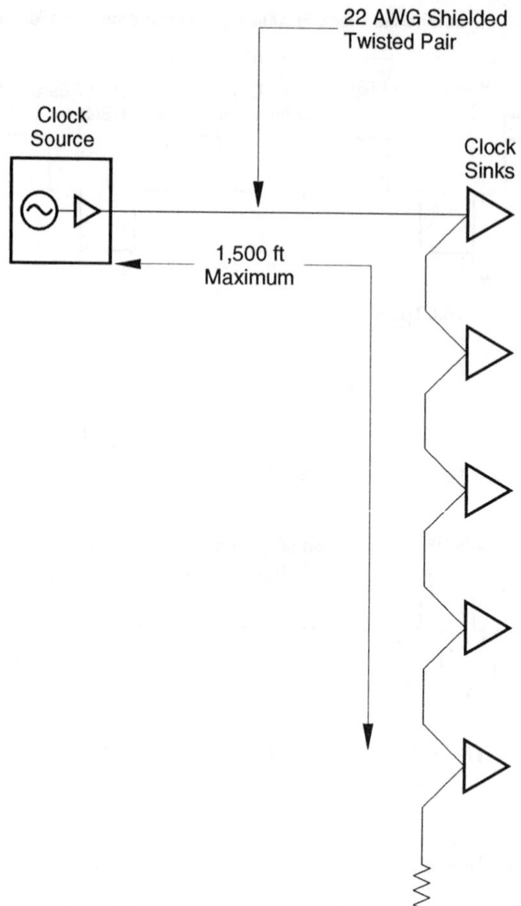

Figure 4-18 Clock distribution chain

reference frequency. In the case of Figure 4-17, this synchronization is via a microprocessor-controlled digital synthesizer, which provides the actual output frequency and drives the distribution circuits. The timing generation and distribution circuits isolate the input from the timing loads to prevent a failure in a single timing load from disrupting the entire building synchronizing system.

A stratum 3 clock is preferred as the TSG clock, although installations requiring high accuracy may be specified as stratum 2 [13]. In some isolated installations it is neither initially necessary nor economical to synchronize the TSG to an external timing source. Therefore, the TSG must be capable of free-run operation while still meeting the slip requirements of isolated installation.

Since the clock signal is a waveform with specific characteristics, its distribution within a building must be treated as a transmission problem. Appropriate care must be taken to ensure that adequate signal levels are available at the inputs to equipment requiring an external clock source. Clock signals are always distributed on dedicated 22 American wire gauge (AWG) twisted pair shielded cables. Some existing installations use 24 AWG shielded cable, but, except for short runs, this should be avoided on new installations. A suitable cable for this purpose is designated ABAM.* The maximum distance between a clock source and the farthest clock sink is 1,500 feet, and the minimum signal level at the sink normally is 1.1 V peak-to-peak, although this latter requirement may vary with the equipment type.

Clock sinks may be chained together, as shown in Figure 4-18. This is the preferable method when individual equipment shelves are stacked in a rack or cabinet. The distance limitation applies from the clock source to the last clock sink in the chain. Also, the clock signal cables must be properly terminated. The clock inputs to many multiplexers and channel banks can be optioned for bridging (high impedance) or terminated connection. The termination impedance can be 100, 120, or 150 Ω, depending on the equipment. Normally, the first and all but the last clock sink in the chain are optioned for bridging, and the last, for terminated. Each output of a centralized clock source normally can drive up to four separate cables, meeting the maximum length and termination requirements.

4.6 Building-Integrated Timing Supply (BITS)

A building-integrated timings supply, or BITS, is a somewhat awkward term that describes a system which provides a single master timing supply for all digital equipment or digital elements within a physical building structure. The BITS is the highest-quality clock in the building and therefore establishes the building synchronizing reference. The BITS drives all other clocks in the building. It usually is found in large installations where many different timing signal generators are used. In many installations the BITS is not distinguishable from a TSG.

The BITS consists of a high-accuracy clock (low stratum number) that is synchronized from an external synchronizing source traceable to a PRS. The BITS reference synchronizing source is always an equal or lower-numbered stratum. The facility by which that reference is brought into the building almost always is a DS-1 rate link. The BITS can, but does not have to, be part of the TSG. The TSG, as previously explained, is primarily used to distribute the timing within the building to a large number of loads.

Figure 4-19 shows a number of ways the BITS can be configured [13]. Figure 4-19(a) shows a BITS clock synchronized to an incoming DS-1 synchronizing signal, which, in turn, drives the TSG, which provides both a composite clock and a DS-1 rate clock. The TSG provides the synchronizing signal to all other clocks in the building. The BITS clock can be part of a digital switching system or digital cross-connect system having a clock output. Figure 4-19(b) is similar but, in this case, the incoming DS-1 synchronizing signal drives a composite clock generator

---

*See Appendix A for more about ABAM, an air core 83 nF/mile, solid polyolefin-insulated, 22 AWG, aluminum-shielded cable.

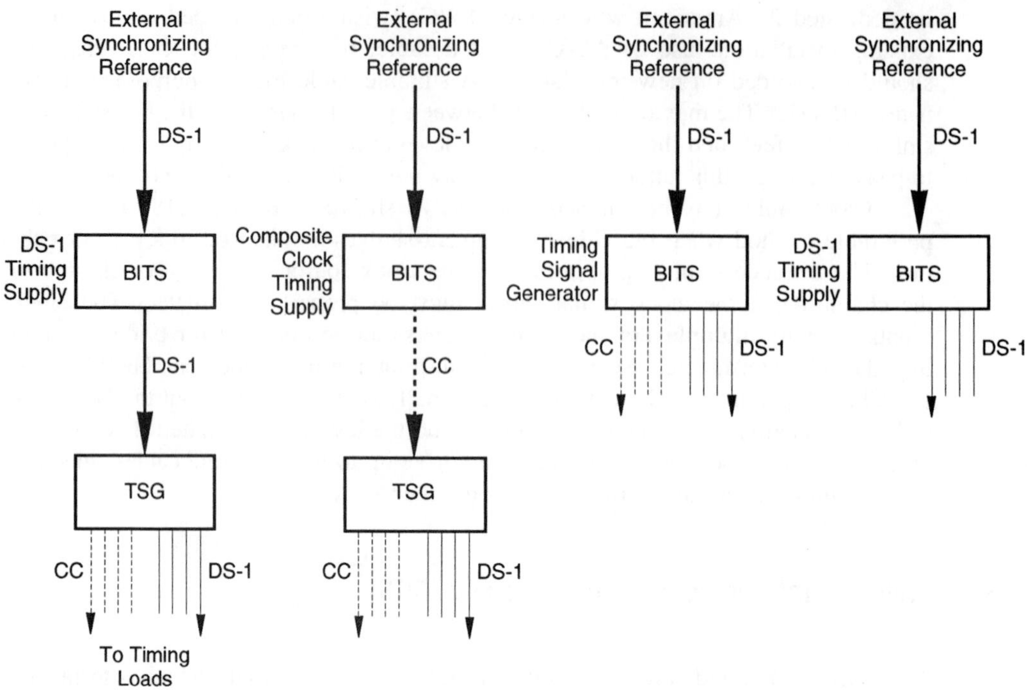

Figure 4-19 BITS configurations

(CCG). The CCG may be built into a subrate digital data system timing supply, which generates and uses both 8 kHz and 64 kHz timings signals. A separate DS-1 timing supply, such as a TSG, is synchronized to it.

In Figure 4-19(c), external timing requirements are low and the BITS and TSG are combined. Figure 4-19(d) shows a situation where the BITS clock is part of a DS-1 timing source, which is synchronized to an incoming DS-1 synchronizing signal. Again, a separate TSG is not used if the external timing requirements are low, such as would be the case where most equipment uses loop timing or the equipment installation is small.

4.7 Primary Reference Sources (PRSs)

There are several types of PRSs from which to choose. Many PRS clocks are based on cesium beam technology. The costs of these systems are high, both initially and operationally, so many networks are based on LORAN-C, global positioning system (GPS), or a combination of both. Some systems use a cesium beam clock with backup to LORAN-C or GPS. Although much cheaper to the user than cesium beam technology, both LORAN-C and GPS are federal government-maintained systems and subject to budget constraints and government vagaries over the long term.

The basic characteristics of LORAN-C and GPS are compared in Table 4-8. In this table, the important measure for network synchronization is frequency comparison accuracy. This defines the expected stability of the time source over a 24-hour period. The time transfer accuracy and ambiguity refer to the transfer of real time with respect to UTC and the expected ambiguity of that transfer. Once a clock is within about 30 mseconds of UTC, the LORAN-C transmission can be used to resolve any further discrepancy. GPS transmissions do not have ambiguity problems.

The GPS is a navigation system used in vehicle location, military, aviation, and marine systems and consists of a constellation of mid-altitude (approximately 11,000 miles) satellites from which high-precision clock signals are transmitted in the 1.2 and 1.5 GHz frequency ranges. For timing reference, unlike navigation, only a single satellite needs to be received. Both the antenna and receiving equipment can be very compact. The GPS can achieve an accuracy in the 1E-13 range under ideal conditions [19].

PRSs usually incorporate both GPS and LORAN-C receivers for signal acquisition reasons. The GPS is only received when a satellite is about 20 degrees above the horizon, and there can be dead-time in mountainous areas when no satellites are in view. Therefore, to ensure reliable operation of a GPS system, the following antenna site guidelines should be followed:

- Mount the GPS antenna at the highest possible point, with a 90% view of the spherical area 20 degrees above the horizon and a 100% view of 40 degrees above the horizon
- Avoid locations that are subject to lightning strikes
- Adjacent buildings or hills may cause reflection or shadow problems
- The best antenna location might have to be determined experimentally

LORAN-C was originally designed as a maritime and aircraft navigation system. It consists of high-power transmitters operating at 100 kHz. Ground wave coverage extends to about 1,500 miles from a transmitter but depends somewhat on transmitter power and receiver location. Sky wave coverage, which is of lower accuracy, can extend to 5,000 miles. As with GPS, more than one LORAN-C station

Table 4-8 Comparison of GPS and LORAN-C Timing Characteristics[a]

Type	Frequency Comparison Accuracy[b]	Time Transfer Accuracy[c]	Ambiguity	Coverage
LORAN-C	5×10^{-12}	1 (Ground wave) 50 (Sky wave)	30 to 50 ms	Most of North American hemisphere
GPS	5×10^{-12}	1	Not applicable	Global

[a]Source: [18]

[b]Under normal conditions

[c]Measured in μsecond(s)

is required for navigation, but only one is required for timing reference. Of course, being able to receive more than one station at any given time improves the long-term reliability of the timing source. The present timing accuracy of LORAN-C is approximately 1E-12/day [19].

The LORAN-C signal can be received 24 hours/day in most areas of the United States, but the signals are subject to fading and shifting during sunrise and sunset, and to signal loss because of precipitation static. However, these effects normally do not affect timing accuracy to any extent.

Typical LORAN-C receivers require only a short, 8-foot whip antenna, which can be mounted almost anywhere outside. The location of the LORAN-C antenna is not as critical as the GPS antenna, but common sense should be used in any case. For example, avoid locations subject to lightning strikes, and physically separate the antenna from low-frequency noise sources such as motors and power transformers.

The reliability of high-precision clock sources was studied between 1978 and 1983; results are shown in Table 4-9. Since this study, clock mean times between failure (MTBFs) have approximately doubled due to improvements in equipment design and construction [5].

Table 4-9 Clock Mean Time Between Failure (MTBF)[a]

TYPE	MTBF
Cesium beam	3 to 4 years
Rubidium atomic	7 to 10 years
Crystal-controlled oscillator	50 years

[a]Source: [5]

4.8 Synchronization Management

Synchronization management thus far in the digital network evolution only takes into account the hierarchical arrangement of clocks connected to various network elements. Limited decision making by a particular clock is allowed during some failure conditions (for example, identification of primary frequency source failure and switching to the secondary source). It is desirable that more sophisticated synchronization management is provided as the network evolves. Some proposed requirements are [20]:

- Each network element should be equipped with the means to select the best possible reference source
- Synchronization source selection algorithms should be compatible with all network topologies (ring, linear, point-to-point, star, mesh)
- Synchronization strategy should be compatible with all timing methods (line, loop, through, external)
- Synchronization messages between network elements should identify the synchronization source quality level

- Message sets should be compatible with all present and foreseen digital transmission environments
- Provisioning rules should prevent timing loops
- Methodologies should be provided that require minimum setup and administration

Of particular significance in the above management strategy are synchronization messages. Synchronization messages are not presently a part of the performance information that is passed between network elements. It is desirable to provide this information for synchronization provisioning when networks are set up or rearranged and to allow network elements to respond to timing source failures from upstream and downstream timing sources.

REFERENCES

[1] Ingle, J. "Identification of Digital Impairments in a Voiceband Channel." 1989 International Communications Conference (ICC'89), Boston, June 1989.

[2] *Synchronization: LATA Switching Systems Generic Requirements (LSSGR): Section 18*. BELLCORE Technical Reference TR-TSY-000518, July 1987.

[3] *Private Digital Network Synchronization*. ANSI/EIA/TIA-594-1991. Available from Electronic Industries Association.

[4] Bridge, R.F. et al. *Jitter Attenuation in T1 Networks*. Paper 314.1.1. Conference Record from the IEEE International Communications Conference 1990 (ICC'90), April 1990.

[5] Kartaschoff, P. *Synchronization in Digital Communications Networks*. Proceedings of the IEEE, Vol. 79, No. 7, July 1991.

[6] *Digital Synchronization Network Plan*. BELLCORE Technical Reference TA-NPL-000436, Nov. 1986.

[7] *Clocks for the Synchronized Network: Common Generic Criteria*. BELLCORE Technical Reference TA-NWT-001244, Nov. 1992.

[8] American National Standard for Telecommunications. *Synchronization Interface Standard*. ANSI T1.101-1994.

[9] Mills, D. "Internet Time Synchronization: The Network Time Protocol." IEEE Transactions on Communications, Vol. 39, No. 10, Oct. 1991.

[10] *Digital Networks, Digital Sections and Digital Line Systems*. CCITT Blue Book, Vol. III, Fascicle III.5, Recommendations G.801–G.956. International Telecommunication Union, Nov. 1988. Available from National Technical Information Service as PB89-143895.

[11] Hawley, G., Nabavi, F. "Timing is Everything." Telephony magazine, Oct. 28, 1991.

[12] *Timing Jitter and Wander Effects on Transmission*. Application Note 404. Telecom Solutions, May 1988.

[13] *Timing Signal Generator (TSG) Requirements and Objectives*. BELLCORE Technical Reference TA-TSY-000378, April 1986.

[14] Thomas, L. *Network Synchronization, DS1 Line Codes and Frame Formats*. Supercomm '89, May 1989.

[15] *Accunet® T1.5 Service Description and Interface Specifications*. AT&T Technical Reference TR62411, Dec. 1988.

[16] *Functional Criteria for Digital Loop Carrier Systems*. BELLCORE Technical Reference TR-TSY-000057, Nov. 1988.

[17] American National Standard for Telecommunications. *Carrier-to-Customer Installation—DS1 Metallic Interface*. ANSI T1.403-1989.

[18] Smith, D.R. *Digital Transmission Systems*. Van Nostrand Reinhold, 1985.

[19] *LORAN-C vs. GPS: Which Primary Reference Source for the Telecommunications Industry?* Engineering Planning Letter 098-41000-02. Telecom Solutions, Dec. 1990.

[20] Horn, F. *Synchronization Messages Technical Report*. ANSI T1X1.3 Working Group, Aug. 1991.

5 Loop Transmission Impairments

Chapter 5 Acronyms

2B1Q	2-binary, 1-quaternary	ISDN	integrated services digital network
ac	alternating current	ISI	intersymbol interference
ALBO	automatic line build-out	Nepers	a dimensionless unit: equal to the natural logarithm of the scalar ratio of two voltages. One Neper = 8.686 dB
AM	amplitude modulation	NEXT	near-end crosstalk
AMI	alternate mark inversion	Ng	noise-to-ground
AWG	American wire gauge	PCM	pulse code modulation
AWGN	additive white Gaussian noise	PI	power influence
BER	bit error rate	PSK	phase-shift keying
CO	central office	Q	quality factor: 2π times the ratio of the maximum stored energy to the energy dissipated per cycle at a given frequency
dBrnC	decibels with respect to reference noise, C-message weighted	QAM	quadrature-amplitude modulation
dc	direct current	RMS, rms	root-mean-square
DCS	electronic digital cross-connect system	RST	remote switching terminal
DLC	digital loop carrier	SM	singlemode
DOV	data-over-voice	SNR	signal-to-noise ratio
DSL	digital subscriber line	SRDL	subrate digital loop
EDD	envelope delay distortion	SSB	single sideband
EO	end-office	TLP	transmission level point
ESF	extended superframe format	VSB	vestigial sideband
FCC	Federal Communications Commission		
FEXT	far-end crosstalk		
FSK	frequency-shift keying		
HDSL	high bit-rate digital subscriber line		

There are a variety of transmission impairments that must be considered when designing and using digital loops. These are:

- Attenuation (loss)
- Background and longitudinal noise
- Impulse noise
- Crosstalk interference
- Delay and delay distortion
- Jitter and wander

This is not an exhaustive list of transmission impairments that may be encountered on any given telecommunication *path*. Rather, it is a list of only those impairments that may be attributed to the loop itself. The effects of each are illustrated in Figure 5-1. Some impairments—for example, differential and longitudinal noise—are specific to twisted pair metallic cables and do not apply to fiber optic cables.

There are other impairments that are not the direct result of the characteristics of loops, but rather, the characteristics of digital transmission itself. For example, signal regeneration can result in a finite probability of error even with no line impairments. Similarly, digital pads, used to adjust the level of analog source signals after they have been converted to digital, can lead to lower signal-to-noise ratios of those signals. The first part of this chapter concentrates on loop-induced impairments; in the latter part, signal regeneration and digital pads are discussed.

Attenuation (or *loss*) is the natural consequence of signal propagation through any practical transmission media. Its effects are controlled by the basic design of digital loop systems, and attenuation is discussed in greater detail in later chapers. Perhaps the next most important transmission impairment is noise (including crosstalk), followed by distortions, and jitter and wander. These are discussed in the following sections.

5.1 Noise

Noise is any interfering signal on the telecommunication channel, whatever its cause. Noise can lead to errors by introducing jitter in timing extraction circuits, by causing pulse detection circuits to detect pulses when none are present, or by preventing pulse detection circuits from detecting a valid pulse. Early experiments and analysis of prototype T1-carrier systems found that errors primarily were due to additive noise at the decision point of the pulse detection circuits in signal regenerators (repeaters) rather than timing jitter [2]. This was true of early systems where path lengths were seldom more than 50 miles. As the length of the T1 lines increased, which required more regenerators, it was found that jitter and wander became a significant problem. All digitial loops in present use that use twisted pair cables are limited by noise, in particular crosstalk, at the pulse detection/decision point.

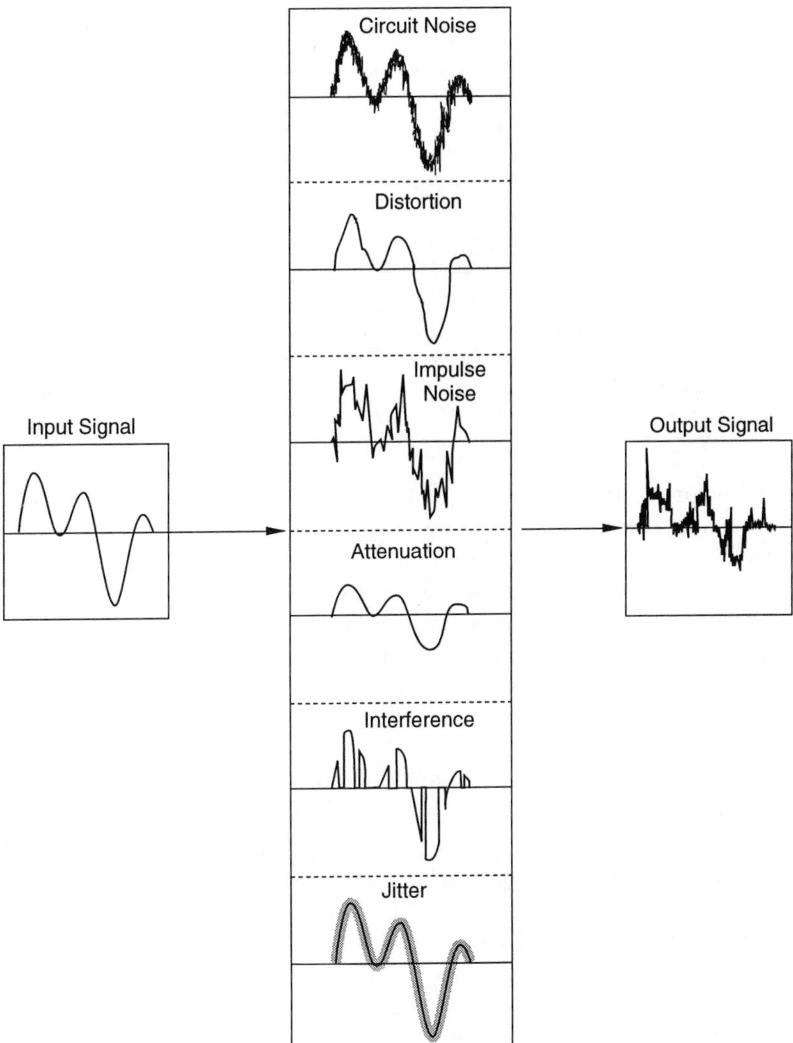

Figure 5-1 Transmission impairments, [1]
Reeve, © 1992 Institute of Electrical and Electronics Engineers. Reprinted with permission

Noise in pulse detection circuits can come from internal and external sources. Noise generated by the regenerator power supply and thermal noise, while principal internal noise sources are small compared to noise from external sources. An example will illustrate this point.

EXAMPLE 5-1

Find the thermal noise voltage, power, and signal-to-thermal noise ratio at the input of a T1 regenerator for a worse-case line section. Assume the section loss is 35 dB, the terminating resistor is at 38 °C (100 °F), and the noise bandwidth is 2 MHz.

$$\text{Root-Mean-Square (RMS) Noise Power} = E^2/R = 4kTB \text{ W}$$

where E = RMS noise voltage in V
R = Resistance in Ω = 100
k = Boltzmann's constant, 1.38E-23 joules per Kelvin
T = Temperature in Kelvins = 273 + 38 = 311
B = Bandwidth in Hz = 2E6

Therefore, noise power = 4 × 1.38E-23 × 311 × 2E6 = 0.0343 pW and rms noise voltage = $\sqrt{3.43\text{E-}14 \times 100}$ = 1.85 μV.

The transmitted signal level for a T1-carrier system is approximately 2.7 V base-to-peak. The peak signal power is then 70 mW into a 100 Ω terminating resistor. The received signal level at the input of a regenerator at the end of a 35 dB section (this would represent a worse-case line) is in the order of 50 mV, or 23 μW. The signal-to-thermal noise ratio is approximately 23E-6/0.0343E-12 = 672E+6, or 88 dB.

For external noise there must be a *noise source*, a *coupling mechanism*, and a *receptor*. The relationship among the three is illustrated in Figure 5-2. Noise sources are either man-made or natural. Practically any piece of electrical or electronic equipment can be a man-made noise source. This includes radio broadcast transmitters, powerlines, personal computers, and noise coupled via crosstalk from other nearby loop transmission systems. Natural noise comes from lightning and other atmospheric conditions, random thermal motion of electrons, and galactic sources, as well as electrostatic discharges.

External noise sources can inject noise into a digital loop directly through conductive, inductive, and capacitive coupling. Crosstalk is introduced through inductive and capacitive coupling from other pairs in twisted pair cables. The predominant external noise source at low frequencies, generally no higher than several kilohertz, is through induction from nearby powerlines. While the fundamental and all significant power harmonic frequencies are too low to cause signal interference with digital loop systems, the actual voltage levels may be high enough to cause equipment malfunction or damage. For example, T1 line regenerators are designed to cope with longitudinal currents from powerline induction up to around 50 mA rms or 300 mA peak-to-peak. Levels above this may saturate coupling transformers, damage input or power supply components, or cause erratic operation due to line repeater power supply upset.

At the higher signal frequencies used by digital loops (tens of kHz to several MHz), the cable shields effectively isolate the pairs from external interference, including most powerline harmonics. At these higher frequencies, however, capacitive and inductive coupling in the form of crosstalk from nearby pairs in the same cable predominates as the interfering noise source.

Conductive coupling is more of a maintenance problem than an operational problem. There are no intentional conductive paths for interfering noise in digital loop systems. Nonetheless, noise can be conducted into the loop through insulation faults or poor or faulty grounding methods.

Electromechanical central office (CO) switching systems are significant sources of noise. In these offices, electromagnetic switches and relays create transient and

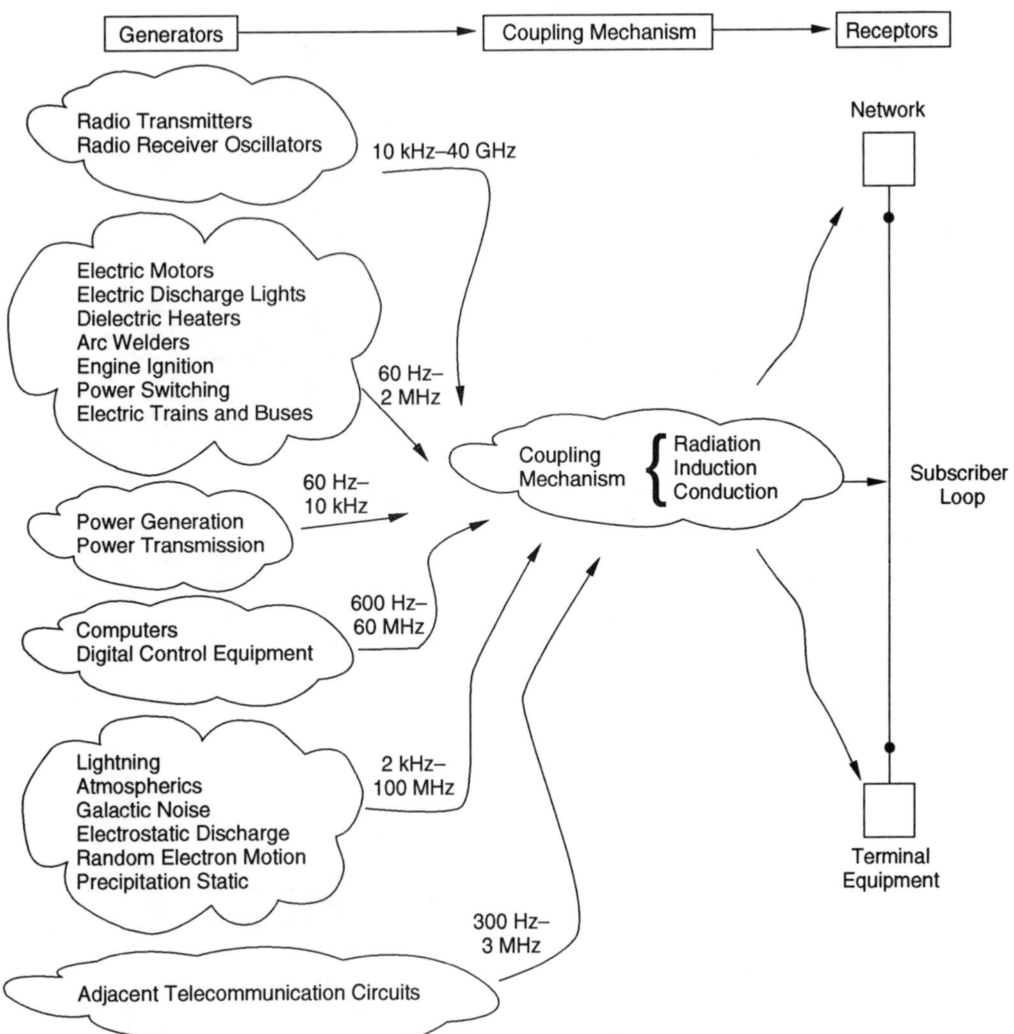

Figure 5-2 Noise diagram, [1]
Reeve, © 1992 Institute of Electrical and Electronics Engineers. Reprinted with permission

repetitive currents with wide frequency spectrums, greater than 2 MHz [2]. Typical impulse noise waveshapes are shown in Figure 5-3.

Digital transmission terminal equipment and systems installed in COs usually are separated or shielded from other equipment and therefore from direct noise coupling; however, noise currents induced into voicegrade lines entering the CO may be crosstalk coupled into digital loops that share common cables in the outside plant. Such noise currents rapidly attenuate with distance from the CO. A similar noise coupling situation exists at customer premises—except that equipment installations

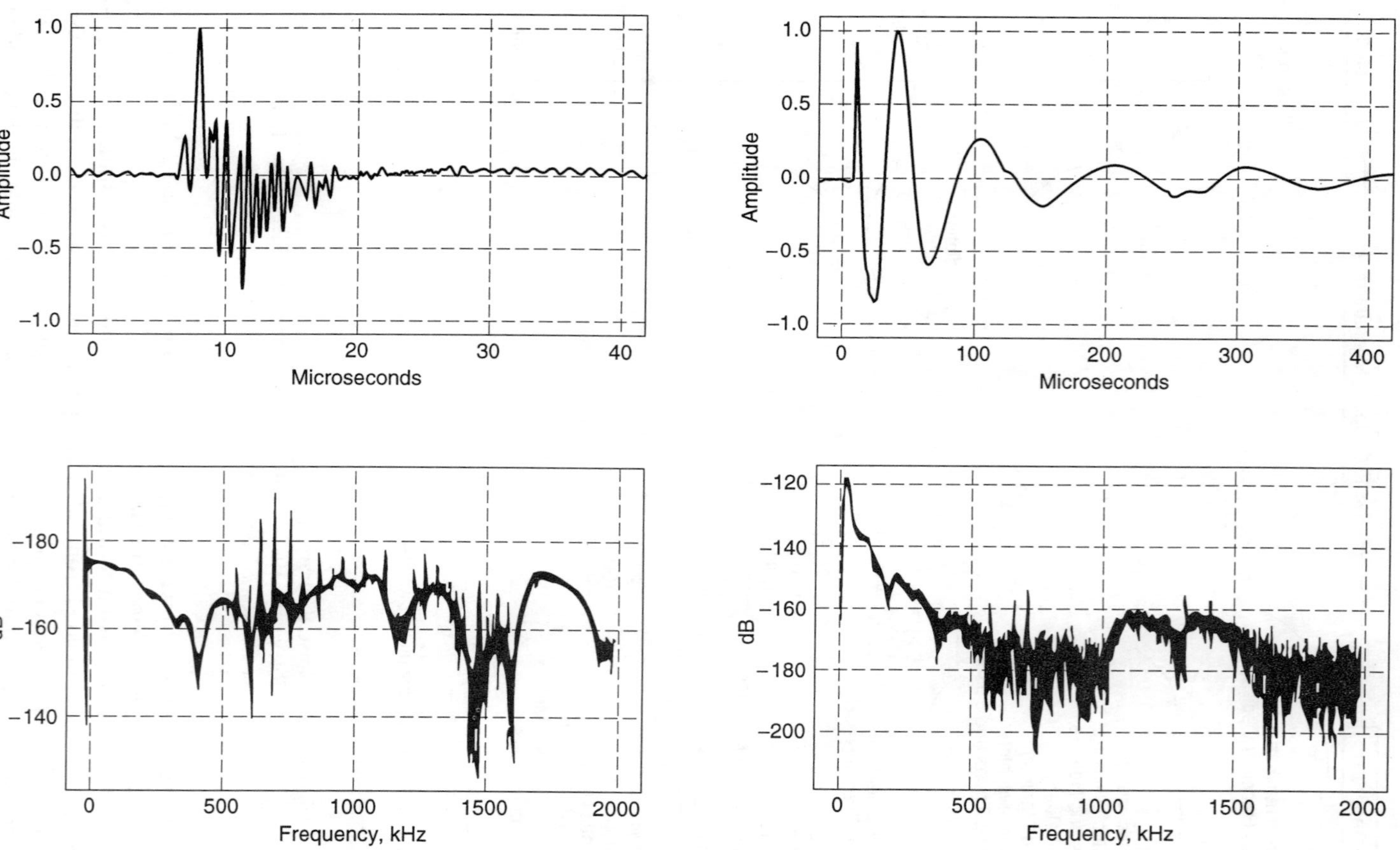

Figure 5-3 Typical impulse noise waveshapes and spectral density, [3]

are not always designed (nor is space always available) to minimize mutual interference.

A number of different baseband and passband telecommunication services may use pairs in a common cable, which can lead to frequency and level coordination problems and mutual interference. For example, a cable may have:

- Analog voicegrade lines with 0 to 3.4 kHz bandwidth
- Analog subscriber carrier lines with 8 kHz amplitude modulation (AM), or 4 kHz single sideband (SSB) bands centered at frequencies from 8 kHz to 150 kHz
- Analog data-over-voice (DOV) lines with 24 or 48 kHz bands centered at frequencies in the range of 50 to 200 kHz
- Integrated services digital network (ISDN) lines with 0 to 80 kHz bandwidth
- Subrate digital lines with 0 to 64 kHz bandwidth
- T1C lines with 0 to 3.1 MHz bandwidth
- Repeatered T1 lines with 0 to 1.5 MHz bandwidth

A number of modulation types are used, including AM, SSB, vestigial sideband (VSB), baseband pulse code modulation (PCM) methods, and others. The potential for mutual interference between different systems is great unless the various signal levels and crosstalk coupling losses are controlled.

Of particular interest in spectrum management are the analog passband systems because they are most vulnerable, especially when operated at their maximum gain. Table 5-1 shows the carrier frequencies used by these systems. Each voicegrade channel in the carrier systems occupies 4 kHz bandwidth on each side of the carrier frequency.

The loop itself, or any individual pair within a cable, can be a noise receptor. The terminal equipment connected to a loop, particularly digital line receivers, and telecommunications network equipment in general are also noise receptors. By their nature, the termination equipment at each end of the loop are subject to conducted noise from the loop. The reverse is also true. Therefore, it is important to specify maximum acceptable noise levels and define how noise affects telecommunication systems.*

On digital loops, noise can be conveniently categorized into four types, each of which are discussed in later sections:

- Differential noise (also called background or circuit noise)
- Longitudinal noise (or power influence)
- Impulse noise
- Crosstalk

*The subjects of noise, its effect on systems, and its reduction fall under the heading electromagnetic compatibility. References [4,5] cover these subjects in detail.

Table 5-1 Carrier Frequencies of Typical Analog Passband Systems Used in the Loop

SYSTEM	TX FROM CO (kHz)	RX AT CO (kHz)
1-Channel subscriber carrier	76	28
1-Channel subscriber carrier	76	32
6-Channel subscriber carrier	84, 92, 100, 108, 116, 124	12, 20, 28, 36, 44, 52
8-Channel subscriber carrier	84, 92, 100, 108, 116, 124, 132, 140	4, 12, 20, 28, 36, 44, 52, 60
8-Channel subscriber carrier	76, 84, 92, 100, 108, 116, 124, 132	8, 16, 24, 32, 40, 48, 56, 64
8-Channel subscriber carrier	88, 96, 104, 112, 120, 128, 136, 144	8, 16, 24, 32, 40, 48, 56, 64, 152
Data-over-voice (9,600 bps)	84/96	36/48
Data-over-voice (19.2 kbps)	168/192	72/96

Tx = transmit; Rx = receive

5.1.1 Noise Objectives

The noise objectives for some digital loop systems are based on the overall noise objectives for the existing voicegrade loop plant. This allows digital line deployment in the existing voicegrade loop plant without special conditioning or measurement. The overall noise objectives of the existing voicegrade loop plant are described later. The following discussion generalizes the basic analog voicegrade objectives so that their effect can be visualized from a digital (and wider bandwidth) perspective. Such a generalization is important because digital and analog loops do, or will, coexist in almost all cables.

For typical analog voicegrade circuits, the overall signal bandwidth is taken to be around 3.1 kHz. When noise on analog loops is measured, however, a C-message weighting filter is used. This filter has an impedance of 600 Ω and an effective bandwidth of about 2.3 kHz. Its special shaping takes into account the interfering effects of various noise frequencies on the human ear. The noise objective for analog loops is 20 decibels with respect to reference noise, C-message weighted (dBrnC).

The C-message filter bandwidth is much less than the spectrum width of almost any digital loop transmission system. Also, the loop impedance at the low frequencies of normal voicegrade analog transmission is somewhat higher than at digital signaling frequencies. Nevertheless, voicegrade noise objectives are well known and easily measured, and they can be used to estimate the noise on digital loops.

First, however, it is necessary to convert voicegrade noise objectives into more general terms of dBm per kHz of bandwidth at digital loop signaling frequencies. This can be done as follows:

- Convert from dBrn to dBm by subtracting 90 dB
- Change bandwidth from 2.3 kHz to 1 kHz by subtracting 10 log(2.3/1)
- Change termination impedance from 600 to 100 (or 135) Ω by adding 10 log(600/100) or 10 log(600/135), depending on the frequency (use 135 Ω for subrate systems; above approximately 150 kHz, use 100 Ω)

The general noise objective becomes

$$\begin{aligned}\text{Noise Objective} &= 20 \text{ dBrnC} - 90 \text{ dB} - 10\log(2.3/1) + 10\log(600/\text{impedance})\\ &= -67.1 \text{ dBm/kHz at } 135\ \Omega \text{ termination impedance, or}\\ &= -65.8 \text{ dBm/kHz at } 100\ \Omega \text{ termination impedance}\end{aligned}$$

EXAMPLE 5-2

Consider a subrate digital loop with 135 Ω impedance operating at 56 kbps. The input filter of the loop receiver has an effective noise bandwidth of 48 kHz. The equalizer in the line receiver that matches the automatic line build-out (ALBO) and loop loss gives a noise gain 3 dB greater than the peak signal gain due to its bandwidth characteristics. This gain is added to the gain at one-half the signaling frequency of 28 kHz, which is 31 dB for a maximum length line. This gain provides a +6 dBm decision level. Assuming loop noise is additive white Gaussian noise (AWGN), find

1. The equivalent noise objective at the receiver decision point (output of the equalizer/input to the 3-level slicing circuit);
2. The total noise power and signal-to-noise ratio (SNR) at the receiver decision point.

Solution

1. The equivalent noise objective is −67.1 dBm/kHz + 31 dB + 3 dB = −33.1 dBm/kHz.
2. To convert from a bandwidth of 1 kHz to 48 kHz, add 10 log(48/1) = 16.8 dB. Therefore, the total noise power in the 48 kHz bandwidth is −33.1 dBm/kHz + 16.8 dB = −16.3 dBm. If the decision point is at +6 dBm, the signal-to-noise ratio is 6 dBm − (−16.3 dBm) = 22.3 dB.

5.1.2 Background Noise and Longitudinal Noise

The following is a general overview of loop noise and balance requirements as they relate to twisted pair loops in an analog voicegrade environment. This is relevant because digital loops operate in the same cables along with voicegrade analog loops, as previously noted. To ensure compatibility, digital loop systems have been designed to provide the specified performance in this loop plant, which originally was designed for analog applications. In some specific installations, notably repeatered T1-carrier and some subrate digital loops, special precautions are required, as will be discussed in later chapters. It is generally true that if a loop meets the analog voicegrade noise requirements, the loop also will meet the digital requirements with minimum adjustments.

The noise that appears across the two conductors (tip and ring) of a loop and that is heard by the user in an analog environment is called *circuit noise*. The same noise appearing at the balanced input of a digital line receiver (digital environment) is called *background noise* (this term will be used throughout this section). This noise also is called message circuit noise, noise metallic, or differential noise. The noise can be due to random thermal motion of electrons (known as white noise or Gaussian noise), or due to static from lightning storms, powerline induction, or crosstalk. On analog subscriber loops, the most likely source of performance-limiting noise is interference from powerline induction. On digital loops, performance is limited by noise from crosstalk sources.

Since different interfering frequencies affect the various services (voice, data, radio studio audio material, digital) differently, a variety of weighting curves have been designed to restrict the frequency response of the noise-measuring sets with which objective tests are made.

Noise is described in terms of dB above a noise reference when measured with a noise meter containing a special filter. There are five common filters that are used to provide the necessary weighting:

- C-message
- 3 kHz Flat
- Program (8 kHz)

- 15 kHz Flat
- 50 kilobit

The most common filter in voice applications, and some digital applications, is called a C-message filter, and measurements are based on dBrnC.* The noise reference is 1 pW (−90 dBm); therefore, a properly calibrated meter will read 0 dBrnC when measuring a 1,004 Hz tone having a power of −90 dBm.

C-message weighting primarily is used to measure noise that affects voice transmission on common telephone instruments, but it also is used to evaluate the effects of noise on both analog and digital data circuits. It weights the various frequencies according to their perceived annoyance such that frequencies below approximately 600 Hz and above 3,000 Hz have less importance.

The *3 kHz Flat* weighting curve is used on voice and digital circuits, too. All frequencies within the 3,000 Hz bandwidth carry equal importance. This filter rolls off above 3,000 Hz and approximates the response of common modems. It generally is used to investigate problems caused by power induction at the lower-power harmonic frequencies, or by higher interfering frequencies (those below several kilohertz). Frequencies in these ranges can affect analog modems as well as voice frequency signaling equipment.

The *Program* weighting curve is used with voice circuits that require bandwidth in the order of 8 kHz or more, such as in the distribution of radio or television program audio material. This curve emphasizes frequencies between 1 and 8 kHz. The *15 kHz Flat* weighting curve also is used on these types of circuits, but it actually carries little weighting over its frequency range (it is essentially "flat" from 20 Hz to 15 kHz).

The *50 kilobit* filter was originally designed for wideband analog data circuits and is used with subrate digital loops (SRDLs) and ISDN digital subscriber lines (DSLs). The 3 dB bandwidth is approximately 50 Hz to 29 kHz. The noise requirements for the SRDL and DSL take into account the limited bandwidth of this filter with respect to the signal spectrum on the loop.

A comparison of the various weighting curves is shown in Figure 5-4. Noise measurements on any particular loop with the 3 kHz, Program, 15 kHz, or 50 kilobit filters generally will be higher than with the C-message filter (because of the higher bandwidth and less overall weighting). For example, if 3 kHz white noise with a total power of 0 dBm is applied to a circuit and then measured with an instrument containing a C-message filter, the meter will read 88.0 dBrnC (−2.0 dBm). With a 3 kHz filter, the meter will read 88.8 dBrn (−1.2 dBm), and with a 15 kHz and 50 kilobit filter, the meter will read 90.0 dBrn (0.0 dBm). All meters will read 90.0 dBrn with a 1,004 Hz, 0 dBm tone.

The maximum acceptable background noise on voicegrade subscriber loops is 20 dBrnC, or 20 dB above the reference noise level. At a −6 transmission level

*The standardized C-message weighting versus frequency characteristic can be found on page 20 of [6].

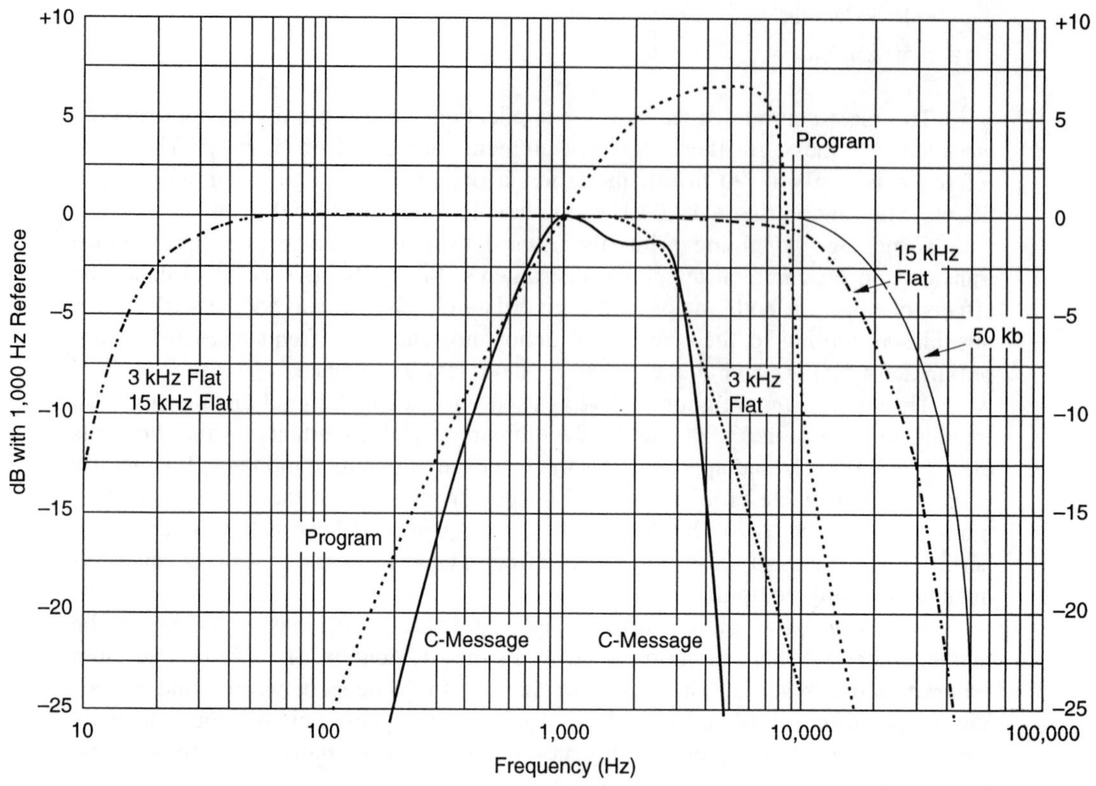

Figure 5-4 Weighting curve comparison, adapted from [1] Reeve, © 1992 Institute of Electrical and Electronics Engineers. Reprinted with permission

point (TLP) [1]—for example, at the receiving end of a loop with 6 dB loss—this noise level would be measured as 26 dBrnC0.

In practice, the average loop will measure around 10 or 11 dBrnC. The lower the noise, in dBrnC, the better. The requirements for noise measurements using 3 kHz Flat weighting have been established at 40 dBrn 3 kHz Flat or less. The noise requirements for wideband (Program) voice circuits are not as well established, but circuits that meet voicegrade requirements can be considered to meet Program requirements.

In general, digital loops are tested with a 50 kilobit filter, which has a bandwidth approaching 30 kHz. Figure 5-5 gives an example of the application of the 50 kilobit filter for measuring background and impulse noise on SRDLs.

In addition to background noise, noise from tip *and* ring to ground, which is called *power influence* (PI), is important. Power influence is similar to *noise-to-ground* (Ng), which is also called longitudinal or common-mode noise, and related by the following expression:

$$\text{Power influence, PI} = \text{Ng} + 40 \text{ dB}$$

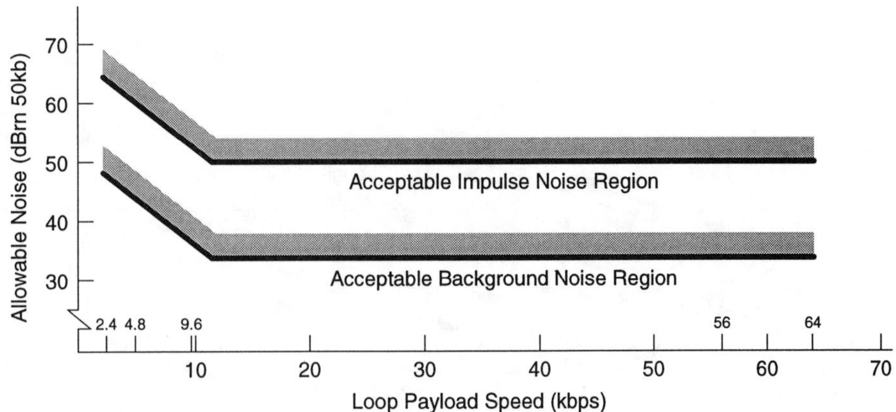

Figure 5-5 Background and impulse noise requirements for SRDLs

The 40 dB factor arises from the attenuator in noise-to-ground measuring sets. It is important to know if PI or Ng is being measured by the transmission or noise test set during loop qualifying tests or troubleshooting. All modern sets built for the telecommunications industry are calibrated to read PI directly, but a number of older sets are not calibrated this way. PI is almost always caused by inductive interference from powerlines.

Normally, the longitudinal currents, which are induced into the loop by nearby powerlines, cannot be heard by the user or are not detected by the balanced digital line receiver because these currents are of equal magnitude and flow in the same direction on both conductors. This gives a zero differential noise voltage (the shield and earth act as return conductors). If the loop becomes unbalanced, however, these longitudinal currents become unbalanced, which causes a differential noise current and, consequently, a differential noise voltage—circuit noise—to develop.

Figure 5-6 illustrates the difference between longitudinal and differential currents on the loop. The maximum acceptable PI level is 80 dBrnC. The lower the PI, in dBrnC, the better. When C-message PI readings exceed 80 dBrnC, additional measurements should be made with the 3 kHz Flat filter to determine if the measurement point is safe.

A measured value exceeding 126 dBrn 3 kHz Flat usually means that the lower powerline harmonic frequencies exceed 50 V, which is considered unsafe. Measurements made with the 3 kHz Flat filter will always exceed C-message readings by as much as 44 dB, due to the higher response of the 3 kHz Flat filter at the lower and higher power frequency harmonics [8].

5.1.3 Balance

The degree of balance indicates the symmetry of the tip and ring conductors with respect to each other, other conductors, and the shield. Sufficient line balance must be maintained for two reasons:

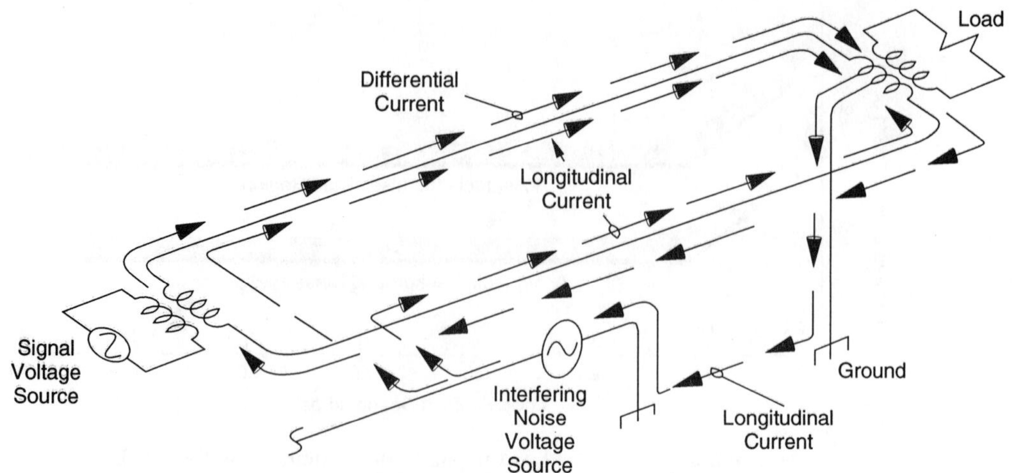

Figure 5-6 Loop noise currents, [1]
Reeve, © 1992 Institute of Electrical and Electronics Engineers. Reprinted with permission

1. To limit the magnitude of longitudinal currents, including crosstalk currents and powerline harmonics, which are converted to circuit noise in terminal equipment; and
2. To limit unbalanced longitudinal currents in cable pairs that may cause crosstalk in adjacent circuits.

The first reason is especially important to digital loops, which usually are crosstalk limited, and to regular telephone instruments, which respond quite well at power frequency harmonics, especially those above 300 Hz. The second reason is equally important to digital loops from a crosstalk standpoint, in that a single noisy circuit could upset a large number of adjacent circuits.

The balance of the loop itself (also called *longitudinal-to-metallic balance* or just *longitudinal balance*) is specified as 60 dB minimum at 1,004 Hz. The balance will vary with frequency; however, in loop applications it is not specified for frequencies other than 1,004 Hz. Minimum terminal equipment balance (in this case called metallic-to-longitudinal balance) is specified in Federal Communications Commission (FCC) Part 68, paragraph 68.310, as shown in Figure 5-7 [12]. Metallic-to-longitudinal balance measures the conversion of metallic currents in the terminal equipment to longitudinal currents that could be injected into the loop and thereby disturb adjacent circuits.

Longitudinal-to-metallic balance requirements specified for various types of digital terminal equipment are given in industry standards [7,9–11]. With longitudinal-to-metallic balance, the concern is for the ever-present longitudinal currents on the loop being injected into the terminal equipment and converted into metallic noise. The higher the balance, in decibels, the better.

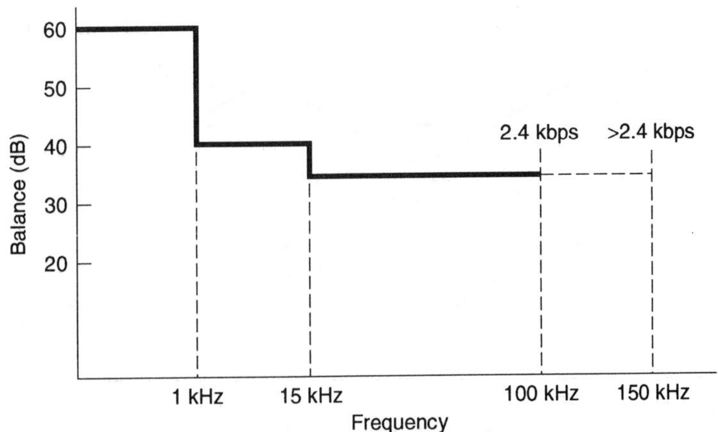

Figure 5-7 FCC terminal equipment balance requirements

If the magnitude of longitudinal noise is great enough (usually because of poor shield continuity, poor grounding practices, or nearby electric utility power factor correction capacitors), even a small circuit unbalance will give interfering circuit noise, especially at higher frequencies. Cable shield continuity and earth grounding provide a path to drain longitudinal noise currents to the earth, where they can be dissipated without disturbing the circuit.

A cable with high PI levels most likely will be improperly grounded, or not at all, and missing shield bonding straps. Once these problems are corrected, any residual circuit noise at voice frequencies almost always is due to unbalance (a poor splice on one or both conductors of the pair, a resistance short from tip or ring to ground, unbalanced equipment connected to the loop) or anomalies in nearby electric utility powerlines (for example, noisy insulator hardware, high harmonic currents from power factor correction capacitors, nonlinear loads, or overexcited distribution transformers).

Crosstalk interference from other loop services is more of a problem at frequencies used by digital loops than is PI, except in extreme cases of equipment upset or line repeater power supply interference in repeatered T1-carrier systems. T1 span line repeaters are line powered by a constant direct current (dc). The induced longitudinal alternating currents (ac) are superimposed on the dc, alternately adding and subtracting from it, as illustrated in Figure 5-8. A large enough opposing swing of the ac will reduce the dc below the operating threshold of the line repeater power supply, causing it to operate erratically. Noise mitigation devices, described in a later section of this chapter, reduce the effects of excessive PI.

Powerline induction also can reduce pulse amplitudes during the crest of the induced currents. Figure 5-9 shows a typical mask for ac powerline induction at the DS-1 rate network interface (demarcation point between network equipment and customer equipment). This figure assumes the longitudinal currents primarily are at 60 Hz. To meet the requirements of this interface, pulse amplitudes must fall inside the shaded area. A full wave of 60 Hz induced current has a length of 16.7 mseconds. At the DS-1 rate, approximately 26,000 pulses will fit in one period of a 60 Hz

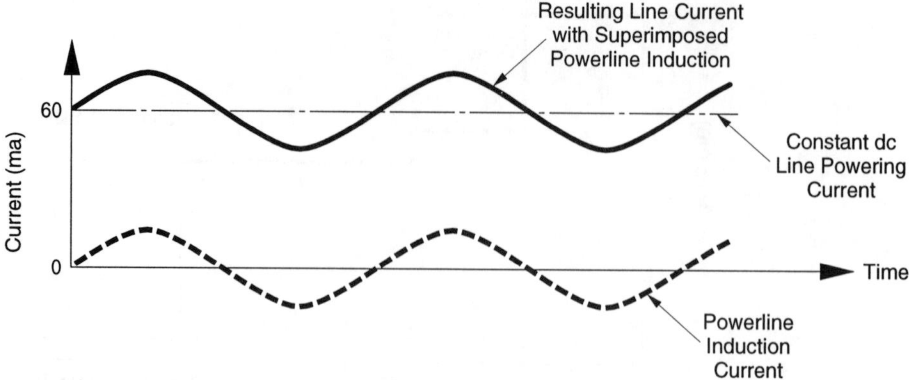

Figure 5-8 DC line current with superimposed ac induction

wave. Generally, ac induction contains many harmonics of the 60 Hz fundamental powerline frequency; the waveshape is more triangular than sinusoidal.

Table 5-2 enumerates loop transmission objectives used throughout the telecommunications industry, as they relate to the previous discussions. Measurements in the "acceptable" range indicate that the loop requires no additional attention. "Marginal" means that the loop requires maintenance attention within a reasonable period, according to the administrative policies of the circuit provider. "Unacceptable" means that the loop requires immediate attention.

These objectives may be approached from a statistical standpoint. For example,

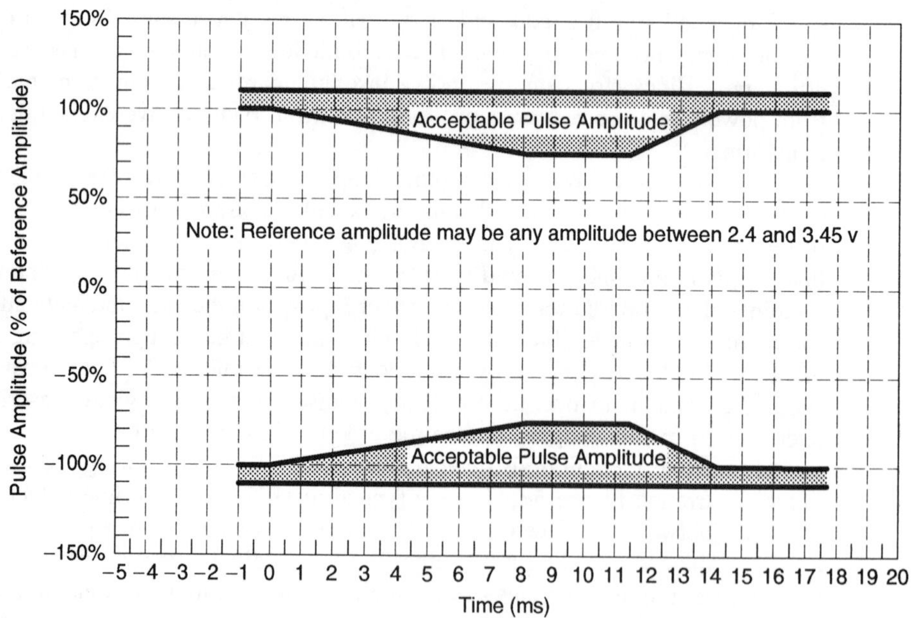

Figure 5-9 Pulse amplitude envelope with ac induction, [13]

Table 5-2 Voicegrade Loop Transmission Objectives[a]

Transmission Quality	Circuit Noise, dBrnC	Circuit Noise, 3 kHz Flat	Power Influence, dBrnC	Circuit Balance, dB
Acceptable	≤20	≤40	≤80	≥60
Marginal	20 to 30	40 to 60	80 to 90	50 to 60
Unacceptable	>30	>60	>90	<50

[a]Source: [6,14]

a company may decide that at least 98% of all loops have characteristics considered acceptable, no more than 2% considered marginal, and 0% considered unacceptable at any given time. The plant is designed, built, and maintained accordingly. See [6,14] for standards related to measuring voicegrade transmission characteristics and loop performance criteria.

5.1.4 Inductive Coordination and Noise Mitigation

As previously explained, powerline induction can affect digital loops by reducing line powering currents and pulse amplitudes. In severe cases, terminating or line equipment can be damaged. Troubleshooting powerline induction problems is difficult because the severity of powerline induction can vary with the weather or seasons. It is possible to predict the minimum required separation between digital loops and powerlines before digital loops are installed; however, this requires some advance knowledge of powerline characteristics [15].

Relatively simple PI tests can be made on digital loops with suspected powerline induction problems. When powerline induction exceeds the specified values, several approaches are available to solve the problem:

- Install shield conductors adjacent and parallel to the affected cable (this can be a very expensive and tedious cure, with a high probability of physical damage during installation to the facilities being shielded).
- Increase the line powering current (this may not be possible unless the line repeaters are designed for higher current).
- Install induction neutralizing transformers in each affected digital loop (this generally is the best solution).

Induction neutralizing transformers can be effective in reducing the effects of longitudinal currents. These devices, inserted in the transmit and receive pairs, as shown in Figure 5-10, increase the longitudinal impedance at powerline frequencies and thus reduce the induced current. The transformers must be specially designed for each type of digital loop.

Compensation must be made for the insertion loss at carrier frequencies and dc resistance introduced by the induction neutralizing transformers. The loss generally is a nonlinear function of pulse amplitude, as shown in Table 5-3. Induction neutralizing transformers should be placed at the approximate center of exposure to

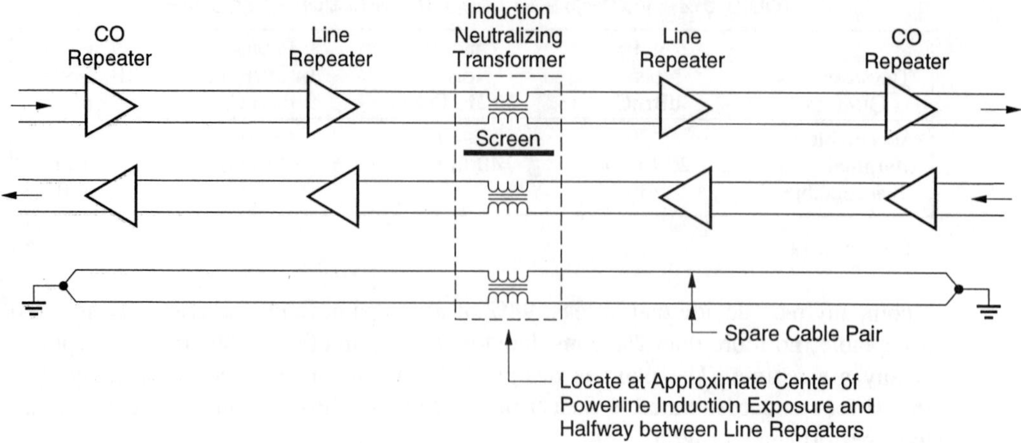

Figure 5-10 Application of induction neutralizing transformers

Table 5-3 Insertion Loss and dc Resistance of Typical Induction Neutralizing Transformers for Repeatered T1-Carrier Applications

	Insertion Loss for Various Signal Levels at Transformer with Respect to Repeater Output		
Frequency	0 dB	−12 dB	−24 dB
0.772 kHz	2.6 dB	2.5 dB	2.5 dB
1.544 MHz	2.9 dB	3.9 dB	3.6 dB
2.370 MHz	4.8 dB	4.9 dB	4.0 dB
3.152 MHz	6.0 dB	5.0 dB	4.5 dB
dc Resistance		10 Ω	

powerline induction and equidistant from the repeaters to reduce reflections caused by impedance discontinuities. Further information on the application of induction neutralizing transformers can be obtained from the manufacturers (for example, [16]).

In order to be effective, the process of solving inductive interference problems must be a cooperative effort between telecommunications and electric utilities serving the area. Standards and procedures have been developed for an *inductive coordination* process [17,18].

5.2 Crosstalk

5.2.1 Spectrum Management

The following discussion of spectrum management on twisted pair cables leads into an analysis of crosstalk and indicates some of the difficulties in designing digital loop systems that are free of mutual interference. In this context, spectrum management on optical fiber cables is not an issue because these are immune from crosstalk effects.

Digital loop spectrum management is the process of controlling the mutual interfering effects of multiple signal sources sharing a common twisted pair cable sheath. The existing cable plant was originally designed and built for baseband analog voicegrade loops. These loops operate in the hertz to low kilohertz range where, for a properly maintained plant, crosstalk interference is negligible. Considerable numbers of analog subscriber carrier systems, which are passband systems, occupy the existing loop plant.

Early digital systems that shared cable sheaths with baseband and passband analog systems were carefully designed to minimize the interfering effect of the transmitted digital signals. This design included limiting signal power levels, filtering to shape the transmitted spectrum, qualification testing of prospective pairs in a cable, and physical segregation to reduce crosstalk. SRDLs are an example of a loop requiring these detailed design considerations. The very high performance and compatibility of SRDLs comes at high cost, both initially and operationally.

The crosstalk coupling function of the imbedded loop plant has been studied extensively, and some of the specific results are discussed in the following sections [19–21,23]. Generally, the crosstalk coupling loss between cable pairs is a function of frequency, with higher coupling loss at lower frequencies. It follows that the transmitted signal spectrum should be centered at the lowest frequency possible and, therefore, be subject to the highest crosstalk coupling loss possible. This is, in fact, done with all current digital loop systems.

A transmitted signal also can cause crosstalk interference by its shape and power level. It is obvious that signals with higher power levels carry with them higher chances for interference on other systems. Also, some transmitted pulse shapes can cause more interference than others. Therefore, all digital loop systems have some constraint on the transmitted power level and signal shape. For all current digital loop systems, these constraints are in the form of pulse templates and maximum pulse voltage specifications.

5.2.2 Manifestations and Types of Crosstalk

Crosstalk manifests itself as errors on digital loops. The types of crosstalk of most concern are near-end crosstalk (NEXT) and far-end crosstalk (FEXT). These are conceptually illustrated in Figure 5-11. With NEXT, shown in Figure 5-11(a), the high-level outputs from digital line transmitters are coupled to the inputs of adjacent digital line receivers through nearby coupling paths.

A different situation exists for FEXT, also shown in Figure 5-11(a), in which the high-level outputs from a line transmitter are coupled into the inputs of line receivers at a different location. Here the coupling paths are all along the same cable section between the two locations.

Figure 5-11(b) shows a route junction where two digital loops in separate cables (cables 1 and 2) converge. If the levels are different at the junction, one of the line receivers will be subjected to higher and possibly destructive crosstalk. Such a situation is avoided by proper design.

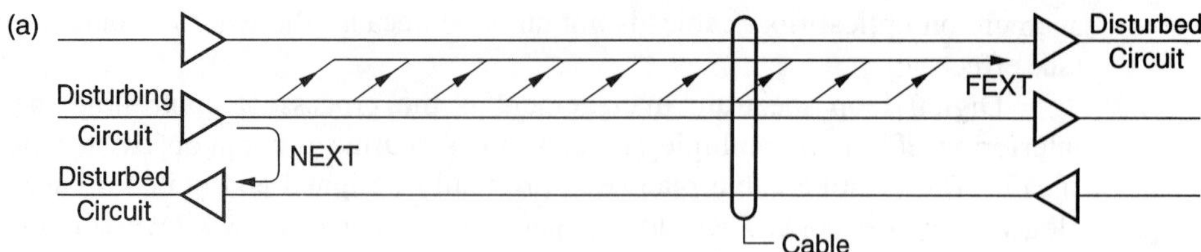

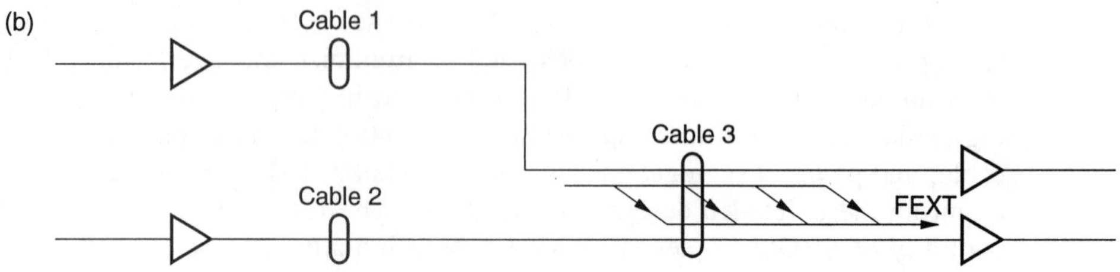

Figure 5-11 Types of crosstalk coupling: (a) NEXT and FEXT in a single cable; (b) FEXT at a route junction

The crosstalk coupling loss in twisted pair cables depends on many factors such as construction, including twist length, lay (physical pattern relationship between pairs), and pair symmetry. The shielding effect of intervening pairs, shields, or core separators (screens) reduce crosstalk coupling. This explains why the crosstalk coupling loss between pairs in the same binder group or unit is lower than between adjacent or nonadjacent groups or units, where physical separation is greater, on average, between the pairs in question. Crosstalk coupling loss due to pair-to-shield capacitance unbalance is lower in smaller cables due to physical constraints in cable constructions.

The coupling loss is highly irregular with frequency, with variations of 15 dB in small frequency bands. Figure 5-12 shows the NEXT loss versus frequency between two pairs in a typical exchange cable. In a cable composed of many twisted pairs, the number of sources and coupling paths is very large. As would be expected for this situation, the coupling loss is a statistical quantity with an assumed normal distribution. For any given pair, the actual coupling loss will be different for each of the other pairs and will have a mean value and standard deviation. Also, due to manufacturing methods, the coupling loss may vary from one reel of cable to the next.

A crosstalk measurement on one pair of a multipair cable is useless. Cable data sheets usually quote a single, statistically worse-case value for crosstalk loss that can be expected in that cable construction. It is recognized that the mean coupling loss between pairs in a given unit or binder group will be less than between pairs in adjacent units or nonadjacent units.

The crosstalk coupling mechanism is idealized in Figure 5-13, in which crosstalk currents i_{cu} and i_{mi} impinge on the disturbed circuit through capacitance unbalances (cu) and mutual inductance (mi). Each circuit shown can be considered a sep-

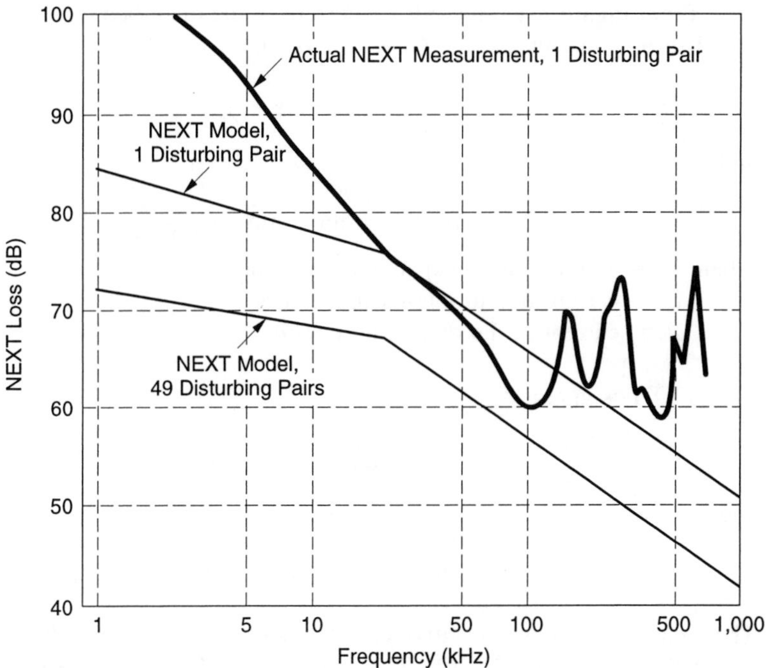

Figure 5-12 Typical NEXT and 1% NEXT loss model, [22] Lechleider, © 1989 Institute of Electrical and Electronics Engineers. Reprinted with permission

arate twisted pair inside a common cable sheath. Only two pairs are shown. In multipair cables, there is a certain level of coupling from every conductor to every other conductor. Crosstalk coupling cannot be eliminated, but it can be reduced by:

- Using balanced terminations
- Twisting the two conductors of a pair and using different twist lengths in a pair group
- Keeping capacitance unbalances to a minimum during cable manufacture

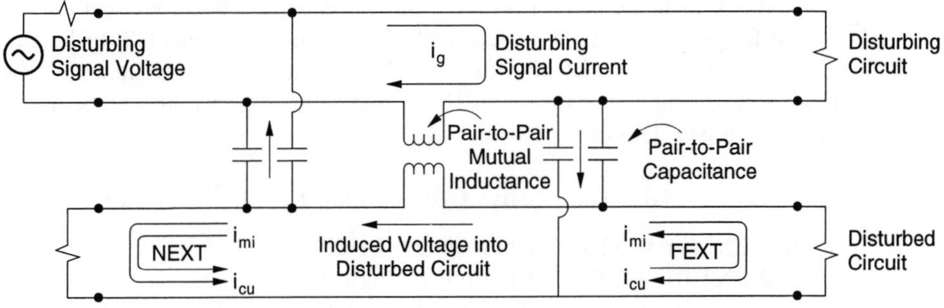

Figure 5-13 Idealized crosstalk coupling

From Figure 5-13, the NEXT current in the disturbed circuit due to current in the disturbing circuit and the NEXT ratio are:

$$i_{NEXT} = i_{cu} + i_{mi}$$

and

$$\text{NEXT current ratio} = i_{NEXT}/i_g$$

where i_{cu} = current due to capacitive unbalance
i_{mi} = current due to mutual inductance
i_g = current in the disturbing circuit

FEXT is given by

$$i_{FEXT} = i_{cu} - i_{mi}$$

and

$$\text{FEXT current ratio} = i_{FEXT}/i_g$$

These equations show capacitive crosstalk and inductive crosstalk oppose each other for FEXT and add for NEXT. The actual levels depend on the relative magnitudes of each component and other circuit characteristics, but for most situations, NEXT is more of a problem than FEXT. This has been verified by field measurements and is taken into account during the design of digital loop systems.

When NEXT is quoted on data sheets, it usually is given as a mean value, m, and standard deviation, σ, both in dB. With this information, NEXT can be determined with 99% confidence from the following:

$$\text{NEXT}(99\%) = m - 2.33 \times \sigma \text{ dB}$$

This provides a NEXT loss value that accounts for approximately 99% of all measurements, and there is only a 1% probability that a pair combination will have worse NEXT. Data sheets frequently show a value based on $(m - \sigma)$. In this case, the value given will account for 84% of all measurements.

NEXT measurements normally are made on a 1,000-foot section of cable at several frequencies. NEXT is sensitive to frequency and sensitive to length in short sections. For lengths that provide ≥ 10 dB loss at the test frequency, NEXT is independent of length. For lengths <10 dB, the following formula can be used to adjust NEXT to a new length:

$$\text{NEXT}(x) = \text{NEXT}(0) - 10 \log\{1 - 10^{-2EL(l_x/10)}\} \text{ dB}$$

where NEXT(x) = NEXT at new length l_x, dB
NEXT(0) = NEXT at reference length l_0, dB
EL = engineering loss of cable pairs in dB per unit length
l_x = length in compatible units

The correction factor part of this equation is plotted in Figure 5-14, with the increase in NEXT coupling loss in dB along the y-axis and equivalent length in dB of loss (that is, EL $\times$ l_x) along the x-axis.

When NEXT is known for one cable gauge, it can be approximated for another gauge of a *similar* cable type by the following formula:

$$\text{NEXT}(x) = \text{NEXT}(0) + 10 \log(\alpha_x/\alpha_0) \text{ dB}$$

where NEXT(x) = NEXT for cable gauge in question, dB
NEXT(0) = NEXT for known cable, dB
α_x = attenuation constant for unknown cable gauge, Nepers* per unit length
α_0 = attenuation constant for known cable gauge, Nepers per unit length

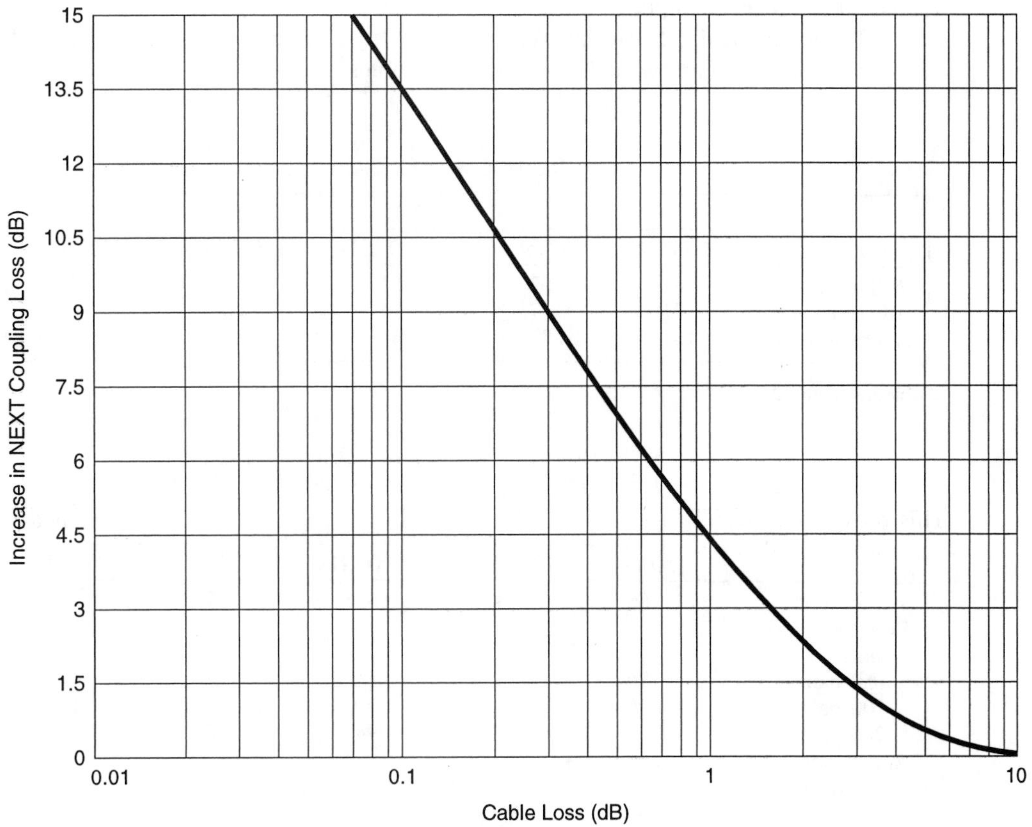

Figure 5-14 NEXT correction for short cables (<10 dB loss)

*Nepers is a dimensionless unit—equal to the natural logarithm of the scalar ratio of two voltages. One Neper = 8.686 dB.

FEXT usually is specified as an rms value of measurements at a test frequency of 150 kHz. The rms value is based on the combined total of all adjacent and alternate pair combinations within the same layer and center to first-layer pair combinations. Once FEXT is determined for a given frequency (f_0) and cable length (l_0), it can be determined for other frequencies (f_x) and lengths (l_x) from the following:

$$\text{FEXT}(x) = \text{FEXT}(0) - 20 \log(f_x/f_0) - 10 \log(l_x/l_0) \text{ dB}$$

where FEXT(x) = FEXT at new frequency f_x or length l_x
FEXT(0) = FEXT at tested frequency f_0 (normally 150 kHz) and length l_0

The effect of FEXT can be controlled by minimizing the level difference between signals traveling in the same direction in the same cable. Said another way, it is important that the received levels are similar at the far-end of a circuit so that crosstalk from the more powerful signal does not interfere with the less powerful signal at the receiver input.

Tables 5-4 and 5-5 can be used as guides in determining FEXT and NEXT objectives for common twisted pair exchange cables.

Table 5-4 FEXT[a]

	FEXT (dB/kft)				
LOCATION	150 kHz	772 kHz	1,000 kHz	1,576 kHz	3,156 kHz
Within unit	81	67	64	60	54
Adjacent unit	91	77	74	70	64
Alternate unit	101	87	84	80	74

[a]Source data: [24]

Table 5-5 NEXT ($m - \sigma$) (assumes cable loss >10 dB)[a]

	M − σ (dB)				
LOCATION	150 kHz	772 kHz	1,000 kHz	1,576 kHz	3,156 kHz
Within unit	66	64	61	57	53
Between opposite units					
50 pairs	86	84	81	77	72
100 pairs	95	93	90	86	82
200 pairs	99	97	94	90	86
Across core separator		Compartmental Core Cables			
25 pairs	96	95	94	91	89
50 pairs	100	99	98	95	93
100 pairs	105	104	103	100	98
200 pairs	111	110	109	106	104

[a]Source data: [24]

EXAMPLE 5-3

Crosstalk is a critical parameter in twisted pair digital transmission, especially at higher frequencies. Therefore, for SRDLs, crosstalk is more severe at a 56 kbps transmission rate than at a 9.6 kbps rate. The worse-case crosstalk in twisted pair cables is NEXT that occurs in pairs within the same unit binder group.

The following is known: (a) tests on a large number of cables pairs show that at 28 kHz (the Nyquist frequency for a 56 kbps SRDL), there is only a 1% chance of NEXT loss being less than 72 dB; (b) the upper bound on signal power expected from unsynchronized 56 kbps loops is +6 dBm, and the signal power expected from synchronized 56 kbps loops is +12 dBm; (c) the decision level referenced to the equalized receiver input (that is, at the loop interface) is −28 dBm; and (d) a decision level-to-crosstalk ratio of 18 dB is needed for a 1E-10 error rate.

Find (1) the expected crosstalk level and signal-to-crosstalk ratio for the unsynchronized and synchronized loops; (2) the operating margin for the synchronized loops if the desired error rate is no worse than 1E-10.

1. For the unsynchronized loops, the expected crosstalk level would exceed +6 dBm − 72 dB = −66 dBm about 1% of the time and the signal-to-crosstalk ratio would be −28 dBm − (−66 dBm) = 38 dB. For the synchronized case, the disturbing power level would be +12 dBm − 72 dB = −60 dBm and the signal-to-crosstalk ratio would be −28 dBm − (−60 dBm) = 32 dB.
2. The margin for the synchronized loops is 32 dB − 18 dB = 14 dB.

5.3 Impulse Noise

Impulse noise (or *impulse event*) is a transient voltage disturbance above some specified threshold separated in time by quiet intervals. The amplitude of the impulse noise bursts or spikes are much higher than the background noise. Impulse noise is characterized by:

- Amplitude
- Duration
- Rate of occurrence or interval between impulses
- Spectrum

Impulse noise is not as disturbing to human listeners as it is to digital receivers used in the subscriber loop. To achieve specified error rates with some systems (for example, repeatered T1-carrier end-sections), the length of loops between regenerators is limited to increase the signal-to-impulse noise ratio. Impulse noise with sufficient magnitude opposite a valid signal pulse will completely mask it, and the line receiver will assume no pulse was present. Similarly, a no-pulse condition can be changed to an invalid pulse. Each such event will result in at least one bit error. Impulse noise is "bursty," so a long impulse compared to the bit interval or a string of impulses could cause many errors.

Much of the impulse noise directly imposed on a loop is thought to be caused by the opening and closing of relay contacts in central offices, telephone set rotary dials, and motor starting and stopping. Other sources are lightning, power system switching activity, and line slaps, and possibly maintenance activity on distribution frames and in outside plant terminals.

Ideally, switched analog loops should be in different cables from digital loops, or at least in physically separated binder groups. Obviously, this is economically impossible where digital loops, such as ISDN lines, are to be widely deployed. Therefore, two of the most serious design issues with digital loops are the effects of impulse noise and crosstalk. A tradeoff is required between the desirability for long loops (to achieve the maximum coverage from the central office) and the acceptable error rate. It follows that, in an impulse noise environment, the error rate can be reduced by using shorter loops.

EXAMPLE 5-4

A string of three impulses is coupled into the input of a repeater of a 1.544 Mbps digital loop carrying a signal with an average ones-density of 70%. Each impulse lasts 100 μseconds. Impulses are spaced 50 mseconds apart and have sufficient amplitude to change a binary zero to a binary one. Assume binary one values are not affected. Find the number of bit errors and the bit error rate for the (1) 1-second and (2) 1-hour interval surrounding the impulse burst.

1. The bit interval of a 1.544 Mbps signal is 1/1.544E06 = 0.65 μsecond/bit. In each 100 μsecond impulse interval, there are 100/0.65 = 153 bits. An average of 30%, or 46 samples, will be binary zero. If each binary zero is changed by the impulse, then there will be 3 × 46 = 138 errors total in the 1-second interval. The bit error rate in the 1-second interval is 138/1.544E06 = 9E-05.
2. In a 1-hour interval, there are still only three impulse bursts, which cause 138 bit errors. The error rate is 138/(1.544E06 × 3,600) = 2E-08.

On a typical twisted pair loop, impulse events with amplitudes in the 5 to 10 mV range occur about one to five times per minute, although excursions in the 10 to 40 mV range occurring ten times per minute have been recorded [25]. A large-scale impulse noise study was conducted in Europe, with the following results [26]:

- Impulse noise activity depends on the time of day, as shown in the histogram of Figure 5-15
- The probability that a particular voltage threshold is exceeded is shown in Figure 5-16
- The pulse interval distribution is shown in Figure 5-17
- The spectral density spectrum is shown in Figure 5-18

Impulse noise has a wide bandwidth, as indicated by the spectral density plot shown in Figure 5-18. The total impulse noise power in a band of frequencies can be found by numerical integration or calculating the area under the curve. Approximately one half of the power is found below approximately 40 kHz. Although there

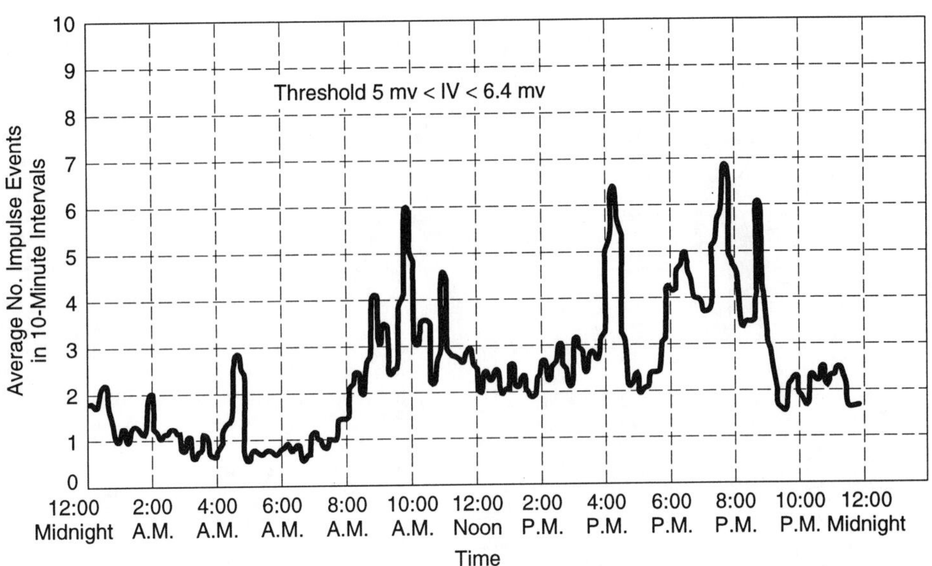

Figure 5-15 Average number of impulse noise events in a ten-minute interval, redrawn from Figure 1 of [26]
Szechenyi, © 1988 Institute of Electrical and Electronics Engineers. Reprinted with permission

are large amplitude peaks above 40 kHz (especially in the 105 to 160 kHz range), the contribution these noise frequencies make is not as significant. Studies in the United States tend to corroborate the European results [25,27].

The European study did not address impulse durations. However, [27] gives durations in the range of 48 to 1,015 μseconds and [28] gives a range of 30 to 150 μseconds.

EXAMPLE 5-5

Find the probability that a 10 mV or greater impulse event will occur on a loop. Assume a 10 mV impulse causes an error on a DS-1 rate digital transmission system. Find the expected interval of 10 mV impulse events and the error rate resulting from it.

From Figure 5-16, $P(V) = (5/V)^2 = (5/10)^2 = 0.25$. Therefore, the threshold of 10 mV will be exceeded 25% of the time. From Figure 5-17, a 10 mV impulse event will occur about every 50 seconds (worse case). In 50 seconds, the transmission system will have passed about 77 million bits, of which one will be in error. Therefore, the error rate is 1/77E6 = 1.3E-8.

Impulse noise counters typically have three separate storage registers, or counters, with thresholds separated by 4 dB (some have selectable thresholds with 1, 2, 4, and 6 dB separation). The maximum counting rate is 8 counts/second (older sets have a 7 counts/second maximum rate). At 8 counts/second, the counter will increment only once in each 125 msecond interval. The 125-msecond blanking (or

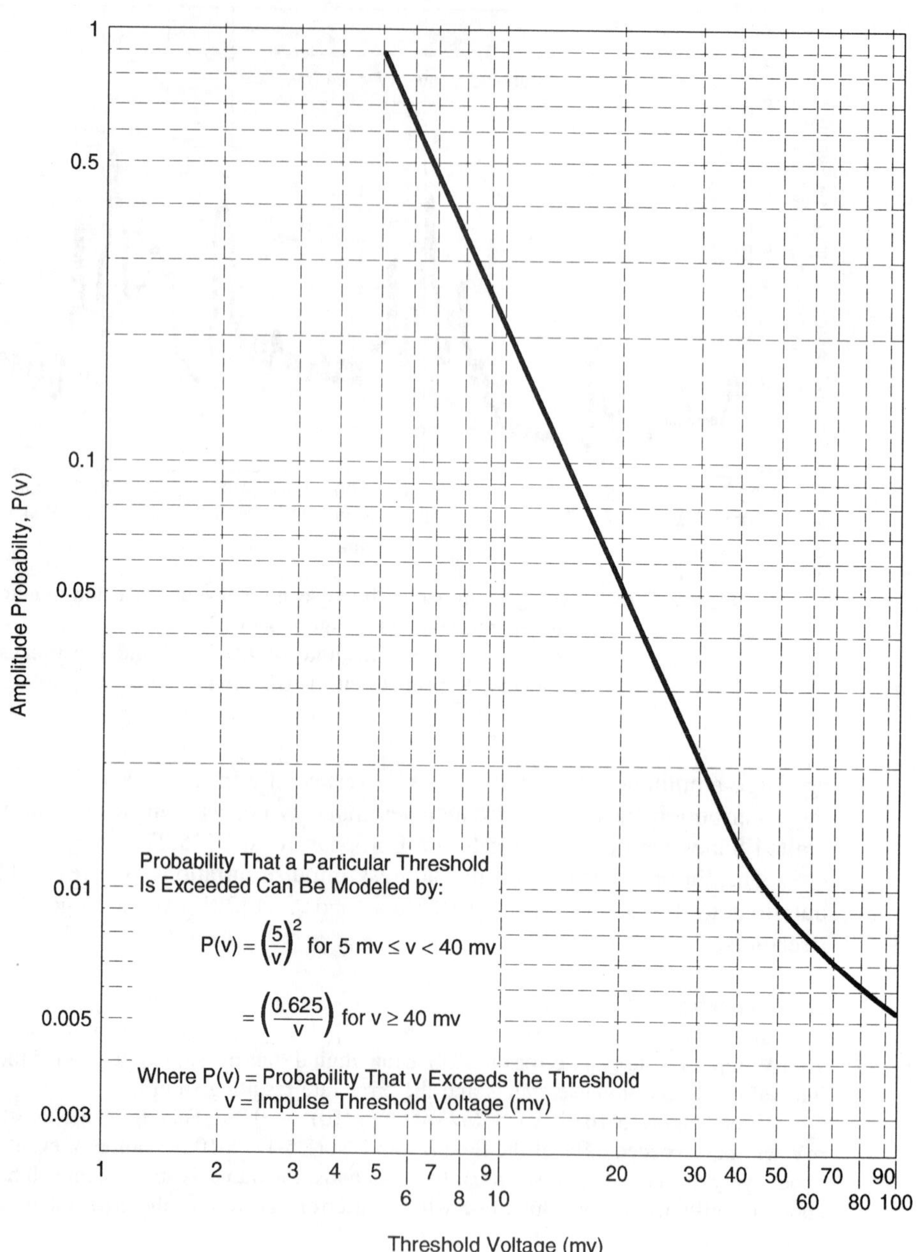

Figure 5-16 Amplitude probability distribution of noise events, redrawn from Figure 2 of [26]
Szechenyi, © 1988 Institute of Electrical and Electronics Engineers. Reprinted with permission

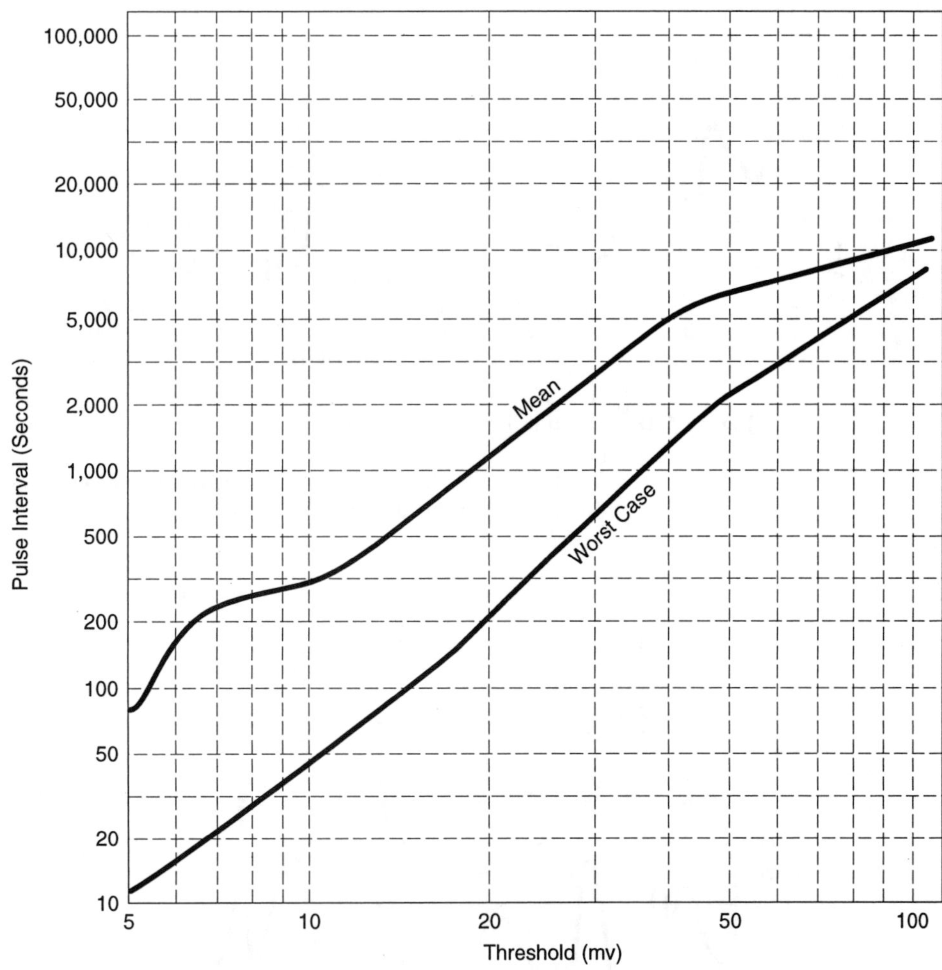

Figure 5-17 Pulse interval distribution of noise events, redrawn from Figure 3 of [26]
Szechenyi, © 1988 Institute of Electrical and Electronics Engineers. Reprinted with permission

"dead-time") interval of each register is independent of the other two registers. Modern test sets have adjustable blanking interval.

With a 125-msecond blanking interval, an impulse exceeding the low threshold, followed 100 mseconds later by another, will be counted as only one low event. An impulse event exceeding the low threshold followed 50 mseconds later by one exceeding the middle threshold will be counted as one low and one middle. An impulse exceeding the low threshold followed 100 mseconds later by one exceeding the middle threshold, followed 50 mseconds later by one exceeding the low threshold, will be counted as two low and one middle. Figure 5-19 shows the various aspects of impulse noise measurements.

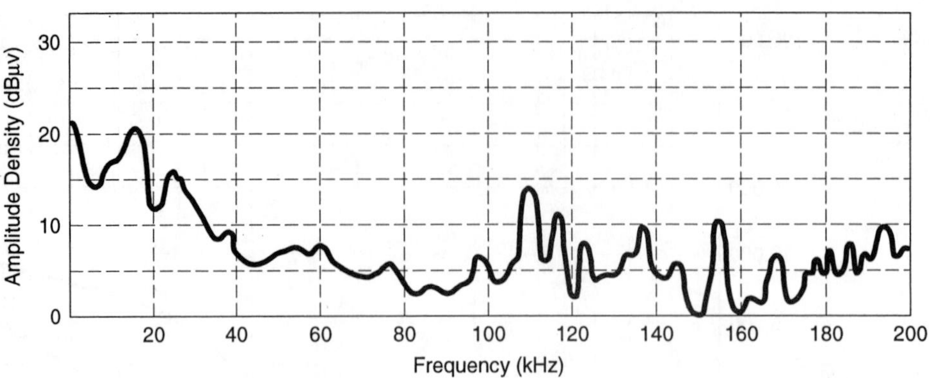

Figure 5-18 Spectral density of noise events, redrawn from Figure 4 of [26]
Szechenyi, © 1988 Institute of Electrical and Electronics Engineers. Reprinted with permission

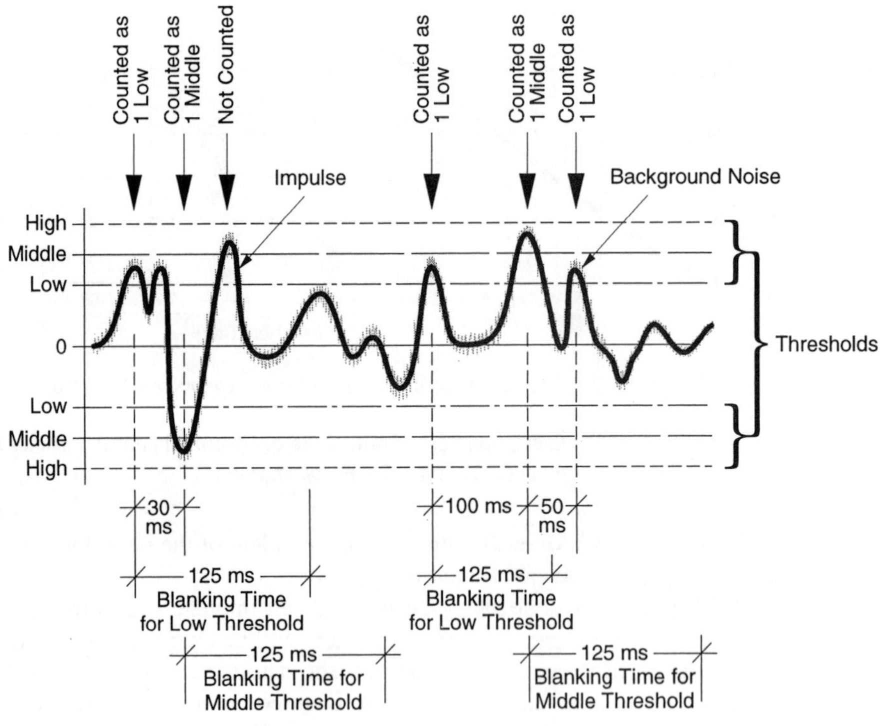

Figure 5-19 Impulse noise counting parameters

Impulse noise occurs in clusters or bursts, many of which last less than 125 mseconds. An instrument with a maximum counting rate of 8 counts/second will obviously miss some impulses in any test interval. In tests performed within the Bell System, the missed counts varied from 0 to as many as 120 counts in a 30-minute interval [29]. Such variability precludes using a simple correction factor with impulse noise measurements. Rather, the stated objectives for the control of impulse noise are designed to account for missed counts.

It has been found that impulse noise counts, as measured on an impulse noise counter, are proportional to the time interval for intervals greater than about 15 minutes, and for every 10 dB decrease in threshold, the number of counts will increase by a factor of ten [30]. This is summarized in the following equation:

$$C(x) = C(0) \times \{[T(x)/T(0) + 10^{\{[L(0)-L(x)]/10\}}\}$$

where C(x) = new counts
C(0) = reference counts during reference test interval
f(x) = new impulse bandwidth, Hz
f(0) = reference impulse bandwidth, Hz
T(x) = new time interval, minutes (≥15)
T(0) = reference time interval, minutes (≥15)
L(x) = new threshold level, dBm
L(0) = reference threshold level, dBm

EXAMPLE 5-6

Assume the allowable impulse noise counts for a digital loop are four in a 30-minute period, at a threshold of 62 dBrn. Find the allowable noise counts if the threshold is set to 50 dBrn for a 15-minute period.

By decreasing the threshold 12 dB to 50 dBrn, the number of counts increases by a factor $10^{1.2} = 15.8$. However, the measurement interval is one half, which gives one half the counts. Therefore, the allowable counts are $15.8 \times 0.5 \times 4 = 32$ counts in a 15-minute period.

The impulse noise counts are affected by the impulse noise bandwidth of the counter. The narrower the bandwidth, the fewer counts at a given threshold. Figure 5-20 can be used to estimate the relative change in counts for different bandwidths. For example, by reducing the bandwidth from 100 kHz to 30 kHz, the allowable number of counts is reduced by (20 − 2 =) 18 from its value at 100 kHz.

Impulse noise thresholds normally are established for a maximum length loop where the SNR is at a minimum. As the loop length and its loss are decreased, the SNR will rise. The impulse noise threshold may be increased on a dB for dB basis, with the decrease in loop loss. For example, if the maximum allowable loss on a digital loop is 42 dB and the impulse noise threshold is 50 dBrn at this loss, the new threshold for a loop loss of 32 dB would be 60 dBrn.

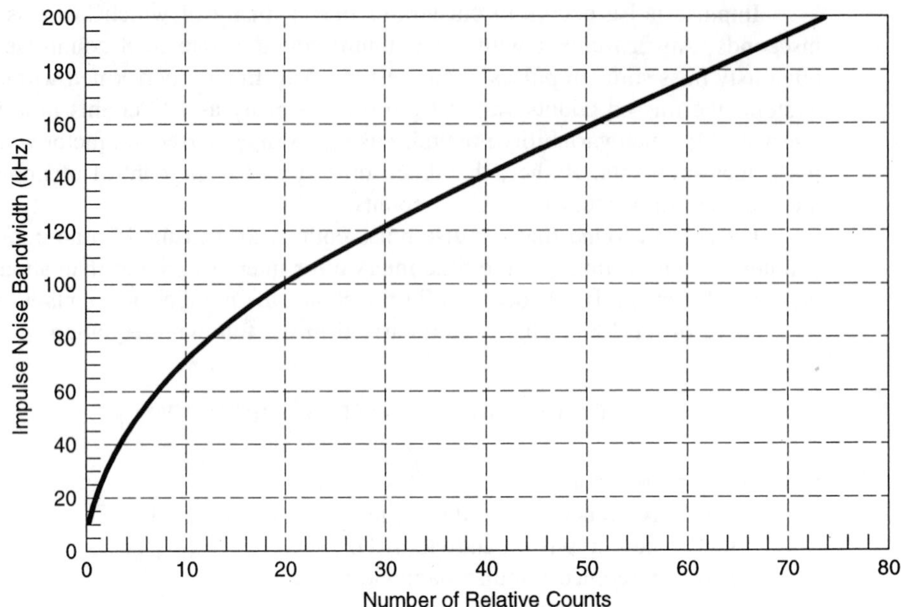

Figure 5-20 Impulse noise bandwidth vs. relative noise counts

5.4 Pulse Detection Errors

Figure 5-21 shows the pulse decoding rules for a typical bipolar alternate mark inversion (AMI) line receiver. If the peak signal amplitude is less than one half the expected received pulse level, the output of the pulse detector is a binary zero. If the signal amplitude is more than one half the expected level, the output is a binary one. It follows that the pulse detection decision point in an ideal regenerator can

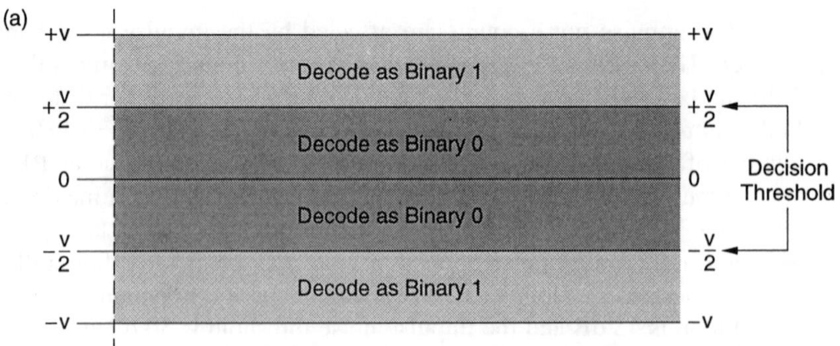

Figure 5-21 Pulse decoding rules: (a) Bipolar pulse decoding; (b) Statistical nature of pulse detection

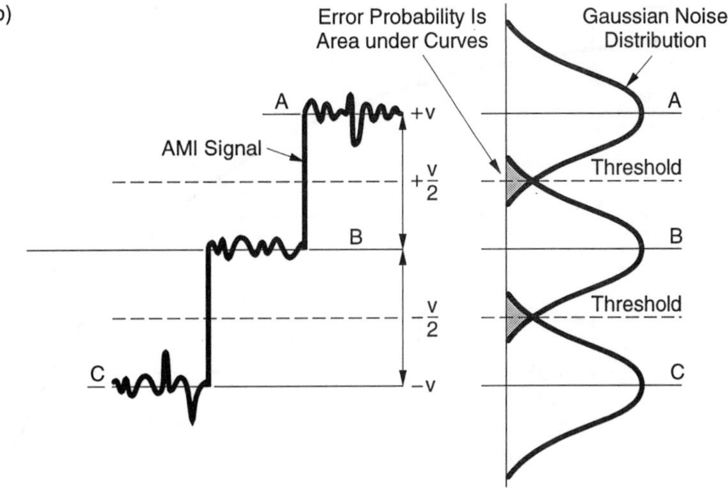

Figure 5-21 Continued

operate with peak noise amplitudes as great as one half the expected level. An error will occur only when the instantaneous noise amplitude at the sampling instant is greater than the decision point value and of the appropriate polarity.

Background noise is assumed to have a Gaussian distribution. Figure 5-21(b) shows how this noise can lead to statistical detection errors for bipolar transmission.

The probability of error for bipolar AMI with a peak signal voltage of V volts is given as [30]:*

$$P(e) = (3/4)\ \mathrm{erfc}\ \{\sqrt{(SNR/2)}\}$$

where P(e) = error probability
erfc(x) = 1 − erf(x), complimentary error function†
SNR = S/N
S = average signal power ($V^2/2$)
N = average noise power (σ^2)

The probability of error versus SNR is plotted in Figure 5-22.

*This probability is one half that given in other literature, which ignore the probability of 1/2 that a noise peak can enhance the signal rather than degrade it.

†Tables to evaluate the error function can be found on page 576 of [32].

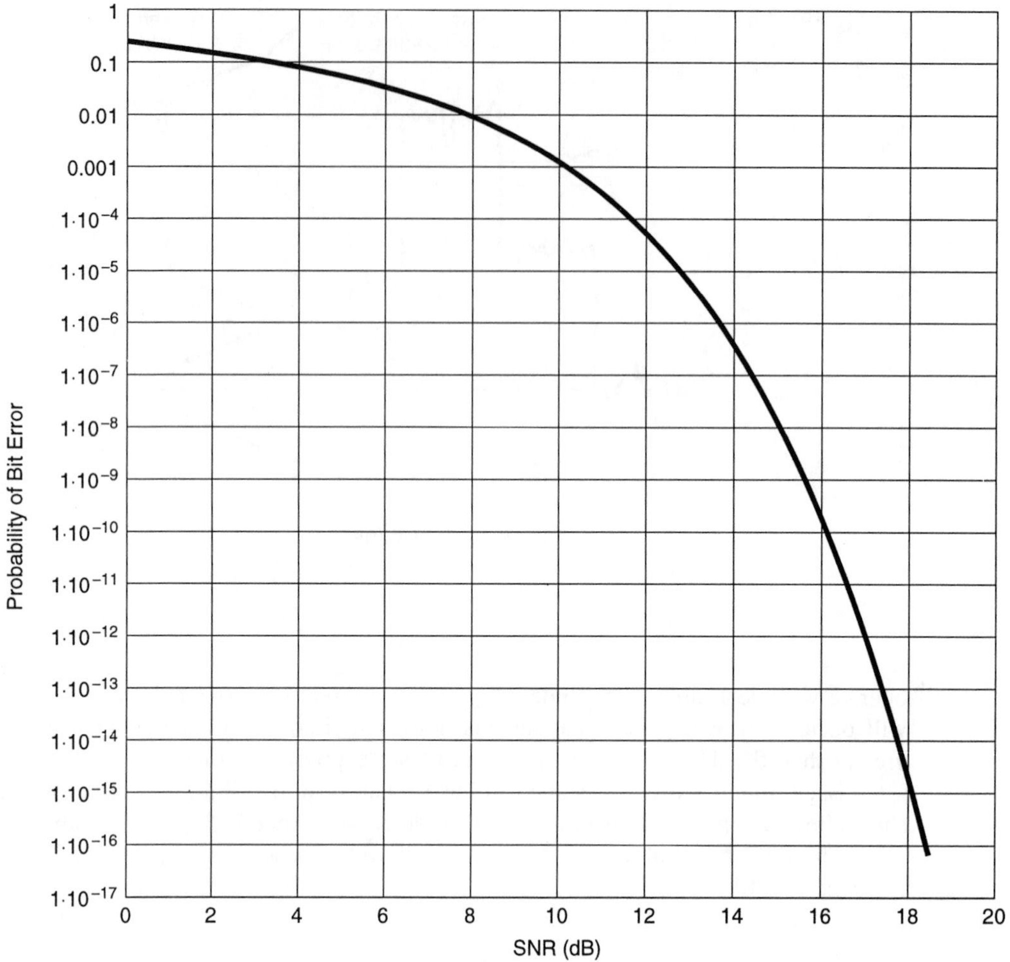

Figure 5-22 Bipolar AMI error probability vs. SNR at decision point

5.5 Distortion

Signals are distorted as they propagate along any real transmission line. Distortions on loops are conveniently categorized into attenuation distortion, impedance distortion, and delay distortion. Each of these are discussed in turn, followed by a brief discussion of equalization, which is used to reduce the negative effects of distortions.

5.5.1 Attenuation Distortion

Attenuation distortion (also called *frequency distortion*) is the departure of a circuit from uniform amplification or attenuation over the frequency range required for transmission [33]. In a twisted pair digital loop, the effect is to attenuate higher frequencies more than lower frequencies, as shown in Figure 5-23.

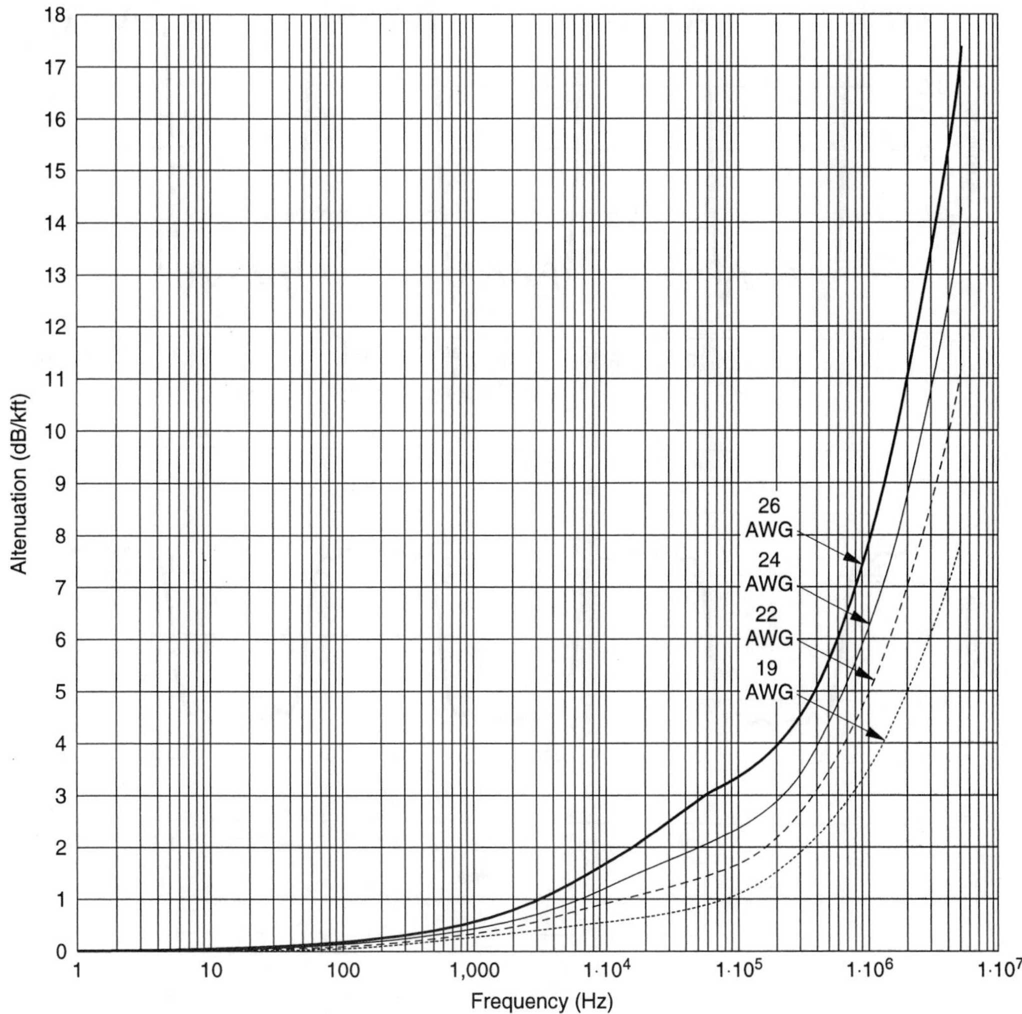

Figure 5-23 Attenuation distortion of typical twisted pair exchange cables, source data from Appendix B

A pulse stream composed of hypothetical pulse shapes is shown in Figure 5-24(a). This shape does not resemble the actual pulse shapes used with digital loop systems and is shown for illustration purposes only. The hypothetical pulses consist of the first five squarewave harmonic frequencies. If the pulse stream is injected into a distortionless transmission medium, only the pulse heights are affected by the medium. The same pulse stream injected into a medium with attenuation distortion, such that higher frequencies are attenuated more than lower frequencies, gives the received pulse shape shown in Figure 5-24(b). The same received pulse stream is shown with nonlinear phase delay in Figure 5-24(c).

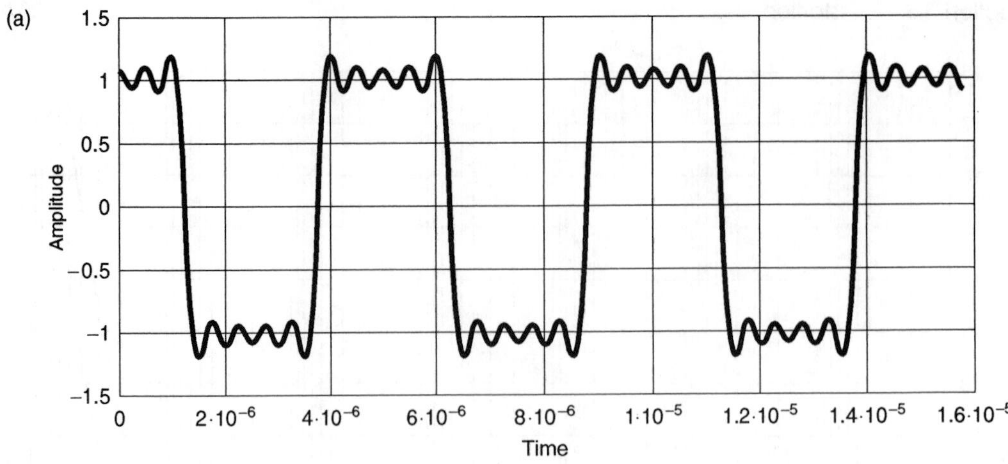

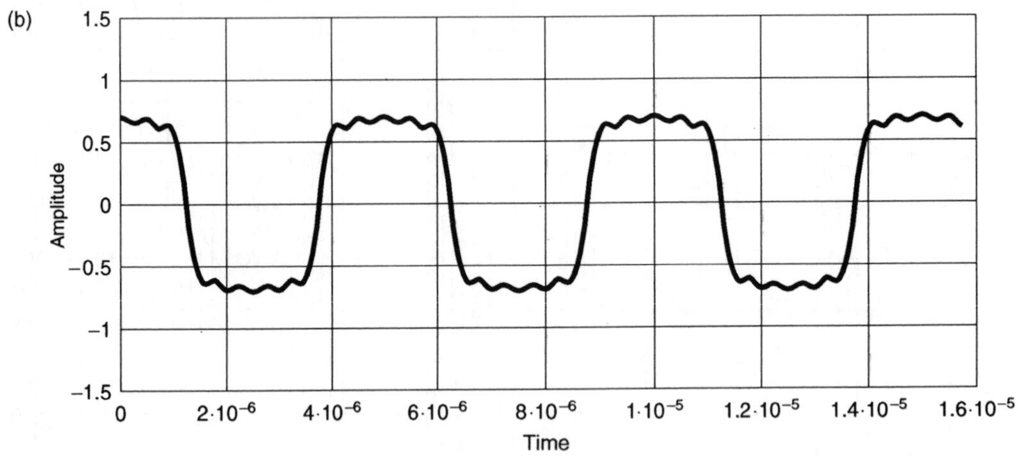

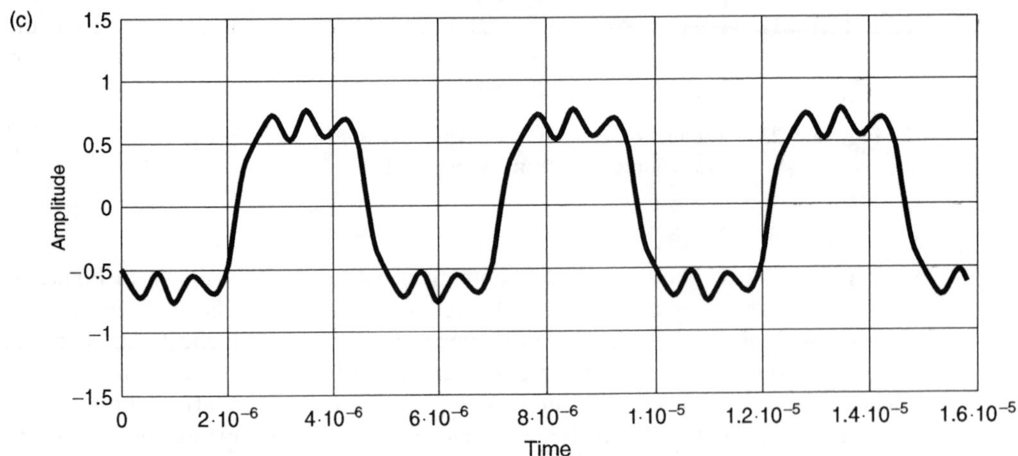

Figure 5-24 Pulse distortion: (a) Waveform with no attenuation or phase delay; (b) Same as (a) with attenuation at each frequency; (c) Same as (b) with nonlinear phase delay at each frequency

To combat attenuation distortion in twisted pair cables, specific line codes, pulse shapes, and equalizers are used. Different pulse shapes will have different spectral characteristics. Pulse shapes used with twisted pair cables are chosen to de-emphasize the importance of higher frequencies. Similarly, line codes used with twisted pair cables are chosen to lower the frequency of signal transmission. For example, a repeatered T1-carrier operates at 1.544 Mbps, but the AMI line code concentrates signal energy in a band centered on 0.772 MHz. Similarly, the 2-binary, 1-quaternary (2B1Q) line code used with the basic-rate ISDN loops lowers the frequency concentration of the 160 kbps source signal to less than 40 kHz.

5.5.2 Impedance Distortion

Impedance distortion is the variation of impedance with frequency over the range of interest. Both the magnitude and angle of twisted-pair-cable characteristic impedance varies with frequency. The impedance magnitude approaches a real value of 100 Ω at around 100 kHz, whereas the impedance angle approaches 0 radians at around 1 MHz.

The input impedance of a loop depends on a combination of the characteristic impedance and termination impedance. The input impedance magnitude and angle of various loops terminated in a 135 Ω resistive impedance is illustrated in Figure 5-25 for exchange cables in the four most common cable gauges.

For maximum power transfer, the loop must be terminated in an impedance that is the complex conjugate of its characteristic impedance. This is difficult to do in practical systems, especially at lower frequencies, where the characteristic impedance has a significant and varying complex component. At higher frequencies, the loop characteristic impedance becomes resistive, and matching is much easier. Any nonmatching termination impedance at one end changes the input impedance to the loop at the other end, further aggravating the matching problem.

Impedance mismatches cause signal reflections, which can lead to two problems. First, signals reflected back to the transmitter output can cause nonlinear distortion in the output circuits, which, in turn, distort the transmitted waveform. The transmitted waveform normally is shaped to minimize interference, so any unplanned distortion could lead to crosstalk problems. Second, reflections introduce *reflection loss*; that is, reflected signal energy is lost to forward transmission and is not available for detection at the line receiver. Digital loops use resistive terminations, as shown in Table 5-6.

Table 5-6 Digital Loop Termination Impedances

Loop Type	Termination Impedance
SRDL	135 Ω resistive
ISDN DSL	135 Ω resistive
Repeatered T1-carrier	100 or 110 Ω resistive
HDSL	135 Ω resistive

HDSL = high bit-rate digital subscriber line

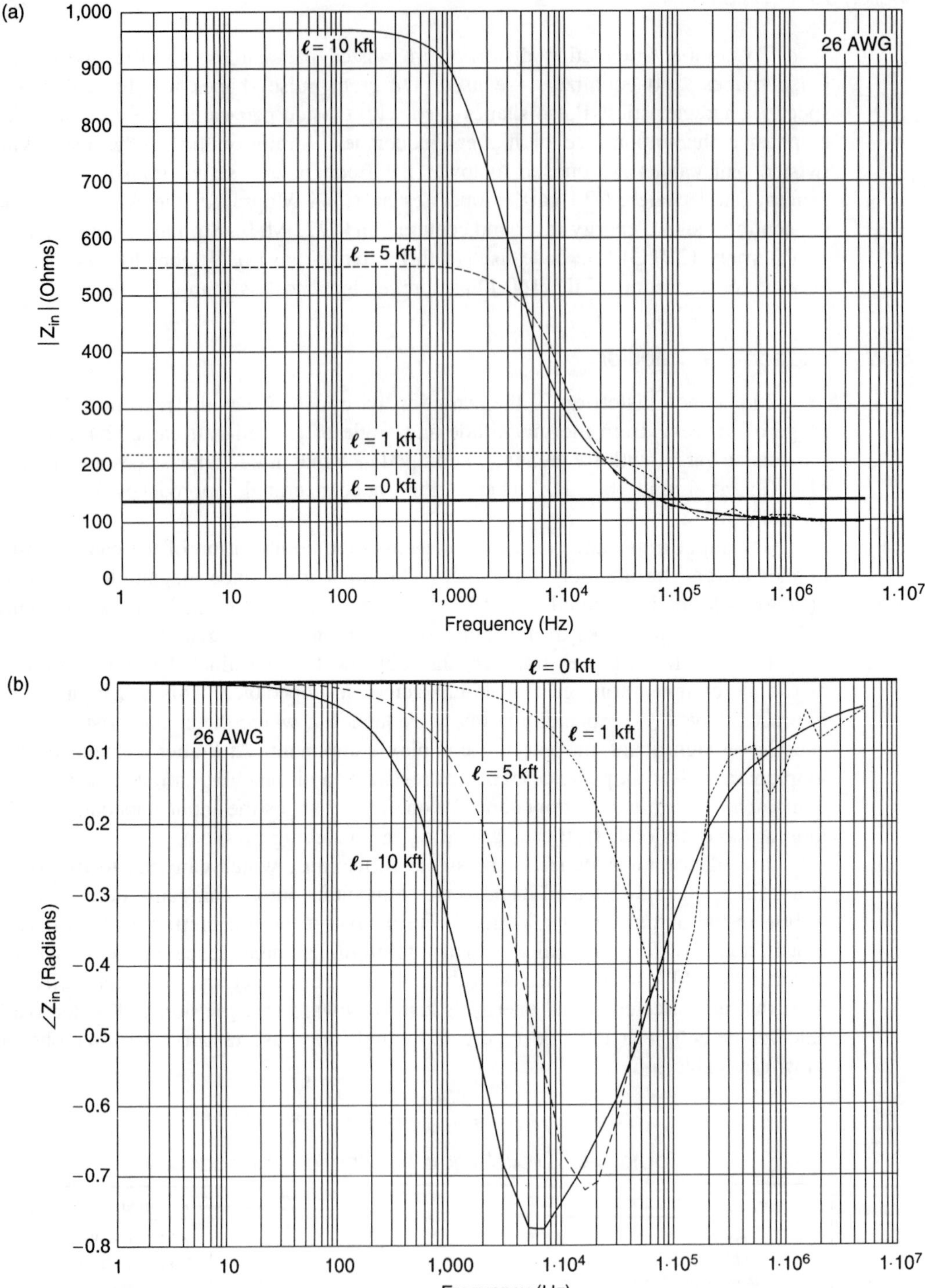

Figure 5-25 Input impedance magnitude and angle of typical twisted pair exchange cables, 135 Ω termination, source data from Appendix B: (a) $|Z|$, 26 American wire gauge (AWG); (b) $\angle Z$, 26 AWG; (c) $|Z|$, 24 AWG; (d) $\angle Z$, 24 AWG; (e) $|Z|$, 22 AWG; (f) $\angle Z$, 22 AWG; (g) $|Z|$, 19 AWG; (h) $\angle Z$, 19 AWG

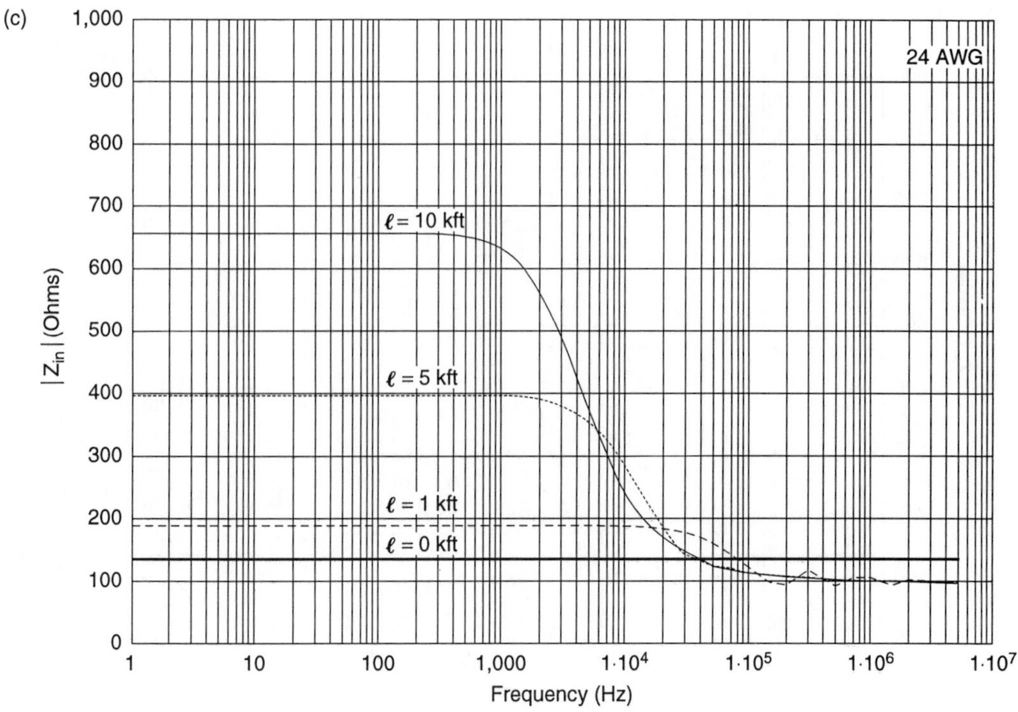

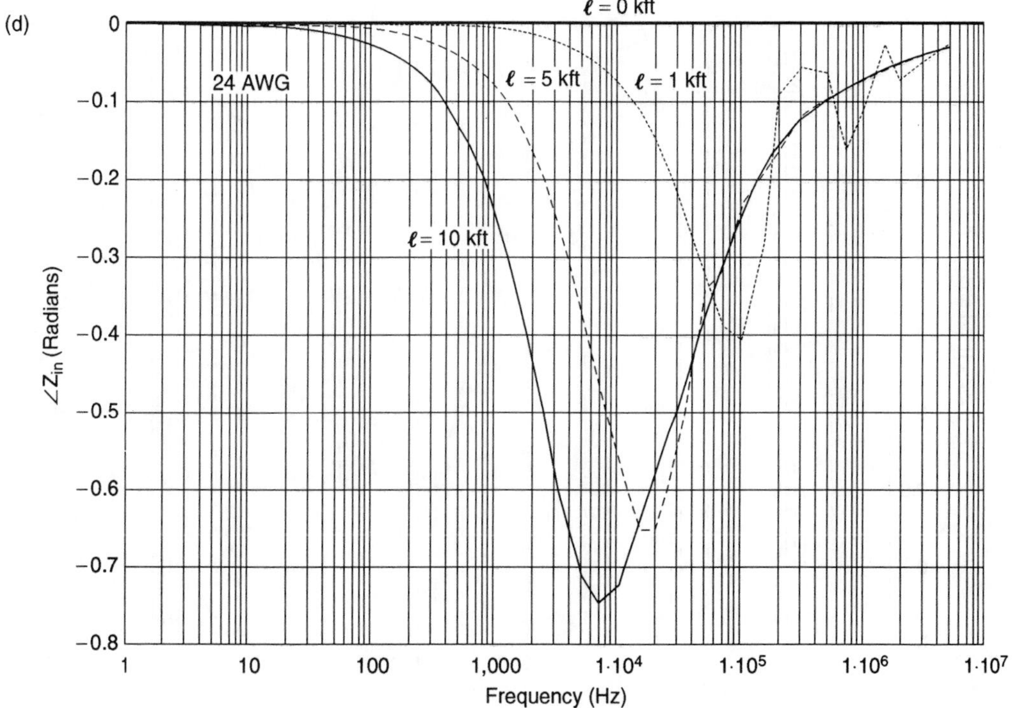

Figure 5-25 Continued

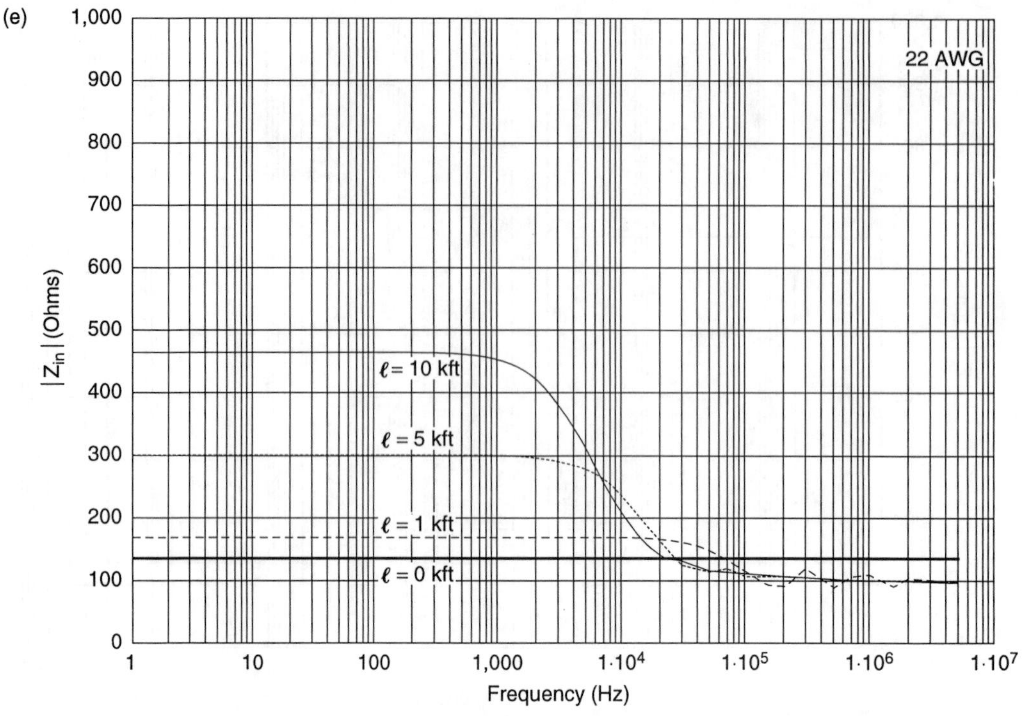

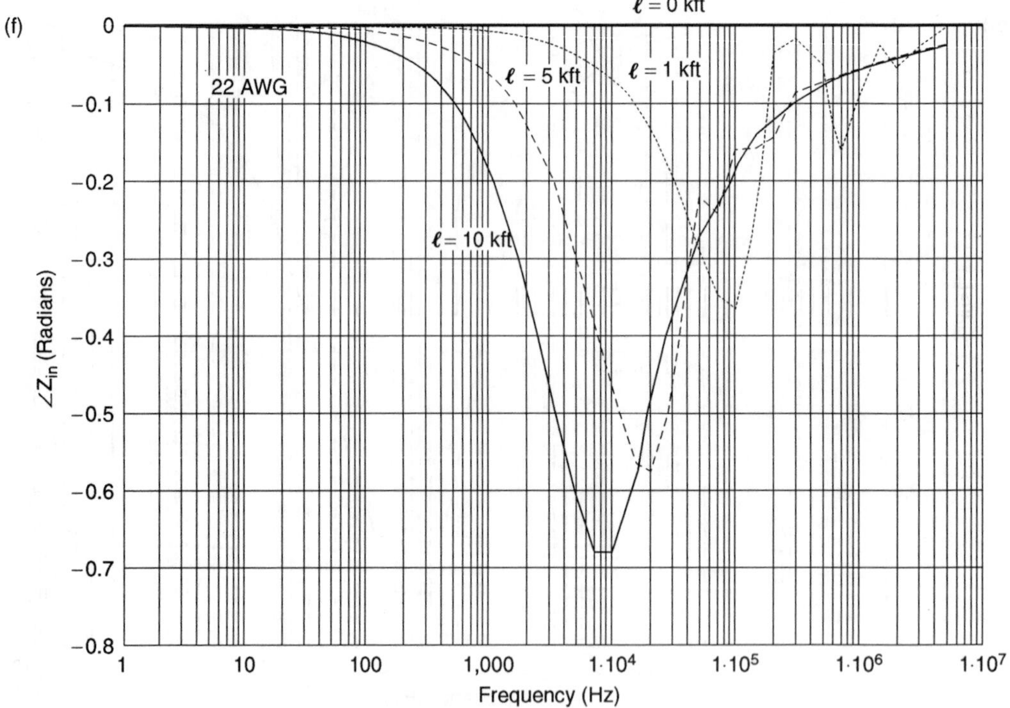

Figure 5-25 Continued

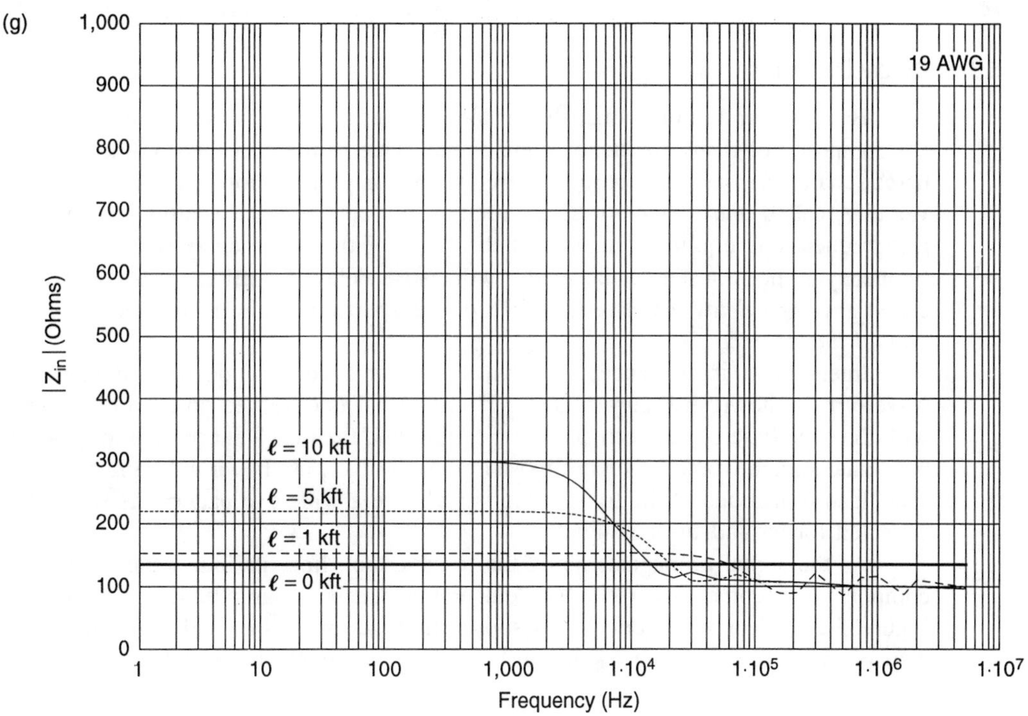

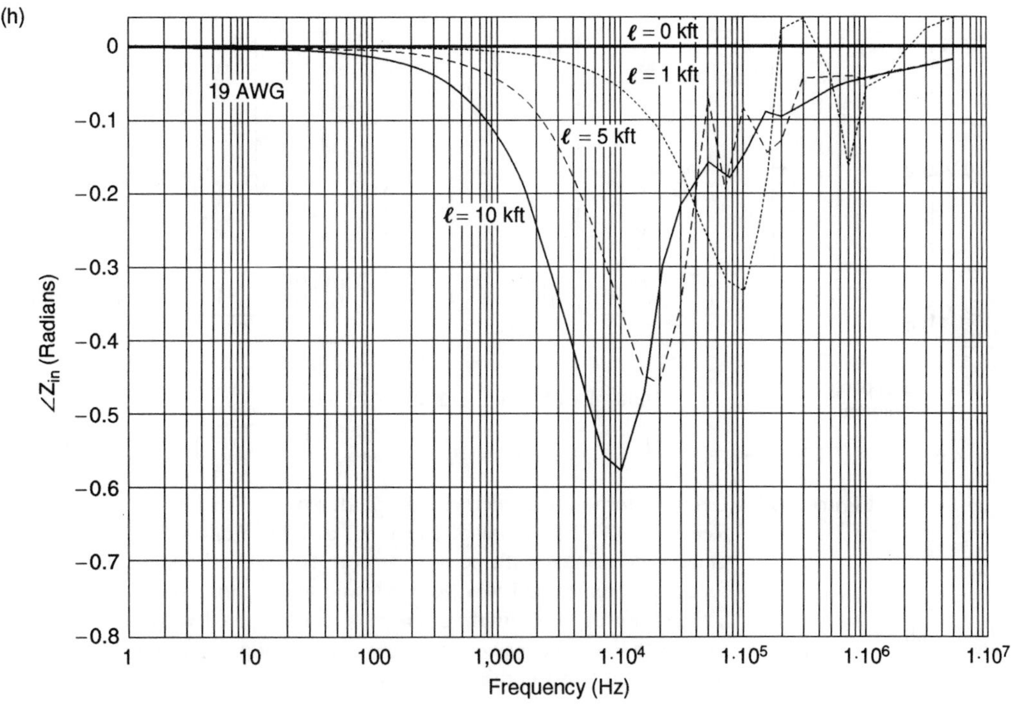

Figure 5-25 Continued

5.5.3 Delay Distortion

Baseband and optical fiber digital loop systems transmit shaped pulses, which are composed of many frequency components. Due to the transmission media's inherent characteristics, each particular frequency component propagates at a different velocity, called *phase velocity*. If the relationship between the resulting phase shift and frequency is nonlinear (that is, if the slope of the phase delay with respect to frequency is not constant), the media will cause delay distortion (also commonly called *envelope delay distortion*, or EDD). Delay distortion is manifested as *intersymbol interference*.

Intersymbol interference is shown in Figure 5-26. An idealized transmitted waveform is shown in Figure 5-26(a). If this signal is observed on an oscilloscope with the time base adjusted to show 2-bit intervals, then each successive interval of the signal stream would be superimposed on each other, as shown in Figure 5-26(b). The clean transitions of the transmitted signal give a wide-eye pattern, as would be expected for an undistorted signal.

As the signal propagates along the transmission line, its individual frequency components become dispersed. The dispersion can be in amplitude, as shown in Figure 5-24(b), or time (delay), as shown in Figures 5-24(c) and 5-26(c). If the dispersion is excessive, the result will be intersymbol interference, whereby the sig-

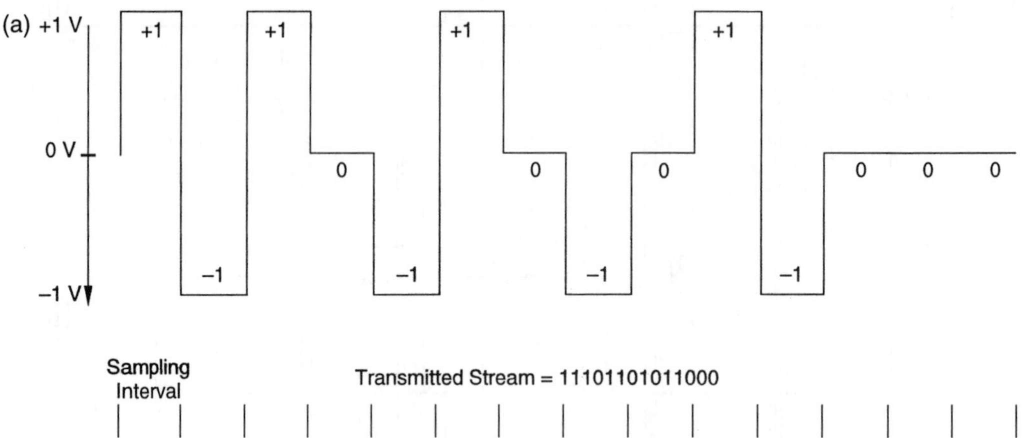

Figure 5-26 Delay distortion and intersymbol interference, [1] Reeve, © 1992 Institute of Electrical and Electronics Engineers. Reprinted with permission: (a) Transmitted waveform; (b) Transmitted waveform viewed by oscilloscope at a time base of two sampling intervals; (c) Received waveform; (d) Received waveform viewed by oscilloscope at a time base of two sampling intervals

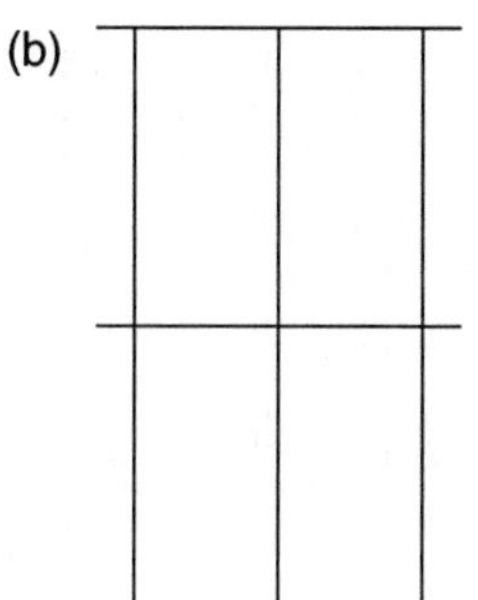

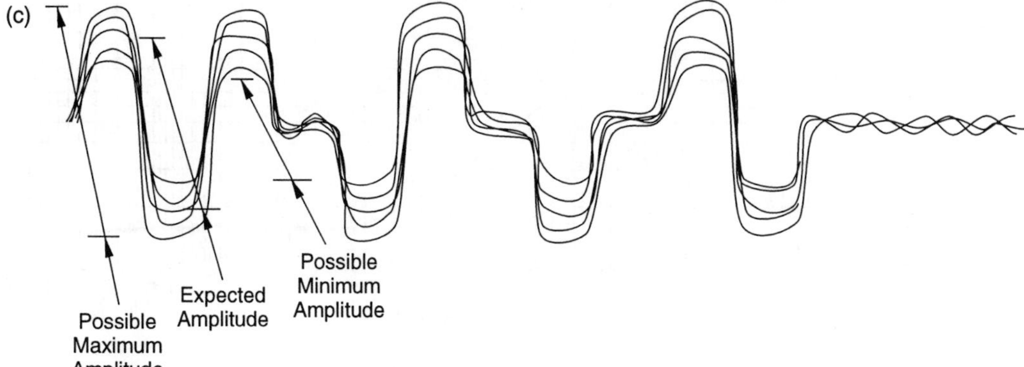

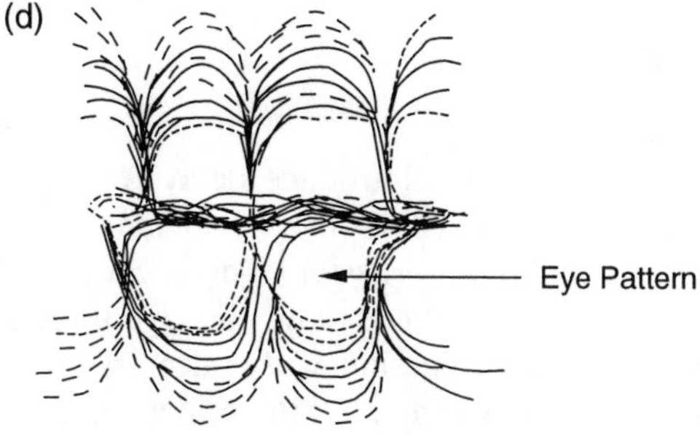

Figure 5-26 Continued

nal energy in a given bit interval is overlapped by signal energy from other bit intervals, thereby degrading it. Excessive intersymbol interference leads to errors in the detection process. The eye pattern for the degraded received signal is shown in Figure 5-26(d). The eye is significantly closed, which indicates delay and amplitude distortion (although, in this case, probably acceptable). Figure 5-27 shows the envelope delay versus frequency for common twisted pair exchange cables.

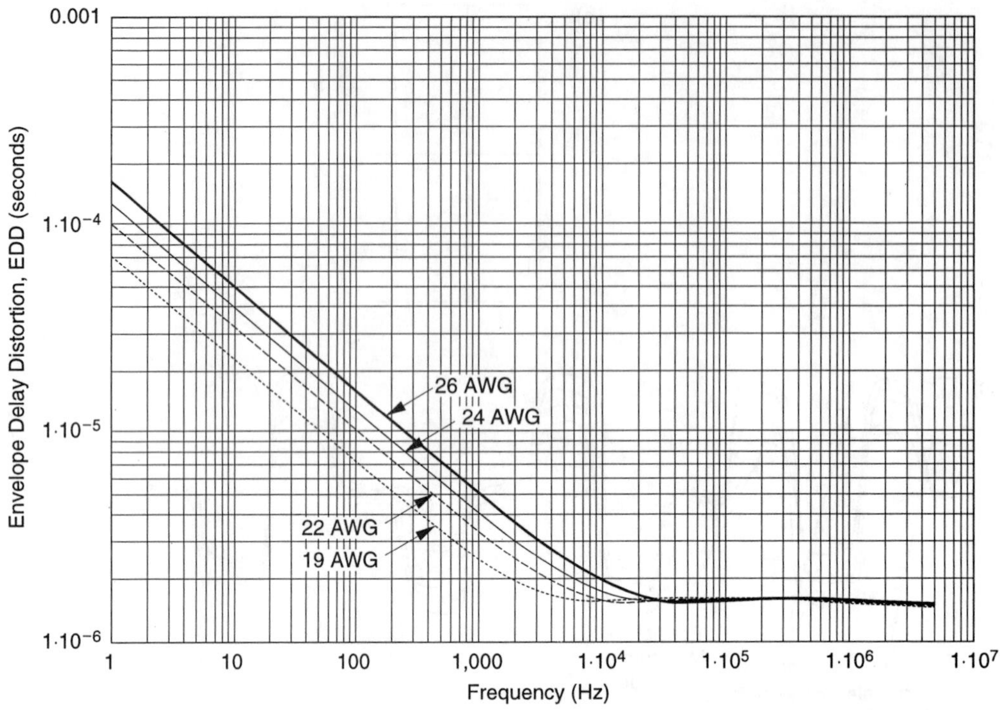

Figure 5-27 Envelope delay for twisted pair exchange cables, source data from Appendix B

5.6 Absolute Delay

All digital transmission facilities introduce delay from propagation and buffering. The delays can affect voice transmissions by making echo more noticeable. Many digital systems depend on the transmission of error check codes and control messages from end-to-end. Excessive delay can slow message throughput or cause equipment malfunction. The following discussion will concentrate on DS-1 rate transmissions as they apply to digital loop systems, but the principles are valid with other systems, as well.

When a DS-1 rate loop connects remote switching terminals (RSTs) with host digital switching systems, propagation delay becomes an important parameter. In order to control the remote system, messages are transmitted between the host and remote location. These messages occupy one or more channels in the bit stream: they may be included in the A- and B-signaling channels, or they may use the facility data link in an extended superframe format (ESF). Since the messages must be received error free, some method of error detection and correction normally is used. One end indicates to the other that a message was received in error, and the message is retransmitted. If the typical message contains 100 data bits and the line error rate is 1E-04, approximately one message in 100 must be repeated. Because the delay introduced may exceed the maximum allowable delay, the call being set up by the delayed message may be mishandled. Typical RSTs are designed to mishandle no more than one call in 50,000 when the line error rate is 1E-04.*

A typical central office error detection protocol will require no more than a 600 μsecond one-way delay. This corresponds to a one-way transmission path length of about 70 miles. Other systems allow no more than a 2 msecond maximum round-trip delay. Assuming this requirement does not include frame buffering at the far-end multiplex equipment, the corresponding maximum one-way transmission path length is about 120 miles. When distributed linear systems are used, as shown in Figure 5-28, the maximum allowable delay is further reduced due to buffering and timeslot interchange in intermediate remote locations.

Listener echo delay is the limiting factor in many systems used for voice transmission. This is illustrated in Figure 5-29, which shows a digital loop carrier (DLC) with a transmission path length of 15 miles. The listener echo path is 45 miles because the voice signal is reflected by the impedance discontinuity in the 2-wire/4-

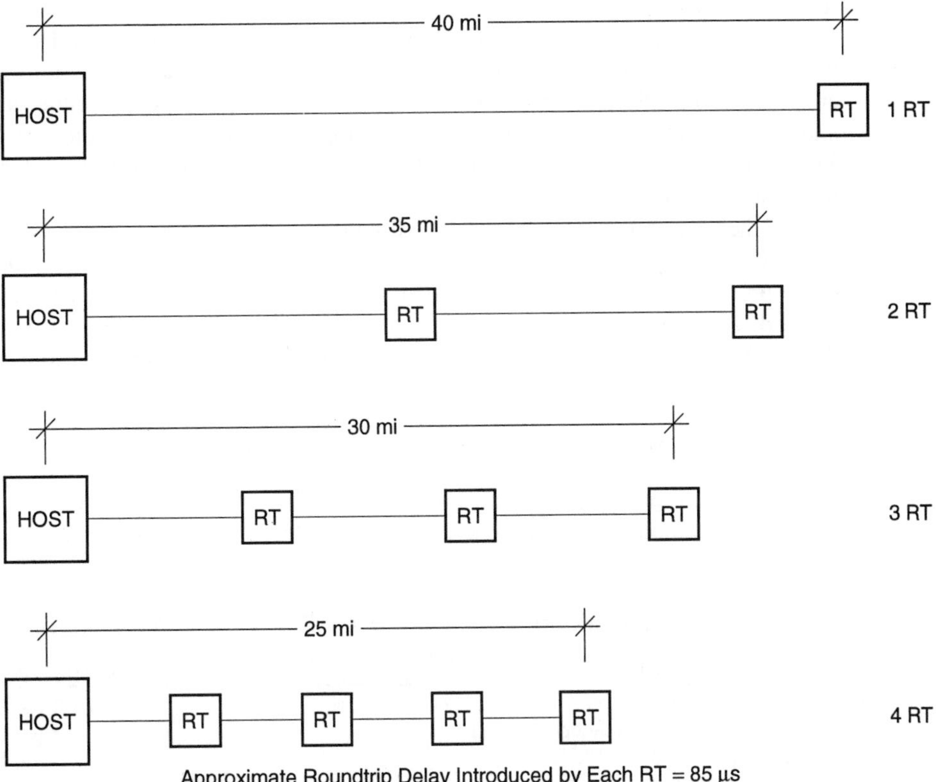

Figure 5-28 Distance limitations due to path and buffering delays in a distributed DLC system

*The actual design specification will vary among manufacturers.

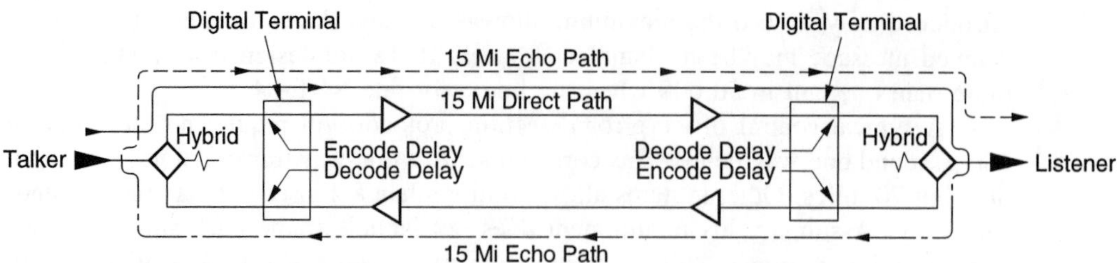

Figure 5-29 Listener echo and voice transmission delay paths

wire hybrid at both the listener and talker ends. The delay introduced by the terminal equipment at each encode/decode point is approximately one frame (125 μseconds) plus buffer hysteresis (up to an additional 125 μseconds) plus propagation delay.

The roundtrip propagation delay can be calculated from Table 5-7 for various media types. The first two entries apply to analog subscriber loops and are provided for comparison.

Electronic digital cross-connect systems (DCSs) operating at the DS-1 rate introduce delay of at least one frame (125 μseconds). The average delay for the digital cross-connect systems is specified as four frames (500 μseconds) in either direction, and the maximum delay of 700 μseconds applies to 99% of all connections [34]. The actual delay is a statistical quantity because of the varied timeslot interchanges that can be made with a number of connections.

Other network elements introduce digital processing delays, as shown in Figure 5-30. The range of values given in this illustration do not include the propagation delay previously described. To calculate the total delay for any particular circuit or network configuration, the processing delay across a network element is added to the propagation delay between the network elements, paying attention to units. That is,

$$\text{Roundtrip Delay} = D1 + P12 + D2 + \ldots \text{ seconds}$$

where D1 = roundtrip processing delay across element D1 in seconds
D2 = roundtrip processing delay across element D2 in seconds

Table 5-7 Roundtrip Propagation Delay for Various Media

Media	Propagation Delay
Twisted pair cable, nonloaded (1 kHz)	16.2 μs/kft
Twisted pair cable, loaded (1 kHz)	32.0 μs/kft
Twisted pair cable (40 kHz)	3.8 μs/kft
Twisted pair cable (200 kHz)	3.3 μs/kft
Twisted pair cable (772 kHz)	3.1 μs/kft
Fiber optic cable (singlemode)	1.6 μs/kft
Terrestrial radio	1.0 μs/kft
Satellite radio	480–530 ms[a]

[a]This delay does not include tail-circuit, encoding, or multiplexer delays.

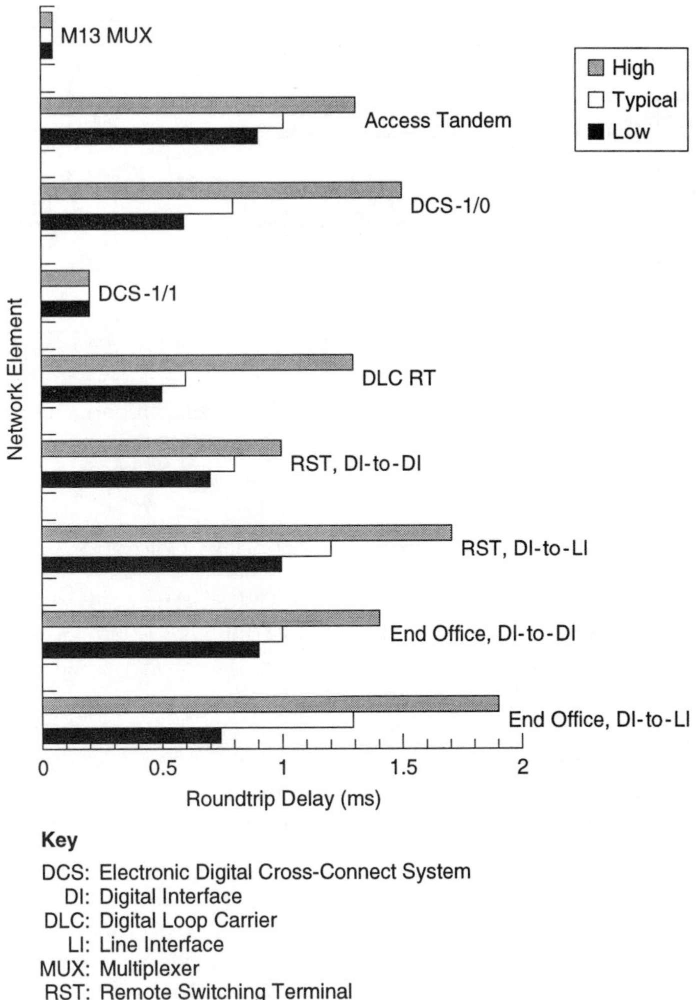

Figure 5-30 Roundtrip processing delays across network elements, source data from [35]

P12 = roundtrip propagation delay between elements D1 and D2 in seconds
= propagation delay from Table 5-7 × circuit length in kilofeet

EXAMPLE 5-7

Figure 5-31 shows two RSTs connected to a host end-office (EO). Find the average roundtrip delay between user A and user B.

The delays are summarized in Table 5-8. A signal from user A travels through eight kilofeet of nonloaded twisted pair cable, an LI-to-DI interface at the RST, and 40 kilofeet of repeatered T1-carrier before entering a DCS in the serving central office. In the central office this signal travels through the DCS, the host EO, and the DCS again. Leaving the central office, the signal travels through 20 kilofeet of repeatered T1-carrier, a DI-to-LI interface at

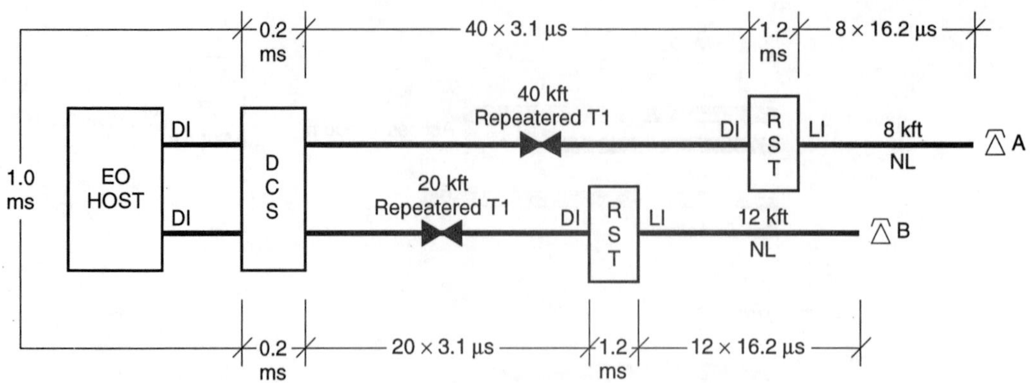

Figure 5-31 For Example 5-7, roundtrip delay between User A and User B

Table 5-8 Summary of Roundtrip Delays for Example 5-7

NETWORK ELEMENT	ROUNDTRIP UNIT DELAY	QUANTITY	TOTAL
User A Loop	16.2 μs/kft	8 kft	0.13 ms
User A RST	1.2 ms	1	1.2 ms
Repeatered T1-carrier	3.1 μs/kft	40 kft	0.12 ms
DCS	0.2 ms	2	0.4 ms
Host EO	1.0 ms	1	1.0 ms
Repeatered T1-carrier	3.1 μs/kft	20 kft	0.06 ms
User B RST	1.2 ms	1	1.2 ms
User B loop	16.2 μs/kft	12 kft	0.19 ms
		TOTAL	4.3 ms

the RST, and finally 12 kilofeet of nonloaded twisted pair cable. The signal is reflected in the hybrid at user B and travels the reverse path back to user A. The total roundtrip delay is approximately 4.3 mseconds, which is 0.7 msecond short of the delay that would require echo cancellers or echo path loss considerations.

5.7 Jitter and Wander

Jitter and wander are variations in the phase position of a pulse with respect to its ideal position. The difference between jitter and wander is the rate at which they are specified; wander is considered to be a timing phase change at a less than 10 Hz rate, and jitter is considered to be a phase change at a greater than 10 Hz rate. Jitter and wander affect digital system synchronization and were discussed in detail in Chapter 4.

5.8 Signal Detection and Regeneration

As digital signals (pulses) are transmitted from the output of the line transmitter (or regenerator) to the input of the line receiver (or next downstream regenerator), they are distorted by the inherent transmission line characteristics and corrupted by noise. The line receiver must decide if a pulse is present or not in each sampling interval. A small fraction of the pulses are not detected, or noise may be detected as a pulse when none was present. The ratio of erroneous bits to actual bits transmitted is called the bit error rate (BER), which is discussed further in the next chapter.

Perhaps one of the most significant problems associated with pulse transmission on twisted pairs is intersymbol interference (ISI). The time dispersive, or nonlinear phase delay, characteristics of twisted pairs cause transmitted symbols (signal elements representing one or more bits) to overlap as they propagate. This effect is a problem with virtually all baseband digital loop systems as well as passband pulse modulation systems, including frequency-shift keying (FSK), phase-shift keying (PSK), and quadrature-amplitude modulation (QAM) modems.

The effects of ISI are reduced by passing the received signal through a filter that approximates the inverse transfer function of the transmission channel, a process called *equalization*. The implementation of an equalizer depends on the symbol rate. At speeds below approximately one kilobaud, nonadaptive or compromise equalization is used in which the filter has fixed coefficients. These coefficients compensate for a wide range of channel characteristics but only on an average basis.

Baseband digital transmission systems, such as ISDN DSL and HDSL, and modems operating at speeds greater than about two kilobaud use adaptive equalization, in which the filter coefficients are adapted to compensate more precisely for the channel characteristics. The coefficients usually are adapted during a training period after the connection is established. This connection may take place upon initial power-up of the digital loop or, if switched, upon establishment of a network call.

Training involves the transmission of a known sequence of bits (or symbols) in both directions. The equalizer automatically adapts the filter coefficients according to a synchronized version of the received training sequence. Upon completion of the training sequence, slight variations in the transmission channel may be tracked by performing additional adaptation based on an estimate of the received symbols. Training periods can last from a few seconds to about one minute in commercial systems.

Signal regeneration is an inherent requirement on a digital loop. As the transmitted pulses propagate down the loop, they suffer attenuation as well as dispersion. A line receiver is used at intermediate locations and at the end of the loop to regenerate the transmitted pulses. Noise due to crosstalk, impulses, foreign voltages, and other interference also appears at the input to the receiver, as shown in Figure 5-32. Among the elements of a line receiver are overvoltage protection, input and output coupling transformers, lowpass filter, preamplifier, ALBO network, equalizer, 3-level slicing circuit, clock circuit, and power supply. A simplified regenerator is shown in Figure 5-33.

In this regenerator, the lowpass filter reduces the higher frequency noise components. The input transformer is terminated with a loop matching impedance on its

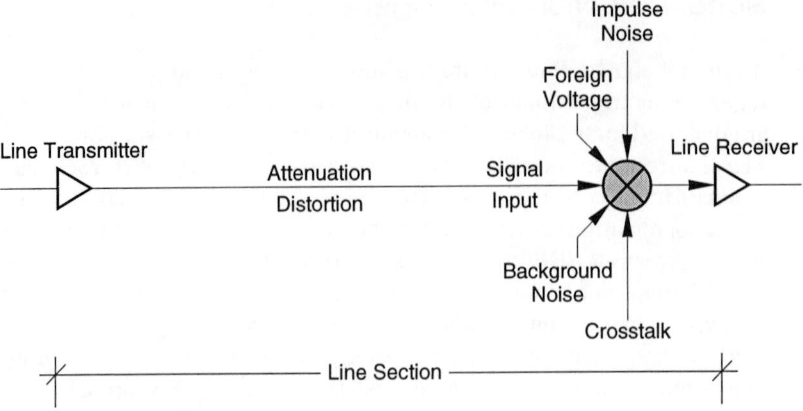

Figure 5-32 Signal and noise inputs to a line receiver

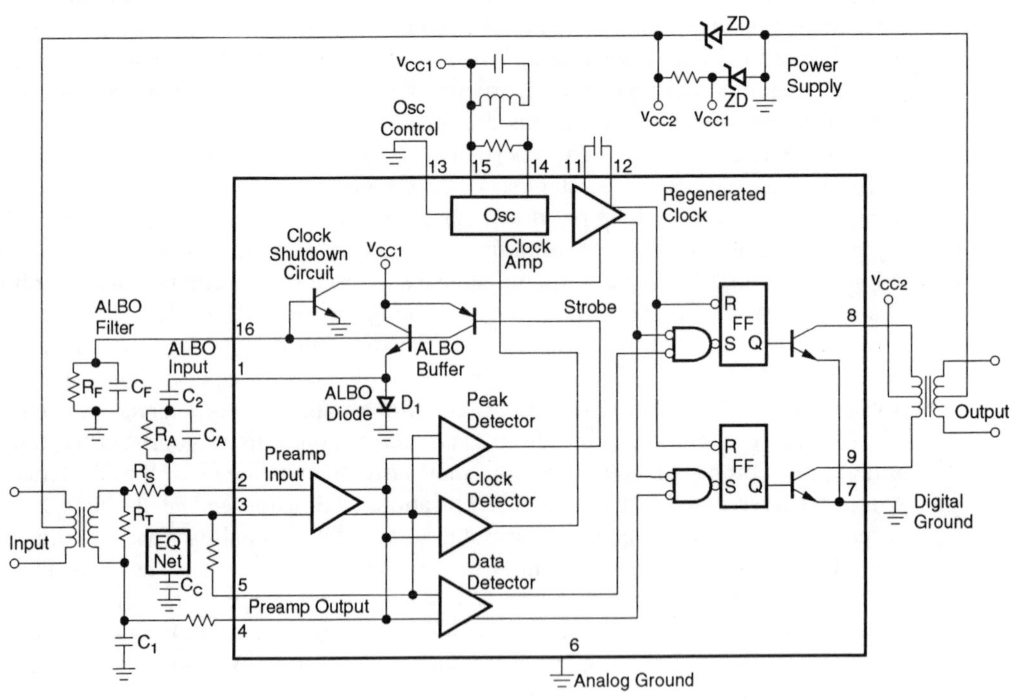

Figure 5-33 Typical digital regenerator, [36]
© 1992 Analog Devices. Reprinted with permission

secondary side. The impedance to ground of the ALBO network (Ra and Ca), which is connected to the transformer through a series resistor Rs, is determined by the current through the ALBO diode. Rs, Ra, and Ca act as a voltage divider and provide attenuation that is proportional to the ALBO diode current. When no current flows through the diode, the ALBO network is isolated from ground and the attenuation is small. Conversely, a high current through the diode shunts the ALBO network to ground, increasing its attenuation.

The preamplifier increases the input signal level and applies it to three comparators (slicers) for data, clock, and peak signal level recovery. Each comparator provides a current pulse output when the signal level exceeds a preset, but different, threshold, as shown in Figure 5-34. The output from the peak detector is integrated by the ALBO network, which causes a relatively constant current to flow in the ALBO diode for a given pulse height. This current is proportional to the incoming signal level, which acts to control the ALBO network attenuation, as previously explained. The clock detector locks the oscillator to the frequency of the incoming pulse stream. The regenerated clock and the data detector outputs drive flip-flops, which, in turn, drive the output coupling transformer. In a T1-carrier regenerator, the preamplifier has a 3 dB bandwidth of approximately 5 MHz.

The clock detector that drives the oscillator tank circuitry has a quality factor (Q) of approximately 10 to 50. The Q is high enough to ensure continued oscillation in the absence of pulses in the incoming data bit stream (for example, a long series of binary zeros). On the other hand, the Q is low enough to allow the incoming pulses to pull the oscillator away from its nominal oscillation frequency. This is necessary because the incoming data bit stream may be timed at a rate slightly different than this nominal frequency, or it may contain jitter where each pulse arrives with a slightly different phase difference from ideal. Therefore, to ensure that no data bits are missed, the oscillator must be able to track the clock frequency of the incoming bit stream.

The final element of the regenerator is the power supply. In a typical digital loop repeater, common mode dc line current is simplexed onto the cable pair. For

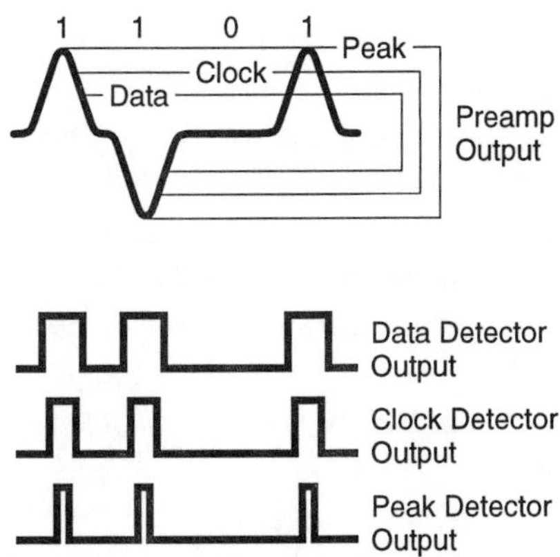

Figure 5-34 Three-level slicer waveforms, [36]
© 1992 Analog Devices. Reprinted with permission

a T1-carrier line repeater, the level usually is a constant 60, 100, or 140 mA (60 mA in all modern designs). The simplex loop current in the ISDN DSL, if provided, is 1.0 to 20 mA and in the HDSL is at least 20 mA. The DSL and HDSL normally do not use line repeaters; the simplex current is a sealing current used for "wetting."

In all digital loop systems, the simplex current is separated from the signal by the center taps of the input and output transformers and converted to voltages necessary for regenerator operation (typically seven to ten V) by zener diodes. When the regenerator is located in terminal equipment (for example, a multiplexer), power is obtained from the equipment power supply. In this situation, the regenerator includes a constant current source that feeds common mode dc current into the loop. The current source can put out voltages of 260 V or more to maintain the specified current under all loop environmental and loop length conditions.

5.9 Digital Pads

Digital pads are not the same as analog (resistive) pads. Digital pads cause a coding translation after an analog source signal has been converted to digital. Analog pads cause no loss in SNR (the signal and noise are reduced by an equal amount), whereas digital pads reduce the SNR equivalent to adding an asynchronous tandem link in which the signal is decoded and then re-encoded. For example, a digital switching system with a SNR of 39 dB will reduce the SNR by 3 dB if a digital pad *of any value* is inserted in the connection [37]. Other disadvantages of digital pads are increased background noise and decreased dynamic range (both change in proportion to the pad value).

References

[1] Reeve, W. *Subscriber Loop Signaling and Transmission Handbook: Analog*. IEEE Press, 1992.

[2] Cravis, H., Crater, T. *Engineering of T1 Carrier System Repeatered Lines*. Bell System Technical Journal, Vol. XLII, No. 2, March 1963.

[3] Kerpez, K., Valenti, C. *Impulse Noise Testing for ADSL Transceiver*. BELLCORE T1E1 Contribution, T1E1.4/93-034, March 10, 1993.

[4] Violette, J. et al. *Electromagnetic Compatibility Handbook*. Van Nostrand Reinhold, 1987.

[5] Ott, H. *Noise Reduction Techniques in Electronic Systems*, 2nd Ed. John Wiley & Sons, 1988.

[6] *IEEE Standard Methods and Equipment for Measuring the Transmission Characteristics of Analog Voice Frequency Circuits*. The Institute of Electrical and Electronics Engineers. IEEE Standard 743-1984. Available from IEEE Service Center.

[7] *Digital Data System Channel Interface Specification*. AT&T Technical Reference TR62310, Nov. 1987. Addendum No. 3, Dec. 1989. Available from AT&T Customer Information Center.

[8] Freeman, R. L. *Telecommunication System Engineering*. John Wiley & Sons, 1989.

[9] *ISDN Basic Access Digital Subscriber Line*. BELLCORE Technical Reference TR-TSY-000393, May 1988. Available from BELLCORE Customer Service.

[10] *Secondary Channel in the Digital Data System: Channel Interface Requirements*. BELLCORE Technical Reference TR-NPL-000157, April 1986. Available from BELLCORE Customer Service.

[11] *Accunet® T1.5 Service, Description and Interface Specifications*. AT&T Technical Reference TR62411, Dec. 1988. Available from AT&T Customer Information Center.

[12] Federal Communications Commission Rules and Regulations, 47 CFR Part 68. Available from Superintendent of Documents.

[13] American National Standard for Telecommunications. *Carrier-to-Customer Installation—DS1 Metallic Interface*. ANSI T1.403-1989.

[14] *IEEE Standard Telephone Loop Performance Characteristics*. IEEE Standard 820-1989. Available from IEEE Service Center.

[15] *T1 Digital Line Transmission and Outside Plant Design Procedures: Carrier Engineering*. AT&T Practice 855-351-101, July 1990. Available from AT&T Network Systems.

[16] SNC Manufacturing Company, Inc., 101 Waukau Ave., Oshkosh, WI 54901.

[17] *IEEE Guide for Inductive Coordination of Electric Supply and Communication Lines*. IEEE Standard 776-1987. The Institute of Electrical and Electronics Engineers. Available from IEEE Service Center.

[18] *Guide for Implementation of Inductive Coordination and Mitigation Techniques*. IEEE Standard 1137-1992. The Institute of Electrical and Electronics Engineers. Available from IEEE Service Center.

[19] Foschini, G. *Crosstalk in Outside Plant Cable Systems*. Bell System Technical Journal, Vol. 50, No. 7, Sept. 1971.

[20] Lechleider, J. *Broad Signal Constraints for Management of the Spectrum in Telephone Loop Cables*. IEEE Transactions on Communications, Vol. COM-34, No. 7, July 1986.

[21] Lechleider, J. *Spectrum Management in Telephone Loop Cables, II: Signal Constraints That Depend on Shape*. IEEE Transactions on Communications, Vol. COM-34, No. 8, Aug. 1986.

[22] Lechleider, J. *Line Codes for Digital Subscriber Lines*. IEEE Communications Magazine, Sept. 1989.

[23] Brosio, A. et al. *A Comparison of Digital Subscriber Line Transmission Systems Employing Different Line Codes*. IEEE Transactions on Communications, Vol. COM-29, No. 11, Nov. 1981.

[24] Chattler, L. *A Guide to Electrical Specification Requirements for Multipair*

Telephone Cable. Paper presented at Wire Asia '82, Singapore. DCM International Corp., 1982.

[25] Werner, J. *Impulse Noise in the Loop Plant*. Paper 348.1, Proceedings of the IEEE International Conference on Communications 1990 (ICC'90).

[26] Szechenyi, K., Bohm, K. *Impulse Noise Limited Transmission Performance of ISDN Subscriber Loops*. Paper 2.3, Proceedings of the International Symposium on Subscriber Loops and Services 1988 (ISSLS'88).

[27] Sistanizadeh, K., Kerpez, K. *A Comparison of Passband and Baseband Transmission Schemes for HDSL*. IEEE Journal on Selected Areas in Communications, Vol. 9, No. 6, Aug. 1991.

[28] Werner, J. *The HDSL Environment*. IEEE Journal on Selected Areas in Communications, Vol. 9, No. 6, Aug. 1991.

[29] *Transmission Parameters Affecting Voiceband Data Transmission: Description of Parameters*. BELLCORE Technical Reference PUB-41008, July 1974. Available from BELLCORE Customer Service.

[30] Bender, E. et al. *Digital Data System: Local Distribution System*. Bell System Technical Journal, Vol. 54, No. 5, May-June 1975.

[31] Bellamy, J. *Digital Telephony*, 2nd Ed. John Wiley & Sons, 1991.

[32] Selby, S. *Standard Mathematical Tables*, 19th Ed. CRC Press, 1991.

[33] *IEEE Standard Dictionary of Electrical and Electronic Terms*. IEEE Std 100-1984.

[34] *Digital Cross-Connect System Requirements and Objectives*. BELLCORE Technical Reference TR-TSY-000170, Nov. 1985. Available from BELLCORE Customer Service.

[35] American National Standard for Telecommunications. *Network Performance—Supplement to Transmission Specifications for Switched Exchange Access Network (Absolute Round-Trip Delay)*. ANSI T1.506a-1992.

[36] 1992 Data Converter Reference Manual, Vol. 1. Analog Devices, Norwood, MA.

[37] Ingle, J. *Identification of Digital Impairments in a Voiceband Channel*. Paper 12.3.1, Proceedings of the International Conference on Communications 1989 (ICC'89), Boston.

6 Transmission Quality and Performance

Chapter 6 Acronyms

ADPCM	adaptive differential pulse code modulation
AIS	alarm indication signal
AM	available minutes
AMI	alternate mark inversion
ANSI	American National Standards Institute
APS	automatic protection switch
AS	available second
BCC	block check character
BER	bit error rate
BPV	bipolar violation
CCIR	The International Radio Consultative Committee
CCITT	The International Telegraph and Telephone Consultative Committee
CEPT	Conference of European Posts and Telecommunications
COFA	change-of-frame alignment
COT	central office terminal
CRC, crc	cyclic redundancy check
CSES	consecutive severely errored second
CSU	channel service unit
DCS	electronic digital cross-connect system
DLC	digital loop carrier
DM	degraded minute
DSS	digital switching system
DSX	digital signal cross-connect
DTMF	dual tone multifrequency
%EFS	percent error-free second
EFS	error-free second
ES	errored second
ESF	extended superframe format
FOT	fiber optic terminal
HRDL	hypothetical reference digital link
HRDP	hypothetical reference digital path
HRDS	hypothetical reference digital section
HRX	hypothetical reference connection
IC	inter-exchange carrier
IDLC	integrated digital loop carrier
ISC	international switching center
ISDN	integrated services digital network
LEC	local exchange carrier
LOF	loss-of-frame
LOFA	loss-of-frame alignment
LOS	loss-of-signal
MTBF	mean time between failure
MUX	multiplexer
OOF	out-of-frame
OSS	operational support system
PCM	pulse code modulation
PEFS	probability of an error-free second
POT	point-of-termination
QRSS	quasi-random signal source
RBER	residual bit error ratio
SES	severely errored second
SF	superframe format
SNR	signal-to-noise ratio
UDLC	universal digital loop carrier
US	unavailable second

6.1 Quality of Service

Quality of service is a broad issue that incorporates the user's often subjective performance requirements, digital loop system performance requirements, and network performance requirements. This is illustrated in Figure 6-1(a) by three concentric circles. The user requirements are the core issue, and the digital loop system that serves the user forms the next performance circle. The outermost circle covers the network performance requirements.

As we may infer from Figure 6-1(a), the performance of digital loop systems can be specified separately from other network parts. To be meaningful, however, these characteristics must be integrated into the overall quality of service requirements of the network. The sum of the parts must meet the user's end-to-end requirements.

Another set of performance circles can be drawn as shown in Figure 6-1(b) if analog services, such as voicegrade or program quality circuits, are carried on the digital loops. The performance of analog circuits is only loosely correlated with the performance requirements for digital transmission, so these are separately specified.

Quality of service in an overall network may be viewed as a tradeoff between value and cost, as seen from the perspectives of the end-user and the carrier. The

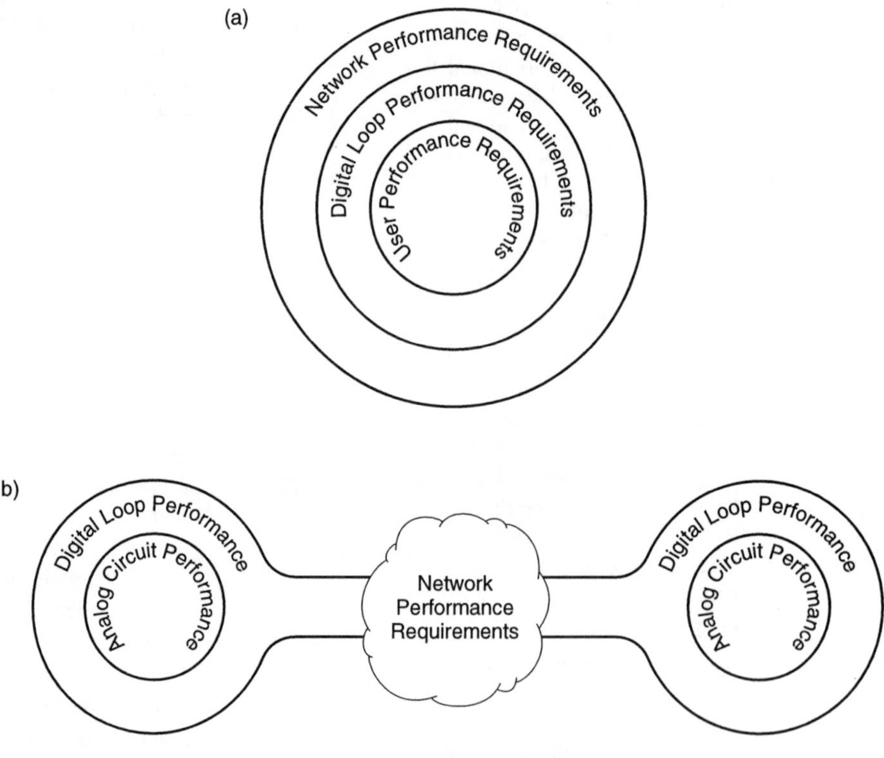

Figure 6-1 Quality of service circles: (a) Digital service performance; (b) Analog service performance

difference in these perspectives is shown in Table 6-1. Yet there are common grounds upon which quality can be measured. This chapter focuses on transmission quality, so the much broader quality of service issues are not discussed.

A matrix can be formed of *communication functions* and *quality criteria,* as seen in Table 6-2 [1]. Such a matrix can be used by users and carriers alike to define service expectations. Since this book concentrates on the technical issues associated with digital loops, this matrix shows only the key technical parameters associated with digital data transmission. The communication functions of a more generalized matrix include [1]:

- *Technical sales and planning*. The ability of the service provider to meet user needs when considering service requests. This includes service conditions, optimization studies, and new applications assistance.
- *Provisioning*. The implementation of network resources and capabilities that allow the user to use fully the service requested.
- *Technical quality*. The requirements or expectations for service use, including error performance and availability.
- *Network and service management by customer*. The ability of the user to monitor and reconfigure network resources in order to fulfill their service needs.
- *Repair*. Responding to and fixing service problems.
- *Technical support*. The attention paid by the service provider to the user's needs and applications after service is implemented.

Table 6-1 Performance Perspectives

From User Perspective	From Serving Carrier Perspective
Confidence in data integrity	Competitive necessity
Backup for billing disputes	Early detection of deteriorating performance
Compliance with performance guarantees	Compliance with tariff or contract requirements
Assessment of service quality vs. carrier	Marketing of premium services

Table 6-2 Technical Quality of Service Framework for Digital Services[a]

Communications Function	Requirement
Speed	Response time; throughput
Accuracy	Error performance
Availability	% Unavailability due to outage and time to restore
Reliability	% Retransmission
Security	Availability and degree of encryption
Simplicity	Ease of use
Flexibility	Reconfiguration ability

[a]Adapted from [1]

The quality criteria are:

- *Speed*. A measure of available data rates and propagation delays.
- *Accuracy*. A measure of the correctness, or fidelity, including error performance.
- *Availability*. The accessibility of a communication function, including the time in any measurement period during which the requested service meets quality-of-service specifications.
- *Reliability*. The dependability or sustainability of a communication function.
- *Security*. The confidentiality of customer information and the protection against fraud or privacy invasion.
- *Simplicity*. The ease of understanding or performing a communication function.
- *Flexibility*. The ability to adapt or customize a communication function to meet individual user needs.

When analyzing a given service, all parameters that affect user satisfaction are identified, with a focus on key items. A systematic approach such as this will ensure that no important service attribute is inadvertently ignored, which is very important during the planning phase for a new service or feature.

For the sake of the following discussion, the technical quality issues will be simply called "transmission quality." Transmission quality in digital transmission systems is measured by error frequency and severity and the effect of these errors on availability.

From the standpoint of user complaints, a digital transmission system devoted exclusively to voice transmission, with 64 kbps encoding, can operate with a higher error level and lower availability than a network devoted exclusively to subrate digital data transmission. A hypothetical voice transmission system, with lower quality and availability, generally would be cheaper to build and operate than a data transmission system, but the better error performance and availability of the data transmission system would be more valuable to the user.

Although this book is concerned primarily with loops, it is necessary to discuss loop performance in perspective of the rest of the network. Therefore, the remainder of this chapter will provide performance information on the local access portion as well as interoffice and long-haul portions of the network.

6.2 Line Performance

Transmission quality is directly related to line performance. All digital loops are designed to provide an error rate below some specified maximum value, which will depend on the type of service, tariff conditions, or contractual requirements. Error rate is controlled by controlling the signal-to-noise ratio (SNR) at the input to the decision circuits in regenerators and terminal and multiplex equipment in the line.

This, in turn, means that signal level, line loss, and noise due to crosstalk and foreign voltages must be kept within predetermined limits. Systems and circuit engineering, described in later chapters, account for these limits.

Digital loops are made up of one or more sections, which are cascaded to provide an overall path between terminals, as shown in Figure 6-2. Some loops contain only one section. In multi-section loops, errors in one section are seldom correlated with errors in another section. Therefore, the overall error rate for a path is equal to the sum of the individual section error rates. That is, if the error rate is e_1 for section 1, e_2 for section 2, and e_n for section n, the overall error rate is $e_1 + e_2 + \ldots + e_n$. Maximum errors normally do not occur in all sections at the same time.

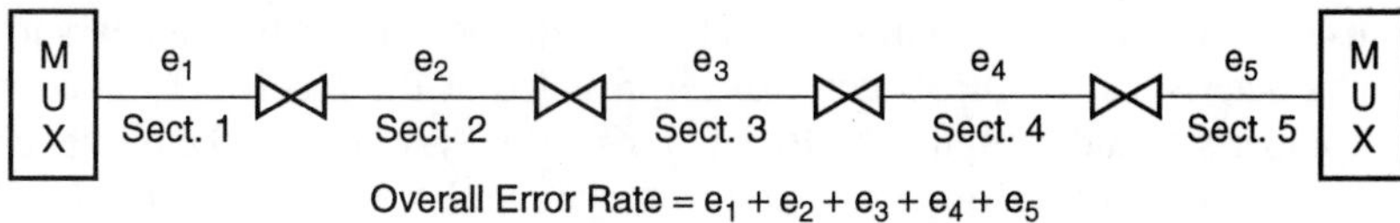

Figure 6-2 Error performance of cascaded digital line sections

6.3 Specifying Error Performance

6.3.1 General Considerations

The performance of digital loop transmission systems can be broadly categorized as *available* or *unavailable*. When a transmission path is unavailable, no useful transmission is possible. Only when the path is available does the quality of that path have meaning. For convenience, and to accommodate the performance of practical systems and the effects of quality on users, the International Telegraph and Telephone Consultative Committee (CCITT) has defined several performance parameters in Recommendation G.821. The parameters are shown in Table 6-3 [2].

CCITT performance definitions generally have been adopted in the United States for circuit-switched digital services [3] and for dedicated digital services [4], but not all industry standards follow the CCITT G.821 requirements. Therefore, in addition to the parameters associated with the CCITT G.821 performance parameters, other industry standard performance parameters are described in this section. A brief tutorial follows this section on digital transmission system performance from an international perspective.

Table 6-3 CCITT Recommendation G.821 Error Performance Requirements

PERFORMANCE	AVAILABLE?	BER	TIME INTERVAL
Acceptable	Yes	≤10E-6	≥1 minute
Degraded	Yes	10E-6 < BER < 10E-3	Any 1-minute interval
Severely errored	Yes	≥10E-3	Any 1-second interval
Unavailable	No	≥10E-3	≥10 consecutive seconds

Bit error rate (BER) traditionally has been used to describe the performance of digital transmission systems. BER states the probability that any given transmitted bit will be erroneously received, or

$$\text{BER} = \frac{\text{Bits received in error}}{\text{Total number of bits transmitted in measurement interval}}$$

The probability of an error has no meaning if it cannot be measured or verified in real systems. This means the receiver would have to know in advance the value of every bit so it can make a comparison with the bit values actually received.

In many digital transmission systems, a quasi-random signal source (QRSS) is used to transmit a known, but repetitive sequence of bits that appears to have random characteristics. A QRSS detector is provided at the receiver to make the necessary comparison and compute the BER. During this type of test, the digital channel cannot be used for live traffic. For live traffic channels, it is possible to infer the BER from the expected values of framing bits and other line code characteristics, such as bipolar violations (BPV) in the case of bipolar alternate mark inversion (AMI).

BER is an average over a selected measurement period and, as such, does not fully describe performance over shorter periods or during periods outside the measurement interval. As an example of its shortcomings, consider a 1.5 Mbps system that transmits data over a period of 5,000 hours (18E6 seconds). During this time, approximately 2.8E13 bits are transmitted. If 1E7 bits are transmitted in error during that time, the average BER is 1 in 2.8E6 (about 3.6E-7). That is an acceptable average in most data applications. What if all those errors are transmitted in one 6-hour (21,600-second) period? During this period, the average BER is about 1 in 3.3E3 (3E-4). This is unacceptable for any modern digital transmission system.

Despite its shortcomings, BER still is frequently cited as a valid performance parameter and usually is specified as a test and measurement parameter during installation and circuit turn-up.

In addition to BER, a number of parameters have been devised to describe performance. These take into account the bursty nature of transmission errors on baseband systems, which are caused by noise impulses or network switching activity. "Bursty" means the errors come in groups, and even though the burst may occur at random intervals, individual errors are not random events. Such error bursts generally last less than one second. The parameters, in addition to BER, usually stated as objectives in performance specifications, are:

- Percent error-free second (%EFS)
- Errored second (ES)
- Severely errored second (SES)
- Consecutive severely errored second (CSES)
- Degraded minute (DM)
- Availability
- Slips

An *error-free second* (EFS) is any second in which no errors are received. The %EFS defines the quality of the service by specifying the percentage of the total seconds in a period (usually one day) that must be completely error-free. Any %EFS specification must consider the length of the circuit and the number of segments that include multiplexing or other bit manipulation, since these affect error performance. Typical specifications are given later and also in industry standards [2,5,6].

The complement of EFS is the ES, such that

$$\text{EFS}(\%) = 100 - \text{ES}(\%)$$

where

$$\text{ES}(\%) = \frac{\text{Number of seconds with at least one error}}{\text{Total seconds in measurement period}} \times 100$$

An ES could contain one error or thousands of errors. From a trouble-reporting standpoint, an ES by itself is of little consequence and no repair action is required. On the other hand, a number of such events (say, several hundred) in a short time (usually 15 minutes) can indicate line or equipment trouble, requiring repair action.

An SES is a parameter that takes into account the bursty nature of errors. Its definition is application specific. A typical specification will set a maximum number of SES in a given time (usually one day). The number of SESs are sensitive to circuit length and multiplexing. According to CCITT G.821 previously cited, an SES is any one-second interval in which the BER is worse than E-3. This definition is independent of data rate, but it implies that BER measurements of the circuit are possible. AT&T defines an SES as any second with 320 or more CRC6 events or one or more out-of-frame event [7]. This definition applies to DS-1 rate systems using the extended superframe format (ESF) and does not necessarily apply to any other digital transmission system.

Although the typical error burst is much shorter than one second, some error bursts can last several seconds, or a number of them can occur over a period exceeding one second. The CSES parameter is used to describe such a situation. The CSES is a consecutive string of between three and ten SESs [3,4]. A typical specification will set the maximum number of CSES in some given period (usually one day). Some service specifications refer to a similar event as a "very severe burst," defining it as a BER of less than 1E-2 lasting for more than two and a half seconds [6].

A degraded minute is a 1-minute interval with 1E-6 $<$ BER $<$ 1E-3. To derive a degraded minute in a measurement interval, remove the SES and then consecutively group the remaining seconds, with BER $>$ 1E-6 into blocks of 60. Each block of 60 such seconds is one DM. The DM is not used presently as a performance parameter in the United States.

Availability is a measure of the percentage of time the circuit is available for use. Availability is the complement of outage, which indicates the circuit is unavailable, or

$$\text{Availability (\%)} = 100 - \text{Outage (\%)}$$
$$= (A/A + U) \times 100$$

where A = Availability time
U = Unavailability time
A + U = Total time in measurement interval

The generally accepted definition of availability (CCITT G.821) specifies that a circuit is considered unavailable at the onset of 10 CSES [2]. The circuit is considered available again at the onset of a ten-second interval with no SES. Availability usually is taken as an average over a one-year period, but other intervals, such as a quarter (-year), can be used. Other definitions are used; for example, a circuit outage is sometimes considered to begin when the circuit has been released to the repair center (for example, a carrier's repair center) and to end when it has been returned to the user [6]. In this context, the terms available and unavailable can have different interpretations, depending on the point of view (carrier or user).

For reference, Table 6-4 quantifies unavailability percentages in terms of time periods. For example, a cumulative outage time of 32 seconds in one year equates to an average unavailability of 0.0001%, or equivalently, 99.9999% availability.

An expression that is sometimes used to relate EFS with BER is

$$\text{PEFS} = (1 - \text{BER})^{\text{R}}$$

where PEFS = Probability of an EFS, given the probability of a bit error and digital rate R in bps

This expression must be used with caution because it assumes a memory-less channel in which a bit error is independent of any other bit errors (that is, errors are not bursty). If this assumption is not valid in a particular case, the expression underestimates PEFS.

Table 6-4 Unavailability in Terms of Time Periods

Availability, %	Unavailability, %	Unavailability		
		Per Year	Per Month	Per Day
0.0	100.0	8,760.0 hours	720.0 hours	24.0 hours
50.0	50.0	4,380.0 hours	360.0 hours	12.0 hours
75.0	25.0	2,190.0 hours	180.0 hours	6.0 hours
90.0	10.0	876.0 hours	72.0 hours	2.4 hours
95.0	5.0	438.0 hours	36.0 hours	1.2 hours
97.5	2.5	219.0 hours	18.0 hours	0.6 hours
98.0	2.0	175.2 hours	14.4 hours	28.8 minutes
99.0	1.0	87.6 hours	7.2 hours	14.4 minutes
99.9	0.1	8.8 hours	43.2 minutes	1.4 minutes
99.99	0.01	52.6 minutes	4.3 minutes	8.6 seconds
99.999	0.001	5.3 minutes	25.9 seconds	0.86 seconds
99.9999	0.0001	31.5 seconds	2.6 seconds	0.09 seconds
99.99999	0.00001	3.2 seconds	0.3 seconds	0.009 seconds

Since error performance is related to segment lengths and multiplexing, each segment would have a set of performance criteria that would have to be higher than the overall end-to-end criteria. Figure 6-3(a) defines the segments in a reference connection. The pre- and post-divestiture situations of Figures 6-3(a) and (b), respectively, refer to the separation of AT&T and the Bell Operating Companies that started in 1982 and was finalized in 1984. The CCITT reference connection of Figure 6-3(c) is seen to fall between the historic and present reference connections used in the United States.

The EFS objective for each segment originally was calculated from

$$\text{Segment EFS Objective (\%)} = 100 - \text{TAP/F}$$

where T = complement of the total EFS objective end-to-end
A = fractional allocation for the segment
P = number of ports demultiplexed on the segment
F = empirical factor relating the average number of ES in a DS-0 port for each one found on the segment

Table 6-5 gives the values used in the original EFS objective calculations. In the present public network, the distinctions between the long-haul and interoffice (and, indeed, local loop access) segments are blurred and may not exist at all in some networks. Therefore, the modern allocations do not include the distinct long-

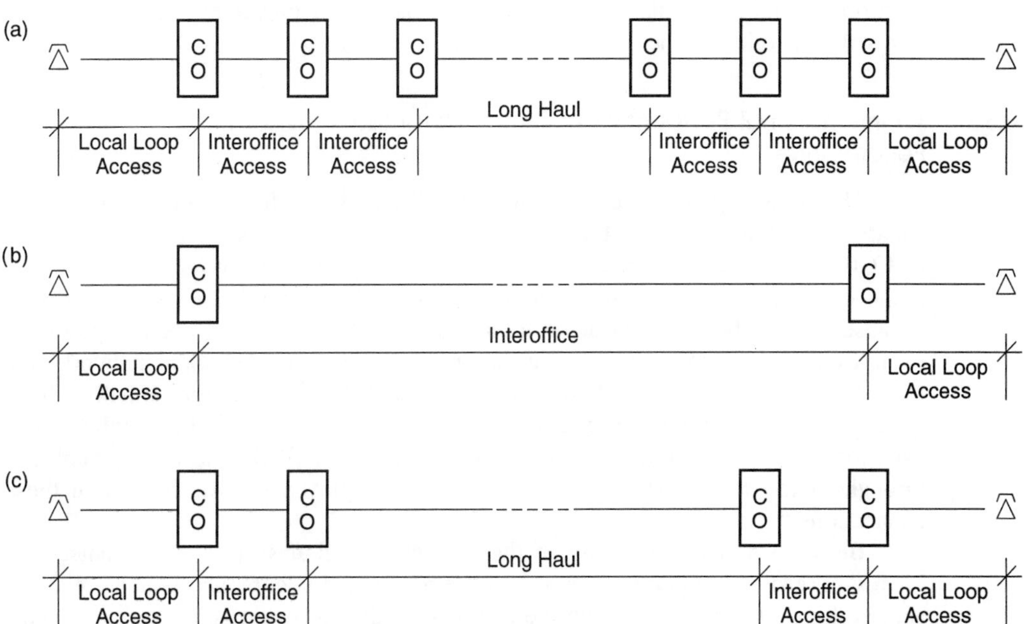

Figure 6-3 Reference connection for subrate digital data system to illustrate EFS allocation: (a) Pre-divestiture; (b) Post-divestiture; (c) CCITT 64 kbps

Table 6-5 Original Components of the EFS Objective Calculations

Segment	EFS	T	A	P	F	No. of Segments
Local loop access	99.95%	0.5%	0.1	1	1.0	2
Interoffice network segment	99.7%	0.5%	0.05	23	2.1	4
Long-haul network segment	98.8%	0.5%	0.6	23	5.75	1

haul network segment shown in Figure 6-3(a). Instead, the generic network access portions shown in Figure 6-3(b) apply. In this case, the local loop access may traverse more than one local exchange carrier (LEC) central office, although that is not shown explicitly in this illustration.

Table 6-6 summarizes the performance specifications for the present subrate digital data service offered by different carriers and for 56/64 kbps service specified in a 1994 industry standard (see references in Table 6-6). The requirements of Table 6-6 are for specific carriers, but they can be considered representative of industry standards for this type of service, the significant differences between carriers notwithstanding.

The performance of other types of digital loops follows a similar format. For example, the performance of commercially available circuits operating at the DS-1 rate is specified in Table 6-7. This table also shows the interoffice and end-to-end performance requirements. The values of SESs shown in parentheses are implicit rather than explicit in the referenced commercial standard.

6.3.2 An International Perspective of Digital Transmission System Performance

The following is a brief tutorial on CCITT and The International Radio Consultative Committee (of the International Telecommunications Union, Switzerland, or CCIR) recommendations and reports that apply to digital service across international connections. Such a discussion is warranted in a book on subscriber loops because, as will be seen, the local loop is allocated a significant percentage of the overall end-to-end performance. Furthermore, in any given national network, the local loop usually defines the overall performance of a connection. Finally, international end-to-end digital connections are more and more common and will become even more so as the integrated services digital network (ISDN) is widely deployed throughout the world. The local loop always will play a significant part in these connections.

Because of the complexity of the subject of digital system performance, the interested reader should obtain copies of the referenced documents, which are available in the United States through the National Technical Information Service.* This

*National Technical Information Service, 5285 Port Royal Road, Springfield, VA 22161, tel. 800-336-4700.

Table 6-6 Subrate Digital Data System Performance in Post-Divestiture Public Network

Connection	Circuit Length	ES/Day	%EFS/Day	SES/Day	Availability	Reference
Local loop access	Any	35	99.96	4	99.75%/year	[8]
Interoffice	<250 mi	25	99.97	10	99.95%/year	[8]
Interoffice	250 to 1,000 mi	30	99.97	25	99.95%/year	[8]
Interoffice	>1,000 mi	35	99.96	30	99.95%/year	[8]
End-to-end	<250 mi	95	99.89	18	99.90%/year	[8]
End-to-end	250 to 1,000 mi	100	99.88	33	99.90%/year	[8]
End-to-end	>1,000 mi	105	99.88	38	99.90%/year	[8]
Local loop access	Any	86	99.9	9	Not specified	[9]
Interoffice	Any	108	99.875	22	Not specified	[9]
End-to-end	Any	216	99.75	30	Not specified	[9]
Local loop access	Any	86	99.90	9	99.900%/month	[29]
Interoffice	Any	86	99.90	22	99.850%/month	[29]
End-to-end	Any	173	99.80	30	99.650%/month	[29]

Table 6-7 DS-1 Rate Services Performance

Connection	Circuit Length	ES/Day	%EFS/Day	SES/Day	Availability	Reference
Local loop access	Any	1,080	98.75	4	99.925%/Qtr.	[6]
Interoffice	<250 miles	86	99.90	10	99.85%/Qtr.	[6]
Interoffice	250 to 1,000 miles	346	99.60	35	99.85%/Qtr.	[6]
Interoffice	>1,000 miles	605	99.30	50	99.85%/Qtr.	[6]
End-to-end	<250 miles	2,246	97.40	(18)[a]	99.7%/Qtr.	[6]
End-to-end	250 to 1,000 miles	2,506	97.10	(43)[a]	99.7%/Qtr.	[6]
End-to-end	>1,000 miles	2,765	96.80	(58)[a]	99.7%/Qtr.	[6]
End-to end (objective)		170	99.80	4	99.975%/Qtr.	[10]
End-to-end (minimum)		363	99.58	4	99.974%/Qtr.	[10]
End-to-end	Any	1,080	98.75	Not specified	99.925%/year	[11]
Local loop access	Any	216	99.75	9	99.925%/month	[29]
Interoffice	Any	216	99.75	22	99.900%/month	[29]
End-to-end	Any	432	99.50	30	99.750%/month	[29]

[a]See text for an explanation of these values.

section serves to supplement and interpret the individual recommendations found in the referenced documents.

6.3.2.1 CCITT and CCIR Reference Models*

CCITT uses three standard hypothetical reference connections (HRXs) in Recommendation G.801 to model (1) the longest envisaged international connection, (2) a moderate international connection, and (3) a moderate international connection with the subscriber near the international switching center (ISC).† These models include switching systems, multiplexing equipment, and transmission systems. See Table 6-8.

The HRX consists of a national part at each end and an international part. The national part consists of two portions: a local part, which covers the facilities between the subscriber terminal, or T-reference point, and the local exchange, and the transit (or interoffice) facilities. The longest HRX and some of its parts are defined in Figure 6-4.

The HRX is further broken down into shorter, media-specific hypothetical models, called hypothetical reference digital path (HRDP) and hypothetical reference digital link (HRDL), which specify the performance objectives appropriate to each media type. Media types included are radio-relay, satellite, optical lines, and twisted pair lines. These models are listed in Table 6-9.

The HRDP and HRDL differ from the HRX in that the HRDP and HRDL are constituents of the HRX and do not include switching systems (but do include multiplexers). In an overall connection, the terms HRDP and HRDL are used interchangeably. The overall connection may be made up of any number of different HRDP and HRDL. CCITT recognizes that real radio-relay and line systems may have longer or shorter lengths than given in Table 6-9. The satellite HRDP consists of a single hop with an equivalent distance (the length of an equivalent terrestrial path) in the range of 10,000 to 13,000 km—nominally 12,500 km.

The HRDP and HRDL are further broken into the hypothetical reference digital sections, or HRDS, to accommodate the performance specification of actual transmission systems that make up the HRDP or HRDL. The HRDS consists of a single media type and associated equipment, but excludes multiplexing and switching. CCITT has tentatively defined the lengths of the HRDS to be 50 and 280 km. These lengths

Table 6-8 HRX Lengths

Type	CCITT Rec. G.801 [2]	Length
Longest length	Figure 1/G.801	27,500 km
Moderate length	Figure 2/G.801	11,000 km
Moderate length with subscriber near the ISC	Figure 3/G.801	10,000 km

*CCITT, the International Telegraph and Telephone Consultative Committee, and CCIR, The International Radio Consultative Committee, were replaced in 1993 by the International Telecommunications Union—Telecommunications Sector (ITU-T) and International Telecommunications Union—Radio Sector (ITU-R), respectively.

†In this section, "Rec." refers to a CCITT or CCIR Recommendation and "Rep." refers to a CCIR Report.

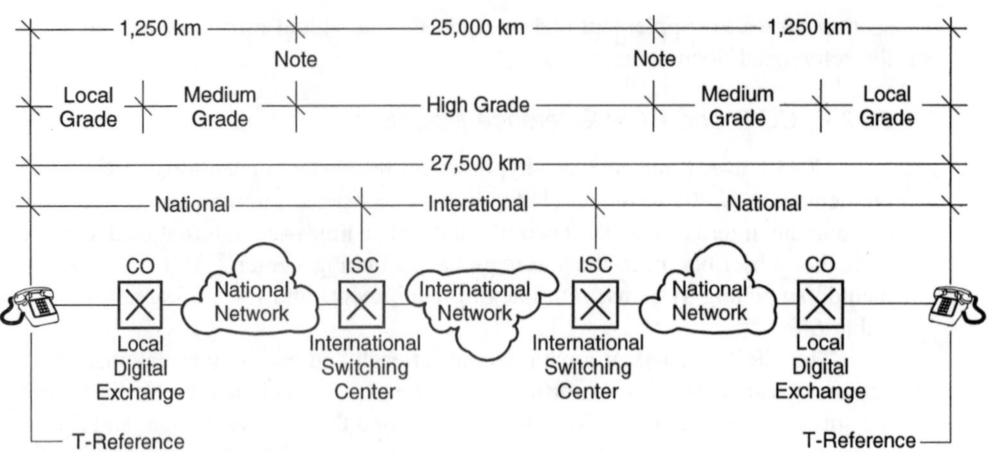

Note: There is No Clearly Defined Demarcation between the Medium- and High-Grade Portions Except for Distance from the T-Reference.

Figure 6-4 HRX

Table 6-9 HRDP and HRDL

Model	Applicability	Recommendation	Length
HRDP	Radio-relay systems	CCIR Rec. 556-1 [13]	2,500 km
HRDP	Satellite systems	CCIR Rec. 521-2 [12]	12,500 km
HRDL	Line systems	CCITT G.801 [2]	2,500 km

represent real systems operating at rates above the second level of the digital hierarchy. The CCITT recommendations do not specifically address the digital hierarchy used in North America but do address the CEPT digital hierarchy. Nevertheless, the concepts apply equally well.

An HRDS is given four different performance quality classifications, depending on its circuit role or application in the HRX. See Figure 6-4 and Table 6-10, which is derived from CCITT Table 2/G.921 [2].

The circuit classifications given in Table 6-10 relate to the allocation of performance objectives in the HRX with respect to the distance of the particular circuit portion from the T-reference. This concept is called circuit quality demarcation and the circuit portions are denoted as local grade, medium grade, and high grade.

The local grade applies to that portion of the HRX between the T-reference

Table 6-10 HRDS

HRDS Quality Classification	Maximum HRDS Length	Circuit Classification
1	280 km	High grade
2	280 km	Medium grade
3	50 km	Medium grade
4	50 km	Medium grade
Not defined	Not defined	Local grade

and the local exchange and is roughly equivalent to the local loop. The medium grade portion applies to that portion between the local exchange and a distance of 1,250 km from the T-reference. Therefore, the local grade and medium grade portions together apply to the first 1,250 km of a circuit extending into the HRX from either end (total of 2,500 km). The high grade portion covers 25,000 km and connects the medium grade portions at each end. The sum of the local grade, medium grade, and high grade portions equals the longest HRX of 27,500 km. There is no clear demarcation between the medium grade and high grade portions except as defined above by distance. This means one portion of a given HRDP or HRDL may be classified medium grade and another portion may be high grade.

It is convenient to consider the national part mentioned above as being the first 1,250 km from the T-reference, although CCITT does not explicitly state this in any recommendation. If this consideration is made, the national part would coincide with the sum of the local grade and medium grade portions of the HRX.

Through various recommendations, CCITT recognizes that the lengths of real transmission networks may differ considerably from the hypothetical models. The main difference is that the real circuit performance objectives relate to actual system lengths, whereas the HRDS objectives are given as block allowances for fixed (usually maximum) reference lengths.

6.3.2.2 Performance Parameters

The two fundamental categories that CCITT uses to describe digital transmission performance are unavailability and quality. Related to quality, but discussed separately, is the number of slips in a network as a result of timing and synchronization differences.

The performance quality parameters specified in CCITT and CCIR recommendations are the degraded minute (DM), severely errored second (SES), errored second (ES), residual bit error ratio (RBER), and unavailability (or, conversely, availability). The RBER applies only to radio-relay systems. Each parameter is expressed in statistical terms and most are based on BERs. All parameters are given at the DS-0 rate, 64 kbps, unless otherwise indicated in the following discussion.

In order to quantify performance, CCITT identifies two threshold BER values. The first threshold of 1E-3 is the point beyond which the connection becomes unacceptable to most services and many processes in a transmission network begin to malfunction (for example, multiplexers lose frame alignment). An integration (or averaging) time of one second applies to the measurement of this BER limit, corresponding to 64 errors per second at the 64 kbps rate. The SES is used to describe a BER of 1E-3 over a one-second period.

The second threshold of 1E-6 is the point at which degradations to voice telephony using pulse code modulation (PCM) become perceptible. An integration time of one minute applies to the measurement of this BER limit. The one-minute time period is a compromise between the length of a typical telephone conversation and a practical measurement period at the 64 kbps rate. The DM is used to describe a BER of 1E-6 over a one-minute period. In practice, a DM at 64 kbps is considered to contain *more than* four errors (four errors corresponds to a BER of 1.04E-6 at the 64 kbps rate). The one-minute intervals used to measure DM are derived by removing unavailable time and SES from the total time and then consecutively grouping the remaining seconds into blocks of 60.

As previously mentioned, an ES is any second that contains at least one error. The ES is considered an alternative to BER in characterizing error performance in systems used primarily for data transmission.

The RBER applies specifically to radio-relay systems and characterizes error performance at the system bit rate during nonfaded propagation conditions. The specified RBER is such that if the DM, SES, and RBER objectives are met at the system bit rate, then the ES objectives should automatically be met at the 64 kbps rate.

Unavailability begins when the BER in each second is worse than 1E-3 for a period of ten consecutive seconds. These ten seconds are considered to be unavailability time. A new period of availability begins with the first second of a period of ten consecutive seconds each of which has a BER better than 1E-3. The ten-second timing criterion is what divides the performance into the two basic categories of quality and unavailability. It is a compromise between the need for immediate action to correct network disruptions and the need to wait for automatic protection switching to take place or fading to subside.

Radio engineers commonly use the term *outage* to describe any period during which the BER exceeds 1E-3 on a digital radio path. Outage in these systems affects both quality and unavailability.

The performance of a circuit is defined by measurement and categorization of the performance parameters. This categorization is performed as follows:

1. Monitor the circuit for S seconds.
2. Count the number of one-second intervals that have more than 64 errors in each second (>1E-3 BER at 64 kbps) for ten consecutive seconds; these are unavailable seconds (US). Note that a one-second interval may have more than 64 errors and still be considered available; ten consecutive seconds, each with more than 64 errors, are required before the circuit is declared unavailable. The available seconds (ASs) are found by subtracting US from the total monitoring time, S; that is, AS = S − US. Available minutes (AMs) = AS/60.
3. Count the number of one-second intervals containing at least one error. These are ES.
4. Count the number of one-second intervals containing more than 64 errors. These are SES.
5. Exclude the one-second intervals with SES found in (4) from the count of AS found in (2). Find groups of 60 consecutive one-second intervals in the remaining AS with more than four total errors (1E-6 < BER < 1E-3 at 64 kbps). These are DM.
6. The allocations for the different HRDP, HRDL, and HRDS given in the next section are applied directly to these values. For example, each local loop access part is allowed 15% of the DM found in (5), 15% of the ES found in (3), and 15% of the SES found in (4).

In practical system design, performance specifications and measurements normally are made at actual system bit rates, while most CCITT recommendations are based on a 64 kbps rate. When an errored signal is demultiplexed from the system bit rate through each hierarchical level to 64 kbps, the BER does not change, provided frame alignment is not lost. This is because the errors are uniformly distributed among all 64 kbps tributaries or channels (assuming a random error distribution). Since the SES and DM are based on BER thresholds, their values remain the same for the different hierarchical levels. Therefore, X% DM and Y% SES at the system bit rate yields X% DM and Y% SES at 64 kbps. This one-to-one relationship also is true of RBER for radio systems, although RBER is defined at the system bit rate and not at the 64 kbps rate.

In the case of loss-of-frame alignment (LOFA), the percentage of SES (Y%) at 64 kbps increases by the percentage of non-SES (Z%) at the system bit rate containing one or more LOFA at the system bit rate. These non-SESs occur during demultiplexing from the system bit rate to the 64 kbps channel rate, when extended error bursts cause LOFA. For this situation, the total SESs at 64 kbps is Y% + Z%. CCIR Rec. 634-2 indicates that Z% falls between 1% and 5% of the measured performance of SES (Y%), which gives a total SES at 64 kbps of 1.01Y% to 1.05Y% [30].

Translation of ES between system bit rate (for example, 1.544 Mbps) and 64 kbps is more difficult because the ES depends on both the number and the distribution of errors in each ES at the system bit rate. CCITT, however, provides guidelines in Annex D of Rec. G.821 for translating ES, but states that "assessment of system error performance by means of these [procedures] does not assure compliance with this Recommendation" at the 64 kbps rate [2]. The guidelines for normalizing ES at the system bit rate to a 64 kbps rate are as follows (note that this relationship requires knowledge of the number of errors at the system bit rate and not the ES at the system bit rate):

$$\%ES_{64} = \frac{1}{j} \sum_{i=1}^{j} \left(\frac{n}{N}\right)_i 100$$

where $\%ES_{64}$ = %ES normalized to the 64 kbps rate

n = number of errors in the *i*th second at the system bit rate

N = ratio of system bit rate to 64 kbps (or, R/64 kbps, where R = system bit rate)

j = integer number of one-second periods (excluding unavailable time) in the measurement period

$(n/N)_i$ = 1, if $n \geq N$, or

= n/N, if $0 < n < N$

The foregoing may be interpreted as follows: If measurements at the system bit rate indicate at least N errors during the measurement period, then $n_i > N$, and there are sufficient errors in each ES at the system bit rate to produce at least one error in each 64 kbps channel. Conversely, if there is just one random error in each ES at the system bit rate, then the number of ES at 64 kbps is smaller by a factor of 1/N. In practice, the actual ES at 64 kbps is better than the value calculated from the foregoing expression.

6.3.2.3 Performance Allocation

The end-to-end objectives for DM, SES, and ES, which apply to any HRX, are given in Table 6-11. All objectives are based on suitable, but unspecified, time periods based on the application. CCITT suggests one month in the notes to Table 1/G.821, but some CCIR recommendations relate performance over one or more one-year periods.

The specific CCITT performance recommendations depend on the location of the measurement point within the HRX. CCITT Table 2/G.821 and Table 3/G.821 break down the overall DM, ES, and SES objectives for the HRX into allocations for each part (local grade, medium grade, and high grade). These allocations are shown in Table 6-12. Any HRX contains two local, two medium, and one high grade portions.

The application of these allocations is shown in the following examples: The total end-to-end DM must be less than 10% over the measurement period. Of these DMs, 15% are allocated to each local grade portion, or (15% × 10% DM =) 1.5% DM. Similarly, 15% of the total end-to-end ES are allocated to the medium grade portion, or (15% × 8% ES =) 1.2% ES.

Notes to the tables in CCITT Rec. G.821 state that administrations may allocate the block allowances for local grade and medium grade portions as necessary within the total allowance of 30% for any one end of the connection (in other words, the allocation does not have to be 15%–15% between local grade and medium grade, but could be, for example, 10%–20%). This would be practical only where a single entity controls both the local grade and medium grade portions or where there is specific agreement between interconnecting entities. As previously mentioned, the sum of the local grade and medium grade portions may be considered the national part of the network.

Although the preceding tables show simple percentage allocations, other notes to the CCITT tables provide several qualifiers. *First*, CCITT has split the overall SES objective of 0.2% into two equal parts. One half of the total allowed SES (that is, 50% × 0.2% SES = 0.1% SES) is divided between local, medium, and high grade portions as shown in Table 6-12. For example, the local grade allowance is (15% × 0.1% SES =) 0.015% SES and the high grade allowance is (40% × 0.1% SES =) 0.04% SES for the first half. The *remaining* half of the total allowed SES (0.1%) is a block allowance to the medium and high grade portions as follows: 50% of the *remaining* one half (that is, 50% × 0.1% = 0.05% SES) is allocated to one 2,500 km radio-relay HRDP, used in the high grade portion of the connection, and

Table 6-11 End-to-End Performance Objectives

Performance Parameter	Objective
DM	<10%
SES	<0.2%
ES	<8%

Table 6-12 Allocation of Performance Objectives

Circuit Classification	DM	ES	SES
Local grade	15% block allowance to each end (no adjustment for length)	15% block allowance to each end (no adjustment for length)	15% block allowance to each end (no adjustment for length; see text)
Medium grade	15% block allowance to each end (no adjustment for length)	15% block allowance to each end (no adjustment for length)	15% block allowance to each end (no adjustment for length; see text)
High grade	40% (equivalent to 0.0016% per km for 25,000 km)	40% (equivalent to 0.0016% per km for 25,000 km)	40% (not more than 0.004% SES in each 2,500 km; see text)

the other 50% of the *remaining* one half (0.05% SES) is allocated to the medium grade portion. Note that this extra allowance is made only in one of the ten possible 2,500 km radio-relay HRDP used in the 25,000 high grade portion of the HRX, and that it can apply to one medium grade portion or can be split between the two medium grade portions. This allowance takes into account occasional adverse propagation conditions occurring during the worst month of the year. If a satellite HRDP is used in the high grade portion, it is allowed 10% of the *remaining* SES in the high grade portion (that is, 10% × 0.1% SES = 0.01% SES).

To confirm that the individual SES allowances add up, consider a hypothetical connection consisting of ten radio-relay 2,500 km HRDP in the high grade portion, two 1,250 km radio-relay HRDP in the medium grade portions, and two local grade portions. The ten high grade HRDP are allowed (10 × 0.004% SES =) 0.04% SES, and one of these HRDP is allowed an additional 0.05% SES. The total is (0.04% + 0.05% =) 0.09% SES. The medium grade portions on each side of the high grade portion are allowed (15% × 0.1% SES =) 0.015% SES, for a subtotal of 0.03% SES. The medium grade portions are allowed an additional 0.05% SES, giving a new total of (0.03% SES + 0.05% SES =) 0.08% SES for both. Finally, the local grade portions are allowed (15% × 0.1% SES =) 0.015% SES each, for a total of 0.03% SES. Summing the local, medium, and high grade portions gives (0.03% SES + 0.08% SES + 0.09% SES =) 0.20% SES, which is the total allowance as given in Table 6-11.

As a second example, consider an HRX consisting of a satellite HRDP in one high grade portion, a radio-relay HRDP in the remaining high grade and medium grade portions, plus two local grade portions. In this case, the satellite HRDP in the high grade portion is allowed 0.02% SES. Assuming the radio-relay HRDP total 12,500 km in the high grade portion, they would be allowed (5 × 0.004% SES =) 0.02% SES. This gives a subtotal of 0.04% SES. In addition, the satellite HRDP is allowed an additional 0.01% SES and the radio-relay portions are allowed an additional 0.05% SES for adverse propagation. However, Par. 3.2 (b) of Rec. G.821 implies that the satellite and radio-relay HRDP are unlikely to experience worst-month conditions during the same month (although Rec. G.921 says this is under study). The total SES for the high grade portion is then (0.04% SES + 0.05% SES =) 0.09% SES, and the extra allowance for the satellite HRDP is ignored. The medium grade portions are allowed 0.08% SES and the local grade portions are allowed 0.03% SES, as computed previously. The total for this HRX is then (0.09% SES + 0.08% SES + 0.03% SES =) 0.20% SES.

To translate the performance percentages to actual values to be measured, simply determine the number of AM or AS during the measurement period. For example, if the measurement is made over a one-month period of 30 days, there is a maximum of (3,600 seconds/hour × 24 hours/day × 30 days/month =) 2,592,000 seconds. Assuming all these seconds are available, the allowable SES for an HRX is (2,592,000 seconds × 0.2% SES =) 5,184. If the unavailability during this measurement period is 0.05%, then the total AS is (2,592,000 seconds × 99.95% =) 2,590,704, and the allowable SES is (2,590,704 seconds × 0.2% =) 5,181.

Although most allocations in Table 6-12 are length independent, CCITT recognizes that the performance of real transmission systems depends on their lengths.

The high grade allocations of DM, SES, and ES are divided on the basis of length to allow the derivation of a block allowance for a defined network model, such as a real radio-relay link or HRDS.

CCIR provides guidance in Rec. 634-1 for real digital radio-relay links used in high grade circuits, which may be significantly shorter than the 2,500 km HRDP [13]. In this case, the various performance objectives are adjusted for links with a length between 280 and 2,500 km by multiplying the performance objective for the 2,500 km HRDP by the factor L/2,500. This is summarized in Table 6-13.

CCITT uses a similar adjustment to derive the performance of 50 km and 280 km HRDS used in medium and high grade circuits.

Finally, if a satellite system is used in the high grade portion, CCIR Rec. 614-1 gives a block allowance of 20% of the allowed DM and 20% of the allowed ES to the satellite HRDP [12]. Therefore, the satellite HRDP is allowed (20% × 10% DM =) 2% DM, while the other systems in the high grade portion are allowed the remaining 2% DM (for a total of 4% DM allowed in the total high grade portion). For ES, the satellite HRDP is allowed (20% × 8% ES =) 1.6% ES, while the other systems in the high grade portion are allowed the remaining 1.6% ES (for a total of 3.2% ES allowed in the total high grade portion).

The DM, SES, and ES requirements of CCITT Rec. G.821 are used to derive BER performance of satellite systems. These requirements are given in CCIR Rec. 614-1 and repeated in Table 6-14. CCIR Report 997-1 describes various methods to derive BER performance from the requirements of CCITT Rec. G.821 [14].

The preceding discussion of satellite system performance applies at the 64 kbps rate for systems forming a part of an international connection in an ISDN. Satellite systems used only for PCM telephony (that is, a voice channel encoded using μ-law or A-law PCM) have slightly different performance requirements, as given in CCIR Rec. 522-3 [12] and repeated in Table 6-15.

Connections consisting of digital lines, such as twisted pair and optical fiber, should meet the overall requirements given for the radio-relay and satellite HRDP already discussed. CCITT does not specifically address the performance of a 2,500 km HRDP consisting entirely of line systems (that is, the HRDL). However, CCITT

Table 6-13 Allocation of Performance to Real HRDP

Performance Parameter	Allocation
DM	0.4% × L/2,500
SES	0.054% × L/2,500
ES	0.32% × L/2,500
RBER	5E-9 × L/2,500

Table 6-14 Satellite System BER Performance When Used in an ISDN

BER	Allowable Time Period
>1E-7	<10% of any month
>1E-6	<2% of any month
>1E-3	<0.03% of any month

Table 6-15 Satellite System BER Performance When Used with PCM Telephony

BER	Allowable Time Period
>1E-6	<20% of any month
>1E-4	<0.3% of any month
>1E-3	<0.05% of any month

Rec. G.801 and G.921 define the HRDS, which is a constituent element of an HRDL and an HRDP. Performance allocations are given for the HRDS in Table 6-16. The HRDS are cascaded as required to make up the HRDL.

The lengths given in Table 6-16 are maximum lengths. If a real HRDS is shorter, there is no reduction of the allocation. If a real HRDS is longer, its overall allocation corresponds to an integer number of HRDS with the same quality classification. A given part of an overall connection may consist of any of the four circuit quality classifications shown in Table 6-16. The choice of classification depends on the propagation characteristics of each HRDS.

A higher quality HRDS, say, class 2, may be used to compensate for the lesser quality of a class 3 HRDS to meet the overall requirements of a medium grade HRDP. The question arises: why not initially design all the HRDSs to the higher quality to guarantee that the medium grade requirements are met? If the goal is simply to meet the minimum requirement, this could result in overdesign. In other cases, it may not be possible to design a class 2 HRDS because of economically insurmountable design or construction problems.

The allocations given are percentages of the overall degradation (at 64 kbps) specified in CCITT Rec. G.821 and previously discussed above; that is, 10% DM, 8% ES, and 0.1% SES. For example, the total high grade portion is allowed (40% × 0.1% SES =) 0.04% SES, as previously mentioned. A 280 km class 1 HRDS, used in a high grade portion, is allowed (0.45% × 0.1% SES =) 0.00045% SES. If the HRDP is a class 2 medium grade, it is allowed (2% × 0.1% SES =) 0.02% DM.

To account for adverse propagation, as detailed in CCITT Rec. G.821 and previously discussed, an additional 0.05% SES is allocated to a 2,500 km HRDP for systems operating in the medium and high grade portions of the HRX. This corresponds to an additional 0.0055% SES in a 280 km HRDS with a quality classification of 1 or 2, and results in an additional allowance of 0.025% SES available for each 1,250 km medium grade portion. Part of this additional allowance may be allocated on a linear basis to HRDS with quality classifications 3 and 4, if necessary.

Table 6-16 Performance Allocation of the HRDS

Quality Classification	Length	Allocation	Circuit Classification
1	280 km	0.45%	High grade
2	280 km	2%	Medium grade
3	50 km	2%	Medium grade
4	50 km	5%	Medium grade

The above allocations for the medium grade portion of a connection are given in CCIR Rec. 696 [13], and the four circuit quality classifications are defined in CCITT G.921 [2]. Table 6-17 shows the performance allocation that results from combining Tables 6-11 and 6-16. For class 1 and 2 in Table 6-17, one half of the allowed SES (0.2%/2) is allocated as a block allowance. The other half of the allowed SES is allocated on the basis of length, as discussed earlier in this paragraph.

The previous discussions focused on systems used in the high and medium grade portions of the HRX. For digital radio-relay systems used in the local grade portion, CCIR Rec. 697 gives the objectives shown in Table 6-18 [13]. These are derived from the basic requirements of CCITT Rec. G.821, as previously discussed.

It should be noted that CCITT Rec. G.921 performance objectives apply specifically to an HRDS based on the 2,048 kbps digital hierarchy. CCITT does not have performance objectives for an HRDS based on the 1,544 kbps digital hierarchy because networks using this hierarchy predominate in North America and Japan, and already are designed to high grade requirements.

Optical lines systems based on the 1,544 kbps digital hierarchy are covered in CCITT Rec. G.955 [2]. Optical lines are not broken into HRDL or HRDS. Instead, CCITT Rec. G.955 provides hypothetical optical lines systems with and without intermediate regenerators, but with no specified maximum length. The only performance parameter recommended is BER, which should be 1E-10 for the maximum regenerator section length. The regenerator length depends on the wavelength, optical transmitter power, optical receiver sensitivity, desired operating margin, and other parameters. CCITT Rec. G.956 provides similar requirements for optical lines based on the 2,048 kbps digital hierarchy.

The CCITT recommendations recognize that some situations may exist where the performance of individual parts of a connection may be below the recommended values for that part. This may require higher performance in other parts to ensure the end-to-end connection meets the recommended performance objectives.

The performance parameters and their recommended values for each part of an HRX are summarized in Table 6-19.

Table 6-17 Performance Allowance for Each Circuit Quality Classification—Radio Relay Systems

Parameter	Class 1, 280 km	Class 2, 280 km	Class 3, 50 km	Class 4, 50 km
DM	0.045%	0.2%	0.2%	0.5%
SES	0.006%	0.0075%	0.002%	0.005%
ES	0.036%	0.16%	0.16%	0.4%
RBER	5.6E-10	Not given	Not given	Not given

Table 6-18 Local Grade Performance Requirements

Parameter	Allowable Time Period
DM	<1.5% of any month
SES	<0.015% of any month
ES	<1.2% of any month

Table 6-19 Summary of CCITT Performance Allocations

Type	Length (km)	Class	Param.	Rate	BER	Objective	Reference
HRX	27,500	Mixed	DM	64 kbps	1E-6	<10%	Rec. G.821 [2]
			SES	64 kbps	1E-3	<0.2%	
			ES	64 kbps	—	<8%	
HRDP (Radio-Relay)	2,500	High	DM	64 kbps	1E-6	<0.4%	Rec. 594-2 [13]
			SES	64 kbps	1E-3	<0.054%	
			ES	64 kbps	—	<0.32%	
HRDP (Radio-Relay)	280–2,500	High	DM	System	1E-6	<0.4% × L/2,500	Rec. 634-1 [13]
			SES	System	1E-3	<0.054% × L/2,500	
			ES	64 kbps	—	<0.32% × L/2,500	
Real (Radio-Relay)	<1,250	Medium	RBER	System	5E-9 × L/2,500	—	Rec. 696 [13]
			DM	64 kbps	1E-6	<1.5%	
			SES	64 kbps	1E-3	<0.04%	
			ES	64 kbps	—	<1.2%	
Real (Radio-Relay)	~10	Local	DM	64 kbps	1E-6	<1.5%	Rec. 697 [13]
			SES	64 kbps	1E-3	<0.015%	
			ES	64 kbps	—	<1.2%	
HRDP (Satellite)	12,500 (equivalent)	High	DM	64 kbps	1E-6	<2%	Rec. 614-1 [12]
			SES	64 kbps	1E-3	<0.03%	
			ES	64 kbps	—	<1.6%	
Optical line	Per section	Any	BER	System	<1E-10	—	Rec. G.955 [2]
HRDS	280	High (1)	DM	64 kbps	1E-6	0.045%	Rec. G.921 [2]
			SES	64 kbps	1E-3	0.006%	
			ES	64 kbps	—	0.036%	
			RBER	System	5.6E-10	—	
HRDS	280	Medium (2)	DM	64 kbps	1E-6	0.2%	Rec. G.921 [2]
			SES	64 kbps	1E-3	0.0075%	
			ES	64 kbps	—	0.16%	
HRDS	50	Medium (3)	DM	64 kbps	1E-6	0.2%	Rec. G.921 [2]
			SES	64 kbps	1E-3	0.002%	
			ES	64 kbps	—	0.16%	
HRDS	50	Medium (4)	DM	64 kbps	1E-6	0.5%	Rec. G.921 [2]
			SES	64 kbps	1E-3	0.005%	
			ES	64 kbps	—	0.4%	

6.3.2.4 Availability and Unavailability

A circuit has measurable quality only when it is available. When the quality of a circuit degrades to a certain point, it becomes unavailable. Table 6-20 lists the CCIR recommendations and corresponding reports that specify the availability and unavailability objectives for the different circuit parts.

Availability and unavailability are related by

$$A\% = 100 - U\%$$

where A% = percentage availability
U% = percentage unavailability

The unavailability (and conversely, availability) varies with the circuit classification. For a satellite HRDP, the unavailability due to propagation should be no more than 0.2% of any month and 0.04% to 0.1% of any year (the annual percentage is still being studied). In these recommendations, "any month" and "any year" are meant to be the worst month over a one-year period and worst year over a four- to five-year period, respectively. CCIR suggests that the value for "any month" corresponds to a period of "any year" by a conversion factor of 5. That is, the average conditions throughout the one-year period of "any year" should be five times better than in the worst month; most months will have very high availability. Unavailability of a satellite HRDP due to equipment failure should be no more than 0.2% of a year.

The foregoing requirements give equipment availability of at least 99.8% and propagation availability of 99.9% to 99.96% during a year. In the United States, public network satellite systems are routinely designed to much higher availability requirements, typically 99.999% or better.

According to the CCIR recommendations previously cited, the availability of a 2,500 km radio-relay HRDP should be at least 99.7% over a period of at least one year. Real radio-relay systems are usually much shorter than 2,500 km. For high grade systems between 280 and 2,500 km long, the following availability, A, is recommended by CCIR:

$$A\% = 100 - (0.3 \times L/2{,}500)$$

CCIR notes that the value of 0.3 in this equation is provisional and, in practice, may fall in the range of 0.1 to 0.5. For medium grade systems, the recommended avail-

Table 6-20 CCIR Availability and Unavailability Recommendations and Reports

RECOMMENDATION	APPLICATION	RELATED REPORT
CCIR 579-1 [12]	HRDP (Satellite)	997-1 [14]
CCIR 557-2 [13]	HRDP (Radio-relay)	784-3 [15]
CCIR 695 [13]	Real (Radio-relay, high grade)	784-3 [15]
CCIR 696 [13]	HRDS (Radio-relay, medium grade)	1052-1 [15]
None	HRDS (Radio-relay, local grade)	1053-1 [15]

ability depends on the circuit classification as shown in Table 6-21. These values are over at least a one-year period.

Table 6-21 Medium Grade Radio-Relay Availability

CLASSIFICATION	LENGTH	AVAILABILITY
1	280 km	99.967%
2	280 km	99.95%
3	50 km	99.95%
4	50 km	99.90%

The availability of radio-relay systems used in the local loop (local grade) is still under study by CCIR. However, a comprehensive CCIR report (see Table 6-20) suggests that the availability should be in the range of 99.99% to 99.00% averaged over one or more years. From a practical standpoint, the actual value would depend on the user requirements and engineering policy for these types of systems.

Radio-relay systems used in the local loop generally are short (less than 10 km), do not use repeaters, and operate at frequencies above 10 GHz. Their availability is determined mainly by equipment reliability and adverse propagation. CCIR suggests that the mean time between failure (MTBF) of local loop radio equipment used in a bidirectional path is at least 25,000 hours. Using typical repair times of six to 48 hours, the annual availability is found to be between 99.99% and 99.8%. The availability can be increased considerably by providing protection switching; however, this doubles the cost of each radio terminal and, in most local loop systems, the extra cost is not justifiable.

Short radio loops are not subject to the same fading conditions that exist for longer paths, but the high frequencies are subject to adverse propagation due to rain attenuation. Rain heavy enough to make the path unavailable may occur several times each year. On the other hand, unavailability due to equipment failure occurs only every three to ten years. Therefore, availability objectives for local loop radio-relay systems are appropriately based on measurements over several years.

Local grade line systems, such as fiber-in-the-loop and twisted pair digital loops, are not subject to the conditions that affect systems based on radio propagation. Instead, these systems are affected by dig-ins (cable damage), line repeater failure, and terminal equipment failure. In unprotected local loop systems, dig-ins may be relatively frequent (on a system basis) and usually lead to the longest outages, whereas equipment failure is very rare. CCITT provides no guidance on these matters. Designers must rely on local or system experience.

6.3.2.5 Slips

Slips are closely related to synchronization problems, as discussed in Chapter 4. Slips, whether controlled or uncontrolled, cause errors. Although it is desirable to minimize slips, they cannot be fully eliminated from interconnected digital networks. Even within a fully synchronized network, slips are inevitable due to equipment failures and environmental factors, including satellite wander or diurnal shifts in propagation velocity in optical fibers, which exceed multiplexer or terminal equipment buffer capacities.

Slip objectives are defined in CCITT Rec. G.822 by equating an average slip rate to each of three proportions of time, as shown in Table 6-22 [2]. An end-to-end connection should not experience more than five slips in 24 hours at least 98.9% of the time during the measurement period (usually one year). For DS-1 rate transmission systems in the United States, controlled slip objectives are ≤ five per day, end-to-end [16].

Table 6-22 CCITT Rec. G.822 End-to-End Slip Performance Objectives

SLIPS	PROPORTION OF TIME PERIOD
≤5 per 24 hours	98.9%
if >5 per 24 hours but ≤30 per hour	1.0%
>30 per hour	0.1%

6.4 Transmission Operational Problems

Transmission problems in digital systems manifest themselves as errors. From a user's point of view, it does not matter what causes the errors; it is only important that the network operator fixes the problem causing the errors—and fixes it quickly! Both private and public network operators need to know what types of problems can occur because that knowledge makes locating and fixing the problem easier. The following sections define those problems. Emphasis is made on DS-1 rate systems because that is where the performance standards are well developed.

6.4.1 Framing Errors

The superframe format (SF) used with repeatered T1-carrier has no inherent performance monitoring method, but limited error detection still is possible. The SF uses a framing bit followed by 192 data bits. The framing bit sequence takes on a fixed pattern, so once synchronization has been achieved, errors in the received frame pattern are easy to detect. However, of the 193 total bits transmitted in a frame, only one erroneous bit is detectable with absolute confidence; therefore, the error detection accuracy is only (1/193 =) 0.52%, at best. The ESF uses a similar framing and framing error detection mechanism, but the ESF also has inherent error detection through the cyclic redundancy check (crc) code. Framing errors, known generically as *frame sync bit errors*, can result in uncontrolled slips of one or more frames.

When the frame pattern is lost, an out-of-frame (OOF) event has occurred. An OOF event is the same as an loss-of-frame alignment (LOFA), or severely errored framing event. Generally, an OOF condition is declared when the terminal or network equipment senses errors in any two-of-four, two-of-five, or three-of-five consecutive terminal framing (F_t) bits. A two-of-four condition is specified by IEEE [17]. An OOF condition clears when reframe occurs. In the absence of framing bit errors, the normal reframe time is less than 50 mseconds for most DS-1 multiplex equipment.

When an OOF event occurs, the framing mechanism may initiate an on-line or off-line framing search. When the on-line search is made, the process causes an OOF condition even if the framing mechanism had correct frame alignment and received inadvertent errors. With off-line framing, the search does not cause an out-of-frame condition if the frame alignment mechanism actually is correctly aligned. When the search reveals a difference in frame alignment, only then is a change-of-frame alignment (COFA) made by choosing a new framing bit. The conditions indicated by a COFA and an OOF condition are summarized in Table 6-23. OOFs and COFAs can be a symptom of severe error bursts, justification (stuffing) errors, high error rates, and synchronization problems. A controlled slip does not result in LOFA.

The slip objectives given previously are to be achieved over a long time. The day-to-day operational requirements of digital loops are slightly different. Slip performance for an integrated digital loop carrier (IDLC) is measured continuously by the host digital switching system (DSS). The DSS issues a maintenance alarm if the number of controlled slips exceed a preset maintenance threshold during the measurement interval. If the controlled slip rate degrades further, the loop is automatically taken out of service. A similar action is taken for COFA events. COFA events are counted, and when the count exceeds the maintenance threshold, a maintenance alarm is issued. When the count exceeds the out-of-service threshold, the loop is automatically taken out of service.

The BER is monitored by digital loop equipment by analyzing the terminal framing (F_t) bits (for superframe formatting) or the CRC6 block (for extended superframe formatting) for errors. For SF, there are 4,000 F_t bits per second. When the estimated error rate exceeds 1E-3, corresponding to four F_t bit errors per second, an *out-of-service* condition is declared. If 1E-6 $\leq$ estimated error rate $<$ 1E-3, a *maintenance limit* condition is declared. For ESF, similar actions are taken, but the error rates are estimated from the CRC6 block rather than from the framing bits. The error rate algorithms are designed to respond to stable, long-term error rate conditions and are not accurate for transient conditions. Table 6-24 shows the approximate detection times for typical digital loop equipment.

Table 6-23 OOF (COFA) Alignment Comparison

EVENT	CONDITION INDICATED
OOF	Percentage of framing bits are in error (for example, two-of-four)
COFA	A new bit has been selected as the framing bit

Table 6-24 BER Detection Times for Typical Digital Loop Equipment Operating at DS-1 Rate

FORMATTING	BER	DETECTION TIME
Superframe (SF)	1E-3	Approx. 10 seconds
	1E-6	Approx. 40 to 60 minutes
Extended superframe (ESF)	1E-3	Approx. 10 seconds
	1E-6	Approx. 10 minutes

Table 6-25 gives typical performance values for IDLC systems connected to digital end-offices. The same values apply to universal digital loop carrier (UDLC) systems, except the digital loop carrier (DLC) central office terminal (COT) takes the appropriate action rather than the DSS.

Table 6-25 Service Thresholds for Digital Loops Operating at the DS-1 Rate

EVENT	CONSEQUENCE	MAINTENANCE THRESHOLD	OUT-OF-SERVICE THRESHOLD
Controlled slip	No LOFA	4 slips in 24-hour period	255 slips in 24-hour period
Uncontrolled slip	LOFA	17 COFA in 24-hour period	511 COFA in 24-hour period
BER	Degraded service	≥1E-6	≥1E-3

6.4.2 BPVs

Digital loop transmission systems follow specific line coding rules. When a line receiver detects a violation of these rules, a line code violation event has occurred. The bipolar AMI line code used in the T1-carrier system is a familiar line code. AMI requires every pulse (binary one) to be transmitted with opposite polarity from the previous pulse. If two pulses in sequence are received with the same polarity, a BPV has occurred, as shown in Figure 6-5.

A BPV indicates at least one error was made. This error detection method is limited because it is possible for several errors to be made during transmission that would not result in a BPV. In systems where errors occur at random and without dependency on other errors, BPV detection provides a good indication of BER [18]. Errors in digital loop systems are bursty, however, and BER measurements using only BPV under these conditions are not always accurate. The operational limits for BPVs on digital loops usually are given on the basis of BPVs per bits transmitted. Table 6-26 shows typical values.

Relying on BPV detection is further limited because BPVs are not passed through multiplexers. This is illustrated in Figure 6-6(a), which shows two repeatered T1-carrier span lines used as digital loops and connected through a high-capacity fiber

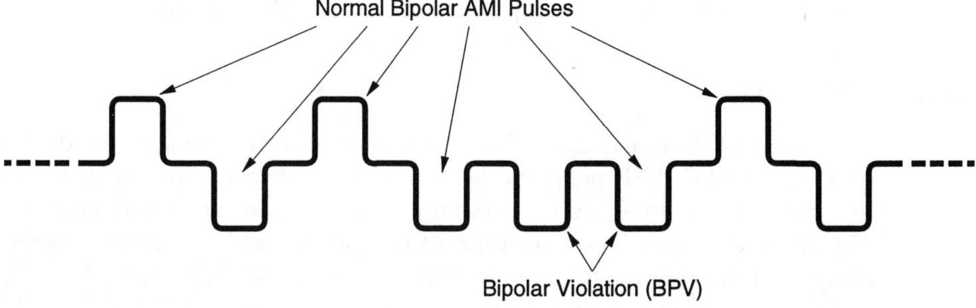

Figure 6-5 Illustration of BPV

Table 6-26 BPVs on Digital Loops Operating at the DS-1 Rate

TYPE	THRESHOLD
Maintenance threshold	1 BPV per 100,000 bits (1E-5)
Out-of-service threshold	1 BPV per 1,000 bits (1E-3)

(a)

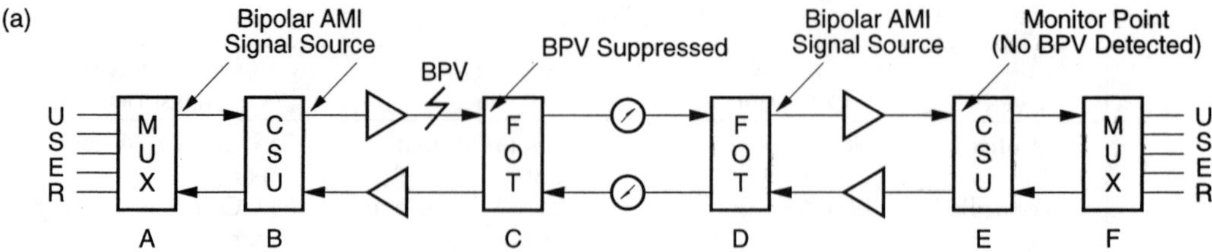

(b)

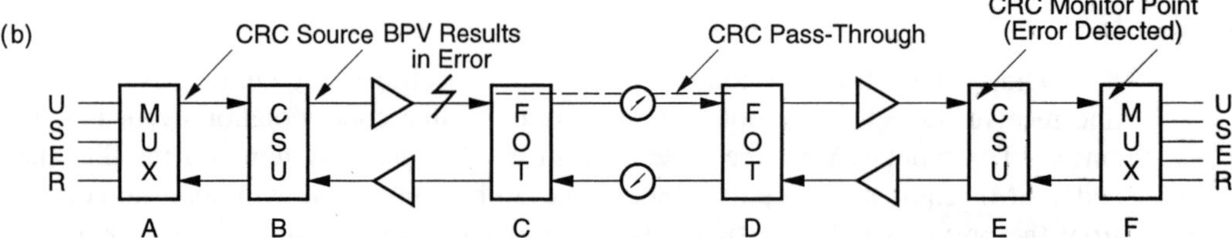

Figure 6-6 Limitations of BPV detection: (a) BPV suppression; (b) CRC pass-through

optic transmission system. Any BPVs occurring between the channel service unit (CSU) associated with T1 terminal multiplexer (MUX) A and the fiber optic terminal (FOT) C are suppressed or filtered by the multiplexer. Regardless of polarity, a pulse always is detected as a binary one, even though it may be a BPV.

The data (including errors) are passed through the fiber optic transmission system to FOT D. The data are re-encoded at the output of FOT D, according to the bipolar AMI line code rules for further transmission on the T1 span line, to MUX F through CSU E. MUX F does not detect the BPV that occurred between B and C. This situation is overcome by using a CRC code, as shown in Figure 6-6(b). With CRC code, an error occurring in the span line is detected at CSU E or MUX F when the received CRC code is compared to the local CRC calculation.

6.4.3 CRC Codes

The ESF is an example of a system used to overcome some of the limitations of T1-carrier SF. ESF provides the capability to detect errors in systems carrying live traffic through the use of a six-bit check code associated with each data block. The check code provides a measure of redundancy and is sent in a cyclic manner along with the data block, so it is called a CRC code. Since the CRC code is part of the framing bits, it is passed through any intervening multiplexers.

The CRC code (called a block check character, or BCC) is generated by a polynomial division process at the transmitting end. The BCC indicates the bits sent in the block immediately preceding it. In ESF, the block is 24 frames in length (4,632 bits), and the BCC resulting from division by a sixth-order polynomial is six bits long. This code is called CRC6 to indicate its length. Industry standards also describe CRC6 in detail [7,19,20].

At the receiver, the bits in the data block are again divided by the same polynomial, and the resulting value is compared to the received CRC6 code. A difference means there has been at least one transmission error. This is called a CRC error event. The CRC6 code will detect 100% of all blocks with fewer than six errors and 98.4% of all blocks with more than six errors [20]. The number of bits in error is not indicated, just that there is at least one. In general,

$$\text{CRC error detection capability (\%)} = 100\,(1 - 2^{-c})$$

where c = number of CRC bits

6.4.4 Loss of Signal

A loss-of-signal (LOS) condition occurs whenever the terminal determines it is not receiving pulses at its network interface. This determination can be made either at the multiplex terminal or CSU. Various industry standards exist for the LOS, as follows:

- 100 (+75, −20) consecutive pulse positions with no pulses [16]
- 175 (+75, −75) consecutive pulse positions with no pulses [19]
- 231,600 successive pulse positions (150 ms) with no pulses [21]

An LOS is caused by the total interruption of the incoming bit stream due to equipment or power failure, or cable cut. Some equipment does not differentiate between an LOS and OOF.

6.4.5 Summary of Error Monitoring Capabilities of SF and ESF

Table 6-27 summarizes the error monitoring capabilities that are generally available with equipment using SF and ESF.

Table 6-27 Summary of Error Monitoring Capabilities of SF and ESF

Error Event Type	SF	ESF
Line code violation	Yes	Yes
CRC	No	Yes
Controlled slip	Yes	Yes
OOF	Yes	Yes
COFA	Yes	Yes
Frame Sync Bit	Yes	Yes
LOS	Yes	Yes

6.5 Performance Monitoring and Measurements

6.5.1 Monitoring Systems

In order to achieve the specified performance objectives, any digital transmission system must provide some form of in-service monitoring and service protection through redundancy, manual or automatic protection line switching, and rapid trouble isolation and circuit restoral.

A typical performance monitoring system will have preset thresholds for the various error events described in the previous sections. Events are continuously monitored and reported on demand. If any given error parameter exceeds a preset threshold, the performance monitoring system will generate a message or alarm depending on the sophistication of the operational support system (OSS).

Performance monitoring is a very important aspect of digital network administration and operation. In many cases, multiple circuit providers are involved, including local exchange carriers, inter-exchange carriers, and value-added network providers, which may be in competition with each other. Without performance verification, there would be no way to assess the overall quality provided by various providers.

Methods and standards for performance monitoring of digital networks are brought about through an evolving process. Tariffs, contracts, and state or federal regulations spell out the performance of a circuit or channel. In many cases, the actual measurement of that performance is not standardized. In other cases, for example DS-1 rate services, standards have been developed, as previously described. The following discussion applies to DS-1 rate, which has well-developed performance monitoring methods.

Circuit performance is defined through the use of certain terminology, as shown in Figure 6-7. A *line* is defined as the channel equipment, including transmission medium, between two regenerators. In transmission engineering, this frequently is called a *section*, but this terminology is avoided when discussing performance.

A *path* is the channel equipment between a CSU located on the customer or user premises and the serving central office. A path can include one or more lines. With T1-carrier optioned for ESF, path monitoring equipment uses the CRC6 data to derive path performance.

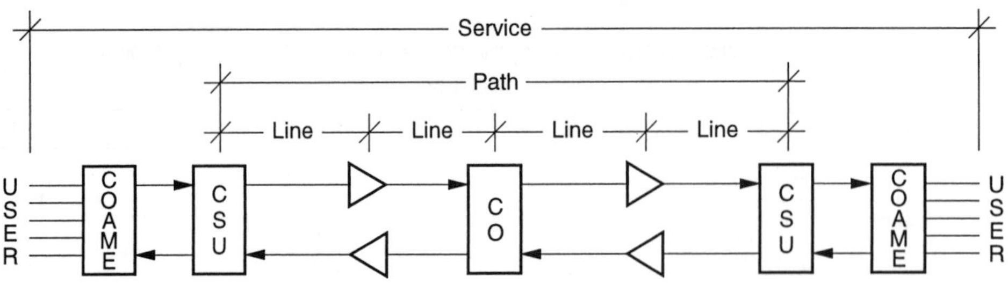

Figure 6-7 Performance terminology

A *service* is the end-to-end channel equipment between two users and includes one or more intermediate central offices, paths, and lines. With T1-carrier, service monitoring equipment monitors the one-second messages in the ESF data link from the user's equipment to derive service performance. Information in the data link can be stripped or read passively at intermediate points.

Equipment used in performance monitoring takes several forms and is connected to digital transmission systems in a variety of ways. Figure 6-8 shows four basic connection methods. A given network will often use all four methods rather than just one, since each has economic or operational advantages and disadvantages that depend on the situation.

The bridging equipment method shown in Figure 6-8(a) uses portable or permanently mounted test equipment designed for temporary connection to the *monitor* jacks of a digital signal cross-connect (DSX) panel through a patch cord. Since the connection is temporary, long-term performance data cannot be collected. Although test equipment can be dedicated to a given circuit, this become prohibitively expensive if more than a few circuits have to be monitored. Also, the performance measured using this method may be between bridging points only, and not from terminal to terminal. Nevertheless, the bridging method is a fast and easy way to obtain performance data where none existed before.

The adjunct equipment method shown in Figure 6-8(b) overcomes some of the disadvantages of the bridging method, but at a greater cost. Adjunct equipment can be designed into the transmission system from the start, or added at a later date. The usual connection point is at the DSX panel. Where a transmission system contains many segments with differing ownership, adjunct equipment can be inserted in each segment. In this case, the performance messages may be exchanged within a given segment only and not traverse the entire path. This precludes collecting end-to-end performance data.

When electronic digital cross-connect systems (DCSs) with performance monitoring capabilities are used, as shown in Figure 6-8(c), the performance of active transmission ports is easily determined. In many networks, however, the DCS is not connected to all lines, and the performance of these lines must be determined by other methods. DCSs are seldom deployed solely to monitor circuit performance.

The final basic method, which uses integral equipment, is shown in Figure 6-8(d). Equipment with this capability usually is a multiplexer, channel bank, or other terminal equipment. The performance monitor is built into the line termination either through retrofit or initial purchase. Terminal equipment with integral performance monitoring capabilities allows the end-to-end performance to be obtained, provided that intervening equipment, such as DCS or adjunct equipment, does not strip the performance messages from the data link.

6.5.2 Alarms

Alarms give quick indications of performance problems. An alarm indication signal (AIS) is an all binary-ones signal (unframed) transmitted to maintain transmission continuity and to notify the receiving terminal (downstream equipment) of an LOS or OOF either in the transmitting terminal or upstream of the transmitting

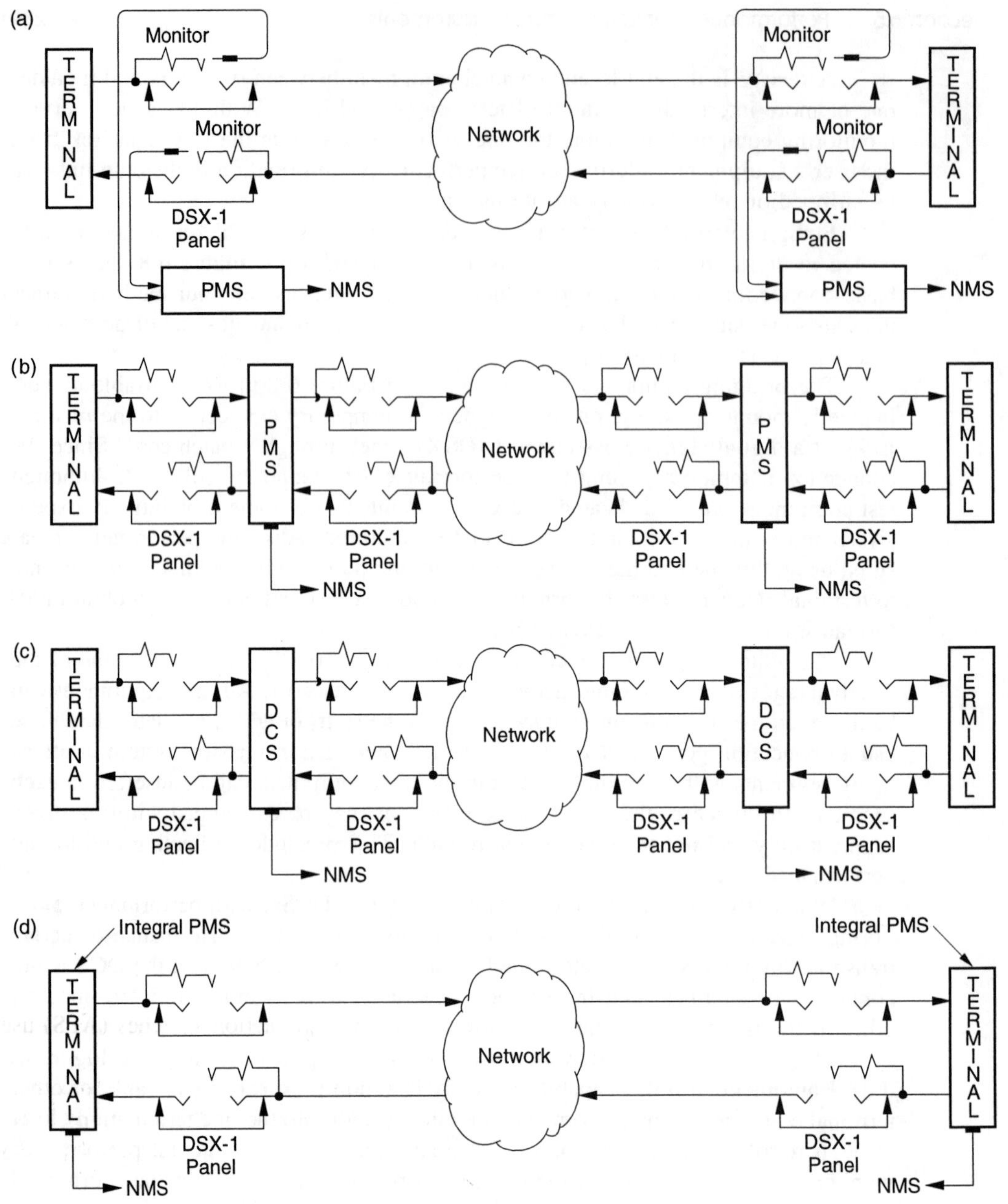

Figure 6-8 Connection of performance monitoring systems: (a) Bridging equipment method; (b) Adjunct equipment method; (c) Digital cross-connect equipment method; (d) Integral equipment method

terminal. The first device to recognize these conditions places the AIS on the network.

At the receiving terminal, an AIS condition is declared whenever an all binary ones and OOF are present at the same time. The AIS also is known as a Blue alarm and is used with both SF and ESF. The CCITT G.704 framing format differs from SF and ESF in that G.704 only has provisions for sending alarm conditions to the other end by changing the S-bit in frame 12 from a binary zero to a binary one.

A Red alarm indicates a locally detected failure. It may be triggered either by a continuous or an intermittent LOS and loss-of-frame (LOF) at the local terminal. An AIS will cause a Red alarm; however, an AIS, normally colored Blue, is declared instead. A Red alarm in a local terminal will cause a Yellow alarm signal to be transmitted to the far-end. When using SF, a Yellow alarm forces bit-2 in all 24 channels to binary zero. With ESF, a repetitive 16-bit pattern of eight binary ones followed by eight binary zeros (1 1 1 1 1 1 1 1 0 0 0 0 0 0 0 0) is continuously transmitted in the data link.

A Yellow alarm is declared at a receiving terminal within one second after the incoming Yellow alarm signal is detected. A Yellow alarm is a remotely detected failure. The colors (Red or Yellow) help to identify the direction in which a failure has occurred: Red, near; and Yellow, far.

In addition to the AIS (Blue) and Red and Yellow alarms, terminal and multiplex equipment usually contain an alarm indicator for LOS. Figure 6-9 shows two typical alarm conditions. In Figure 6-9(a), two terminals, A and B (for example, two channel banks) are connected together by a digital line. The failure of the signal received at B causes a Red alarm due to the LOS. Terminal B then sends the Yellow alarm signal back toward A. Terminal A detects the Yellow alarm signal and activates the Yellow alarm lamp. In Figure 6-9(b), two terminals, A and B, are connected together through a multiplexer or CSU, A′ and B′. The loss of the incoming signal to A′ causes it to activate the Red alarm (LOS) and send the AIS downstream. The CSU (or multiplexer) at B′ detects the AIS and activates the Blue alarm (AIS) and passes the signal to terminal B. Terminal B sends the Yellow alarm signal, which is detected by all upstream equipment.

Central office switching systems and automatic protection switches (APS) use a threshold mechanism to initiate alarm actions. For example, an APS may be set to switch to the protection line if the BER (as determined from BPVs or frame sync bit errors) exceeds a certain value, typically 1E-3 for lines carrying primarily voice and 1E-5 for lines carrying primarily data. Central office switching systems typically are set to annunciate a minor alarm if the BER on an incoming line exceeds 1E-4, and are set to take the line out-of-service and declare a major alarm if the BER exceeds 1E-3. A similar action is taken if frame slips exceed a certain count within a predetermined period.

6.5.3 Measurements

Performance measurements of digital loops can be categorized as

- Error measurements
- Interface measurements

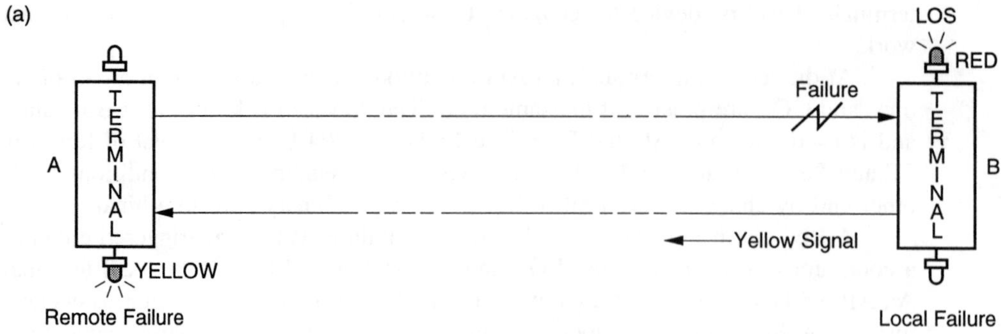

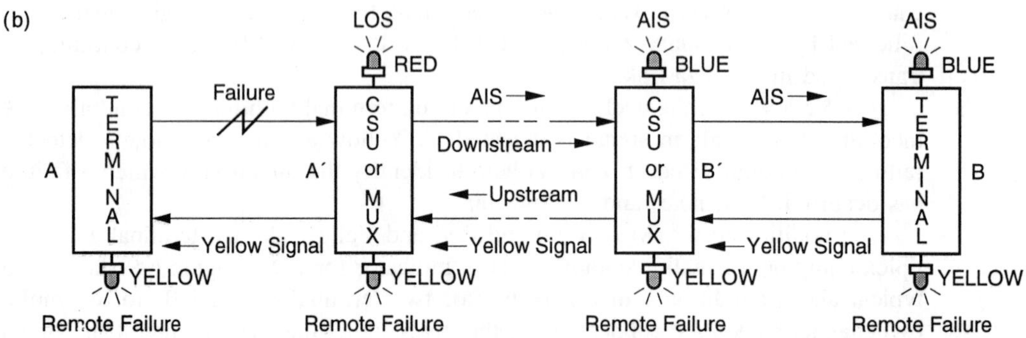

Figure 6-9 Alarm signals: (a) Between terminals; (b) Between terminals with intermediate MUX or CSU

Error measurements determine whether the binary values of the received signal are the same as the values of the transmitted signal, and determine the action taken when these values do not agree. These measurements include BER, EFS, ES, and alarm timing. *Interface measurements* determine whether the transmitted signals meet the requirements for pulse shape (width, height, overshoot, and rise time), jitter, and timing. Interface measurements also include the electrical characteristics of the transmitting and receiving circuits (for example, impedance). Test and measurement methods have been standardized for the most common digital transmission rates [17].

The test time for BER measurements is critical, especially with slow-speed circuits or very low error rates. At any given BER, the time required to detect an error is inversely proportional to the measured bit rate. In other words, if a BER of 1E-8 is required, at least 1E8 bits must be sent. However, for reliable error rate tests, at least ten times the minimum number of bits should be monitored. Figure 6-10 can be used to equate the test time required to detect an error, given the desired BER and measurement bit rates from 300 bps to 45 Mbps (DS-3 rate). This illustration includes the 10× factor for reliable analysis.

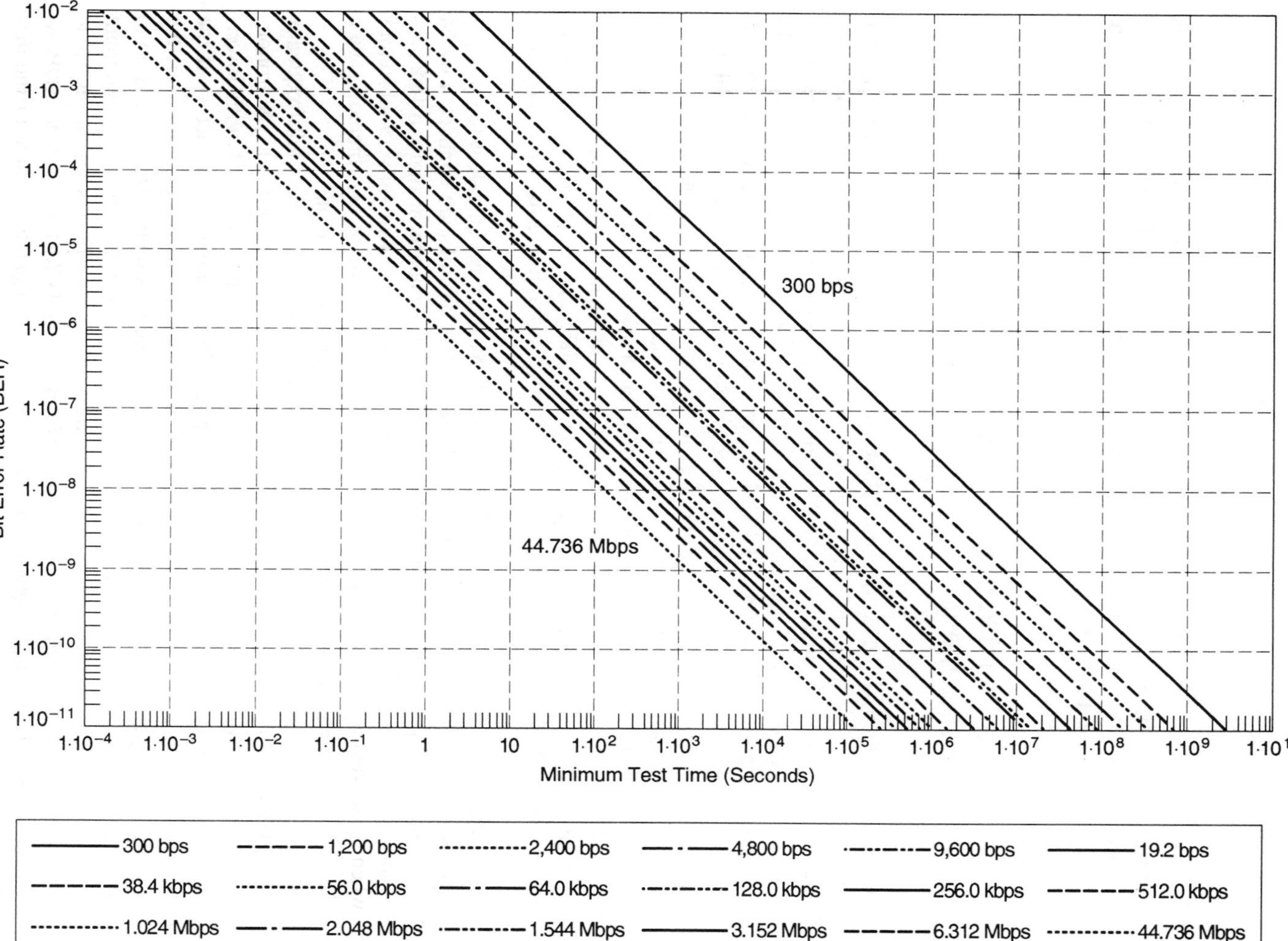

Figure 6-10 BER vs. minimum test time

6.6 Performance Requirements for Analog Circuits on Digital Loops

A generic set of requirements is used to specify the performance of analog circuits transmitted on digital loops. These are summarized in Table 6-28, with specific reference to industry requirements for digital loop carrier systems. A more complete discussion of the analog parameters can be found in [23].

6.7 Performance Limitations of ADPCM Encoding in a DLC Environment

Adaptive differential pulse code modulation (ADPCM) channel units are available with many DLC systems and channel banks. If these are to be used, network design and service planning must take ADPCM limitations into account. These limitations include degraded transmission on tandem ADPCM links, particularly when the links are asynchronous. The use of ADPCM channel units in DLC systems must be administratively controlled to ensure that true *zero-option* (explained later) capability, when required, is available from one user network interface to another.

It is important to distinguish between synchronous and asynchronous tandem links. Referring to Figure 6-11, two ADPCM tandem links are said to be synchronous if the connecting transmission channels are equivalent to 64 kbps PCM and do not have additional encoding or bit processing. Similarly, two links are asynchronous if they are decoded and then re-encoded (converted from digital to analog and back to digital) or processed at some point in the circuit.

The performance of two synchronous ADPCM links is equivalent to one synchronous ADPCM link, except for delay. If two ADPCM links are connected by 64 kbps bit streams but bit integrity is disturbed, the performance of the two links together is better than two asynchronous links but worse than two fully synchronous links [25]. Bit integrity is disturbed by any digital signal processing, including gain or phase equalization, and echo control (such as echo cancellation).

Nonvoice applications such as voiceband data and dual tone multifrequency (DTMF) signaling require special consideration, especially in regard to asynchronous tandem ADPCM connections. The performance of 32 kbps ADPCM with lower-speed modems (up to 2.4 kbps) and of some DTMF signaling is satisfactory if the number of asynchronous tandem connections is not greater than four. However, the performance of asynchronous tandem ADPCM connections degrades at speeds up to 4.8 kbps. Above 4.8 kbps, ADPCM cannot be used except with V.32 compliant modems. Tests have been performed on tandem links, with results shown in Table 6-29. The recommended maximum number of tandem ADPCM links include ADPCM encoders embedded in the network, plus those used on the end-user side of the network interface.

End-to-end signaling schemes (such as DTMF and single frequency) concentrate energy at specific frequencies in the voiceband. In order to accommodate such

Table 6-28 Analog Voicegrade Channel Performance on a DLC[a]

Parameter	Measurement Conditions	Objective	Requirement
Longitudinal balance per IEEE Std 455	200 Hz		>58 dB (>63 dB avg.)
	500 Hz		>58 dB (>63 dB avg.)
	1,000 Hz		>58 dB (>63 dB avg.)
	3,000 Hz		>53 dB (>58 dB avg.)
Echo return loss	600 Ω + 2.16 μF	>22 dB	>19 dB
Singing return loss	600 Ω + 2.16 μF	>14 dB	>11 dB
Maximum system loss	0 dBm0, 1,004 Hz		≤2 dB
System loss tolerance	0 dBm0, 1,004 Hz	±0.5 dB	±1.0 dB
Frequency response	400 Hz to 2,800 Hz	±0.5 dB	+1.0/−0.5 dB
Loss at 60 Hz	60 Hz		20 dB below 1,004 Hz
Amplitude tracking	−37 to +3 dBm0 input		±0.5 dB (±0.25 dB avg.)
	−50 to −37 dBm0 input		±1.0 dB (±0.5 dB avg.)
	−55 to −50 dBm0 input		±3.0 dB (±1.5 dB avg.)
Idle channel noise		<18 dBrnC	<20 dBrnC
Signal-to-distortion ratio	0 to −30 dBm0		>33 dB
	−30 to −40 dBm0		>27 dB
	−40 to −45 dBm0		>22 dB
Impulse noise	47 dBrnC0 threshold		<15 counts in 15 minutes
Intermodulation distortion	−13 dBm0 input per IEEE Std 743	R2 > 48 dB[b]	R2 > 43 dB
		R3 > 49 dB[b]	R3 > 44 dB
Overload compression	+3 dBm0 input		≤0.5 dB increase
	+6 dBm0 input		≤1.8 dB increase
	+9 dBm0 input		≤4.5 dB increase
Single frequency distortion	0 dBm0, 0 Hz to 12 kHz		<−28 dBm0
	0 dBm0, 1,004 Hz		<−40 dBm0
Peak-to-average ratio	−13 dBm0 input		≥90
Channel crosstalk	0 dBm0, 200 Hz to 3,400 Hz		<−65 dBm0
System-generated tones	0 Hz to 16 kHz		<−50 dBm0
Frequency offset			≤0.4 Hz

[a]Source: [24]

[b]R2 is the test tone signal to 2nd-order intermodulation distortion power ratio, and R3 is the test tone signal to 3rd-order intermodulation distortion power ratio.

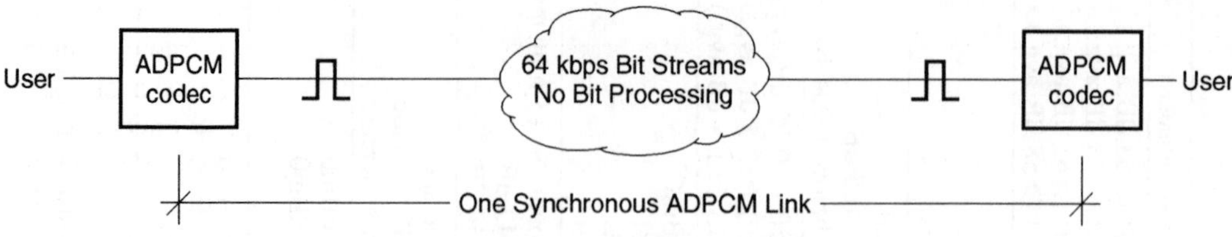

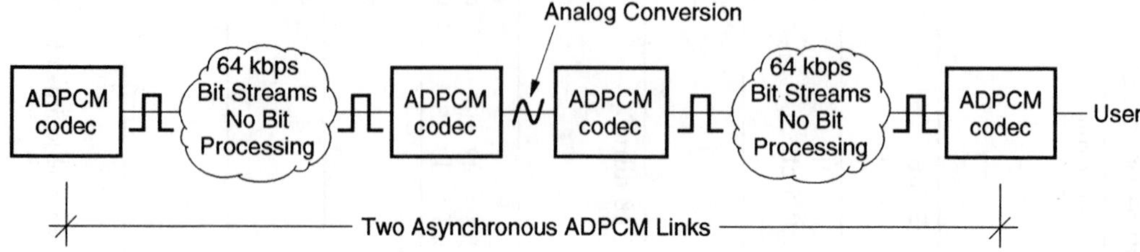

Figure 6-11 Synchronous and asynchronous tandem ADPCM links

Table 6-29 Guide to Number of Tandem ADPCM Links Conforming to ANSI[a] T1.301[b]

Source Signal	Number of Possible Tandem ADPCM Links	Recommended Maximum Tandem ADPCM Links	Note
Voice	3 to 4	3	c
DTMF	2 to 4	3	d
Voiceband data			
<2.4 kbps	4	3	e
4.8 kbps	2 to 4	3	f
9.6 kbps	0	0	
V.32 9.6 kbps	0 to 2	3	
Facsimile groups I & II	2 to 4	3	
Facsimile group III	0	0	

[a]American National Standards Institute

[b]Source: [26]

[c]Total round-trip delay should be less than five mseconds if no echo control devices are used.

[d]Actual performance depends on the type of DTMF receiver; actual tests may be required to determine if operation will be satisfactory with one or more ADPCM links.

[e]Some low-speed (300 and 1,200 bps) modems require at least 20 dB echo return loss at 1,100 and 2,100 Hz in the 2-wire, full-duplex mode. Also, some 2.4 kbps modems will require the scrambling option.

[f]Some modems may not function properly with more than one ADPCM link.

signaling, many ADPCM systems use a signal classifier that detects the tones and encodes them into eight-bit words for transmission in two successive frames (since each frame only uses four bits per channel per frame). The encoding rate is still 32 kbps. Not all ADPCM encoders have this capability, particularly early encoders.

With DTMF, in particular, the success of using an ADPCM connection is highly dependent on the type of DTMF receiver used. It has been found experimentally that filter type DTMF receivers have fewer restrictions than zero-crossing type receivers. The following rules of thumb apply [27]:

- Filter type DTMF receivers can be supported on up to four ADPCM links
- Zero-crossing receivers are sensitive to the number of links and circuit characteristics and may not work at all with ADPCM links
- Satisfactory operation of end-to-end DTMF signaling may not be possible with ADPCM; pretesting is recommended

Although informative, these pointers may be of little avail to the user, who has little, if any, prior knowledge of the type of DTMF receiver at the other end of the call. Furthermore, there is little chance that the user has any idea of the number of tandem connections, or whether the connections are synchronous, asynchronous, or composed of 32 or 64 kbps channels. This means the user has no prior assurance that high-speed modems or end-to-end signaling will work. It is possible to identify ADPCM links by using a 21-tone test, described in [28], but such tests are administratively difficult on a per-call basis. In private networks, the problem can be reduced by proper design and by avoiding asynchronous tandem connections.

ADPCM limitations have led many service providers to make available so-called *zero-option* ADPCM links. Most inter-exchange carriers (IC) have zero-option links between the point-of-termination (POT) and each end-office, and include zero-option links within their own networks. In addition, many local exchange carriers have similar zero-option links between the end-office and network interfaces located on customer premises, including links served by a DLC.

References

[1] Richters, J., Dvorak, C. "A Framework for Defining the Quality of Communications Services." IEEE Communications Magazine, Oct. 1988.

[2] *Digital Networks, Digital Sections and Digital Line Systems.* Recommendations G.801–G.956, Vol. III, Fascicle III.5, Blue Book, Nov. 1988. Available from National Technical Information Service, Order No. PB89-143895.

[3] American National Standard for Telecommunications. *Network Performance Parameters for Circuit-Switched Digital Services—Definitions and Measurements.* ANSI T1.507-1990.

[4] American National Standard for Telecommunications. *Network Performance Parameters for Dedicated Digital Services—Definitions and Measurements.* ANSI T1.503-1989.

[5] *Subrate Data Multiplexing, A Service Function of Dataphone© Digital Service*. AT&T Technical Reference TR54075, Nov. 1988. Available from AT&T Customer Information Center.

[6] *Accunet® T1.5 Service, Description and Interface Specifications*. AT&T Technical Reference TR62411, Dec. 1988. Available from AT&T Customer Information Center.

[7] *Requirements for Interfacing Digital Terminal Equipment to Services Employing the Extended Superframe Format*. AT&T Technical Reference TR54016, Sept. 1989. Available from AT&T Customer Information Center.

[8] *Digital Data System Channel Interface Specification*. AT&T Technical Reference TR62310, Nov. 1987; Addendum No. 1, Jan. 1988; Addendum No. 2, Oct. 1989; Addendum No. 3, Dec. 1989. Available from AT&T Customer Information Center.

[9] *Digital Data Special Access Service: Transmission Parameter Limits and Interface Combinations*. BELLCORE Technical Reference TR-NWT-000341, Feb. 1993.

[10] *Access Specification for High Capacity (DS1/DS3) Dedicated Digital Services*. AT&T Technical Reference TR62415, June 1989. Available from AT&T Customer Information Center.

[11] *High-Capacity Digital Special Access Service: Transmission Parameter Limits and Interface Combinations*. BELLCORE Technical Reference TR-INS-000342, Feb. 1991.

[12] *Fixed Satellite Service*. Recommendations of the CCIR, Vol. IV—Part 1, 1990. Available from National Technical Information Service, Order No. PB91-200048.

[13] *Fixed Service Using Radio-Relay Systems*. Recommendations of the CCIR, Vol. IX—Part 1, 1990. Available from National Technical Information Service, Order No. PB91-200105.

[14] *Fixed Satellite Service*. Reports of the CCIR, Vol. IV—Part 1, 1990. Available from National Technical Information Service, Order No. PB91-200048.

[15] *Fixed Service Using Radio-Relay Systems*. Reports of the CCIR, Vol. IX—Part 1, 1990. Available from National Technical Information Service, Order No. PB91-200105.

[16] *Digital Synchronization Network Plan*. BELLCORE Technical Advisory TA-NPL-000436, Nov. 1986. Available from BELLCORE Document Registrar.

[17] *IEEE Standard Methods and Equipment for Measuring the Transmission Characteristics of PCM Telecommunications Circuits and Systems*. IEEE Standard 1007-1992. Institute of Electrical and Electronics Engineers. Available from IEEE Service Center.

[18] Bellamy, J. *Digital Telephony,* 2nd Ed. John Wiley & Sons, 1991.

[19] American National Standard for Telecommunications. *Carrier-to-Customer Installation—DS1 Metallic Interface*. ANSI T1.403-1989. Available from American National Standards Institute.

[20] *Extended Superframe Format (ESF) Interface Specification*. BELLCORE Tech-

nical Reference TR-TSY-000194, Dec. 1987. Available from BELLCORE Customer Service.

[21] EIA/TIA Standard—*Network Channel Terminal Equipment for DS1 Service*. EIA/TIA-547-1989. Available from Global Engineering Documents.

[22] Kenepp, G. "Guarantee Digital Services—Now." Telephone Engineer & Management, Nov. 1991.

[23] Reeve, W. *Subscriber Loop Signaling and Transmission Handbook: Analog*. IEEE Press, 1992.

[24] *Functional Criteria for Digital Loop Carrier Systems*. BELLCORE Technical Reference TR-TSY-000057, Nov. 1988.

[25] Woinsky, M. "National Performance Standards for Telecommunication Services." IEEE Communications Magazine, Oct. 1988.

[26] American National Standard for Telecommunications: *Network Performance—Tandem Encoding Limits for 32 kbit/s Adaptive Differential Pulse-Code Modulation (ADPCM)*. ANSI T1.501-1988.

[27] *DE-4 Enhanced PCM Channel Bank,* Practice 368-5151-120. *Line Interface Unit (LIU) Description, Installation, and Maintenance,* QPP562A 32KB/S. Northern Telecom Inc., March 1984.

[28] Ingle, J. *Identification of Digital Impairments in a Voiceband Channel*. 1989 International Communications Conference (ICC'89). Boston: June 1989.

[29] American National Standard for Telecommunications. *Network Performance Parameters for Dedicated Digital Services—Specifications*. ANSI T1.510-1994.

[30] 1992—CCIR Recommendations. RF Series, Fixed Service. International Telecommunications Union. Geneva: March, 1992.

7 Loop Transmission Media, Part I—Twisted Pair Metallic Cables

Chapter 7 Acronyms

AL	acceptance limit	ML	maintenance limit
AMI	alternate mark inversion	Neper	One Neper = 8.686 dB
AWG	American wire gauge	NESC	National Electrical Safety Code
DSL	digital subscriber line	PIC	polyolefin insulated cable
ECCS	electrolytic chrome-coated steel	PSTN	public switched telephone network
EHS	extra high strength	PVC	polyvinyl chloride
EMI	electromagnetic interference	REA	Rural Electrification Administration
FAC	facility area connector	RMS, rms	root-mean-square
FCC	Federal Communications Commission	RZ	return-to-zero
HDSL	high bit-rate digital subscriber line	SAI	serving area interface
I/O	inside/outside wire	SNR	signal-to-noise ratio
IAL	immediate action limit	UG	utilities grade
ICEA	Insulated Cable Engineers Association	XC	cross-connect
ISDN	integrated services digital network		

7.1 General Considerations

The most commonly used transmission media in present-day digital loop applications is twisted pair metallic (copper) cable, but fiber optic cables are seeing significantly more and more use. Although both media transmit electromagnetic waves, their spectra are very widely separated, and the transmission and reception of signals depend on completely different propagation mechanisms. Because of the breadth of information, the descriptions of transmission media are broken into two parts. Part I (this chapter) covers twisted pair metallic cables and Part II (next chapter) covers fiber optic cables.

In spite of the lack of similarity between the two transmission media, the materials used in the outer (protective) layers of both cable types have many things in common and are identical in many respects. In many cases, the cables are installed under identical service conditions, and the associated loops and networks serve identical purposes.

There are cases, however, in which each cable type (twisted pair or fiber) can serve purposes the other cannot either because of economy or service requirements. For example, current use of twisted pair cable is exclusively for the final portion of the subscriber loop, commonly called "the last mile", because of its economy. On the other hand, fiber optic cable is used as a replacement for twisted pair feeder cables between the central office and remote switching terminals or digital loop carrier systems because of its economy in backbone and feeder applications. Work is under way by several large telephone companies to find economical methods to replace twisted pair cables altogether, but this is not expected to happen for many more years.

An economic trigger point is widely considered the best rationale for mass deployment of fiber optics in the loop. This point already has been reached for backbone and feeder plants. This point will be reached for the remaining portions of the loop (distribution and service drop) when the incremental cost of fiber over twisted pair is equal or close to zero, and when users demand services that can only be provided by fiber.

Economic considerations aside, the application of either twisted pair or fiber optic cables is predicated on their

- Physical characteristics
- Mechanical performance
- Electrical performance

Physical characteristics have to do with the geometry of the cable and transmission media. For example, the individual conductors of a twisted pair have a certain geometry, so they conform to a specified gauge and provide the necessary transmission characteristics. Similarly, optical fibers have a particular geometry associated with core and cladding, as well as dimensional requirements, which provide the necessary transmission characteristics.

Mechanical performance is related to the strength, abrasion resistance, and fatigue resistance of cables and the enclosed transmission media. This performance is specified in different ways for twisted pair and fiber optic cables. The same is true for electrical performance, which is due to their radically different transmission characteristics. The general physical characteristics and mechanical performance of both cable types and the specific physical and electrical characteristics of twisted pair cables are discussed in this chapter. Chapter 8 discusses the similar physical and electrical attributes of fiber optic cables.

7.2 Environment

Outside plant cables are subjected to a number of environmental hazards and conditions throughout their life. Some of these affect twisted pair cables differently than fiber optic cables. Each cable design attempts to mitigate any adverse effects over the anticipated cable life. Outside plant cables generally last at least 15 years, but in most regulated utility applications, their usefulness extends 25 to 35 years.

Environmental conditions of concern to the engineer are:

- Moisture
- Animal attack
- Electric fields
- Atmospheric disturbances (lightning)
- Temperature variations
- Absolute temperatures
- Snow and ice loading
- Wind loading
- Vibration
- Ultraviolet radiation (above ground only)

Moisture affects twisted pair and fiber optic cables through different mechanisms. Penetration and permeation of moisture into a twisted pair cable increases the attenuation by increasing the capacitance between conductors of a twisted pair. It also will corrode mechanical splices. When liquid water is present and it freezes, the expansion of the ice can crush the insulation or damage cable jackets from the inside. Such damage may not be apparent until the cable thaws and the pairs become shorted. In fiber optic cables, the crushing forces can fracture the fibers or cause microbending, both of which increase losses. Also, long-term exposure to moisture deteriorates the fiber's tensile strength.

Rodent attacks are a significant problem in some parts of the country.* The rodent's powerful jaws can easily penetrate plastic sheaths and allow moisture penetration. Also, the transmission media can be severed unless armoring is provided against the attack.

Strong electric fields from high voltage powerlines can lead to failure of the plastic sheath materials from corona discharge and arcing. Induction from powerlines can cause damaging current flow on conductive portions of the cables (armor, shields, and conductors in twisted pair cables; conductive strength members and armor in fiber optic cables).

Lightning is particularly troublesome in some parts of the country.† It can cause extremely large currents to flow on conductive portions of a cable. Also, large and

---

*For example, see page 246 of [1] for areas prone to gopher attack.

†See pages 3–5 of [2] or pages 2–5 of [3].

sudden currents from a lightning discharge can cause the cable to buckle as a result of mechanical forces and burn the insulation and sheaths. Rapid thermal expansion of the soil and moisture in it can exert significant forces on nearby buried cables. This damage can occur in completely dielectric fiber optic cables if they are buried near conductive material such as twisted pair or power cables.

Diurnal and seasonal temperature variations cause expansion and contraction of the materials used in cables. In some areas, the temperature can vary annually over a 90 °C range, and daily variations can easily exceed a 35 °C range. In twisted pair cables, temperature variations affect the attenuation but otherwise have no direct effect on modern cables. Older filled cables used a filling compound that flowed under high ambient temperatures; this compound would then drain out of cables used as risers. Modern filling compounds for cables have flow temperatures of 65 °C or 80 °C.

In areas of extremely cold temperatures, aerial twisted pair cables can contract enough to damage splices or supporting structures. Similarly, high temperatures increase the sag in aerial cables and allow accidental contact with vehicles or other moving objects. In fiber optic cables, the mechanical movement due to expansion and contraction can cause excessive bending loss.

Snow and ice loading can damage the cables as well as supporting structures through excessive weight. For example, one-half inch of radial ice on a one-half inch diameter cable weighs approximately 68 lb per 100 feet of length. One inch of radial ice on this same cable weighs about 200 lb per 100 feet. Snow and ice also increase the cross-sectional area of aerial cables and can amplify the adverse effects of wind.

Wind loading normally only affects aerial facilities; however, in especially windy areas, ground-mounted equipment may need special anchoring. The adverse effects of wind are manifested in four ways:

- As a transverse load on the supporting structure (for example, wooden poles)
- As a transverse load on the cable and messengers, which are coupled to the supporting structures
- Through vibration
- Through cable whipping and dancing

The transverse load, if large enough, can damage the supporting structures. Vibration, whipping, and cable dancing cause fatigue in supporting structures and in the cable itself. Even very small winds can cause large-amplitude vibrations on long sections of aerial line. Another source of vibration is road traffic, and over a long period, this source of vibration can lead to catastrophic failures of supporting structures. It is difficult to build vibration damping into cables, so external means are frequently employed, including specially made sails, spiraling the lashed cable, additional strands above the supporting strand and attached to it, and shorter spans between supports. In extreme cases of cable dancing, the only alternative is to bury the cable or change the route.

Sunlight can damage cable sheaths through breakdown of the plastic materials from ultraviolet radiation. Any facility exposed to the sun is subject to this problem, and special materials or coatings are used to combat it.

7.3 Service Classifications

All telecommunication cables are categorized according to their service classifications, as shown in Table 7-1. These classifications are universal, but general, in nature. Most cables are not specifically made for a single service classification, so a given cable type may be used wherever it is suitable. For example, filled twisted pair cable normally is considered for direct buried applications, but it is also used in aerial and underground service. Armored cable may be used in aerial as well as direct buried service. The selection of general cable types is aided by the flow chart in Figure 7-1.

As previously noted, cables used in an outside environment normally last 25 to 35 years. Although indoor cables are manufactured for a similar life, technological advances of equipment connected to indoor cables and user requirements for more and more bandwidth may render the cables obsolete long before their physical life is exhausted. This is especially true of most existing twisted pair cables. With the widespread deployment of fiber optic cables and high-performance twisted pair cables in buildings, the threat of premature obsolescence is somewhat eased.

Table 7-1 Cable Service Classifications

Classification	Application
Indoor	Plenums in buildings (CMP)[a] Risers in buildings (CMR) General use in buildings (CM)
Aerial	Lashed to separate messenger or self-supporting on above-ground structures
Buried	Buried directly in earth
Underground	Installed underground in duct or raceway

[a]The designations in parentheses for indoor cable indicate classifications defined in Article 800 of the National Electrical Code.

7.4 General Cable Construction Attributes

Both twisted pair and optical fiber cables have similar external physical attributes, which are illustrated in Figure 7-2. All cables have an outer jacket, which for twisted pair and fiber optic cables is identical. This also is true of the armor (with an optional inner jacket), although fiber optic cables are equipped with armor in more applications than twisted pair cables. The buffering, strength member, and transmission media are radically different.

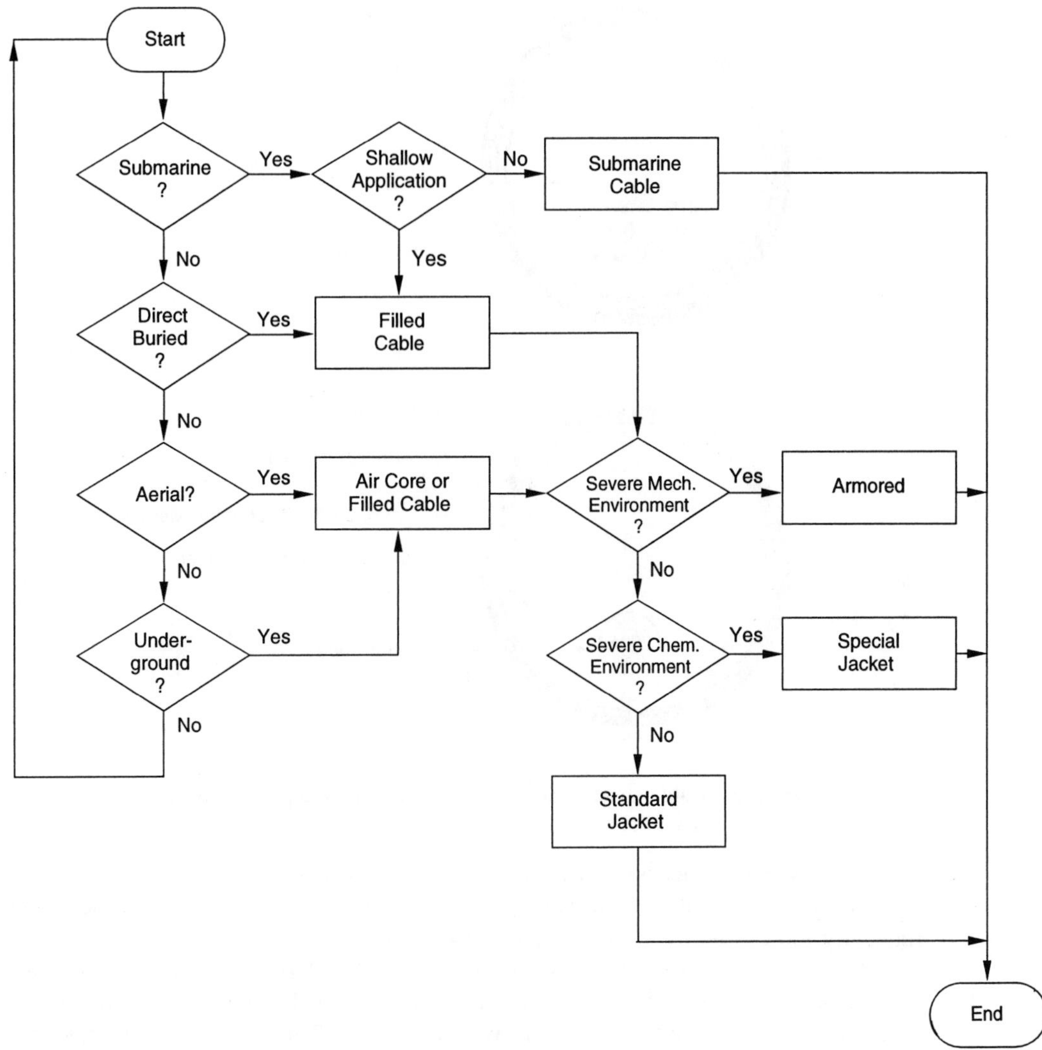

Figure 7-1 Cable selection flow chart [4]
Reeve, © 1992 Institute of Electrical and Electronics Engineers. Reprinted with permission

7.5 Other Outside Plant Equipment

Although twisted pair cables represent the highest proportion of outside plant investment, they are not the only transmission equipment found outdoors. The cables and other facilities used with digital transmission systems or digital loops are frequently shared with analog loops. Except for digital loop carrier, remote switching systems, and all present optical fiber transmission systems, digital and analog sub-

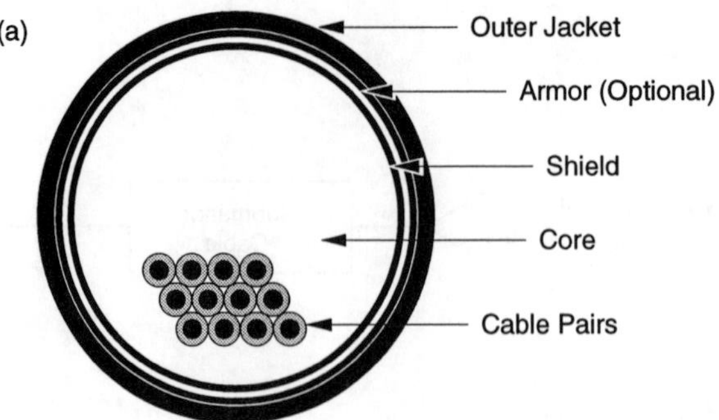

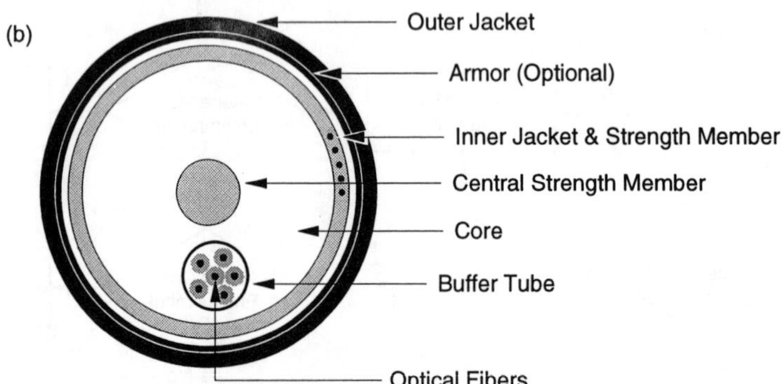

Figure 7-2 Cable construction attributes: (a) Twisted pair cable; (b) Fiber optic cable

scriber loops require access at frequent intervals. Access is typically at every pole in aerial systems installed in urban areas, or at least once in every city block (and frequently more often) for buried or underground systems.

The equipment used for access to subscriber loops depends on the type of plant. Aerial plant is accessed through "ready access" or sealed fixed-count terminals located adjacent to the pole and in-line with the cable. The aerial terminal has provisions for shield bonding and grounding, and connection of service wires to distribution pairs. In buried or underground plant, access is at pedestal or outdoor housings. The pedestals sometimes contain loop carrier equipment, loading coils, cross-connect blocks, or other devices, including protectors, for connecting distribution plant to service wires.

Digital loops frequently are used as backbone facilities to feed digital loop carrier and remote switching systems, which may be separated from the central office by several miles. Access to the digital loop in these cases is not necessary except at repeater points. Repeaters on twisted pair cables typically are about five to seven kilofeet apart for repeatered T1 systems. Fiber optic transmission systems used in digital loop applications do not require repeaters because of their relatively short lengths, so access is not required anywhere along the route except at reel-end splice points.

Repeatered T1 repeater housings used with twisted pair cables are available in a variety of sizes and forms, from one repeater to several dozen. The housings have printed wiring board connectors and mounting provisions for one or more repeater cards, grounding and bonding, and lightning protection. Some housings also may have order wire and fault location equipment, which are used for maintenance and troubleshooting. Repeater housings can be pressurized with an inert gas or with dry air through cable pressurization systems to keep out moisture. Sealed units have provisions for venting in case of excessive pressure build-up.

7.6 Twisted Pair Transmission Media: Physical and Electrical Characteristics

7.6.1 Introduction

The twisted pairs in all modern polyolefin insulated cables (PICs) are enclosed in a heavy-duty plastic sheath. Although pulp-insulated cables are still manufactured for in-kind replacements, this chapter is concerned with PIC exclusively. First, the physical aspects and non–transmission-related electrical characteristics are described in detail. Then the electrical transmission characteristics are described for a frequency range extending to 5 MHz, which covers all present digital loop applications.

7.6.2 Physical Characteristics

7.6.2.1 Twisting

Each cable pair consists of two twisted insulated conductors. The twist is used to decrease electromagnetic interference due to crosstalk between adjacent pairs in a cable. The twist also helps keep the conductors of the pair balanced with respect to the shield and other pairs.

Neighboring pairs in a given 25-pair group have a different twist length to further reduce crosstalk between them. The twist length (or pitch) is usually left-hand (counterclockwise), and it varies from two to six inches. It is optimized to reduce crosstalk at higher frequencies (hundreds of kilohertz and above), since it is in this region where crosstalk can become a problem.

The requirement for adequate crosstalk performance in all telecommunication cables is highly stressed during their design and construction. Crosstalk due to cable pair irregularities is managed by tightly controlling capacitance unbalance (pair-to-pair, pair-to-shield, and pair-to-ground) and resistance unbalance during the manufacturing process.

7.6.2.2 Feeder and Distribution Cables

Feeder cables comprise the loop from the central office to main junction points called serving area interfaces (SAIs), facility area connectors (FACs), cross-connects (XCs), or control points. The actual name depends on the company, but the function is to provide a point at which feeder cable pairs can be administered and connected to individual distribution cable pairs.

Distribution cables comprise the loop from the main junctions to distribution terminals such as pedestals or aerial terminals. In any given geographic area, there are generally more distribution pairs than feeder pairs. This allows efficient use of larger and more expensive feeder cables through easier administration and assignment of subscribers.

The most widely used feeder and distribution cable size is 24 American wire gauge (AWG). This is due to economic tradeoffs that take into account the typical loop length. Relatively short loops in metropolitan environments are typically 26 AWG.

In the smaller cable sizes, good design practice calls for using cables that are multiples of the basic 25-pair unit (25 and 50). Pair counts of 6, 12, 18, and 37 are avoided unless other factors make their use clearly more economical than 25- or 50-pair sizes. There is very little difference between the installed cost of 6- or 12- and 25-pair cables or 37- and 50-pair cables in most applications.

The extra pairs not only give extra insurance to accommodate unforeseen growth, but also allow pairs in the field to be allocated and used more logically due to the multiple 5-pair and 25-pair color coding scheme. The long-term result, of course, is easier administration and maintenance.

With twisted pair cables, each pair may be made available at many different locations in a "preferred count" splicing scheme, or at predetermined single locations in a "dedicated count" splicing scheme. Generally, feeder cables serve a given area on an express basis, while distribution cables serve the individual subscribers in that area. Large cables cannot be provided in lengths of more than 1,000 or 2,000 feet and so must be frequently spliced along long routes. The splice, connector, and access points are required to use the facility properly. Splices and connectors add very little to the overall transmission loss in the cable route.*

7.6.2.3 Service Wires

Service wires (also called *service drops* or *service cables*) make up the final portion of the metallic loop. They are installed between the distribution terminal and network interface device. This can be at an office building or residential establishment. Beyond the network interface device are premises wiring systems, which are described in Chapter 11.

Service wires can be one pair or up to hundreds or thousands of pairs, depending on the establishment to be served. In many situations there is nothing to distinguish a service cable from a distribution cable, except for the point of termination. Exactly the same materials may be used for the cables themselves.

Service wires in the smaller sizes are not always twisted because they are usually short and do not measurably degrade transmission quality. Aerial service wires are usually 18 or 19 gauge, whereas buried service wires are usually 19, 22, or 24

*The loss of an individual twisted pair splice connector is not easily measured because it is so small; therefore, splice loss is not considered a factor in feeder and distribution design.

gauge. Highly pure copper is the most widely used conductor material for feeder and distribution cables and underground service wires, whereas copper with special alloying (or other means for strength) is used for aerial service wires.

7.6.2.4 Current Limitations

The maximum continuous current-carrying capacity of any given conductor gauge depends on many factors, such as ambient temperature, allowable temperature rise, number of conductors in a bundle, and insulation type. Table 7-2 shows the values for various copper conductor gauges, assuming a 25 °C ambient temperature and 40 °C rise. These are provided to give an idea of current magnitudes and not as design values.

In analog applications, the maximum loop current available from an analog central office line circuit is in the range of 45 to 145 mA, although some older systems can supply at least 175 mA. Modern digital equipment limits loop current to approximately 50 mA. The current magnitude normally drops off as loop length increases beyond a certain limit. Many digital central offices use current-limiting type line circuits to limit loop current on short loops.

Digital loop circuits may be arranged to not carry any dc loop current at all, but this is to be avoided. Sealing current of 10 to 20 mA is used to keep mechanical splices and connections free of high-resistance metal oxides. Repeatered lines for digital loop carrier systems are usually loop powered with 60 to 140 mA dc. The lower value is preferred in order to reduce power consumption and dissipation in repeaters and digital line terminal equipment.

For signaling and control applications outside of the digital transmission arena, loop current normally is limited to 350 mA. In some special applications, a loop current up to 1.3 amps is sometimes allowed. This higher value coordinates with 26 AWG cable. In public switched telephone network (PSTN) applications, the local exchange carrier always will limit current to this value.

In premises wiring applications, the Federal Communications Commission (FCC) regulates the maximum allowed currents on cable pairs connected to the public telephone network. The FCC limits currents, as shown in the right-hand column of Table 7-2. The FCC does not list a value for 19 AWG conductors but does give the allowable currents of 7.5 A and 10 A for 20 AWG and 18 AWG, respectively. Also, a 45 °C rise is used for 22, 20, and 18 AWG, and de-rating is required where ambient temperatures exceed 25 °C.

Also shown in Table 7-2, for reference, is the approximate fusing current for each gauge, which is the point at which the conductor will fuse open. These are

Table 7-2 Exchange Cable Pair Current Ratings

AWG	Approx. Fusing Current (0.1 Seconds)	Max. Current Capacity (40 °C Rise)	Max. Current per FCC Rule 68.215 [5]
19	600 A	8 A	See text
22	300 A	5 A	5.0 A
24	190 A	2 A	2.1 A
26	110 A	1.3 A	1.3 A

approximations; actual values depend on many variables and published information varies widely. Fusing currents and fusing times are inversely related. For example, if the current is increased by a factor of 3, the fusing time will decrease by a factor of about 1/10.

7.6.2.5 Twisted Pair Cable Specifications

Telecommunication cables are manufactured for an intended application according to a specification prepared by the cable customer. In many cases, these specifications are generic and are in wide use throughout a segment of the telecommunications industry. Of particular interest are the telecommunication cable standards published by the Insulated Cable Engineers Association (ICEA) [6–9], Rural Electrification Administration (REA) [10–17], and BELLCORE [18–25].*

Generally, the specifications include, among other requirements, an insulation breakdown voltage, a high-voltage proof test, and insulation resistance. The insulation breakdown voltage (or dielectric strength) is the voltage at which the insulation will fail. It is usually specified in volts per mil† of insulation thickness. Typical insulation thickness is around 10 to 14 mils, but this differs among cable types and manufacturers. Cable specifications normally only show an insulation breakdown voltage, and the manufacturer then applies enough of the particular material to satisfy the requirement.‡

The high-voltage proof test is at a somewhat lower voltage than the insulation breakdown voltage. It is the voltage at which cables are tested after manufacture and before shipment to the customer. The actual values depend on such factors as cable gauge, whether filled or air core, and solid or expanded foam insulation. All proof tests are made for three seconds. The cable is not expected to withstand these voltages indefinitely.

Insulation resistance is a measure, in MΩ-miles, of the leakage between a given conductor and the other conductors and shield in the cable. By inspecting the units, it is clear that insulation resistance decreases with increasing length. For example, an air core cable five miles long will measure at least 2,000 MΩ (10,000 MΩ-miles/five miles). Insulation resistance tests are usually made at around 550 V for new cables at the factory, or 250 V for cables in the field. The higher the resistance, the better. Table 7-3 shows the ratings for typical outside plant telecommunication cables.

Unlike power cables, telecommunication cables usually do not have a specified maximum working voltage. Nevertheless, most telecommunication cables are considered to have a working voltage across the tip and ring of any given pair, and between pairs of 300 V.§

---

*The listed BELLCORE documents are representative of the available outside-plant cable technical references. Additional specifications can be found in [26].

†One mil is one one-thousandth of an inch.

‡This is not an arbitrary process because the insulation thickness affects cable capacitance, which is very tightly controlled, and conductance.

§The working voltage considered here is root-mean-square (rms) or dc.

Table 7-3 Exchange Cable Voltage and Insulation Ratings

PARAMETER	RATING
Insulation breakdown voltage	300 to 1,200 volts/mil
High-voltage proof test (3 seconds)	
Conductor-to-conductor	2 to 7 kV
Conductor-to-shield	10 to 20 kV
Insulation resistance (@ 100 to 550 V)	
Air core cable	10,000 MΩ-miles
Filled cable	1,000 MΩ-miles

This does not mean the user can arbitrarily apply 300 V to the cable pairs. As expected, for equipment connected to the PSTN, FCC Part 68 has certain maximum voltage and current limitations for terminal equipment in various configurations that preclude such operation.* Basically, the terminal equipment cannot apply more than 56.5 Vdc to a loop connected to an off-premises station, or supply more than 140 mA to the loop under any circumstances. During ringing, no more than 300 V peak-to-peak can be applied to the loop.† The reasons for these limitations are obvious: safety and network harm. Some telephone company tariffs may specify more stringent voltage and current limitations.

The voltage and current limitations discussed above, as imposed by the FCC on equipment connected to the loop, do not apply to the telephone companies themselves. Although the working voltage of cable pairs is considered to be 300 V, telephone companies will use higher voltages under certain conditions. These are always associated with special equipment located outside the central office. For example, the line repeaters in long digital loop carrier span lines may be powered by up to 360 V at the central office. This voltage dwindles to much lower values along the span line due to voltage drop across the repeaters, and resistive voltage drop in the cable pairs.

Cable specifications also include requirements for resistance, capacitance, crosstalk, and loss at particular frequencies. Although the exact details of industry specifications may vary with the market segment (that is, REA, BELLCORE, AT&T, GTE, and others), the overall effect on transmission performance is not significant except in the most demanding applications (for example, at high bit rates where specially designed cables might be used).

Other basic requirements of virtually all specifications are the marking requirements for outer jackets. These include sequential marks at two-foot intervals, cable type, pair count, gauge, and manufacturer. Some cables have two sets of interval markings when shipped from the factory. Normal marking usually is white, but if this is unusable for any reason, a second interval marking is made in yellow. Also, as required in 1993 by the National Electrical Safety Code (NESC), the outer jacket must be marked with a telephone handset symbol to differentiate the communications cable from power cables (NESC Sec. 35, par. 350.6) [27].

*Par. 68.306 of [5].

†300 V peak-to-peak corresponds to about 106 V RMS, assuming a sinusoidal ringing voltage.

7.6.2.6 Splices

Copper conductors are easily spliced by a number of means. Because of requirements for long life, reliability, and fast installation, all splicing in the telecommunications industry is by specially designed insulation displacement mechanical connectors. Single-pair or bulk (25 pair at a time) splice connectors are available. All require special tools or jigs to assemble the splice properly and to ensure a long and reliable life. Twisting together of bare conductors is never used to splice exchange cables.

Of special concern in outside plant applications is the need for a weatherproof splice connector. Even when splices are enclosed in a sealed splice case, enough moisture can enter to corrode the splice connections and introduce high loss or noise. All splice connectors are available with filling compounds that prevent moisture problems. Although these type of connectors find the widest use in direct buried cable applications, they are also recommended for use with pressurized plant.

The splice connectors used in the telecommunications industry have negligible loss at all frequencies encountered in twisted pair transmission systems. Nevertheless, proper care must be taken in choosing splice connectors and the splice cases that enclose them. Without this care, the expected service reliability and availability will never be obtained over any significant period.

7.6.2.7 Summary of Twisted Pair Cable Physical Characteristics

The previously described physical characteristics are summarized in Table 7-4 for modern copper telecommunication cables. Figure 7-3 illustrates the construction of a few cable types.

Table 7-4 Outdoor Twisted Pair Cable Physical Characteristics

Parameter	Characteristic
Insulation	Polyethylene Polypropylene Dual expanded polyethylene Expanded polypropylene
Conductors	19, 22, 24, 26 AWG, copper
Pair Count	6, 12, 18, 25, 37, 50, 75, 100, 150, 300, 400, 600, 900, 1,200, 1,500, 1,800, 2,100, 2,400, 2,700, 3,000, 3,600, 4,200[a]
Core	Air or filled
Shield	Bonded aluminum (8 mils), copper (5 or 10 mils) or copper alloy (6 mils)
Armor	Bonded steel (6 mils) when separate from the shield
Sheath (jacket)	Polyethylene jacket Special high-temperature jacket
Special features	Compartmental core Low capacitance External wire armor

[a]Not all pair counts are available in all gauges.

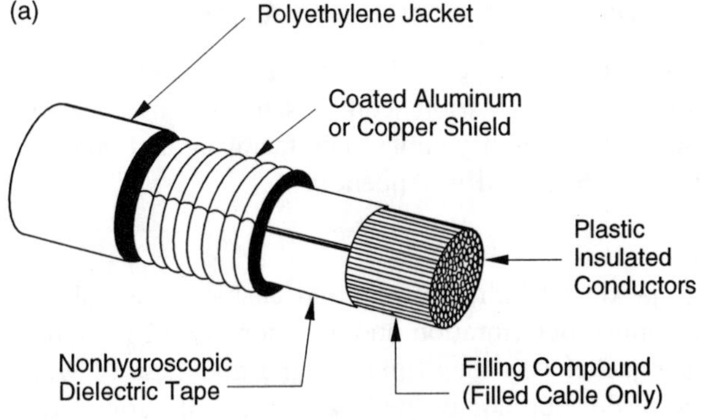

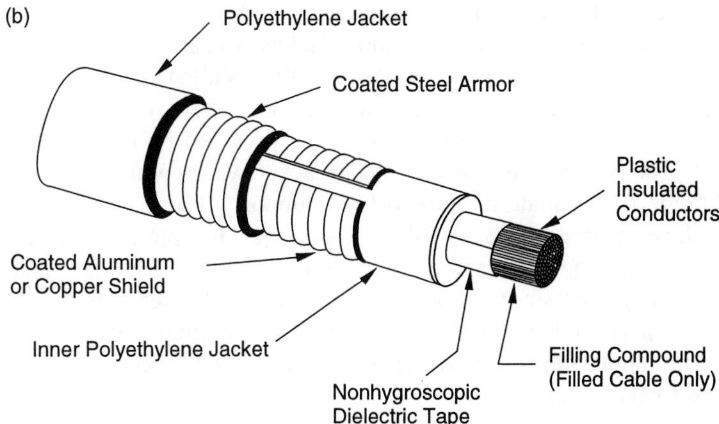

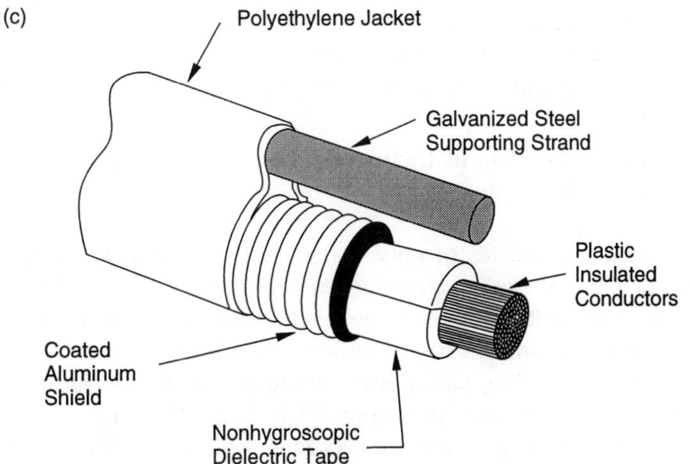

Figure 7-3 Twisted pair cable constructions, [4]
Reeve, © 1992 Institute of Electrical and Electronics Engineers. Reprinted with permission

7.6.2.8 Description of Each Construction Attribute

This section describes the cable constructions used in subscriber loop applications; refer to Figure 7-3. Various coding schemes have been used in the telecommunications industry to simplify cable manufacturing and ordering. One of the more common schemes is described in Appendix A.

Outer Jacket

The outer jacket (sheath) protects and cushions the cable core. It also offers resistance to chemical deterioration and abrasion. In all modern telecommunication cables, the outer jacket is a tough plastic, the thickness of which varies with size and type of cable. For outdoor twisted pair cables, it varies from approximately 60 mils on cables with diameter of 0.75 inch, to 110 mils or more on cables with a two-inch or larger diameter.

The sheath thickness for indoor cables typically is less than comparable outdoor cables, because indoor cables are usually installed in a relatively benign environment. In those cases in which an indoor cable is installed in an abusive environment, special outer jacket constructions can be provided by cable manufacturers.

The type of plastic used in outer jackets varies with the application. A polyvinyl chloride (PVC) material that will not support a flame and has nontoxic combustion products is used for riser cables in buildings. For plenum applications in buildings, which is the most demanding service in terms of fire rating, cables typically have a fluoropolymer resin (for example, TEFLON™*) jacket. Low-density, high molecular weight polyethylene (or ethylene copolymer) or high-density polyethylene is used in outside applications. These types of polyethylene are typical choices for the outer jacket because they have:

- Relatively low humidity permeability
- Good mechanical properties, such as high resistance to abrasion, weathering, and environmental stress cracking
- Good chemical properties, including high resistance to deterioration from sunlight
- Low coefficient of friction, which is important when pulling cables through conduits and duct systems

The disadvantages of polyethylene are its stiffness and flammability, both of which preclude its use in an indoor environment.

In aerial applications, the cable is lashed to a supporting messenger (commonly called *suspension strand* or just *strand*), which varies from 3/16 inch to 7/16 inch in diameter, depending on the strength required and the material used. Lashed aerial construction is illustrated in Figure 7-4(b). Table 7-5 shows the breaking strengths for messengers used in telecommunications applications. The messenger tension never

---

*"Teflon" is a trademark of DuPont-Nemours, Delaware.

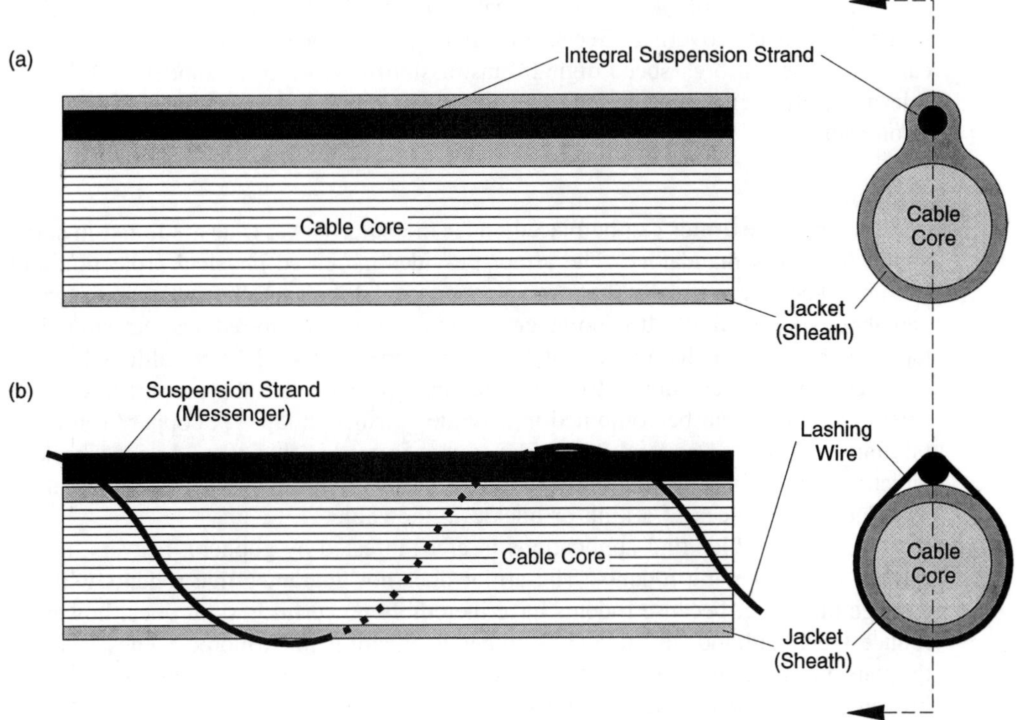

Figure 7-4 Cross-sections of aerial construction: (a) Figure 8; (b) Lashed

Table 7-5 Messenger Strand Breaking Strengths

Designation	Strand Size	Breaking Strength (lb)
6M	3/16 inch extra high strength (EHS)	6,000
6M	1/4 inch utilities grade (UG)	6,000
10M	5/16 inch EHS	10,000
10M	3/8 inch UG	10,000
16M	3/8 inch EHS	16,000
16M	7/16 inch UG	16,000

exceeds 60% of the breaking strength, under all mechanical loading conditions, in a properly designed aerial plant.*

In some aerial applications, a self-supporting cable is used, as shown in Figure 7-4(a). This cable incorporates a built-in supporting messenger along with the outer sheath. This cable is called "figure 8" because of its shape. Due to strength limitations, figure 8 cable generally is not available in cable pair counts above 300 pairs (26 AWG), 200 pairs (24 AWG), 100 pairs (22 AWG), or 50 pairs (19 AWG).

*Mechanical loading conditions and design criteria are covered in [27–29].

Many older outdoor twisted pair cables have a lead sheath, which acts as both a shield and outer covering. Because of their age and construction, few lead sheath cables are used in high-speed digital transmission service. Some modern cables use a lead sheath over a polyolefin jacket for added protection in highly corrosive environments.

Armor

A steel tape armor can be placed under the outer jacket if desired, and in some applications it is mandatory. The armor provides increased physical protection and strength. The tape is usually (but not always, especially on smaller cables) corrugated and about six mils thick. It is polymer coated on both sides to reduce corrosion. The tape material is typically an electrolytic chrome-coated steel (ECCS), although stainless steel and copper alloys also are used. In copper cables, the properties of both shield and armor can be combined in a single corrugated tape. A copper/iron alloy (six mils thick) or extra thick (ten mils) copper is used in this case. An inner jacket, which is similar in construction to the outer jacket, may be used in armored cables to reduce the chances of a kink or buckle in the cable.

Armored cables find applications in areas infested by gophers, beavers, squirrels, or other chewing rodents. The armor does not guarantee that rodents will not damage the cable, because rodents have incredibly powerful jaws; it does, however, reduce the likelihood of damage. Another application for armored cables is when they are installed in a severe physical environment, such as direct burial in sharp, broken rocks, where bedding sand is not available (or economical); or in lakes or stream crossings, where anchor or other boat-caused damage is possible.

In lakes and river crossings, a cable with a lead antimony alloy armor can be used to offset the natural buoyancy and weight the cable down. Otherwise, the cable can be installed with a submersible plow or weighted at intervals. Freshwater has a density of 62.4 lb/feet^3 and seawater has a density of 64 lb/feet^3. The volume of water displaced is that of a cylinder equal in dimensions to the cable. If the cable weight exceeds the weight of the water displaced, the cable will not float.

Specially made armored cables are available for true submarine service in oceans and deep lakes. Figure 7-5 is a photograph of heavily armored twisted pair and fiber optic cables for submarine service such as lakes and river crossings, and Figure 7-6 is a drawing of the cable components (in this case, a heavily armored fiber optic cable for water crossings). Special jacket materials can be provided for highly corrosive, potentially damaging, or chemically contaminated environments. The added protective layers are usually applied over standard cables.

Shielding

Outside plant twisted pair cables are most commonly shielded with corrugated aluminum or copper tape. Copper alloys and copper-clad stainless steel also are used. To increase corrosion resistance, the aluminum shield is bonded to a thin plastic (copolymer). This coating is not necessary with copper shields. A corrugated rather than smooth surface is used to increase shield strength and flexibility. The shield isolates the enclosed conductors from outside sources of electromagnetic interference (EMI), such as induction from powerline harmonic frequencies. The ability of a

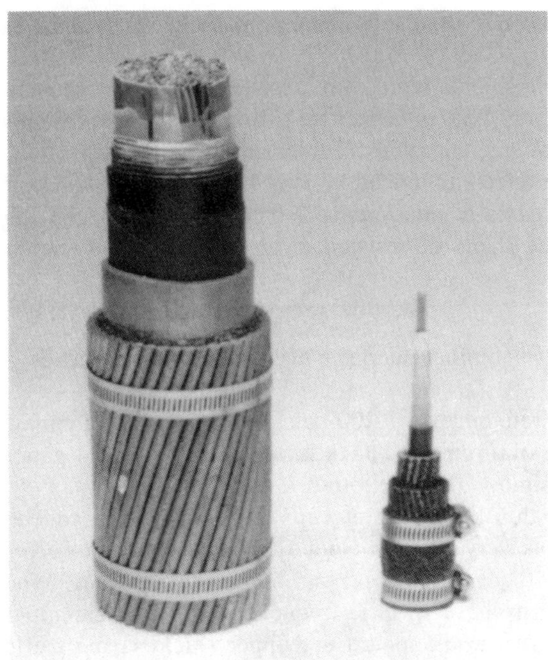

Figure 7-5 Comparison of heavily armored twisted pair and fiber optic cables
Taken from [30], with permission of AT&T Network Systems

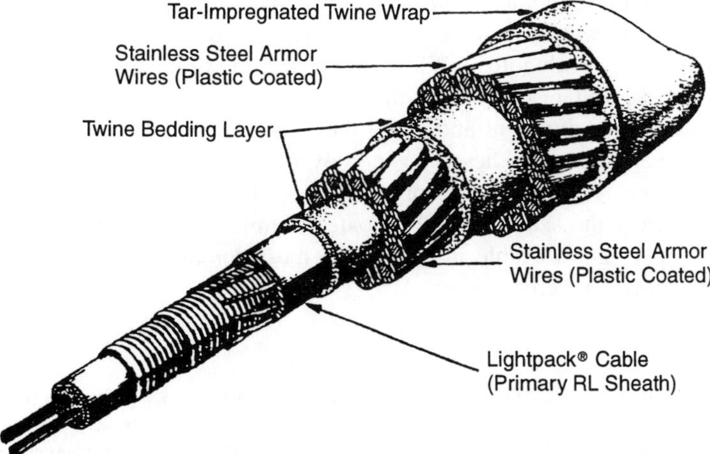

Figure 7-6 Components of heavily armored fiber optic cable
Taken from [30], with permission of AT&T Network Systems

shield to isolate the cable pairs from such interference is determined by its shield factor.

Cable shields with a low resistance and low mutual impedance have a higher shield factor. Larger cables have both these features, and so offer a better shield factor than smaller cables. Tables are available to simplify the calculation of shield factor for telecommunication cables, but these calculations are only needed in exceptional cases of interference.* If the diameter of the cable over the outer shield is known, the shield dc resistance can be determined from:

$$\text{dc shield resistance} = 0.75/\text{D}\ \Omega/\text{kilofoot}$$

where D = nominal diameter over the shield, in inches

For reference, a 1,200-pair, 24 gauge, aluminum shielded cable has a shield resistance of around 0.2 Ω/kilofoot; a 100-pair, 24 gauge cable has a shield resistance of almost 0.7 Ω/kilofoot.

The thin aluminum or copper shield used in conventional telecommunication cables does very little to isolate the pairs from powerline interference at the fundamental frequency (60 Hz) and first few harmonics. When it is necessary to shield a cable from these specific frequencies, an iron shield (for example, iron conduit) works best. The aluminum or copper shield is most effective at higher harmonic frequencies, which are annoying to the subscriber in analog service or can cause problems with digital transmission, due mostly to equipment upset. Since the telephone instrument and other terminal equipment—digital or analog—are designed to have little response at 60 Hz, induction at the fundamental powerline frequency usually is not a problem.

The shield in a given cable provides almost twice the shielding (induced voltage cut almost in half) at 540 Hz than it does at 60 Hz. Also, a cable with four times the diameter of another cable will provide twice the shielding. These low frequency-induced voltages have little direct interference on digital signals because of the relatively wide bandwidths used in digital transmission systems. If large enough, however, these voltages can upset the electronic circuitry used in digital regenerators or transmission equipment and cause the system to fail or produce excessive errors.

The shield thickness is generally 0.008 inch (eight mils) for aluminum and 0.005 inch (five mils) for copper. The different thicknesses are used to give approximately the same strength. Coated aluminum is the preferred shield because it is cheaper and is usable in all but the most demanding applications (such as some highly corrosive environments).†

---

*See [31] for calculation methods. If a particular interference problem is severe enough to require these calculations, it may be easier or more economical simply to reroute the cable or use a different transmission method. For example, fiber optic transmission systems are inherently immune to all forms of EMI in the outside plant environment.

†The cost of twisted pair cables naturally varies with the costs of the materials used to make them. The biggest variables are the costs of aluminum and copper. There may be times when a copper-shielded cable is cheaper than the same cable with an aluminum shield.

The shield is made continuous through splice points, such as vaults, manholes, aerial terminals, and pedestals, and connected to earth ground at regular intervals. Aerial messengers must be grounded at least four times per mile to comply with National Electrical Safety Code requirements.*

Shield continuity and grounding are very important; otherwise, the cable pairs are susceptible not only to noise but to damaging overvoltages during lightning activity, power faults, and power system switching transients.

For digital carrier applications, an additional shield may be used that separates the cable pairs into two compartments. This is called compartmental core or screened cable. "Z-Screen," "D-Screen," or "T-Screen" are well-known trade names. The last indicates the primary application, which is T-carrier, whereas the first two indicate the shape of the core separator. These types are interchangeable in almost all applications. Some special designs are available that are optimized for the digital rate to be carried on the cable. For example, the core separator in cables designed for T1C (DS-1C rate of 3.152 Mbps) service has more overlap, and provides more immunity to crosstalk at the higher frequencies, than cables optimized for repeatered T1 (DS-1 rate of 1.544 Mbps) service.

In a given cable, the transmit pairs are put in one compartment and the receive pairs are put in the other compartment. The separation is necessary to reduce crosstalk between the transmit and receive circuits of the transmission system.

The cross-section of a compartmental core cable is illustrated in Figure 7-7. The core separator, or screen, is made electrically continuous at all splice locations, but it is not deliberately grounded or connected to the cable shield (this is an important, but often overlooked, requirement).

Wrapped between the shield and the cable core of modern cables is a nonhygroscopic dielectric tape that provides additional insulation to increase the pair-to-shield withstand voltage.

Buffering

The term *buffering*, as it is used here, is borrowed from fiber optic technology, where it describes the protective measures used in fiber optic cable constructions.

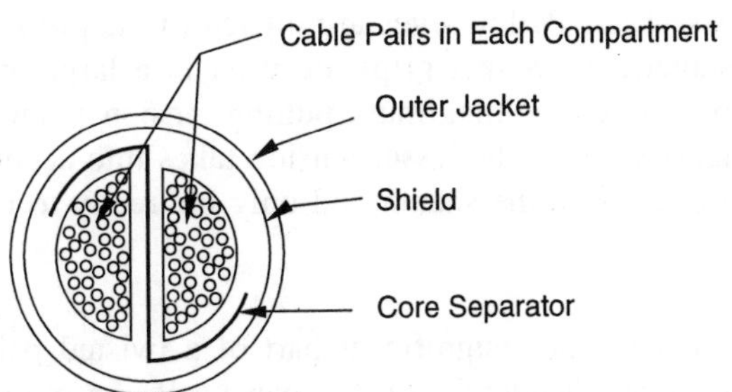

Figure 7-7 Compartmental core cable cross-section, [4] Reeve, © 1992 Institute of Electrical and Electronics Engineers. Reprinted with permission

*See section 92C of [27].

The term is easy to envision for any cable construction, however, and it is used here for that reason. In twisted pair cables, buffering is naturally provided by the outer jacket, armor (if used), and the conductor insulation itself, and no additional specific measures are required (unlike optical fibers as explained in the next chapter).

Strength Member

Unlike fiber optic cables, twisted pair telecommunication cables normally do not employ a separate strength member (although one can be specified if needed). Instead, the strength mostly comes from the conductors themselves and the jacket/shield/armor assembly.

When calculations are made to determine the maximum pulling tension for underground copper twisted pair cables, the following formula is used:

$$\text{Tmax} = 0.008 \times \text{CM} \times \text{N} \times 80\% \text{ lb}$$

where Tmax = maximum pulling tension in lb
CM = conductor area in circular mils
N = number of conductors (N = 2 × number of cable pairs)

The maximum tension formula can be rewritten for the common cable gauges, as follows:

$$\text{Tmax}(19) = 16.5 \times \text{no. pairs}$$

$$\text{Tmax}(22) = 8.2 \times \text{no. pairs}$$

$$\text{Tmax}(24) = 5.2 \times \text{no. pairs}$$

$$\text{Tmax}(26) = 3.3 \times \text{no. pairs}$$

where Tmax(AWG) is the maximum tension in lb for the gauge indicated

The foregoing formulas assume a pulling eye is used. A pulling eye is soldered to each individual conductor to transfer the pulling strain uniformly throughout the cable cross-section. Pulling eyes are difficult to apply in the field (they are usually factory installed), so basket grips are used in a large number of installations. If a basket grip is used, the maximum pulling tension is the calculated Tmax or 1,000 lbs, whichever is less. The lesser tension takes into account that the pulling force is transferred directly to the sheath and only indirectly to the conductor itself.

Cable Pairs

Probably the most significant part of a twisted pair digital loop, in terms of investment, is the outside plant cable pair itself. An individual pair comprised of tip and ring conductors is inherently bidirectional; that is, it will work just as well for transmission or reception in either direction. In common digital loop applications, however, with the exception of basic rate integrated services digital network (ISDN) digital subscriber line (DSL), and the high bit-rate digital subscriber line (HDSL), two cable pairs (four conductors) are used to provide independent transmit and receive paths. Associated with each cable pair of either the 2-wire or 4-wire digital loop are ancillary devices for protection and termination, and repeaters for regenerating the digital signal on long loops.

Twisted pair copper telecommunication cables can be obtained in any of four gauges, as shown in Table 7-6. Aluminum conductors have seen limited use in telecommunications. Aluminum conductors provide approximately the same conductivity as copper when they are two gauge sizes larger. For example, a 20 AWG aluminum conductor provides approximately the same conductivity as a 22 AWG copper conductor.

Table 7-7 gives the dc loop resistance per unit length for the different copper cable pair gauges at different temperatures. The loop resistance at temperatures not shown in the table can be found from:

$$R_T = R_{T0}[1 + \rho_{T0}(T - T_0)]$$

where R_T = resistance at temperature T
R_{T0} = resistance at reference temperature T_0
ρ_{T0} = temperature coefficient of resistivity per °C at temperature T_0

For copper, ρ_{T0} = 0.00401/°C at 15 °C and 0.00393/°C at 20 °C. Figure 7-8 shows the loop resistance as a function of loop length for each gauge at typical design temperatures.

For economic reasons, analog and digital signals may be carried on the same cable. When combined service is used, the cables are commonly referred to as exchange cables. When a cable is dedicated to digital service between central offices, it is sometimes referred to as a trunk cable.* Exchange cables are available in stan-

Table 7-6 Twisted Pair Copper Conductor Gauges[a]

American Wire Gauge (AWG)	Diameter (inch)	Cross-Section (CM)[b]	Equivalent Diameter (mm)
19	0.03589	1,288	0.912
22	0.02535	642.4	0.643
24	0.02010	404.0	0.511
26	0.01594	254.1	0.404

[a]Source: [32]
[b]CM = circular mils = (diameter in mils)^2

Table 7-7 DC Loop Resistance of Copper Twisted Pairs (Ω/kilofoot)

Design Temperature	19 AWG	22 AWG	24 AWG	26 AWG
13 °C (55 °F)	15.9	31.7	50.3	81.1
38 °C (100 °F)	17.4	34.8	55.3	89.2
60 °C (140 °F)	18.9	37.7	59.8	96.5

*A trunk cable is more strictly defined as a cable not providing exchange service to individual subscribers, but instead providing trunk service between exchanges. In the digital transmission environment the difference becomes less clear.

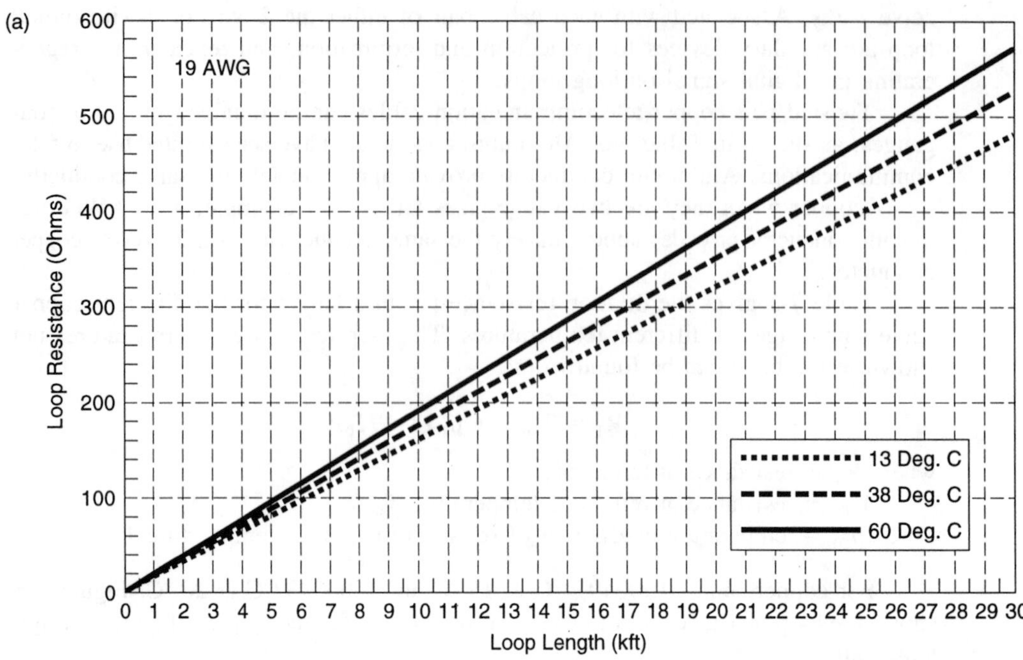

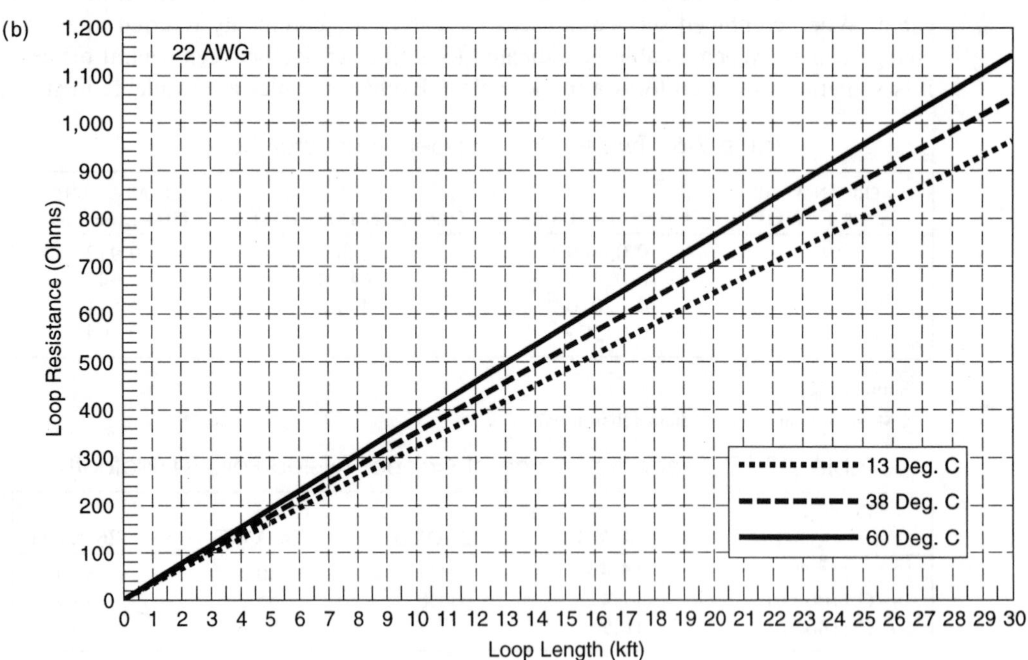

Figure 7-8 dc loop resistance vs. loop length: (a) 19 AWG; (b) 22 AWG; (c) 24 AWG; (d) 26 AWG

(c)

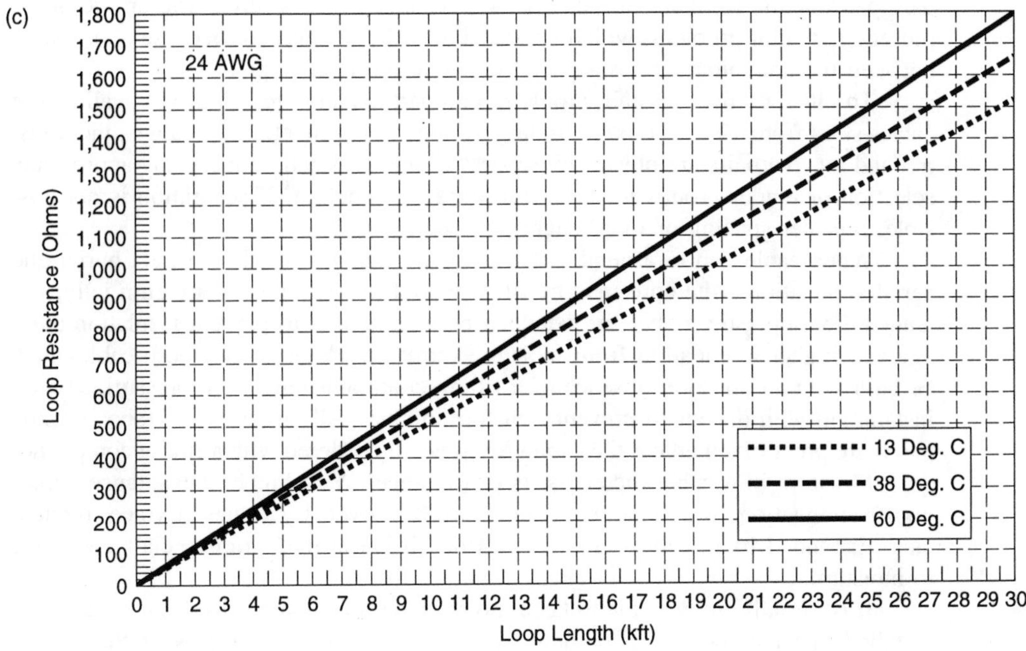

(d)

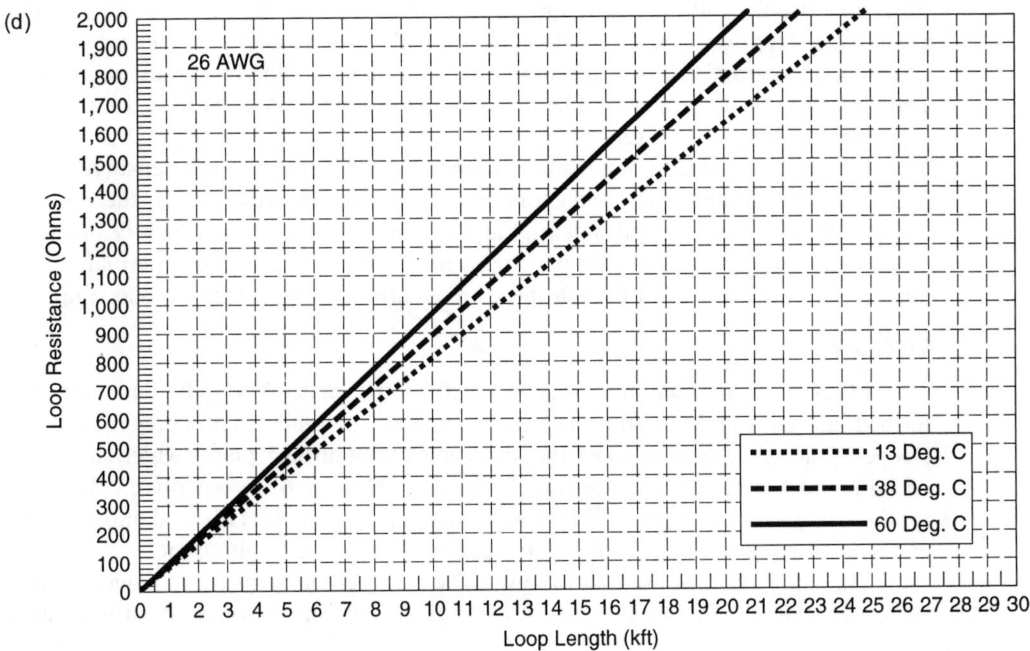

Figure 7-8 Continued

dard pair counts, from six to 4,200 pairs. The smaller sizes (up to 150 or 200 pairs) present particular problems with high-speed digital transmissions because of crosstalk between the unavoidably close-spaced transmit and receive pairs.

To alleviate the crosstalk problem, compartmental core cables are available in cables sizes from 12 to 200 pairs. In larger, noncompartmental core cables, the transmit and receive pairs are physically separated, and crosstalk reduced, by appropriate selection of binder groups for each transmission direction. If separation is not possible, repeater spacing or loop length must be reduced.

When cables up to around 1,200 pairs are manufactured for direct burial and aerial applications, the core may be filled with a polyethylene/petroleum jelly, extended thermoplastic rubber, or extended petroleum wax to prevent insulation damage and higher attenuation from moisture intrusion. Alternately, these cables may be built with an air core (absence of a filling compound) and then pressurized after installation with dry air to prevent moisture ingress. All larger cables (above 1,200 pairs) are pressurized unless the cables are specially ordered with a filling compound.

Moisture in a cable increases its capacitance. With higher capacitance comes higher attenuation, which can be as much as 40% over dry values at voice frequencies. Expanded foam cables are always filled because they are especially sensitive to moisture.

It is virtually impossible to keep moisture out of unpressurized, air core cables installed in an outdoor environment over a long period, regardless of the service classification. This applies to aerial, direct buried, and underground cables. If pressurization is not available, or if the cable is relatively small, filled cables provide much better long-term service than unpressurized, air core cables in all service classifications. Contrary to fairly common practice among some companies, unpressurized air core cables never should be placed underground or directly buried if the goal is long-term, trouble-free service.

Historically, telecommunications terminal equipment has been sensitive to the polarity of the tip and ring conductors. For switched subscriber loops, the tip conductor normally is at ground potential and the ring conductor is at battery potential (−48 V nominal). Conductors in pairs used for dedicated digital loops normally have no specified polarity, even though these frequently are designated tip and ring.

Color Coding

Many pairs are combined together to form the cable. The individual pairs are formed into units and the units are then assembled, or cabled, into cores (much like a rope). The basic unit of most modern cables contains 25 pairs, which are color coded for identification, as shown in Table 7-8.* Note also the repetitive color code for the tracer every five pairs. When necessary, the basic unit is divided into subunits of 12 or 13 pairs to attain cable roundness and the desired pair count.

The color code is used as follows. The basic color of the tip conductor insulation material is the first color shown in the table. The insulation has a striped tracer

*Appendix D provides cable core layout details for most cable types.

Table 7-8 Telephone Pair Color Code—25-Pair Unit Binder Group

Pair No.	Color (Tip-Ring)	Pair No.	Color (Tip-Ring)	Pair No.	Color (Tip-Ring)
1	White-blue	11	Black-blue	21	Violet-blue
2	White-orange	12	Black-orange	22	Violet-orange
3	White-green	13	Black-green	23	Violet-green
4	White-brown	14	Black-brown	24	Violet-brown
5	White-slate	15	Black-slate	25	Violet-slate
6	Red-blue	16	Yellow-blue		
7	Red-orange	17	Yellow-orange		
8	Red-green	18	Yellow-green		
9	Red-brown	19	Yellow-brown		
10	Red-slate	20	Yellow-slate		

of the second color. The reverse is true of the ring conductor; that is, the basic color is the second color shown in the table and the tracer is the first color. For example, the basic color of pair 1 tip conductor is white and the tracer is blue, and the basic color of the ring conductor is blue and the tracer is white.

In cables larger than 25 pairs, the groups (or units) are wrapped with a colored tracer ribbon (unit or group binder) according to a similar color coding scheme, as shown in Table 7-9. For example, a 100-pair cable contains four groups of 25 pairs each. The first group (pairs 1 to 25) is wrapped with a white-blue binder, the second group (pairs 26 to 50) is wrapped with a white-orange binder, and so on. Since there are 25 basic colors in the color code, up to twenty-five 25-pair groups can be identified this way. Only 24 are used, however, which takes care of cables up to 600 pairs. Beyond 600 pairs, "super unit binders" are used. Each 600-pair super unit is wrapped with an appropriate binder. This color code is shown in Table 7-10. Figure 7-9 shows the unit and super unit relationships in a 3,000-pair exchange cable.

Table 7-9 Group Binder Color Code[a]

Pair Group	Group No.	Color (Tip-Ring)	Pair Group	Group No.	Color (Tip-Ring)
1–25	1	White-blue	301–325	13	Black-green
26–50	2	White-orange	326–350	14	Black-brown
51–75	3	White-green	351–375	15	Black-slate
76–100	4	White-brown	376–400	16	Yellow-blue
101–125	5	White-slate	401–425	17	Yellow-orange
126–150	6	Red-blue	426–450	18	Yellow-green
151–175	7	Red-orange	451–475	19	Yellow-brown
176–200	8	Red-green	476–500	20	Yellow-slate
201–225	9	Red-brown	501–525	21	Violet-blue
226–250	10	Red-slate	526–550	22	Violet-orange
251–275	11	Black-blue	551–575	23	Violet-green
276–300	12	Black-orange	576–600	24	Violet-brown

[a]In cables of 100 pairs (4 groups) or fewer, the white binder is optional.

Table 7-10 Super Unit Binder Color Code

Super Unit	Binder Color
Pairs 1 to 600	White
Pairs 601 to 1,200	Red
Pairs 1,201 to 1,800	Black
Pairs 1,801 to 2,400	Yellow
Pairs 2,401 to 3,000	Violet
Pairs 3,001 to 3,600	Blue
Pairs 3,601 to 4,200	Orange

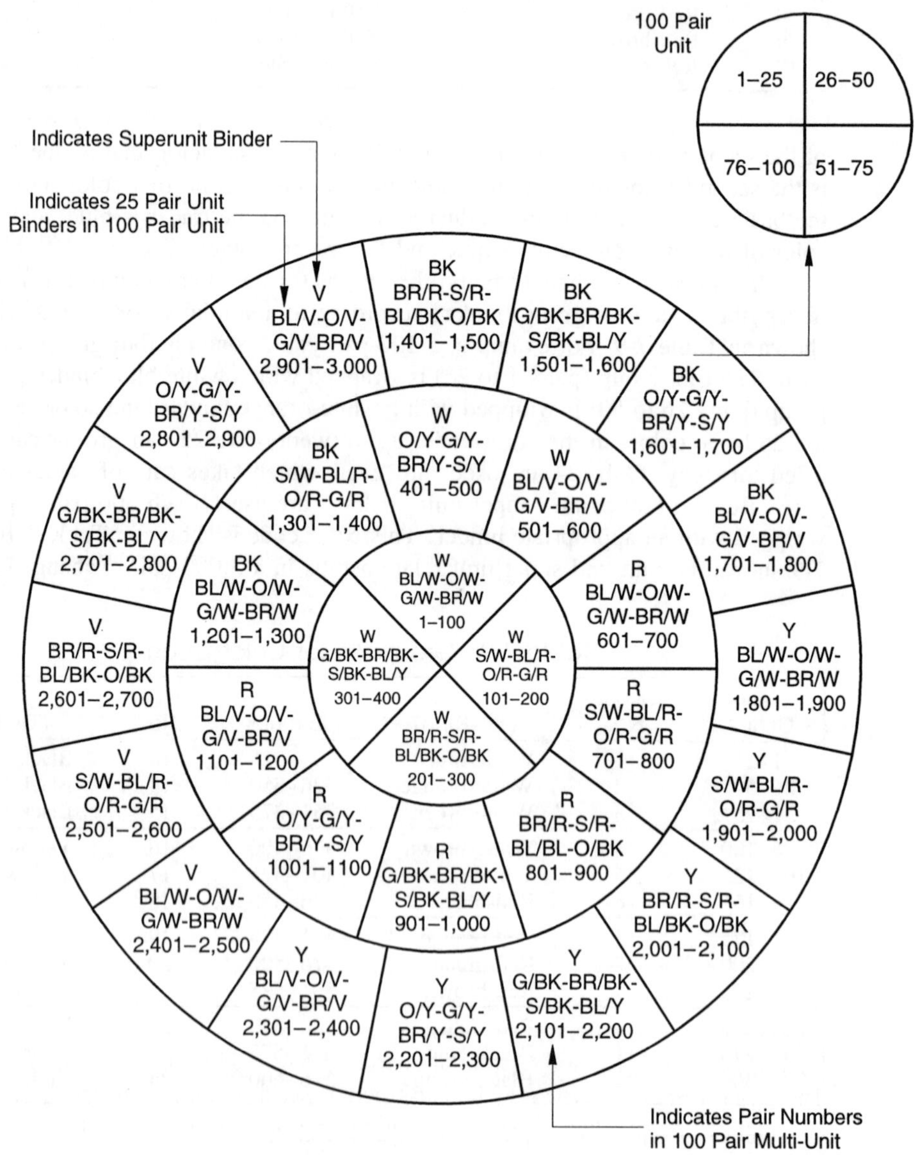

Figure 7-9 Super unit cable layout

On large exchange cables consisting of 1,200 pairs or more, the mirror-image color code and unit arrangement can be obtained, if desired. This color code is sometimes called the "Bell Design" and differs from the unit and super unit binder groups described above. For mirror-image cables, the 25-pair units are assembled into 100-pair multi-units. Each unit is identified by a standard color sequence binder: blue (pairs 1 to 25); orange (pairs 26 to 50); green (pairs 51 to 75); and brown (pairs 76 to 100).

Multi-units in a mirror-image cable are identified by a two-color binder and layering scheme. The first multi-unit in each layer has a green binder, which designates the "marker" unit. Marker units are situated radially, and a red and a blue unit are placed on each side, as shown in Figure 7-10, for a 3,000-pair mirror-image cable. The second color in the multi-unit binder indicates the layer and alternates between yellow and black. The outermost layer is yellow, second layer in is black, third layer in is yellow, and so on. Therefore, the outermost marker will have a green-yellow binder, and the marker for the "second layer in" will have a green-black binder.

With one's back to the central office, the pair counts of mirror-image cables start with the centermost marker multi-unit, with pairs 1 to 100, and continue clockwise to pairs 101 to 200 in the second multi-unit (red), and so on. The count continues through subsequent layers by following the same counting rules, always starting with the marker unit.

The most common residential type station wire (also known as "quad" or inside/outside wire, I/O) consists of four conductors, with the color code shown in Table 7-11. Quad normally is not recommended for digital service because of its limited twisting (and higher crosstalk), but it can be used on short runs (<50 feet) at subrates when nothing else is available.

Other types of multipair station wire will conform to the color code shown in Table 7-8. In addition, various other color codes are used for wiring electronic equipment, some of which may be used in telecommunication systems.

Table 7-11 Quad Station Wire Color Code[a]

PAIR NO.	COLOR (TIP-RING)
1	Green-red
2	Black-yellow

[a]The yellow conductor (or yellow and black conductors together) is used as a ground lead, if required, by the terminal equipment.

Conductor Insulation

All modern cable conductors are insulated with a polyolefin (plastic) formed by an extrusion process. Older cables use a paper (or pulp) insulation. PIC, as defined earlier, is a general term used to describe all modern plastic insulated cables. Such cables may use solid, cellular foam, or foam skin (skinned foam) insulation. The actual insulation type may be classified as solid (polyethylene or polypropylene) or expanded foam (dual expanded polyethylene or expanded polyethylene, PVC).

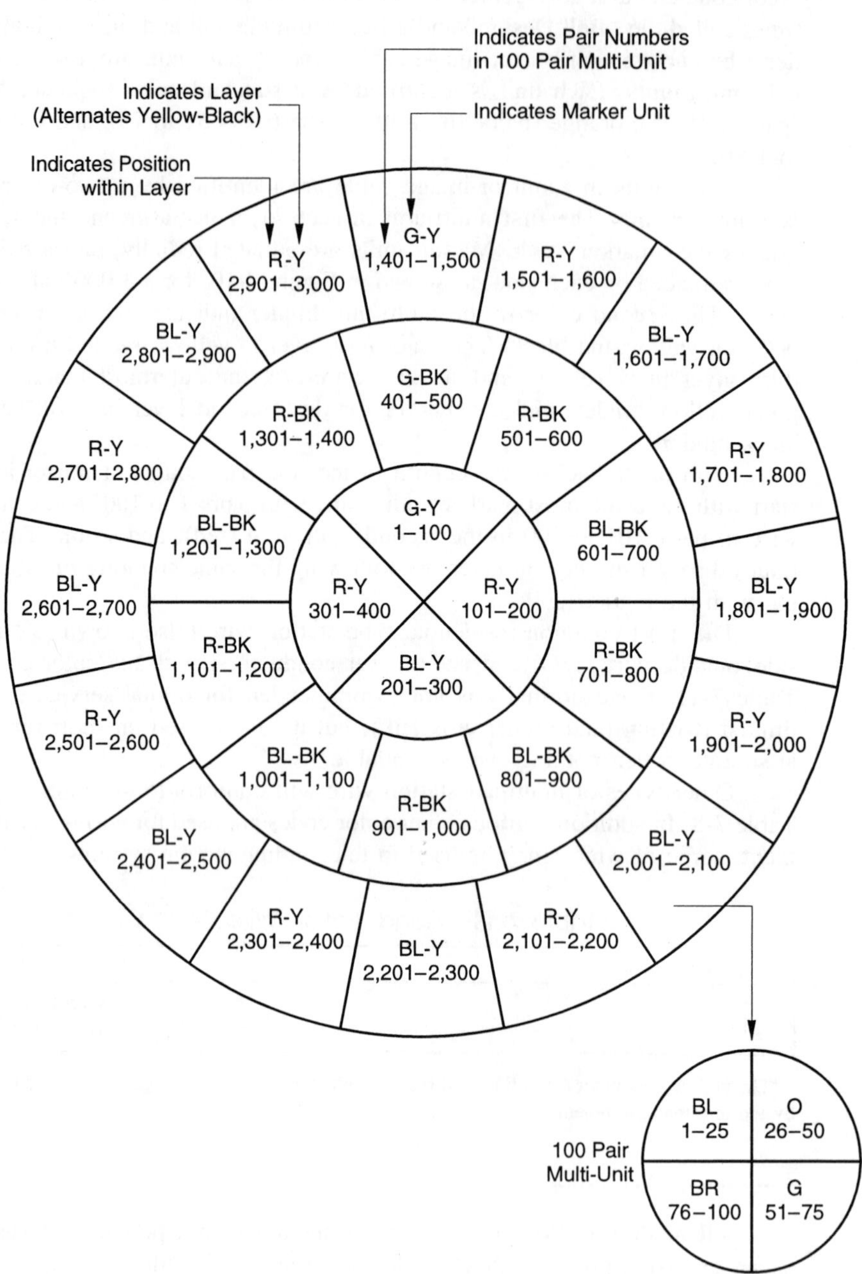

Figure 7-10 Mirror-image cable pair layout

A combination of solid and foam insulation is used in foam skin cables. In these cables, the regular foam insulation is covered by a solid plastic skin of about 2-mil thickness for added strength. Expanded foam cables have found widespread use since about 1980.

The dimensions and dielectric constant of the insulation and surrounding material determine the capacitance of twisted pair cables. The dimensions are shown in Figure 7-11 for each of the common cable gauges. The dielectric constant of solid plastic insulations used in exchange cables is about 2.3, whereas the dielectric constant of the expanded foam is about 1.6. Insulation thicknesses vary from about 10 mils for solid insulations, used in air core cables and filled foam cables, to about 14 mils for solid insulations used in filled cables. The relationship for twisted pair capacitance is

$$C = \frac{K\varepsilon}{\log\left(\frac{1.5D}{d}\right)} \text{ nF/length}$$

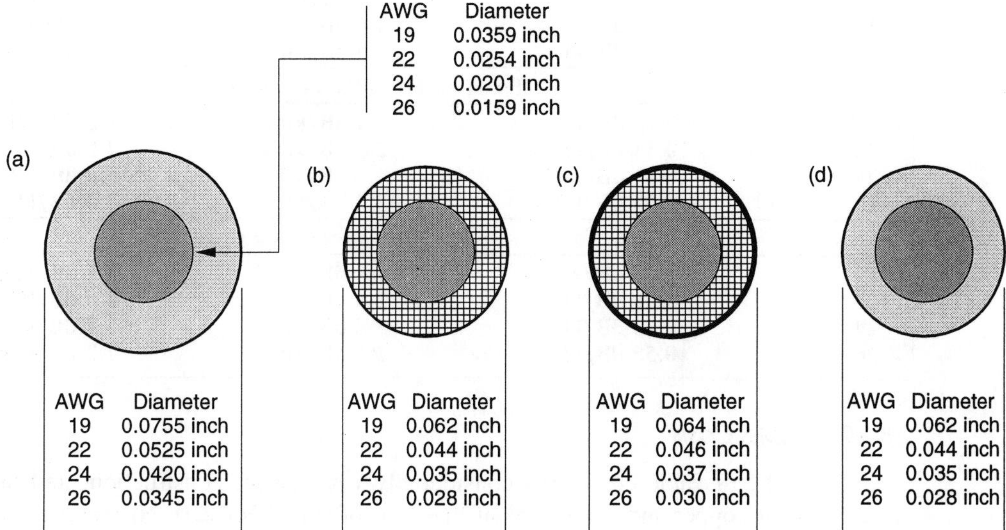

Figure 7-11 Twisted pair insulation and conductor dimensions: (a) Filled core/solid insulation; (b) Filled core/foam insulation; (c) Filled core/skinned-foam insulation; (d) Air core/solid insulation

where K = constant of proportionality, which depends on the dielectric properties of the medium surrounding the insulation (either air or filling compound)
ε = dielectric constant of the insulating material
D = diameter of the insulation
d = diameter of the conductor (units of D and d must be the same)

The transmission performance of the different conductor insulations is the same at voiceband frequencies, but this is not the case at the higher frequencies. The transmission performance of expanded foam cables at digital carrier frequencies falls between solid insulated air core and solid filled core cables at 772 kHz, which is the Nyquist frequency for repeatered T1-carrier.

A key indicator of performance is attenuation, which is given in Table 7-12 for various insulation types and frequencies. The values shown are averages for a wide variety of cables. Significant deviations exist for specific cable types at 772 kHz. A more complete list is given for this frequency in Chapter 9.

Table 7-12 Average Attenuation of PICs at 20 °C (68 °F)

AWG	1 kHz	150 kHz	772 kHz
		SOLID INSULATION—FILLED CORE	
19	0.23 dB/kft	1.3 dB/kft	2.8 dB/kft
22	0.34 dB/kft	1.8 dB/kft	4.0 dB/kft
24	0.43 dB/kft	2.4 dB/kft	5.0 dB/kft
26	0.55 dB/kft	3.5 dB/kft	6.4 dB/kft
		FOAM OR FOAM-SKIN—FILLED CORE	
19	0.23 dB/kft	1.4 dB/kft	3.2 dB/kft
22	0.34 dB/kft	2.0 dB/kft	4.5 dB/kft
24	0.43 dB/kft	2.7 dB/kft	5.6 dB/kft
26	0.55 dB/kft	3.7 dB/kft	7.0 dB/kft
		SOLID INSULATION—AIR CORE	
19	0.23 dB/kft	1.4 dB/kft	3.3 dB/kft
22	0.34 dB/kft	2.0 dB/kft	4.6 dB/kft
24	0.43 dB/kft	2.7 dB/kft	5.7 dB/kft
26	0.55 dB/kft	3.6 dB/kft	7.2 dB/kft

7.6.2.9 Cable Costs

Twisted pair cable costs vary considerably with gauge, construction, and the current cost of copper and aluminum (if that type of shield is used). Nineteen gauge cable can be twice as expensive as 24 gauge cable for a given pair count. Filled cable is 10 to 25% more expensive than air core cable, depending on the gauge and pair count. Expanded foam insulation cables are less expensive than comparable solid insulation cables by 7 to 20%, depending on the gauge and pair count.

Tables 7-13 and 7-14 give multipliers to find the relative material costs for various cable gauges and sizes. Multipliers for adding a compartmental core separator (screen) to exchange cables are shown in Table 7-15. The use of each cable is explained in the note below it.

Table 7-13 Relative Costs of Twisted Pair Cable Gauges and Sizes Compared to 24 AWG

	PAIRS							
AWG	12	25	50	100	200	400	900	1,200
26	0.72	0.73	0.82	0.78	0.77	0.76	0.77	0.76
24	1.00	1.00	1.00	1.00	1.00	1.00	1.00	1.00
22	1.18	1.21	1.41	1.31	1.43	1.40	1.36	1.37
19	1.94	2.33	2.59	2.59	2.56	2.57	N/A	N/A

This is a vertically oriented matrix where costs are shown relative to 24 gauge, filled PIC, 8-mil aluminum shield. For example, a 26 AWG, 100-pair cable is 78% of the cost of a 24 AWG, 100-pair cable, and a 22 AWG, 400-pair cable is 140% of the cost of a 24 AWG, 400-pair cable.

Table 7-14 Relative Costs of Twisted Pair Cable Sizes for a Given Gauge

	PAIRS							
AWG	12	25	50	100	200	400	900	1,200
26	0.22	0.34	0.57	1.00	1.86	3.72	8.18	10.44
24	0.23	0.35	0.54	1.00	1.91	3.71	8.31	10.73
22	0.20	0.32	0.58	1.00	2.08	3.96	8.62	11.22
19	0.17	0.31	0.54	1.00	1.88	3.68	N/A	N/A

This is a horizontally oriented matrix where costs are shown relative to a 100-pair cable of a given gauge. For example, a 26 AWG, 50-pair cable is 57% of the cost of a 26 AWG, 100-pair cable, and a 22 AWG, 400-pair cable is 396% of the cost of a 22 AWG, 100-pair cable.

Table 7-15 Multiplier for Adding Compartmental Core Separator to a Cable

	PAIRS							
AWG	12	25	50	100	200	400	900	1,200
26	1.19	1.16	1.12	1.10	1.06	1.08	1.07	N/A
24	1.15	1.13	1.10	1.08	1.05	1.06	1.06	N/A
22	1.23	1.22	1.14	1.13	1.08	1.10	1.10	N/A
19	1.18	1.16	1.11	1.09	1.06	N/A	N/A	N/A

This matrix gives multipliers for adding a compartmental core separator (screen) to a cable of given size and gauge. For example, to determine the cost of adding a screen to a 26 AWG, 50-pair exchange cable, multiply its cost by 1.12, and to determine the cost of adding a screen to a 22 AWG, 200-pair exchange cable, multiply its cost by 1.08.

7.6.3 Electrical Characteristics

7.6.3.1 Bandwidth

Digital loops consist of one or two cable pairs. The cable pair is a transmission line, and it has predictable electrical characteristics. Twisted pair cables are not used presently with digital transmission systems requiring more than several megahertz of bandwidth for several reasons, as will be discussed. For comparison with voiceband applications, the usable bandwidth on an analog loop is considered to be limited to around 3,100 Hz (300 to 3,400 Hz) [4]. The limitation is due to the coupling transformers and other devices connected to voiceband loops.

In digital loops, a much wider bandwidth is needed for information transmission. This point is often missed in preliminary considerations of the advantages of digital transmission and cannot be overemphasized. For example, a digitally encoded voicegrade (3,100 Hz) circuit could actually require a transmission line bandwidth of as much as 64 kHz.

It is important to realize the difference between circuit bandwidth and transmission bandwidth. Figure 7-12 shows the power spectral density of a hypothetical voiceband signal that has been encoded at an 8 kHz rate into 8-bit samples (64 kbps), filtered by a first-order Butterworth lowpass filter, and transmitted using a return-to-zero (RZ), bipolar, alternate mark inversion (AMI) line code. The main lobe of the spectrum is centered at 32 kHz and extends out to at least 64 kHz.

The frequency response of a loop derived entirely from twisted pairs extends down to dc (zero frequency). Digital loop transmission systems avoid using dc for information carrying, so the loops can be transformer-coupled to regenerators. DC continuity is still required in the metallic loop circuit, however, for loop powering schemes and sealing current. When signals with frequency content approaching zero frequency are placed on the loop, the transmission is considered to be in the baseband.

A twisted pair digital loop has no precise upper frequency limit. Transmission systems using the loop are designed to accommodate its attenuation characteristics over the required bandwidth. For example, the T1C system, which operates at 3.152

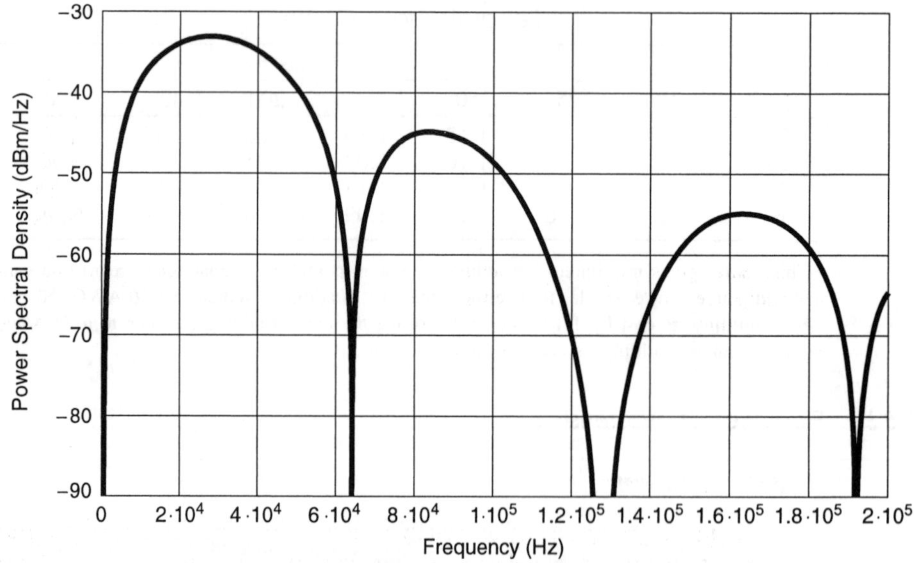

Figure 7-12 Power spectral density of a 64 kbps digitally encoded voiceband signal using bipolar AMI transmission

Mbps, has a bandwidth requirement of about 3.152 MHz. This one-for-one bit rate to bandwidth correspondence does not hold for all digital transmission systems. The T1C has the highest bandwidth requirements (and bit rate) of digital loop systems using twisted pair cable considered in this book.*

7.6.3.2 Transmission Line Primary Constants

Like all transmission lines, twisted pairs can be described by their fundamental electrical characteristics, or primary constants.† The primary constants are series inductance L, shunt capacitance C, shunt conductance G, and series resistance R. These are described in customary electrical units per unit length. Figure 7-13 shows a T-equivalent circuit commonly used to represent telecommunication transmission lines.

Inductance L of the typical cable pair depends on the geometry of the conductors and is affected by the magnetic flux distribution within and around them. It is around 0.19 mH/kilofoot at voice frequencies for air core PIC and about 7 to 10% higher for filled PIC. Due to skin effect, inductance starts to decrease at about 20 kHz and continues to decrease to at least 5 MHz.‡

Capacitance C depends on the geometry of and dielectric medium between the two conductors of the twisted pair. It is standardized during manufacture at 12.5 or 15.7 nF/kilofoot (66 or 83 nF per mile), although cables with different values are available. Cable with 15.7 nF/kilofoot capacitance is the most common in digital loop applications, but a special low-capacitance cable, with 12.5 nF/kilofoot, is preferred for higher bit-rate systems such as the 3.152 Mbps T1C system. For all practical purposes, capacitance is independent of frequency.

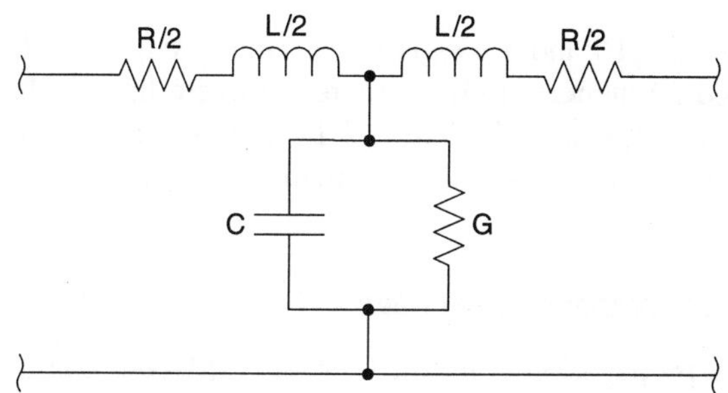

Figure 7-13 Transmission line T-equivalent circuit in terms of primary constants

---

*The differences between digital loop systems, digital backbones, and digital trunk transmission systems are becoming less clear as these systems become more prevalent. It is arguable whether a digital system operating at the DS-3 rate (45 MBPS) can be considered a digital loop system.

†As will be seen, the primary constants are not really constant, but the terminology is traditional and will be used here.

‡Skin effect is the tendency for higher-frequency currents to travel near the outside surface of a conductor rather than be distributed evenly throughout the entire cross-section.

Conductance G depends on the dielectric medium between the conductors and is negligible at about 0.04 microsiemen (μS) per kilofoot at 1 kHz for modern cables. It increases by a factor of ten for each decade (factor of ten) increase in frequency. The value can vary considerably from those given here, due to different insulation types and core filling compounds.

Resistance R, measured in Ω/kilofoot, depends on the resistivity of the conductors and the current distribution within them. It is constant at lower frequencies (less than a few hundred hertz) but is approximately proportional to the square root of frequency at higher frequencies (above several tens of kilohertz) where skin effect and proximity effect dominate. The resistance of each conductor gauge is different due to the different cross-sectional area.

The primary constants are affected by temperature in different ways. Some primary constants depend on whether the cable has a filled or air core. The following summarizes the temperature effects:

- Inductance varies only slightly with temperature (1.8% from −18 to +49 °C).
- Capacitance of filled cables decreases linearly at a rate of 0.016% per °C over the frequency range of 1 kHz to 3 MHz. This is attributed to a decrease in insulation density, cable expansion, and a decrease in dielectric constant with temperature increase. In air core cables, the capacitance varies only slightly with temperature over the same frequency range [38].
- Conductance varies only negligibly with temperature.
- Resistance is highly temperature dependent, increasing by about 4% for each 10 °C increase in conductor temperature at voice frequencies and 2 to 3% at frequencies higher than 10 kHz.

Tables 7-16 and 7-17 give the primary constants for typical cable pair gauges at selected frequencies [33]. A more complete listing of outside plant characteristics is given in Appendix C. Figure 7-14 shows the variation of the primary constants with frequency in graphical form (capacitance is not shown because it is frequency independent).

7.6.3.3 Secondary Parameters

The propagation constant ($\rho = \alpha + j\beta$) and characteristic impedance (Z_0) are the secondary parameters used to describe transmission line characteristics.* These parameters, which are functions of frequency, can be found from the primary constants from

$$\rho = \alpha + j\beta = \sqrt{(R + j\omega L)(G + j\omega C)}$$

where ω is the radian frequency, $2\pi f$.

---

*The secondary parameters are mathematically derived from the primary constants in [4]. Simplified analyses of twisted pair cable secondary parameters when the primary constants are known, and primary constants when secondary parameters are known, are covered rather well in [34]. There are numerous other sources of information on twisted pair transmission characteristics; see [33–37].

Table 7-16 Air Core PIC, Exchange Cable Pair Primary Constants, 20 °C

FREQUENCY (KHZ)	R (Ω/KFT)	L (MH/KFT)	C (NF/KFT)	G (μS/KFT)
		19 AWG		
1	16.1	0.182	15.7	0.0398
40	18.7	0.173	15.7	1.58
192	35.3	0.156	15.7	7.58
772	70.2	0.144	15.7	30.4
1,576	98.9	0.140	15.7	62.0
		22 AWG		
1	32.4	0.184	15.7	0.0398
40	34.2	0.177	15.7	1.58
192	50.4	0.167	15.7	7.58
772	100.8	0.151	15.7	30.5
1,576	141.2	0.146	15.6	62.0
		24 AWG		
1	51.3	0.181	15.7	0.0398
40	53.0	0.175	15.7	1.58
192	67.0	0.167	15.7	7.58
772	127.7	0.151	15.7	30.5
1,576	182.1	0.145	15.6	62.0
		26 AWG		
1	82.0	0.183	15.7	0.0398
40	83.6	0.177	15.7	1.58
192	94.4	0.172	15.7	7.59
772	159.8	0.159	15.7	30.5
1,576	230.3	0.152	15.7	62.0

The propagation constant is a complex number consisting of real and imaginary parts. In general, it describes the attenuation and phase characteristics of any transmission line. The real portion of the propagation constant, denoted Re(ρ), is called the attenuation constant α, and the imaginary portion, denoted Im(ρ), is called the phase constant β. The attenuation constant is expressed in Nepers per unit length and the phase constant, in radians per unit length.*

The characteristic impedance is independent of length. It, too, can be found from the primary constants:

$$Z_0 = \sqrt{\frac{(R + j\omega L)}{(G + j\omega C)}}$$

The equations for the secondary parameters can be simplified somewhat for modern outside plant PIC, as follows:

$$\rho = \sqrt{-\omega^2 LC + j\omega RC} \text{ for frequency } <500 \text{ kHz, and}$$

*A more convenient unit than the Neper is the decibel such that 1 Neper = 8.686 dB.

Table 7-17 Filled Core PIC, Exchange Cable Pair Primary Constants, 20 °C

FREQUENCY (KHZ)	R (Ω/KFT)	L (MH/KFT)	C (NF/KFT)	G (μS/KFT)
		19 AWG		
1	16.1	0.212	15.7	0.0398
40	18.2	0.202	15.7	1.58
192	32.8	0.189	15.7	7.58
772	63.8	0.178	15.7	30.5
1,576	89.6	0.174	15.6	62.0
		22 AWG		
1	32.4	0.212	15.7	0.0398
40	33.9	0.205	15.7	1.58
192	47.7	0.196	15.7	7.59
772	93.0	0.182	15.7	30.5
1,576	129.4	0.178	15.7	62.0
		24 AWG		
1	51.3	0.212	15.7	0.0398
40	52.7	0.206	15.7	1.58
192	64.1	0.199	15.7	7.58
772	117.4	0.186	15.7	30.5
1,576	165.5	0.180	15.6	62.0
		26 AWG		
1	82.0	0.212	15.7	0.0398
40	83.3	0.205	15.7	1.58
192	92.1	0.200	15.7	7.59
772	149.4	0.190	15.7	30.5
1,576	212.7	0.183	15.7	62.0

See text for discussions of R, L, C, and G.

$$\rho = \frac{R}{2}\sqrt{\frac{C}{L}} + j\omega\sqrt{LC} \text{ at higher frequencies}$$

with

$$\alpha = \mathrm{Re}(\rho) \text{ and } \beta = \mathrm{Im}(\rho) \text{ as before}$$

$$Z_0 = \sqrt{\frac{L}{C} - j\frac{R}{\omega C}} \text{ for frequency } <500 \text{ kHz}$$

$$Z_0 = \sqrt{\frac{L}{C}} \text{ at higher frequencies}$$

The characteristic impedance also is a complex number. At low frequencies (around 1 kHz), the real and imaginary parts are about equal in magnitude. At higher frequencies (above 100 kHz), the imaginary part becomes small, and the impedance is essentially constant and real at approximately 100 Ω.

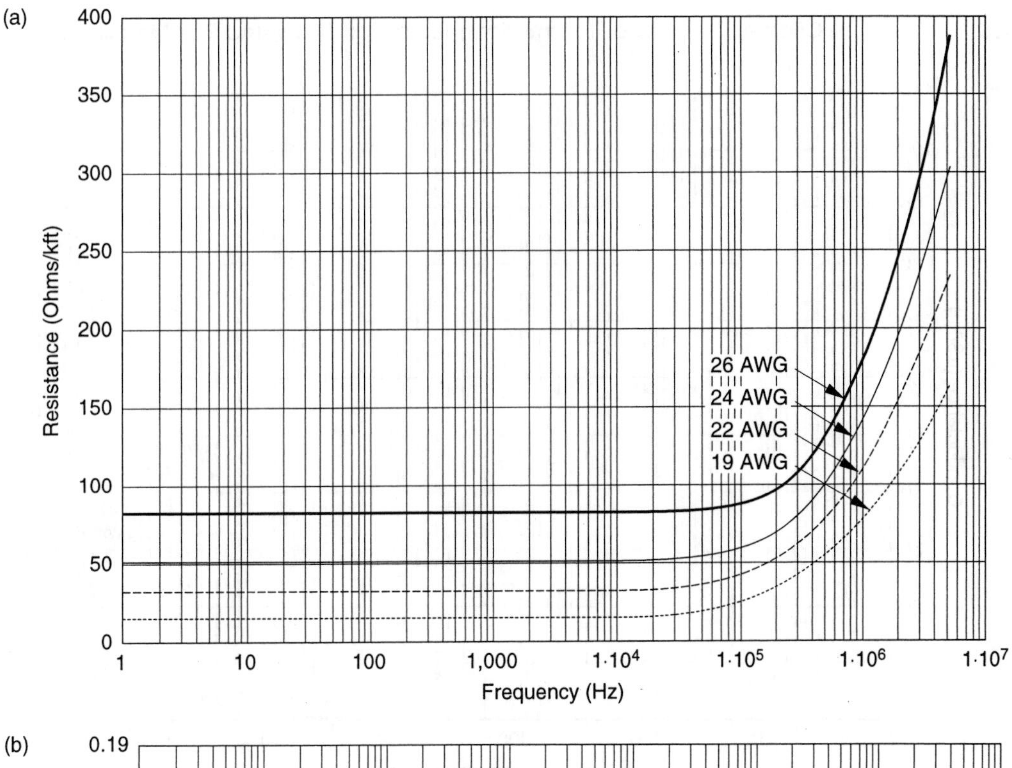

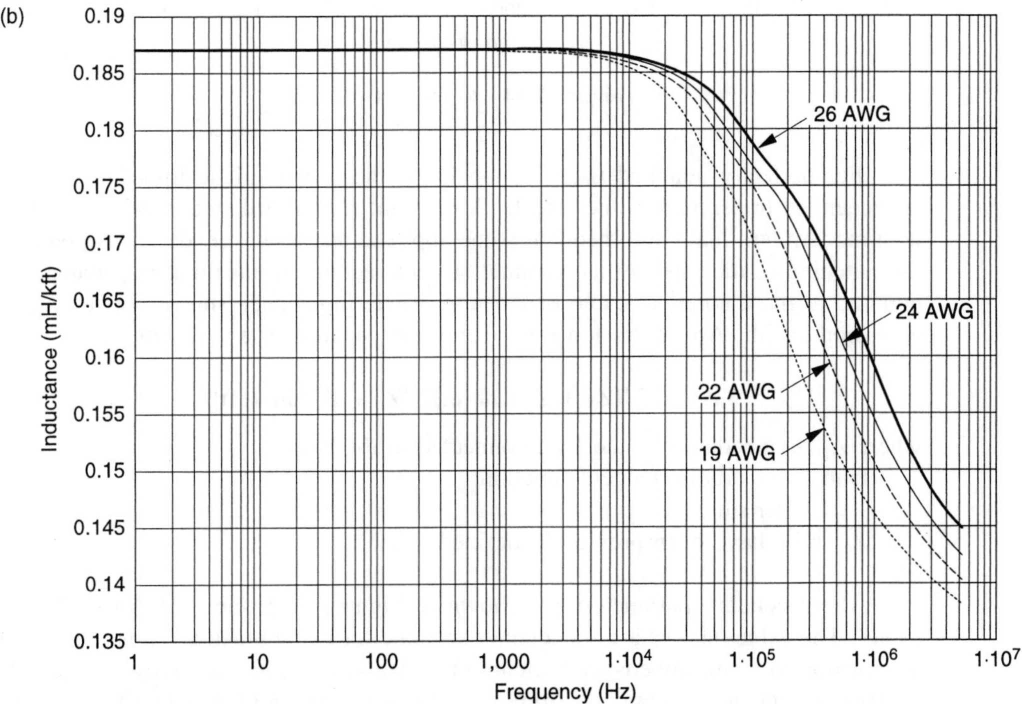

Figure 7-14 Twisted pair primary constants for typical PIC at 21 °C: (a) Resistance vs. frequency; (b) Inductance vs. frequency; (c) Conductance vs. frequency

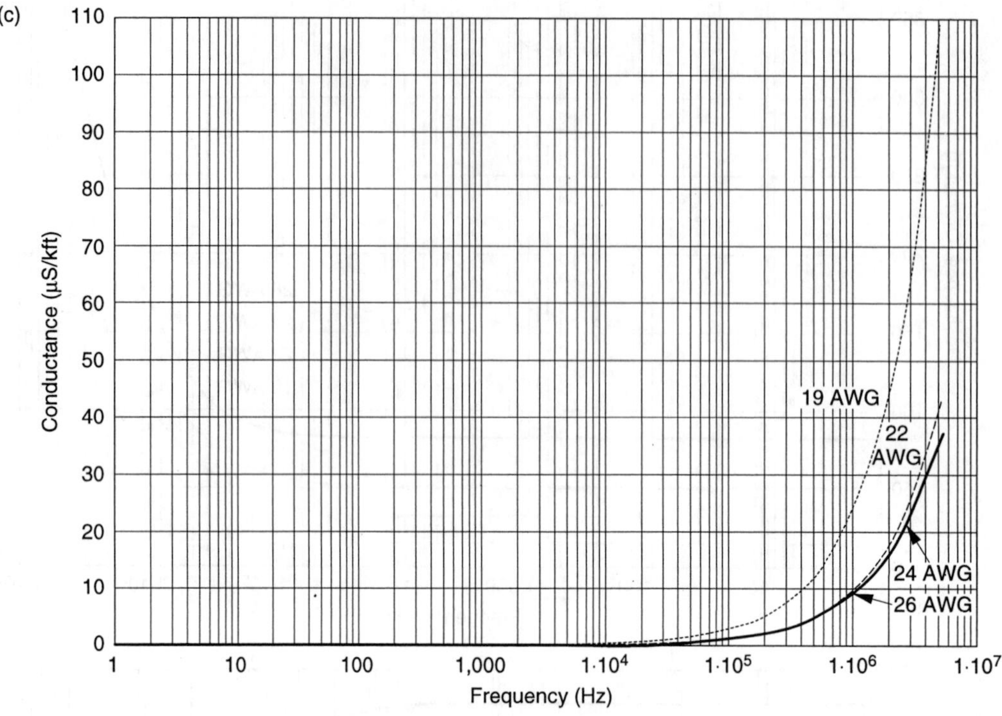

Figure 7-14 Continued

For perfectly matched transmission lines, the characteristic impedance is independent of transmission line length. A constant real impedance is easy to match with modern terminal equipment. In some applications, however, due to impedance mismatching of the loop by its terminations, the actual circuit input impedance (as opposed to the theoretical cable pair characteristic impedance shown above) varies with loop length. An expression for the input impedance of a real loop is:

$$Z_{in} = Z_0\{[Z_T + Z_0 \tanh(\rho l)]/[Z_0 + Z_T \tanh(\rho l)]\}$$

where Z_T = impedance of the load connected to the loop
tanh = hyperbolic tangent function
l = length
Z_0 and ρ have been previously defined

The secondary parameters are shown in Figures 7-15 and 7-16 for exchange cables. The values shown are sufficiently accurate for many practical situations, such as insertion loss and impedance matching problems. If greater accuracy is required, actual test data can be obtained from the cable manufacturers or actual field installations.

As with the primary constants, secondary parameters are temperature dependent. Attenuation is related to temperature such that the higher the temperature, the

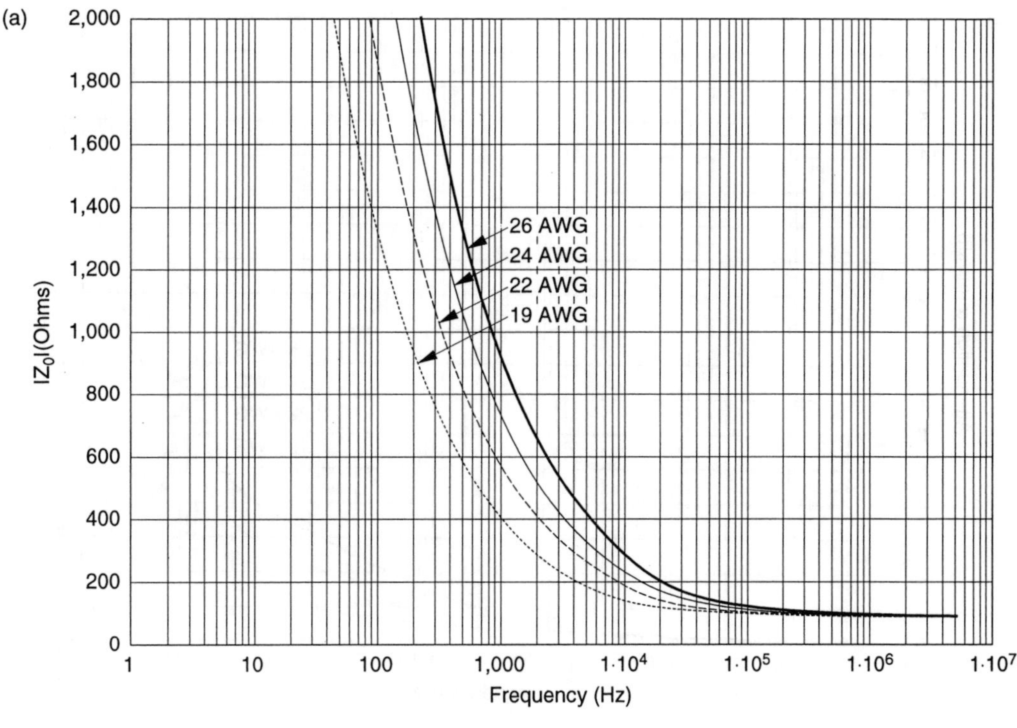

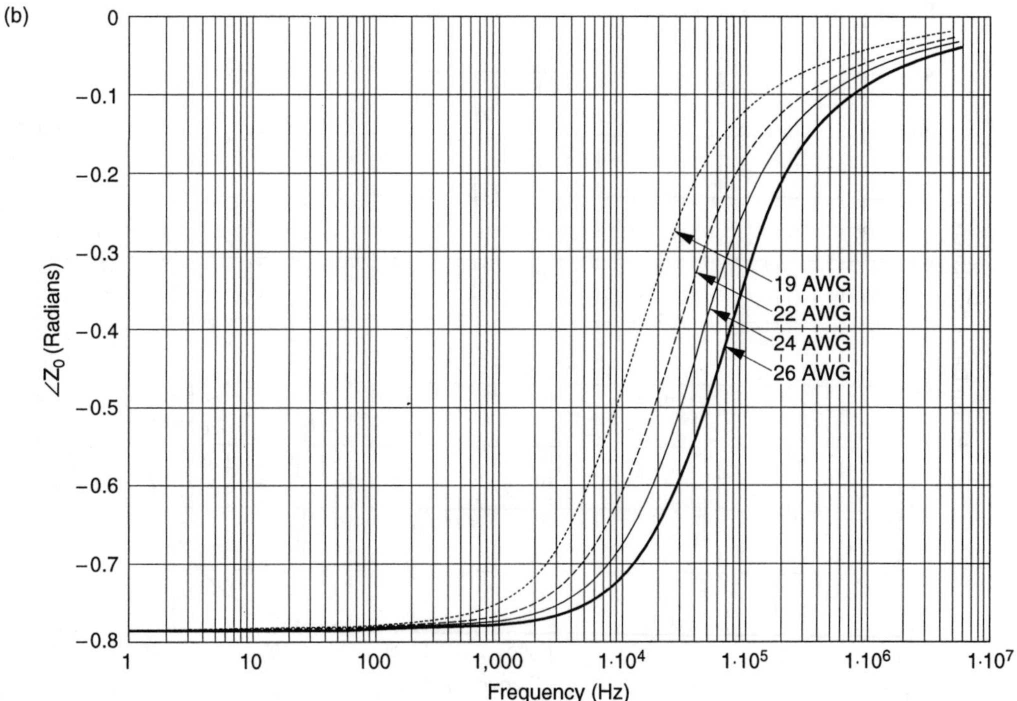

Figure 7-15 Twisted pair secondary parameters, characteristic impedance: (a) Impedance magnitude; (b) Impedance angle

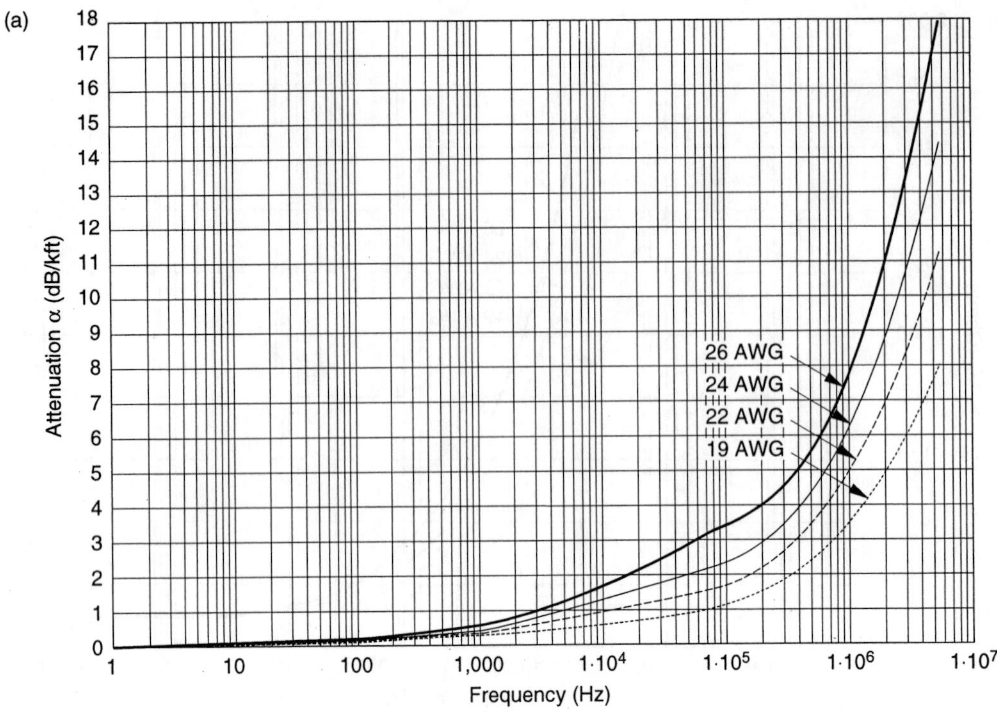

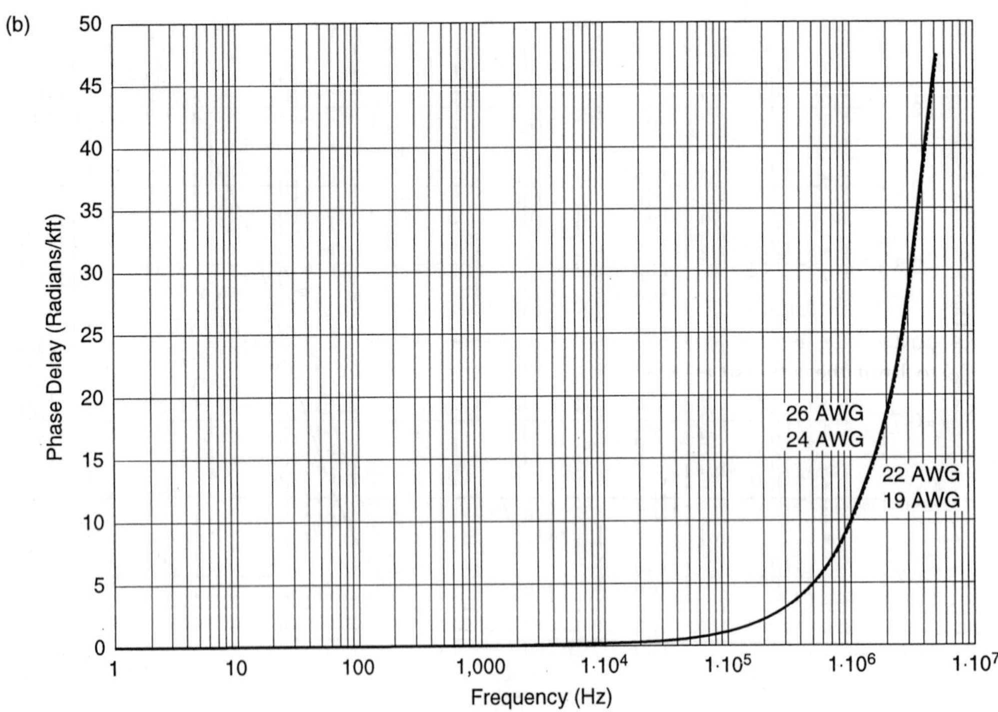

Figure 7-16 Twisted pair secondary parameters: (a) Attenuation vs. frequency; (b) Phase delay vs. frequency

higher the attenuation. The relationship is nonlinear and due primarily to changes in conductor resistance. Transmission parameters provided on cable manufacturer's data sheets usually are given at 55 °F (12.8 °C) or 68 °F (20 °C). Current digital loop design methods require the quoted values to be adjusted (corrected) to the maximum expected temperature of operation.

The attenuation characteristics of PIC frequently are modeled using a single parameter. The single-parameter model assumes attenuation is proportional to $\sqrt{f}$. Therefore, if the attenuation $\alpha(f_0)$ is known at frequency f_0, the attenuation $\alpha(f)$ at any other frequency f can be obtained from:

$$\alpha(f) = \alpha(f_0)\sqrt{\frac{f}{f_0}}$$

This model assumes that skin effect causes the predominant impact on attenuation, which is not true at low frequencies. Therefore, although fairly accurate above 200 kHz, the single-parameter model is useless in many digital loop applications, particularly the ISDN DSL and subrate digital loops.

7.6.3.4 Transmission Requirements

The fundamental transmission requirements of digital loops follow the same principles as analog loops.* On any digital loop:

- The signal must have adequate level so the presence of a signal element may be reliably detected, and
- The signal level must be sufficiently higher than the noise to ensure reliable differentiation between the signal elements and noise (high enough signal-to-noise ratio, SNR).

Adequate level is provided by controlling circuit loss for a given transmitter power output. Adequate SNR is provided by controlling crosstalk and taking care to limit the adverse effects of impulse and other noise through proper shielding.

For reference, it is informative to know what the acceptable SNRs are for various types of circuits. Digital loops operate with acceptable error performance if the SNR is above approximately 15 dB. As a rule of thumb, each dB of change in SNR will change the error rate by a factor of ten for error rates in the vicinity of 10E-6. For example, if the error rate is 10E-6 with 14 dB SNR, it will be 10E-5 at 13 dB SNR. Generally, the SNR for individual analog circuits should be at least 30 dB for voice calls (based on subscriber satisfaction) and 15 to 20 dB on data calls using analog modems (based on the specified error rate, coding, and modulation method). These figures are end-to-end. Each portion of the transmission system (loops, switching systems, terrestrial and satellite radio systems, and so on) contribute to lowering

*See [4] for analog transmission requirements.

the SNR. It follows that the SNR for any given part of the overall connection must be higher than the overall requirement.

For an analog voice circuit being transmitted on a digital loop carrier system, poor error performance of the digital loop manifests itself indirectly to the listener in terms of distortion, dropouts, and other annoyances. A degraded channel in this case may still be usable for voice transmission. Digital data circuits, on the other hand, are directly affected by error performance in that bit integrity is a requirement of digital transmission. Tolerable error levels on a digital data circuit are always much less than on a voice circuit. It should be noted there is no correlation between the attenuation of an analog signal and attenuation of the digital signal that represents it.

7.6.3.5 Loop Transmission Loss

Loop transmission loss is highly dependent on the frequency of operation. Equipment design compensates for this dependency by employing a number of equalization techniques. Digital loops have frequency components over a wide frequency range. To simplify the loop design process, a nominal frequency is chosen based on the frequency of peak energy or an upper frequency limit of significant energy for a particular loop type and the loss specified for that frequency. This nominal frequency depends on the transmission format, pulse shape, and a number of other factors.

A typical maximum loss of 32 to 42 dB at the nominal frequency is used, but the actual value depends somewhat on the technology. Also, the maximum quoted loss value may have to be reduced to counteract the adverse effects of crosstalk or other transmission impairments.

Some digital loop types require a minimum loss, which depends on the ability of the loop interface electronics to equalize the received signal automatically. For repeatered T1-carrier, the minimum loss is 7.5 dB unless the repeaters have built-in pads. If necessary, this minimum loss can be provided by external attenuators placed in the line at the receive end.

From this point forward, attenuation, as determined from the secondary parameters, is used interchangeably with insertion loss (or just loss). This is only strictly correct when the digital loop is terminated in the complex conjugate of its characteristic impedance (Z_0). In practice, this does not always happen (if ever), but for practical problems it is sufficiently accurate.

7.6.3.6 Transmission Levels

Adequate signal power at the receive end of a circuit is essential to providing satisfactory error rates in any digital transmission system. It is obvious this is not possible unless loop loss is thoroughly controlled and enough signal level is provided at the loop input.

Too much signal power, on the other hand, can overload the transmission facilities or terminal equipment and cause distortion and crosstalk. Because digital loop systems frequently share transmission facilities with other systems, such as voice and analog carrier systems, mutual interference is a potential problem. Therefore, the signal powers in these systems are coordinated with each other. This requires close attention to power levels at particular frequencies or particular frequency bands.

In digital systems, the signal power is not specified in the same way as with analog systems, where power levels (for example, in dBm) are referenced to a standard impedance (for example, 600 Ω). Instead, the digital levels are specified by pulse shape and maximum voltage level into a standard load impedance (typically 100 or 135 Ω). These two parameters serve to determine the spectral limits of the digital pulse train, and crosstalk considerations ensure noninterference.

7.6.3.7 Presentation of Transmission Parameters

Loss and noise directly affect regenerator and equipment error performance. The loss and noise parameters can be presented in a variety of ways. Typically, a range, variability, or other distribution of a particular parameter is specified. There should be some predetermined agreement between the design value and the measured value, but the measured value always will show some variability due to:

- Test equipment calibration
- Environmental variations that affect the circuit components as well as the measuring equipment
- Human factors associated with reading meters or displays
- Position of the test equipment or lead lengths
- Cable splicing techniques
- Cable and equipment manufacturing tolerances

These effects will influence the measurements of a large number of loops so that the results show some type of distribution. Therefore, when the transmission parameters of a digital loop are determined, they really can only be specified on a statistical basis.

For voicegrade (analog) applications, the Institute of Electrical and Electronic Engineers (IEEE) has identified three forms for the results of transmission parameter calculations or estimations [39]:

- A cumulative distribution
- A histogram
- An analytical distribution function

Although these forms are proposed by IEEE for voicegrade applications, they are generic in nature and apply just as well to digital applications.

Other forms, or variations of the above forms, can be used. The most typical, of course, is the single-value specification; for example, "The digital loop loss shall be 23.2 dB." More typically, in digital applications, the specification might read, "The equipment shall function with a digital loop loss between 7.5 and 35.0 dB." This provides a lower and upper bound only. The exact loss value for any loop does not matter as long as it falls within the specified bounds. Similarly, a cumulative distribution would specify "95% of the loop losses shall fall between 7.5 and 33.0 dB." This specification allows some loops to have a loss outside of the boundaries, but no specification is given as to how far outside. A better specification would state "95% of the loop losses shall fall within 7.5 and 33.0 dB, and no loop shall have a loss less than 5.5 dB nor greater than 35.0 dB." If a normal distribution curve is used to specify the parameters, a mean value and a standard deviation are given. For example, "mean loop loss shall be 22.0 dB with standard deviation of 7.0 dB."

The normal distribution curve is shown in Figure 7-17. The various points labeled on the curve illustrate how transmission parameters are typically quantified. The average, or mean, value for the parameter is at the peak of the normal distribution curve. This is the value used for most field transmission calculations. At the outer edges, representing a small percentage of circuits, are the immediate action limits (IALs). If a parameter measurement gives an IAL or worse, immediate remedial action is required. Closer toward the mean are the maintenance limits (MLs). A parameter measurement falling between the ML and the IAL will receive remedial action during routine maintenance. The acceptance limits (ALs) are used when a new circuit or system is installed; these are necessarily tighter than the other limits to give allowance for equipment aging and other variations.

Because of this statistical nature, it would seem that individual parameters for several loop segments in tandem could not be combined by simple addition. This is

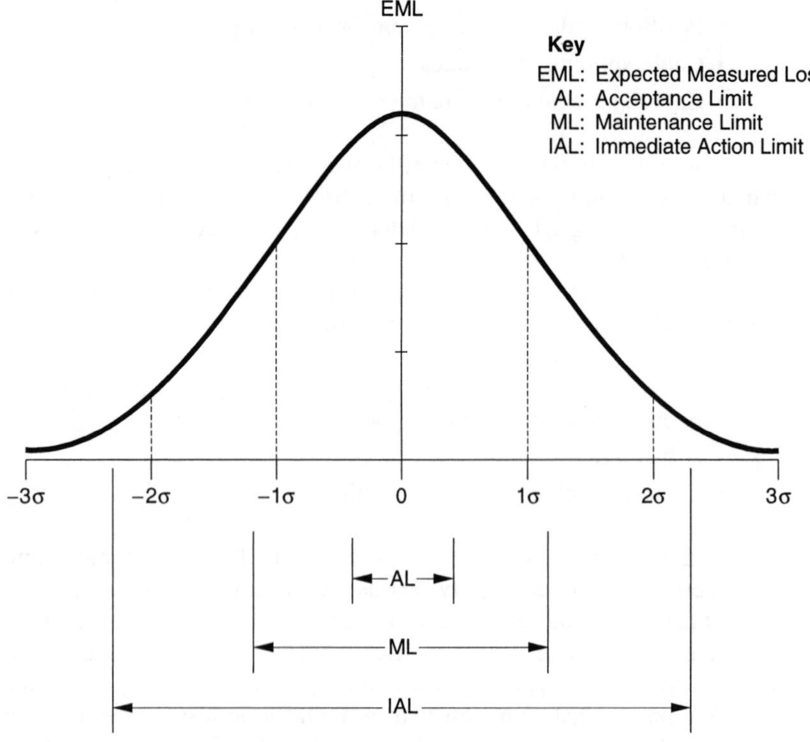

Figure 7-17 Normal distribution curve with circuit performance limits

indeed true, and various mathematical techniques are used to describe the entire connection.* For most field problems, however, parameter distributions are ignored, and simple addition is frequently used with satisfactory results.

7.6.3.8 Twisted Pair Cable Purchase Checklist

The checklist below is useful during the development of twisted pair exchange cable purchase specifications.

1. Specification? (ICEA, REA, BELLCORE) ☐
2. Deviations from the referenced specification? (specify) ☐
3. Pair gauge? (19, 22, 24, 26 AWG) ☐
4. Pair count? (not all pair counts are available with all cable gauges) ☐
5. Screened cable required? (T1, T1C, or T-Screen, D-Screen, Z-Screen) ☐
6. If screened, service pair requirements? (specify) ☐
7. Use? (direct buried, underground, aerial, riser, aerial self-supporting, service wire) ☐
8. Jacket environment? (severe, normal, submarine, rodent infested) ☐
9. Shield? (coated aluminum, copper) ☐
10. Armor? (heavy copper, steel, copper alloy) ☐
11. Factory pulling eyes required? (yes, no) ☐
12. Lengths required? (standard, custom; specify) ☐
13. Reel size or weight limitations? (specify) ☐
14. Reel requirements? (standard, wood, steel) ☐
15. Reel protection requirements? (none, wood lagging, pressboard) ☐
16. Shipping instructions? (specify) ☐

References

[1] Miller, S., Kaminow, I. *Optical Fiber Telecommunications II*. Academic Press, 1988.

[2] *Grounding, Bonding, and Shielding for Electronic Equipments and Facilities,* Vol. 1 of MIL-HDBK-419, Jan. 1982.

[3] *Grounding, Bonding, and Shielding Practices and Procedures for Electronic Equipments and Facilities,* Vol. 1: Fundamental Considerations. Federal Aviation Administration Report FAA-RD-75-215, NTIS Order No. ADA-022332.

[4] Reeve, W. *Subscriber Loop Signaling and Transmission Handbook: Analog*. IEEE Press, 1992.

*See, for example, Appendix B of [39].

[5] *Connection of Terminal Equipment to the Telephone Network.* 47 CFR Part 68. US Government Printing Office.

[6] Insulated Cable Engineers Association. *Standard for Polyolefin Insulated Communication Cables for Outdoor Use.* ANSI/ICEA S-56-434-1983. Available from ICEA.

[7] Insulated Cable Engineers Association. *Standard for Telecommunications Cable, Filled, Polyolefin Insulated, Copper Conductor Technical Requirements.* ANSI/ICEA S-84-608-1988. Available from ICEA.

[8] Insulated Cable Engineers Association. *Standard for Telecommunications Cable, Aircore, Polyolefin Insulated, Copper Conductor Technical Requirements.* ANSI/ICEA S-85-625-1989. Available from ICEA.

[9] Insulated Cable Engineers Association. *Standard for Telecommunications Cable, Buried Distribution and Service Wire Technical Requirements.* ANSI/ICEA S-86-634-1991. Available from ICEA.

[10] *REA Specification for Filled Telephone Cables,* Bulletin 345-67. PE-39, Jan. 1987. Available from Rural Electrification Administration (REA).

[11] *REA Specification for Filled Telephone Cables with Expanded Insulation,* Bulletin 1753F-208, June 1993.

[12] *REA Specification for Aerial and Underground Telephone Cable,* Bulletin 345-13. PE-22, Jan. 1983.

[13] *REA Specification for Self-Supporting Cable,* Bulletin 345-29. PE-38, Feb. 1982.

[14] *REA Specification for Filled Buried Wire,* Bulletin 345-70. PE-54. (Note: This specification was rescinded by REA in March 1987, but cables meeting it still are available from some suppliers. Specification PE-86 replaces PE-54.)

[15] *REA Specification for Filled Buried Service Wire,* Bulletin 345-86. PE-86, Oct. 1982.

[16] *REA Specification for Parallel Conductor Drop Wire,* Bulletin 345-36. PE-7, Jan. 1983.

[17] *REA Specification for Terminating (TIP) Cables,* Bulletin 345-87. PE-87, Dec. 1983.

[18] *Generic Requirements for Metallic Telecommunications Cables.* BELLCORE Technical Reference TR-TSY-000421, Sept. 1991.

[19] *PIC Filled ASP Cable.* BELLCORE Technical Reference TR-TSY-000100, June 1988.

[20] *Aircore PIC ALPETH Cable.* BELLCORE Technical Reference TR-TSY-000101, June 1988.

[21] *PIC Self Support Cable.* BELLCORE Technical Reference TR-TSY-000102, June 1988.

[22] *Underground Foam-Skin PIC Bonded STALPETH Cable.* BELLCORE Technical Reference TR-TSY-000106, June 1988.

[23] *PIC Filled Screened ASP Cable.* BELLCORE Technical Reference TR-TSY-000109, June 1988.

[24] *PIC Riser Cable*. BELLCORE Technical Reference TR-TSY-000111, June 1988.

[25] *Generic Requirements for Multiple Pair Buried Wire*. BELLCORE Technical Reference TR-TSY-000124, Sept. 1991.

[26] *Catalog of Technical Information*. BELLCORE Special Report SR-NWT-000264, July 1992–July 1993.

[27] *National Electrical Safety Code,* ANSI C2-1993. Available from IEEE.

[28] REA TE&CM Section 635: *Construction of Aerial Cable Plant*. Feb. 1962, with 1966 and 1979 addenda. Available from REA.

[29] *Outside Plant Engineering Handbook*. AT&T Document Development Organization, 1990.

[30] "When Even More Protection Is Needed." Lightguide Digest, No. 1, 1992, AT&T Network Systems.

[31] REA TE&CM Section 451.2: *Shield Continuity,* April 1984. Available from REA.

[32] Westman, H., ed. *Reference Data for Radio Engineers,* 4th Ed. ITT, 1964.

[33] *Understanding Transmission,* Vol. 7. Lee's ABC of the Telephone, 1976. Available from ABC Teletraining, Inc.

[34] Hamsher, D. *Communication System Engineering Handbook*. McGraw-Hill Book Company, 1967.

[35] *Understanding Transmission,* Vol. 8. Lee's ABC of the Telephone, 1976. Available from ABC Teletraining, Inc.

[36] *Transmission Systems for Communications*. Bell Telephone Laboratories, Inc. 1982. Available from AT&T Customer Information Center.

[37] REA TE&CM Section 406: *Transmission Facility Data*. Aug. 1977. Available from REA.

[38] Goldberg, B., ed. "Transmission properties of filled thermoplastic insulated and jacketed telephone cables at voice and carrier frequencies," in *Communication Channels: Characterization and Behavior*. IEEE Press, 1976.

[39] *IEEE Standard Methodologies for Specifying Voicegrade Channel Transmission Parameters and Evaluating Connection Transmission Performance for Speech Telephony*. IEEE Std 823-1989. Available from IEEE.

8 Loop Transmission Media, Part II—Optical Fiber Cables

Chapter 8 Acronyms

APD	avalanche photodiode
BER	bit error rate
CCITT	The International Telegraph and Telephone Consultative Committee
EIA	Electronic Industries Association
FC	screw-on fiber optic connector
FEXT	far-end crosstalk
FITL	fiber-in-the-loop
GaInAsP, Ge, InGaAs	laser diode and photodiode materials and configurations
GRIN	graded index
ICEA	Insulated Cable Engineers Association
LD	laser diode
LED	light-emitting diode
MDP	minimum detectable power
MFD	mode field diameter
MIFL	maximum individual fiber loss
MRP	minimum required power
NESC	National Electric Safety Code
n factor	stress corrosion susceptibility factor
ODTR	optical time domain reflectometer
OH	hydroxyl ion
PIN	positive-intrinsic-negative
PON	passive optical network
psi	pounds per square inch
PVC	polyvinyl chloride
REA	Rural Electrification Administration
SC	push-pull fiber optic connector
SMA	fiber optic connector
ST	bayonet fiber optic connector
WDDM	wavelength division demultiplexer
WDM	wavelength division multiplexer

8.1 Introduction, Physical and Electrical Characteristics

From a mechanical standpoint, fiber optic cables present only a slightly different set of engineering and installation problems than do twisted pair cables. The mechanical performance requirements of twisted pair cables are well understood, and cable designs have evolved to solve virtually all installation and operation problems. Based on this experience, cable manufacturers were able to solve many initial problems with fiber optic cables as well. However, for the optical fibers themselves, the resolution of strength, abrasion resistance, and fatigue resistance issues are still under way.

As will be seen in the following discussion, the physical characteristics of the outer portions of twisted pair and fiber optic cables are very similar. Since the mechanical performance of fiber optic cables goes beyond the physical characteristics of outer cable layers, additional detailed information is provided on optical fibers themselves.

The telecommunications industry inconsistently uses metric and English measurements and units when *specifying* and *constructing* facilities. For example, fiber optic cable length is specified in kilometers or kilofeet, and the burial depth is specified in inches or feet rather than centimeters or meters. Installation tensions are specified in either newtons or pounds-force (abbreviated lb in this book). In this chapter, metric units are used in length measurements (kilometers, meters, centimeters, millimeters, and nanometers) and English units are used elsewhere.

As would be expected, the electrical characteristics of optical fibers in no way resemble metallic twisted pairs, and fiber loop design issues differ markedly from their twisted pair counterpart. Optical components, such as splices and connectors, play a more visible role in fiber loop design because of their higher losses relative to media transmission losses. Later sections of this chapter discuss the electrical characteristics and design data for optical fibers and components; fiber loop design procedures are given in Chapter 10. The emphasis is on the 1,300-nm region of singlemode fibers, which is the recommended optical region for telecommunication services in fiber loops [1]. The 1,550-nm region is touched on briefly in this chapter. This region is used for currently nontraditional telecommunication services, such as video, and for some long-distance applications.

This chapter avoids introductory and theoretical material. Introductory information may be found in [2] and theoretical information in [3] and [4].

8.2 Physical Characteristics

8.2.1 Fiber Optic Cable Specifications

Like twisted pair cables, fiber optic cables used in telecommunications are manufactured for an intended application, according to a specification prepared by industry organizations or the cable customer. Specifications have been prepared by the Insulated Cable Engineers Association, Inc. (ICEA) [5], Rural Electrification

Administration (REA) [6], BELLCORE [7–10], and Electronic Industries Association (EIA) [11–15].

Generally, the specification includes, among other requirements, jacketing requirements, strength member, buffering, armoring, fiber number and types, maximum attenuation at nominal wavelengths, and test requirements and methods. The various construction attributes of fiber optic cables will be described in a format similar to the previous chapter.

8.2.2 Outer Jacket

The outer jackets on fiber optic cables serve the same purposes as on twisted pair cables, and identical materials are used (see Chapter 7). The outer jacket is a nominal 50 mils thick for all cable sizes, which are generally less than 3/4 inch in diameter. Fiber optic cables are available in the figure-8 configuration; the built-in supporting messenger can be either dielectric or nondielectric (conducting). Lead sheaths are not used with fiber optic cables.

8.2.3 Armor and Inner Jacket

When armor is required for fiber optic cables, its construction is identical to twisted pair cables. In many armored optical fiber cables another layer of plastic, or inner jacket, is installed to further protect the fiber and its surrounding structure. Armored cables are always used where there is a chance of damage from dragging anchors and fishing activities in lakes and rivers, or if there is a chance of rodent

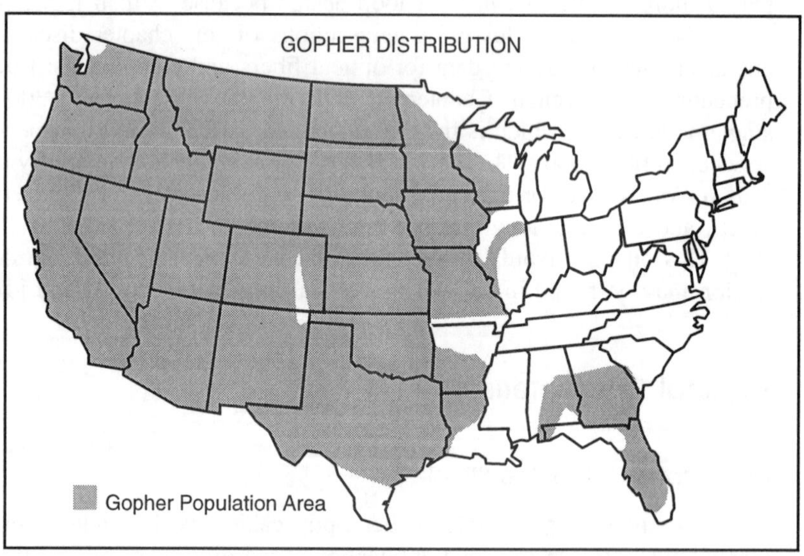

Figure 8-1 Gopher distribution in the continental United States
From [16] with permission

damage. Fiber optic cables used in direct burial applications are almost always armored regardless of the chance of rodent damage. This provides an added measure of service protection against dig-ups and rough handling. For extra severe submarine service, special armoring methods are used (see Chapter 7).

Gophers and similar chewing rodents probably are the biggest threat to buried fiber optic cables. Figure 8-1 shows the areas in the United States where gophers are known to live. According to work done by AT&T Network Systems, there are only two known methods to prevent gopher damage. One is to place the cables in rigid polyvinyl chloride (PVC) conduit (schedule 40 or 80), between 2-1/4 and 4 inches in diameter, and the other is with stainless-steel armoring under the cable jacket [16]. The large diameter prevents the rodent from getting a leveraged bite. A typical armored cable is shown in Figure 8-2.

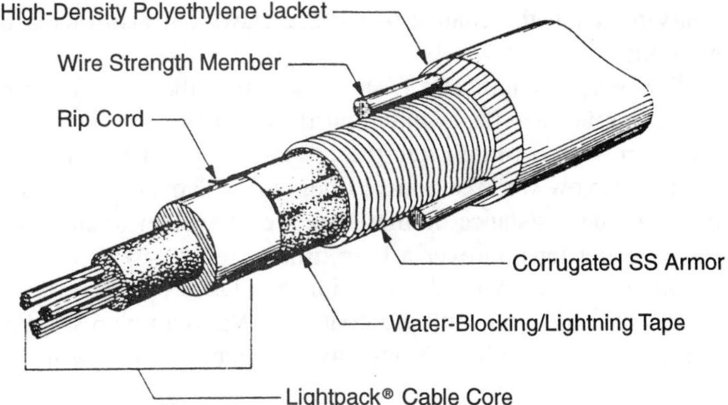

Figure 8-2 Typical gopher-resistant armored fiber optic cable
From [16] with permission

8.2.4 Buffering

The long-term reliability of optical fibers used in exchange carrier applications is determined by three properties related to fiber strength [17]:

- The size and distribution of flaws along the fiber's length
- The growth behavior of flaws when the fiber is stressed
- The aging behavior of the flaws without stress

Optical fibers are in many ways considerably more fragile than insulated copper conductors and must be protected from any actions that might cause microbending, fracturing, or media damage of any type. The manufacturing process is highly developed to reduce the chance of surface flaws in the fiber. The core of the fibers used in telecommunications applications is a pure silica glass, which has considerable tensile strength but little resistance to abrasion, microbending, or handling and installation forces.

The theoretical tensile strength of pure glass is about 2.5E6 pounds per square inch (psi). Due to molecular-size material inclusions and surface flaws around 0.02 μm deep, however, the actual strength is reduced to around 500 to 650 kpsi in short lengths, and 60 kpsi in long lengths for standard fibers. For a fiber 125 μm in diameter, this is equivalent to about 12.4 lb and 1.1 lb, respectively. Special high-strength fibers are available with 90 kpsi tensile strength.

The reason the strength is so much lower in long lengths is because the probability of strength-limiting surface flaws is greater. All standard fibers are proof-tested at 50 kpsi before shipment from the factory. This process weeds out fiber lengths with flaws greater than approximately 2 to 3 μm deep.

Even at long lengths, an optical fiber has considerably more tensile strength than copper conductors.* However, the risk of a fiber break in a typical cable has more serious consequences than a single conductor break in a twisted pair cable of comparable capacity.† Therefore, fiber manufacturers dedicate considerable effort to not only reducing the chance of surface flaws but also increasing the strength of a fiber strand.

The basic strength of a fiber comes from the manufacturing process, which is designed to minimize flaws and impurities. Abrasion resistance and fatigue resistance are two terms used to describe the longevity of that basic strength. *Abrasion resistance* defines how well the optical fiber resists damage from installation and handling forces. Fatigue resistance defines how well the fiber strength is maintained when it is placed under tension over a long time.

One way to increase the abrasion resistance of a fiber is to provide special and multiple claddings and multiple coatings. This can almost double the strength. An example of a fiber with high abrasion resistance is shown in Figure 8-3.

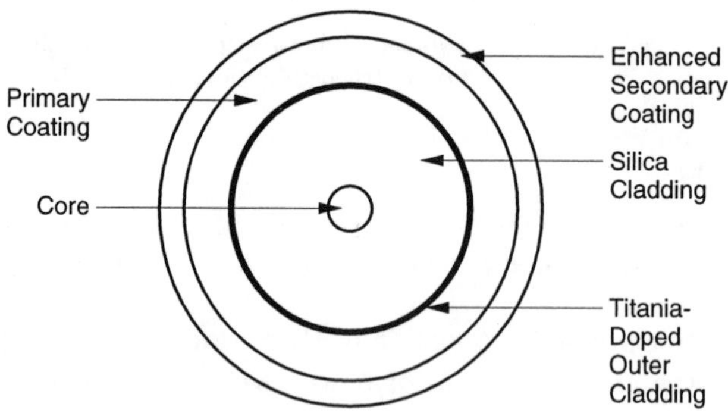

Figure 8-3 Abrasion-resistance properties of commercially available optical fibers
Adapted with permission from [19]

---

*Annealed soft copper wire has a tensile strength of about 38,000 psi, and overall elongation at fracture is 25% in short lengths. See pages 2–20 of [18]. Glass has very little elongation before breaking.

†Until the 1980s, most twisted pair cables were made with extra pairs just for this eventuality. For example, a 200-pair cable actually had 202 pairs.

Fatigue resistance is important for long life. It is impossible to eliminate completely the tension in a fiber of any significant length either during or after installation. When a fiber is under tension, and especially in the presence of moisture, the inherent surface flaws will slowly deepen. This weakens the fiber, and it may eventually break. Fatigue resistance is characterized by the stress corrosion susceptibility factor, or simply n factor. The n factor for normal production fibers is in the range of 20 to 25, but fibers with higher fatigue resistance have n factors in the range of 28 to 30 [19].

Microbending is any random bend or deformation in the fiber core that causes radiation losses of the optical signal. It is not normally visible to the unaided eye. Typical causes of microbending are strains on the fiber during manufacturing or installation, and dimensional variations due to operational temperature changes.

To reduce the likelihood of problems, fibers are protected by at least two buffering mechanisms. First, the cladded fiber core is coated with a colored polymer (plastic) that typically brings the diameter of the fiber plus coating to 250 to 900 μm, as shown in Figure 8-4. This is sometimes called the primary coating, and it serves several purposes:

- Abrasion protection
- Microbending loss protection
- Static fatigue protection
- Identification
- Crush resistance

The second buffering mechanism is a tube or ribbon assembly, which surrounds the coated fiber and protects the fiber core from abuse during manufacturing and

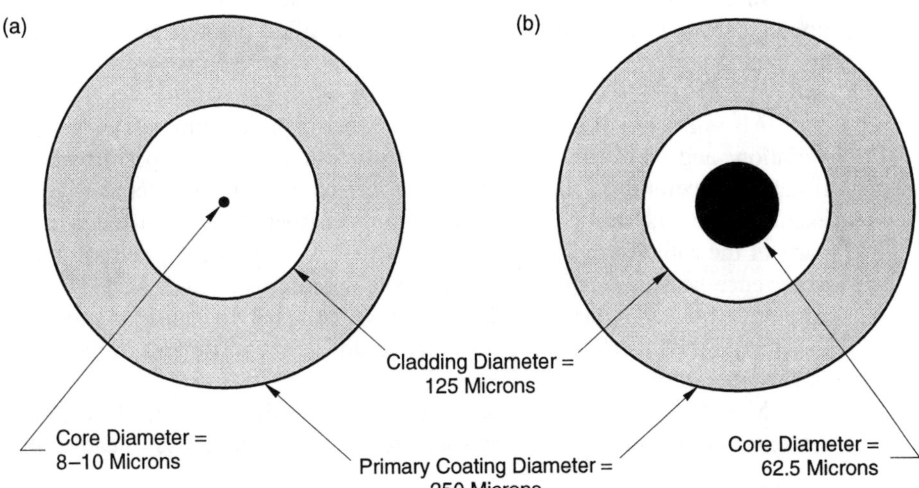

Figure 8-4 Coated optical fibers: (a) Singlemode; (b) Multimode

installation. Two tube constructions are used: loose tube and tight tube. Loose tube constructions are illustrated in Figures 8-5(a)–(c). The slotted core cable in Figure 8-5(c) does not have a separate tube for each group of fibers, but still is a form of loose tube construction. The ribbon assembly in Figure 8-5(d) can be considered a hybrid loose tube and tight tube design.

Loose tube or ribbon constructions are universally recommended for outside plant applications, whereas the tight tube is widely used in buildings because tight tube cables are smaller and a little easier to handle.* Tight tube cables are cheaper than loose tube cables by about 15%, all else being equal.

Except in the bundled loose tube (also called a loose-fiber bundle), the tube itself is typically 2 to 3 mm in diameter. The bundled loose tube is typically 6 to 8 mm in diameter. These relatively large diameters allow free movement of the optical fibers within the tube during installation or service. The ability to move is illustrated in Figure 8-6. Ribbon cables have movement ability, too, but in a different way. The ribbon is twisted inside the cable, and any cable length change is taken up in the ribbon's twist.

The tubes are constructed of nylon, various polymers, or polypropylene. Each loose tube holds as little as two fibers or as many as six or 12, with an entire cable holding upwards of 200 fibers. When multiple tubes are used, they are stranded (wrapped in a helical fashion) around the cable core. In the bundled tube, only one large tube is used, and the fibers are wrapped with a color-coded binder for identification. This construction holds upwards of 100 fibers.

As would be expected in a loose tube design, the fiber is longer than the tube. The amount of excess length and how it is distributed in the tube is a critical cable design parameter because of the potential for high bending losses in the fiber. The excess length is in the order of 0.1%, the loss from which can be neglected in subscriber loop applications. The exact percentage will vary with the cable construction and can be obtained from the manufacturer if needed.

In all outdoor applications the tube is filled with a moisture-resistant compound, which is very similar to that used in filled twisted pair cables. The filling compound protects the fiber from long-term adverse effects of moisture on the glass core.

Also of concern in outside facilities are absolute temperatures and temperature variation, and their effects on the mechanical and electrical performance of cables. In cold temperatures, the cables will contract, and in high temperatures they will expand. The amount of length change for the fibers will be different than the other parts of the cable (armor, jackets, and tubes). All cables designed for outside service experience no significant loss change at temperatures as high as 80 °C. At low temperatures, however, there is additional loss, primarily because of excess bending of the fiber in the tubes (bending loss) due to different coefficients of thermal expansion of the fiber and tube.

Most fiber manufacturers have standard cables that perform adequately to −40 °C. At least two manufacturers have standard (that is, catalog listed) cables that perform

*At least one manufacturer produces tight tube cables for outdoor applications. These are derived from cables designed for military applications.

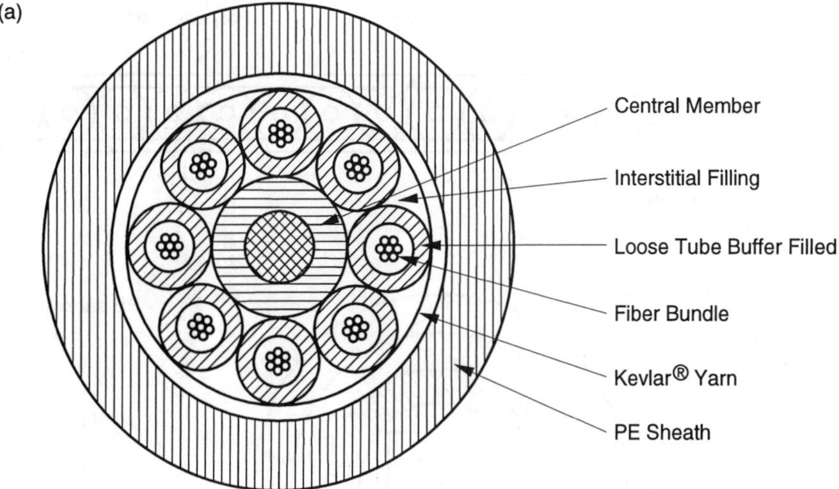

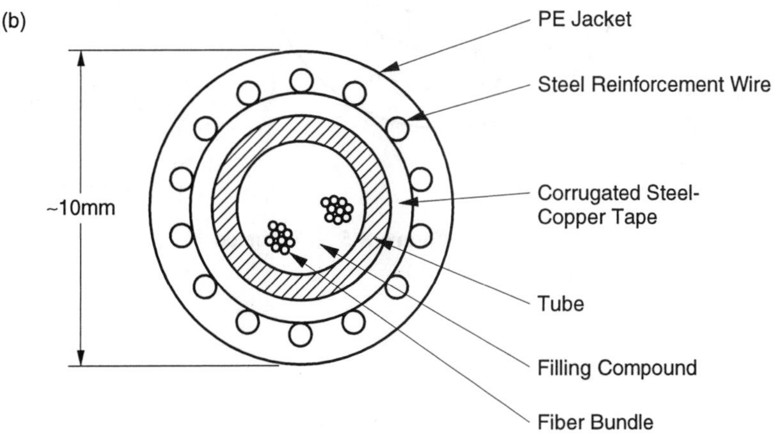

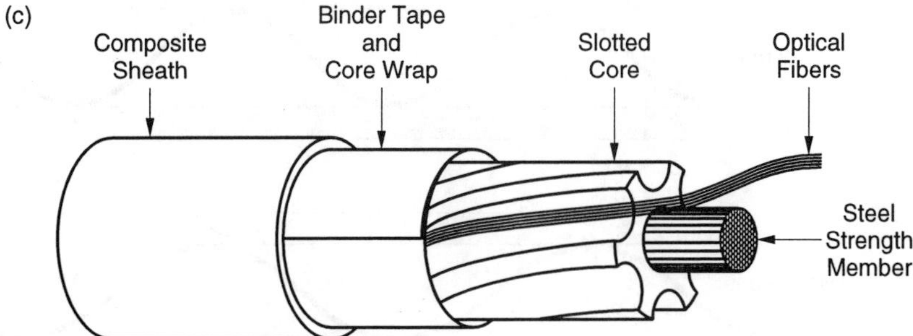

Figure 8-5 Optical fiber constructions: (a) Loose tube; (b) Bundled loose tube; (c) Slotted core; (d) Ribbon assembly
Adapted with permission from [3]

(d)

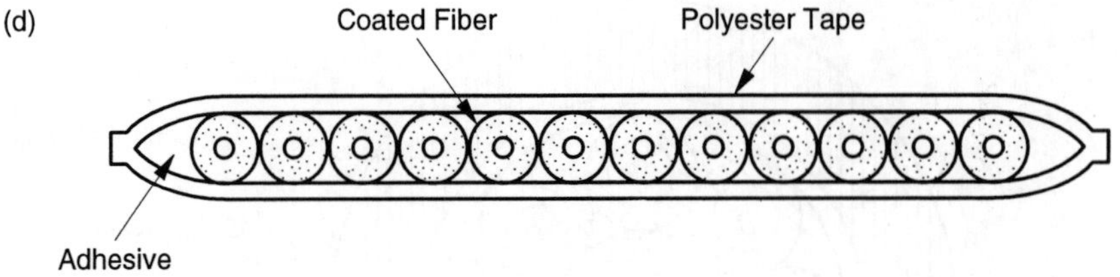

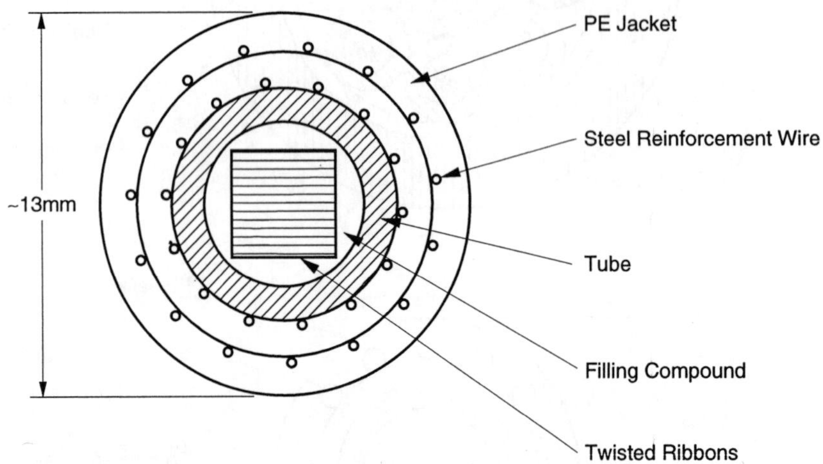

Figure 8-5 Continued

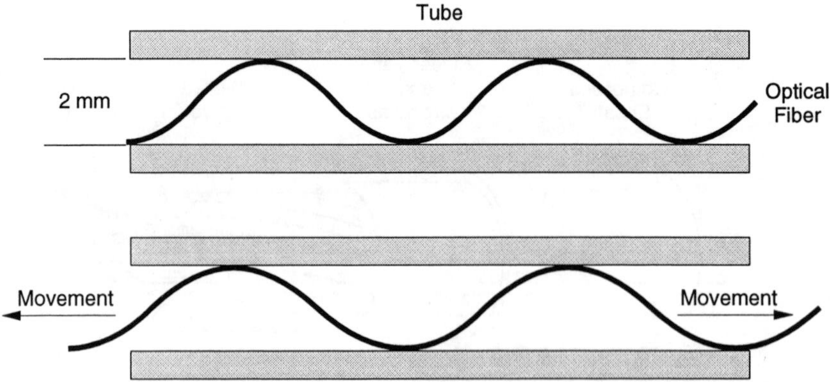

Figure 8-6 Loose tube fiber movement

down to -55 °C. Manufacturers can provide special cables if lower-temperature performance is needed. In low-temperature cables, the excess fiber length is very closely controlled (within $\pm 0.1\%$), and low-density polyethylene is used in the inner and outer jackets. Special low-temperature cables cost 5 to 7% more than standard cables.

In tight tube construction, an additional nylon (or polymer) coating is applied directly to the fiber primary coating. It increases the overall fiber diameter to around 900 μm, as shown in Figure 8-7. Except in special constructions, tight tube buffered cables do not have the low-temperature performance of loose tube cables; they are typically useful to 0 °C, which precludes their use in most outside applications. Tight tube cables are more sensitive to outside forces, but they are more flexible, easier to handle (for low fiber counts), and easier to connectorize than loose tube cables.

The tight tube fibers are stranded directly around the cable core and then jacketed (stranded design, Figure 8-8(a)) or are individually jacketed (with an integral strength member) before the stranding and outer jacketing operation (subunit design, Figure 8-8(b)). Tight tube fiber cables are available with 144 or more fibers.

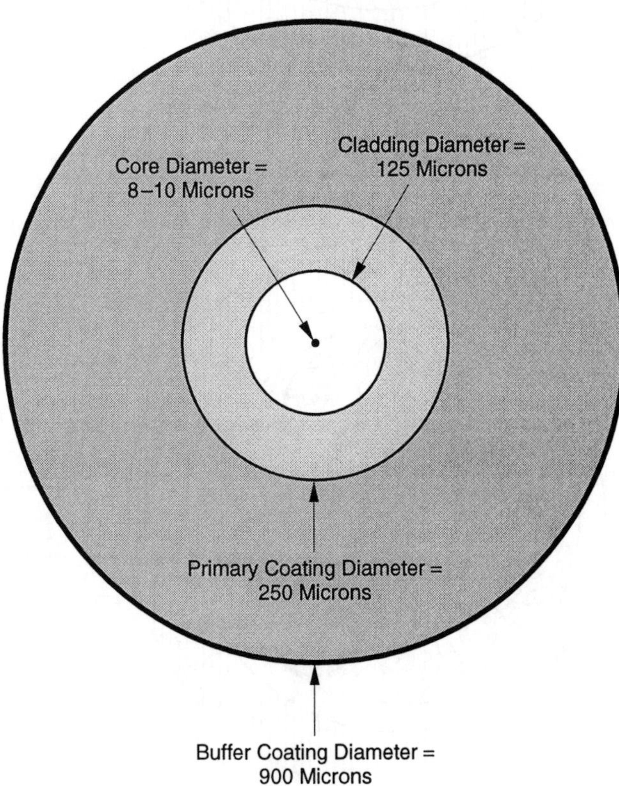

Figure 8-7 Buffered optical fiber for tight tube cables

(a)

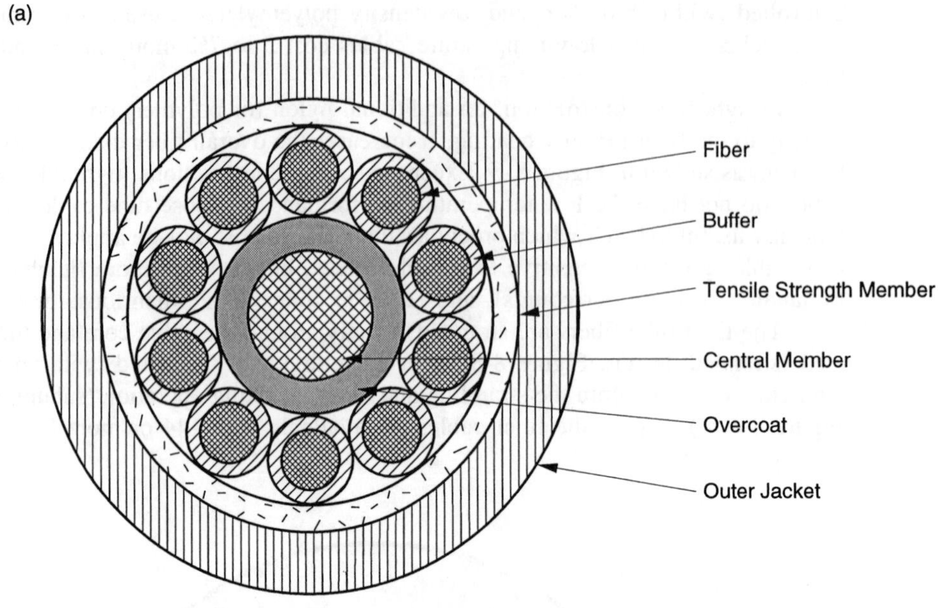

(b)

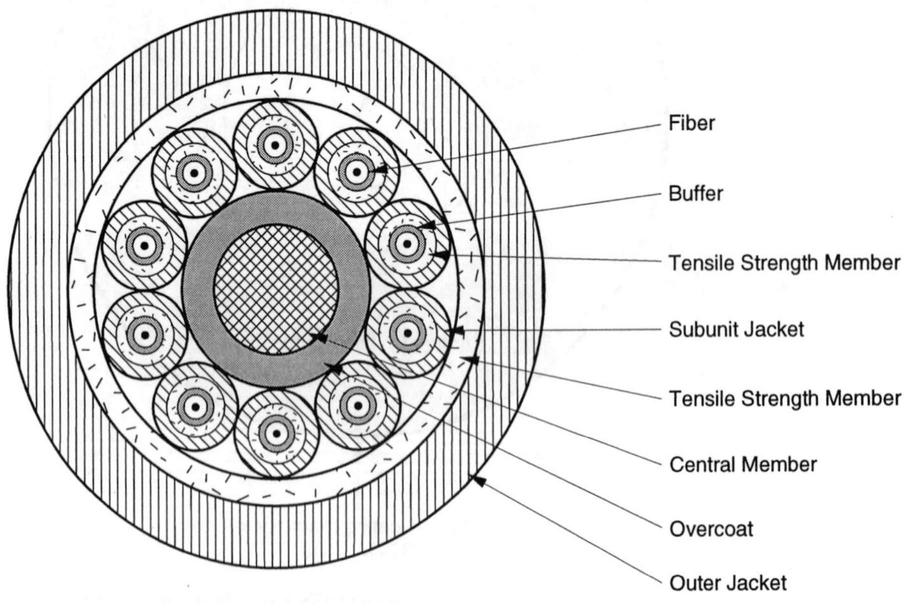

Figure 8-8 Tight tube cable construction: (a) Stranded design; (b) Subunit design

8.2.5 Strength Member

Optical fibers themselves are not used as a strength member, for obvious reasons. Instead, the cable is manufactured with separate strength members. These are available in several configurations, as listed below. The strength member is designed to minimize fiber stress during installation or handling.

- Central core strength member
- Filler material between the fiber tubes
- Spun glass material woven around the periphery of the buffering material
- Steel wires helically wound around the cable core and imbedded in the jacket
- Any combination of the above

Regardless of the type of strength member, the cable will stretch when placed under tension during pulling or stringing operations. Some increased length will remain under static conditions after installation. In loose tube cables, the fiber is intentionally longer than the cable. The strength member provides the required strain relief when the cable is under tension. The fibers themselves are mechanically decoupled from the strength member, which serves to minimize tension transmitted to the fiber.

The structural design considerations for aerial fiber cable are essentially the same as for twisted pair cables. Since fiber cable is so light, much longer spans are possible and may be attractive from an economic standpoint. However, longer spans can experience severe dancing in even light wind conditions and may require special dampers or twisting between supports. Dancing can greatly fatigue the strength member and may abrade the fiber at structure supports under extreme conditions.

Aerial fiber plant has been shown to perform well when exposed to lightning. In tests by AT&T Network Systems, there was no difference between the performance of cables with conducting and dielectric strength members as long as the cable was lashed to a metallic messenger strand [20]. Even though the strand carried the bulk of the lightning current, in almost all the tests, the metallic lashing wire fused open. The only other damage during these tests was to the fiber cable outer jacket when the heated lashing wire melted the plastic. As a result, AT&T recommends dielectric lashing wire in high lightning areas. Further results of the tests were as follows: There was no evidence of arcing between the metallic strength members within the outer jacket; no evidence of arcing between the metallic strength members and the lashing wire or supporting messenger strand; and no substantiation of the common belief that cables with dielectric strength members are better suited for aerial plant than armored cables or cables with conducting strength members.

The bending radius of fiber optic cables must be limited to minimize bending loss. A central strength member helps to limit the bending radius and also prevents the cable from buckling (kinking) when it is bent. A typical minimum bending radius is 10 to 16 times the cable diameter for unarmored cables and 20 times the cable diameter for armored cables.

Unless the cable data sheet or manufacturer indicates that a smaller radius can be used, 16 to 20 times the cable diameter provides a measure of safety and should be observed in all installations. For example, if the unarmored cable diameter is one half inch, the minimum bend radius is eight inches. Some manufacturers specify a

16× (read "sixteen times") bending radius for unarmored cables during installation (when the cable is under tension) and 10× bending radius while in operation. Corresponding values for armored cables are 20× and 16×. Table 8-1 summarizes these important requirements.

The strength member either can be dielectric (insulating) or nondielectric (conducting). Nondielectric members are usually a central core steel strand or periphery steel wires helically wrapped and imbedded in the jackets. Cables with conducting strength members are typically designed to withstand 15 kVdc between the core and the armor for three seconds. The conducting strength members in these cables should be grounded wherever the conducting surface is exposed. The National Electrical Safety Code (NESC), which normally applies to utility-type installations outside of buildings, requires at least four grounds per mile of telecommunication cable [21].

Dielectric strength members are of spun glasslike materials such as aramid yarn, which is a spun glass fiber with considerable strength. Kevlar™ is an example of such a yarn.*

Typical strength members are designed to give the tensile strengths shown in Table 8-2. Two values are shown: one for installation only and one for long-term service. These values are typical for fiber optic cables used in the telecommunications industry, but the manufacturer should always be consulted, if possible, to confirm the actual values to be used.

Table 8-1 Fiber Optic Cable Bending Radius

Condition	Bending Radius (× Cable Diameter) Unarmored	Armored
Installation	16×	20×
Operation	10×	16×[a]
When in doubt	16×	20×

[a]Some wrinkling of jacket may occur but is not harmful.

Table 8-2 Typical Fiber Optic Cable Tensile Strengths

Size/Buffering	Installation[a]	Service[a]
Small/tight tube	900 (200)	450 (100)
Large/tight tube	2,700 (600)	900 (200)
Small/loose tube	2,700 (600)	585 (130)
Large/loose tube	2,700 (600)	585 (130)

[a]Measured in units of newton (pound)
Small ~ <1/2 inch; large ~ >1/2 inch

8.2.6 Media

All fibers used in outside telecommunication applications are made of a pure silica glass.† The fiber consists of a glass core and a glass cladding. The index of

*Kevlar is a trademark of DuPont-Nemours, Delaware.

†"Today, glass fiber purity is so high that if the ocean were as pure as optical fiber glass, then the bottom could be seen at its deepest point." [3]

refraction of the core and cladding are very slightly different, which gives the fiber its outstanding light-guiding properties. Typical values for index of refraction are 1.47 to 1.48 for the core and 1.46 for the cladding.

Figure 8-9 illustrates the light trajectories in optical fibers. Fibers are available

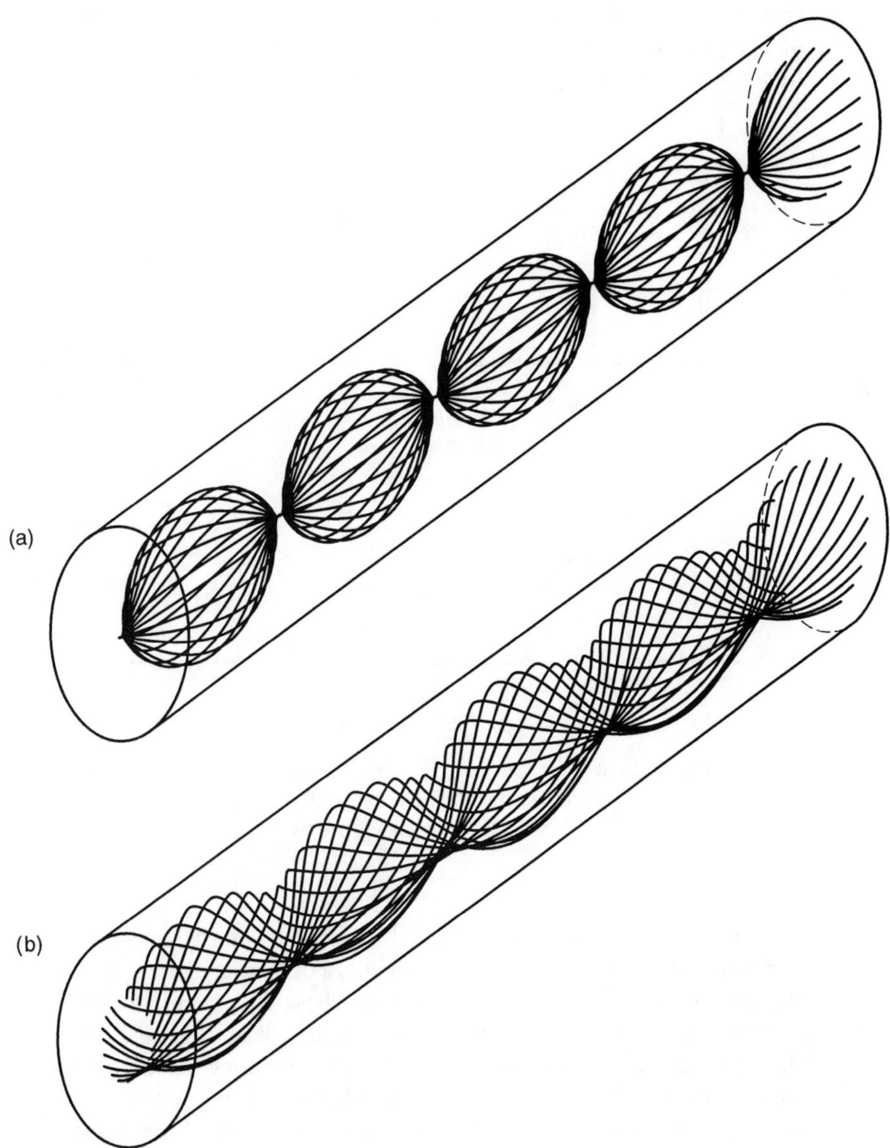

Figure 8-9 Ray trajectories in optical fibers, [4]: (a) Ray passing through axis; (b) Skewed ray
© 1990 TAB Professional and Reference Books. Reprinted with permission

in two basic configurations, singlemode and multimode, which indicate the light propagation characteristics. Guided light energy in singlemode fibers is concentrated in one, and only one, mode.* In multimode fibers, the energy is distributed throughout hundreds or even thousands of modes. In this context, a mode is a path or trajectory followed by the propagating light energy.

In singlemode fibers, the cladded core diameter is between 8 and 10 μm (microns). The cladding brings the fiber diameter to 125 μm, as was seen in Figure 8-4. The principal core/cladding relationships or profiles presently used in singlemode fibers are step-index matched clad and depressed clad, as illustrated in Figure 8-10(a).

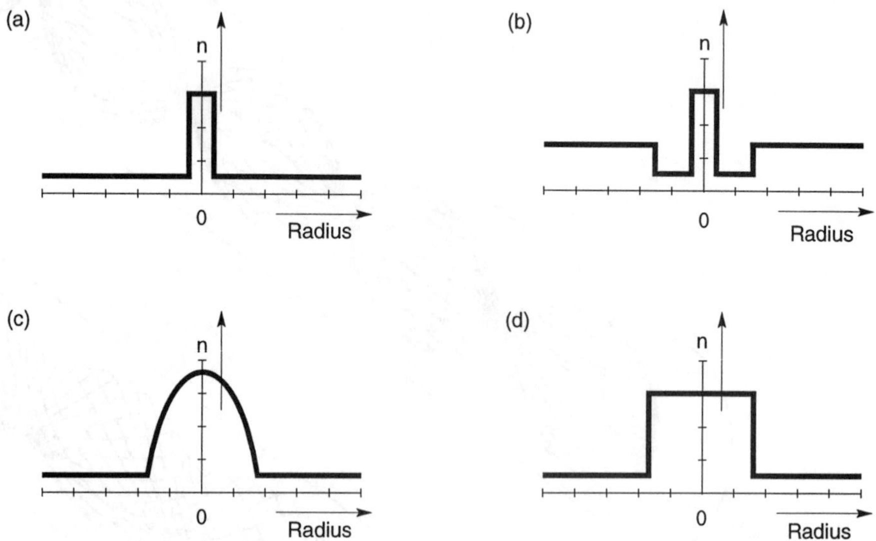

Figure 8-10 Typical index profiles for optical fibers: (a) Singlemode step-index matched cladding; (b) Singlemode depressed cladding; (c) Multimode graded-index; (d) Multimode step-index

In multimode fibers, the cladded core is between 50 and 100 μm, with 50 and 62.5 being the most popular at this time. A 62.5-μm fiber is shown in Figure 8-4. The cladded diameter is 125 or 140 μm. These fibers are commonly referred to as 50/125, 62.5/125, or 100/140 fibers, which indicate the core/cladding diameters. The step index and graded index (GRIN) are commonly used profiles for multimode fibers. Figure 8-10(b) shows multimode fiber index profiles. The coated singlemode fiber is indistinguishable to the unaided eye from a coated multimode fiber; both are typically 250 μm in diameter.

---

*There are actually two orthogonal modes, but no attempt is made to distinguish between them in digital loop applications.

8.2.7 Color Coding

The fiber primary coating and tube are colored according to a color code adapted from the color code for twisted pair cables. Although industry standards exist for fiber color coding [5,6], there are variations among manufacturers, especially beyond ten fibers in a bundle. The variations are subtle but, for the most part, inconsequential if care is taken when trying to identify a particular fiber.

One industry standard color code, this one specified by the REA, is shown in Table 8-3. This scheme will adequately identify up to 50 fibers in a cable, with five tubes of ten fibers each. The color code used by Siecor Corporation, a large supplier of fiber cables, is shown in Table 8-4. This scheme will adequately identify up to 144 fibers (12 tubes of 12 fibers each). The color code specified by ICEA is shown in Table 8-5. Finally, AT&T, which is another large supplier, uses the scheme shown in Table 8-6 for its bundled fiber constructions; it will identify up to 96 fibers. The AT&T ribbon cables contain 12 fibers in a ribbon, which are identified by the same color code, but each 12-fiber ribbon assembly has a number printed on it for identification. Where tracers are used in a color code, they always will be smaller than the width of base color hatching, stripes, or circumferential marks.

Table 8-3 REA Optical Fiber Color Code

FIBER NO.	COLOR	BUNDLE NO.	COLOR
1	Blue	1	Blue
2	Orange	2	Orange
3	Green	3	Green
4	Brown	4	Brown
5	Slate	5	Slate
6	White		
7	Red		
8	Black		
9	Yellow		
10	Violet		

Table 8-4 Siecor Corporation Optical Fiber Color Code: Loose Tube Fibers

FIBER NO.	COLOR (BASE; TRACER)	BUNDLE NO.	COLOR (BASE; TRACER)
1	Blue	1	Blue
2	Orange	2	Orange
3	Green	3	Green
4	Brown	4	Brown
5	Slate	5	Slate
6	White	6	White
7	Red	7	Red
8	Black	8	Black
9	Yellow	9	Yellow
10	Violet	10	Violet
11	Blue; slate	11	Blue; slate
12	Orange; slate	12	Orange; slate

Table 8-5 Insulated Cable Engineers Association Optical Fiber Color Code

Fiber No.	Color (Base; Tracer)	Bundle No.	Color (Base; Tracer)
1	Blue	1	Blue
2	Orange	2	Orange
3	Green	3	Green
4	Brown	4	Brown
5	Slate	5	Slate
6	White	6	White
7	Red	7	Red
8	Black	8	Black
9	Yellow	9	Yellow
10	Violet	10	Violet
11	Rose	11	Rose
12	Aqua	12	Aqua
13	Dashed blue; black	13	Dashed blue; black
14	Dashed orange; black	14	Dashed orange; black
15	Dashed green; black	15	Dashed green; black
16	Dashed brown; black	16	Dashed brown; black
17	Dashed slate; black	17	Dashed slate; black
18	Dashed white; black	18	Dashed white; black
19	Dashed red; black	19	Dashed red; black
20	Dashed black; yellow	20	Dashed black; yellow
21	Dashed yellow; black	21	Dashed yellow; black
22	Dashed violet; black	22	Dashed violet; black
23	Dashed rose; black	23	Dashed rose; black
24	Dashed aqua; black	24	Dashed aqua; black

Table 8-6 AT&T Optical Fiber Color Code: Bundled Fibers

Fiber No.	Color	Bundle No.	Color
1	Blue	1	Blue
2	Orange	2	Orange
3	Green	3	Green
4	Brown	4	Brown
5	Slate	5	Slate
6	White	6	White
7	Red	7	Red
8	Black	8	Black
9	Yellow		
10	Violet		
11	Dashed blue		
12	Dashed orange		

It is apparent from these schemes that any fiber can be identified by its bundle (or unit) and fiber number. For example, under the REA color code, fiber 2 in a 30-fiber cable would be orange, and it would be contained in a blue tube. Fiber 30, in a cable with six fibers per tube, would be white in a slate tube. Under the Siecor color code, for a 96-fiber cable with 12 fibers in a tube, fiber 71 would be blue-slate in a white tube. Some care must taken in identifying fibers unless the manufacturer and its color code is known. In most situations, examination of the fiber bundles and fiber count per bundle should yield enough information to allow proper identification of each fiber.

8.3 Feeder and Distribution Cables

The application of fiber optic cables in digital loops has not yet fully matured, although industry standards have been developed for fiber-in-the-loop (FITL) architectures and other requirements [1]. As a consequence, various digital loop architectures are still being developed, and the feeder and distribution nomenclature is not as easily applied as it is with twisted pair cables.

At the present, fiber optic cables used in interoffice and feeder applications are primarily arranged in two basic architectures. One is a physical star configuration, which is shown in Figure 8-11(a). This configuration provides the most flexibility and is the easiest to administer with the present technology.

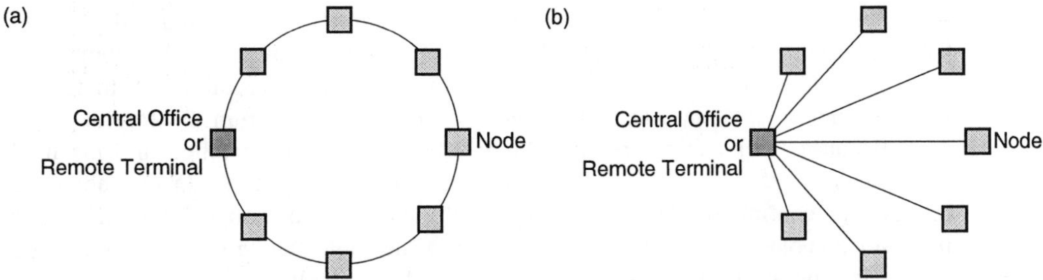

Figure 8-11 Fiber network architectures: (a) Physical loop; (b) Physical star

When a user or application requires especially reliable service, the loop configuration shown in Figure 8-11(b) commonly is used. This architecture frequently is found in high-capacity metropolitan environments and between central offices. Upon close inspection, the looped configuration is an outgrowth of the star.

8.4 Fiber Access

The access philosophy used with fiber optic cables differs from twisted pair cables. In fiber optic cable routes, splice and connector losses can account for a significant amount of the overall loss budget. This is why fiber access and splicing are always kept to a minimum. Therefore, fiber optic cables are not made accessible at frequent intervals along a route, as opposed to twisted pair cables.

In digital loop carrier applications, each fiber pair serves a large number of subscribers in a carrier serving area. The fibers are dedicated to feeding that carrier serving area and are not accessed between the central office and the digital loop carrier remote terminal. Industry standards are available for splice closures and access terminals [22,23].

8.5 Direct Buried and Underground Installations

Underground fiber optic cables are encased in buried conduits, whereas *buried* cables are placed directly in the ground. Underground cables are protected by the conduit and normally are not armored. Buried cables are protected by integral armor and additionally by a direct buried duct liner in some installations. Additional protection of underground and buried installations is provided by their physical separation from foreign structures (other cables and buried installations) and burial depth. Generally, fiber optic cables should be separated by at least 12 inches from foreign structures.

The depth at which cables are placed depends on their relative importance. Table 8-7 gives the minimum and typical depths from finished grade to the top of the cable or conduit for generalized use classifications. The minimum depths are based on industry standard [24]. Where cables are buried in public easements, the agency granting the right-of-way usually has minimum depth requirements that override any industry standard and that depend on the location within the easement.

Because fiber cables are small and light, they can be handled and installed quite easily in very long lengths. Manufacturers can supply up to 25 km (about 80 kilofeet) of continuous length for singlemode and 2.5 km (about 7.5 kilofeet) for multimode cables.* Splicing is not necessary if the cable can be installed in a single length and the associated splice losses are therefore avoided.

Although fiber optic cables can be easily installed directly in a duct, this is not always done. By itself, a 1/2 inch cable could easily snarl a 4 inch duct and prevent

Table 8-7 Burial Depth for Underground and Direct Buried Fiber Optic Cables

Cable Use or Location	Minimum Depth (Inches)	Typical Specification (Inches)
Interoffice and trunk cable	30	36
Feeder and distribution cable	24	30
Service or drop cable	18	24
Underground (all uses)	30	36
Ditch lines or public roads (all uses)	36	48

*Multimode cables are not used in long route applications anymore, so long reel lengths are not needed.

any future cables from being pulled through it. The smaller cable is also subject to damage during installation of later cables. Therefore, fiber optic cables are installed in a 1 or 1-1/4 inch inside diameter plastic inner-duct or duct liner.

The duct liner can be pulled into the duct with or without the fiber cable already installed. Four such duct liners can be installed in a 4 inch duct, which is the most widely used size. This is illustrated in Figure 8-12. If only a single fiber cable is initially required, but more are anticipated in the future, a bundle of four duct liners can be installed with only one containing a cable. The duct liner also provides additional protection, although in most duct installations this is not necessary.

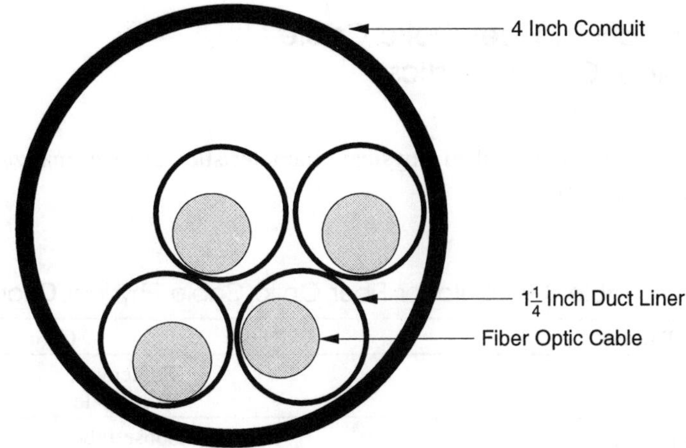

Figure 8-12 Duct liner application

Duct liner is frequently used in direct buried applications. The fiber optic cable can be pre-installed in the duct liner (called cable-in-conduit) and the assembly directly buried by plowing or trenching. Attempts have been made to place empty duct liner and later pull fiber optic cable into it. This method does not work well with certain types of duct liner except in short lengths (a few hundred feet), because the installed duct liner has many bends that accumulate to increase greatly the side-wall pressure and pulling tension. Certain types and sizes of duct liner perform better in direct-buried installations, as shown in Table 8-8 [25]. Basic duct liner specifications are given in [26].

Table 8-8 Direct Buried Duct Liner Application Guidelines

Duct Liner Type	Diameter (Inches)	Application
Corrugated wall	1 and 1-1/4	Avoid
Corrugated wall	1-1/2 and 2	Satisfactory if ≤ 2 kilofeet
Smooth wall	1 through 2	Very good, low pulling tension
Ribbed wall	1 through 2	Very good, low pulling tension

To maintain a low coefficient of friction, the cross-sectional area of cables installed in duct liner should not exceed 50% of the duct liner cross-sectional area. The percentage fill can be found by simply taking the ratio of the respective diameters, squared, as follows:

$$\frac{d^2}{D^2} \leq 50\%$$

where d = cable outside diameter
D = duct liner inner diameter

8.6 Summary of Fiber Optic Cable Physical Characteristics

The previously described physical characteristics are summarized in Table 8-9 for fiber optic cables.

Table 8-9 Outdoor Fiber Optic Cable Physical Characteristics

PARAMETER	CHARACTERISTIC
Type	Singlemode Multimode
Buffering	Loose tube Ribbon assembly Slotted core Tight tube
Size	8 to 10/125 μm (singlemode) 50/125 to 100/140 μm (multimode)
Primary coating Fiber count Cable core	250 to 900 μm outer diameter 2 to 200 Filled
Strength member Armor Sheath	Dielectric (nonconducting) Nondielectric (conducting) Bonded steel (6 mils) Polyethylene

8.7 Fiber Optic Cable Costs

Tables 8-10 and 8-11 can be used to estimate the costs, per meter, of singlemode cables in various sizes. The actual costs at any given time will vary with market conditions and raw material costs.

Table 8-10 Comparative Costs for Loose-Tube Singlemode Fiber Optic Cable

SIZE	6-FIBER	12-FIBER	24-FIBER	48-FIBER	96-FIBER	192-FIBER
$/m	1.550	2.231	3.625	6.419	12.385	24.426
$/fiber-m	0.258	0.186	0.151	0.134	0.129	0.127

Costs in this table are current as of January 1994 for quantities of 10 kilofeet. Cable has the following characteristics: Dual-window (1,310/1,550 nm), 0.4/0.25 dB/km MIFL, dielectric strength member, no armor, 6 fibers/tube (<48 total fibers) or 12 fibers/tube (≥48 total fibers), 2,700 N tensile rating.

Table 8-11 Comparative Costs for Armored, Loose-Tube Singlemode Fiber Optic Cable

SIZE	6-FIBER	12-FIBER	24-FIBER	48-FIBER	96-FIBER	192-FIBER
$/m	1.870	2.592	4.011	6.996	13.222	25.394
$/fiber-m	0.312	0.216	0.167	0.146	0.138	0.132

Costs in this table are current as of January 1994 for quantities of 10 kilofeet. Cable has the following characteristics: Dual-window (1,310/1,550 nm), 0.4/0.25 dB/km MIFL, dielectric strength member, armor, 6 fibers/tube (<48 total fibers) or 12 fibers/tube (≥48 total fibers), 2,700 N tensile rating.

8.8 Optical Fiber Electrical Characteristics

8.8.1 Fiber Refractive Index and Cladding

Singlemode fibers used in fiber loop applications generally are matched-clad with simple step-index profile or depressed-clad, as previously mentioned. The effective group refractive index of the core in a common 1,310 nm, optimized fiber is nominally 1.470, with a fractional index change of 0.36%. The cladding always has a lower refractive index, which can be determined from:

$$\text{Fractional index change} = (n1 - n2)/n1$$

where n1 = core refractive index
n2 = cladding refractive index

The group velocity in an optical fiber can be determined from

$$\text{Group velocity} = c/n$$

where c = velocity of light in a vacuum (3E8 m/second)
n = refractive index

The velocity of propagation for the typical singlemode fiber is approximately (3E8/1.47 =) 204E6 m/second.

8.8.2 Transmission Characteristics

8.8.2.1 Attenuation Characteristics

In fiber loop applications, the most important transmission parameter is the optical fiber loss per unit length.

Mean unit fiber loss and unit loss standard deviation is given by:

$$\text{Mean unit fiber loss} = U_c + U_{cT} + U_\lambda \text{ dB per unit length}$$

$$\text{Unit fiber loss standard deviation} = \sqrt{(\sigma_c^2 + \sigma_m^2 + \sigma_t^2 + \sigma_T^2)}$$

where U_c = mean fiber loss, dB
U_{cT} = loss due to temperature effects
U_λ = loss due to wavelength dependence
σ_c = standard deviation of fiber loss related to manufacturing tolerances
σ_m = standard deviation of fiber loss related to measurement uncertainty
σ_t = standard deviation of fiber loss related to aging
σ_T = standard deviation of fiber loss related to temperature effects

Optical fiber unit loss is conveniently specified as a maximum individual value for all fibers in a given cable. This is called the *maximum individual fiber loss* (MIFL), and the unit loss of all fibers will be ≤MIFL. This type of specification removes the complications associated with obtaining statistical data from suppliers and manufacturers. MIFL is given in manufacturer's data sheets as attenuation cells. For example, standard attenuation cells of 0.35 and 0.40 dB/km at 1,310 nm, and 0.25 and 0.3 dB/km at 1,550 nm are common attenuation cells; as are 0.3 and 0.4 dB/km at 1,310 nm, and 0.25 and 0.3 dB/km at 1,550 nm.

For a given fiber route, the maximum total loss will be

$$\text{Total fiber loss} = \text{MIFL} \times \text{length dB}$$

If mean losses and standard deviations are to be used in the fiber section design, the total loss and total standard deviation are:

$$\text{Total fiber loss} = \text{Mean unit fiber loss} \times \text{length dB}$$

$$\text{Total fiber loss standard deviation} = \text{Unit fiber loss standard deviation} \times \text{length dB}$$

The loss curve for a typical fiber is given in Figure 8-13(a) [27]. Loss in singlemode fibers is caused by the following phenomena, each of which is discussed in turn:

- Scattering (primarily Rayleigh scattering)
- Absorption (primarily due to impurities in the glass)
- Geometric effects (due to microbending and macrobending)

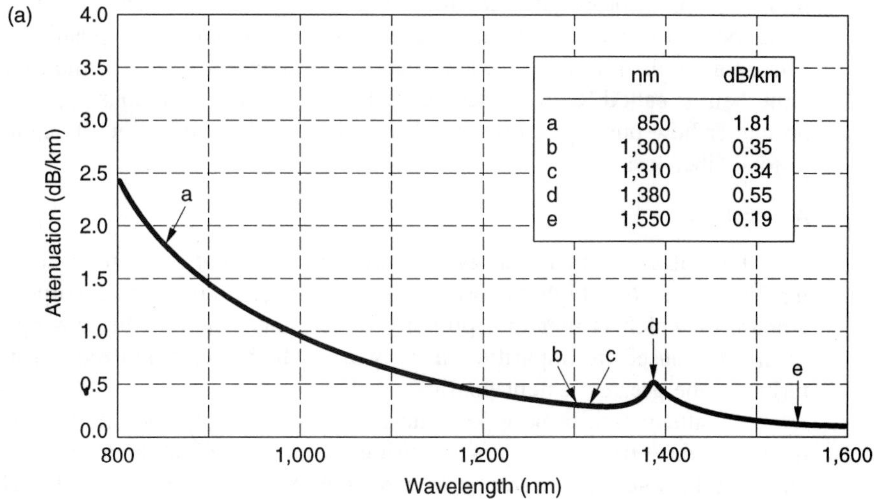

	nm	dB/km
a	850	1.81
b	1,300	0.35
c	1,310	0.34
d	1,380	0.55
e	1,550	0.19

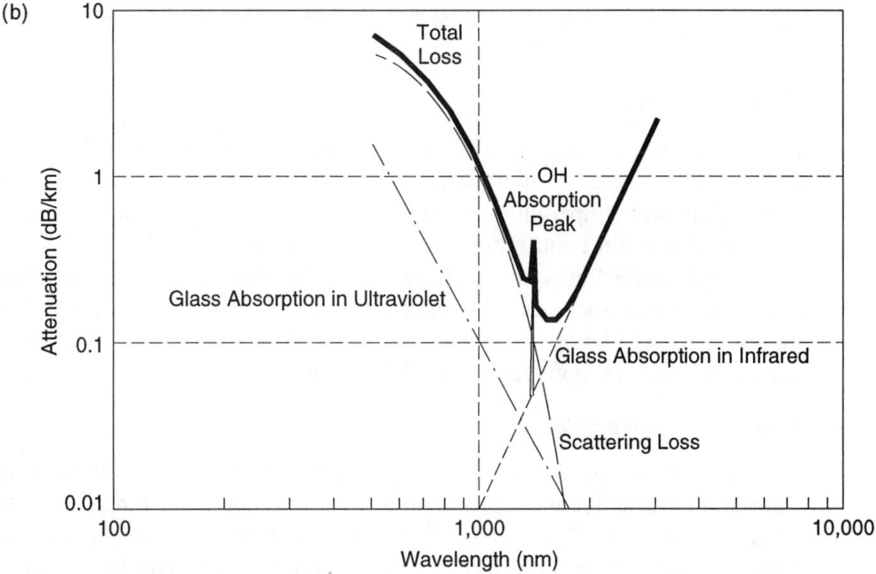

Figure 8-13 Loss in optical fibers: (a) Spectral attenuation curve for commercially available fiber cable, [27] © 1990 Corning, Inc. Reprinted with permission; (b) Loss mechanisms in optical fibers

Scattering

Scattering is an effect of the interaction between electromagnetic radiation and matter. As light propagates down an optical fiber, a certain amount of the energy is scattered by local variations in the glass refractive index or compositional materials. These variations normally are less than a wavelength across, and the loss associated with them is called Rayleigh scattering loss [2]. These variations appear in the glass no matter how pure it is. Figure 8-13(b) shows the contribution of scattering losses to total fiber loss.

Absorption

Part of the optical energy is converted into heat, or causes electrons in the material to move to higher energy states. This process is called absorption, a certain amount of which is a natural property of the glass itself. Other absorption mechanisms are caused by impurities in the glass, which are minimized during manufacturing to the greatest extent possible.

An attenuation peak appears near a wavelength of 1,380 nm and is due to the hydroxyl ion (OH) from water impurities. In a typical singlemode fiber, the attenuation at the so-called "water peak" does not exceed 2.1 dB/km [27]. OH impurities also contribute to other minor peaks near 1,230 and 950 nm. Figure 8-13(b) shows the contribution of absorption and other loss mechanisms to total fiber loss. This graph is representative of the loss mechanisms but does not show actual values for any particular fiber.

Geometric Effects

As already discussed, the primary geometric effects on fiber loss are from physical distortions due to microbending and macrobending. Optical fiber packaging, installation, and temperature stresses contribute to microbending losses; and cladding and core distortion contribute to macrobending losses.

Singlemode fibers are packaged so as to minimize geometric effects. For example, a typical fiber can be wrapped around a 75-mm mandrel 100 times, and the induced loss will be less than 0.05 dB in the 1,300-nm region. Bending effects and losses are more pronounced in the 1,550-nm region.

Temperature Dependence

The transmission performance of standard singlemode fibers used in outside plant applications is typically specified over a temperature range of −40 °C to +55 °C or −60 °C to +85 °C. At 1,300 nm, the induced loss over the temperature range typically is less than 0.05 dB/km, but in some designs can be as high as 0.2 dB/km at 1,300 nm. The loss specified by design processes normally is an MIFL under any environmental condition, so special design considerations are not required for cables installed within their rated temperature range.

Wavelength Dependence

As seen in Figure 8-13, the loss characteristics of optical fibers depend on the wavelength. This illustration shows only the gross spectral dependence. There also are variations within narrow wavelength regions due to manufacturing tolerances. Typical values are given in Table 8-12.

Table 8-12 Attenuation Variation within Narrow Wavelength Regions for Typical Singlemode Fibers[a]

Wavelength Range	Nominal Wavelength	Additional Loss
1,285 to 1,310 nm	1,310 nm	0.10 dB/km
1,310 to 1,330 nm	1,310 nm	0.05 dB/km
1,525 to 1,575 nm	1,550 nm	0.05 dB/km

[a]Data from [27].

8.8.2.2 Transmission Modes

Modern fiber loop applications use singlemode fibers exclusively. These fibers are designed to support one and only one propagation mode. All singlemode fibers have a cutoff wavelength, above which the optical fiber supports only one propagation mode, and below which multiple modes are supported. Operation below this cutoff wavelength may result in modal noise, modal distortion (increased pulse broadening), and improper operation of connectors, splices, and wavelength division multiplexer (WDM) couplers.

For a given fiber, it is necessary to ensure that the maximum cutoff wavelength is less than the optical transmitter wavelength, to guarantee that the system operates entirely within the optical fiber's singlemode region. Pigtail, station cables, and outside plant cables may have different cutoff wavelengths, so this parameter should be confirmed for each component. This is especially critical for short fibers used in repair because cutoff wavelength increases with decreased fiber length. This also applies to pigtails, which are short, pre-connectorized lengths of fiber.

For a typical singlemode fiber meeting CCITT recommendations, the uncabled fiber cutoff wavelength will lie between 1,100 and 1,280 nm [28]. Commercially available fibers normally have a cutoff wavelength less than 1,250 nm [27], but cutoff wavelengths as high as 1,330 nm may be encountered in practice [3]. Bends in the cabled fiber reduce the cutoff wavelength. The difference between the cabled and uncabled cutoff wavelength is a matter of definition and measurement [4].

8.8.2.3 Dispersion

While multimode fibers are subject to a number of significant dispersive mechanisms, singlemode fibers generally experience only material and waveguide dispersion, with material dispersion being the most significant. Material dispersion, also called chromatic dispersion, is the name given to the variation of optical wave velocity with wavelength that is due to the properties of the optical fiber [2]. Dispersion depends on length and is measured in picoseconds/nm-km.

The optical pulse launched into an optical fiber has a finite spectral width. Each frequency component propagates with a slightly different velocity, although only a single propagation mode is supported at any given wavelength. The variation of velocity with wavelength causes the shape and width of the pulse to change. The transmitted pulse broadens, as shown in Figure 8-14. This ultimately limits the transmission rate because the pulses overlap, resulting in intersymbol interference. Conversely, at a given transmission rate, the overlap limits the distance. The maximum distance at which the incoming pulses are still separated enough to be detected with the specified bit error probability is the point at which the link becomes dispersion limited.

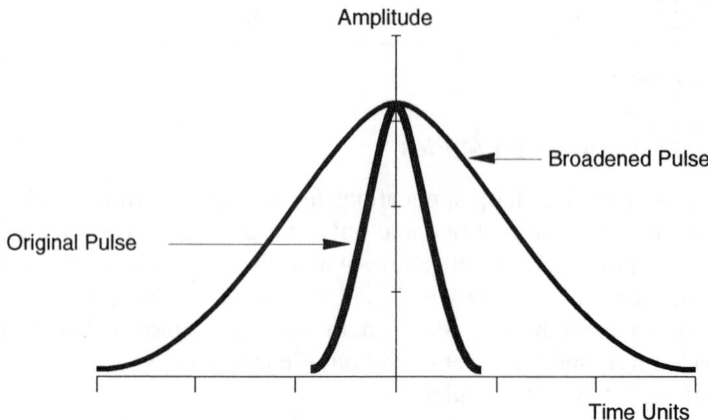

Figure 8-14 Pulse broadening in optical fiber transmission

An infinite transmission rate would be possible with a spectrally pure pulse launched into a fiber at a wavelength in which the dispersion is zero and the dispersion slope is zero. In this case, the maximum distance would be limited only by the system gain. In practical singlemode fiber systems, the bandwidth or distance is determined by the dispersion slope at the zero dispersion wavelength. At other wavelengths, these limitations are determined by the dispersion [3].

Standard singlemode fibers fortuitously have zero dispersion, typically in the wavelength band from 1,300 to 1,321 nm, and the dispersion slope is low enough to allow wide bandwidth transmission without extraordinary fiber manufacturing processes [27]. However, as mentioned, any practical pulse has a finite spectral width, and fibers do not have zero dispersion slope, so infinite bandwidths are not possible, nor are they presently required, in practical fiber loop systems.

Even though the dispersion is zero at a particular wavelength, it is nonzero at other wavelengths. In a typical singlemode fiber, the dispersion has a positive slope of 0.090 ps/nm^2-km, as illustrated in Figure 8-15. The dispersion at a given wavelength can be found from [27]:

$$\text{Dispersion}(\lambda) = (S/4)(\lambda - \lambda_0^4/\lambda^3) \text{ ps/nm-km}$$

where S = dispersion slope = 0.090 ps/nm^2-km
λ_0 = zero dispersion wavelength
λ = wavelength in question (valid for a range of 1,200 to 1,600 nm)

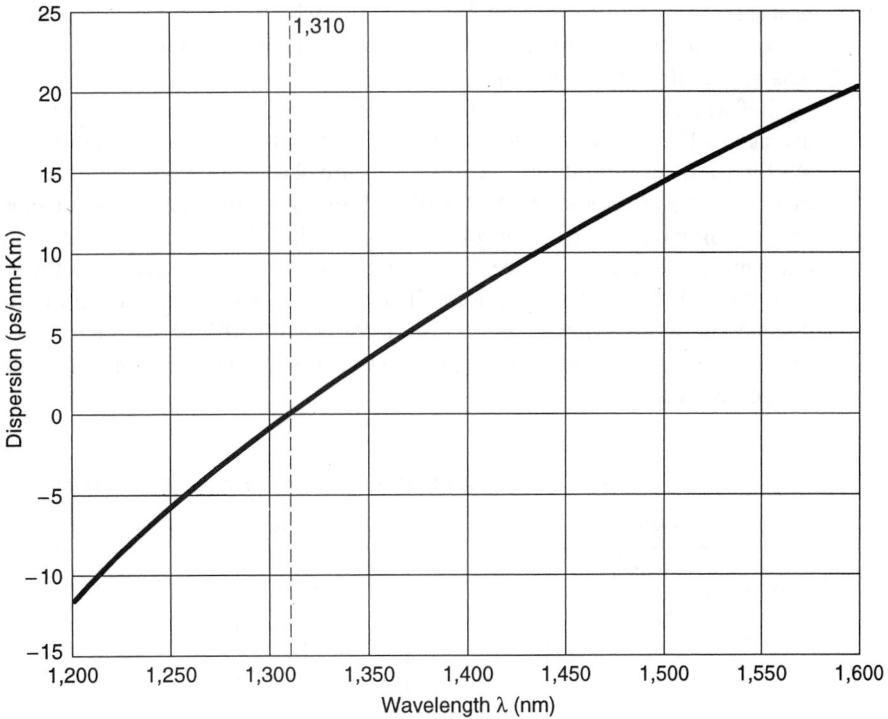

Figure 8-15 Material dispersion in a typical singlemode fiber

The dispersion slope is given in units of ps/nm^2-km, which indicates broadening of the transmitted pulse in picoseconds per nanometer of transmitted pulse spectral width per nanometer of displacement of the zero dispersion wavelength from the transmitted wavelength, per kilometer of length.

The dispersion for a given singlemode fiber usually is given as a maximum value in ps/nm-km, which indicates broadening of the transmitted pulse in picoseconds per nanometer of spectral width, per kilometer of length. This takes into account the maximum displacement of the zero dispersion wavelength. The dispersion in singlemode fibers varies from 0.9 to 6.0 ps/nm-km, with typical values of 2.8 to 4.5 ps/nm-km. The lower number indicates a better fiber. The effect of pulse dispersion on system design is discussed in Chapter 10.

8.9 Splices and Connectors

Unlike twisted pair cable systems, splice and connector loss is a significant factor in fiber optic systems design. For example, in a fiber system with a 10 dB section loss budget, five splices with 0.5 dB loss each account for 25% (2.5 dB) of the budget. If two connectors are used, each with 0.5 dB loss, another 10% of the budget is used. This is illustrated in Figure 8-16, which shows the allocation of loss for various combinations of splices and splice losses in a 10-km section, with 4 dB fiber loss and 1 dB connector loss.

Three different configurations are shown in Figure 8-16, each differing by the average distance between splices (one km/splice, two km/splice, and four km/splice). Within each configuration, two situations are shown: one with an average splice loss of 0.15 dB and another with 0.5 dB. High splice loss and short distance between splices can have a large impact on the loss budget. For example, where there is only one km/splice and the splice loss is 0.15 dB, splice loss accounts for a little more than 20% of the total section loss. If the average splice loss is 0.5 dB for the same distance between splices, splice loss accounts for 50% of the total. As the number of splices decreases (greater distance between splices), their impact on the total loss budget decreases.

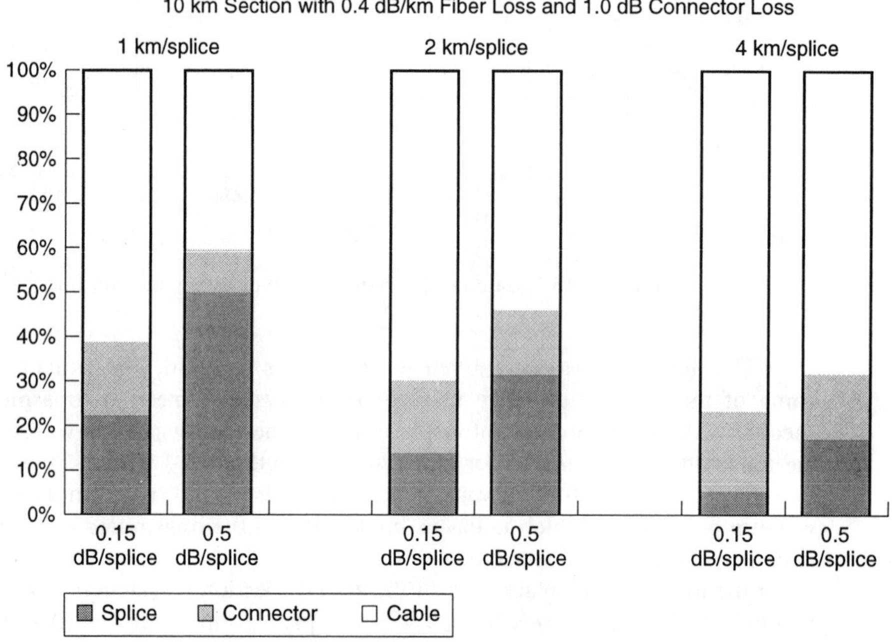

Figure 8-16 Fiber optic system loss allocation

In the above example, a splice with 0.5 dB loss was used. This is considered high by today's standards, where 0.2 to 0.3 dB or less are more typical, depending on the type of splice used. In any case, the number of splices and connectors must be minimized, which explains why all companies strive to install fiber optic cables in the longest unspliced lengths possible.

One industry standard for splicing systems is [29]. Splice loss, reflection, and tensile strength are the most important parameters associated with any splice method.

These parameters affect the type of splice chosen for field installations. Splices can be categorized as:

- Fusion
- Mechanical, passive
- Mechanical, active

The fusion splice is made by holding the accurately cleaved fiber ends together in precise alignment and permanently welding the ends with a high-voltage arc. Fusion splicing machines either automatically or manually align the fibers. The effects of face quality are minimized by the fusion process, and end-gaps are nonexistent after the splice is fused.

Mechanical splices longitudinally clamp the fibers to hold the ends together. Various techniques are used to minimize offset and tilt, and fiber end-polishing and index-matching gels or adhesives help minimize the effects of end-gap reflections. An active mechanical splice is "tuned" by rotating the fibers to minimize splice loss, as viewed on a test set in real-time. Some passive mechanical splices can be tuned by disassembling and remaking the splice until an acceptably low splice loss is attained.

All types are acceptable in fiber loop applications. The mechanical types are especially attractive for restoration of damaged facilities because they can be installed quickly and easily without special preparation or environmental concerns. Also, most mechanical splices can be reused, and some actually exhibit better splice loss performance as the splice is remade [30]. Table 8-13 compares the salient characteristics of each splice type.

Regardless of manufacturer or type, splices are designed to minimize losses caused by the extrinsic and intrinsic mechanisms shown in Table 8-14. These mechanisms also apply to connectors, as discussed later. Extrinsic loss mechanisms are mechanical and splice process-related.

The intrinsic splice loss mechanisms are determined when the optical fiber to be spliced is manufactured. Fusion and active mechanical splicing can minimize loss due to core/cladding concentricity; otherwise, the intrinsic splice loss mechanisms are not affected by the splice type.

For singlemode fibers, the dominant intrinsic loss mechanism is mode field diameter (MFD) mismatch [31]. Figure 8-17 shows the loss due to MFD mismatch based on the ratio of one fiber MFD to another. Typical fibers will have a mode field diameter of 8 to 10 μm. MFD has a normal distribution with a two or three standard deviation (σ) tolerance of ± 0.5 μm.

Each splice type will show a statistical variation in splice loss. While this variation is traditionally assumed to have a normal distribution, in reality this is not the case. Table 8-15 summarizes the normal distribution parameters for each splice type, as taken from information available from the manufacturers. Figure 8-18 shows typical histograms for each splice type.

For the fusion splice, two situations are given in the accompanying figure and table. The first is for splicing two fibers with the same cladding profile (in this case, step-index matched clad); the second is for splicing a step-index matched clad profile

Table 8-13 Comparison of Splice Types

Characteristic	Fusion	Active Mechanical	Passive Mechanical
Splice loss	0.01 to 0.10 dB	0.01 to 0.3 dB	0.1 to 0.3 dB
Splice loss standard deviation	0.04 dB	0.15 dB	0.1 dB
Return loss	60 to 90 dB	40 to 50 dB	35 to 45 dB
Configuration	Permanent	Temporary or permanent	Temporary or permanent
Initial cost	\$5,000 to \$50,000	\$1,600	\$2,000[a]
Splice incremental cost	\$0.50 to \$1.50	\$20 to \$35	\$10 to \$15
Test equipment cost	\$0 to \$15,000	\$8,000 to \$15,000	\$8,000 to \$15,000
Operator skill	Highest	Low	Lowest
Crew size[b]	1 person	2 persons	2 persons
Splice time[c]	2 to 3 minutes	7 to 15 minutes	3 minutes
Tensile strength	High	Low	Low
Tunable?	No	Yes	No
Restoration?	No	Yes	Yes
Splicing in manhole?	No	Yes	Yes
Adhesives required?	No	Yes	No
Precise cleaving?	Yes	No	Yes
Angled end-polishing required?	No	Yes[d]	No
Performance over fiber life	Excellent	Good	Fair to good
Temperature performance	Excellent	Good	Good
Splice test equipment	Built-in on more expensive models	External	External
Test equipment annual maintenance	Calibration and cleaning	Cleaning	Cleaning
Test equipment annual maintenance cost	5% of initial	1 to 5%	1 to 5%
Controlled environmental conditions during splicing?	Yes	No	No

[a]Includes high-quality cleaver.
[b]Two-person crew includes one person to splice, one person to test.
[c]Does not include set-up time or splice loss testing (unless splicing machine is equipped with local injection and detection).
[d]Required to achieve >40 dB return loss.

Table 8-14 Splice and Connector Loss Mechanisms

Extrinsic Mechanisms	Intrinsic Mechanisms
Core misalignment (offset) Angular misalignment (tilt) Longitudinal misalignment (end-gap)	Fiber mode field diameter mismatch Fiber index mismatch Core eccentricity (out-of-roundness)
Face quality (polish) Reflection Contamination Core deformation	Core/cladding concentricity

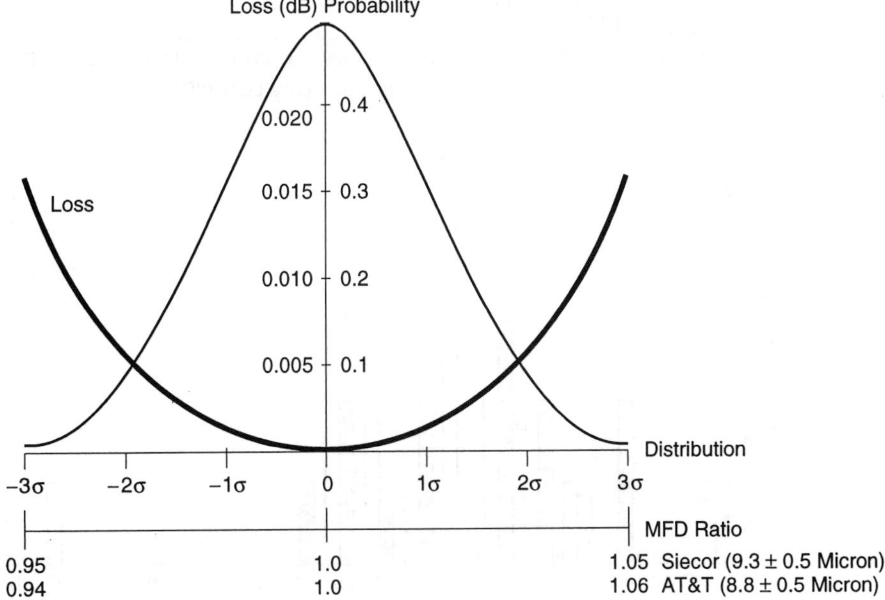

Figure 8-17 Splice loss due to mode field diameter mismatch

Table 8-15 Normal Distribution Parameters for Basic Splice Types

Splice Type	Mean Splice Loss	Standard Deviation	Maximum Splice Loss	Source
Fusion (step-index to step-index)	0.05 dB	0.04 dB	0.16 dB	[32]
Fusion (step-index to depressed clad)	0.07 dB	0.03 dB	0.19 dB	[32]
Passive mechanical	0.13 dB	0.09 dB	[a]	[30]
Active mechanical	0.18 dB	0.15 dB	[b]	[33]

[a] 1% of splices had loss >0.5 dB.

[b] 4% had loss >1 dB and were remade.

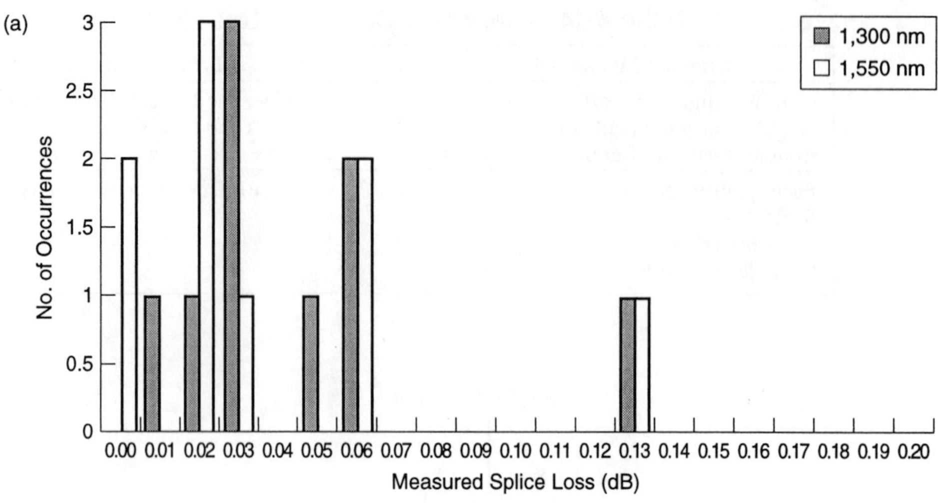

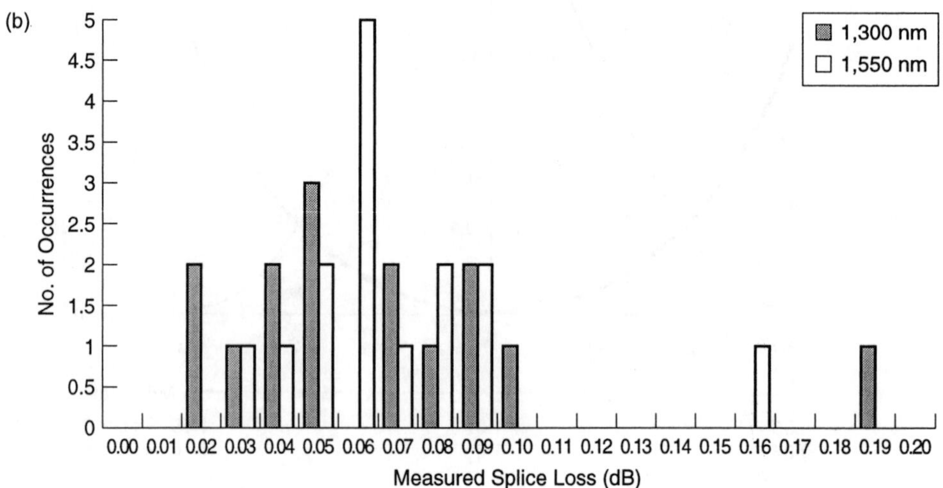

Figure 8-18 Splice loss distribution: (a) Fusion splice—step-index to step-index; (b) Fusion splice—step-index to depressed clad; (c) Passive mechanical splice; (d) Active mechanical splice

to a depressed clad profile. The latter configuration provides a slightly higher splice loss, but the loss still is quite low. It may be concluded that no significant loss penalty is incurred when singlemode fibers from different manufacturers with different cladding profiles are spliced together.

Real splices always will have a wider distribution of loss above the mean than below. Also, most splice loss specifications require any splice over 0.5 dB to be remade. Therefore, although any splice type is capable of having a loss greater than 0.5 dB, splices actually placed into service always will have less.

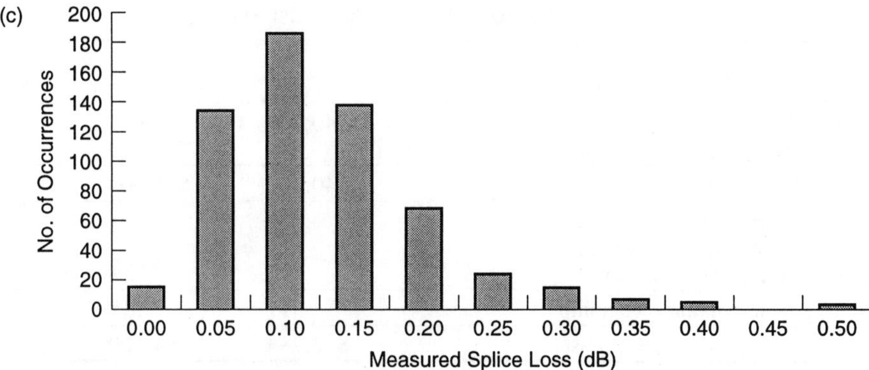

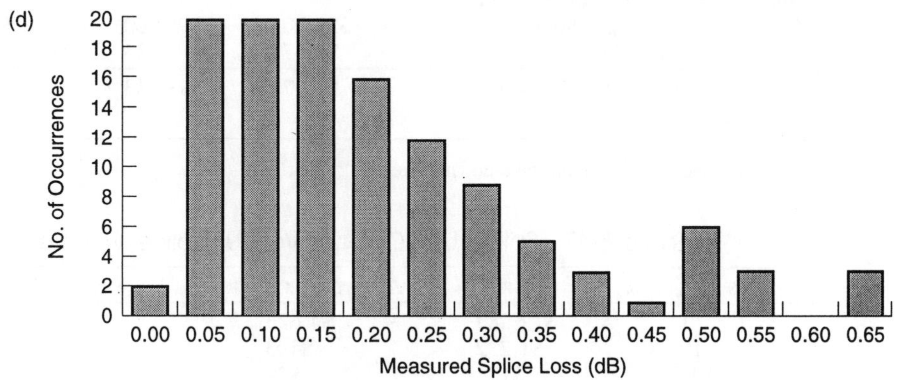

Figure 8-18 Continued

Fiber preparation varies with the splice. Fusion and passive mechanical splices require the fiber to be precisely cleaved. Some fusion splicing machines have a built-in cleaver. An active mechanical splice does not require precise cleaving but does require end-face polishing to improve splice loss. Some mechanical splices require adhesives to hold the fibers together. The addition of adhesives and polishing both increase the time to make a splice.

The cost of a particular splice type depends on several factors. Table 8-16 shows an evaluation based on typical equipment, splice materials, and labor costs. The unit splice cost, including ownership costs of the splicing machine, depends highly on the volume of splices to be made throughout the life of the splicing machine. Table 8-17 shows these costs spread over one, ten, 100, 1,000, and 10,000 splices per year. A splicing machine life of five years is used in these tables.

Figure 8-19 represents the same information in graph format. Almost all analyses will show the fusion splicer to be most economical above a few hundred splices per year. However, specific requirements, including noneconomic considerations, may shift this threshold either way by a wide margin.

Table 8-16 Splice Cost Evaluation

Item	Fusion	Active Mechanical	Passive Mechanical
Splicing machine	$40,000	$1,600	$2,000
Splice test equipment	$0	$15,000	$15,000
Total investment	$40,000	$16,600	$17,000
Capital recovery factor (at 9% for 5 years)	26%	26%	26%
Maintenance factor	5%	2%	2%
Total annual ownership	31%	28%	28%
Annual ownership cost	$12,284	$4,600	$4,711
Consumable material cost per splice	$1.50	$25.00	15.00
Splice and test labor time (minutes)[a]	3	10	5
Loaded labor rate per hour	$50.00	$50.00	$50.00
Labor cost per splice	$1.67	$8.33	$4.17
Total labor and material cost per splice	$3.17	$33.33	$19.17

[a]Does not include set-up and take-down time.

Table 8-17 Splice Unit Cost for Various Splice Volumes

Splices per Year	Fusion	Active Mechanical	Passive Mechanical
1	$12,287	$4,633	$4,730
10	$1,232	$493	$490
100	$126	$79	$66
1,000	$15	$38	$24
10,000	$4	$34	$20

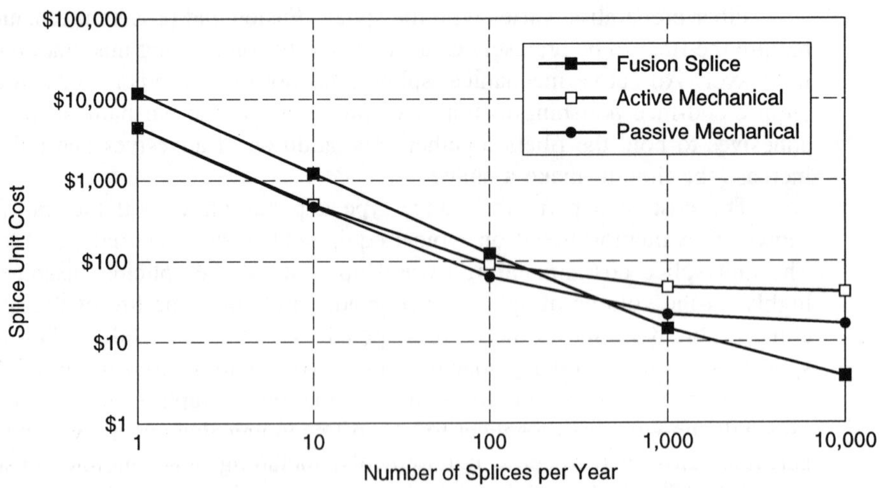

Figure 8-19 Splice cost comparison curve

The quality and performance of any splice type depends on the care taken during the splice process. Dust and precipitation as well as the effects of corrosive atmospheres must be controlled at the splice site to avoid problems with fiber alignment and contamination. Both misalignment and contamination can increase splice loss and reduce tensile strength.

The splice procedures or scenarios to be used must be determined in advance. For fusion splicing, in which the splicing machine has local injection and detection for splice loss estimation measurements, each splice is tested as it is completed. With mechanical splices and fusion splices without local injection and detection, three scenarios currently are used.

In the first scenario, a splice point (for example, reel-end splice or manhole splice) is visited by a two-person crew and the cables are prepared for splicing. The splices at each splice point are made blind; that is, all fiber splices are finished before any individual splice loss tests are made. When each splice point is almost completed, one person in the crew moves to another point where an optical time domain reflectometer (OTDR) is connected and measurements made. Meanwhile, the person remaining at the splice point finishes splicing and is available to remake any unacceptable splices. Since an OTDR is used, tests are required from both sides of the splice. Travel time between measurement points can add significantly to the average splice cost.

In the second scenario, all splice points are visited in succession and no splice loss tests are made until all splice points are completed, end-to-end. This method allows each splice site to be completed much faster, but additional time is required to return to each point that has unacceptable splices, to open the splice closure, and to remake the splice.

The third scenario is similar to fusion splicing with local injection and detection. However, since the splicing machine (either mechanical or fusion) in this case does not have local injection and detection capabilities, external local measurement equipment is used at each splice point. Generally, all splices at a splice point are completed before loss measurements are made so as not to disturb the splicing "assembly line." Then, each splice is measured, remade as necessary, and the splice closure sealed; the crew then moves to the next splice point.

Each scenario described above requires some amount of time to make splice loss measurements using external equipment. One estimate, for twenty 50-fiber splice points (1,000 splices total), places the time and cost shown in Table 8-18 [34]. This estimate is based on a 5% remake and a $60/hour loaded labor rate for the 1,000 splices.

Once the splice site is set up, the optical fiber cable must be properly prepared. In addition to stripping the outer sheaths, a sufficient amount of slack fiber must

Table 8-18 Comparison of Splicing Scenarios for Twenty 50-Fiber Splice Points

ITEM	SCENARIO 1	SCENARIO 2	SCENARIO 3
Additional time	142.5 minutes	200 minutes	52.5 minutes
Cost at $60/hour loaded labor rate	$2,850	$4,000	$1,050

always be provided in case an unacceptable splice needs to be redone; the outer jacket and armor, if any, is stripped back one or two meters from each cable end for this purpose. All splice organizers and closures have space for coiling excess fiber. Additional unstripped slack is required if the splice is made in a specially equipped vehicle parked adjacent to the splice site. Generally, an additional ten m is provided for each cable end. Slack length must be included in loss calculations. When the splice is completed, the slack is coiled and buried below the splice pedestal (for buried plant), coiled in an enclosure mounted on the supporting structure (aerial plant), or coiled and attached to the wall in a manhole (underground plant).

The optical fiber itself is prepared by stripping the fiber primary coating with a mechanical, thermal, or chemical stripper. The mechanical stripper is preferred because it is safe, inexpensive, and creates a well-defined coating termination [31]. Mechanical stripping force typically is one-half to one lb. The fiber surface is cleaned after stripping to remove any coating residue, dust, and body oils. Wiping the fiber surface clean should be done with proper materials designed specifically for that purpose; if improperly done, wiping can introduce surface scratches, which tend to weaken the fiber.

Fiber ends are prepared by proper cleaving. End angles less than two degrees are acceptable; high-quality cleavers provide end angles approximately one-half degree [31]. The fusion and active mechanical splice require fiber alignment, which can be done by either power monitoring, using an external light source and detector, or by using an OTDR. Modern fusion splicing machines allow two additional methods for fiber alignment: power monitoring by local injection and detection, and profile alignment.

Power monitoring by external source and detector allows the fiber alignment to be optimized by maximizing the amount of power transferred through the splice from the light source to the detector. This method requires craft personnel at each end of the fiber to operate the test equipment. An *OTDR* can be used at one end to optimize fiber alignment, and only one extra craft person is required with this method.

Local injection and testing is the preferred fiber alignment method with fusion splicing because it does not require any extra craft people, nor does it require time-consuming termination of fibers at one or both ends of the fiber being spliced. With local injection and testing, the fibers on either side of the splice are bent around mandrels on the splicing machine, and light is injected on one side of the splice and detected on the other. Some splicing machines can measure the splice loss directly by this method.

The *profile alignment* method directs collimated light through the fibers at right angles to the fiber axis. The machine then automatically aligns the fibers, or the fibers can be manually aligned using a microscope. The remaining steps in the splice process depend on the splice type and manufacturer and will not be covered here.

Once the splices are made, they must be evaluated. All splice types are evaluated by a loss test and a tensile strength test. Strength can be measured by manual pull (this requires practice to keep from overstressing the splice) or by attaching a weight about two or three inches below the splice for one or two seconds. If the splice does not break, its tensile strength is acceptable. The weight varies from one-half to one lb (one to two kg) for fusion splices on 50 kpsi fiber [35]. Mechanical splices are tested with a one lb weight.

Loss is measured using an external source and detector, local injection and detection (with fusion splicing machines so equipped), or an OTDR. However, the OTDR requires averaged bidirectional measurements, described later.

Fusion splices are covered with heat shrink tubing; mechanical splices are self-protecting. All splice types are further protected by placing them in an organizer. Splices are never installed on a cable and then pulled into a duct.

Splice loss and splice loss standard deviation is given by:

$$\text{Splice unit loss} = U_S + U_{ST} \text{ dB per splice}$$

$$\text{Splice unit loss standard deviation} = \sqrt{(\sigma_S^2 + \sigma_T^2)}$$

where U_S = average splice loss, dB
U_{ST} = splice loss due to temperature effects
σ_S = standard deviation related to splice tolerances
σ_T = standard deviation related to temperature effects

Total splice loss and total splice loss standard deviation in a given fiber section are:

$$\text{Total splice loss} = \text{Splice unit loss} \times (Ns + Nr)$$

$$\text{Total standard deviation} = \sqrt{(Ns + Nr)} \times \text{splice unit loss standard deviation}$$

where Ns = number of engineered splices
Nr = number of projected repair splices

The specification of splice loss is a very important consideration in fiber loop design. Several approaches are used. The most conservative is to specify a maximum allowable splice loss value, typically 0.25 to 0.5 dB, and to use this value in the design process. This approach does not account for the statistical variation of splice losses and can result in gross overdesign, especially on long routes.

Another approach is to specify a mean and maximum splice loss value. The mean value is used for design purposes, and the maximum value provides a measure of control over individual splice quality. A typical specification would require a mean splice loss of 0.1 to 0.15 dB for all splices and a maximum individual splice loss of 0.2 to 0.5 dB. A maximum individual splice loss of 0.2 dB may not be possible on large volume installations, but it should be used as an objective.

Yet another approach is to specify a mean splice loss with a margin of 2σ (97.7%) or 3σ (99.9%) variation of 0.5 dB. A mean splice loss of 0.1 to 0.15 dB is specified for all splices, as before. However, in the case of a 0.15 dB mean, any splice over 0.27 dB (1σ) is required to be remade. If a splice loss less than 0.27 dB cannot be achieved after three or four tries, the splice is left alone unless it exceeds 0.5 dB. If it exceeds 0.5 dB, splice tooling or splicing techniques are checked and work continues on this splice until it is made less than 0.5 dB. Other variations on this theme are possible or may be desirable, based on splicer skills or equipment and type of splice used.

Any splice method requires a certain number of remakes to keep all splices within the 0.5 dB maximum. Depending on the splice type, manufacturer, and splicer skill, as many as 5 to 10% of all splices may have to be remade. If accurate estimates are to be made of the time and cost required, remakes must be taken into account.

To determine if splices meet specifications, they must be measured. This is most conveniently done with a fusion splicer that has an integral splice loss estimation feature, as previously described. If local injection/detection is not available, two loss measurement methods are used. Loss may be measured by injecting a known optical signal on one side of the splice and measuring the detected signal on the other side. Alternatively, loss may be measured by an OTDR. This second method has specific limitations due to differences in the MFD of optical fibers.

MFD characterizes the width of optical power distribution across the fiber cross-section. The MFD can vary along the fiber's length, but this normally does not affect measurements. Significant differences, however, can arise at reel-end splices or repair splices. Loss can be affected by differences in core diameters and core eccentricities, but most splice types compensate for these characteristics. MFD, on the other hand, affects not only *actual* splice loss but also splice loss as *measured* by an OTDR.

The *actual* splice loss due to MFD differences is the same in each direction across a singlemode fiber splice. The *measured* loss by an OTDR, however, is different in each direction. This can lead to the phenomenon of splice "gain," in which the OTDR shows a negative loss in one direction across the splice. The other direction will show a higher-than-actual loss.

One major manufacturer of singlemode fiber (AT&T Network Systems) claims a normal distribution for MFD, with a mean of 8.8 ± 0.5 μm [36]. Assuming 0.5 μm = 3σ, almost 98% of all splices made with this fiber will have a loss less than 0.02 dB due to MFD difference. This is a negligible value. Even when two fibers at the extreme (8.3 and 9.3 μm) are spliced, the maximum loss is 0.056 dB. Nonetheless, the one-way measurement error with an OTDR is 0.5 dB. Such an error magnitude could lead to unnecessary splice remakes. Therefore, one-way OTDR measurements should not be used to qualify splices.

> The only reliable OTDR method for measuring splice loss is to take the algebraic average of two measurements, one in each direction, across the splice.

The number of splices have a significant effect on the operating margin obtained from a working system, but the number is easy to underestimate. At least one splice is required at each facility entrance, and one splice is needed at each reel-end. If the cable must be cut between reel-ends due to construction requirements and obstacles, each cut will require one splice.

Repair splices are difficult to predict because the route typically will have a 20- to 40-year life, and it is impossible to predict where and when damage may occur. An estimate, perhaps based on historical records, is required of how vulnerable the route is to damage from dig-ins or other disasters. Repair splices always are counted in pairs, as shown in Figure 8-20. The damaged cable ends cannot be spliced directly together because they are too short. A piece of repair cable is required to connect the damaged ends. All fiber splices preferably should be above ground, as shown, to allow future access. Splices may be directly buried in a suitable splice closure, but this should be avoided wherever possible; if access is required, locating the splice will be difficult.

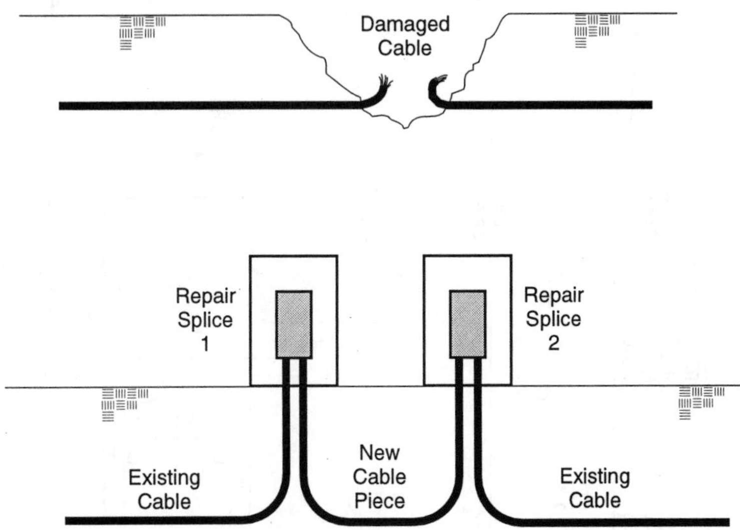

Figure 8-20 Counting repair splices

Connectors are another important element of fiber loop systems, and industry standards are available [37–39]. Three general types of fiber optic connectors are used in loop applications. These are:*

- push/pull (SC), shown in Figure 8-21(a)
- bayonet (ST), shown in Figure 8-21(b)
- screw-on (FC), shown in Figure 8-21(c)

These connectors all use a 2.5-mm ferrule, as shown in the dimensional drawings. The drawings show both the connector and the adapter (or coupler) to which two connectors attach in patching applications.

Although manufacturer's published specifications may indicate one type has slightly better performance than another, in field applications they are essentially

*Other types have been used, in particular the SMA and biconic connector.

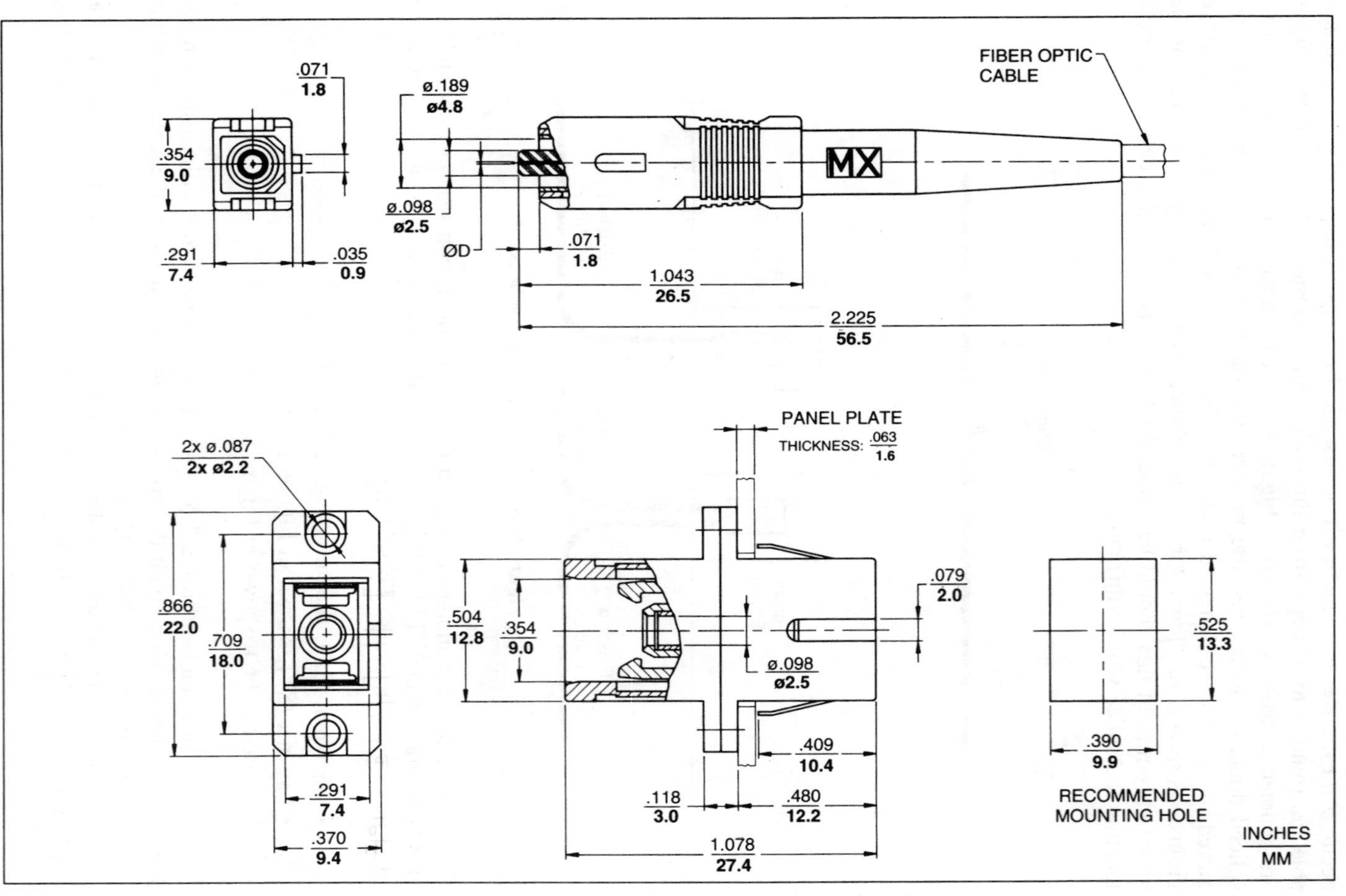

Figure 8-21(a) Mechanical dimensional drawing of SC fiber optic connector © 1993 Molex® Fiber Optic Interconnect Technologies, Inc.

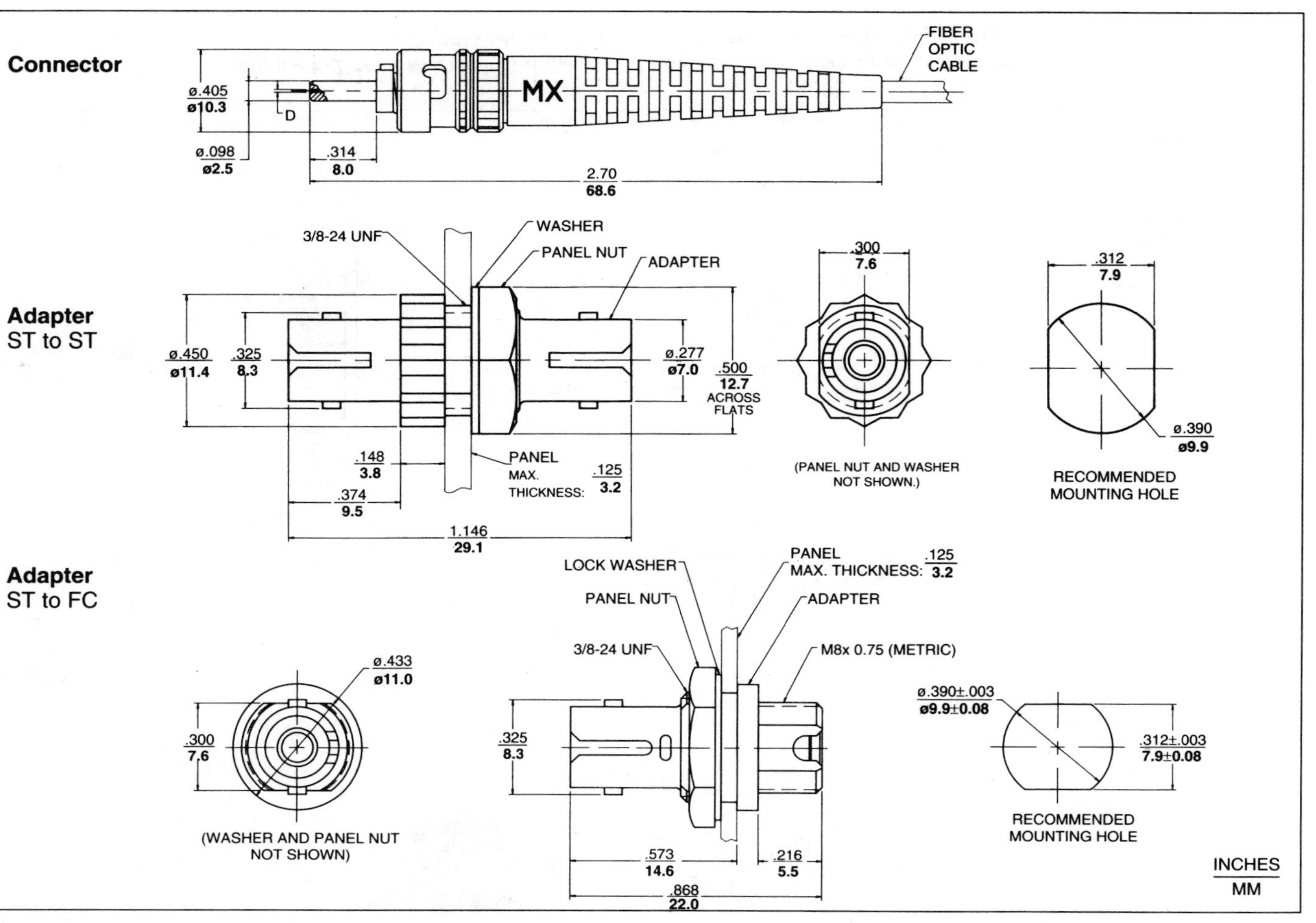

Figure 8-21(b) Mechanical dimensional drawing of ST fiber optic connector
© 1993 Molex® Fiber Optic Interconnect Technologies, Inc.

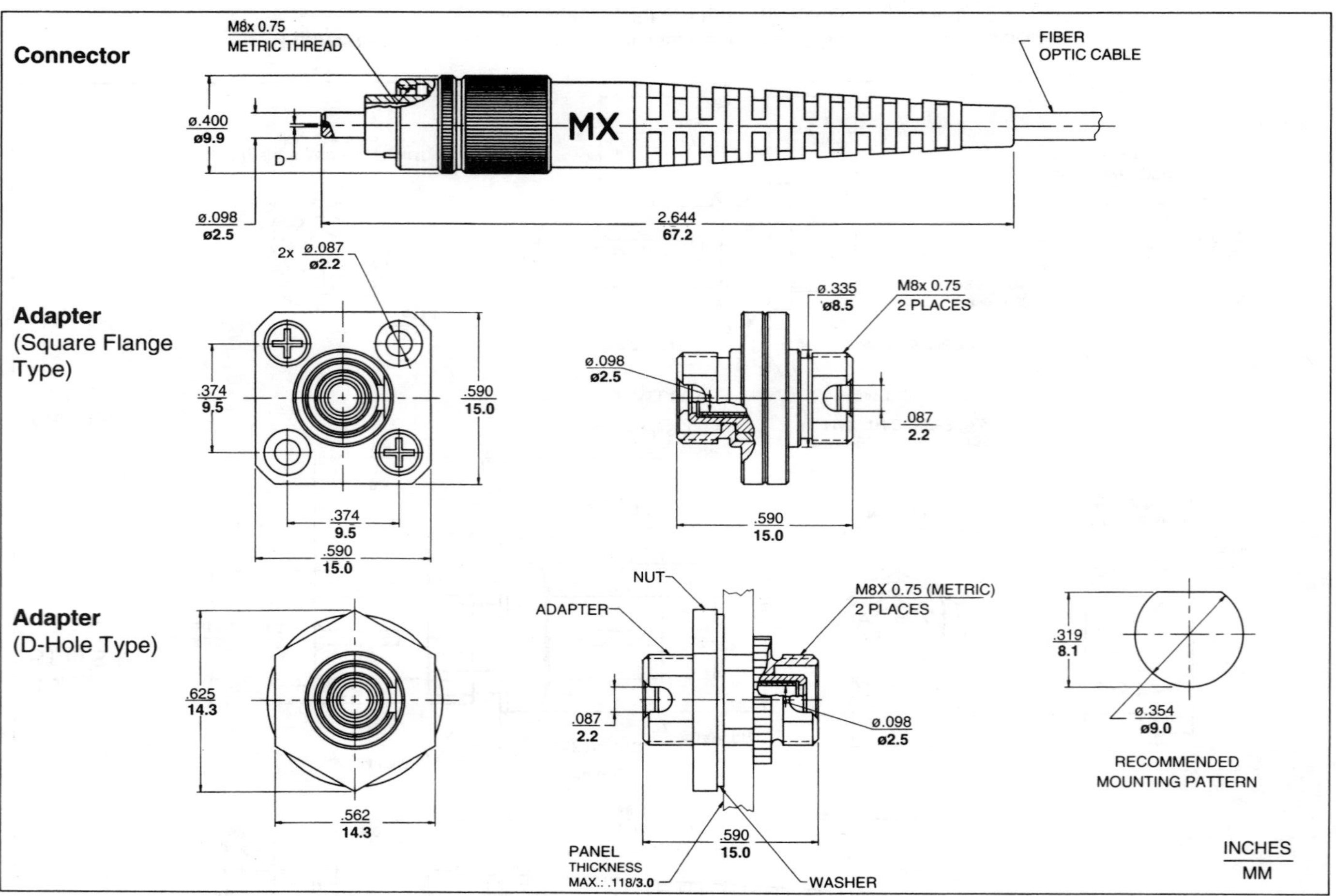

Figure 8-21(c) Mechanical dimensional drawing of FC fiber optic connector
© 1993 Molex® Fiber Optic Interconnect Technologies, Inc.

equivalent. Table 8-19 summarizes the relevant statistical insertion loss information for the connectors, assuming a normal distribution. Figure 8-22 shows the insertion loss histograms for connectors from one manufacturer (Molex® Fiber Optic Interconnect Technologies, Inc.).

Table 8-19 Connector Insertion Loss Statistics

Type[a]	Mean (dB)	σ(dB)	Typical (dB)
SC	0.18	0.09	0.20
ST	0.20	0.09	0.25
FC	0.13	0.05	0.15

[a]See text for explanation of types of connectors.

When choosing connectors, important considerations are:

- Optical performance
- Repeatability
- Durability
- Pull-out resistance
- Multiple connection capability
- Ease of use
- Field assembly and installation
- Optical performance

The three connector types have comparable costs. The SC is popular for new fiber loop installations. Connectors installed on optical fiber terminal equipment line interface units usually are selected by the manufacturer and usually are beyond the control of the link designer.

With respect to optical performance, connectors suffer from many of the same loss mechanisms as splices. The SC and some FC connectors can be factory tuned for lower insertion loss. The ST is a keyed connector and is not normally tunable. The insertion loss depends on the mechanical accuracy of the ferrule. This accuracy can be improved by orienting the ferrule to a preferred position during connector assembly. High return loss depends, among other things, on high-quality end-face finish or polish. High-performance connectors are angle-polished to improve return loss, but this tends to increase insertion loss. The angle varies from eight to twelve degrees from perpendicular. Typical return loss values for singlemode connectors are 31 to 32 dB (normal polish) and 40 to 42 dB (polish using factory processes).

Connector loss and loss standard deviation are given by:

$$\text{Connector unit loss} = (U_{con} + U_{cont}) \text{ dB per unit}$$

$$\text{Connector unit loss standard deviation} = \sqrt{(\sigma_{con}^2 + \sigma_t^2 + \sigma_T^2)} \text{ dB}$$

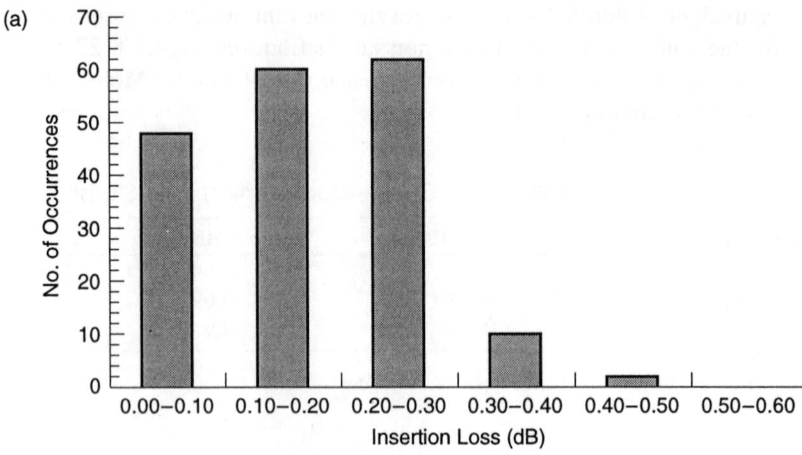

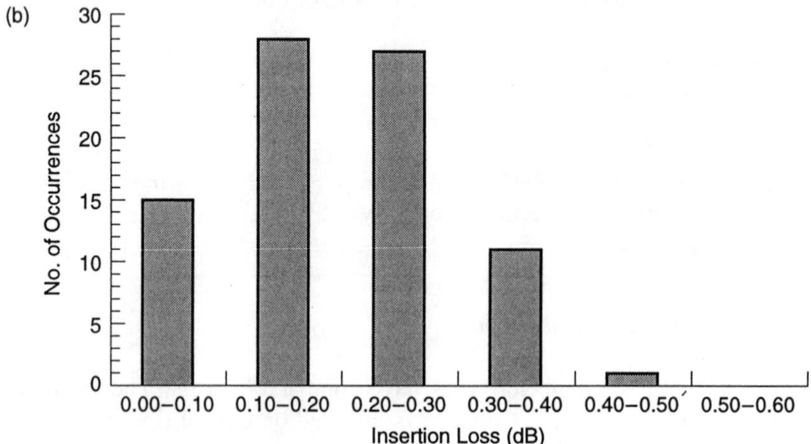

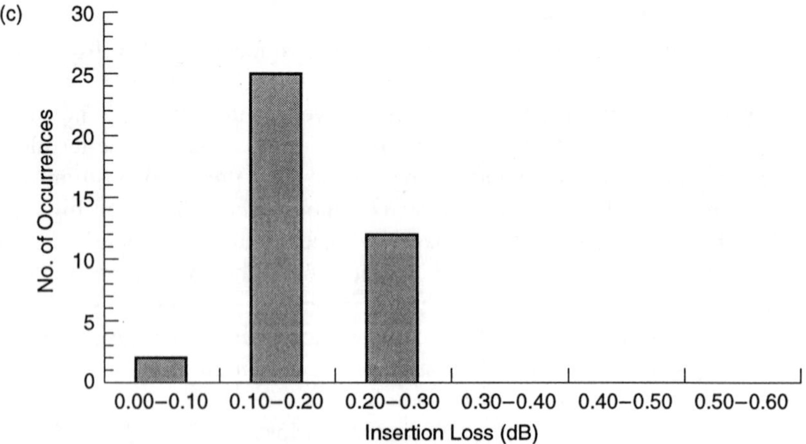

Figure 8-22 Fiber optic connector insertion loss histograms (a) SC; (b) ST; (c) FC

where U_{con} = average connector loss
U_{cont} = loss due to temperature variations
σ_{con} = standard deviation related to connector tolerance
σ_t = standard deviation related to mating cycles and age
σ_T = standard deviation related to temperature

The total connector loss and loss standard deviation in any given fiber section are:

$$\text{Total connector loss} = N_{con} \times \text{connector unit loss dB}$$

$$\text{Total standard deviation} = \sqrt{(N_{con})} \times \text{connector unit loss standard deviation dB}$$

All connectors exhibit increased loss after aging and many mating cycles. Connector data do not always take into account optical fiber tolerances. The end result is to increase the connector loss. With singlemode fibers, the increase normally is less than 0.1 dB. Unless more specific data are available, all connector types should be assumed to have 0.5 dB average loss and 1 to 2 dB maximum loss.

Overall system loss can be reduced slightly by using connectors with factory-installed optical fiber pigtails. The pigtails are installed by the manufacturer under controlled conditions and with tested, repeatable results. The only variable is where the pigtail is spliced to the station or entrance cable. As indicated above, splice loss can be made arbitrarily small with modern splicing machines. It is easy to reduce system loss by 0.2 dB for each connector when factory-installed connector pigtails are used and the pigtail splice loss is made very small.

Connectors must have high repeatability and durability; that is, the optical parameters should be within specifications after many connector operations. Repeatability and durability generally are a function of the most accurate part of a connector: the ferrule. The ferrule must have a highly accurate outside diameter, concentricity, and hole diameter. This minimizes loss due to offset and tilt. Most connectors are available with ferrules made of alumina ceramic, zirconia ceramic, metal alloy, and plastic.

Alumina ceramic holds tight tolerances quite well but is perceived as being less rugged than other materials. Zirconia ceramic is somewhat more rugged while still holding tight tolerances. Both ceramics, however, have the potential for allowing glass undercutting during the final polishing stages because they are harder than glass. If the undercut is not too extensive, certain polishing papers can be used to repair the connector. If the glass is polished too far, it will wear away and create an air gap when two connectors are mated. The air gap may cause wide variations in both loss and return loss. Also, if loss is unacceptable due to inadequate polishing of the glass, no further work is possible and the connector must be completely replaced.

Ferrules made of metal alloys, such as stainless steel, do not have the same thermal properties as ceramics but generally perform as well, over a −40 °C to + 55 °C temperature range. For colder or warmer temperatures, ceramic ferrules generally are specified. Metal alloy ferrules do not normally undercut; the metal wears faster than the glass as it is polished.

Plastic ferrules presently are not used in high-performance connectors in loop applications. As materials technology improves, this should change.

In a pull-proof connector, the ferrule is decoupled from the connector body. Since the fiber optic cable is anchored to the body, the ferrule maintains optical contact when the cable is pulled outward or sideways. In a connector without adequate pull-out resistance, optical contact is broken or degraded when the cable is pulled. This can happen whenever work is being performed in a fiber optic patch or connector panel, especially with closely spaced connectors. Pull-out resistance becomes increasingly important when connectors are in outside plant equipment and not specially protected. The SC and FC connectors are designed to have high pull-out resistance, whereas the standard ST connector is not. Some versions of the ST connector are pull-proof.

The SC push/pull connector is available in single and dual configurations. In the dual configuration, one connector body holds both the transmit and receive fibers. The standard ST and FC connectors are available only in the single configuration; therefore, two separate connectors are required for a duplex connection. Some ST connectors are available in duplex configurations.

Ease of use is an obvious requirement, and the SC connector is perceived to be superior in all respects. Its push/pull operation allows easy connection and disconnection in close quarters. The ST and FC connectors require a twisting action, which is difficult in close quarters or with cold hands. Nevertheless, the ST and FC are popular and widely used connector types.

8.10 Couplers

The bandwidth of any FITL transmission system is inherently very high and is much higher than any single user requires at the present time. This bandwidth is allocated to many users. The tradeoff between signal transmission to a number of users and the complexity of transmission and power loss to those users is a typical engineering problem. Couplers play a key role in passive optical networks (PONs) such as are proposed in the industry standard for FITL systems [1]. Figure 8-23 shows a typical application in a PON.

A coupler couples multiple optical signal inputs or outputs, or both. A coupler normally is bidirectional and can be used either as a splitter or combiner. In the splitter mode, the coupler splits one incoming signal into two or more paths at its output. The reverse is true of a combiner. The combiner combines two or more optical inputs into a single output. The coupler also can be built with more than one input (for a splitter) or more than one output (for a combiner). The splitting or combining can be done on a wavelength basis or signal basis. When a coupler splits or combines on a wavelength basis, it is called a wavelength division multiplexer (WDM) or wavelength division demultiplexer (WDDM).

The *tree coupler* splits a single optical input into multiple outputs, as shown in Figure 8-24(a), or combines multiple inputs into one, as shown in Figure 8-24(b). Tree couplers normally split the power equally among all output ports. Therefore,

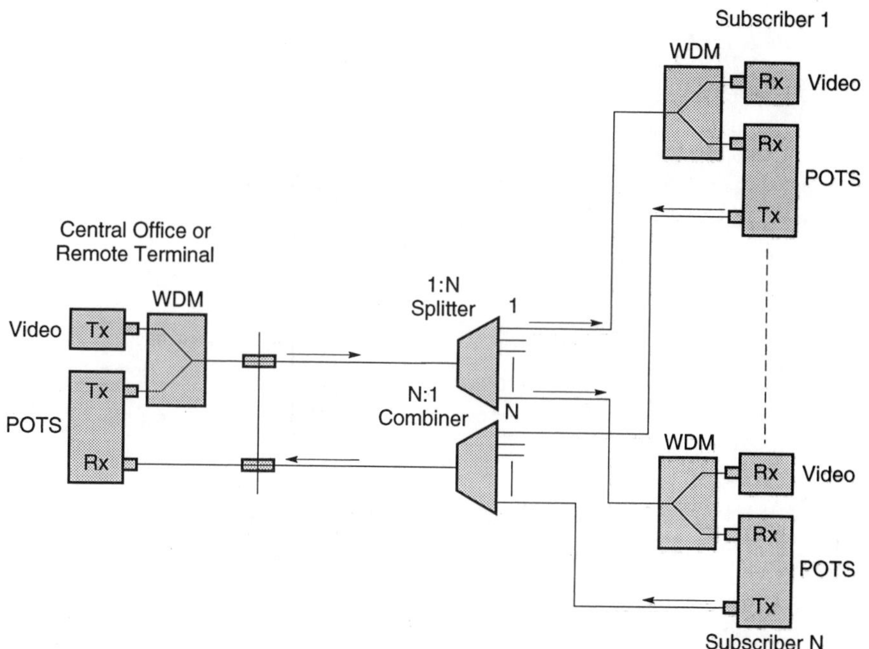

Figure 8-23 Typical coupler application in a passive optical network

for a 1:n coupler (one input port to n output ports) with input power P_i, the output power at any output port will be P_i/n. Typical values for n are powers of two, between 2 and 32, although the normal maximum value for n is eight. An alternate designation can be used; instead of 1:n the configuration can be indicated as $1 \times n$ (read "1 by n"). In a 1:2 tree coupler, 50% of the power is coupled to each output port.

Tap couplers, shown in Figure 8-24(c), normally are configured with one input port and two output ports (1:2) but have an unequal splitting ratio between the output ports. This means that more of the input power is coupled to one output port than the other. Common splitting ratios are in 5% increments, between 5 and 95% to the tap port. For example, with a 5/95 splitting ratio, 5% of the input power is coupled to one output port and 95% is coupled to the other port.

The *star coupler* shown in Figure 8-24(d) is similar to a tree coupler, except that it usually has an equal number of input and output ports (n:n or $n \times n$); but this is not necessary. When the input and output ports are not equal, the designation is m:n or $m \times n$. Star couplers are bidirectional, but to establish a consistent reference, the input ports are labeled alphabetically and the output ports are labeled numerically, as shown in the illustration.

The WDM splits or combines signals of different wavelengths, as shown in Figure 8-24(e), (f), and (g). In (e), two wavelengths on two separate input ports are combined onto a single output port. This is used, for example, to combine the output from a system operating at 1,310 nm with another system operating at 1,550 nm

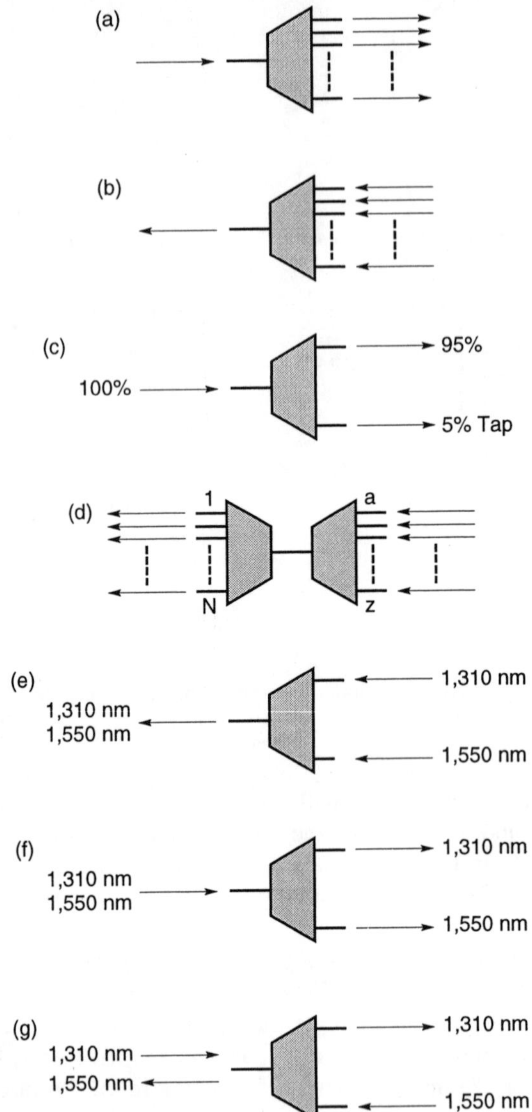

Figure 8-24 Passive couplers: (a) Tree coupler as a splitter; (b) Tree coupler as a combiner; (c) Tap; (d) Star coupler; (e), (f), (g) Wavelength division multiplexer

onto a common fiber. The WDM coupler shown in (f) performs the inverse function of splitting the two wavelengths on one input port, to individual wavelengths on each of two output ports. A combination WDM coupler, shown in (g), provides bidirectional wavelength coupling. In this configuration, one wavelength is coupled into one port while the other wavelength is coupled out of the other port. Both wavelengths appear at the common port.

Couplers are not ideal devices, which means the coupling parameters (ratios and losses) are not precise. The parameters of interest in coupler specifications are:

- Passband
- Insertion loss
- Uniformity
- Directivity
- Far-end crosstalk isolation
- Back-reflection

The passband is the wavelength range over which the coupler should meet specified requirements and usually is specified as a range within the wavelength window or windows of interest. In telecommunications applications, either the second or third window (1,310 or 1,550 nm), or both, are used. The actual wavelength range within the window needed for any given system depends on the spectral widths of optical sources, but usually is ±50 nm for either of the windows, as shown in Table 8-20. A coupler that shows little dependence on wavelength is said to be *achromatic*.

Coupler insertion loss is a measure of the ratio between the total input power to the total output power. Since a coupler has more than just one input and one output port, consideration must be given to the loss through to each port. The coupler loss mechanisms are given in Table 8-21 along with typical values.

The basic power splitting loss is simple to determine for any coupler configuration, as previously mentioned. For a 1:2 coupler, the basic power split is 50%

Table 8-20 Typical Coupler Performance (Dual-Window Type)

Item	1:2 Splitting Ratio	1:4 Splitting Ratio	1:8 Splitting Ratio
Passband	1,310 ± 50 nm 1,550 ± 50 nm	1,310 ± 50 nm 1,550 ± 50 nm	1,310 ± 50 nm 1,550 ± 50 nm
Basic splitting loss	3 dB	6 dB	9 dB
Insertion loss at 1,310 nm	≤3.8 dB	≤7.8 dB	≤11.5 dB
Insertion loss at 1,550 nm	≤4.0 dB	≤7.8 dB	≤11.5 dB
Maximum insertion loss at 1,310 nm	3.9 dB	7.8 dB	11.7 dB
Uniformity	≤0.4 dB	≤1.0 dB	≤1.5 dB
Directivity	40 to 60 dB	40 to 60 dB	40 to 60 dB
Back-reflection loss	Typically 45 dB	Typically 45 dB	Typically 45 dB

Table 8-21 Optical Coupler Loss Mechanisms

Loss Mechanism	Typical Value
Basic power splitting loss	10 log (1/n) dB
Excess loss	0.2 dB
Power-splitting nonuniformity	0.5 dB
Wavelength dependencies	1 dB

(3 dB loss) into each port; for a 1:4 coupler, the basic power split is 25% (6 dB loss) into each port; and so on. However, excess losses, nonuniformity, and dependencies on wavelength and polarization increase the basic power splitting loss through to each port. Table 8-20 shows typical insertion loss values for commercially available couplers, and the maximum allowable insertion loss, according to industry standards [1].

Coupler uniformity indicates the insertion loss difference between output ports. It is difficult to make couplers with high splitting ratios and high uniformity. This is shown in Table 8-20, where the uniformity for a 1:2 coupler is ≤0.4 dB but for a 1:8 coupler is ≤1.5 dB. This means that for the 1:2 coupler, the insertion loss to one output port could be as high as a 0.4 dB difference from the loss to the other port. Similarly, for the 1:8 coupler, the loss difference could be as high as 1.5 dB.

Coupler directivity measures how well the coupler transmits the optical input power to the desired output port. In bidirectional systems, misdirected signals can lead to crosstalk or other undesirable interference effects. Directivity also can be considered in the context of back-reflections (back scatter) from one port interfering with another port. Figure 8-25 shows what is meant by directivity and how it can be affected by reflection. Laser diodes are particularly sensitive to reflection.

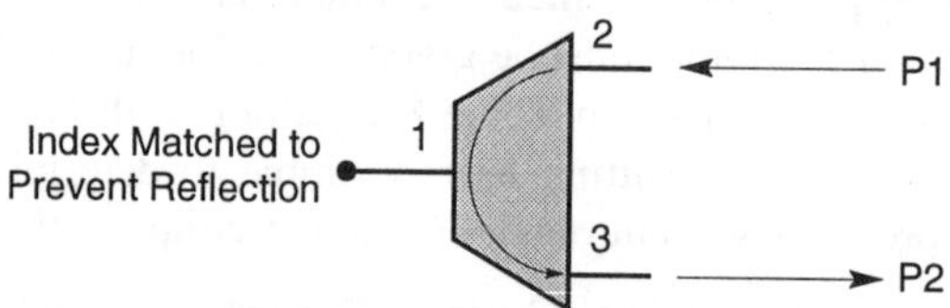

Figure 8-25 Coupler directivity

A signal coming in port 2 should pass through to port 1, and (ideally) none of the signal from port 2 should be coupled to port 3. In reality, the directivity typically is 50 dB between ports 2 and 3, which means the coupling loss between ports 2 and 3 is 50 dB. However, due to back-reflection from port 1, part of the signal from port 2 is reflected back to port 3. If the back-reflection—say, due to deformities in the coupler or a reflective splice or connector—is 30 dB (a not uncommon value), the actual signal from port 2 coupled to port 3 is attenuated by only 36 dB. The extra 6 dB comes from the basic splitting loss from port 2 to port 1 and from port 3 to port 1. This illustrates the importance of minimizing the reflections at coupler ports. Coupler directivity is given by:

$$\text{Directivity} = 10 \log \left| \frac{P_1}{P_2} \right| \text{ dB}$$

Far-end crosstalk (FEXT) isolation is a significant parameter with WDM couplers. It measures the isolation between two wavelengths at one of the output ports. This is illustrated in Figure 8-26 for a 1:2 WDM. Here, signals composed of two wavelengths enter the input port and are split, one wavelength to each output port. The FEXT isolation is the amount of the undesignated wavelength at one of the output ports. In the figure, 1,310 nm is the designated wavelength at port 2 and 1,550 nm,

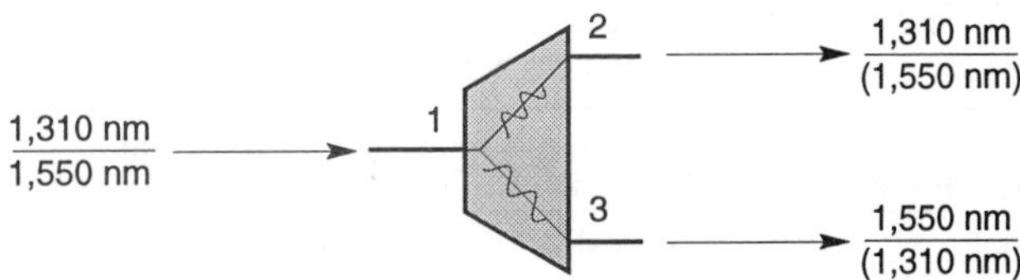

Figure 8-26 FEXT coupling in a 1:2 wavelength division multiplexer

at port 3; however, some energy at 1,550 nm appears at port 2. The FEXT coupling loss for typical WDM couplers is 25 dB; that is, the signal at 1,550 nm will be attenuated by 25 dB at port 2. FEXT coupling loss is given by:

$$\text{FEXT} = 10 \log \left| \frac{P_{1,1550}}{P_{2,1550}} \right| \text{ dB}$$

where $P_{1,1550}$ = power with wavelength of 1,550 nm at port 1
$P_{2,1550}$ = power with wavelength 1,550 nm at port 2

A similar expression may be written for the FEXT at port 2 due to the wavelength 1,310 nm.

8.11 Optical Sources and Detectors

Optical transmitters and receivers are built into line interface units on fiber optic terminals and multiplexers. These devices are covered here because a fiber loop system cannot be designed without full coordination between the line interfaces and transmission media previously discussed. Although there are a wide variety of commercial devices, the exact types or configurations usually are determined by the manufacturer of the line interface unit. With some equipment, two choices are given; for example, a high- or low-power transmitter. When choices are given, they usually are to allow the designer to optimize the system gain or life based on route length, bit rate, or temperature performance.

Optical transmitters for singlemode fiber loop applications can be categorized as light-emitting diode (LED) and laser diode (LD). Typical data are summarized in Table 8-22. The wavelengths given are nominal values. Optical transmitters used in fiber loop applications can operate over a range of wavelengths, but this range is restricted to 1,260 to 1,360 nm for the 1,310-nm region, and to 1,430 to 1,580 nm for the 1,550-nm region [1]. As would be inferred from the power output shown in Table 8-22, the LED is restricted to relatively short fiber loops when compared to the laser diodes. The signal output from an LED transmitter has such a wide beamwidth that relatively little power is coupled to the fiber. LD transmitters have a narrow beamwidth and high coupling efficiency.

In general, the power output and other electrical/optical characteristics of an optical transmitter are specified at the output connector or pigtail and should include

Table 8-22 Optical Transmitters Used in Singlemode Fiber Loop Applications[a]

Parameter	1,310 nm LED	1,310 nm Laser Diode	1,550 nm Laser Diode
Material	GaInAsP	GaInAsP	GaInAsP
Spectral width	50–150 nm	—	—
Line width	—	0.002–150 MHz	0.002–150 MHz
Output power	0.4–0.6 mW	0.5–8 mW	0.5–8 mW
Coupled power	0.003–0.03 mW	—	—
Extinction ratio	—	25:1	25:1
Drive current	100–150 mA	25–130 mA	—
Rise time	2.5–10 nseconds	0.3–0.7 nseconds	0.3–0.7 nseconds
Temperature drift of wavelength	0.6 nm/°C	0.3 nm/°C	0.3 nm/°C
Temperature drift of power	−0.9 %/°C	—	—
Lifetime	5E7–1E9 hours	5E5–5E7 hours	5E5–5E7 hours
Degradation over life	1–3 dB	1–3 dB	1–3 dB

[a]Adapted from [41]

the effects of aging and temperature. Aging and temperature can be accounted for separately, if necessary, provided these effects are known. When not otherwise specified, 3 dB power penalty usually is used for aging and temperature. Industry standards are available for optical transmitters used in fiber loop applications [40].

Optical receivers are either of PIN (positive-intrinsic-negative) photodiode or APD (avalanche photodiode) construction. The main difference of concern to the designer is that the PIN detector does not have gain built into it, whereas the APD does. As a result, the APD generally is more sensitive than the PIN diode. Table 8-23 compares the attributes of typical optical detectors, and Figure 8-27 shows the sensitivities of various optical receivers for a BER of 1E-9. Industry standards are available for optical detectors used in fiber loop applications [42].

The most important receiver parameter is minimum *detectable* power (MDP) for a given bit error rate, or BER. This is the power required at the semiconductor

Table 8-23 Optical Detectors Used in Singlemode Fiber Loop Applications

Parameter	PIN Photodiode	Avalanche Photodiode
Material	Ge or InGaAs	Ge or InGaAs
Nominal wavelength	1,310 or 1,550 nm	1,310 or 1,550 nm
Minimum sensitivity at BER = 1E-9	See Figure 8-27	See Figure 8-27
Dynamic range	30 dB	30 to 40 dB
Sensitivity de-rating for temperature	+0.05 dB/°C	+0.05 dB/°C
Degradation over life	None	None

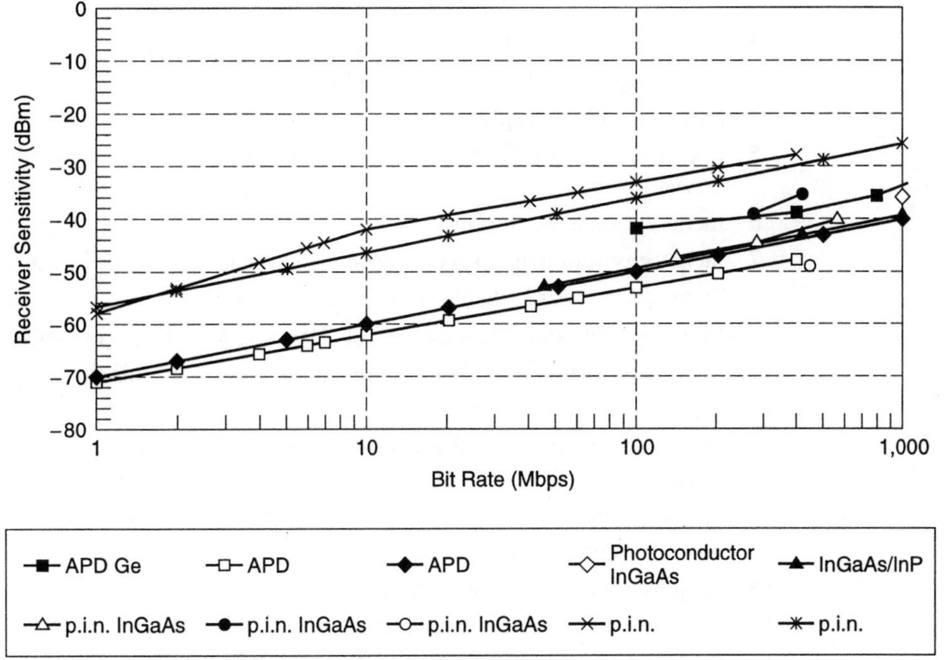

Figure 8-27 Optical receiver sensitivity at BER = 1E-9
Data taken from [3,4,43]

device. The minimum required *receiver* power (MRP), or sensitivity, always is higher by an amount equal to the fiber-to-detector coupling loss L_c, as given by:

$$MRP(dBm) = MDP(dBm) + L_c(dB)$$

Coupling losses include [41]:

- Connector and pigtail loss
- Loss due to the coupling geometry, including relative diameters, distance, and numerical aperture
- Fresnel reflections

8.12 Fiber Optic Cable Purchase Checklist

The checklist that follows is useful during the development of fiber optic cable purchase specifications.

1. Specification? (ICEA, REA, BELLCORE, EIA) ☐
2. Deviations from the referenced specification? (specify) ☐
3. Buffering type? (loose tube or tight tube) ☐
4. Fiber count? (multiples of 2) ☐
5. Use? (direct buried, under ground, aerial, riser, aerial self-supporting, service drop) ☐
6. Jacket? (single or double) ☐
7. Jacket environment? (severe, normal, submarine, rodent infested) ☐
8. Strength member? (dielectric or nondielectric) ☐
9. Nominal operating wavelength? (1,310 nm, 1,550 nm, or both) ☐
10. Maximum individual fiber loss or attenuation cell? (dB/km) ☐
11. Dispersion shifted? (yes, no) ☐
12. Mode field diameter and tolerance? (specify manufacturer's standard or custom) ☐
13. Special requirements? (specify: for example, tests, temperature range) ☐
14. Armor? (heavy copper, steel, copper alloy) ☐
15. Lengths required? (standard, custom; specify) ☐
16. Reel size or weight limitations? (specify) ☐
17. Reel requirements? (standard, wood, steel) ☐
18. Reel protection requirements? (none, wood lagging, pressboard) ☐
19. Shipping instructions? (specify) ☐

References

[1] *Generic Requirements and Objectives for Fiber in the Loop Systems*. BELLCORE Technical Reference TR-NWT-000909, Dec. 1991. Available from BELLCORE Customer Service.

[2] Palais, J. *Fiber Optic Communications*, 2nd Ed. Prentice Hall, 1988.

[3] Miller, S., Kaminow, I. *Optical Fiber Telecommunications II*. Academic Press, 1988.

[4] Tosco, F., ed. *Fiber Optic Communications Handbook*, 2nd Ed. Technical Staff of CSELT, TAB Professional and Reference Books, 1990.

[5] Insulated Cable Engineers Association. *Standard for Fiber Optic Outside Plant Communications Cable*. ANSI/ICEA S-87-640-1992.

[6] Rural Electrification Administration. *REA Specification for Totally Filled Fiber Optic Cable*, REA Bulletin 345-90. PE-90, May 1986.

[7] *Generic Requirements for Optical Fiber and Optical Fiber Cable*. BELLCORE Technical Reference TR-TSY-000020, March 1989.

[8] *Generic Requirements for Intrabuilding Optical Fiber Cable*. BELLCORE Technical Reference TR-TSY-000409, Sept. 1990.

[9] *Generic Requirements for Optical and Optical/Metallic Buried Service Cable.* BELLCORE Technical Reference TR-TSY-000843, Jan. 1989.

[10] *Generic Requirements for Optical Distribution Cable.* BELLCORE Technical Reference TR-TSY-000944, July 1990.

[11] Electronic Industries Association. *Generic Specification for Fiber Optic Cables.* ANSI/EIA-472-85, March 1985.

[12] Electronic Industries Association. *Blank Detail Specification for Fiber Optic Communication Cable for Outside Plant Use—All Dielectric.* ANSI/EIA-472DA00, June 1992.

[13] Electronic Industries Association. *Sectional Specification for Fiber Optic Communication Cables for Outside Aerial Use.* ANSI/EIA-472A-85, Oct. 1985. (Note: This sectional specification is associated with Blank Detail Specification ANSI/EIA-472AXX0.)

[14] Electronic Industries Association. *Sectional Specification for Fiber Optic Communication Cables for Underground and Buried Use.* ANSI/EIA-472B-85, Oct. 1985. (Note: This sectional specification is associated with Blank Detail Specification ANSI/EIA-472BXX0.)

[15] Electronic Industries Association. *Sectional Specification for Fiber Optic Communication Cables for Outside Telephone Plant Use.* ANSI/EIA-472D-85, Oct. 1985. (Note: This sectional specification is associated with Blank Detail Specification ANSI/EIA-472DXX0.)

[16] "Rodent Protective Sheath Is Necessary For Most Fiber Optic Cables." Lightguide Digest, No. 2. AT&T Network Systems, 1992.

[17] Gulati, S. *Large Flaws: The Culprit in Fiber Reliability.* Photonics Spectra magazine, Dec. 1992.

[18] Pender, H., Del Mar, W. *Electrical Engineers' Handbook—Electric Power.* John Wiley & Sons, 1949.

[19] *Fiber Toughens Up.* Corning, Inc. Guidelines magazine, Vol. 6, No. 1, 1991.

[20] "Tests Show Aerial Plant Fiber Performs Well In Lightning." Lightguide Digest, No. 3. AT&T Network Systems, 1992.

[21] Institute of Electrical and Electronics Engineers. *National Electrical Safety Code.* ANSI C2.1990. Available from IEEE Service Center.

[22] *Generic Requirements for Universal Splice Closures for Fiber Optic Cables.* BELLCORE Technical Reference TR-TSY-000771, April 1991.

[23] *Generic Requirements for Distribution/Service Closures Used with Fiber Optic Cables.* BELLCORE Technical Reference TR-TSY-000950, Sept. 1990.

[24] *Standard for Physical Location and Protection of Below-Ground Fiber Optic Cable Plant.* ANSI/EIA/TIA-590-1991, June 1991.

[25] Pope, D. *Duct Liner As A Buried Conduit? Be Picky About What You Bury.* Telephony, Aug. 26, 1991.

[26] *Optical Fiber Ductliner.* BELLCORE Technical Reference TR-TSY-000356, Dec. 1986.

[27] SMF-28™ CPC3 Single-Mode Optical Fiber. Product Information PL-11, Corning, Inc., Nov. 1990.

[28] *Transmission Media—Characteristics*, Vol. III, Fascicle III.3. Recommendations G.601–G.654 (Study Group XV), CCITT Blue Book, 1989.

[29] *Splicing Systems for Single-Mode Optical Fibers*. BELLCORE Technical Reference TR-TSY-000765, Dec. 1989.

[30] Bradley, S. "The Role of Fiberoptic Mechanical Splices—Mechanical Splicing May Dominate Short Distance Networks." Fiberoptic News, July 1992.

[31] *Single Fiber Fusion Splicing*. Corning™ Optical Fiber Application Note AN-121, Corning, Inc., June 1990.

[32] Robinson, S. M. *CORGUIDE® SMF28™ Fiber Splice Characterization Study*, Feb. 1987.

[33] "AT&T CSL LightSplice System Goes Afield." Lightguide Digest, reissued No. 2. AT&T Network Systems, 1990.

[34] Midkiff, J. "Mechanical Splicing Drives Need for Local Loss Measurements." Lightwave magazine, Aug. 1991.

[35] Personal communication with Dan Davila, Siecor Corp.

[36] "Effects of Mode Field Diameter Mismatch on Single Mode Fiber Splice Loss and OTDR Splice Loss Measurement Error." Lightguide Digest, No. 2. AT&T Network Systems, 1991.

[37] Electronic Industries Association. *Generic Requirements for Fiber Optic Connectors*. ANSI/EIA/TIA-4750000-B-89, Aug. 1989.

[38] Electronic Industries Association. *Sectional Specification for Fiber Optic Connectors—Type BFOC/2.5*. ANSI/TIA/EIA-475E000-92, June 1992. (Note: This specification is associated with Blank Detail Specification ANSI/TIA/EIA-475EA00-92, -475EB00-92, and -475EC00-92.)

[39] *Generic Requirements for Optical Fiber Connectors and Connectorized Jumper Cables*. BELLCORE Technical Reference TR-TSY-000326, March 1991.

[40] *Optical Source Module for Subscriber Loop Distribution*. BELLCORE Technical Reference TR-TSY-000786, Dec. 1989. Available from BELLCORE Customer Service.

[41] Hoss, R. *Fiber Optic Communications Design Handbook*. Prentice-Hall, 1990.

[42] *Optical Detector Module for Fiber in the Loop Systems*. BELLCORE Technical Reference TR-TSY-001092, Sept. 1990. Available from BELLCORE Customer Service.

[43] Killen, H. *Digital Communications with Fiber Optics and Satellite Applications*. Prentice-Hall, 1988.

9 Transmission Engineering, Part I—Repeatered T1-Carrier

Chapter 9 Acronyms

A/D	analog-to-digital
ABAM	cable type: air core 83 nF/mile, solid polyolefin insulated, 22 AWG, aluminum-shield cable
ABMM	Cable type: air core 83 nF/mile, solid polyolefin insulated, 24 AWG, aluminum-shield cable
AIS	alarm indication signal
ALBO	automatic line build-out
AMI	alternate mark inversion
AML	actual measured loss
APIC	air core PIC
APS	automatic protection switch, -ing
AWG	American wire gauge
B8ZS	bipolar with 8-zero substitution
BER	bit error rate
BPV	bipolar violation
CDF	combined distribution frame
COR	central office repeater
CRC6	6-bit cyclic redundancy check
CSU	channel service unit
D/A	digital-to-analog
DACS	digital access and cross-connect system (*see* DCS)
dBrnC	decibels with respect to reference noise, C-message weighted
DCS	digital cross-connect system (*see* DACS)
DEPIC	dual-expanded foam PIC
DLC	digital loop carrier
DSX	digital signal cross-connect
DTE	data terminal equipment
ESF	extended superframe format
FEXT	far-end crosstalk
FLBO	fixed line build-out
FPIC	filled PIC
HDSL	high bit-rate digital subscriber line
ICOT®	intercity and outstate trunk
ISDN	integrated services digital network
LOS	loss-of-signal
MAT	metropolitan area trunk
MDF	main distribution frame
MON	monitor (jack)
MUX	multiplexer
NEXT	near-end crosstalk
NPL	noise path loss
OOF	out-of-frame
ORB	office repeater bay
PIC	polyolefin insulated cable
QRSS	quasi-random signal source
REA	Rural Electrification Administration
RST	remote switching terminal
Rx	receive
SF	superframe format
SNR	signal-to-noise ratio
TSI	timeslot interchange
Tx	transmit
VF	voice frequency
ZBTSI	zero-byte timeslot interchange

Transmission engineering consists of interference considerations, loop design, and loop qualification. To make the description of these processes manageable, the transmission engineering topic is broken into two parts. The first part (this chapter) is singularly devoted to twisted metallic pair repeatered T1-carrier span lines used as digital loops. It will become apparent that designing repeatered T1 lines is not a trivial task if it is to be done properly. The second part (next chapter) covers the design of subrate digital loops, integrated services digital network (ISDN) digital subscriber lines, the high bit-rate digital subscriber lines (HDSLs), and fiber-in-the-loop.

9.1 Repeatered T1-Carrier Span Lines*

Although T1-carrier originally was designed for interoffice applications, it soon found widespread use with digital pair gain devices associated directly with the subscriber or user. As will become apparent from the following sections, the design process for repeatered T1-carrier span lines is rather involved, which makes deployment of this technology expensive. Nevertheless, repeatered T1-carrier will be used in digital loop applications well into the future.

It is interesting that the transmission design requirements for repeatered T1-carrier span lines have remained virtually unchanged since their introduction in the 1960s, even in the face of vast technological advancements in repeaters and multiplexers. This illustrates one of the most fundamental requirements in telecommunication systems design: backward compatibility. The terms *span* and *span line* are introduced in Figure 9-1. No effort is made in this chapter to differentiate between the two.

9.1.1 Interference Considerations

Interference in repeatered T1-carrier systems is a complex issue and is highly dependent on cable types. There is a large variety of exchange and trunk cable types being used by the telecommunications industry. The characteristics of each type must be considered during the span line design process. This book emphasizes loop systems; therefore, cables for trunk systems will only be mentioned below and in some design tables to be given later for completeness. It is unlikely that trunk cables will be encountered in digital loop carrier (DLC) design.

In most DLC applications, exchange or compartmental core cables will be used. Exchange cables consist of pulp/paper insulated cables and polyolefin insulated ca-

---

*Fractional T1 may be considered a subset of repeatered T1 and HDSL systems. However, using fractional T1 in the loop presently is a bandwidth allocation issue and not a transmission design issue. Between the network termination on the user's premises and the serving central office, fractional T1 services are transported on conventional T1-carrier span lines or HDSL. The user is allocated only that part of the full DS-1 rate bandwidth, as desired. Therefore, transmission design of the loop portion of fractional T1 is the same as for repeatered T1-carrier described in this chapter and HDSL described in the next chapter.

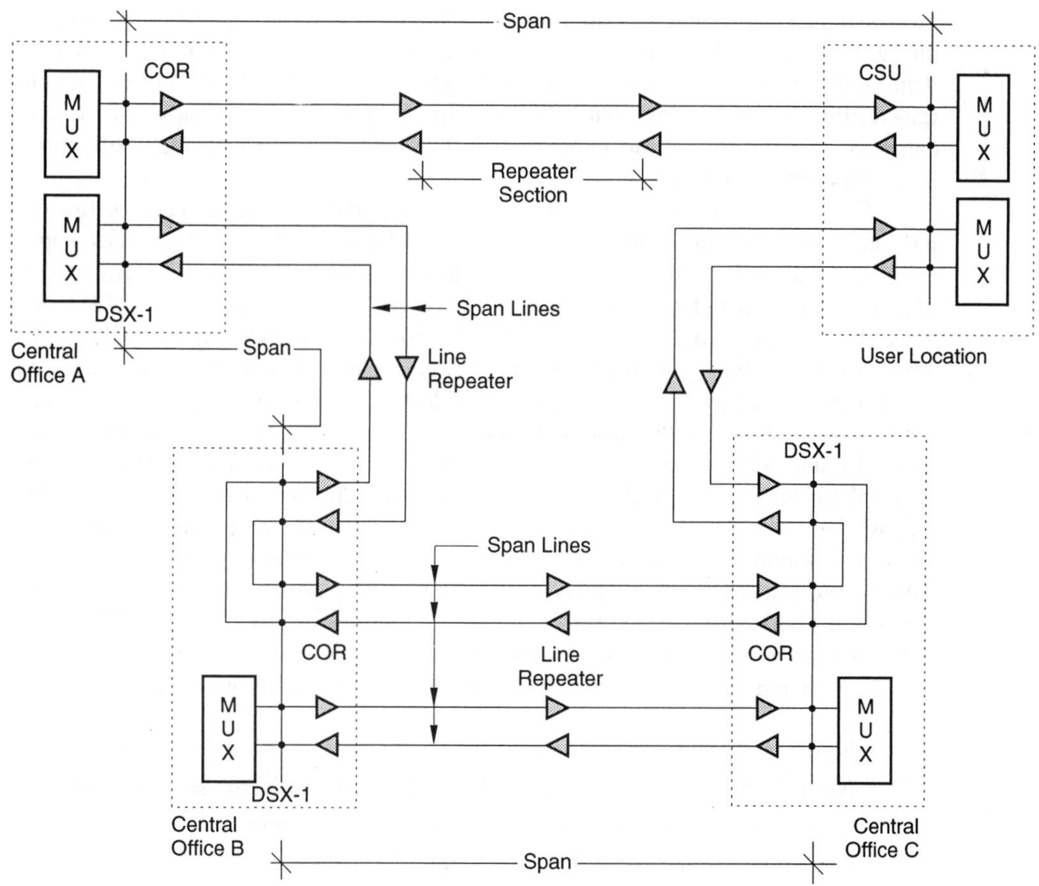

Figure 9-1 Repeatered T1-carrier spans and span lines

bles (PIC). To differentiate the various PICs further, the following nomenclature will be used in this chapter: cables with solid insulation will be simply referred to as PIC while cables with expanded foam plastic (foam skin) insulation will be referred to as DEPIC (dual-expanded foam PIC).

All DEPIC is filled (waterproof). PIC can be either air core PIC (APIC) or filled PIC (FPIC).* Compartmental core cables are called by various names, such as D-screen or T-screen as discussed in Chapter 7. Compartmental core cables are always a type of PIC and, except for the core separator and pair counts, are identical to exchange cables.

Pair capacitance varies with the cable type. All exchange PICs are nominal 83 nanofarad (nF) per mile (15.7 nF/kilofoot), whereas the capacitance of paper and

*The APIC and FPIC nomenclature is unique to this book. It is used to avoid confusion arising from the many different trade names and coding systems used by different manufacturers and large users. Appendix A includes information that may be used to compare one of the larger manufacturers' codes with the nomenclature used in this book.

pulp insulated cables can be a nominal 66, 72, 73, 79, 83, or 84 nF/mile, depending on the type. The actual capacitance of installed paper/pulp cables is highly dependent on the amount of moisture in them. Moisture also affects APIC, but well-maintained FPIC is virtually impervious to moisture. In general, it is not advisable to mix cables of widely different capacitances on a span line; this is so as to avoid reflections from impedance mismatches.

The trunk cables mentioned above are compartmental core (screened), low capacitance, twisted pair cables specially designed for interoffice digital carrier applications, including T1, T1C, T1D, and T1G (this nomenclature indicates general classes of digital transmission systems; T1D and T1G are little used outside of AT&T and the Bell Operating Companies). AT&T refers to these cables as Metropolitan Area Trunk (MAT) cables and InterCity and Outstate Trunk (ICOT®*) cables, but they are generically called "Lo-Cap" cables. The MAT cable is an air core 25 American wire gauge (AWG) DEPIC designed specifically for metropolitan trunk applications using T1 and T1C. Available sizes for MAT cables are 400 to 1,800 pairs, and nominal capacitance is 64 nF/mile. ICOT® is either air core or filled core 24 AWG DEPIC designed for suburban and rural trunk systems. Available sizes are 50 to 900 pairs, and nominal capacitance is 52 nF/mile (air core) or 60 nF/mile (filled). Both cables have lower loss than 24 AWG exchange cables because of the lower capacitance, and both allow 100% fill of T1 or T1C systems. MAT and ICOT® should not be mixed with exchange cables on T1 or T1C span lines.

Repeatered T1-carrier span line design requires the determination of:

- Number of span lines ultimately installed
- Number, types, and compatibility of other services in the same cables
- Choice between bidirectional or unidirectional repeaters
- Choice between 1- or 2-cable operation
- Suitability and year-round accessibility of repeater locations
- Repeater section length
- Number, type, age, and condition of cables, including entrance cables and premises wiring used to terminate span lines
- Uniformity of cable types from terminal to terminal
- Locations of branches and route junctions
- Location of manholes, pedestals, and control points in terms of accessibility for construction and maintenance
- Splicing configurations and integrity for one-cable operation
- Exposure of span lines and repeaters to electrical and mechanical hazards

Where an existing route consists of several cables, the newer ones should be chosen for repeatered T1-carrier, and PIC should be chosen over pulp or paper insulated cables. Older cables generally should be avoided because they have a history

*ICOT is a registered trademark of Western Electric.

of more trouble, and lower performance and reliability. Cables selected for T1-carrier should:

- Not have paraffin or twisted splices
- Not be more than 25 years old
- Extend from distribution frame to distribution frame (that is, should avoid intermediate terminations)
- Be free of hazards to service continuity (for example, vehicle damage, flooding, earthquake damage, vandalism)
- Be chosen after examining cable maintenance records, if available, and after discussing cable condition with construction and maintenance forces
- If air core, be pressurized and completely dry and have no history of flooding or moisture in the core

If an existing cable is chosen, all bridged taps, build-out capacitors, load coils, and unterminated and unused cable stubs must be removed. Older cables (installed prior to the 1970s) generally will require resplicing to provide the required integrity for one-cable operation. At the very least, a sample of the splices in older cables should be opened to ascertain splicing consistency and arrangement in accordance with current practice. It is important that, if resplicing is necessary, all splices be redone before the first T1-carrier installation to avoid subsequent service disruptions. Even in 2-cable operations, resplicing may be necessary if existing splices are located close to repeater locations (within approximately 1,000 feet). This will help to eliminate marginal repeater operation due to impedance irregularities and the resulting reflections.

Where the luxury of multiple cables and cable routes is available, the most permanent and accessible route should be chosen. It is desirable, but not absolutely necessary, that each of the two cables in 2-cable operation follow the same route. If not, additional maintenance costs will be incurred because the repeaters for each direction will be in different locations. There are no diversity advantages in having the two transmission directions follow different routes (a span line is useless unless both transmission directions are functional).

Where many cables follow the same route and are potential candidates for T1-carrier span lines, rearranging some of the cables to obtain uniform size and gauge from termination to termination should be considered. If these same cables already have some T1-carrier span lines or other important services, rearrangements should be made only after considering the need for service reliability and continuity on existing services.

Entrance and protector terminating cables (also called stub or tip cables) can present a particular problem because of the common practice of bringing both the transmit and receive directions into the central office in a single cable of relatively fine gauge. Using a single cable increases the chances for harmful crosstalk, and the fine gauge complicates the calculation of end-section loss. However, if these cables are short (less than a few hundred feet), they will have negligible effect. To avoid future problems, new entrance and terminating (tip) cables should be spliced to contain only one transmission direction.

Where the installation of new cables is contemplated, compartmental core PIC should be used. The cables should not be terminated anywhere except at each end of the span lines. It is the practice of some companies to directly splice exchange cable feeder pairs assigned to T1-carrier service that pass through cross-connect cabinets (serving area interfaces). This means the T1-carrier pairs are neither cross-connected nor available for cross-connect. This reduces disruptions by maintenance personnel but complicates feeder pair administration over the life of the cable. Other companies use specially marked or color-coded jumpers in cross-connect cabinets for the T1-carrier pairs. This alerts maintenance personnel to the special nature of the cable pairs.

Along new routes, even if the requirement for T1-carrier is not foreseen, repeater locations should be planned and provisions made. This requires very little extra engineering and construction effort in most cases, and the slightly increased costs will be small compared to later rearrangement and resplicing costs. New cables must be installed without any bridge taps or line conditioning devices on the pairs to be used for T1-carrier.

Due to crosstalk coupling, T1-carrier span line design is highly dependent on whether both transmission directions are in one cable (1-cable operation as shown in Figure 9-2(a)) or one direction in each of two cables (2-cable operation, as shown in Figure 9-2(b)). Therefore, one of the early decisions in span line engineering is whether to use one or two cables.

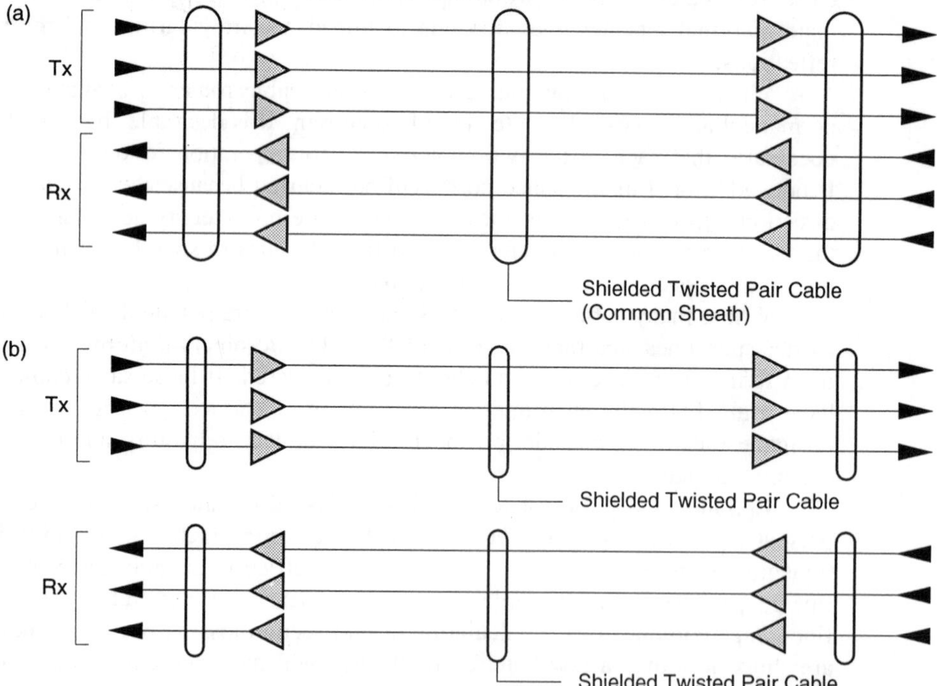

Figure 9-2 Cable operation: (a) 1-cable operation; (b) 2-cable operation

In 1-cable operation, both transmission directions are in one cable. In 2-cable operation, the transmit direction is in one cable and the receive direction in another, physically separate cable. Operation in compartmental core (screened) cables is equivalent to 2-cable operation. The design of repeater sections using compartmental core cables is identical to design based on 2-cable operation.

Crosstalk interference, specifically near-end crosstalk (NEXT) between T1 lines, generally is a significant problem only with 1-cable operation. The design considerations used to control NEXT are described later. If two cables, or a compartmental core (screened) cable are used, NEXT does not enter into line design process.

Far-end crosstalk (FEXT) always is present in cable pairs used for the same transmission direction, and this can occur in both 1- and 2-cable operations. Three FEXT conditions may exist in an outside plant, as shown in Figure 9-3. In Figure

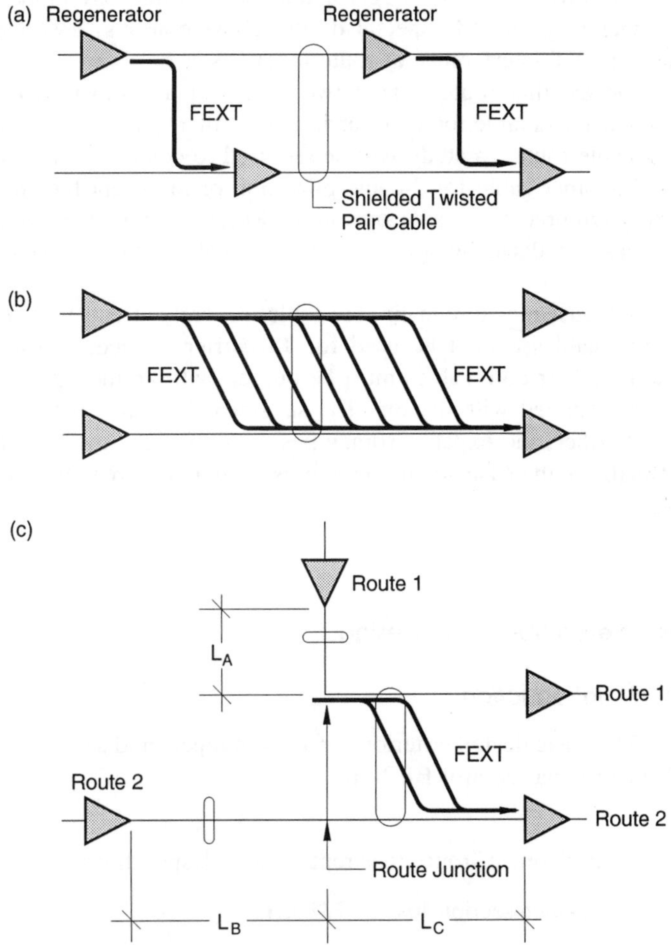

Figure 9-3 FEXT coupling in digital lines: (a) FEXT with misaligned repeater housings; (b) FEXT over entire section length; (c) FEXT at route junctions

9-3(a), the output of a regenerator in one housing is coupled into the input of another regenerator in another housing that is misaligned with the first. This situation occurs when regenerators are added at slightly different locations as growth occurs and there is not enough space at the original location for additional repeater housings. In such cases, it is necessary to use the next or prior pole, pedestal, or manhole along the cable route for access.

In Figure 9-3(b), FEXT is coupled from the output of a digital line transmitter (for example, a regenerator) at one end of a line section into the input of a digital line receiver at the other end.

A slightly different set of circumstances gives rise to FEXT, as shown in Figure 9-3(c), where a route junction exists between two T1 lines. A route junction exists wherever two T1 lines follow different routes for part of the section. If the loss of the section in route B is greater than the loss of the section in route A ($L_B > L_A$), the signals from A have a larger magnitude and interfere more seriously with B than the other way around. Special design considerations, described later, are used to reduce FEXT interference at route junctions.

On existing routes where two cables or a compartmental core cable already exist and are suitable for a T1-carrier, no further special considerations are necessary. On 1-cable routes, a study will be required to determine if it is more economical to install another cable for 2-cable operation or to accept the shorter repeater spacing usually required in 1-cable operation. Refer to Table 9-1, which compares the advantages and disadvantages of 1-cable, 2-cable, and compartmental core cable operation.

Because moisture in air core cables severely affects transmission, unpressurized air core cables cannot be used for T1-carrier service. Moisture in previously unpressurized air core cables must be purged with an inert gas, such as nitrogen, and then pressurized with dry air. The satisfactory results of this effort are not guaranteed. Further, the expense sometimes far outweighs the installation of new cables, especially in the case where the unpressurized air core cables are more than 25 years old.

9.1.2 Repeatered Span Line Design

9.1.2.1 Introduction

The basic design criterion for a given repeatered section is bit error rate (BER). The factors that control BER are:

- Interference from other repeatered T1 span lines
- Repeater section loss at 772 kHz
- Central office impulse noise
- Powerline induction

Table 9-1 Comparison of Cable Operation Types[a]

Cable Operation	Advantages	Disadvantages
1-cable	• Only one cable required. • Existing cable plant may be used, providing a large measure of economy.	• Section lengths are shorter to compensate for increased NEXT, which requires more repeater sections for a given route. • 100% cable fill may produce erratic or unreliable operation due to crosstalk. • Cable-count integrity and pair separation must be known. • Cable splices may have to be redone to ensure count integrity.
2-cable	• NEXT is not a design factor permitting longer section lengths and fewer repeater sections for a given route. • 100% cable fill is possible. • Cable count integrity and pair separation is not a design factor except as it affects FEXT.	• Greater cost both initially and if second cable is added later. • Repeater failure can result in the loss of two span lines, because side 1 and side 2 of a given repeater provide the same transmission direction on two span lines. Two spare lines may be needed to provide complete protection.
Compartmental core cable	• NEXT is not a design factor permitting longer section lengths and fewer repeater sections for a given route. • 100% cable fill is possible. • Cable count integrity and pair separation is not a design factor except as it affects FEXT. • One-cable repeater configuration (repeaters are operated bidirectionally), resulting in the loss of only one span line with repeater failure.	• Compartmental core cable costs more than regular exchange cable.

[a]Data from [1]

Section lengths can be adjusted to give the desired BER, as will be shown. In early tests of T1-carrier systems used for voice transmissions, a BER of 1E-6 was found to be difficult to detect by listening, and a BER of 1E-5 did not seriously impair voice quality [2]. As a result of these tests, the terminal-to-terminal BER objective was set to 1E-6. Since errors on any given section usually are not related to errors on other sections, the error rates of individual sections add arithmetically. For example, if the error rate on two sections are 0.6E-6 and 0.4E-6, respectively, the total error rate is 1.0E-6.

To achieve an overall error rate of 1E-6, the objective span BER originally was set to 3E-7. This allowed three spans (each consisting of multiple sections) to be connected in tandem and still achieve the overall terminal-to-terminal specification. Tandem span lines are shown in Figure 9-1 for the route from central office A through central offices B and C to the user location.

The basic repeatered T1 line engineering rules are derived from a statistical requirement that 95% of all such lines will meet this minimum BER; however, many modern repeatered T1 lines carry a large proportion of data traffic that requires better error performance—at least 1E-7 but more typically 1E-9. The line engineering rules are modified to account for this, as explained later.

From the 95% probability requirement, the probability of a line meeting the 1E-6 BER due to one section alone is less than (5% ÷ N), where N is the number of repeater sections in the terminal-to-terminal span lines [3]. In the engineering rules discussed in this section, N is taken to be 50 or less. This results in span lines at least 50 miles long, which covers most practical DLC systems. The design of span lines with more than 50 repeatered sections only requires a closer look at potential crosstalk problems and their resolution; this is discussed briefly for completeness. Due to jitter accumulation, repeatered T1-carrier on twisted pair cable is normally restricted to no more than 200 repeatered sections. Transmission delay may restrict span line length before jitter from regenerators restricts the length.

System error performance is related to the signal-to-noise ratio (SNR) at the input to the repeater. The most significant noise is due to NEXT. The SNR is controlled by controlling the loss in the cable between repeaters. With bipolar alternate mark inversion (AMI), the frequency spectrum peaks at a frequency equal to one half the bit rate, or 772 kHz with T1-carrier. Even though significant spectrum components exist at frequencies widely separated from this value, the design of line sections is based on cable pair loss at 772 kHz.

Span line sections consist of *end-sections* and *intermediate sections*, as shown in Figure 9-4. The allowable loss in each section is limited either by the maximum loss range of the repeater or crosstalk and noise, whichever requires the lowest transmission loss. Usually all intermediate sections will be designed for the same loss; however, in situations in which central office noise and crosstalk are expected to be especially severe because of relatively low crosstalk coupling loss in central office entrance cables or exchange cables, the sections adjacent to end-sections (called *adjacent section* and shown in Figure 9-4) may have to be reduced. Where special considerations are not necessary, adjacent sections are designed the same as intermediate sections.

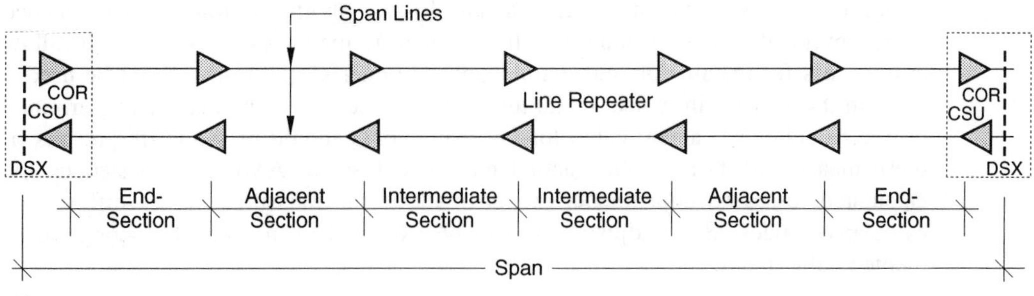

Key
DSX: Digital Signal Cross-Connect
CSU: Channel Service Unit
COR: Central Office Repeater

Figure 9-4 Repeatered T1-carrier sections

When the repeatered T1-carrier was developed originally, section lengths were designed to coincide with H88* load points on 22 AWG cables, resulting in end-sections 3,000 feet long and intermediate sections 6,000 feet long. This greatly simplified and reduced the cost of installing line repeaters, since the cables were readily accessible at each load point (the load coils, of course, were removed from the pairs used for the span line). As will be seen, the repeater locations are not necessarily restricted to these intervals. Section lengths, and resulting repeater spacings—which are determined by the interaction between loss and noise requirements and cable types and gauges—can vary significantly from the relatively rigid load coil spacings.

9.1.2.2 Transmission Design Objectives

Of paramount importance in repeatered T1-carrier span line design is determining repeater section loss. Section loss in a 1-cable operation is to some degree governed by the maximum number of span lines, including spares, to be installed in a given cable. In a 2-cable operation, the number of span lines is not a design parameter.

The length of a section, based on the maximum allowed section loss of the cable, must take into account the design temperature, cable types and gauges, and the expected spread in insertion loss values based on a mean and standard deviation. The maximum section length is then determined by dividing the maximum section design loss by the per unit engineering loss. Engineering loss is the cable loss in decibels per unit length at some standard temperature.

The design of intermediate sections is covered first. The designs differ for 1- and 2-cable operations due to the complicating effect of NEXT on 1-cable operations. Also, there are several ways to calculate the section parameters. Two methods will be given in this book. From a practical standpoint, both give equivalent results. The

*"H" indicates 6,000-foot spacing, and "88" indicates 88 mH load coils.

design methods give maximum lengths and losses. Shorter sections may be needed to accommodate existing plant structures, such as manholes or poles, or to allow easy access for installation and maintenance. Longer sections should not be used.

In those cases in which *maximum* section losses are to be used, as determined by the calculations, actual cable loss measurements should be made (the results of these measurements are called actual measured loss, or AML) to cross-check the calculated losses so adjustments can be made if the measured values differ from calculated values. Such adjustments can be expected in an older or poorly documented cable plant.

9.1.3 Intermediate Section Loss: 2-Cable Operation (and Compartmental Core Cable), *Method 1*

For a 2-cable operation, the basic design equation for intermediate section loss is based on the allowable loss range of the line repeater. Modern line repeaters function with a cable loss range of 7.5 to 35 dB at 772 kHz. Some repeaters have a loss range of 0 to 27 dB, and others are switch selectable between the two ranges. The following assumes repeaters with a loss range of 7.5 to 35 dB; however, other loss ranges can be accommodated by simple arithmetic changes to the equations given.

Adjustments must be made to repeater loss range limits to account for cable pair, splicing, and temperature variations. The maximum allowable loss limits in any intermediate section are:

$$(7.5 + V)/CF_T \text{ dB} \le ISL_T \le (35 - V)/CF_T \text{ dB}$$

where ISL_T = intermediate section loss at design temperature due to repeater loss limitations (dB)

V = largest expected increase in loss due to cable pair and splicing variations (dB)

CF_T = compensation factor for the design temperature (dimensionless)

The value of CF_T varies with the cable type and is given in Table 9-2, along with the engineering loss at 12.8 °C (55 °F) and the resulting maximum ISL_T for buried and aerial cables. For most applications, the following average values can be used for CF_T and V:

$$CF_{T(buried)} = 1.040$$
$$CF_{T(aerial)} = 1.077$$
$$V = 1.5 \text{ dB}$$

With the foregoing values for V and CF_T, the equation for intermediate section loss reduces to

General case $9/CF_T \text{ dB} \le ISL_T \le 33.5/CF_T \text{ dB}$

Buried $8.7 < ISL_T < 32.2 \text{ dB}$

Aerial $8.4 < ISL_T < 31.1 \text{ dB}$

Table 9-2 Cable Characteristics at 772 kHz and 12.8 °C (55 °F)[a]

Type	Capacitance (nF/mile)	Engineering Loss (dB/kft)	Buried $CF_{T(38)}$[b]	Aerial $CF_{T(60)}$[b]	Buried $ISL_{T(38)}$ dB[c]	Aerial $ISL_{T(60)}$ dB[c]	Loss Variation (%)[d]	TC (dB/kft/10 °C)
				19 AWG				
FPIC	83	2.94	1.031	1.059	32.5	31.6	1.5	0.036
APIC	83	3.30	1.046	1.088	32.0	30.8	2.5	0.061
DEPIC	83	3.20	1.038	1.072	32.3	31.3	1.5	0.049
FPIC T-screen	83	2.94	1.050	1.095	31.9	30.6	1.5	0.036
APIC T-screen	83	3.28	1.043	1.080	32.1	31.0	2.5	0.061
FPIC D-screen	83	2.94	1.050	1.095	31.9	30.6	1.5	0.036
APIC D-screen	83	3.18	1.043	1.080	32.1	31.0	2.5	0.061
Paper unit or layer	84	3.80	1.033	1.063	32.4	31.5	2.8	[e]
Pulp	83	3.90	1.051	1.077	31.9	31.1	2.8	[e]
Paper unit or layer	66	3.00	1.038	1.071	32.3	31.3	3.6	0.045
				22 AWG				
FPIC	83	3.99	1.039	1.074	32.2	31.2	1.5	0.063
APIC	83	4.60	1.044	1.083	32.1	30.9	2.5	0.081
DEPIC	83	4.40	1.041	1.077	32.2	31.1	1.5	0.072
FPIC T-screen	83	3.99	1.050	1.095	31.9	30.6	1.5	0.063
APIC T-screen	83	4.47	1.044	1.083	32.1	30.9	2.5	0.081
FPIC D-screen	83	3.99	1.050	1.095	31.9	30.6	1.5	0.063
APIC D-screen	83	4.39	1.044	1.083	32.1	30.9	2.5	0.081
Paper unit or layer	83	5.10	1.042	1.078	32.1	31.1	2.1	[e]
Pulp	83	5.20	1.038	1.058	32.3	31.7	2.1	[e]
Paper unit or layer	73	4.60	1.039	1.074	32.2	31.2	2.3	0.072

continued

Table 9-2 Continued

Type	Capacitance (nF/mile)	Engineering Loss (dB/kft)	Buried $CF_{T(38)}$[b]	Aerial $CF_{T(60)}$[b]	Buried $ISL_{T(38)}$ dB[c]	Aerial $ISL_{T(60)}$ dB[c]	Loss Variation (%)[d]	TC (dB/kft/10 °C)
24 AWG								
FPIC	83	5.00	1.043	1.082	32.1	31.0	1.5	0.086
APIC	83	5.80	1.043	1.081	32.1	31.0	2.5	0.099
DEPIC	83	5.50	1.043	1.081	32.1	31.0	1.5	0.094
FPIC T-screen	83	4.92	1.050	1.095	31.9	30.6	1.5	0.086
APIC T-screen	83	5.52	1.026	1.052	32.7	31.8	2.5	0.099
FPIC D-screen	83	4.92	1.050	1.095	31.9	30.6	1.5	0.086
APIC D-screen	83	5.58	1.026	1.052	32.7	31.8	2.5	0.099
Paper unit or layer	84	6.80	1.044	1.083	32.1	30.9	1.6	[e]
Pulp	83	6.30	1.048	1.063	32.0	31.5	1.7	[e]
Paper unit or layer	72	5.85	1.044	1.083	32.1	30.9	1.8	[e]
ICOT, air core	52	3.60	1.055	1.104	31.8	30.3	1.5	0.079
ICOT, filled core	60	3.90	1.046	1.087	32.0	30.8	2.0	0.072
26 AWG								
FPIC	83	6.30	1.046	1.088	32.0	30.8	1.5	0.117
APIC	83	7.30	1.042	1.079	32.1	31.1	2.5	0.122
DEPIC	83	6.90	1.044	1.083	32.1	30.9	1.5	0.121
Paper unit or layer	83	8.17	1.053	1.100	31.8	30.5	1.3	[e]
Pulp	69	6.79	1.054	1.101	31.8	30.4	1.6	[e]
Paper unit or layer	79	7.70	1.054	1.103	31.8	30.4	1.4	[e]
25 AWG								
MAT, air core	64	5.10	1.049	1.092	31.9	30.7	3.0	0.099

[a]Data from [4] with additions from [3]

[b]$CF_{T(\ )}$ is compensation factor for the design temperature shown in parentheses (°C).

[c]$ISL_{T(\ )}$ is maximum intermediate section loss at the design temperature shown in parentheses (°C).

[d]Loss variation is expressed as a percentage in this table. Divide by 100 before using in equations.

[e]Use linear interpolation to calculate CF_T for temperatures other than 38 and 60 °C. If the temperature is between 12.8 and 38 °C, use $CF_{T(12.8)} = 1.00$ and $CF_{T(38)}$ from the table. If the temperature is between 38 and 60 °C, use $CF_{T(38)}$ and $CF_{T(60)}$, respectively, from the table.

The equations give the ISL_T corrected for the design temperature, using average temperature coefficients. Higher accuracy can be obtained by using the values for ISL_T and CF_T given in Table 9-2.

The intermediate section length is found from

$$ISD_T = ISL_T/EL \text{ kilofeet}$$

where ISD_T = intermediate section length (kilofeet) corrected for design temperature
ISL_T = intermediate section loss due to repeater loss limitations, corrected for design temperature (dB)
EL = engineering loss at 12.8 °C (55 °F) (dB/kilofoot), from Table 9-2

The maximum section lengths for predominantly buried and predominantly aerial cables are

$$ISD_T = 32.2/EL \text{ kilofeet (buried)}$$
$$ISD_T = 31.1/EL \text{ kilofeet (aerial)}$$

If a particular section contains more than one cable type or gauge, three methods can be used to determine the overall section loss:

1. The losses for each cable can be arithmetically combined to determine the overall section loss.
2. A more conservative design is to include a junction loss to account for mismatches between cable gauges. Junction losses for typical cables are given in Table 9-3. This loss is added at each change in cable type or gauge.
3. The maximum section loss of each cable gauge, assuming the entire section consists of that gauge, is found, and the higher of the maximum losses is used. This is the most conservative method.

It is only necessary to use (2) or (3) when repeaters are to be spaced at their absolute maximum spacings, and when cable and splicing characteristics are uncertain. Each method is illustrated in the following example.

EXAMPLE 9-1

Using Figure 9-5, find the total section loss at 55 °F. All cables are DEPIC.

Using Table 9-2, the losses of segments L1, L2, L3, and L4 are:

L1 (24 AWG) = 0.200 kilofeet × 5.5 dB/kilofoot = 1.1 dB

L2 (22 AWG) = 1.000 kilofeet × 4.4 dB/kilofoot = 4.4 dB

L3 (24 AWG) = 0.500 kilofeet × 5.5 dB/kilofoot = 2.8 dB

L4 (26 AWG) = 3.000 kilofeet × 6.9 dB/kilofoot = 20.7 dB

1. Total section loss = L1 + L2 + L3 + L4 = 29.0 dB

Table 9-3[a] Junction Loss between Cables[b]

AWG	nF/MILE	19 AWG 66	19 AWG 83	22 AWG 73	22 AWG 83	24 AWG 72	24 AWG 83	26 AWG 69	26 AWG 79	26 AWG 83
19	66	0.0	0.2	0.2	0.3	0.2	0.3	0.2	0.3	0.3
19	83	0.2	0.0	0.2	0.1	0.2	0.1	0.3	0.2	0.1
22	73	0.2	0.2	0.0	0.1	0.2	0.2	0.2	0.2	0.2
22	83	0.3	0.1	0.1	0.0	0.2	0.1	0.3	0.2	0.1
24	72	0.2	0.2	0.1	0.2	0.0	0.1	0.2	0.2	0.2
24	83	0.3	0.1	0.2	0.1	0.1	0.0	0.3	0.2	0.1
26	69	0.2	0.3	0.2	0.3	0.2	0.3	0.0	0.1	0.2
26	79	0.3	0.2	0.2	0.2	0.2	0.2	0.1	0.0	0.1
26	83	0.3	0.1	0.2	0.1	0.2	0.1	0.2	0.1	0.0

[a]Data from [1]

[b]For short lengths of inserted cable, the junction losses given in this table apply *to each end* as follows: For lengths <50 feet or less, do not include any junction loss; for lengths between 50 and 150 feet, use 1/2 of the junction loss given in the table; for lengths >150 feet, use full junction loss given in the table.

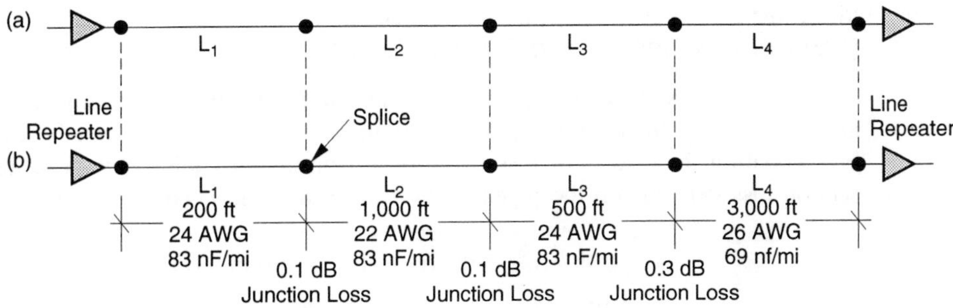

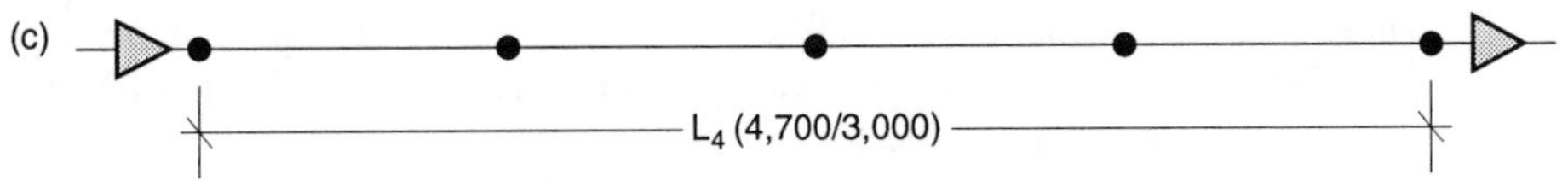

Figure 9-5 Illustration for Example 9-1

2. Total section loss = L1 + 0.1 + L2 + 0.1 + L3 + 0.3 + L4 = 29.5 dB
3. Total section loss = L4 (4,700/3,000) = 32.4 dB

When values for CF_T are required for various PIC at temperatures other than 38 or 60 °C, as given in the table, the following equation applies:

$$CF_T = 1 + [(TC\ \Delta T/10)/EL_{T0}]$$

where CF_T = compensation factor at temperature T
TC = temperature coefficient of loss, from Table 9-2 (dB/kilofoot/10 °C)
ΔT = $(T - T_0)$, or difference between the desired temperature, T, and reference temperature, T_0, or 12.8 °C (°C)
EL_{T0} = engineering loss at 12.8 °C, from Table 9-2 (dB/kilofoot)

The temperature coefficient of loss for pulp and paper insulated cables is nonlinear, and a simple relationship, such as for PIC above, does not exist. For pulp and paper insulated cables, values for CF_T can be interpolated from the values for 38 and 60 °C with sufficient accuracy. The value for $CF_{T(12.8)} = 1.00$.

EXAMPLE 9-2

Find the compensation factor for a 24 AWG DEPIC to be operated at 40 °C.

From Table 9-2, engineering loss at 12.8 °C is 5.5 dB/kilofoot, and the temperature coefficient TC is 0.094 dB/kilofoot/10 °C. Therefore:

$$CF_{T(40)} = 1 + \frac{0.094\left(\dfrac{40 - 12.8}{10}\right)}{5.5} = 1.046$$

9.1.4 Intermediate Section Loss: 2-Cable Operation (and Compartmental Core Cable), *Method 2*

This method gives a slightly less conservative result than method 1 described above. For a 2-cable operation, the maximum intermediate section loss is determined by evaluating the effects of loss variation for particular cable types. The minimum section loss depends on the repeater, as before, and the maximum section loss is:

$$\text{ISL} \leq 33.5 \text{ dB (repeater loss limitation)}$$

$$\text{ISL}_\text{T} = \text{ISL}/(2.33\ \text{LV} + \text{CF}_\text{T})$$

where ISL = maximum intermediate section loss due to repeater loss limitations (dB)
ISL_T = maximum intermediate section loss corrected for the design temperature and cable loss variation (dB)
LV = loss variation, from Table 9-2 (%/100)
CF_T = compensation factor for design temperature (dimensionless), from Table 9-2

The maximum temperature-corrected intermediate section length is then found from:

$$\text{ISD}_\text{T} = \text{ISL}_\text{T}/\text{EL kilofeet}$$

where ISD_T = intermediate section length (kilofeet) corrected for design temperature and cable loss variation
ISL_T = intermediate section loss due to repeater loss limitations, corrected for design temperature and cable loss variation (dB)
EL = engineering loss at 12.8 °C (55 °F), from Table 9-2 (dB/kilofeet).

EXAMPLE 9-3

An existing route between a central office and remote switching terminal (RST) is to be converted to repeatered T1-carrier. The route consists of a 600-pair, 24 AWG exchange cable and 25-pair, 22 AWG compartmental core (screened) cable, as shown in Figure 9-6. Both cables are filled. Assume the end-sections have a loss of 18 dB each. Also, assume section length limitations are due to loss and not crosstalk. Determine the minimum number of intermediate sections.

The lengths of the end-sections are found as follows: At the central office end, the engineering loss of 24 AWG FPIC is 5.0 dB/kilofoot and $\text{CF}_\text{T} = 1.043$ for buried cable (both quantities from Table 9-2). The length of the central office end-section is 18 dB/(5.0 dB/kilofoot × 1.043) = 3.452 kilofeet. Intermediate sections will be placed in the remaining 24 AWG cable, or 20.950 − 3.452 = 17.498 kilofeet.

Similarly, for the RST end, the engineering loss of 22 AWG FPIC is 3.99 dB/kilofoot and $\text{CF}_\text{T} = 1.039$. The length of the RST end-section is 18 dB/(3.99 dB/kilofoot × 1.039) = 4.342 kilofeet. Intermediate sections will be placed in the remaining 22 AWG cable, or 86.250 − 4.342 = 81.908 kilofeet.

The maximum intermediate section length for the 24 AWG cable segment is 32.1 dB/(5.0 dB/kilofoot × 1.043) = 6.155 kilofeet and for the 22 AWG cable segment is 32.2 dB/(3.99 dB/kilofoot × 1.039) = 7.767 kilofeet. In the 24 AWG cable segment, there can be no more than 17.498 kilofeet/6.155 kilofeet/section = 2.8 intermediate sections. Since there

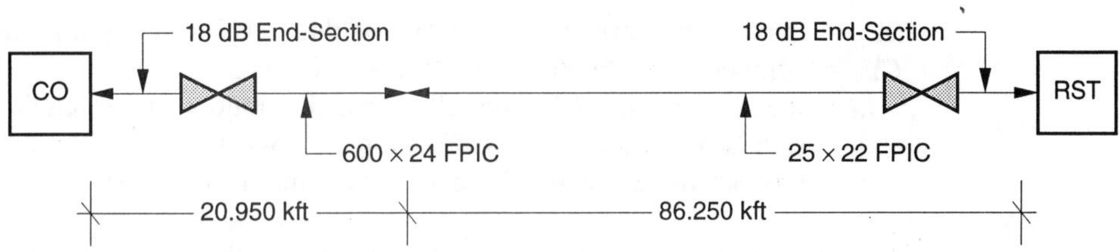

Figure 9-6 Illustration for Example 9-3

cannot be fractional sections, it is necessary to take the next lowest integer. Two sections require 12.310 kilofeet with 17.498 − 12.310 = 5.188 kilofeet left over. In the 22 AWG cable segment, there can be no more than 81.908 kilofeet/7.767 kilofeet/section = 10.5 sections. The next lowest integer is ten, and ten sections require 77.670 kilofeet with 81.908 − 77.670 = 4.238 kilofeet left over.

Considering the leftover cable segments, it is clear at least two additional intermediate sections are required. It is desirable to place the repeater separating these sections at the junction of the two cable gauges.

Therefore, the 24 AWG cable segment requires three intermediate sections, and the 22 AWG segment requires 11 intermediate sections, for a total of 14 intermediate sections.

The above example determined the *minimum* number of intermediate sections from the *maximum* calculated section lengths. It is not considered good engineering practice to use maximum section lengths without field-verifying section losses (being sure to adjust the measurements for temperature). Even if measurements verify the values used, it is possible that parts of either cable segment may be rerouted in the future, extending their length. Therefore, it is appropriate to reduce the calculated maximum section lengths either by several hundred feet or several decibels.

On existing routes, to reduce the amount of construction, it is desirable to place repeaters in existing manholes, pedestals, or aerial terminals. When this is done, some sections may be shorter than others. Since the signal is completely regenerated in each repeater, it is only necessary that the equalization range of the repeater be observed. For example, one section could be 9 dB and the sections on either side of it could be 27.5 dB.

9.1.5 Intermediate Section Loss: 1-Cable Operation

For a 1-cable operation, the maximum intermediate section loss is determined by evaluating the effects of both loss and crosstalk coupling from the following equations. The minimum section loss depends on the repeater, as before, and the maximum section loss is

$$ISL_T \leq 33.5/CF_T \text{ dB (repeater loss limitation)}$$

or

$$ISL_N \leq [(m - s) - 32 - 10 \log n]/CF_T \text{ dB (crosstalk limitation)}$$

where ISL_T = maximum intermediate section loss corrected for design temperature due to repeater loss limitations, as previously shown for 2-cable operation

ISL_N = maximum intermediate section loss due to NEXT coupling limitations
CF_T = compensation factor for design temperature
m = mean value of NEXT coupling loss distribution in the cable
s = standard deviation of NEXT coupling loss distribution in the cable
n = number of repeatered T1-carrier systems in the cable

The smaller of ISL_T and ISL_N is used to determine the section length from

$$ISD = ISL/EL \text{ kilofeet}$$

where ISD = intermediate section length (kilofeet)
ISL = the smaller of ISL_T and ISL_N (dB)
EL = engineering loss at 12.8 °C (55 °F), from Table 9-2 (dB/kilofoot)

The equation for ISL_T is identical to that given for 2-cable operation. The factor 10 log n in the equation for ISL_N arises from assuming that all disturbing signals due to NEXT add up on the disturbed pair on a 10 log n basis. Table 9-4 gives values for 10 log n.

Shortening the section length reduces the effect of NEXT by reducing the loss (increasing the signal level at the receiver). The value 32 in the equation for ISL_N arises from the basic repeater design parameters, error rate, and the probability density of the crosstalk interference amplitude, and includes a 6 dB reduction in the value of m.* This reduction makes ISL_N conservatively low to increase the operational margin, and accounts for

- Cable manufacturing tolerances
- Splicing variations
- The presence of FEXT
- Crosstalk in repeater housings and terminal shelves

Table 9-4 Values of 10 log n

n	10 LOG n	n	10 LOG n	n	10 LOG n	n	10 LOG n	n	10 LOG n
1	0.0	11	10.4	21	13.2	31	14.9	41	16.1
2	3.0	12	10.8	22	13.4	32	15.1	42	16.2
3	4.8	13	11.1	23	13.6	33	15.2	43	16.3
4	6.0	14	11.5	24	13.8	34	15.3	44	16.4
5	7.0	15	11.8	25	14.0	35	15.4	45	16.5
6	7.8	16	12.0	26	14.2	36	15.6	46	16.6
7	8.5	17	12.3	27	14.3	37	15.7	47	16.7
8	9.0	18	12.6	28	14.5	38	15.8	48	16.8
9	9.5	19	12.8	29	14.6	39	15.9	49	16.9
10	10.0	20	13.0	30	14.8	40	16.0	50	17.0

*A derivation of maximum section loss due to NEXT can be found in [2].

The values of m and (m − s), used for the two transmission directions (labeled Tx for transmit and Rx for receive), are average values based on a large number of measurements at 772 kHz and apply to cable segments with a loss greater than 10 dB at 772 kHz. For shorter lengths, the coupling loss is greater due to the shorter crosstalk exposure length. Figure 9-7 shows the coupling loss increase versus the loss of the segment in question. The increase is added to the NEXT coupling loss found in Table 9-5, which provides the values of m and (m − s) for various cable types and pair orientations.

If a particular cable is not shown in the table, the value of (m − s) can be approximated from the data for a similar cable, as follows:

$$(m - s)_x = (m - s) + 10 \log (EL_x/EL) \text{ dB}$$

where $(m - s)_x$ = value for the cable in question
$(m - s)$ = value for the nearest cable with known data
EL_x = engineering loss for cable in question
EL = engineering loss for the nearest cable with known data

The factor (m − s) in the above equation and the equations for ISL_N reduce the mean

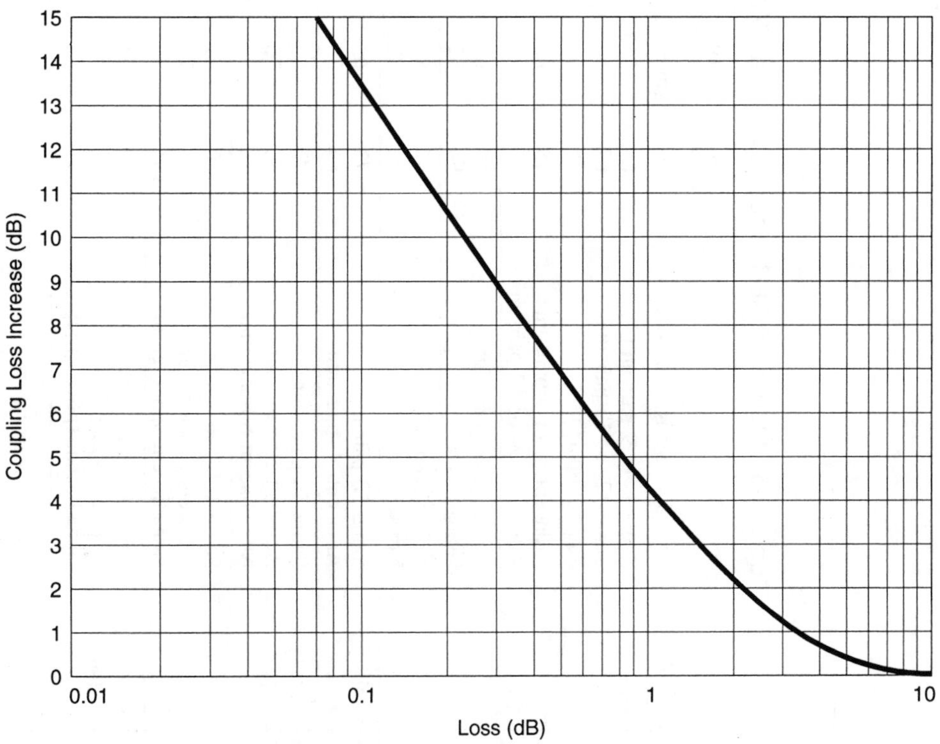

Figure 9-7 Coupling loss increase vs. cable segment loss (the equation for this curve is given in Chapter 5)

Table 9-5 Measured NEXT Coupling Loss at 772 kHz[a] Valid for Section Losses ≥10 dB

Cable Type	Tx and Rx Location	19 AWG m	19 AWG (m − s)	22 AWG m	22 AWG (m − s)	24 AWG m	24 AWG (m − s)	26 AWG m	26 AWG (m − s)
PIC, All	Same 8- or 9-pair unit	64	55	66	57	67	58	68	59
	Same 12- or 13-pair unit	67	57	69	59	70	60	71	61
	Same 25-pair unit	74	61	76	63	77	64	78	65
	Adjacent 8- or 9-pair unit	75	66	77	68	78	69	79	70
	Adjacent 12- or 13-pair unit	75	66	77	68	78	69	79	70
	Adjacent 25-pair unit	80	67	82	69	83	70	84	71
	Nonadjacent 8- or 9-pair unit	90	82	92	84	93	85	94	86
	Nonadjacent 12- or 13-pair unit	90	82	92	84	93	85	94	86
	Nonadjacent 25-pair unit	90	82	92	84	93	85	94	86
Paper unit-type ≤200-pair	Same 50-pair unit	73	63	75	65	76	66	78	68
	Adjacent 50-pair unit	86	78	88	80	89	81	90	82
Paper unit-type ≥200-pair	Same 50-pair unit	73	63	75	65	76	66	77	67
	Adjacent 50-pair unit	86	78	88	80	89	81	90	82
	Nonadjacent 50-pair unit	98	92	100	94	101	95	102	96
	Same 100-pair splice group	80	69	82	71	83	72	84	73
	Adjacent splice group	88	79	90	81	91	82	92	83
	Nonadjacent splice group	101	94	103	96	104	97	105	98
Paper layer-type ≥200-pair	Same 100-pair splice group	73	64	75	66	76	67	77	68
	Adjacent splice group	81	73	83	75	84	76	85	77
	Nonadjacent splice group	91	82	93	84	94	85	95	86

[a]Data from [4]
Tx = transmit; Rx = receive

NEXT coupling loss by one standard deviation, thus giving a value that will be exceeded by approximately 84% of all pair combinations.

Alternately, Table 9-6 can be used with PIC manufactured to industry standards, including Rural Electrification Administration (REA) specifications [5–7]. The presentation of the information in this table is slightly different from Table 9-5 in that it shows only the NEXT mean, less one standard deviation (m − s). Also, this table shows specified minimum requirements rather than average values from measurements. Either one of the tables is satisfactory for field applications, but Table 9-6 will give more conservative results.

Table 9-6 Specified NEXT Coupling Loss at 772 kHz

Cable Type	Tx and Rx Location	19 AWG (m − s)	22 AWG (m − s)	24 AWG (m − s)	26 AWG (m − s)
PIC, All	Same 12- or 13-pair unit	56	56	56	56
	Same 25-pair unit	60	60	60	60
	Adjacent 13-pair unit or less	65	65	65	65
	Adjacent 25-pair unit	66	66	66	66
	Nonadjacent (all)	81	81	81	81

The following list and examples illustrated in Figure 9-8 (see page 428) show, in descending order of desirability (first, most desirable; last, least desirable), the choice of cable pairs for the Tx and Rx circuits of a span line when a 1-cable operation is used:

- Nonadjacent groups or units
- Adjacent 25-pair groups or units
- Adjacent 8- or 9-pair, or 12- or 13-pair units
- Same 25-pair unit
- Same 12- or 13-pair unit
- Same 8- or 9-pair unit

Figure 9-9 shows how pairs should be chosen for the Tx and Rx circuits with different cable sizes to obtain the maximum physical separation. Only cables up to 150 pairs are shown. Larger cables will allow Tx and Rx circuits to be assigned in nonadjacent, 25-pair binder groups similar to the 150-pair cable. It is worth repeating that, to maximize section length, it is important that the Tx and Rx pairs be physically separated when a 1-cable operation is used.

EXAMPLE 9-4

Using Table 9-5, find the NEXT coupling loss that will be exceeded for 84% of all Tx and Rx pair combinations if the two pairs are located in the (a) same 25-pair unit, (b) adjacent 25-pair unit, and (c) nonadjacent 25-pair unit of a 22 AWG PIC. Assume losses are >10 dB.

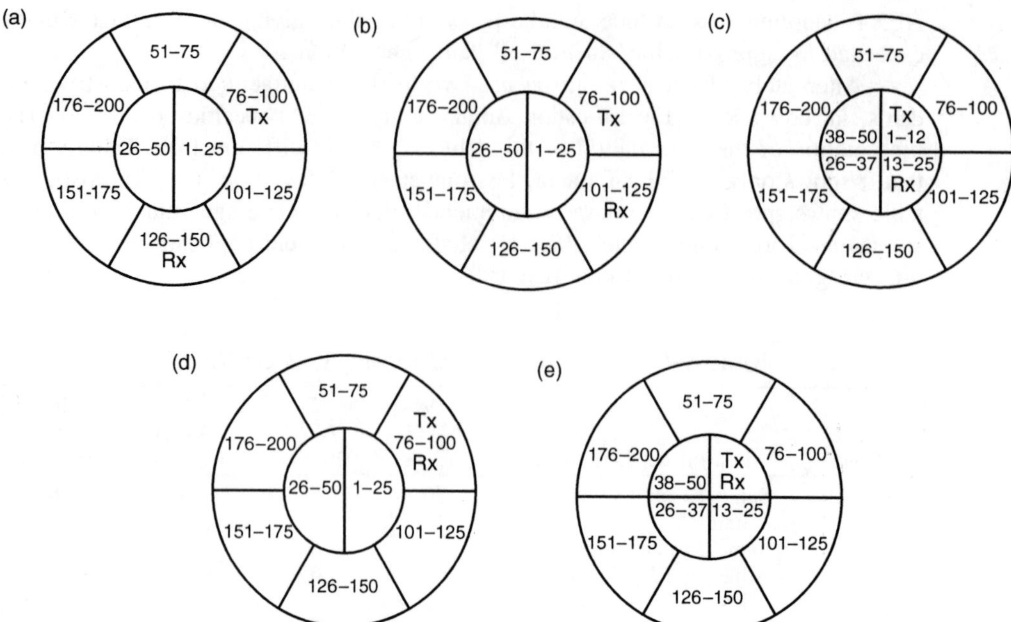

Figure 9-8 Examples of circuit locations in a 200-pair cable operation: (a) Example of Tx and Rx pairs in nonadjacent binder groups; (b) Example of Tx and Rx pairs in adjacent binder groups; (c) Example of Tx and Rx pairs in adjacent 12- or 13-pair units; (d) Example of Tx and Rx pairs in same 25-pair binder group; (e) Example of Tx and Rx pairs in same 12- or 13-pair unit

From Table 9-5,

(a) (m − s) = 63 dB,
(b) (m − s) = 69 dB,
(c) (m − s) = 84 dB.

EXAMPLE 9-5

Find the NEXT coupling loss that will be exceeded for 84% of all pair combinations where the Tx and Rx pairs are located in adjacent 25-pair units for three 22 AWG DEPIC segments that are 500, 1,500, and 3,000 feet long, respectively.

From Table 9-5, the engineering loss for 22 AWG DEPIC is 4.40 dB/kilofoot, giving the loss of each segment as follows:

500-foot segment: 4.40 dB/kilofoot × 0.5 kilofoot = 2.20 dB

1,500-foot segment: 4.40 dB/kilofoot × 1.5 kilofeet = 6.60 dB

3,000-foot segment: 4.40 dB/kilofoot × 3.0 kilofeet = 13.20 dB

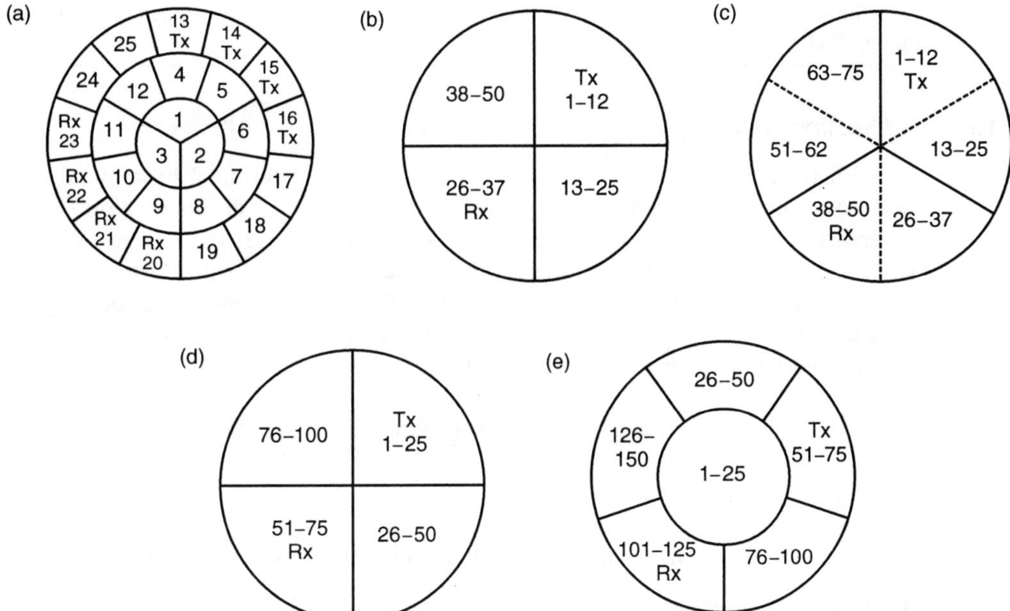

Figure 9-9 Examples of Tx and Rx pair assignments for 1-cable operation: (a) 25-pair cable; (b) 50-pair cable; (c) 75-pair cable; (d) 100-pair cable; (e) 150-pair cable

Since the losses of the two shorter segments are <10 dB, the NEXT coupling loss will be higher than given in the tables. The added coupling losses from Figure 9-7 corresponding to the segment losses are 2.0, 0.2, and 0.0 dB, respectively. Therefore, the (m − s) values (rounded) for 22 AWG DEPIC are:

500-foot segment: 69 dB + 2.0 dB = 71 dB

1,500-foot segment: 69 dB + 0.2 dB = 69 dB

3,000-foot segment: 69 dB + 0.0 dB = 69 dB

EXAMPLE 9-6

Confirm the assumption in Example 9-3 that the intermediate section lengths are not crosstalk limited. Assume that 12 systems are to be installed, and that nonadjacent 25-pair binder groups are used in the 24 AWG exchange cable to separate the Tx and Rx circuits. The Tx and Rx circuits are separated by the core separator in the 22 AWG compartmental core cable.

Since crosstalk is not considered an issue in compartmental core cable, it is only necessary to evaluate the 24 AWG exchange cable. From Table 9-5, the mean, less one standard deviation, crosstalk coupling loss (m − s) for nonadjacent 25-pair binder groups, is 85 dB. Also, from Table 9-4, log(12) = 10.8. Therefore:

$$ISL_N \leq [(m - s) - 32 - 10 \log n]/CF_T = (85 - 32 - 10.8)/1.043 = 40.5 \text{ dB}$$

Since $ISL_T = 33.5/CF_T = 33.5/1.043 = 32.1$ dB, it is clear that repeater loss is the limiting factor, and the assumption is correct.

9.1.6 End-Section Loss

Where cables serve *only* repeatered T1-carrier, from terminal to terminal, impulse noise in end-sections normally is not a problem. Such an assumption should be verified by measurement. In the case of no impulse noise, the end-section loss can be as high as 32.5 dB, less the artificial line loss of the central office repeater (COR) (usually 7.5 dB). However, the end-section in a 1-cable operation may be limited by NEXT. Therefore, where impulse noise is not a problem, the end-section loss must be the smaller of

$$\text{ESL} \leq (32.5 - \text{COR artificial line loss}) \text{ dB}$$

or

$$\text{ESL} \leq (\text{ISL}_N - \text{COR artificial line loss}) \text{ dB}$$

where ESL = end-section loss (dB)

ISL_N = maximum intermediate section loss due to NEXT coupling, as found in the previous section

Impulse noise in end-sections is a common situation. Impulse noise is coupled from voice frequency pairs to repeatered T1-carrier pairs, where cables serve mixed voice frequency services and T1-carrier. To combat the effects of impulse noise, it is necessary to reduce end-section loss, as compared to an intermediate section. The difference between the impulse noise path loss (NPL), and the signal loss (that is, from the disturbing pair to the disturbed pair connected to the repeater input) must be at least 52 dB. A conservative value for NPL is 75 dB. Therefore, the maximum end-section loss is $75 - 52 = 23$ dB, which is the first design constraint.

The end-section loss may be further limited by the temperature effects (1- and 2-cable operations) or NEXT effects (1-cable operation). It is common to use a proportional reduction (that is, $23/35 \doteq 2/3$) to account for these effects. This leads to the second and third constraints, which reduce the end-section loss from temperature and NEXT considerations by a factor of 2/3. Therefore, where impulse noise is present, the end-section loss must be the smaller of:

$$\text{ESL} \leq (2/3)\, 35 = 23 \text{ dB}$$

or

$$\text{ESL} \leq (2/3)\, \text{ISL}_N \text{ dB}$$

or

$$\text{ESL} \leq (2/3)\, \text{ISL}_T \text{ dB}$$

where ESL = end-section loss (dB)
ISL_N = maximum intermediate section loss due to NEXT coupling, as found in the previous sections (dB)
ISL_T = maximum intermediate section loss due to repeater loss limitations, corrected for design temperature, as found in the previous sections (dB)

EXAMPLE 9-7

The end-section losses in Example 9-3 are given as 18 dB. Determine if these losses are appropriate.

It was found in Example 9-6 that the intermediate sections were repeater loss limited rather than crosstalk limited. Therefore, the maximum end-section loss for the 24 AWG cable is $(2/3)\ ISL_T = (2/3)\ 32.1 = 21.4$ dB.

Similarly, the maximum end-section loss for the 22 AWG cable is $(2/3)\ 32.2 = 21.5$ dB. Both quantities are greater than the assumed 18 dB, so the value used is conservative and appropriate.

Where span lines are connected in tandem through a central office, as was shown in Figure 9-1, the end-section loss may need to be further reduced, as shown in Table 9-7. This table is used when a conservative design is needed for span lines passing through noisy central offices. The reduction value is subtracted directly from the end-section loss, found above, before the end-section length is calculated (this calculation is described below).

The calculation of end-section losses must include losses in entrance cables and central office interbay (switchboard) cables. The loss of the inside cable is subtracted from the end-section loss calculated above to determine the length of the outside portion of the end-section. Table 9-8 shows the average unit loss values (per 100 feet) for commonly used inside cables.

The most frequently cited inside cable type for connection between the entrance and digital signal cross-connect (DSX) panels and between multiplex equipment is the AT&T ABAM multipair cable. It is not necessary to use this cable; any high-quality, shielded, twisted pair cable with 100 Ω impedance at 772 kHz may be used

Table 9-7 End-Section Loss Reduction for Tandem Span Lines[a]

No. of Tandem Span Lines	Reduction in ESL
2	0.0
3	0.0
4	0.9
5	1.6
6	2.3
7	2.8
8	3.3
9	3.7
10	4.0

[a]Data from [4]
ESL = end-section loss

Table 9-8 Shielded Entrance and Switchboard Cable Average Losses at 772 kHz and 21 °C[a]

Cable Type	Configuration	Loss (dB/100 ft)	AWG	Impedance
AT&T ABAM[b]	Multipair	0.46	22	100
AT&T ABMM[b]	Multipair	0.52	24	100
AT&T 600 series	Multipair	0.41	22	100
AT&T 750 series	Multipair	0.61	22	85
AT&T 761 series	Single-pair	0.62	24	100
AT&T 1249 and 2249 series	Multipair	0.65	26	100
Alcatel P/N 600015-100-241	Single-pair	0.95	24	100
Protector stub (tip) cable	Multipair	0.52	22	100
Protector stub (tip) cable	Multipair	0.62	24	100

[a]Data from [3] and [4]

[b]See Appendix A for a description of ABAM and ABMM cable types.

(85 Ω cables also may be used); however, adjustments must be made to the maximum length of this cable if its unit loss is different from ABAM cable.

The environment in central offices is controlled, which means high-temperature effects on entrance and switchboard cables normally are not considered; however, cables inside commercial office buildings or industrial establishments may be run in walls or air handling plenums, in which case the temperature effects must be considered. Without more accurate information, a temperature compensation factor of 1.077 may be used for cables installed where temperature is not tightly controlled. The new loss value for the inside cable can be found from:

$$AL_T = CF_T AL_{(21)}$$

where AL_T = temperature-corrected average loss per unit (dB/100 feet)
$AL_{(21)}$ = average loss per unit at 21 °C, from Table 9-8
CF_T = temperature compensation factor = 1.077

Where span lines are terminated in an office repeater bay (ORB) or DSX-1 panel, the cable length from the ORB or DSX-1 through the main distribution frame (MDF), if used, to the entrance cable must be known. If an ORB or DSX-1 panel is not used, the distance from the span line termination equipment to the entrance cable must be known. Inside cabling loss is determined from:

$$ICL_T = L\ AL_T$$

where ICL_T = temperature-corrected loss of inside cabling (dB)
L = length of inside cabling (in multiples of 100 feet)
AL_T = temperature-corrected average loss per unit (dB/100 feet)

The maximum length of the outside portion of an end-section cable is found from:

$$ESD_o = (ESL - ICL_T)/EL$$

where ESD_o = maximum end-section length of outside cable (kilofeet)
ESL = maximum end-section loss (dB)
ICL_T = temperature-corrected inside cabling loss (dB)
EL = engineering loss of outside cable at 12.8 °C (55 °F), from Table 9-2 (dB/kilofeet)

Inside cabling used with repeatered T1-carrier span lines should bypass the MDF if possible and run directly to the ORB, DSX-1 panel, or span line terminating equipment. Separate cables should be used for each transmission direction. If termination on the MDF is unavoidable (as it is in most central offices), the first choice would be to run the inside cabling directly from the MDF terminations to the ORB, DSX-1, or span line termination equipment. The last choice would be to use jumpers between the inside and outside cabling terminations on the MDF. In this case, shielded jumpers are recommended, one shielded pair for each transmission direction; however, unshielded jumpers may be used on small MDFs and combined distribution frames (CDFs).

9.1.7 Route Junctions

A route junction occurs anywhere span lines from different systems enter the same cable at different signal levels. Harmful FEXT may occur if the level difference is too great. A hypothetical route junction is shown in Figure 9-3(c). For simplicity, only one transmission direction is shown. A more complete illustration is given in Figure 9-10(a) for a 1-cable operation and in Figure 9-10(b) for a 2-cable operation. Through span lines are shown in Figure 9-10(c).

If the losses are such that $L_B > L_A$, the stronger signals from system 1 will interfere with the weaker signals from system 2. The route junction places limits on the loss difference $|L_B - L_A|$ with respect to the exposure segment loss, L_C, such that:

$$|L_B - L_A| \leq (\text{FEXT} - \text{EF})\ \text{dB}$$

where $|L_B - L_A|$ = magnitude of loss difference between segment A and segment B (dB)
FEXT = equal-level FEXT coupling loss at 772 kHz between system 1 and system 2 in the exposure segment C (dB), from Table 9-9
EF = exposure factor (dB), from Table 9-10

FEXT for various cable types is given in Table 9-9. The exposure factor EF varies with the exposure segment loss L_C and is given in Table 9-10 for 25 interfering systems. For 50 interfering systems, increase EF by 2 dB; for 12 systems, reduce EF by 2 dB.

Route junction interference problems can be avoided by installing a repeater at the junction point, as shown in Figure 9-11(a) on page 436, or by using separate cables for each system. If a repeater is installed at the junction, and it is too close to another repeater in one of the sections on either side of it, an attenuator can be

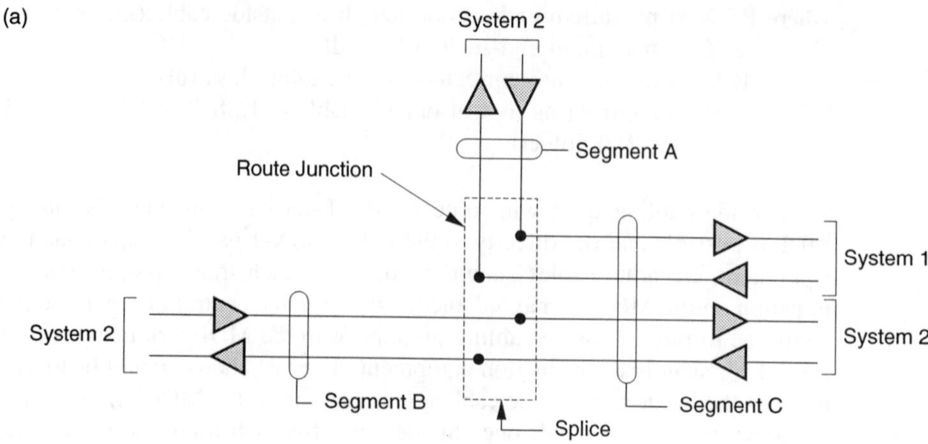

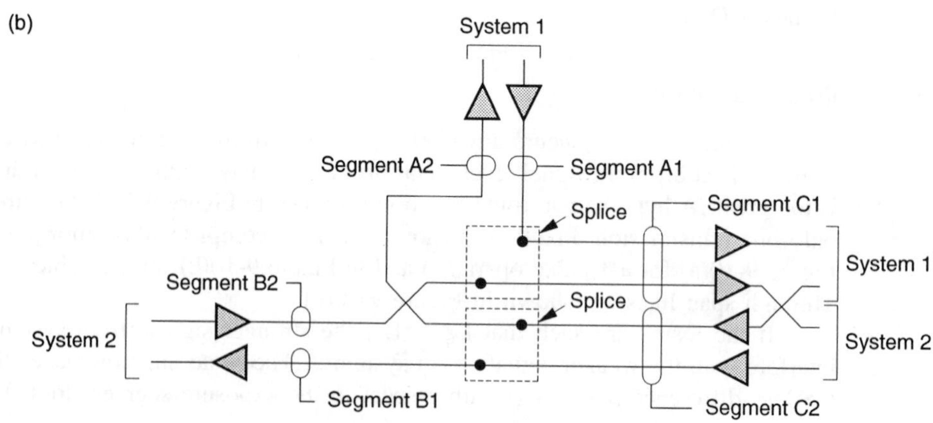

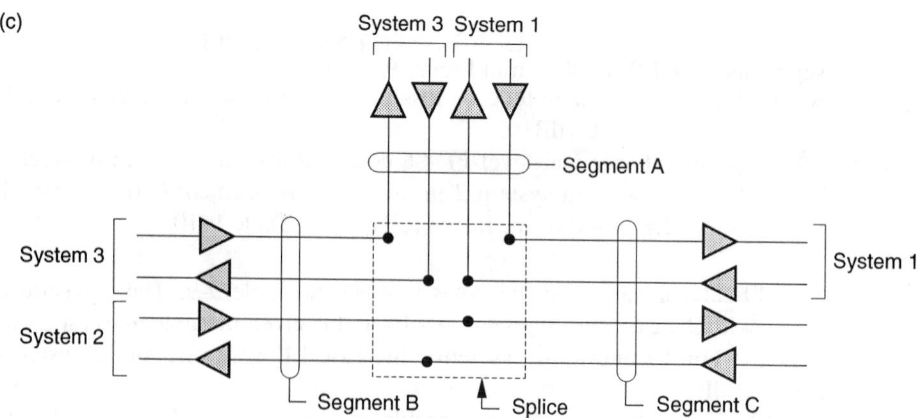

Figure 9-10 T1-carrier route junction: (a) Route junction with 1-cable operation; (b) Route junction with 2-cable operation; (c) Route junction with through systems

Table 9-9 Average Equal-Level FEXT at 772 kHz

LOCATION	ELFEXT (dB)
Same unit	71
Adjacent unit	81
Alternate unit	91

Table 9-10 Exposure Segment Loss vs. Exposure Factor (EF) for 25 Interfering Systems

L_C (dB)	EF (dB)
0.5	46
1	50
2	53
4	56
6	57
8	58
10	58
12	59
14	60
16	60
18	60
20	61
22	61
24	62
26	62
28	62
30	62
32	63

installed, as shown in Figure 9-11(b), to reduce the level to within the equalization or line build-out range of the new repeater.

For example, if a repeater has a specified loss range of 7.5 to 35 dB, and the loss between the junction and one of the adjacent repeaters is less than 7.5 dB, an attenuator (pad) can be installed to increase the new section loss to something greater than 7.5 dB. The attenuation should be adjusted to provide a total loss equal to the loss in the noninterfering segment. Pads, if used with older repeaters, should be located at the line receiver rather than the line transmitter to reduce reflection problems.

Pads also may be used at route junctions to reduce loss difference between segments A and B, as shown in Figure 9-12. Ideally, the difference $|L_A - L_B|$ should be zero, but it is only necessary to reduce the difference such that the relationship $|L_A - L_B| \leq (\text{FEXT} - \text{EF})$ is satisfied. For example, if $|L_A - L_B| = 10$ dB, FEXT $= 81$, and EF $= 63$ dB, the route junction will not cause interference.

9.1.8 Designing for Improved Error Performance

As indicated earlier, the original design procedures for repeatered T1-carrier used the characteristics of NEXT and SNR and assumed margins to give the required error performance for line sections. The bit error probability on a section can be improved to any practical, but arbitrarily small, value by increasing the SNR.

In a 1-cable operation, the limiting design factor usually is NEXT, so the SNR is increased by increasing the NEXT margin. In 2-cable and compartmental core cable operations, NEXT is not a design issue, so the SNR is increased by simply increasing the received signal level. In both cases, the SNR is increased by short-

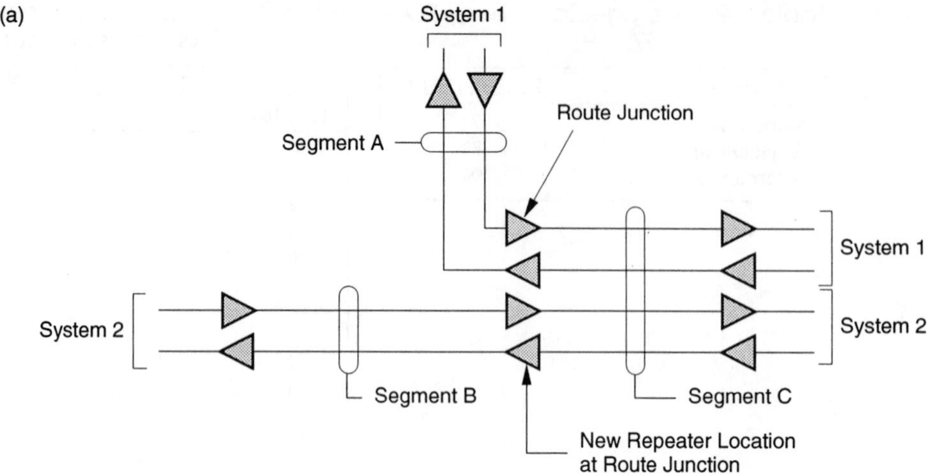

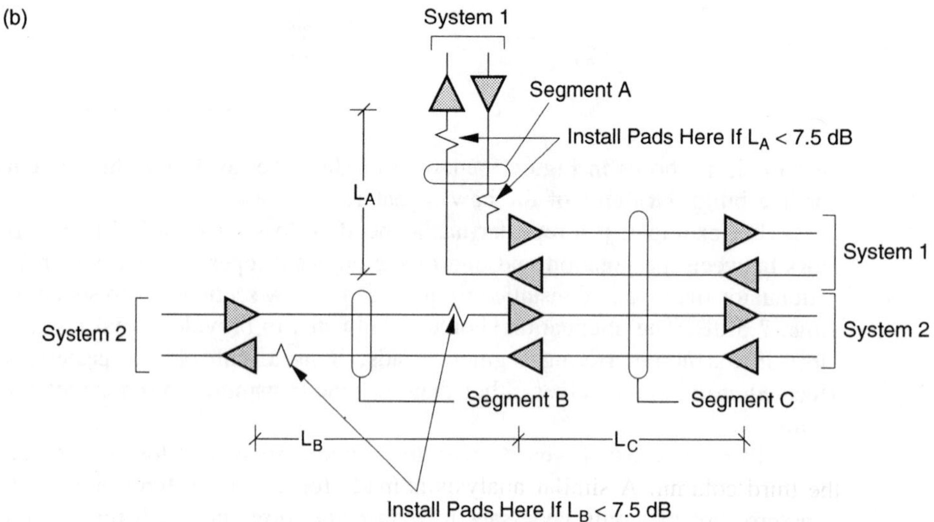

Figure 9-11 Reducing route junction interference problems: (a) Installation of a repeater to eliminate route junction interference problems; (b) Application of pads to accommodate repeater equalizer range

ening the section length. For a nominal error probability of 1E-6, the section loss is limited to the lesser of:

$$ISL_T \leq 33.5/CF_T \text{ dB (repeater loss limitation)}$$

or

$$ISL_N \leq [(m - s) - 32 - 10 \log n]/CF_T \text{ dB (crosstalk limitation)}$$

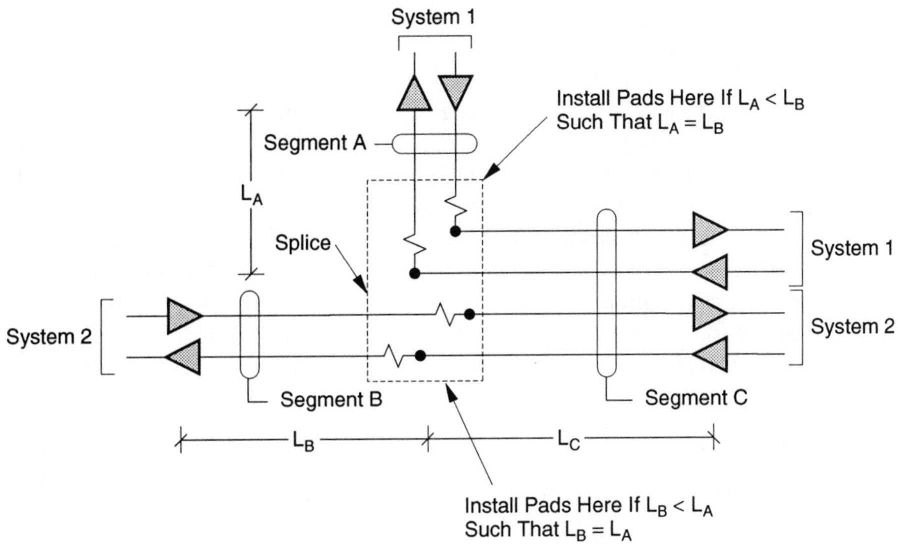

Figure 9-12 Using pads to reduce the loss difference between segments A and B

The constant (32) in the crosstalk limited loss equation accounts for the statistical nature of noise and signal variations. A new constant for any bit error probability can be determined by using the techniques described in [2]. Table 9-11 gives new constants for error probabilities between 1E-6 and 1E-12. To use this table, simply replace the constant 32 in the section loss equation with the new constant for the desired BER. For example, for BER = 1E-9,

$$ISL_N \leq [(m - s) - 33.3 - 10 \log n]/CF_T \text{ dB (crosstalk limitation for BER = 1E-9)}$$

As seen from the table and the equations above, the section loss is reduced by Δ in the third column. A similar analysis is made for a 1-cable operation, in which loss is not limited by NEXT, and for a 2-cable or compartmental core cable operation. Here, the error probability on the section is improved by reducing the section loss by a similar amount. Therefore, for these situations, the new loss equation becomes

$$ISL_T \leq (33.5 - \Delta)/CF_T \text{ dB (repeater loss limitation for BER = 1E-9)}$$

Table 9-11 Loss Equation Constant for Various Error Probabilities

Bit Error Rate	Constant	Δ (dB)
1E-7	32.0	0.0
1E-8	32.7	0.7
1E-9	33.3	1.3
1E-10	33.8	1.8
1E-11	34.2	2.2
1E-12	34.6	2.6

The above changes in the section loss equation are equivalent to increasing the chances that any given section will have an error probability <1E-7. The original line design procedures used a 95% probability that the BER on an entire line, having of up to 50 sections, would be less than 1E-7. This means that some sections would have an error rate of 1E-8 and some, 1E-9, and so on, due to the statistical distribution of section characteristics. Increasing the constant in the loss equation is the same as increasing 95% to some higher percentage.

In the AT&T repeatered T1 outstate line design, the constant is increased by 5 dB, from 32 to 37, to account for the allocation of error to up to 200 repeater sections rather than 50 in the normal design [8]. In other words, the chance of any one section exceeding a 1E-7 error rate is reduced from 5% ÷ 50 = 0.1% to 5% ÷ 200 = 0.025%.

9.1.9 Span Line Powering

Span line repeaters are powered from the central office through the simplex resistances of the directional cable pairs, as shown in the simplified diagram of Figure 9-13. The illustration shows a common powering scheme, but it is only one of several possible schemes. For example,

- The COR could be separately powered rather than loop powered.
- The powering scheme is different when the line repeaters are arranged for unidirectional operation.
- The powering scheme can be different where the span line terminates on a customer's premises rather than another central office or DLC remote terminal.

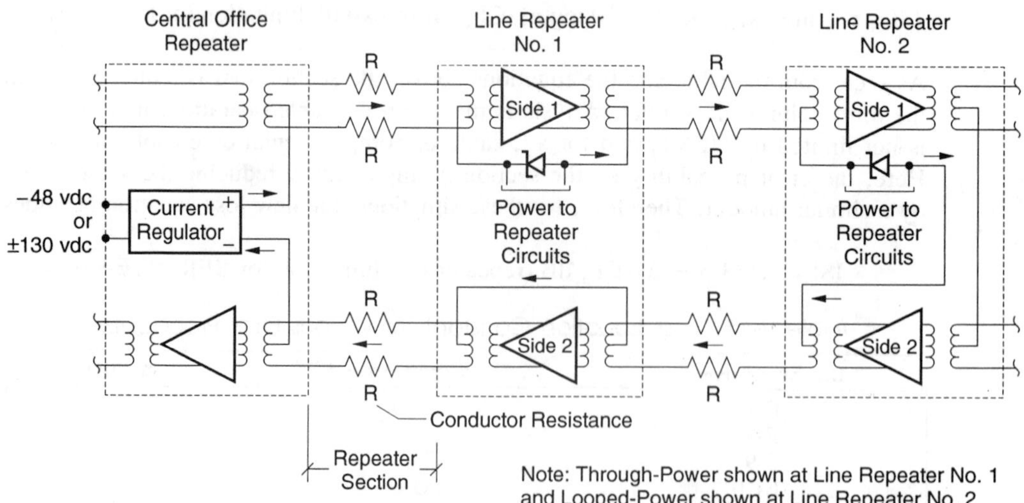

Figure 9-13 Simplex configuration for span line repeater powering

- A span line could be powered from each end and looped at an intermediate line repeater.

The line current regulator in the central office can accept various input voltages, as required, to support a constant current over different line lengths and line repeater powering schemes. Most line powering equipment can be set to provide 60, 100, or 140 mA line currents. The 60 mA current is preferred for all new systems, although higher line currents may be needed to offset high levels of powerline induction in special situations. Older line repeaters required 100 or 140 mA. In a given span, all line repeaters must be capable of accepting the same current. Some line repeaters can accept, without adjustment, any of the three "standard" currents noted above. To avoid administrative and maintenance problems, repeaters with different current requirements should not be mixed in a repeater housing. If this is done, field swapping may not work, may lead to confusion, and may prolong line maintenance activities.

In a simplex circuit, the two conductors of a pair are connected in parallel. In Figure 9-13, each *conductor* in a given section has dc resistance R Ω. The simplex resistance from one side 1 repeater to the next side 1 repeater is $R/2\ \Omega$. With loop powering, the two simplex circuits are connected in series, so the loop resistance of the simplex cable pairs between repeaters is $R/2 + R/2 = R\ \Omega$, or the same as the resistance of one conductor. The operating voltage for each line repeater (both side 1 and side 2) is derived from loop current flowing through a zener diode in side 1. The side 2 repeater is identical in all respects except for the zener diode in series with the loop. Instead, side 2 has a dc short, as shown.

The minimum voltage required to power the span line is determined by the voltage drop across the simplex loop. This voltage drop is the sum of the voltage drops across the resistance of each section, the line repeater power supplies (zener diode), and the central office repeater power supplies, or

$$\text{Line powering voltage} = V_{corn} + V_{s1} + V_{s2} + \ldots + V_{sn} + V_{corf}$$

where V_{corn} = voltage drop across the near-end central office repeater
V_{s1} = voltage drop across section 1
V_{s2} = voltage drop across section 2
V_{sn} = voltage drop across section n
V_{corf} = voltage drop across the far-end central office repeater

Two procedures are given below to calculate the line powering voltage. The choice of which procedure to use will depend on the information available. The first procedure, called the "resistance method," is used when the equivalent resistances of the central office and line repeaters are known. The second procedure, called the "voltage method," is used when the voltage drops of the repeaters, rather than the equivalent resistances, are known. It is possible to convert from one method to the other. If the voltage drop across a line repeater is known, the equivalent resistance can be found from a simple application of Ohm's Law. Similarly, if the voltage drops are known, the equivalent resistances can be found.

For example, say the equivalent resistance of a line repeater is given in its specification sheet as 128 Ω when the line current is 60 mA. The voltage drop across this repeater is IR = 60 mA × 128 Ω = 7.7 V. Similarly, the voltage drop across a central office repeater is given as 11 V at a line current of 60 mA. The equivalent resistance of this repeater is E/R = 11 V/60 mA = 183 Ω. When these types of calculations are made prior to determining the line powering voltage, it is important that the maximum equivalent resistances and maximum voltage drops are used, respectively, to ensure that the worst case is being considered.

The simplex loop resistance of twisted pair cables is shown in Table 9-12 at various temperatures and conditions. The average value generally is used in voltage drop calculations, but the maximum values are given if worse-case voltage drops need to be found. Resistance values also are given for temperatures of predominantly buried plant (38 °C) and predominantly aerial plant (60 °C). The resistance can be found at any temperature from the following relationship:

$$R_T = R_{20}\,[1 + \rho_{20}\,(T - 20)]$$

where R_T = resistance at temperature T (Ω)
R_{20} = resistance at 20 °C (Ω)
ρ_{20} = temperature coefficient of resistance = 0.00393 per °C
T = temperature (°C)

EXAMPLE 9-8

A span line between a central office and an RST consists of seven line repeaters and has a route length of 37,253 feet of 24 AWG buried cable. The line repeaters have a voltage drop of 7.7 V at 60 mA line current. The COR in the central office is locally powered and the COR in the RST is loop powered. The resistance of the RST COR is 183 Ω at 60 mA line current. Find the line powering voltage.

The simplex resistance of the buried cable (at 38 °C) is 27.6 Ω/kft × 37.253 kilofeet = 1,028 Ω. The simplex voltage drop is 1,028 Ω × 60 mA = 61.7 V. The voltage drop across the seven line repeaters is 7 × 7.7 V = 53.9 V. The voltage drop across the RST COR is 183 Ω × 60 mA = 11.0 V. The total voltage drop in this line is 61.7 V + 53.9 V + 11.0 V = 126.6 V. Therefore, the minimum powering voltage is 127 V. Typical span line power supplies have output voltages of 48 V, 130 V, 178 V, 260 V, and 356 V. In this case, the 130 V powering voltage would be selected.

Table 9-12 DC Resistance of Simplex Cable Pairs

AWG	Maximum Resistance[a] at 20 °C	Average Resistance[a] at 20 °C	Average Resistance[a] at 38 °C	Average Resistance[a] at 60 °C
19	8.7	8.2	8.8	9.5
22	17.2	16.2	17.4	18.8
24	27.3	25.8	27.6	29.9
26	43.9	41.4	44.3	47.9

[a]Simplex loop resistance is measured in units of Ω/kilofeet.

9.1.10 Repeatered T1-Carrier Loop Qualification

One of the very necessary requirements for repeatered T1-carrier span lines is the qualification of the cable pairs. Normally, each section is separately qualified, and variations from the specified requirements are investigated and resolved before the span line is turned up for acceptance testing. Tests are best performed after all line repeater housings have been installed so that splicing errors and defective stub cables can be found during the first round of tests.

The qualification tests are listed in Table 9-13. The insertion loss at 772 kHz and dc resistance are the most important tests. The capacitance test is quick and simple and will indicate water-flooded air core cable by abnormally high readings.

The ac induction tests will indicate excessive power influence. Older style repeaters (indicated as *standard* in Table 9-13) are more resistant to the effects of power influence; however, virtually all new systems use the so-called mini-T1 repeaters, which have a lower ac induction tolerance. The ac induction test is made as shown in Figure 9-14. If the voltmeter is to be used, the pair to be tested is first shorted and grounded at one end. The pair is then shorted at the other end and connected to ground through a 300 Ω, 5 W resistor. Any induced ac from nearby powerlines will cause current to flow in this resistor. The voltage developed across the resistor must not exceed 15 Vac (equivalent to 50 mA of ac current) for the mini-T1 repeater and 21 Vac (equivalent to 70 mA) for the standard repeater.

It is more convenient to measure power influence with a regular transmission noise test set. Set the tester to measure power influence with the 3 kHz flat filter and short the cable pair at both ends. The far-end is grounded as in the voltmeter test above. At the near end, connect the test set to ground and measure the power influence of the shorted pair. The 300 Ω resistor is not required. The power influence

Table 9-13 Repeatered T1-Carrier Loop Qualification Limits (per Section)

TEST	LIMITS
ac Induction	≤15 Vac or 119 dBrn3kHz (mini-T1 repeater) ≤21 Vac or 122 dBrn3kHz (standard T1 repeater)[a]
Insulation resistance at 250 Vdc test voltage	>450 MΩ[b]
Loop dc resistance, temperature corrected	Calculated value ± 3 Ω[c]
Insertion loss at 772 kHz, temperature corrected	Calculated value ± 1.0 dB[d]
Capacitance	Calculated value ± 2 nF

[a]See text for test methods using a voltmeter or transmission noise test set; see Chapter 5 glossary for definition of dBrn.

[b]Insulation resistance is measured tip to ring, tip to ground, and ring to ground.

[c]Loop dc resistance is measured with the far-end shorted.

[d]Insertion loss is made with 100 Ω termination in signal generator and signal detector.

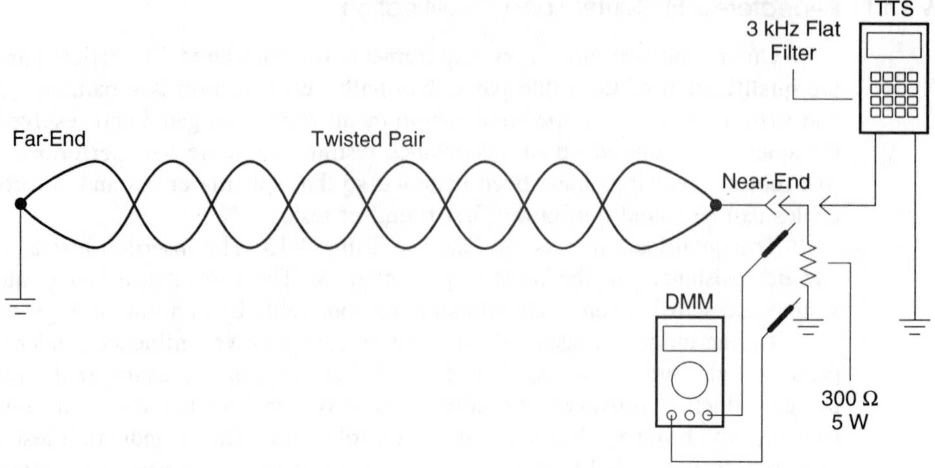

Figure 9-14 AC induction test

should be <119 dBrn3kHz for mini-T1 repeaters and <122 dBrn3kHz for standard repeaters.

Calculated values for insertion loss can be determined by multiplying the engineering losses given in this chapter by the section lengths. Similarly, the loop dc resistance may be determined by multiplying twice the dc simplex resistances given in Table 9-12 by the section lengths. The calculated and measured values of insertion loss and resistance must be temperature corrected to the same temperature. Capacitance can be measured by any digital multimeter with a capacitance scale. The calculated capacitance will be 15.7 nF/kilofoot times the section length for all exchange cables. If cables with other capacitance values are to be used, the capacitance must be accurately known for the test to be meaningful.

Values of capacitance, loop dc resistance, and insertion loss are given in Table 9-14 for typical section lengths. The actual measured values may vary from those shown, depending on the section lengths.

Table 9-14 Typical Values for Repeatered T1 Loop Parameters

PARAMETER	22 AWG	24 AWG
Capacitance	120 nF (intermediate section) 75 nF (end-section)	80 nF (intermediate section) 50 nF (end-section)
Loop dc resistance	250 Ω (intermediate section) 150 Ω (end-section)	250 Ω (intermediate section) 150 Ω (end-section)
Insertion loss at 772 kHz	29 dB (intermediate section)[a] 15 dB (end-section)[a]	24 dB (intermediate section)[b] 15 dB (end-section)[b]

[a]Values are for compartmental core cable.
[b]Values are for exchange cable (without compartmental core).

9.1.11 Fault Locating Systems

Span line fault locating equipment allows the span lines to be tested from one or both ends and a fault pinpointed to a particular repeater location before a maintenance dispatch is made. All line repeaters have a tertiary fault winding built into the output coupling transformers of the regenerator. This winding can be connected to either a passive or active fault locating filter, which is mounted in the housing with the repeater. A test signal is injected into the interrogation unit in the central office line termination shelf, and from there, coupled to the span line. The interrogation signal passes through a good repeater and is coupled to the output transformer fault winding. Each repeater location has a fault filter tuned to a different frequency. A typical fault locating system is shown in Figure 9-15.

Fault locating is performed by transmitting an interrogation signal from a pulse generator over the transmit side of the span line to be tested. At a selected repeater location, a voice frequency (VF) component of the interrogation signal is extracted by a bandpass filter and returned on the fault-locate pair to the interrogating end. A frequency-selective level meter is used to detect the incoming signal on the fault-locate pair. If the received signal is not returned at the expected level, the repeater with a fault filter corresponding to the interrogating frequency is bad.

The interrogation signal is a 1.544 Mbps bit stream, but the pulses are transmitted in trios with a large number of intentional bipolar violations (BPVs), as shown in Figure 9-16. The specific pattern of positive and negative trios determines the VF components of the signal. When this signal is passed through the fault locating filter, the 1.544 Mbps bit stream is filtered out and the VF components are passed through to the fault-locate pair. A good repeater will pass the interrogation signal through to the next repeater location.

The fault windings in each repeater in a given housing are connected in parallel (bridged) at the fault locating filter. This way, one filter and fault locating frequency can be used at each repeater housing to test all the span lines served by that housing. A polarized dc voltage normally is placed on the fault-locate pair at the central office to select either side 1 or side 2 of a given repeater. Where a repeater is wired for bidirectional operation, side 1 transmits in one direction and side 2 transmits in the other direction. In unidirectional operation, side 1 and side 2 transmit in the same direction.

Both passive and active filters are used in repeatered T1-carrier applications, but only active filters are used in new systems. Different filter frequencies, designated by an alphabetical code A through M, are used to select the repeater location. There are 12 filter frequencies in the fault locating system, which allows up to 12 repeater locations to be tested with one fault-locate pair. If more than 12 locations are to be tested, an additional fault-locate pair is required for each group of 12 filters. If a span line route splits, a fault-locate pair is required for each route beyond the splitting point. One of the pairs can be used for the portion of the route shared by the span lines and for one of the lines beyond the splitting point. The second, and subsequent pairs, are used along each of the other routes after the split. Filter frequencies and typical filter output levels are shown in Table 9-15.

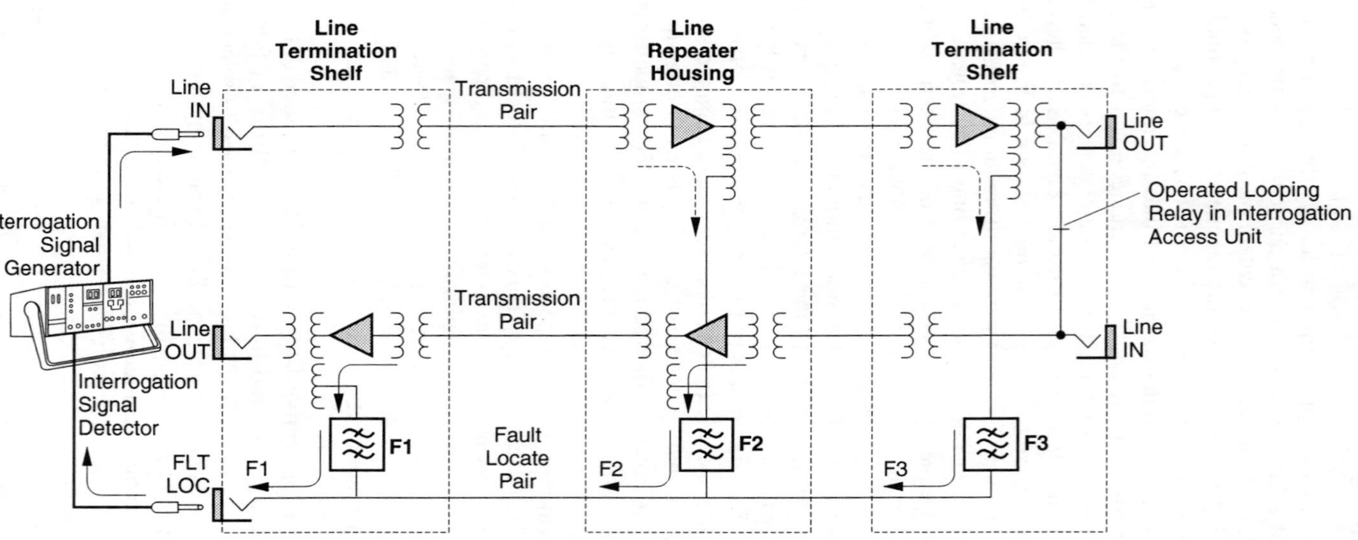

Figure 9-15 Typical T1-carrier fault locating system

(a)
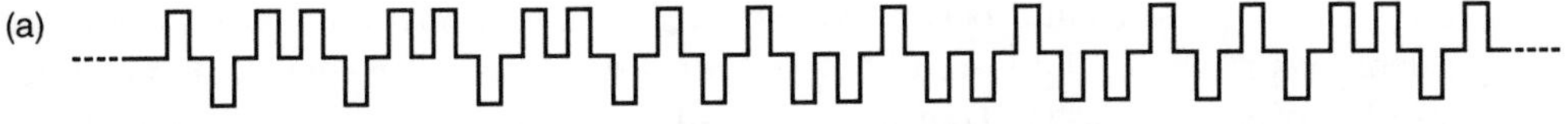

(b)
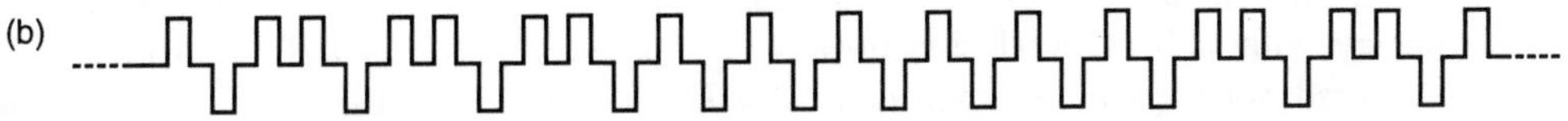

(c)
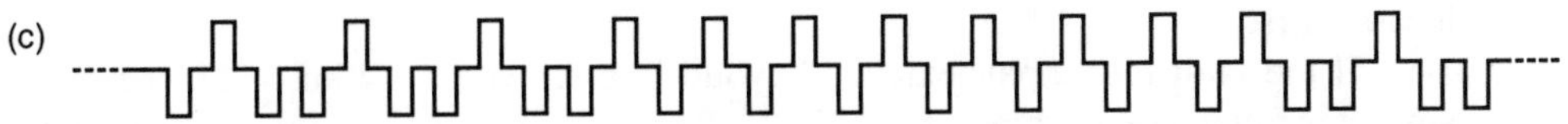

(d)
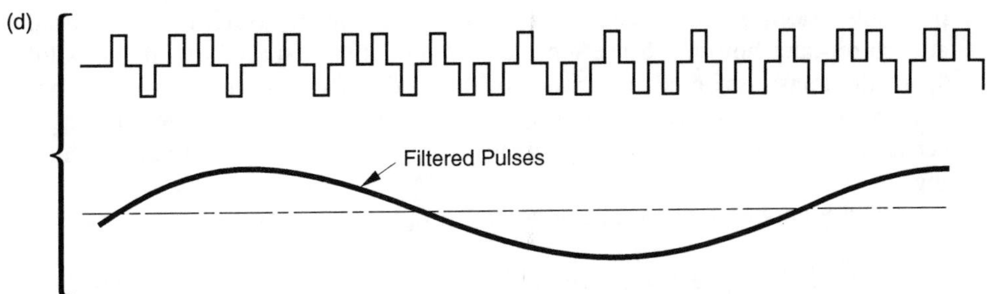

Figure 9-16 Interrogation pulses: (a) Positive and negative trios; (b) Positive trios and bipolar AMI; (c) Negative trios and bipolar AMI; (d) Positive and negative trios compared to filtered pulses

Table 9-15 Typical Fault-Locate Filter Characteristics

		Nominal Filter Output Level (dBm)[a]	
Filter Code	Filter Frequency (Hz)	Passive Filter	Active Filter
A	832	−56.0	−29.0
B	928	−54.5	−29.0
C	1,048	−52.5	−29.0
D	1,206	−51.5	−29.0
E	1,340	−51.0	−29.0
F	1,508	−50.0	−29.0
G	1,722	−48.0	−29.0
H	2,008	−46.5	−29.0
J	2,193	−46.0	−29.0
K	2,413	−45.5	−29.0
L	2,680	−44.5	−29.0
M	3,017	−43.0	−29.0

[a]The output levels can decrease by up to 6 dB depending on the number of repeaters connected to filter. Active filter output may vary by up to ±4 dB with the temperature and impedance of the fault-locate pair.

The design of fault locating systems requires transmission and powering considerations if active filters are to be used, or just transmission considerations if passive filters are to be used. Transmission design uses analog VF techniques; however, the VF transmission requirements differ considerably from analog subscriber lines. The maximum design loss of a fault-locate pair is 20 dB at the highest filter frequency, although higher losses may be acceptable if noise is well controlled. The measured noise on fault-locate pairs should not exceed 20 dBrnC = decibels with respect to reference noise, C-message weighted (dBrnC). In addition to the information given below, [11] can be used to determine the VF characteristics of fault-locate pairs.

If the fault-locate pair extends beyond five repeater locations or if the loss at any filter frequency is greater than 20 dB, the fault-locate pairs generally are loaded. Unlike subscriber loops, however, load coils on fault-locate pairs do not have to be precisely spaced along the route. Normally, 88 millihenry (mH) coils are installed at each repeater housing. Since repeaters are not always spaced exactly six kilofeet apart, the transmission characteristics of the loaded fault-locate pair will not be the same as loaded subscriber loops. Where the distance between repeater housings exceeds six kilofeet, the load coils generally are installed halfway between each housing.

The following guidelines can be used to place the fault filters:

- Place the highest frequency filters at central office locations, if possible.
- Place the highest frequency filter farthest from the interrogation point, and progressively lower frequencies closer to the interrogation point.
- Do not place two filters with the same frequency on the same fault-locate pair.

Since the signal levels available at the filter output are relatively low, it is important that the noise on the fault-locate pair be low, also. A design goal is to obtain a 10 dB SNR and a 20 dB signal-to-interference ratio. Noise can come from many sources, most likely powerline induction. Interference is also possible from filters other than the one being interrogated. Fault-locate pairs should be tested for circuit noise using C-message weighting. Table 9-16 can be used as a guide after tests are made to determine if a fault-locate pair is suitable. In general, if the fault-locate pairs provide the required 10 dB SNR, the fault locating system will work acceptably.

Active filters are of the dual-amplified type. This means two amplifiers are included in each filter module in a repeater housing. One amplifier is connected to

Table 9-16 Fault-Locate Pair Noise Guidelines

Fault-Locate Pair Length (kft)	Maximum Circuit Noise (dBrnC)
$0 \leq \text{length} \leq 40$	14
$40 < \text{length} \leq 80$	33
$80 < \text{length} \leq 160$	45
$160 < \text{length} \leq 320$	57

the side 1 regenerators and the other is connected to the side 2 regenerators. In 1-cable and compartmental core cable operation, the side 1 regenerator is for the Tx direction and side 2, for the Rx direction. The particular amplifier, and therefore the direction of the repeater being tested, is selected by the polarity of the dc voltage placed on the fault-locate pair. This voltage not only selects the appropriate direction but also powers the filter. For 2-cable operation, side 1 is for one span line and side 2 is for another span line, both in the same direction. Therefore, the polarity selects the filter for one of two spans. Side selection is not possible, and powering is not required, with passive filters.

The required filter powering voltage depends on the filter manufacturer. Typical minimum values are 10 to 17 Vdc at the filter. This means the last filter on the fault-locate pair must have 10 to 17 Vdc across it. The typical filter draws 0.5 to 0.8 mA at 60 Vdc, but the current-voltage characteristic curve is nonlinear. For design purposes, however, these values can be used to determine the fault-locate pair powering voltage.

The fault-locate pair must provide adequate voltage support at the last fault filter on the fault-locate pair. If this is done, all other filters will have sufficient operating voltage. A simplified diagram of the fault-locate pair and filters is shown in Figure 9-17. The circuit analysis is straightforward but tedious. The problem can be simplified somewhat by assuming that each repeater section has equal length (and equal dc resistance, R_s) and that each fault filter draws equal current (i_{ff}). The required fault-locate pair powering voltage is then

$$V_{fp} \geq V_{ff} + (N \times i_{ff}) [R_{fd} \times l_{fd} + R_{fl} \times l_{fl} (N - 1)/2]$$

where V_{fp} = minimum fault-locate pair powering voltage (V)
V_{ff} = minimum filter powering voltage (V)
N = number of filters outside the central office connected to the fault-locate pair
i_{ff} = current draw of each filter (A)
R_{fd} = unit loop resistance of the fault-locate pair feeder (Ω/kilofoot)
l_{fd} = length of fault-locate pair feeder (kilofeet)
R_{fl} = unit loop resistance of the average repeater section (Ω/kilofoot)
l_{fl} = length of the average repeater section (kilofeet)

A repeater housing can hold up to 25 repeaters (or 50 regenerators, since each repeater has a side 1 and a side 2 regenerator). The filters in any given housing serve all the repeaters in that housing. The signal outputs from all passive and most active filters decrease with each additional repeater installed due to input loading. The output decreases by 6 dB when the number of repeaters increases from 1 to 25 (50 regenerators). The outputs from some active filters do not vary with input loading.

The received signal level can be found from:

$$\text{Received signal level (dBm)} = P_f - 0.24 \times N - L_c(f) - L_s \text{ dB}$$

where P_f = output power from the filter for one repeater (dBm)
N = number of repeaters loading the filter input

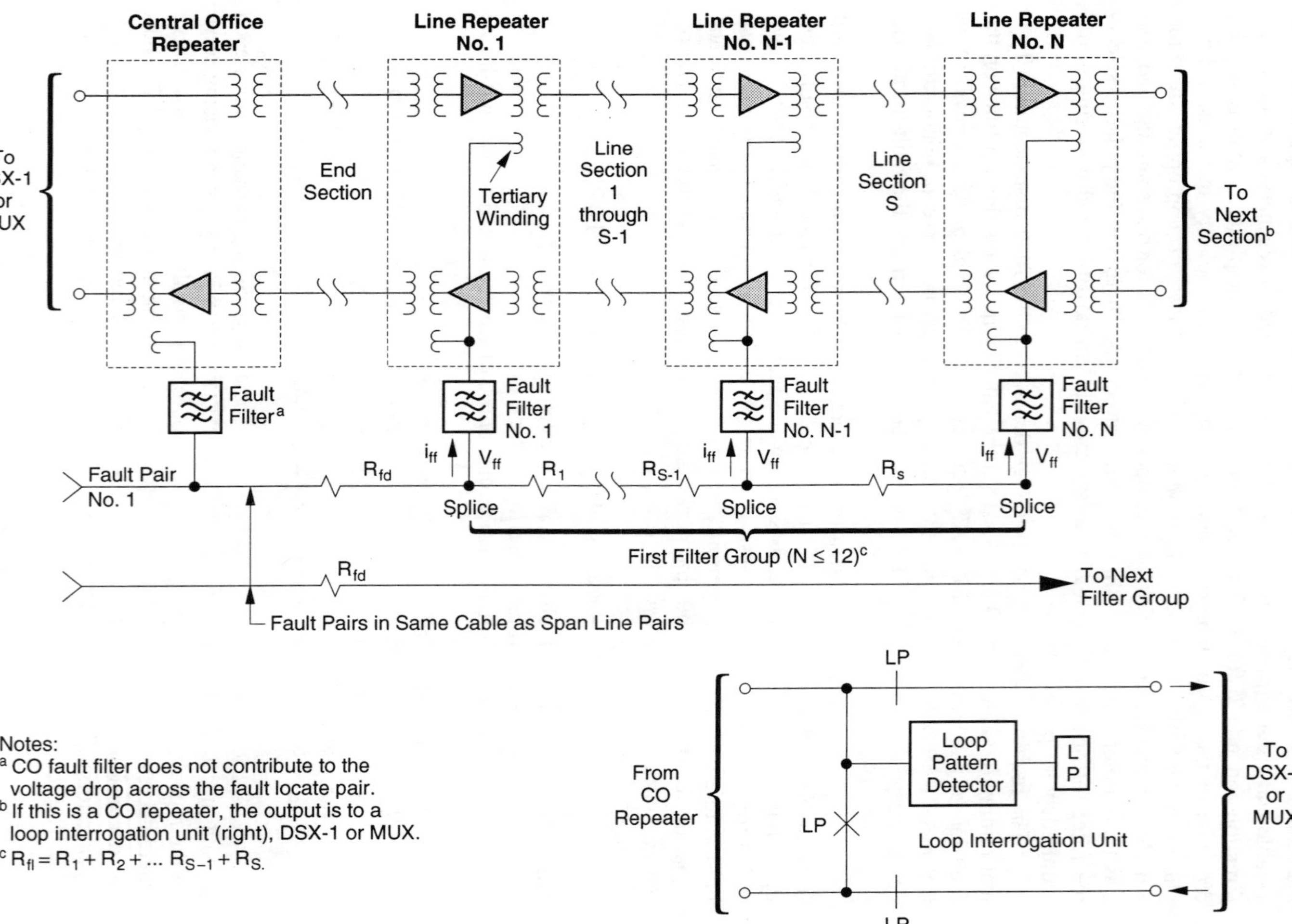

Notes:
[a] CO fault filter does not contribute to the voltage drop across the fault locate pair.
[b] If this is a CO repeater, the output is to a loop interrogation unit (right), DSX-1 or MUX.
[c] $R_{fl} = R_1 + R_2 + \ldots R_{S-1} + R_S$.

Figure 9-17 Fault locating system

$L_c(f)$ = cable loss as a function of frequency, from Table 9-17, adjusted for length (dB)

L_s = splitting loss = 3.0 dB

The splitting loss is 3.0 dB because the fault locating signal is split at the repeater housing; half the power can travel back toward the interrogator and half can travel toward the other end of the line. The cable loss varies with frequency and is different for each cable gauge. Table 9-17 gives the unit loss for common cable types and can be used to estimate the loss on fault-locate pairs. The loss is given at 68 °F, but fault locating system transmission design must be made at the worse-case temperature expected on the route. These temperatures are 100 °F for predominantly buried cable and 140 °F for predominantly aerial cable. Therefore, Table 9-17 gives an adder to adjust the cable loss at 68 °F to the higher temperatures. For example, the loss for 22 AWG buried cable at 2,193 Hz at 68 °F is 0.16 dB/kilofoot; the loss at 100 °F is 0.16 + 0.01 = 0.17 dB/kilofoot.

Table 9-17 Unit Loss of H88 Loaded Cables at 20 °C (68 °F)

Fault Filter Frequency (Hz)	Filter Code	Loss (dB/kft) 19 AWG	Loss (dB/kft) 22 AWG	Loss (dB/kft) 24 AWG	Loss (dB/kft) 26 AWG
Coil Spacing→		5.7 kft	6.0 kft	6.0 kft	4.0 kft
832	A	0.083	0.16	0.24	0.35
928	B	0.083	0.16	0.24	0.36
1,049	C	0.083	0.16	0.24	0.36
1,206	D	0.083	0.16	0.24	0.36
1,340	E	0.083	0.16	0.24	0.36
1,508	F	0.083	0.16	0.24	0.36
1,722	G	0.083	0.16	0.24	0.37
2,008	H	0.084	0.16	0.24	0.37
2,193	J	0.085	0.16	0.24	0.37
2,413	K	0.087	0.17	0.25	0.38
2,680	L	0.091	0.17	0.25	0.38
3,017	M	0.10	0.19	0.28	0.40
Adder for 38 °C (dB/kft)		0.005	0.01	0.02	0.02
Adder for 60 °C (dB/kft)		0.011	0.02	0.04	0.05

9.1.12 Order Wire

The order wire unit allows technicians at a repeater location to communicate with another technician at the central office or remote terminal. A separate cable pair is used for the order wire, and it frequently is connected to a central office line circuit. This allows the technician in the field to obtain a dialtone and call a central service or test center. With this feature, a technician does not have to be at the terminal location. Only one order wire is needed per route regardless of the number of individual span lines or repeaters along the route.

Where the order wire is connected to a central office line circuit, it is restricted to the signaling and supervision range of the line circuit. This range can be extended with loop extenders and VF repeaters, if necessary. If the order wire units are not connected to a central office line circuit, the loop resistance is limited to approximately 2,200 Ω. This range can be extended by powering the loop from both ends or boosting the battery feed voltage. When the order wire is powered from both ends, blocking capacitors are required to isolate the battery feeds from each other. These capacitors are usually located halfway between the two ends.

When the order wire length exceeds 18 kilofeet, it is loaded according to normal practice with 66 mH or 88 mH load coils spaced at 4.5 or 6.0 kilofoot intervals, respectively. When the repeater housings are spaced exactly 6 kilofeet apart, the order wire load coil is conveniently installed in the repeater housing (usually on the order wire card itself).

9.2 Line Terminating Equipment

The fault locating and order wire systems already discussed are part of both the span line and line termination equipment. Additional termination components include the CORs, test signal sources and detectors, DSX, digital cross-connect system (DCS), and automatic protection switches (APSs).

Span lines do not have to terminate in a central office. The lines may terminate on a user's premises or in a DLC remote terminal unit. In the case of the DLC, the CORs may be identical to those used in the central office, or they may be specially designed units that are integral to the DLC. In the case of a user's premises, the COR functions usually are included in the channel service unit, or CSU, located on the premises.

The CSU serves as a complete DS-1 rate interface, loopback, and demarcation point. The CSU contains two basic functional blocks: a network interface block and a feature block. A functional block diagram of a digital network terminated in CSUs at each end is shown in Figure 9-18(a). The interface and feature blocks are detailed in Figure 9-18(b). Referring to this figure, the CSU network interface provides:

- DC isolation between the network and the user's installation
- DC power looping or power-through of the simplex powering current from the network
- Loopback of the network signal and loopback indication signal to the user's installation
- Signal regeneration of the network signal through the looped back transmission path, but no regeneration of the user's signal
- Selectable attenuation (line build-out) of the user signal outgoing to the line
- Surge protection on the network side
- Resistive terminations for all signals

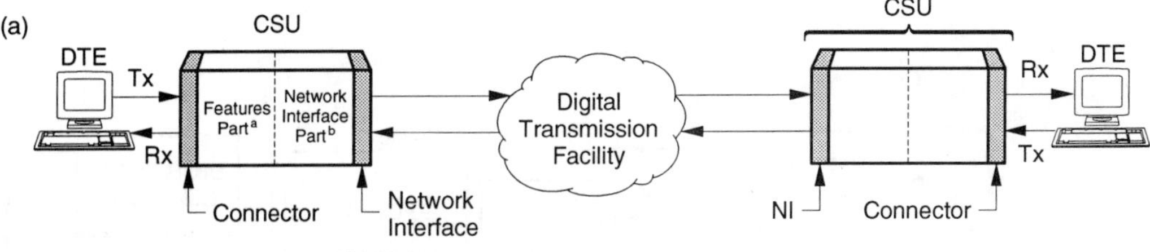

[a] Figures 9-18(c) and (d)
[b] Figure 9-18(b)

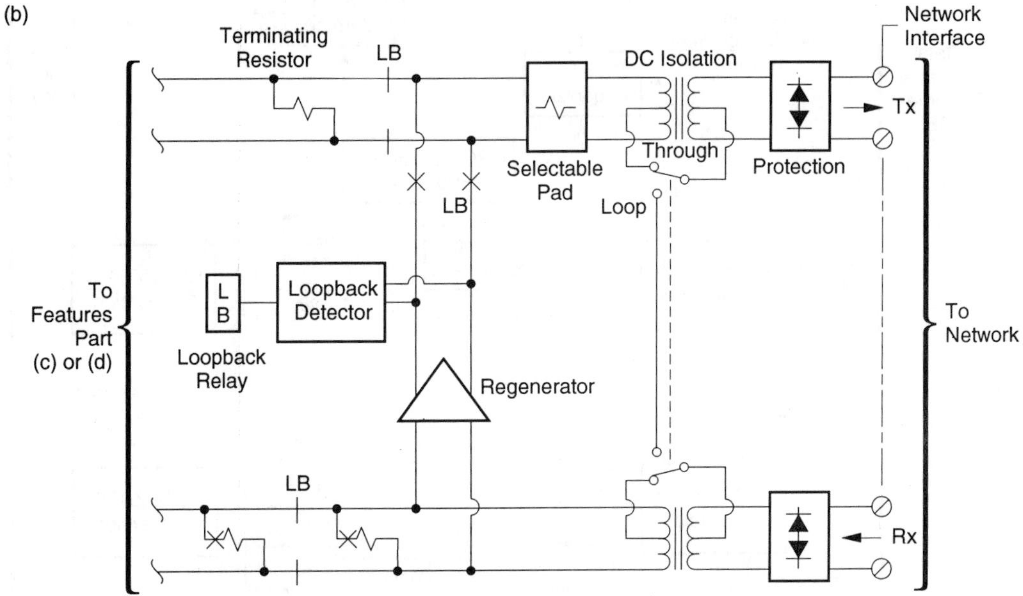

Figure 9-18 CSU block diagram: (a) Overview diagram of end-to-end network diagram; (b) Network interface block; (c) CSU feature part for SF to ESF interface; (d) CSU feature part for ESF to ESF interface

Other specifications and details of the network interface can be found in [9,10]. The features of a CSU are manufacturer specific. A typical CSU can terminate DS-1 rate data terminal equipment (DTE) and span lines with either superframe formatting (SF) or extended superframe formatting (ESF). Figure 9-18(c) applies when the DTE signal is configured for SF and the lines are configured for ESF.

For SF to ESF interfaces, the incoming signal from the local DTE or central office terminal is first converted from B8ZS or ZBTSI, if necessary, to the internal signal required by the CSU, usually a binary unipolar signal. The 1.536 Mbps payload is then separated from the 8 kbps SF framing by an internal multiplexer. The payload is passed through the CSU while the SF framing is converted to ESF in the

(c)

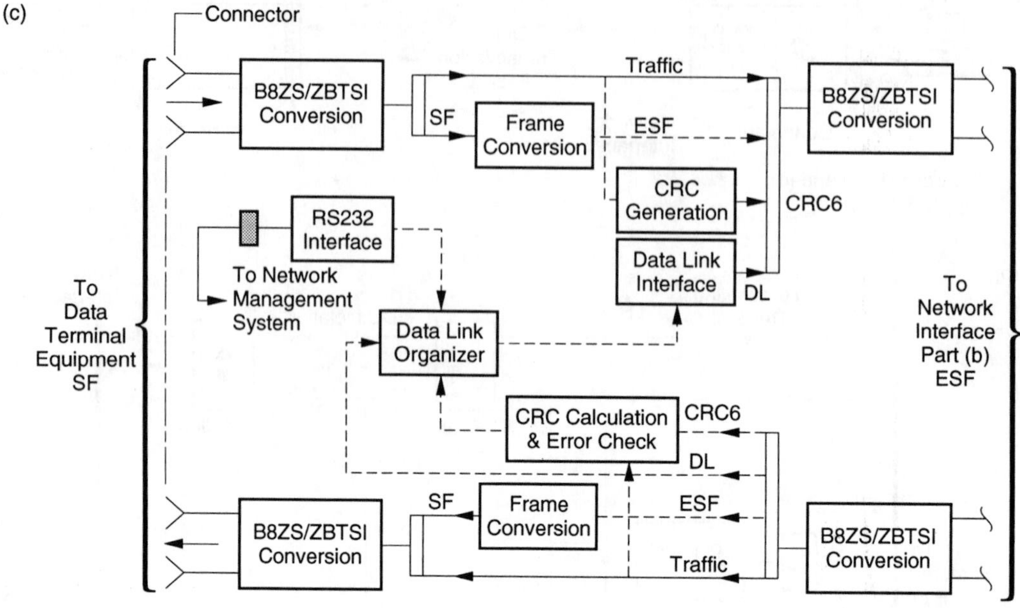

(d)

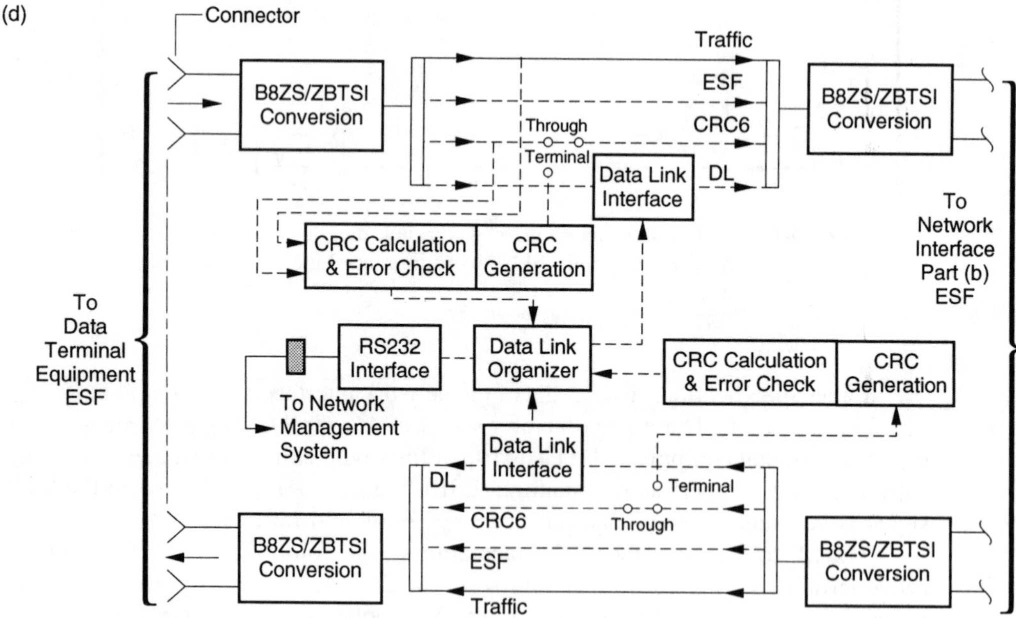

Figure 9-18 Continued

frame converter. Meanwhile, a 6-bit cyclic redundancy check (CRC6) block is generated from the payload and combined with the payload, ESF framing, and data link bit streams. The data link is a network management control channel that provides network management functions and performance data. The aggregate signal at 1.544 Mbps is converted back to B8ZS or ZBTSI, if necessary, equalized, and sent out on the line.

For a signal coming into the CSU from the line, similar actions take place in terms of framing and B8ZS or ZBTSI conversion; however, instead of generating a CRC6 block for transmission on the line, a CRC6 block is received from the line and compared to a local CRC6 calculation. An error report is made to the data link for transmission on the other side of the CSU. Also, the incoming data link is presented to an external network management system.

The same CSU can be optioned for ESF to ESF interface, as shown in Figure 9-18(d). In this case, however, the signal from the local terminal already is in the ESF format. A CRC6 calculation is made on the incoming payload from both directions and regenerated if necessary. Otherwise, the CSU is very similar to the SF to ESF interface described above. CSUs usually use one of the RJ-48 series jacks for user connection.

In central office applications, the DSX normally is considered the demarcation point between the line and equipment. Where a DSX does not exist, other types of jack arrangements can be used, such as a line access unit or the jacks on the COR.

APS can be provided as a part of line terminations. The APS automatically switches out a failed span and switches in a protection (spare) span. APSs with one-for-one (1:1) capability usually are installed on interoffice T1-carrier lines that carry embedded data from subrate digital loops. In a 1:1 configuration, each active service line is protected with a backup protection line. The switches give high availability (and high cost) to the digital data service.

A more distributed protection configuration normally is provided for the DLC that transports only voiceband traffic. In this case, a 1:2 protection configuration, in which one protection (spare) span backs up two active spans, is relatively common. In typical interoffice applications, a 1:4 to 1:14 protection configuration is common. A 1:2 protection configuration is shown in Figure 9-19.

APS continuously monitors the service (active) lines for loss-of-signal (LOS), out-of-frame (OOF), BER, and alarm indication signal (AIS) and compares these parameters to preset performance criteria. See Chapter 6 for further discussions of these terms. The APS instantly switches from the service line to the protection line whenever an LOS, OOF, or AIS occurs. It also switches whenever the measured BER degrades below the preset values. Some systems can be set to switch only on LOS, OOF, or AIS and to ignore BER on the service line; however, high BER usually leads to loss of framing, and switching takes place when the OOF event occurs.

The length of the span lines will influence the choice of protection. For example, a system requiring two span lines over a distance of two miles in a benign residential environment may not be equipped with any protection capability. A similar application over a distance of 20 miles in a rural environment may be equipped with 1:2 protection to account for the much larger number of line repeaters and

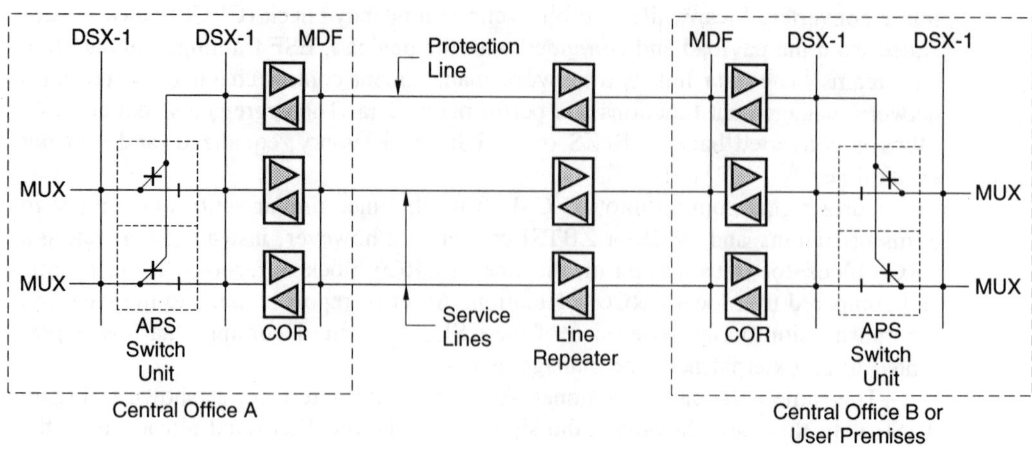

Figure 9-19 1:2 Automatic protection switch configuration

damage exposure in a potentially hostile environment. A DLC system serving a high-density business center with a high proportion of digital data may be equipped with 1:1 protection.

In 1:N protection schemes, where N > 1, once the switch has been made to a protection line, the APS is locked out to preclude another service line from switching to the protection line. The typical APS allows certain service lines to have higher priority than others. This means if a low priority service line switches to a protection line and a higher priority service line subsequently fails, the lower priority line is dropped back and the higher priority line is switched to the protection line. If another, lower priority service line fails after the higher priority line, the APS takes no action.

The detection time for BER depends on the BER threshold setting. Detection times are discussed in Chapter 6. All protection systems can be set to switch on BER values between 1E-3 and 1E-7, and some go as low as 1E-9. The APS normally is set to switch back automatically to the service line when trouble on the service line has cleared. The switchback normally does not take place until 20 seconds after the trouble clears and the BER has decreased below the protection threshold.

Some switching hysteresis is used. For example, if the switching threshold is set to 1E-5, the APS will switch as soon as the BER on the service line degrades to 3E-6. In the case of switching caused by high BER, the switch does not take place unless the protection line has a BER better than the service line. Once the switch has been made, the switchback to the service line will not take place until the service line improves to 1E-5, or better. In the case of LOS, OOF, or AIS on the service line, a switch to the protection line is made even if its BER is worse than the BER threshold.

BPVs and terminal framing bit errors are used to calculate the BER for lines using the SF. In the case of ESF with B8ZS encoding, BPVs are intentionally introduced to suppress long strings of binary zeros. Therefore, an APS that protects service lines using B8ZS must be compatible with it. Early protection systems are not B8ZS capable but most late model systems are.

A quasi-random signal source (QRSS) frequently is part of the span line terminating equipment. It is used to test central office and line repeaters and is used as a "keep-alive" signal on the protection line of an APS. The QRSS is an integral part of most automatic protection systems but is a separate plug-in card or external test set when used for testing. The QRSS appears to be a random pattern of binary ones and zeros, but it actually is a known, repeating sequence. For testing, the QRSS is injected into the line and either received or looped back at the far-end. A QRSS detector receives the signal. If the received bit pattern is exactly as expected, no errors have occurred. If there are errors, they are detected and can be used to drive an error counter. The QRSS cannot be used on a span line carrying live traffic.

Bridging repeaters are installed in line terminating bays and are used for the following applications, shown in Figure 9-20:

- Obtain a synchronization signal from an active span line
- Control signal levels of span lines connected in tandem at a central office
- Connect error detectors and other test interfaces

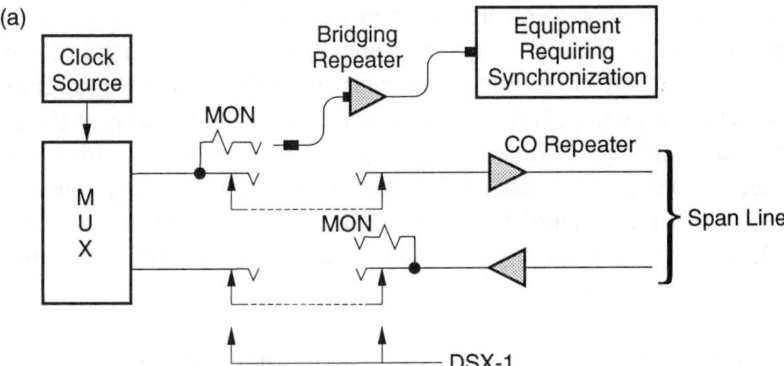

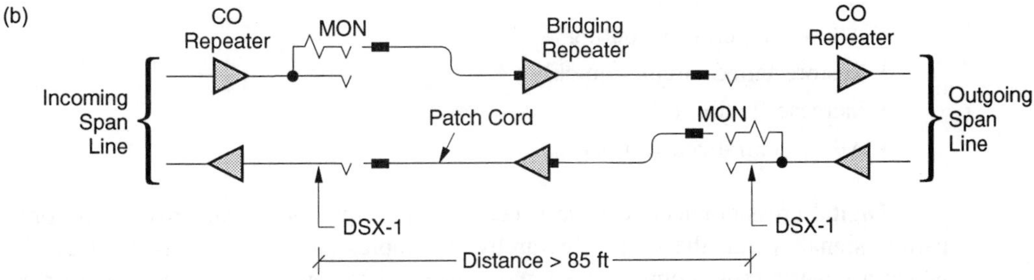

Figure 9-20 Bridging repeater applications: (a) Synchronization; (b) Tandem span lines; (c) Connecting test equipment

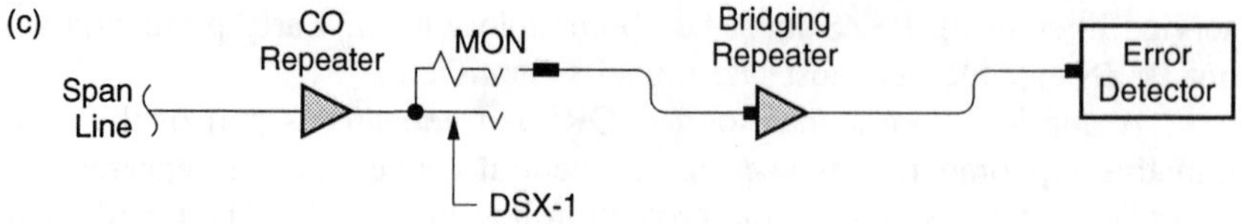

(c) Span Line — CO Repeater — MON — DSX-1 — Bridging Repeater — Error Detector

Figure 9-20 Continued

A bridging repeater is specially designed to accept low signal levels and retransmit them at the DSX level. It normally is connected to the MON (monitor) jack or a special isolated jack that allows a nonintrusive tap. This nonintrusive connection allows bridging repeaters to be used for in-service maintenance patching.

9.3 Digital Cross-Connect Equipment

Digital cross-connect systems are used to access and rearrange digital channels and circuits and are an integral part of any DS-1 transmission system. Digital cross-connect systems range in sophistication from simple panel-mounted electromechanical jacks and plugs to electronic multiprocessor controlled systems. Jack and plug systems operate at the DS-1, DS-1C, and DS-3 rates, whereas electronic systems operate at DS-3 and DS-1 aggregate rates and DS-0 and subrate channel rates. In this context, a digital *circuit* can be considered to be working at the primary transmission rate, and the *channels* working at subsidiary rates within the circuit. For example, a repeatered T1-carrier operates at the DS-1 primary rate of 1.544 Mbps, while each of its 24 channels operates at the DS-0, 64.0 kbps subsidiary rate.

Digital cross-connect systems offer significant advantages to digital transmission and network systems:

- Speed rearrangement of digital channels and circuits
- Allow hubbing, grooming, and consolidation of individual channels and circuits
- Allow temporary rerouting
- Restore failed circuits quickly
- Increase flexibility
- Allow centralized test access

Digital cross-connects can be used with any equipment that provides a compatible signal within the digital hierarchy. Examples are given in Figure 9-21, in which a generic cross-connect is shown connected to a variety of equipment operating at the DS-1 rate. With this arrangement, any circuit can be interconnected with any other circuit. In the following discussion, the term DSX refers to the electromechanical devices, and DCS refers to the electronic systems.

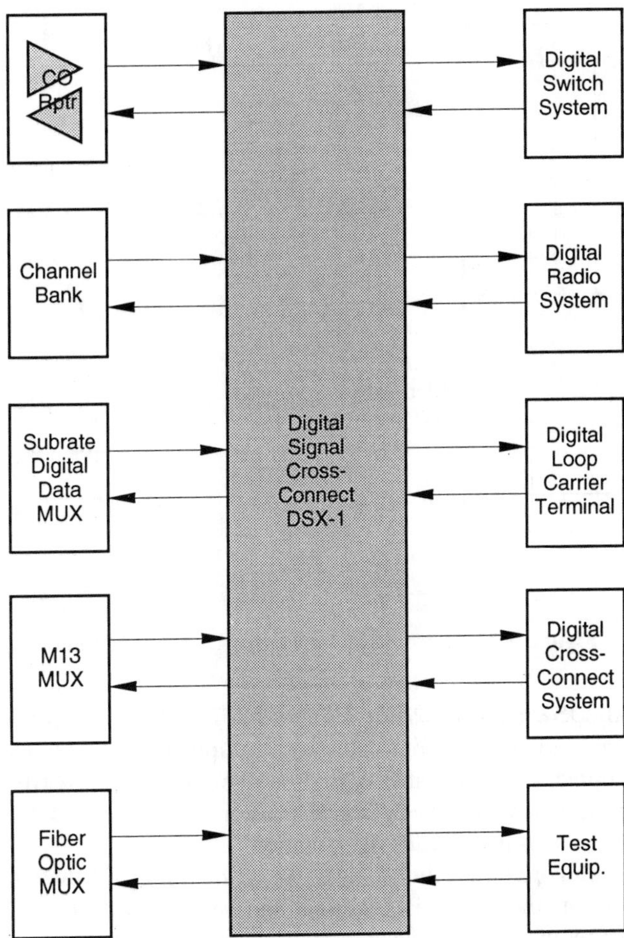

Figure 9-21 Typical digital cross-connect application

9.3.1 Manual DSX

The following discussion is limited to DSX panels operating at the DS-1, DS-1C, and DS-2 rates. The DSX is primarily an electromechanical (plug and jack) system that functions as a digital signal distribution frame. A DSX allows manual rearrangement of digital circuits at a given primary rate, such as DS-1 or DS-1C, but does not allow access to individual channels at subsidiary rates, such as DS-0 or subrates. The DSX is transparent to the network and has negligible effect on transmission performance. Detailed specifications for manual DSX panels are given in [12–14].

DSX panels are wired to multiplex equipment and digital lines as shown schematically in Figure 9-22. Labeling for line and equipment connections are shown based on industry standards [8]. Signal directional sense and circuit labeling is described in greater detail later.

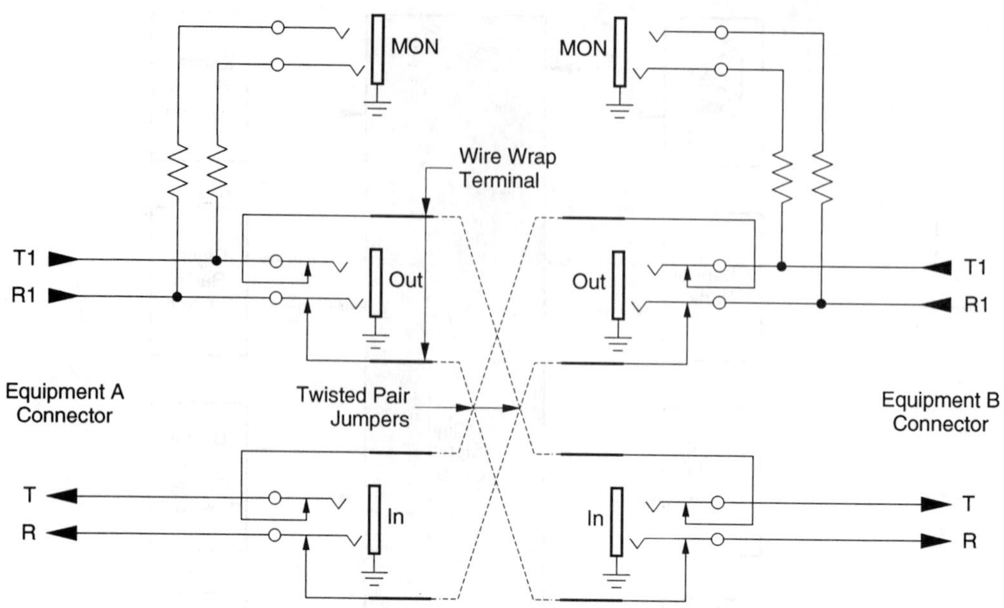

Figure 9-22 Schematic drawing of DSX-1, -1C, and -2 panels

All DSX panels provide IN and OUT jacks and most have MON jacks. The MON jack allows a nonintrusive test equipment connection. With the appropriate test equipment, a live traffic signal can be monitored for BPVs or other errors. As seen in the schematic, the MON jack is connected to the OUT jack through isolating resistors, which provide 20 dB isolation.

Patch cord connections to the line and equipment IN/OUT jacks are intrusive, which means the circuit is disturbed and any live traffic is disrupted. Such connections are required when tests are made to and from the line or equipment side of the digital circuit, or when temporary connections are made to other circuits. As the patch cord plug is inserted into the jack, the circuit is broken and one side is isolated. For example, when a plug is inserted into the equipment OUT jack, the patch cord is connected to the equipment transmit port and the line port is isolated. This feature allows pieces of equipment or lines to be quickly connected or disconnected. Dummy plugs or termination plugs also are used to disconnect equipment. Termination plugs have a 100 or 110 Ω resistor connected across the tip-and-ring leads.

The DSX panel provides interconnections through semipermanent jumpers or temporary patch cords. Such an arrangement is illustrated in Figure 9-23, where DS-1 port 1 on a digital switching system is permanently connected to span line 1 by a 3-pair jumper wire. The same port is temporarily connected to span line 2 by a 4-wire patch cord. A commonly made mistake when patching is to connect an OUT jack to another OUT jack. Normally this will not harm equipment, but the patch will not work, for obvious reasons. A patch always is made from one OUT jack to another IN jack. Patch cords should be shielded and have a 100 or 110 Ω characteristic impedance.

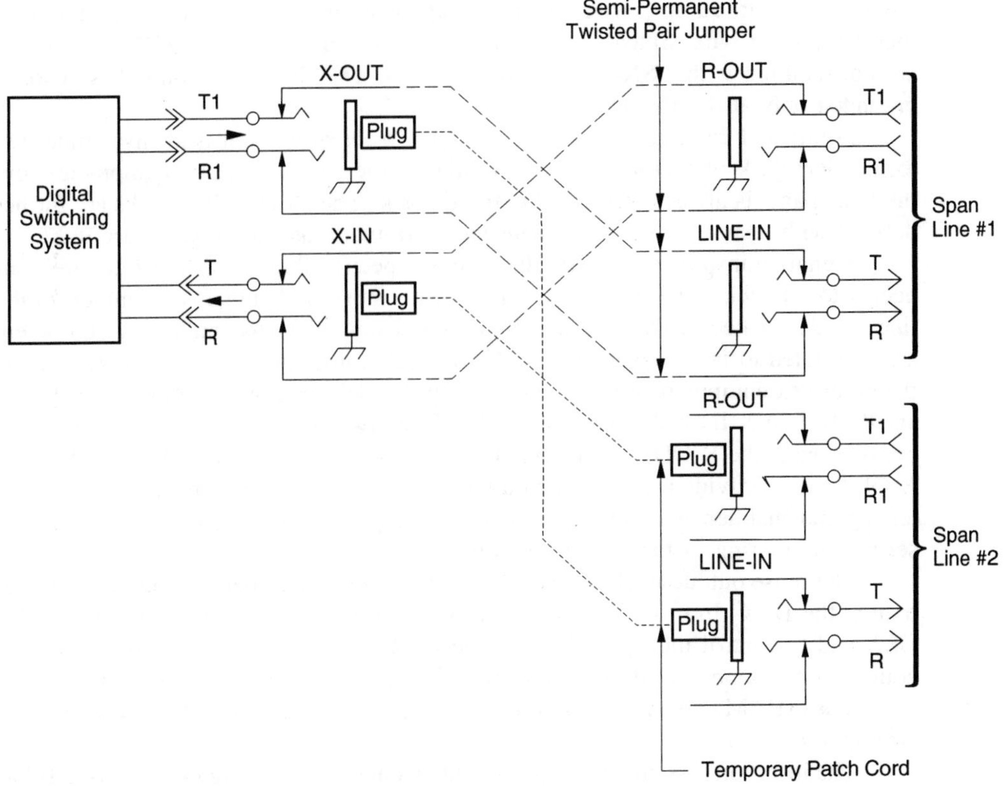

Figure 9-23 DSX patching

Patch cords serve four basic purposes:

- *Restoration*. When a line fails, operation of the equipment connected to it may be restored by patching to a spare or maintenance line.
- *Rerouting*. A working system can be temporarily rerouted to a new line while permanent jumpers are installed.
- *Looping*. The OUT and IN jacks of a given jack set are patched together to connect the transmitter to the receiver for loopback testing during maintenance and installation.
- *Cutovers*. Patch cords are connected in parallel with existing jumpered circuits. This isolates the existing jumpered connections. The existing jumpers are then removed and new jumpers installed for the new facility. At cutover, the patch cords are removed to transfer service.

A semipermanent cross-connection is made by jumpering the IN and OUT jack set of one circuit to the OUT and IN jack set of another circuit (note the IN to OUT transposition). The jumper wire used to make these connections must be twisted to reduce crosstalk. Jumper wire for digital applications generally has a tighter twist than VF jumper wire. Shielded twisted pair jumpers are used with DSX-2 panels but are not required with DSX-1 or DSX-1C panels. The shield, if required, should be grounded only at the IN jack.

Equipment cabling that carries the signal *from* equipment is always connected to the OUT jack of a DSX jack set. Similarly, the signal going *to* equipment from the DSX panel is always connected to the IN jack. The IN and OUT jacks are further defined with respect to the equipment they terminate, as shown in Figure 9-22.

Signals transmitted *from* the line (more specifically, from the COR) are connected to the jacks designated R OUT (repeater out), and signals transmitted *to* the line are designated L IN (line in). Signals transmitted *to* the multiplex equipment are connected to jacks designated X IN (multiplex in), and signals transmitted *from* the multiplex equipment are designated X OUT (multiplex out). Jumper connections are R OUT to X IN and X OUT to L IN. When two lines are patched together, the cords connect the R OUT of one line to the L IN of the other line. The R OUT and X OUT jacks provide access to the output of the CORs and multiplex equipment for testing and maintenance purposes. Also, the respective MON jacks provide for in-service monitoring of these same signals.

The pulse output level of multiplex and repeater equipment is nominally higher than at the DSX. In order to ensure interoperability of all systems connected to the DSX and to control the signal level at the DSX, the output from equipment is pre-equalized and then sent through a cable with specific characteristics. This cable is known as ABAM, which is an AT&T designation for 22 AWG, twisted pair cable with an overall shield.

The insertion loss and phase per 1,000 feet are shown in Figure 9-24 for ABAM cable. The chart values must be adjusted to the actual length used. The reference length that applies to the DSX-1 and DSX-1C normally is 655 feet, so values in the chart must be reduced by the ratio 655/1,000 to obtain reference loss and phase. Other cable types can be substituted for ABAM cable. The maximum lengths given in the following discussion should be adjusted (increased or decreased) by the ratio of unit loss in the new cable to the unit loss of ABAM cable at 772 kHz.

Using Figure 9-25 for reference, the maximum cable lengths for the various equipment connected to the DSX and the maximum cross-connect jumper lengths at the DSX are shown as a matrix in Table 9-18. In this matrix, the different combinations are shown in the rows and columns. The nominal voltage at the DSX-1 is ±3 V. Most equipment operating at the DS-1 rate has an output pulse voltage of either ±3 or ±6 V, and some are switch selectable. When the output voltage is ±3 V, the connecting cable and jumper lengths are restricted to 15 feet rather than the 85 feet shown in Figure 9-25. This does not cause any problems in small installations, such as those on user premises and small central offices or remote terminals, but can greatly restrict the expansion of larger installations. Figure 9-25 also shows the pulse voltages at different points in the cabling between a digital terminal and the distribution frame in a central office. This illustration shows a terminal and COR with ±6 V outputs.

CORs with ±3 V output are treated differently, depending on the type of line build-out in the first span line repeater. If the first line repeater has a fixed line build-

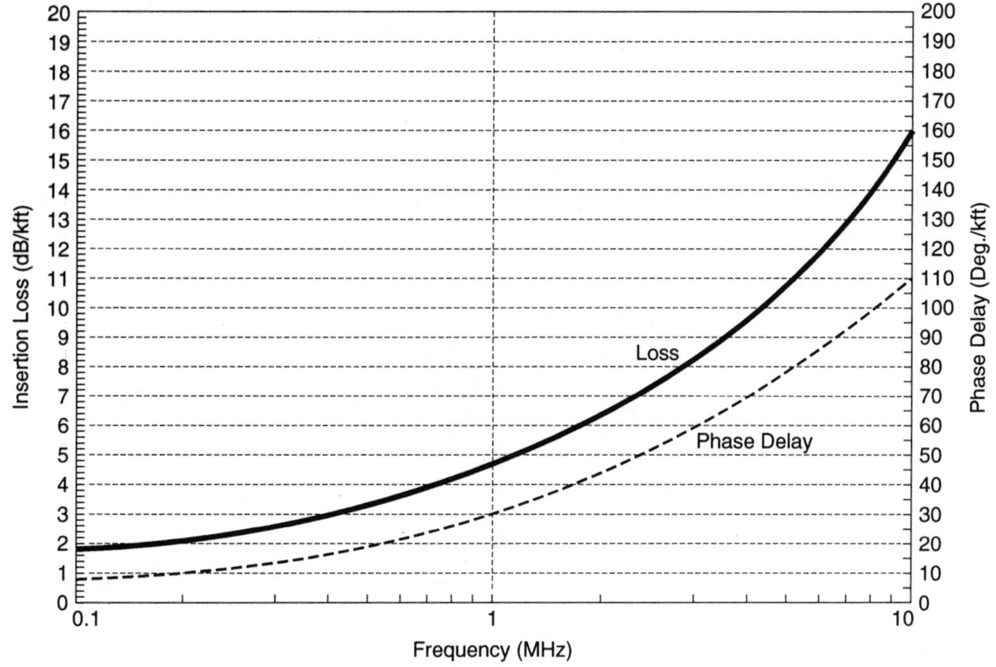

Figure 9-24 Reference cable for equipment-to-DSX connection

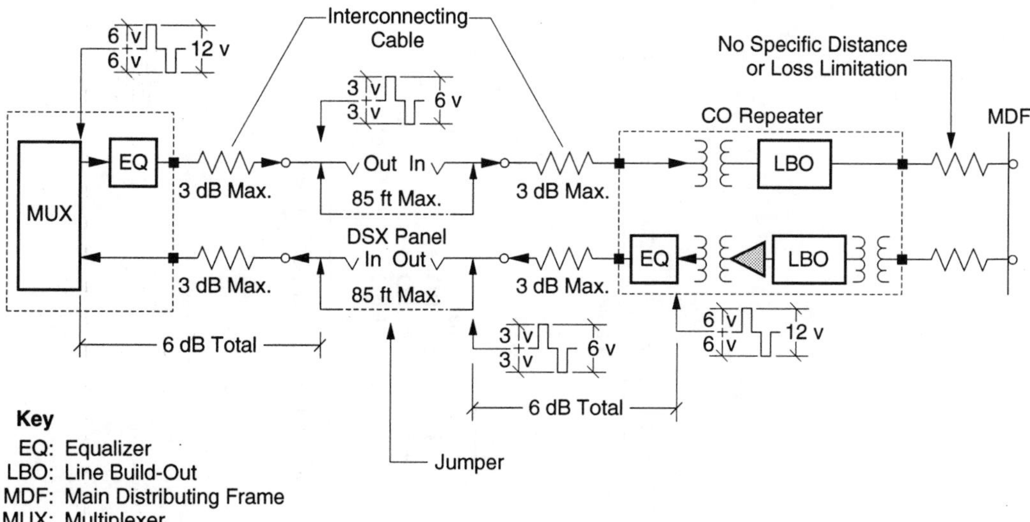

Figure 9-25 Voltage levels between a typical terminal and a distribution frame (assuming equipment output of 6 V p-p)

Table 9-18 Maximum Equipment-to-DSX-1 Interconnecting Cable[a] Lengths and Maximum DSX-1 Jumper Lengths

To ↓	From →	Maximum ABAM Cable Length to DSX-1 (ft)	1	2	3	4	5
			Maximum Cross-Connect Jumper Length at DSX-1 (ft)				
			±3 V Terminal	±3 V COR FLBO	±3 V COR ALBO	±6 V COR ALBO	±6 V Terminal
1	±3 V Terminal	85	85	85	85	85	85
2	±3 V COR FLBO	85	85	15	15	85	85
3	±3 V COR ALBO	85	85	15	85	85	85
4	±6 V COR ALBO	655	85	85	85	85	85
5	±6 V Terminal	655	85	85	85	85	85

[a]If other than ABAM cable is used, adjust lengths to provide equivalent loss (see Table 9-8).

out (FLBO), the maximum cross-connect jumper length at the DSX is only 15 feet. This is to ensure adequate signal at the input to the first line repeater, since the COR transmit circuit is passive and does not regenerate the signal output from a terminal or another COR. If this same COR is connected to a more modern line repeater with automatic line build-out (ALBO), the maximum jumper length is increased to 85 feet because the ALBO can compensate for more loss.

The DSX is a fundamental reference point in all digital transmission systems. The requirements for signals at the DSX are very important for network and equipment interoperability. The electrical requirements for DSX-1 interfaces that apply to field engineering work are given in Chapter 3. Additional requirements for the DSX-1 (and DSX-1C, DSX-2, and DSX-3) reference signals are given in [15].

Some restrictions apply to equipment cabling for DSX panels: each cable should carry only one direction of transmission (all Tx or all Rx, but not both). Since the cable is shielded, the shield must be grounded, but at one end only, usually at the equipment end (repeater bay or multiplex bay). No switched VF circuits are allowed to share a cable with digital services of any type, and the cables should not be run in cable racks or trays with switched VF pairs. The latter requirement largely is ignored in small installations.

Tracer lamps are not provided on all DSX panels but are recommended when DSX panels are installed for more than just a few circuits. The purpose of the tracer lamp is to allow a circuit to be traced by inserting a patch cord into the MON jack. When one lamp lights, the lamps associated with all other MON jacks cross-connected to the first also light. This provides a quick method for locating interconnected jacks and circuits.

Most DSX panels have a trough at the bottom to allow neat jumpering and housekeeping. This trough also provides support for interbay cords and jumpers. When more than approximately four bays or more than one lineup are installed, interbay patch panels are used to avoid long patch cords.

9.3.2 Electronic DCS

The electronic DCS (also called digital access and cross-connect system, DACS) terminates primary rate (DS-1 or DS-3) transmission lines, extracts the DS-0 or subrate channels from the primary rate stream, and electronically cross-connects these channels to other primary rate lines. The DCS replaces the need for back-to-back multiplexers, which, until the introduction of the electronic cross-connect system, were required to allow rearrangement of individual analog or subrate digital channels. Figure 9-26(a) shows an example of a group of channels from one source being transmitted to two locations via back-to-back multiplex terminals (channel banks). The channel mapping is shown in Table 9-19. Figure 9-26(b) shows a timeslot interchange (TSI), which electronically connects the individual channels in the DCS to the required aggregate output.

The DCS provides benefits not available from the DSX. The DCS provides buffering, timing regeneration, circuit grooming, and consolidation at the DS-0 rate, and eliminates the need for A/D (analog-to-digital) and D/A (digital-to-analog) con-

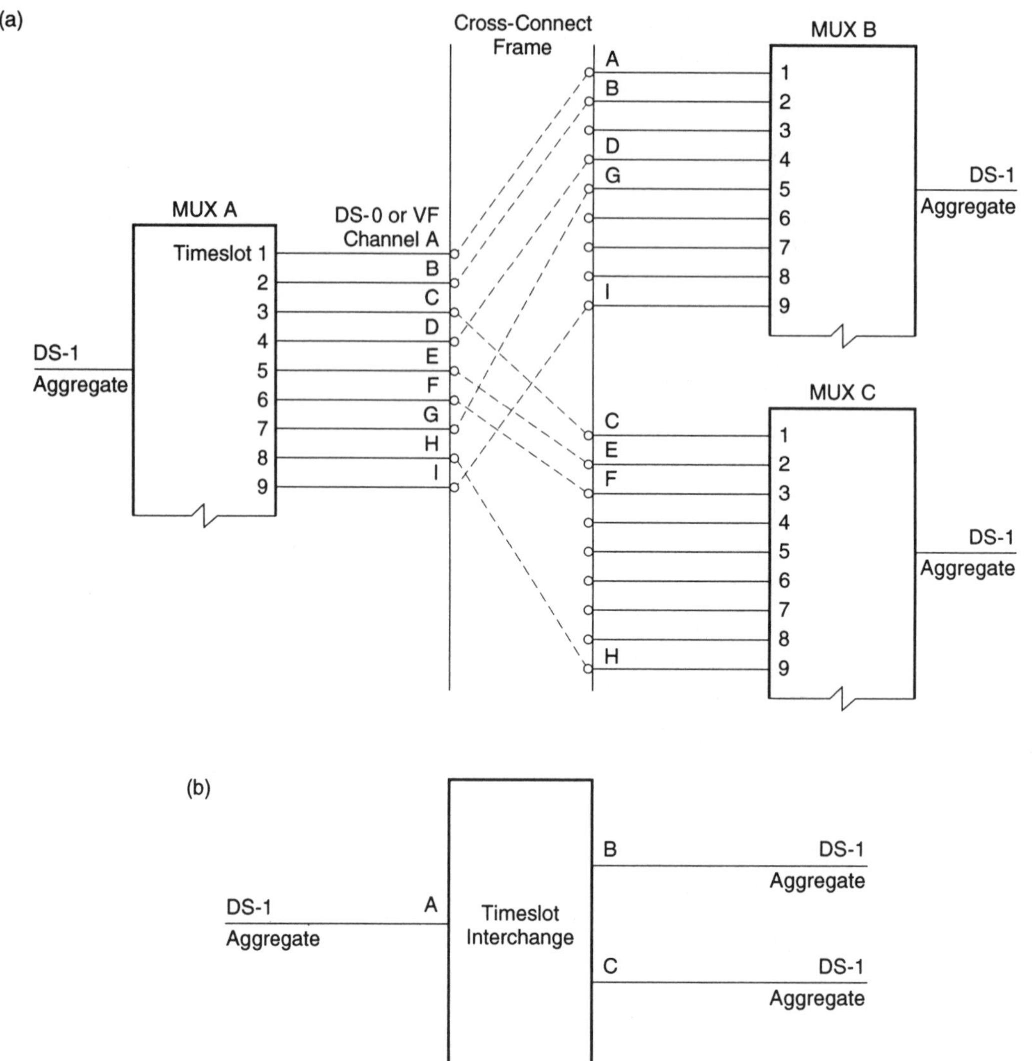

Figure 9-26 Examples of channel connections: (a) Back-to-back channel banks; (b) TSI in DCS

version. As a result, it provides increased transmission performance for both analog and digital source signals.

Modern electronic DCSs can be considered true network management systems, especially in large network applications, rather than just another transmission component. The network management aspect is beyond the scope of this book; the electronic DCS will be described only as it affects transmission performance and bandwidth management.

Table 9-19 Example Mapping of Channels in Back-to-Back Channel Banks

MUX A		MUX B		MUX C	
TIMESLOT	CHANNEL	TIMESLOT	CHANNEL	TIMESLOT	CHANNEL
1	A	1	A	—	—
2	B	2	B	—	—
3	C	—	—	1	C
4	D	4	D	—	—
5	E	—	—	2	E
6	F	—	—	3	F
7	G	5	G	—	—
8	H	—	—	9	H
9	I	9	I	—	—

MUX = multiplexer.

Bandwidth management is the allocation of digital channel and circuit bandwidth to achieve some goal. An example of this is channel grooming at the DS-0 rate, whereby a number of individual 64 kbps channels from different sources are combined into a single DS-1 stream for bulk transmission. More sophisticated DCSs provide analog circuit interfaces and perform A/D and D/A conversion, thereby combining multiplexer and DCS functions in one unit.

Since all DCS equipment uses time division multiplexing techniques, the switching of bits associated with one channel to another channel requires bit storage (buffering) until the timeslot for the other channel becomes available. Any buffering introduces delay. Table 9-20 shows the delay limits for electronically cross-connecting various circuits.

The DCS does not eliminate the need for DSX panels. In fact, the preferred method for connecting a DCS to digital line and multiplex equipment is through DSX panels. A comparison of DCS and DSX is provided in Table 9-21. Detailed specifications for an electronic DCS are given in [16,17,18]. Additional descriptive information can be found in [19].

Table 9-20 Delay Limits for DCS Cross-Connect Times[a]

CONNECTION TYPE	MAXIMUM CONNECTION OR DISCONNECTION TIME
Connect (1) DS-0 channel	3 seconds
Connect (6) DS-0 channels	6 seconds
Connect (24) DS-0 channels	14 seconds
Disconnect (1) DS-0 channel	3 seconds
Disconnect (6) DS-0 channels	6 seconds
Disconnect (24) DS-0 channels	19 seconds

[a]Data from [16]

Table 9-21 Comparison of Digital Cross-Connects

Type	Digital Signal Cross-Connect (DSX)	Digital Cross-Connect System (DCS)
Signal rate	DS-1, DS-1C, DS-2, DS-3	Subrate, DS-0, DS-1, DS-3
Cross-connect method	Patch cord or jumper	Electronic, software controlled
Circuit access and rearrangement method	Patch cord	Centralized video display terminal for control
Circuit capacity	24 to 64 per panel	16 to 4,096 per system
Testing and performance monitoring	No built-in capability, but allows nonintrusive access for monitoring. Uses metallic access for manually patched external test or performance monitoring equipment	Simple systems have limited built-in capability for monitoring preset performance thresholds. More sophisticated systems provide built-in test functions in addition to allowing metallic access for external test equipment
Maintenance capability	Usually physically grouped to allow centralized maintenance, but requires on-site personnel	Control terminal can be located at local or remote site. Remote allows centralized control of numerous sites.
Delay	None	Average 500 μseconds Maximum 750 μseconds (99%)
Signal regeneration	None	Signal fully regenerated at each primary rate port
Floor space requirements	Approximately 280 DS-1 circuits (in and out) per 7-foot bay	Approximately 160 DS-1 circuits (in and out) per 7-foot bay
Power requirements	No intrinsic power requirements except for lighted patch cord trace indicators (LED)	Varies with the manufacturer and equipment vintage

LED = light-emitting diode

REFERENCES

[1] *Carrier Outside Line Engineering Considerations and Design Procedures.* Alcatel Network Systems B303 T1, Oct. 1985.

[2] Cravis, H., Crater, T. "Engineering of T1 Carrier System Repeatered Lines." Bell System Technical Journal, Vol. XLII, No. 2, March 1963.

[3] *T1 Digital Line, Transmission and Outside Plant Design Procedures: Carrier Engineering.* AT&T 855-351-101, July 1990.

[4] *T1 Digital PCM Span Line Engineering Design Considerations.* Alcatel Section ITTR-20-700, March 1987.

[5] *REA Specification for Aerial and Underground Telephone Cable.* Bulletin 345-13 PE-22, Jan. 1983.

[6] *REA Specification for Filled Telephone Cables.* Bulletin 345-67 PE-39, Jan. 1987.

[7] *REA Specification for Filled Telephone Cables with Expanded Insulation.* Bulletin 1753F-208, June 1993.

[8] *T1 Outstate Digital Line, Transmission and Outside Plant Design Procedures: Carrier Engineering.* AT&T 855-351-200, April 1981.

[9] American National Standard for Telecommunications. *Carrier-to-Customer Installation—DS1 Metallic Interface.* ANSI T1.403-1989.

[10] *Functional Criteria for the DS1 Interface Connector.* BELLCORE Technical Reference TR-TSY-000312, March 1988.

[11] Reeve, W. *Subscriber Loop Signaling and Transmission: Analog.* IEEE Press, 1992.

[12] *Fundamental Generic Requirements for Metallic Digital Signal Cross-Connect Systems DSX-1, -1C, -2, -3.* BELLCORE Technical Reference TR-NPL-000320, April 1988.

[13] *Generic Requirements for Manual Digital Signal Cross-Connect Frames DSX-1, -1C, -2.* BELLCORE Technical Reference TR-NPL-000321, April 1989.

[14] *Interconnection Specification for Digital Cross-Connects.* AT&T Compatibility Bulletin No. 119, Oct. 1979.

[15] American National Standard for Telecommunications. *Digital Hierarchy: Electrical Interfaces.* ANSI T1.102-1987.

[16] *Digital Cross-Connect System Requirements and Objectives.* BELLCORE Technical Reference TR-TSY-000170, Nov. 1985.

[17] *Electronic Digital Signal Cross-Connect (EDSX) System Generic Requirements and Objectives.* BELLCORE Technical Advisory TA-TSY-000241, July 1989.

[18] *Digital Access and Cross-Connect System Technical Reference and Compatibility Specification.* AT&T Compatibility Bulletin No. 143, Jan. 1983.

[19] *Digital Cross-Connect and Drop and Insert Systems.* REA TE&CM Section 958, Sept. 1988.

10 Transmission Engineering, Part II—Nonrepeatered Digital Loops Including Fiber Optics

Chapter 10 Acroynms

AGC	automatic gain control	LT	line termination
AMI	alternate mark inversion	MACAL	maximum allowable cable loss
AML	actual measured loss	MIFL	maximum individual fiber loss
APD	avalanche photodiode	NEXT	near-end crosstalk
AWG	American wire gauge	NMS	noise measuring set
BER	bit error rate	NT	network termination
CPE	customer premises equipment	OSP	outside plant
CSA	carrier serving area	PIC	polyolefin insulated cable
DBRP	dispersion bit-rate product	PIN	positive-intrinsic-negative
DFB	decision feedback	POTS	"plain old telephone service"
DSL	digital subscriber line	RFI	radio frequency interference
EIA	Electronic Industries Association	Rx	receive
EIP	excess interfering power	SAI	serving area interface
EMC	electromagnetic compatibility	SM	separation margin
EML	expected measured loss	SNR	signal-to-noise ratio
FET	field effect transistor	SRDL	subrate digital loop
FEXT	far-end crosstalk	TC	temperature compensation
FITL	fiber-in-the-loop	TCM	time compression multiplexing
FLBO	fixed line build-out	TMS	transmission measuring test set
FWHM	full width, half maximum	Tx	transmit
HDSL	high bit-rate digital subscriber line	VF	voice frequency
IL	insertion loss	WDM	wavelength division multiplexer
ISDN	integrated services digital network		
LD	laser diode		
LED	light-emitting diode		

The previous chapter covered the extensive design details for repeatered T1-carrier span lines used in digital loop applications. This chapter is concerned with other digital loop types, including the subrate digital loop (SRDL), switched 56 loops, integrated services digital network (ISDN) digital subscriber line (DSL), high bit-rate digital subscriber line (HDSL), and fiber-in-the-loop (FITL). For the loop types that use twisted pairs, the engineering considerations are broken down into three areas: interference, design, and loop qualification procedures. It is well known that optical fiber transmission systems are immune to interference such as crosstalk and impulse noise.* In the section on FITL design, only design is considered. Fiber loop qualification is normally limited to simple loss or optical time domain reflectomer tests to ensure continuity and splice integrity (see Chapter 8).

10.1 Compatibility

The interference matrix given in Table 10-1 shows the general electromagnetic compatibility (EMC) of the various twisted pair loop types that may share a cable sheath. Even though a particular loop type may not interfere with other loop types under normal conditions, as indicated by the "√" in the *from* loop type *to* loop type column, some combinations may cause problems under certain conditions of relatively high crosstalk coupling or low signal level. The "C" indicates conditional EMC, which means it may be possible to reduce the interference to an acceptable level by shortening the affected loop length or using other methods to reduce its loss (and to increase the signal-to-noise ratio, SNR), or rerouting to reduce the exposed section. Unfortunately, in most practical situations, this is not possible. Additional information is provided in the following sections to aid in interference decisions.

10.2 Subrate Digital Loops

10.2.1 Interference Considerations

The SRDL normally will share cable facilities with a variety of loop types, including other SRDLs, repeatered T1 span lines, ISDN DSL, HDSL, analog subscriber carrier lines, 2-wire and 4-wire voicegrade lines, and "plain old telephone service" (POTS) lines. Because of crosstalk coupling between adjacent pairs in the cable, the potential exists for mutual interference between these different services. The frequency spectra of SRDLs extend well beyond the voice frequency (VF) range. Bridged taps and VF conditioning devices such as powerline induction neutralizing transformers, loading coils, and build-out capacitors adversely affect transmission at these higher frequencies when connected to the SRDL.

---

*FITL architectures that use couplers can experience crosstalk. The problems are not as extensive as with twisted pair cables and are more easily controlled because the crosstalk exposure is at limited, well-defined points.

Table 10-1 Loop Compatibility Matrix ⊕

From → / To ↓	Repeatered T1-Carrier	Subrate Digital Loop[a]	ISDN DSL	HDSL	Program Audio Channel	Single-Channel Analog Carrier	Multichannel Analog Carrier	POTS
Repeatered T1-Carrier	√	√	√	√	√	√	√	√
Subrate digital loop[a]	√	√	√	√	√	√	√	√
ISDN DSL	√	√	√	√	√	√	√	√
HDSL	√	√	√	√	√	√	√	√
Program audio channel	√	C	C	C	√	√	√	√
Single-channel analog carrier	√	C	C	X	√	√	√	√
Multichannel analog carrier	√	C	C	X	√	√	√	√
POTS	√	√	√	√	√	√	√	√

[a]Includes switched 56 loop

√ = Compatible; C = Conditionally Compatible; X = Incompatible

To limit mutual interference and negative effects of VF conditioning, the assignments of the different loop types to pairs in a cable must be evaluated. This activity is part of the loop design process. Once the loop design, including pair assignments, has been completed, qualification tests are needed to ensure the SRDL meets the design requirements.

The crosstalk coupling loss of exchange cable pairs is sufficiently high at SRDL frequencies to minimize interference from other services into the SRDL in most cases; however, any cables with pairs carrying dial-pulse trunks, dc alarm loops, or high unbalanced signaling currents should be avoided to limit data errors from impulse noise. If avoidance is not possible, the pairs to be used as SRDLs will require extensive impulse noise testing to ensure suitability. Further, the SRDL pairs must be physically isolated from the noisy pairs by being assigned to nonadjacent binder groups.

Different types of noise can affect the SRDL. Thermal noise is negligible compared to the expected SRDL signal levels, and further consideration of it is not needed. Generally, noise due to powerline induction and radio frequency interference (RFI) will have no effect on the SRDL if the cable pair meets the basic requirements for voicegrade services. These requirements are described later in this chapter; a more thorough discussion appears in [1]. The qualification tests described later provide additional specific noise tests that are required for SRDL loops.

SRDL signal levels have been set to limit the potential for interference on other services, and no restrictions generally are needed when assigning pairs in cables with mixed services, except as previously indicated. In most situations, the SRDL can be assigned to the same binder group as any other service, with the exceptions noted above and one additional exception. The latter occurs when an SRDL operating at 56 or 64 kbps is assigned to the same binder group with private line services or analog subscriber carrier systems that operate in the same frequency range. In this case, one or the other service may have to be reassigned to a different binder group. Nonadjacent binder groups normally will provide the necessary separation and additional crosstalk coupling loss.

For almost all other situations, interference from the SRDL is only likely if the crosstalk coupling loss of the cable is unusually low. This can occur if improper splicing techniques have been used or if there has been unrepaired facility damage. It may be necessary to assign program audio channels (8 and 15 kHz) in binder groups that are not adjacent to 4.8 and 9.6 kbps SRDLs to ensure reliable operation under all conditions. SRDLs may be mixed in the same binder group as T1-carrier span lines, limited distance data channels (using limited distance modems), and all POTS and voicegrade private line services.

If SRDL interference does exists, it is generally due to near-end crosstalk (NEXT) coupling of signals from the SRDL transmit (Tx) circuit into the receive (Rx) circuit of another loop type. This can occur where:

- The Tx pairs of the disturbing service (SRDL) and the Rx pairs of the disturbed service are exposed to each other at some common point in the outside cable plant

- The Tx pairs of the disturbing service (SRDL) and the Rx pairs of the disturbed service are in the same binder group or, in some situations, in adjacent binder groups
- The receiver of the disturbed service is operating at a relatively low SNR due to maximum length loop or low Tx levels

In order to determine if the potential for NEXT exists, it is necessary to evaluate the amount by which the disturbing signal power exceeds the maximum allowable interfering signal input to the disturbed circuit. This evaluation is quantified by the *separation margin*. The separation margin is a measure of the signal loss between the output of the disturbing transmitter and the exposure point ($L_A + L_B$) plus the loss between the exposure point and the disturbed receiver ($L_B + L_C$). The exposure point is illustrated in Figure 10-1. If there is no binder group separation anywhere in cable B, $L_B = 0$.

The separation margin can be found from [2]:

$$SM = L_A F_A + 2L_B F_B + L_C F_C$$

where SM = separation margin
F_A, F_B, and F_C = loss factors from Table 10-2 for each section of cable plant
L_A, L_B, and L_C = lengths of each section of cable plant from Figure 10-1

If SM exceeds the excess interfering power (that is, SM ≥ EIP) shown in Table 10-2 for the type of service being evaluated, the cable pair assignments are generally acceptable. If SM < EIP, the disturbed circuit may or may not work. In some marginal cases, it may be possible to increase the SNR of the disturbed receiver by increasing the signal level at its corresponding transmitter or by reducing the circuit

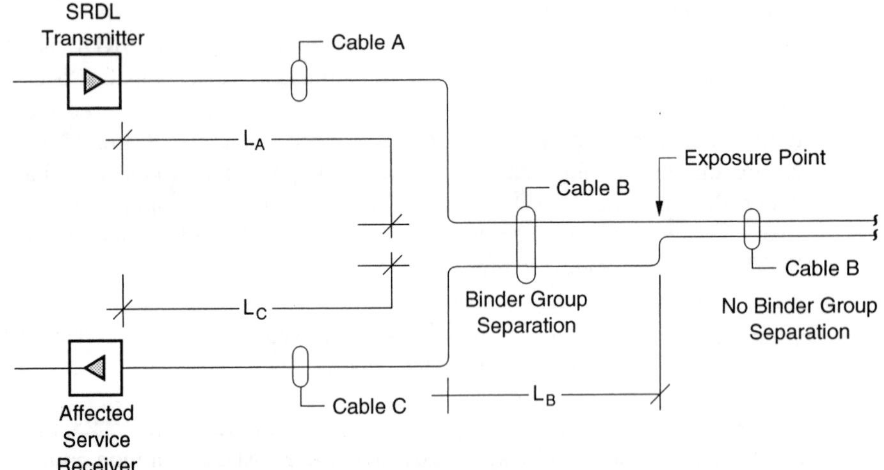

Figure 10-1 NEXT exposure point for SRDLs

Table 10-2 Cable Loss Factors and Excess Interfering Power[a]

Affected Service	Cable Loss Factor, F (dB/kft)				EIP[b] (dB)
	19 AWG	22 AWG	24 AWG	26 AWG	
15 kHz Program channel	0.5	0.9	1.2	1.6	5
Single-channel analog subscriber carrier system, adjacent binder group	0.7	1.2	1.6	2.2	9
Single-channel analog subscriber carrier system, same binder group	0.7	1.2	1.6	2.2	36
Multichannel analog subscriber carrier system	0.7	1.2	1.6	2.2	8

[a]Data from [2]

[b]EIP is excess interfering power from a 56 or 64 kbps SRDL

loss. Reassignment of one or the other circuits into nonadjacent binder groups usually will solve the problem in other cases.

The EIP used in the above calculations is based on a 1% probability that the SRDL will interfere with the other service. If interference does indeed occur in this case, relatively simple circuit changes, such as pair reassignment, usually solve the problem. The probability of interference increases with the difference between EIP and SM. Table 10-3 can be used to determine the chance of interference for different EIP values. For example, if EIP is greater than SM by 10 dB, there is a 10% chance of interference. As the probability of interference increases, binder group separation or a repeater should be considered in the initial loop design process. The foregoing quantification of interference probability is somewhat general and should only be used to estimate compatibility.

Table 10-3 Chance of Interference

EIP − SM	Chance of Interference
0 dB	1%
3 dB	2%
6 dB	5%
10 dB	10%
15 dB	20%
20 dB	40%

EXAMPLE 10-1

An existing single-channel analog subscriber carrier is assigned to pair 152 in feeder cable 1 at the central office, as shown in Figure 10-2. A 64 kbps SRDL is to be assigned to pair 1,139 in feeder cable 2, which is at the same central office and follows approximately the same route. The two feeders merge at serving area interface SAI-1. Both the carrier system and SRDL continue from SAI-1 to SAI-2 in cable SAI-12 with the carrier system on pair SAI-1–201. The SRDL is to be assigned to pair SAI-1–238. Because of cable congestion, the SRDL must be assigned to pair SAI-22–215 beyond SAI-2, whereas the carrier system

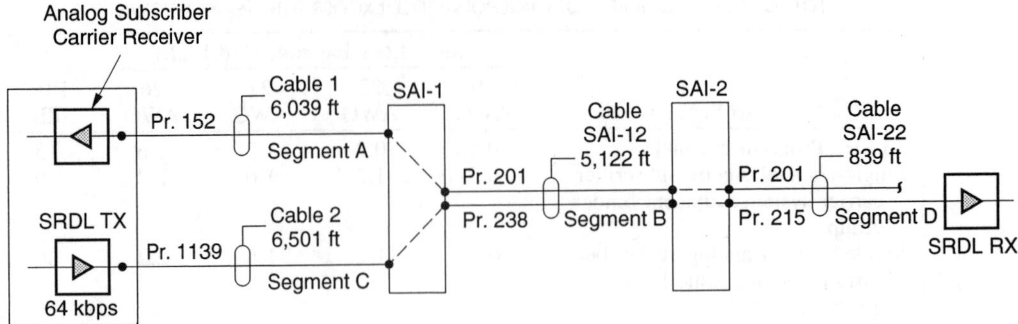

Figure 10-2 SRDL separation margin example

remains assigned to pair SAI-22–201. Cables 1 and 2 are aerial 26 AWG polyolefin insulated cable (PIC), whereas cables SAI-12 and SAI-22 are direct buried 24 AWG PIC. Find the separation margin and determine if the SRDL will function as assigned.

Between SAI-1 and SAI-2, the carrier system and SRDL will be in adjacent binder groups [the carrier in binder group 9 (pairs 201–225) and the SRDL in binder group 10 (pairs 226–250)]. The point of exposure is SAI-2, where both the carrier system and SRDL enter the same binder group [binder group 9 (pairs 201–225)]. Therefore, the exposure point is SAI-2, and it is necessary to apply the separation margin formula from the central office to SAI-2. Using Figure 10-2 and Table 10-2:

$$L_A = 6.039 \text{ kilofeet} \qquad F_A = 2.2 \text{ dB/kilofoot} \qquad F_A L_A = 13.3 \text{ dB}$$

$$L_B = 5.122 \text{ kilofeet} \qquad F_B = 1.6 \text{ dB/kilofoot} \qquad F_B L_B = 8.2 \text{ dB}$$

$$L_C = 6.501 \text{ kilofeet} \qquad F_C = 2.2 \text{ dB/kilofoot} \qquad F_C L_C = 14.3 \text{ dB}$$

Therefore, $SM = 13.3 + 2 \times 8.2 + 14.3 = 44.0$ dB. Since the carrier system and SRDL are assigned to the same binder group for at least part of the route, the worse-case EIP will apply. From Table 10-2, the EIP for this situation is 36 dB. Therefore, SM > EIP, and the pair assignments are suitable. The 8 dB margin between SM and EIP will allow proper operation under worse-case conditions.

10.2.2 Subrate Digital Loop Design

SRDL design starts by determining:

- Cable route distance from the central office to the loop termination at the user premises
- Length and gauge of each cable making up the proposed SRDL
- Length and gauge of bridged taps
- Cable makeup (capacitance, conditioning)
- Type of plant (underground, buried, or aerial)
- Types of services and loops existing in the same cables and the physical relationship of the pair assignments with respect to the proposed SRDL (same, adjacent, or nonadjacent binder group)

The basic requirements for exchange cable pairs used for SRDLs are:

- Actual measured loss (AML) ≤ 34 dB (including bridged taps) at the Nyquist frequency (1/2 the bit rate; see Table 10-4)*
- Limited bridged tap lengths (see Table 10-4)
- Nominal 69 to 83 nanofarad (nF) per mile, 19, 22, 24, and 26 AWG PIC or pulp exchange cable pairs
- No open wire
- Nonloaded cable pairs only
- No build-out capacitors
- No VF repeaters or loop extenders

The conditioning requirements preclude loaded cable pairs and build-out capacitors, which means these devices must be removed from the cable pairs before the SRDL is installed. Bridged taps increase the attenuation and present impedance discontinuities on a loop. The higher attenuation lowers the available SNR at the receiver, whereas impedance discontinuities cause signal reflections that can adversely affect the transmitter and also increase the SNR at the receiver.

The SRDL will function with bridged taps provided the lengths are limited. The intent is to limit any bridged tap to one-quarter wavelength at the highest frequency of the SRDL. Since bridged tap lengths seldom are shown accurately in plant

Table 10-4 SRDL Design Data[a]

Payload Data Rate (kbps)	Loop Bit Rate (kbps)	Secondary Channel	Nyquist Frequency (Hz)	Longest Single Bridged Tap (ft)	Total Bridged Tap (ft)
2.4	2.4	No	1.2	6,000	6,000
2.4	3.2	Yes	1.6	6,000	6,000
4.8	4.8	No	2.4	6,000	6,000
4.8	6.4	Yes	3.2	6,000	6,000
9.6	9.6	No	4.8	6,000	6,000
9.6	12.8	Yes	6.4	6,000	6,000
19.2	19.2	No	9.6	6,000	6,000
19.2	25.6	Yes	12.8	6,000	6,000
38.4	38.4	No	19.2	2,000	2,500
38.4	51.2	Yes	25.6	2,000	2,500
56.0	56.0	No	28.0	2,000	2,500
56.0	72.0	Yes	36.0	2,000	2,500
64.0	72.0	No	36.0	2,000	2,500

[a]Source: [3]

*The 34 dB maximum AML is according to the loop rates given in the original SRDL specification and subsequent industry technical references [3,4]. Other intermediate rates and losses are given by specific manufacturers for their dataport equipment. These are discussed later in the loop qualification section.

records, all bridged taps should be removed wherever it is economically possible to do so. Where plant records indicate the presence of bridged taps, but their extent cannot be determined, they should be removed.

The maximum AML of 34 dB should not be used as a design value. Instead, a maximum expected measured loss (EML) of 31 dB should be used as a design value. The EML gives an additional 3 dB of margin. Some systems (or more accurately, specific equipment by some manufacturers) can meet the error rate specifications with AML = 45 dB at 56.0 and 64.0 kbps rates; however, without specific information indicating otherwise, an AML of 34 dB should be used at all rates unless other values are specified for a particular equipment set.

Certain differences between AML and EML are normal due to variations in splicing, records accuracy, and measurement techniques and equipment. Measurements falling within requirements of Table 10-5 indicate the AML is within acceptable tolerances. Readings outside the range indicate plant or record problems that must be corrected before the SRDL is installed. In no case should the AML exceed 34 dB for a proposed SRDL except as noted above.

Table 10-5 AML Tolerance

AML	EML + 5 dB − 2.5 dB
AML	≤34 dB
EML	≤31 dB

The EML is calculated at the highest anticipated temperature of the outside plant cable pairs. Therefore, measured values must be adjusted for temperature before they can be compared to design values. Generally, the design temperature for predominantly buried and underground plant is 38 °C (100 °F) and for predominantly aerial plant is 60 °C (140 °F).

The EML is determined as follows:

- Multiply each segment length by the per-unit insertion loss for that segment
- Multiply the length of each bridged tap by the per-unit bridged tap loss for that bridged tap
- Add the resulting segment losses and bridge tap losses to get the total EML

In terms of an equation:

$$EML_T = L_i IL_i + L_{i+1} IL_{i+1} + \ldots + L_n IL_n + BTL_k\, BT_k + BTL_{k+1}\, BT_{k+1} + \ldots + BTL_m\, BT_m$$

where EML_T = total EML

L_i = length of segment i for i = 1 to n

IL_i = per-unit insertion loss of segment i for i = 1 to n (Tables 10-6 and 10-8)

BT_i =length of bridged tap segment k for k = 1 to m

BTL_i = per-unit bridged tap loss of bridged tap segment k for k = 1 to m (Table 10-10 or Figure 10-3)

Table 10-6 can be used to determine the per-unit insertion loss of SRDLs using various PIC pair gauges in both underground and aerial environments. The temperature compensation percentages allow adjustment of the nominal values given for 21 °C to any design temperature. Table 10-7 is provided to simplify calculations for the standard design temperatures of 38 and 60 °C. Tables 10-8 and 10-9 give the same information for pulp insulated cables.

Table 10-10 gives the bridged tap loss for short lengths of the various cable gauges. Bridged tap loss is a nonlinear function of length for longer lengths, as shown in Figure 10-3 on pages 483–488. Bridged tap loss also varies slightly with the point at which the tap is connected to the main loop; however, for practical problems, this dependence can be ignored. Finally, bridged tap loss is assumed to be independent of temperature.

Tables 10-11 and 10-12 give the maximum allowable loop lengths at 31 dB EML for PIC and pulp insulated cables, respectively, at standard design temperatures of 38 and 60 °C. These tables assume that the loops consist of straight cable with no bridged taps. Any loop less than the lengths shown will satisfy the EML requirements provided that the loop is a single gauge and has no bridged taps. For loops with mixed gauges, each segment will require a separate calculation, as previously described. The data provided in these tables are sufficiently accurate for all engineering applications. Some equipment may require fixed line build-outs (FLBOs) if the SRDL has loss <10 dB.

The losses at 38 and 60 °C are emphasized because these are the worse-case temperatures at which predominantly buried and predominantly aerial loops are presently designed. In other words, if the maximum desired loss under all reasonable temperature conditions is 31 dB, the loop length giving this loss would be calculated at 38 °C (100 °F) for predominantly buried plant, and so on.

When both buried and aerial facilities are used along a given route, the dominant facility type (buried or aerial) will determine the design temperature to be used, although it is acceptable to determine the contribution of each type to the total. Additional tables and formulas for temperature correcting transmission parameters are available from cable manufacturers. See also [1] and [5].

Obviously, loops will be built where the temperature of the earth (and consequently of the copper conductors) at common burial depths never reaches, say, more than 10 °C (50 °F) and the ambient air temperature, never more than 21 °C (70 °F). In these cases, it would seem that designing for 38 °C (100 °F) and 60 °C (140 °F) temperatures would be overdesign. Good design practice, however, requires some amount of conservatism, and there should be no economic penalty in designing for the higher temperatures in most applications.

Where it can be shown that a significant economic penalty does exist, the temperature requirements can be adjusted to account for local conditions, as follows [6]:

1. Calculate the average ambient temperature for 60 consecutive days during the high-temperature season in the local area. This information is available from local National Weather Service offices in a number of useful forms.*

---

*The National Weather Service has local offices in most cities.

Table 10-6 Nominal PIC Per Unit Insertion Loss at 21 °C and Temperature Compensation

Nyquist Frequency (kHz)	19 AWG IL[a] (dB/kft)	19 AWG TC[b] (%)	22 AWG IL[a] (dB/kft)	22 AWG TC[b] (%)	24 AWG IL[a] (dB/kft)	24 AWG TC[b] (%)	26 AWG IL[a] (dB/kft)	26 AWG TC[b] (%)
1.2	0.259	1.9	0.374	2.0	0.476	1.9	0.603	1.9
1.6	0.296	2.1	0.430	2.1	0.549	2.0	0.696	2.0
2.4	0.351	2.1	0.518	2.0	0.663	2.0	0.845	1.9
3.2	0.395	2.3	0.590	2.1	0.761	2.0	0.972	2.0
4.8	0.459	2.5	0.703	2.3	0.916	2.1	1.179	2.0
6.4	0.504	2.6	0.791	2.3	1.040	2.1	1.345	2.1
9.6	0.564	2.9	0.921	2.4	1.232	2.3	1.614	2.1
12.8	0.598	3.2	1.007	2.6	1.369	2.4	1.816	2.2
19.2	0.648	3.5	1.132	2.9	1.581	2.6	2.141	2.4
25.6	0.688	3.4	1.204	3.1	1.712	2.7	2.362	2.5
28.0	0.702	3.4	1.229	3.1	1.758	2.8	2.440	2.5
36.0	0.747	3.2	1.290	3.2	1.861	3.0	2.622	2.7

[a]Insertion loss (IL) values apply at 21 °C to solid or foam skin insulated, filled or air core PIC with nominal 83 nF/mile capacitance. These data have been extracted from Appendix I of [3] with corrections and adjustments for frequency.

[b]Increase (decrease) the nominal loss at 21 °C by the temperature compensation (TC) percentage for each 10 °C increase (decrease) in temperature.

Table 10-7 Nominal PIC Per Unit Insertion Loss at Design Temperatures

NYQUIST FREQUENCY (kHz)	19 AWG IL (dB/kft) 38 °C	19 AWG IL (dB/kft) 60 °C	22 AWG IL (dB/kft) 38 °C	22 AWG IL (dB/kft) 60 °C	24 AWG IL (dB/kft) 38 °C	24 AWG IL (dB/kft) 60 °C	26 AWG IL (dB/kft) 38 °C	26 AWG IL (dB/kft) 60 °C
1.2	0.27	0.28	0.39	0.40	0.49	0.51	0.62	0.65
1.6	0.31	0.32	0.45	0.46	0.57	0.59	0.72	0.75
2.4	0.36	0.38	0.54	0.56	0.69	0.72	0.87	0.91
3.2	0.41	0.43	0.61	0.64	0.79	0.82	1.00	1.05
4.8	0.48	0.50	0.73	0.77	0.95	0.99	1.22	1.27
6.4	0.53	0.56	0.82	0.86	1.08	1.13	1.39	1.45
9.6	0.59	0.63	0.96	1.01	1.28	1.34	1.67	1.75
12.8	0.63	0.67	1.05	1.11	1.42	1.50	1.88	1.97
19.2	0.69	0.74	1.19	1.26	1.65	1.74	2.23	2.34
25.6	0.73	0.78	1.27	1.35	1.79	1.89	2.46	2.59
28.0	0.74	0.80	1.29	1.38	1.84	1.95	2.55	2.68
36.0	0.79	0.84	1.36	1.45	1.96	2.08	2.74	2.90

Table 10-8 Nominal Pulp Insulated Cable Characteristics at 21 °C and Temperature Compensation

Nyquist Frequency (kHz)	19 AWG IL[a] (dB/kft)	19 AWG TC[b] (%)	22 AWG IL[a] (dB/kft)	22 AWG TC[b] (%)	24 AWG IL[a] (dB/kft)	24 AWG TC[b] (%)	26 AWG IL[a] (dB/kft)	26 AWG TC[b] (%)
1.2	0.262	2.0	0.380	2.0	0.483	1.9	0.599	1.9
1.6	0.300	2.0	0.437	2.0	0.557	1.9	0.692	1.9
2.4	0.357	2.1	0.527	2.0	0.674	2.0	0.839	2.0
3.2	0.403	2.2	0.601	2.1	0.773	2.0	0.965	2.0
4.8	0.471	2.4	0.719	2.2	0.932	2.1	1.171	2.0
6.4	0.520	2.5	0.810	2.2	1.059	2.1	1.337	2.0
9.6	0.589	2.7	0.947	2.4	1.258	2.2	1.604	2.1
12.8	0.632	2.9	1.040	2.6	1.402	2.3	1.806	2.2
19.2	0.697	3.2	1.180	2.8	1.628	2.5	2.132	2.3
25.6	0.745	3.2	1.265	3.0	1.773	2.7	2.355	2.5
28.0	0.762	3.2	1.296	3.0	1.824	2.7	2.435	2.5
36.0	0.816	3.1	1.369	3.2	1.942	2.9	2.623	2.6

[a]Insertion loss (IL) values apply at 21 °C to pulp insulated, air core cable with nominal 83 nF/mile capacitance. These data have been extracted from Appendix I of [3] with corrections and adjustments for frequency.

[b]Increase (decrease) the nominal loss at 21 °C by the temperature compensation (TC) percentage for each 10 °C increase (decrease) in temperature.

Table 10-9 Nominal Pulp Insulated Cable Per Unit Insertion Loss at Design Temperatures

Nyquist Frequency (kHz)	19 AWG IL (dB/kft) 38 °C	19 AWG IL (dB/kft) 60 °C	22 AWG IL (dB/kft) 38 °C	22 AWG IL (dB/kft) 60 °C	24 AWG IL (dB/kft) 38 °C	24 AWG IL (dB/kft) 60 °C	26 AWG IL (dB/kft) 38 °C	26 AWG IL (dB/kft) 60 °C
1.2	0.27	0.28	0.39	0.41	0.50	0.52	0.62	0.64
1.6	0.31	0.32	0.45	0.47	0.58	0.60	0.71	0.74
2.4	0.37	0.39	0.55	0.57	0.70	0.73	0.87	0.90
3.2	0.42	0.44	0.62	0.65	0.80	0.83	1.00	1.04
4.8	0.49	0.52	0.75	0.78	0.96	1.01	1.21	1.26
6.4	0.54	0.57	0.84	0.88	1.10	1.15	1.38	1.44
9.6	0.62	0.65	0.99	1.04	1.31	1.37	1.66	1.74
12.8	0.66	0.70	1.09	1.15	1.46	1.53	1.87	1.96
19.2	0.73	0.78	1.24	1.31	1.70	1.79	2.22	2.32
25.6	0.79	0.84	1.33	1.41	1.85	1.96	2.46	2.58
28.0	0.80	0.86	1.36	1.45	1.91	2.02	2.54	2.67
36.0	0.86	0.92	1.44	1.54	2.04	2.16	2.74	2.89

Table 10-10 Bridged Tap Loss at Nyquist Frequency

Nyquist Frequency (Hz)	Loss (dB/kft)			
	19 AWG	22 AWG	24 AWG	26 AWG
1.2	0.055	0.078	0.104	0.142
1.6	0.072	0.103	0.135	0.184
2.4	0.100	0.145	0.188	0.255
3.2	0.127	0.182	0.241	0.320
4.8	0.173	0.247	0.324	0.425
6.4	0.206	0.299	0.390	0.506
9.6	0.253	0.377	0.489	0.627
12.8	0.280	0.428	0.557	0.708
19.2	0.318	0.496	0.651	0.824
25.6	0.339	0.536	0.703	0.889
28.0	0.349	0.545	0.720	0.910
36.0	0.371	0.573	0.758	0.957

Valid for bridged taps ≤2.5 kilofeet

2. Convert the ambient temperature to cable core temperature for different cable types:
   a. Aerial cables: add 17 °C (30 °F) to the ambient temperature to obtain cable core temperature
   b. Buried cables: subtract 6 °C (10 °F) from the ambient temperature to obtain cable core temperature
   c. Underground cables: use 20 °C (68 °F) for cable core temperature

EXAMPLE 10-2

Find the EML for the SRDL in Example 10-1 and determine if it meets SRDL engineering requirements.

Because all cables are PIC, Table 10-6 is used to obtain the per-unit insertion losses (ILs) for each segment. The table values are for 21 °C, so they must be adjusted for the design temperatures that apply to each segment. The design temperature for segment A is 60 °C because it is aerial. The design temperature for segments B and D is 38 °C because they both are buried. For segment A, the difference between the baseline temperature and the design temperature is 39 °C, or 3.9 units of 10° each. Therefore, the baseline per-unit insertion loss for 26 AWG PIC at the Nyquist frequency of 36.0 kHz is increased by 3.9 × 2.7% = 10.5%; hence IL_C = 2.622 × 1.105 = 2.90 dB/kilofoot. Similarly, for segments B and D, the baseline for 24 AWG PIC is increased by 1.7 units of 10° each, or 1.7 × 3.0% = 5.1%, or 1.96 dB/kilofoot. Applying these values to the respective segment lengths yields

L_A = 6.039 kilofeet	IL_C = 2.90 dB/kilofoot	$IL_C L_C$ = 17.5 dB
L_B = 5.122 kilofeet	IL_B = 1.96 dB/kilofoot	$IL_B L_B$ = 10.0 dB
L_D = 0.839 kilofoot	IL_D = 1.96 dB/kilofoot	$IL_D L_D$ = 1.6 dB
L_T = 12.462 kilofeet		EML_T = 29.1 dB

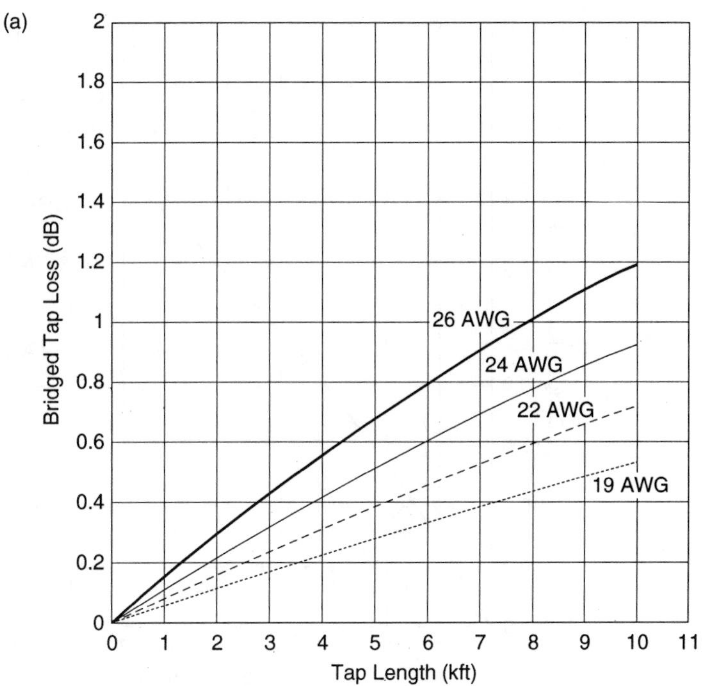

Figure 10-3(a) Bridged tap loss vs. bridged tap length, 1.2 kHz

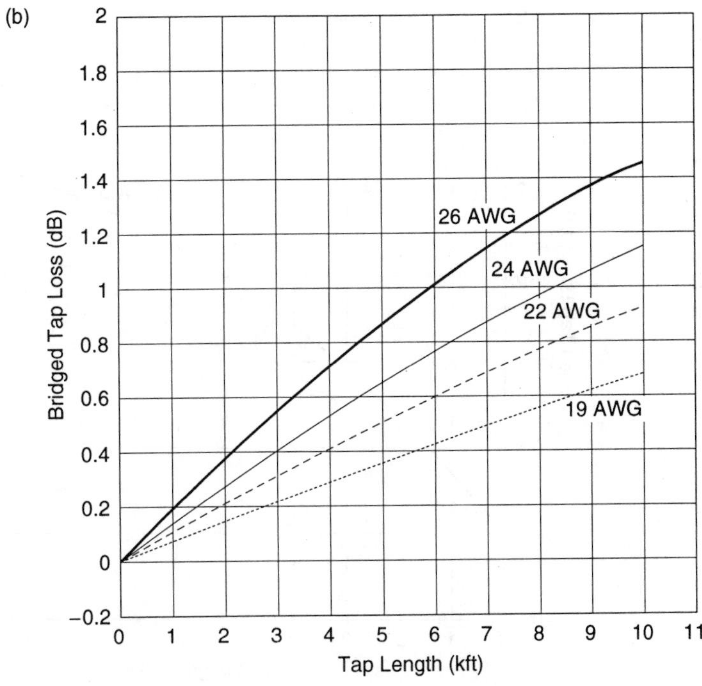

Figure 10-3(b) Bridged tap loss vs. bridged tap length, 1.6 kHz

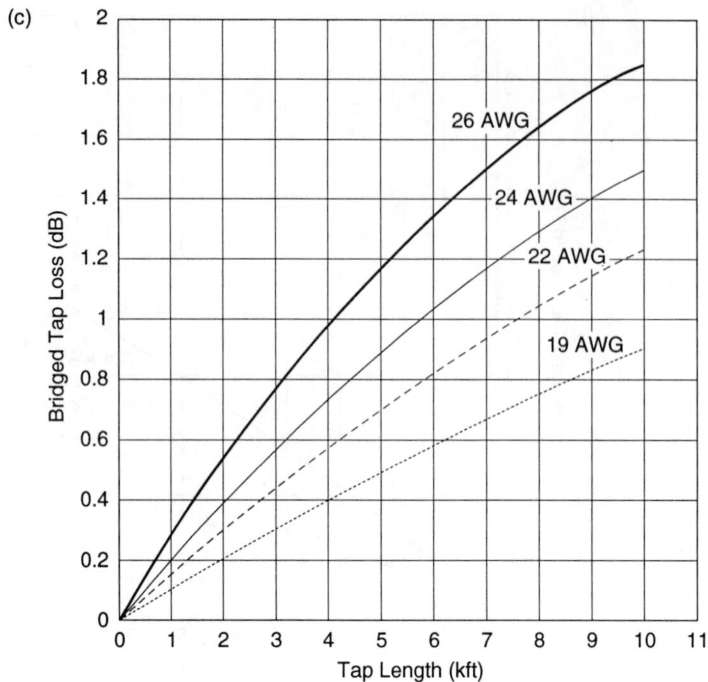

Figure 10-3(c) Bridged tap loss vs. bridged tap length, 2.4 kHz

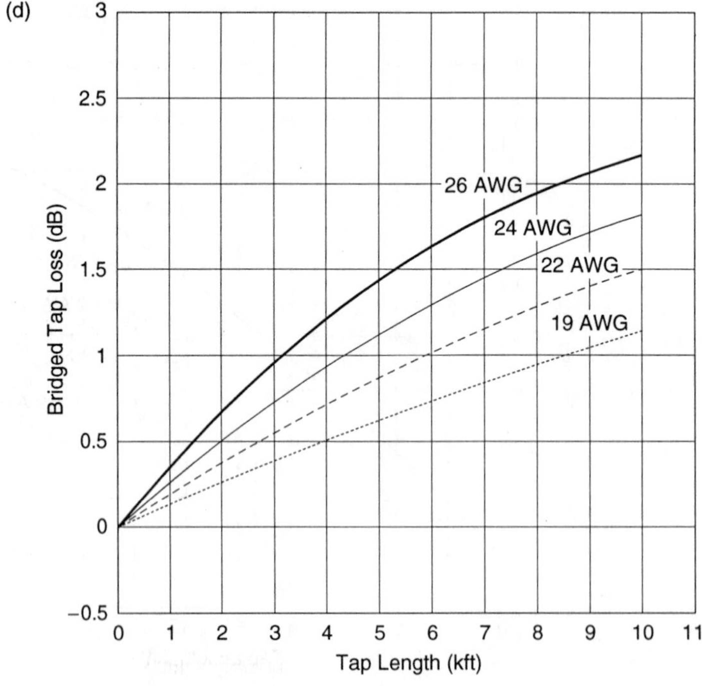

Figure 10-3(d) Bridged tap loss vs. bridged tap length, 3.2 kHz

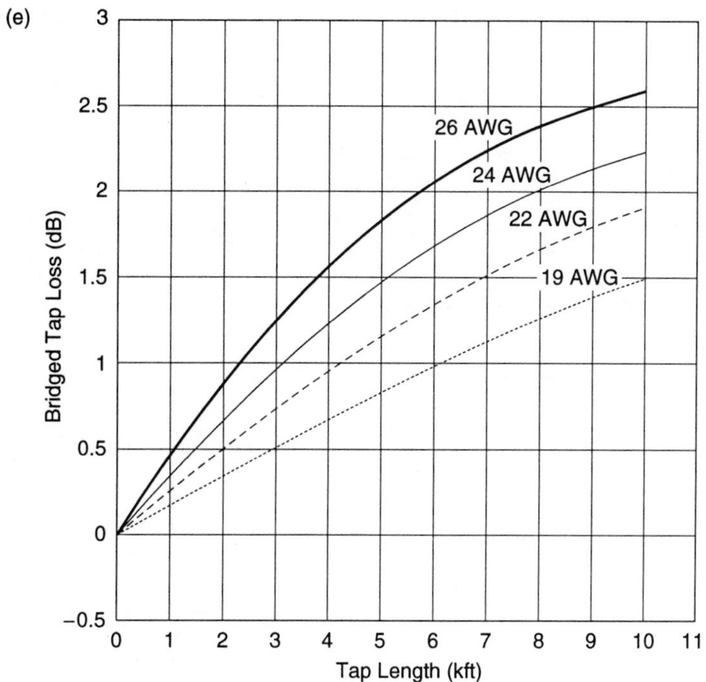

Figure 10-3(e) Bridged tap loss vs. bridged tap length, 4.8 kHz

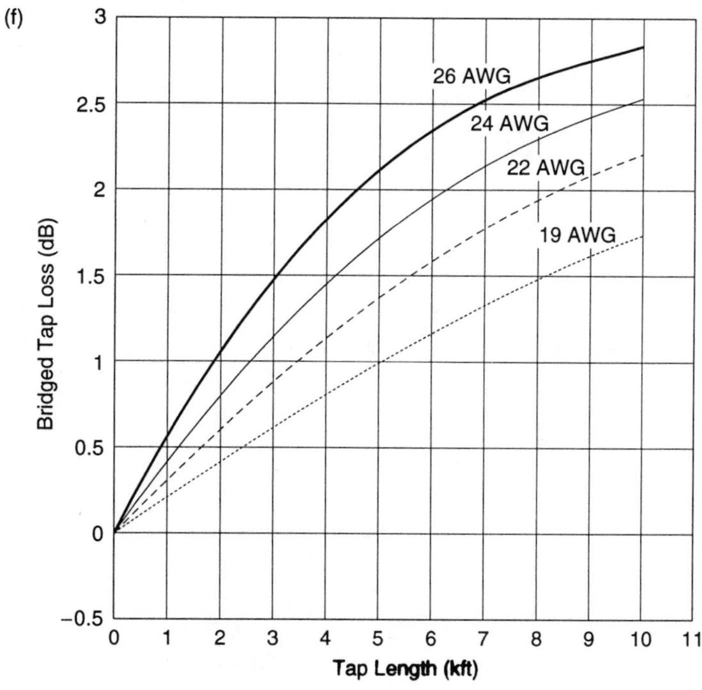

Figure 10-3(f) Bridged tap loss vs. bridged tap length, 6.4 kHz

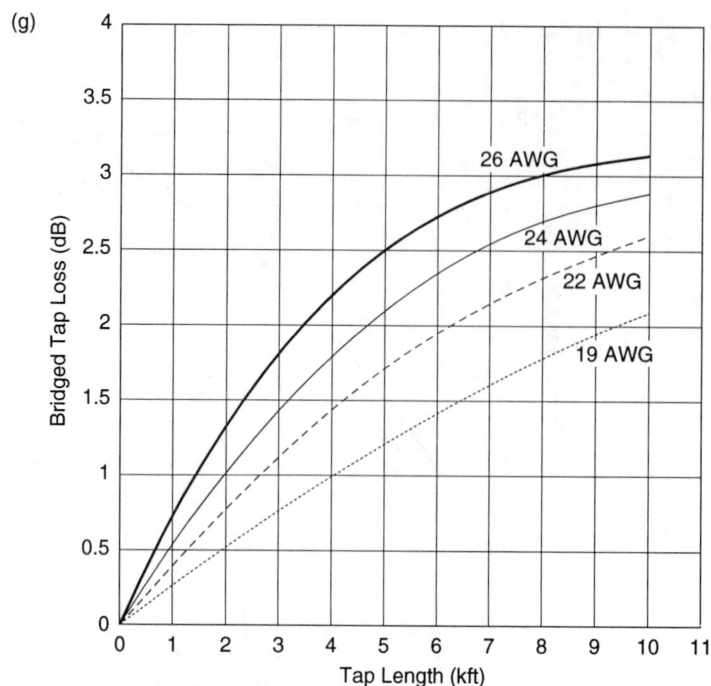

Figure 10-3(g) Bridged tap loss vs. bridged tap length, 9.6 kHz

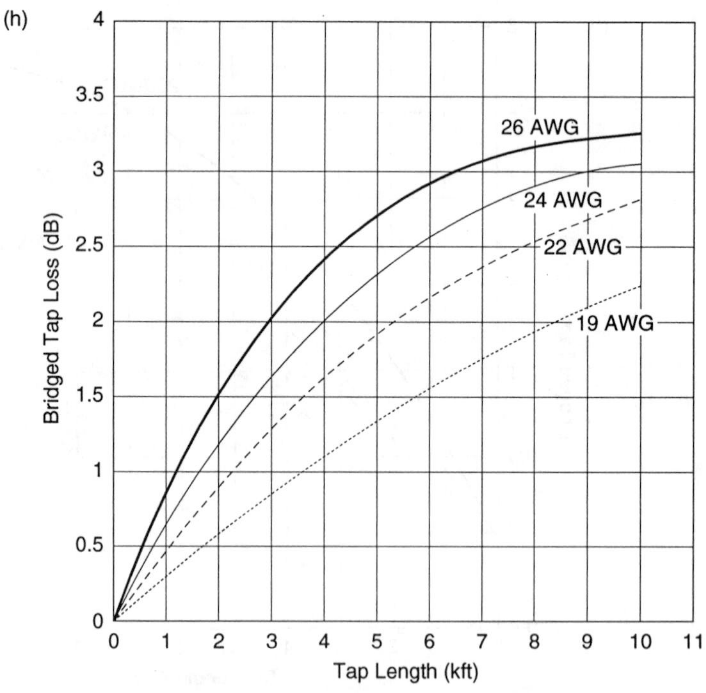

Figure 10-3(h) Bridged tap loss vs. bridged tap length, 12.8 kHz

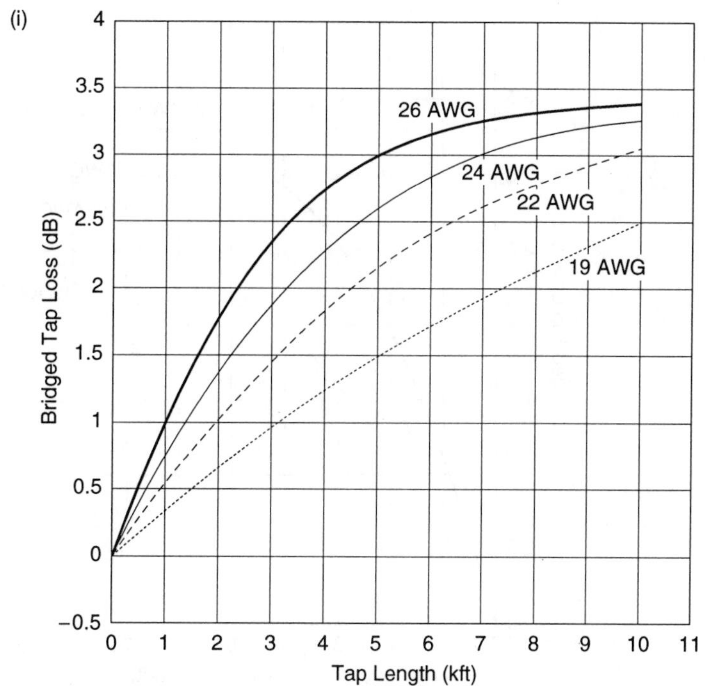

Figure 10-3(i) Bridged tap loss vs. bridged tap length, 19.2 kHz

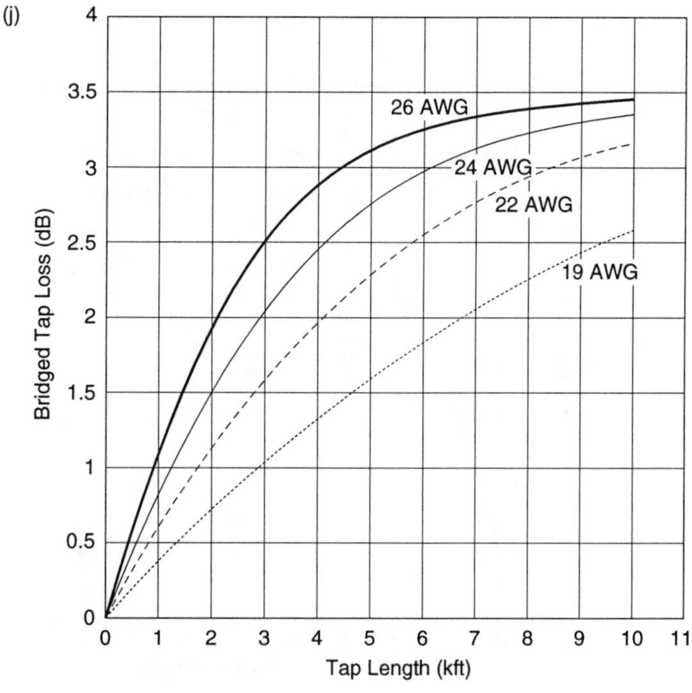

Figure 10-3(j) Bridged tap loss vs. bridged tap length, 25.6 kHz

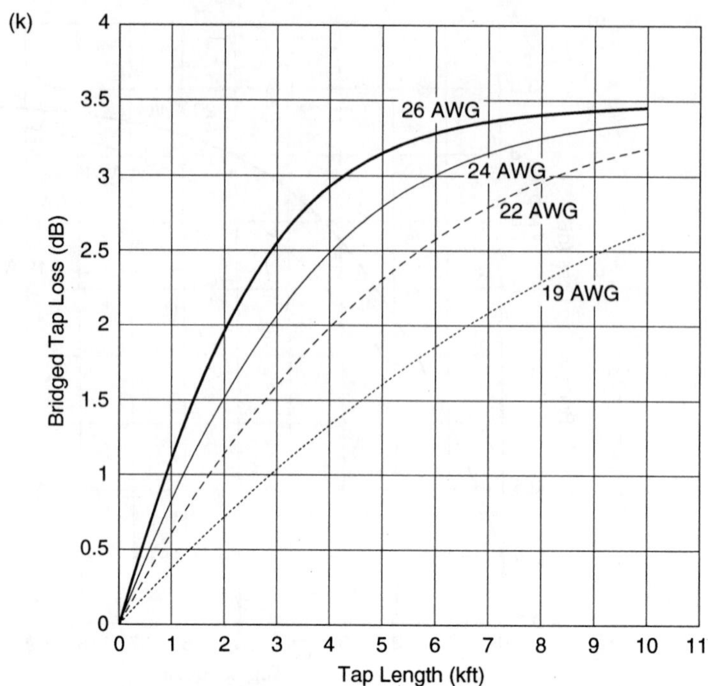

Figure 10-3(k) Bridged tap loss vs. bridged tap length, 28.0 kHz

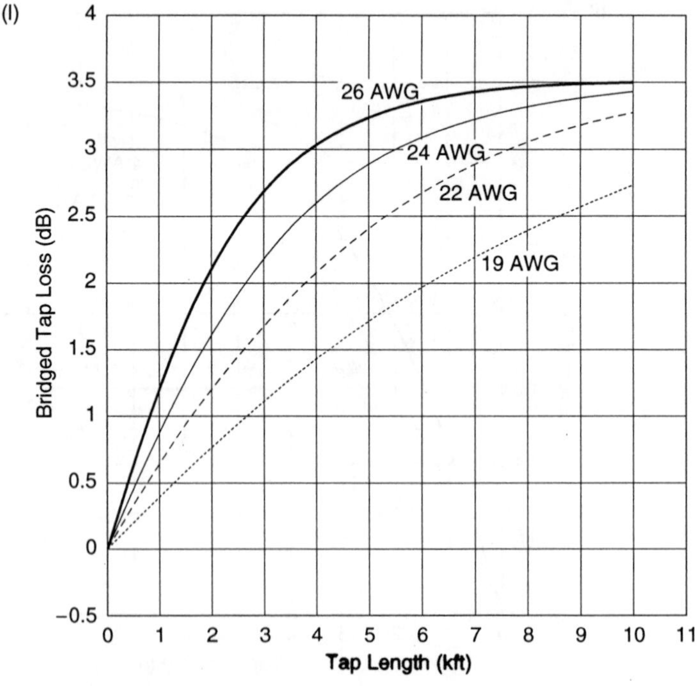

Figure 10-3(l) Bridged tap loss vs. bridged tap length, 36.0 kHz

Table 10-11 Maximum Allowable SRDL Lengths for PIC at Design Temperatures (No Bridged Taps)

Nyquist Frequency (kHz)	Single-Gauge Length at 31 dB Maximum EML (kft)							
	19 AWG 38 °C	19 AWG 60 °C	22 AWG 38 °C	22 AWG 60 °C	24 AWG 38 °C	24 AWG 60 °C	26 AWG 38 °C	26 AWG 60 °C
1.2	115.9	111.3	80.2	76.9	63.1	60.7	49.8	47.9
1.6	101.2	97.0	69.6	66.7	54.7	52.5	43.1	41.4
2.4	85.2	81.5	57.9	55.5	45.2	43.3	35.5	34.1
3.2	75.6	72.1	50.7	48.5	39.4	37.8	30.9	29.6
4.8	64.8	61.6	42.4	40.5	32.7	31.3	25.4	24.4
6.4	58.9	55.8	37.7	36.0	28.8	27.5	22.3	21.3
9.6	52.5	49.2	32.3	30.7	24.2	23.1	18.6	17.7
12.8	49.2	46.3	29.5	27.9	21.8	20.7	16.5	15.7
19.2	44.9	41.9	26.1	24.6	18.8	17.8	13.9	13.2
25.6	42.5	39.7	24.4	23.0	17.3	16.4	12.6	12.0
28.0	41.9	38.9	24.0	22.5	16.8	15.9	12.2	11.6
36.0	39.4	37.0	22.8	21.3	15.9	14.9	11.3	10.7

Table 10-12 Maximum Allowable SRDL Lengths for Pulp Insulated Cable at Design Temperatures (No Bridged Taps)

Nyquist Frequency (kHz)	Single-Gauge Length at 31 dB Maximum EML (kft)							
	19 AWG 38 °C	19 AWG 60 °C	22 AWG 38 °C	22 AWG 60 °C	24 AWG 38 °C	24 AWG 60 °C	26 AWG 38 °C	26 AWG 60 °C
1.2	114.3	109.6	78.9	75.7	61.9	59.7	50.1	48.2
1.6	99.9	95.8	68.6	65.7	53.9	51.8	43.4	41.7
2.4	83.8	80.3	56.9	54.5	44.5	42.7	35.8	34.3
3.2	74.1	70.8	49.8	47.7	38.8	37.2	31.1	29.8
4.8	63.2	60.1	41.6	39.7	32.1	30.8	25.6	24.5
6.4	57.1	54.2	36.9	35.2	28.3	27.0	22.4	21.5
9.6	50.0	47.7	31.3	29.8	23.7	22.6	18.7	17.8
12.8	47.0	44.3	28.4	27.0	21.2	20.3	16.6	15.8
19.2	42.5	39.7	25.0	23.7	18.2	17.3	14.0	13.4
25.6	39.2	36.9	23.3	22.0	16.8	15.8	12.6	12.0
28.0	38.6	36.1	22.7	21.4	16.2	15.4	12.2	11.6
36.0	36.1	33.9	21.5	20.1	15.2	14.3	11.3	10.7

The total EML is less than 31 dB, and therefore meets the engineering requirements. This problem could have been solved more quickly by obtaining the adjusted IL for each design temperature directly from Table 10-7.

EXAMPLE 10-3

An SRDL operating at 9.6 kbps without a secondary channel is to be installed on a 24 AWG PIC that is entirely underground. The length from the central office to the customer premises equipment (CPE) is 34.3 kilofeet. Determine the EML with and without a 24 AWG bridged tap 2.1 kilofeet long.

A 9.6 kbps SRDL without secondary channel operates at a Nyquist frequency of 4.8 kHz. Since the cable is underground, the design temperature is 38 °C. Referring to Table 10-7 at a Nyquist frequency of 4.8 kHz and a temperature of 38 °C, the unit IL for a 24 AWG cable is 0.95 dB/kilofoot. The total EML = 0.95 × 34.3 = 32.6 dB. Since the EML exceeds 31 dB, the circuit may not work during hot weather, even without considering the bridged tap. This can be confirmed by referring to Table 10-7, which gives the maximum length for a 24 AWG loop of 32.7 kilofeet at 38 °C. The bridged tap adds 2.1 kilofeet × 0.324 dB/kilofoot = 0.68 dB. The total loss for the loop with bridged tap is 33.3 dB.

The background noise and impulse noise requirements of SRDLs are shown in Table 10-13. The most restrictive background noise requirements are at the higher bit rates. The 34 dBrn50kb noise level for a 64 kbps SRDL equates to a voicegrade

Table 10-13 Maximum SRDL Background Noise and Impulse Noise Levels[a]

Payload Data Rate (kbps)	Secondary Channel	Background Noise (dBrn50kb)[b]	Impulse Noise Threshold (dBrn50kb)[b,c]
		Loop Loss ≤34 dB	
2.4	Yes or no	48	64
4.8	Yes or no	45	61
9.6	Yes or no	37	53
19.2	Yes or no	34	50
38.4	Yes or no	34	50
56.0	Yes or no	34	50
64.0	No	34	50
		Loop Loss ≤40 dB	
19.2	No	30	51
19.2	Yes	30	55
38.4	Yes or no	30	55
		Loop Loss ≤45 dB	
56.0	No	28	53
56.0	Yes	31	56
64.0	No	31	56

[a]Source: [3]

[b]For noise in dBm, subtract 90.

[c]All measurements made with a filter having a passband of 40 Hz to 30 kHz (50 kb filter). Maximum of 7 counts in any 15-minute measurement interval.

loop with about a 16 dBrnC noise level. At 9.6 kbps and lower rates, higher noise levels are allowed.

Impulse noise measurements are made at a fixed threshold. Any impulse that crosses that threshold is counted as an impulse event. Any further crossings of the threshold within the deadtime of the noise counter are ignored. Impulse noise tests should be made at both ends of the SRDL to ensure that local impulse noise sources do not lead to excessive errors.

10.2.3 SRDL Qualification

Because the SRDL has relatively high performance requirements, extensive pre-service tests are used to qualify an SRDL. All tests are made before the loop is terminated in any equipment. Tests to be performed and their limits are shown in Table 10-14. As each test described below is completed, the results should be compared to the test limits. Any SRDL that does not fall within the specified limits does not qualify and must be rejected until repaired or reassigned to a qualifying pair.

Table 10-14 SRDL Qualification Test Limits

Test	Limits
Foreign voltage	±1 Vdc maximum
Insulation resistance	≥300 kΩ
Loop dc resistance	≤4,200 Ω (initial) Initial test value ±30% Ω (operating)
Insertion loss (AML)[a]	≤34 dB (2.4 to 9.6 kbps) ≤40 dB (19.2 to 38.4 kbps) ≤45 dB (56.0 to 64.0 kbps)
Background noise (135 Ω impedance)	see Table 10-13
Impulse noise (135 Ω impedance)	see Table 10-13

[a]Not all equipment can meet the 40 and 45 dB actual measured loss (AML) as shown in this table for rates above 9.6 kbps. The original subtrate digital loop (SRDL) specification only allows 34 dB AML.

Foreign dc voltages must be measured on all conductors of the 4-wire SRDL. A high-impedance voltmeter is used to test for foreign voltages, as follows (T and R are *transmit* circuit conductors; T1 and R1 are *receive* circuit conductors):

Test	Results
T to R	≤ ±1 Vdc
T to ground	≤ ±1 Vdc
R to ground	≤ ±1 Vdc
T1 to R1	≤ ±1 Vdc
T1 to ground	≤ ±1 Vdc
R1 to ground	≤ ±1 Vdc

The insulation resistance tests are made with an analog or digital ohmmeter, as follows (set the analog meter to ohms × 10,000 range; set the digital meter to the 2 MΩ range):

TEST	RESULTS
T to R	≥300 kΩ
T to ground	≥300 kΩ
R to ground	≥300 kΩ
T to T1	≥300 kΩ
R to R1	≥300 kΩ
T1 to R1	≥300 kΩ
T1 to ground	≥300 kΩ
R1 to ground	≥300 kΩ
T1 to R	≥300 kΩ
R1 to T	≥300 kΩ

Loop resistance is measured with the T and R conductors shorted together at one end of the SRDL. A separate test is made with the T1 and R1 conductors shorted together. The dc resistance is measured at the other end with an analog or digital ohmmeter, as follows (set the analog meter to ohms × 100 range; set the digital meter to the 20 kΩ range):

TEST	RESULTS
T to R	≤4,200 Ω
T1 to R1	≤4,200 Ω

The AML qualification tests include not only the loss at the Nyquist frequency but other frequencies, as well, to give an indication of rolloff and to detect excessive bridged taps. Bridged taps will cause unexplained differences between AML and EML. Tests must be made on both the Tx and Rx circuits of the 4-wire loop. Two transmission measuring test sets (TMSs) capable of transmitting and receiving at frequencies up to 82 kHz will be needed. Termination impedances should be set to 135 Ω. If the TMS does not have 135 Ω impedance, it can be set to make a bridging measurement and the loop terminated with an external 135 Ω, noninductive resistor at both ends. Table 10-15 provides the test limits.

Background noise is measured on the Tx and Rx circuits of the 4-wire loop using a noise measuring set (NMS) with a 50 kb filter and 135 Ω termination impedance at both ends. The 50 kb filter has a 3 dB passband of 40 Hz to 30 kHz; the response is down approximately 22 dB at 50 kHz.*

Impulse noise tests are made with an impulse noise counter on both the Tx and Rx circuits of the 4-wire loop over a 15-minute interval with the appropriate thresholds and a 50 kb filter. The noise counter should be set to 135 Ω impedance and the other end of the circuit terminated in a 135 Ω resistor. Readings in dBrn can be converted to dBm by subtracting 90 dB; that is, dBm = dBrn − 90 dB.

---

*The specified characteristics of a 50 kb filter are given in [8].

Table 10-15 Initial Installation SRDL AML[a]

Payload Data Rate (kbps)	Test Frequency (kHz)	AML Limits
2.4	1.2	AML(1.2 kHz) ≤ 34 dB and AML = EML + 5 dB, −2.5 dB
	4.8	AML(4.8 kHz) ≤ 2 × AML(1.2 kHz)
4.8	2.4	AML(2.4 kHz) ≤ 34 dB and AML = EML + 5 dB, −2.5 dB
	4.8	AML(4.8 kHz) ≤ AML(2.4 kHz) + 15 dB
9.6	4.8	AML(4.8 kHz) ≤ 34 dB and AML = EML + 5 dB, −2.5 dB
19.2	9.6	AML(9.6 kHz) ≤ 40 dB and AML = EML + 5 dB, −2.5 dB
38.4	19.2	AML(19.2 kHz) ≤ 40 dB and AML = EML + 5 dB, −2.5 dB
56.0	28.0	AML(28.0 kHz) < 34 dB and AML = EML + 5 dB, −2.5 dB
	82.0	AML(82.0 kHz) ≤ AML(28.0 kHz) + 20 dB
	48.0	AML(48.0 kHz) = [AML(28.0 kHz) + AML(82.0 kHz)]/2 ± 2.5 dB
64.0	28.0	AML(28.0 kHz) ≤ 34 dB and AML = EML + 5 dB, −2.5 dB
	82.0	AML(82.0 kHz) ≤ AML(28.0 kHz) + 20 dB
	48.0	AML(48.0 kHz) = [AML(28.0 kHz) + AML(82.0 kHz)]/2 ± 2.5 dB

[a]Tests are valid for loops with or without secondary channel.

10.3 Switched 56 Loops

From a loop engineering perspective, the 4-wire switched 56 loop is identical to the dedicated SRDL operating at 56 kbps. Refer to the previous section for details. Switched 56 loops based on 2-wire time compression multiplexing (TCM) technology use basically the same loop engineering techniques as the 4-wire SRDL, but the design frequency and noise requirements are different, as will be described in this section. The incompatibility problems with wideband analog services and analog subscriber carrier, as described in the previous section, also apply to 2-wire TCM switched 56 loops.

The 2-wire TCM switched 56 loop uses a bipolar alternate mark inversion (AMI) line code, which peaks the power at a frequency equal to one-half the line transmission rate (Nyquist frequency). Since the line rate is 160 kbps, the Nyquist frequency is 80 kHz. The peak transmitted voltage is 2.4 V nominal, which provides a maximum reach of 45 dB loop loss (at 80 kHz). Due to propagation delay considerations, however, the loop length is limited to 18 kilofeet (24, 22, and 19 AWG) and 14 kilofeet (26 AWG), even though the loss may be less than 45 dB for certain loops longer than these values. There may be situations in which a loop complies with the loss and length requirements but still does not work due to impulse noise or NEXT.

The loss of switched 56 loops consists of three basic components: cable loss, junction loss at the point where two different cable gauges (or cable types) are joined, and bridged tap loss. The loss components are tabulated for the loop in question, and if the total is less than 45 dB, the loop is acceptable from an engineering standpoint. The procedure described in the previous section may be used with the addition of junction losses. The loop requires qualification, which may cause it to be rejected, requiring selection of another cable pair.

Cable loss is given in Table 10-16 for each common cable gauge. Figure 10-4 shows the maximum loop length at 38 °C for PIC assuming a single gauge, no junction losses, and no bridged tap losses. Junction losses are given in Table 10-17.

Bridged tap loss can be determined using techniques previously described for the SRDLs; however, to simplify the design process, it is acceptable to use a flat 3.5 dB of bridging loss for each bridged tap on the loop. For example, two bridged taps would contribute a total of 7 dB of additional loss. The 3.5 dB bridging loss per bridged tap is a conservative value, which applies to most practical situations.

The impulse noise requirements for a maximum loss 2-wire switched 56 loop (45 dB at 80 kHz) are zero counts in 15 minutes at a threshold of 49 dBrn50kb. If a weighting filter with a noise bandwidth of 100 kHz is available, the threshold remains the same, but the counts may be increased to 44 counts in any 15-minute test period.

The impulse noise threshold may be increased by 1 dB for every 1 dB that the loop loss is less than 45 dB. This increase in threshold applies to loop losses down to 25 dB, where the threshold is 69 dBrn. The threshold remains fixed at 69 dBrn for loop losses between 0 and 25 dB. Background noise should be less than 32 dBrn with either a 50 kb filter or 100 kHz filter. All transmission tests are made using a 135 Ω termination impedance. The design requirements for switched 56 loops are summarized in Table 10-18.

Table 10-16 80 kHz Loss at Design Temperatures, Nonloaded Twisted Pair Exchange Cable

80 kHz Loss	19 AWG (dB/kft)	22 AWG (dB/kft)	24 AWG (dB/kft)	26 AWG (dB/kft)
		APIC, FPIC, DEPIC		
21 °C	0.986	1.545	2.206	3.204
38 °C	1.025	1.625	2.330	3.375
60 °C	1.077	1.731	2.494	3.603
TC (%/10 °C)	2.4%	3.1%	3.3%	3.2%
		PULP INSULATED CABLE		
21 °C	1.108	1.683	2.353	3.237
38 °C	1.151	1.768	2.482	3.407
60 °C	1.209	1.882	2.654	3.635
TC (%/10 °C)	2.3%	3.0%	3.3%	3.1%

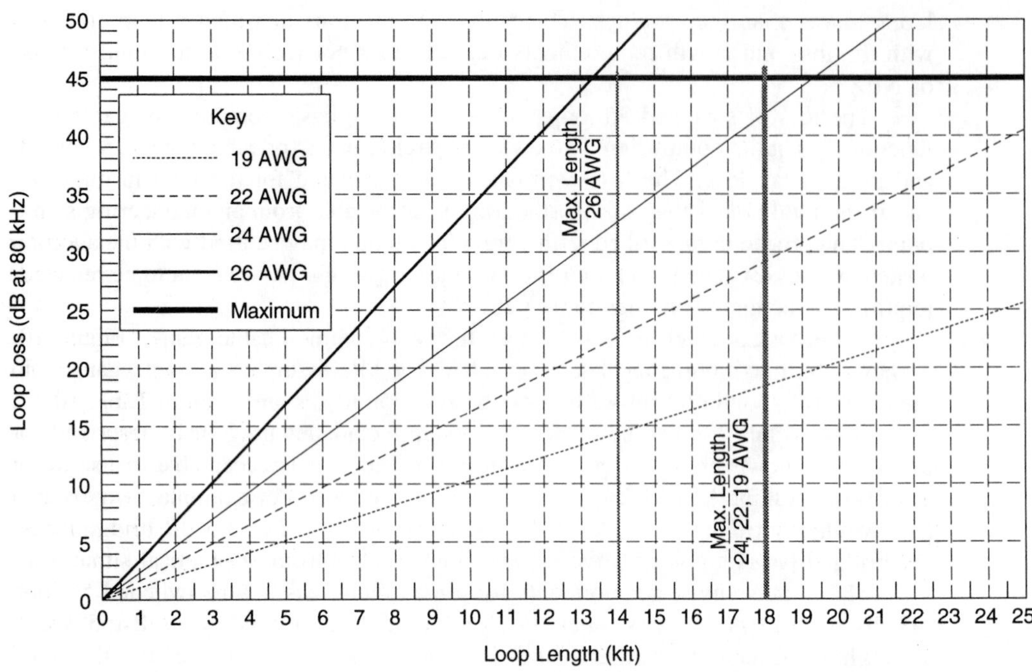

Figure 10-4 Loop loss at 80 kHz vs. loop length at 38 °C

Table 10-17 Junction Loss for 2-Wire Switched 56 Loops

CONDITION	JUNCTION LOSS (dB)
Length of both cables > 400 ft, adjacent gauge	0.25
Length of both cables > 400 ft, nonadjacent gauge	0.5
Length of one cable < 400 ft	0.0

Table 10-18 Switched 56 Loop Design Summary

PARAMETER	LIMITS
Loop loss at 80 kHz	45 dB
Loop length	18 kft (19, 22, 24 AWG) 14 kft (26 AWG)
Background noise (135 Ω impedance)	32 dBrn50kb
Impulse noise (135 Ω impedance)	0 counts in 15 minutes at 49 dBrn50kb threshold

10.4 ISDN DSL

10.4.1 Interference Considerations

For the ISDN DSL to be widely deployed, it must coexist with other types of services in the loop plant. This means that energy coupled from the DSL into other loops must not degrade their transmission performance, nor must energy from other loops degrade the DSL performance. Studies have shown that NEXT and impulse noise are the primary sources of interference to a DSL; far-end crosstalk (FEXT) is not perceived as a problem [6,7]. Generally, the most significant interference problem in the loop plant is *from* the DSL *to* other loop types.

The design and application of the DSL is one of constraints. In order to minimize the crosstalk induced by the DSL into other services within the same cable, the DSL transmit power must be constrained. This, in turn, constrains the distance over which the DSL can operate, because less power means less operational range. Similarly, crosstalk induced into the DSL by other services constrains the minimum Rx power of the DSL. Self-interference caused by one DSL onto another DSL also constrains the range. Any range limitations translate into limitations on the percentage of loops served by the DSL.

Self-interference is not a problem with DSLs meeting the design requirements described in the next section. That is, DSLs coexist with other DSLs without problems. Similarly, DSLs coexist with other digital loop systems, such as SRDLs and repeatered T1-carrier span lines. However, the DSL is known to interfere with some existing analog services, particularly analog subscriber carrier and program audio channels. The DSL has significant energy components at the same frequency as these services, and crosstalk within the same binder group and within adjacent binder groups will degrade their performance.

Performance degradation of other services in the same cable may not be noticed when only a few DSLs are operated. However, as more DSLs are assigned to the cable, the cumulative crosstalk power will increase, and at some point, destructive interference will result. This can be appreciated by considering that 49 DSLs in a 50-pair cable will induce approximately 10 dB more crosstalk power than one DSL. The point at which destructive interference occurs is not predictable.

Any loop that operates within the band between approximately 4 kHz and 40 kHz is a candidate for interference from DSLs. Single-channel and multichannel analog subscriber carriers are such candidates. This is particularly the case when the analog carrier is operated at its maximum range; that is, where the receiver is operating at its maximum gain. This is the operational condition in which the SNR is the lowest and the receiver is most susceptible to interference. In some cases, the analog carrier can be made to operate properly by reducing the exposure distance between the central office and the remote subscriber channel unit, which effectively increases the SNR; however, this may not be effective if the receiver is equipped with automatic gain control (AGC), or may not be possible due to plant limitations.

10.4.2 DSL Design

The DSL is designed for service on twisted pair loops that meet the following basic rules:

- ≤18 kilofeet for 19, 22, and 24 AWG; 15 kilofeet for 26 AWG
- ≤42 dB loss (including bridged tap loss) at 40 kHz
- ≤1,300 Ω dc loop resistance
- Nonloaded loops only

It has been estimated that approximately 99% of all existing nonloaded loops in North America meet these requirements [9]; however, due to interference considerations, additional engineering provisioning rules are needed to minimize performance degradation on the 1% of loops that do not meet these requirements, and on other loops in the same cable. These will be discussed later.

Subscriber loops designed according to the resistance design method form the cornerstone of plant suitable for the DSL. Resistance design originally was conceived for analog VF loops, and because of its ubiquity, the DSL was made to fit within its confines.

The general rules for resistance design are listed in Table 10-19. The most important transmission requirement from the DSL standpoint is the loss, and not resistance or length. DSLs of 24 AWG and larger with lengths ≤18 kilofeet and 26 AWG with lengths ≤15 kilofeet implicitly meet the basic loss constraint at 40 kHz. The DSL may work on longer loops under some conditions and may not work on some shorter loops, particularly those with long bridged taps.

Loop powering is an important consideration, and this is where the dc resistance of the loop comes into play. Limiting the dc resistance of the loop to 1,300 Ω meets the DSL design requirements.

Implicit in the resistance design method is the flexibility to use single or mixed cable gauges depending on the situation. An objective is to use a single gauge in each distribution area identified in the planning process. This is quite easy to do

Table 10-19 Resistance Design Rules[a]

RULE	DESCRIPTION
1	Limit conductor loop resistance to 1,300 Ω maximum. Mixed gauges are permissible
2	Limit length of nonloaded loops to 18 kilofeet maximum (15 kilofeet for 26 gauge)
3	Load all loops over 18 kilofeet but less than 24 kilofeet
4	Use digital loop carrier (DLC) on loops over 24 kilofeet

[a]The Bell System revised the resistance design rules slightly in 1983 for their own purposes, although the principles are applicable anywhere [10]. The revised rules are essentially identical to those previously described, but the conductor loop resistance for loops between 18,000 and 24,000 feet is limited to 1,500 Ω (rather than 1,300 Ω), and H88 loading is used for loops over 18,000 feet (15,000 feet for 26 gauge). Rules (3) and (4) above and the Bell modifications just described are not relevant to DSL design but are included here for completeness.

when a particular distribution area is compact (equivalent radius approximately three miles). If a single gauge is not feasible, adjacent gauges should be used (for example, 26 and 24, or 24 and 22).

With the resistance design method, the use of mixed gauges is handled in a straightforward manner: the 40 kHz loss and resistance of each gauge are each tabulated and added to determine if their totals to a particular end-user location meet the transmission and powering criteria. In mixed-gauge applications, the smaller gauges normally are placed closer to the central office, with progressively heavier gauges farther from the central office.

When loops are designed according to the foregoing method, it is helpful to have detailed cable pair data at various temperatures. For practical high-frequency problems, loss variation with temperature is considered to be due to resistance and inductance variation.

Table 10-20 shows the 40 kHz loss for various cable gauges at 38 °C (100 °F) and 60 °C (140 °F). Loop losses versus loop length for PIC are given in Figure 10-5 at a temperature of 38 °C. This illustration assumes no bridged taps. Other design temperatures can be used as discussed in the design section for the SRDL.

Figure 10-6 shows the bridged tap loss for the four cable gauges at 40 kHz. Beyond approximately two kilofeet, the bridged tap loss is a nonlinear function of length. Table 10-21 can be used for short bridged taps. Bridged taps made of 19 AWG cable should not be used with the DSL because of reflection problems from impedance mismatch.

A rule of thumb is to increase the loss by 3% for every 10 °C increase in temperature. The dc loop resistance increases by about 4% for every 10 °C increase in temperature. Table 10-22 gives the dc loop resistance of twisted pairs at various temperatures, and Figure 10-7 shows the resistance as a function of length at the same temperatures. Where applicable, the maximum resistance of 1,300 Ω is shown.

Other engineering provisioning rules apply to DSL design, as previously mentioned. The application of engineering provisioning rules will have a somewhat different tone in existing plant versus new plant. That is, new plant design will take

Table 10-20 40 kHz Loss at Design Temperatures, Nonloaded Twisted Pair Exchange Cable

40 kHz Loss	19 AWG (dB/kft)	22 AWG (dB/kft)	24 AWG (dB/kft)	26 AWG (dB/kft)
	APIC, FPIC, DEPIC			
21 °C	0.768	1.316	1.904	2.700
38 °C	0.808	1.390	2.001	2.824
60 °C	0.861	1.485	2.127	2.984
TC (%/10 °C)	3.1%	3.3%	3.0%	2.7%
	Pulp Insulated Cable			
21 °C	0.843	1.402	1.992	2.704
38 °C	0.886	1.478	2.094	2.828
60 °C	0.942	1.577	2.225	2.989
TC (%/10 °C)	3.0%	3.2%	3.0%	2.7%

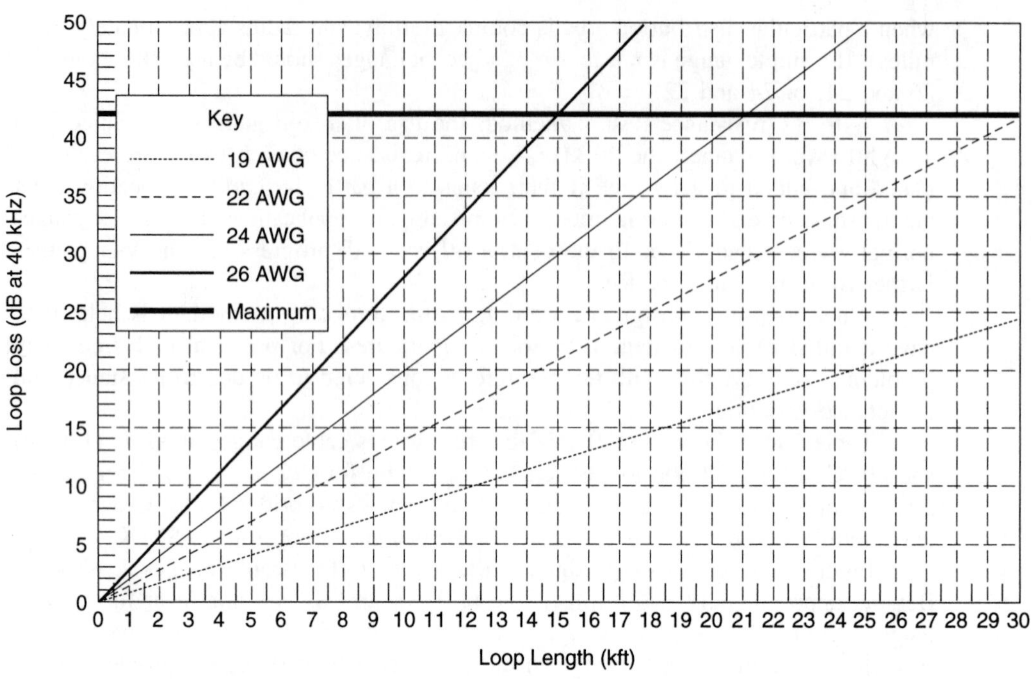

Figure 10-5 Loop loss at 40 kHz vs. loop length at 38 °C

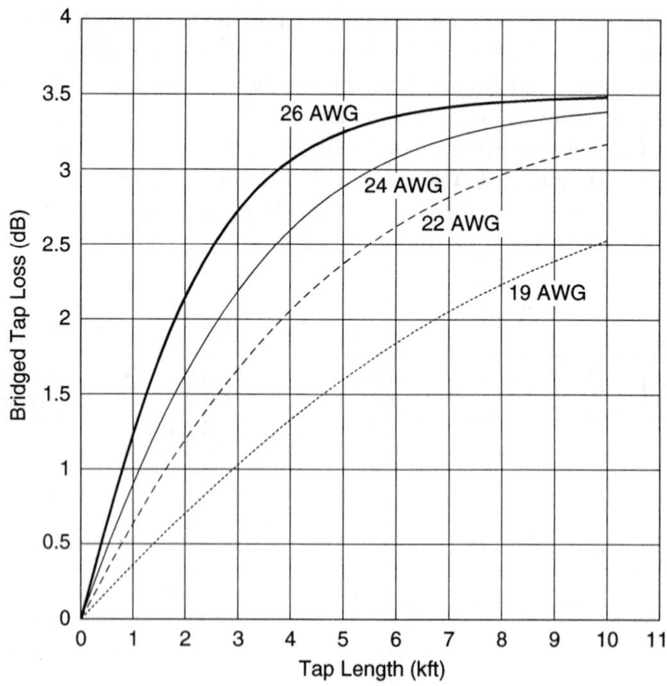

Figure 10-6 Bridged tap loss at 40 kHz vs. length

Table 10-21 Bridged Tap Loss at 40 kHz[a]

Bridged Tap Loss (dB/kft)			
19 AWG	22 AWG	24 AWG	26 AWG
0.343	0.561	0.757	0.969

[a]Valid for bridged taps ≤2.5 kft

Table 10-22 DC Loop Resistance of Copper Twisted Pairs (Ω/kft)

Design Temperature	19 AWG	22 AWG	24 AWG	26 AWG
13 °C (55 °F)	15.9	31.7	50.3	81.1
38 °C (100 °F)	17.4	34.8	55.3	89.2
60 °C (140 °F)	18.9	37.7	59.8	96.5

into account the DSL requirements when initially designed and built. The constraints on existing plant are not as serious, for economic reasons. This is particularly true of bridged taps. Whereas it is not necessary to undertake wholesale removal of bridged taps in existing plant used for DSL service, the design of new plant will strive to eliminate bridged taps altogether.

Other engineering considerations are:

- Removal of load coils and induction neutralizing transformers
- Assignment of non-DSL services in the same cable
- Resolving transmission constraints due to bad cable and poor splicing practices
- Application of fixed count terminals and sealed plant concept
- Elimination of ready access plant in favor of fixed or dedicated count plant
- Elimination of party lines

EXAMPLE 10-4

The makeup of a DSL between the line termination (LT) and network termination (NT) is shown in Figure 10-8; all cables are direct buried. Find the total loss at 40 kHz.

The loop consists of 7.5 kilofeet of 26 AWG, 7.5 kilofeet of 24 AWG, and 1.0 kilofeet of 22 AWG. Bridged tap lengths are 2.0 kilofeet of 24 AWG and 1.0 kilofeet of 22 AWG. The loss of each segment is summarized in the table below.

AWG	Unit Loss	Segment Length	Segment Loss
		Main Cable	
26	2.824 dB/kft	7.5 kft	21.2 dB
24	2.001 dB/kft	7.5 kft	15.0 dB
22	1.390 dB/kft	1.0 kft	1.4 dB
		Bridged Taps	
24	0.757 dB/kft	2.0 kft	1.5 dB
22	0.561 dB/kft	1.0 kft	0.6 dB
		Total	39.7 dB

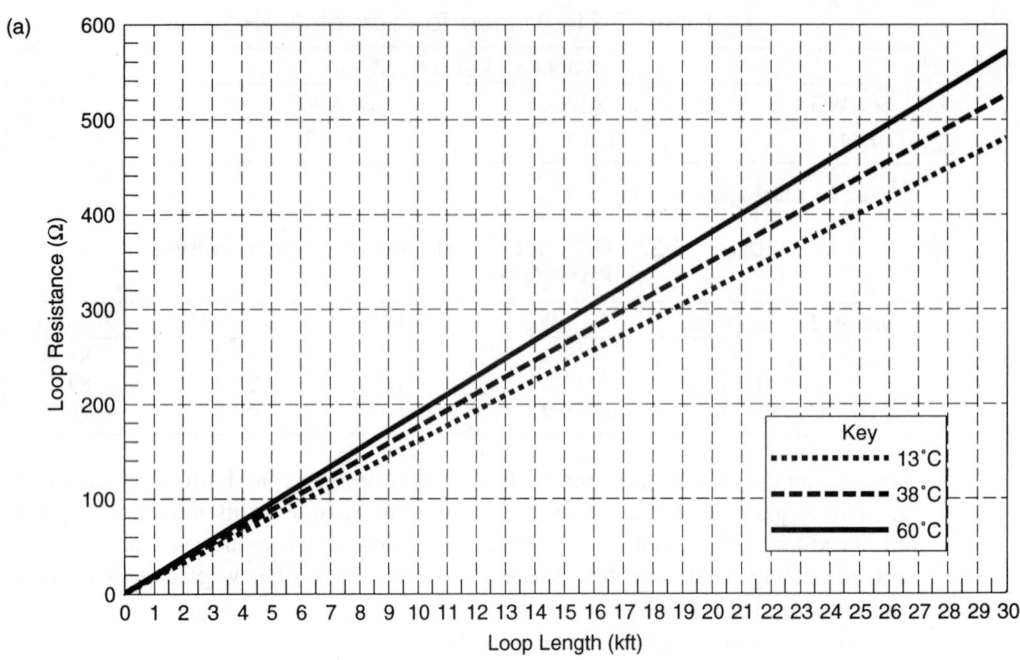

Figure 10-7(a) DC loop resistance vs. loop length, 19 AWG

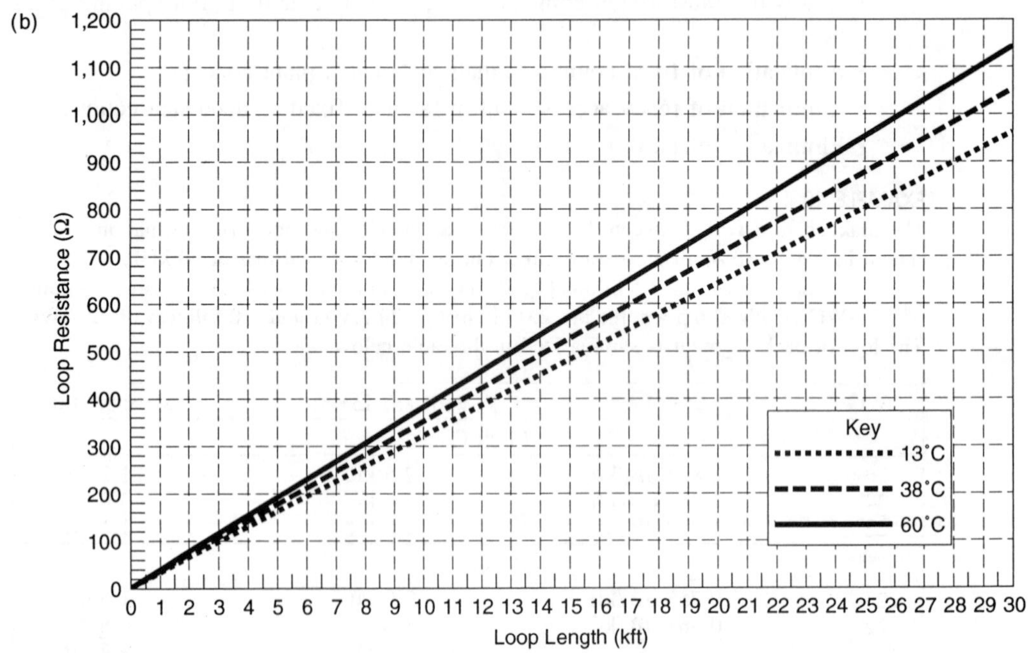

Figure 10-7(b) DC loop resistance vs. loop length, 22 AWG

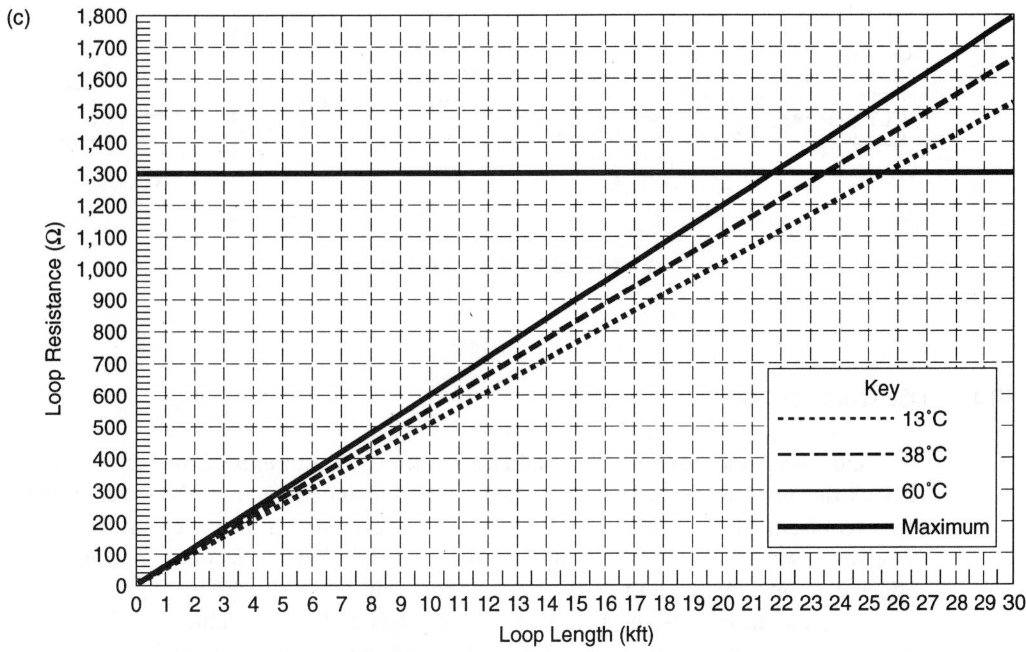

Figure 10-7(c) DC loop resistance vs. loop length, 24 AWG

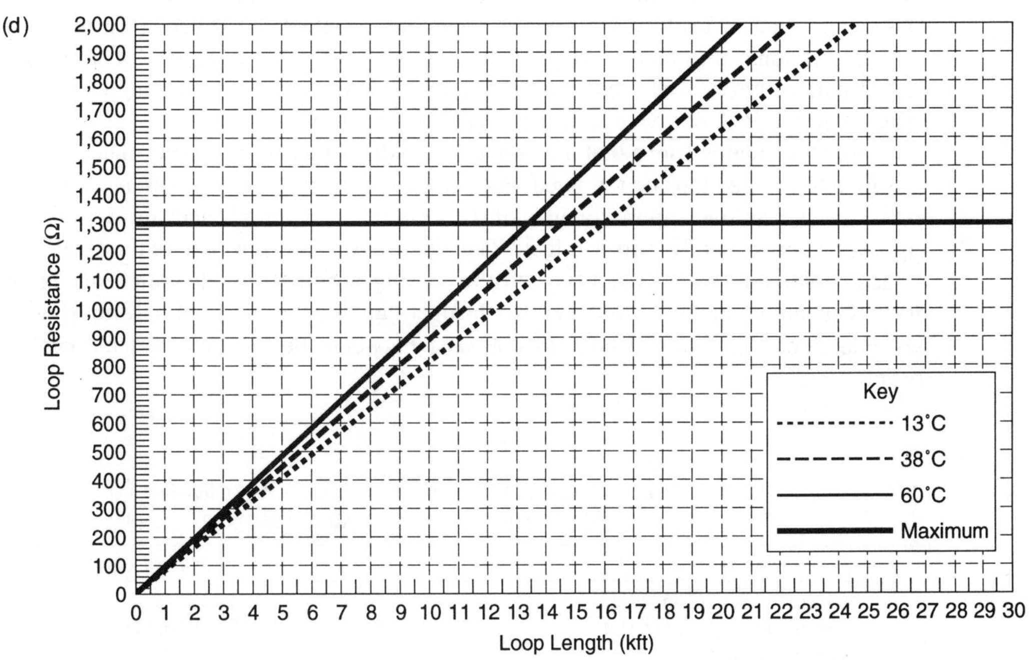

Figure 10-7(d) DC loop resistance vs. loop length, 26 AWG

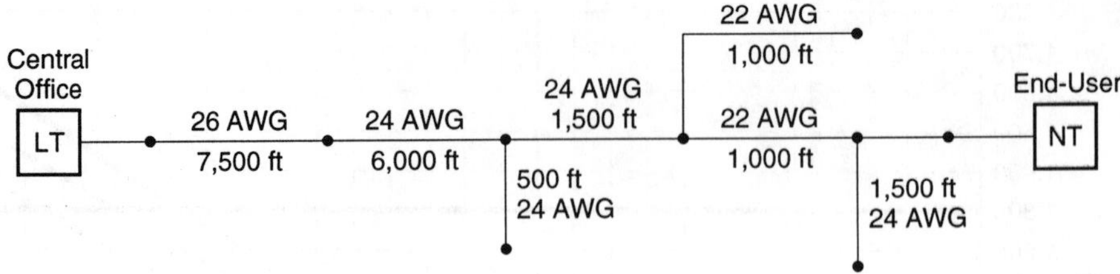

Figure 10-8 DSL for Example 10-4

10.4.3 DSL Qualification

Industry standards for DSL qualification limits do not exist; however, since the DSL is designed to work on any twisted pair that meets analog VF requirements, these requirements are given in Table 10-23 for reference in this section. Table 10-24 gives practical qualification information for the DSL. The noise requirements reflect the wider bandwidth of the noise measuring set 50 kb filter.

Consideration must be given to bridged taps and line conditioning. Although there are no specific restrictions on DSL bridged taps as there are with other digital loops, the DSL will not work with virtually all bridged taps in the field. Other loops (specifically the SRDL) basically require that bridged taps be kept shorter than 1/4 wavelength at the frequency of maximum power. Using this same basic rule for the DSL could lead to bridged tap lengths over three kilofeet; however, a bridged tap or taps this long are considered poor practice in new plant designed for universal digital connectivity. It is more reasonable to limit total bridged tap lengths to one kilofoot where it is not economical to avoid them altogether.

Existing analog loops that are converted to digital loop service normally do not require that bridged taps be removed unless they are known to be "excessive," where excessive means over approximately three kilofeet total length and any individual tap over approximately two kilofeet. The dedicated plant concept requires all assigned cable pairs be cut dead beyond the serving terminal, as illustrated in Figure 10-9. This automatically eliminates any bridged tap beyond the terminal, but it does not guarantee that bridged taps do not exist between the terminal and the central office.

Table 10-23 Analog Loop Transmission Objectives for Reference in DSL Qualification

Parameter	Acceptable Limits
Circuit noise, C-message	≤20 dBrnC
Circuit noise, 3 kHz flat	≤40 dBrn3kHz
Power influence	≤80 dBrnC
Circuit balance	≥60 dB

Table 10-24 DSL Qualification Test Limits

Test	Limits
Foreign voltage	±1 Vdc maximum
Insulation resistance	≥300 kΩ
Loop dc resistance	≤1,300 Ω, initial test value ± 30% Ω (operating)
Insertion loss (AML) at 40 kHz	≤42 dB
Background noise (135 Ω impedance)	≤37 dBrn50kb
Impulse noise (135 Ω impedance)	0 counts in 30-minute period at 57 dBrn50kb threshold

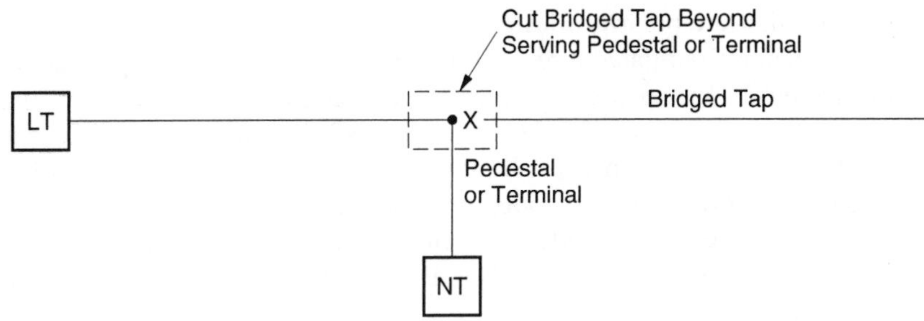

Figure 10-9 Elimination of bridged taps beyond the serving pedestal or terminal

Loops over 18 kilofeet generally are not suitable for use with the DSL. Such loops, when used with analog VF services, are always loaded or otherwise treated according to the design rules applicable to the service involved.* There will be situations in which plant rearrangements will make these cable pairs suitable and available for the DSL. If so, any loop treatment, such as load coils, build-out capacitors, equalizers, induction neutralizing transformers designed for VF service, VF repeaters, and loop extenders, must be removed before the pair can be used for the DSL.

The sealed plant concept yields significant economic benefits in all loop plant. Therefore, existing ready access terminals and exposed splicing in pedestals should be replaced with sealed fixed count terminals during routine or rehabilitation work, and new plant should be constructed with the sealed fixed count terminals from the very beginning.

Any digital loop system is sensitive to impulse noise. Impulse noise can be minimized by removing any dc signaling systems, such as alarm systems and loop current-operated teletype systems, from the cable. Also, shield integrity and proper grounding will help minimize performance degradation from impulse noise as well as from powerline induction. In particularly severe environments, DSL-compatible induction neutralizing transformers will be needed.†

*See [1].

†One source of induction neutralizing transformers for digital loop applications is [11].

The final conditioning requirement is to remove all analog subscriber carrier pair gain systems from cables to be used for the DSL in particular, and universal digital connectivity in general. In cases in which pair gain is required due to congested plant, the digital loop carrier should be used in place of the analog carrier.

One significant deployment advantage of DSL is its ability to work properly without regard to tip-ring polarity. The tip and ring conductors in the outside plant or premises wiring between the network interface and the ISDN network termination (NT1) can be reversed with virtually no effect on the DSL. Other digital loops, namely 4-wire loops such as repeatered T1-carrier and the SRDL, require proper polarity for voltage-operated loopback control. Even though field reversals are avoided in properly installed plant, they do exist to a limited extent. With polarity independence, the DSL provides yet lower installation and maintenance cost when compared to loops that are polarity sensitive.

Technically related to the ISDN DSL are the digital pair gain devices described in Chapter 2. The loop design requirements are identical to the DSL just described; that is, the loop loss at 40 kHz should be ≤42 dB and the dc loop resistance ≤1,300 Ω. These constraints usually limit the loop to ≤18 kilofeet, but digital pair gain loops may work on longer loops that meet the loss and resistance criteria. The interference issues also are identical: the digital pair gain systems are not compatible with analog subscriber carrier. Figure 10-10 shows a typical digital pair gain system application.

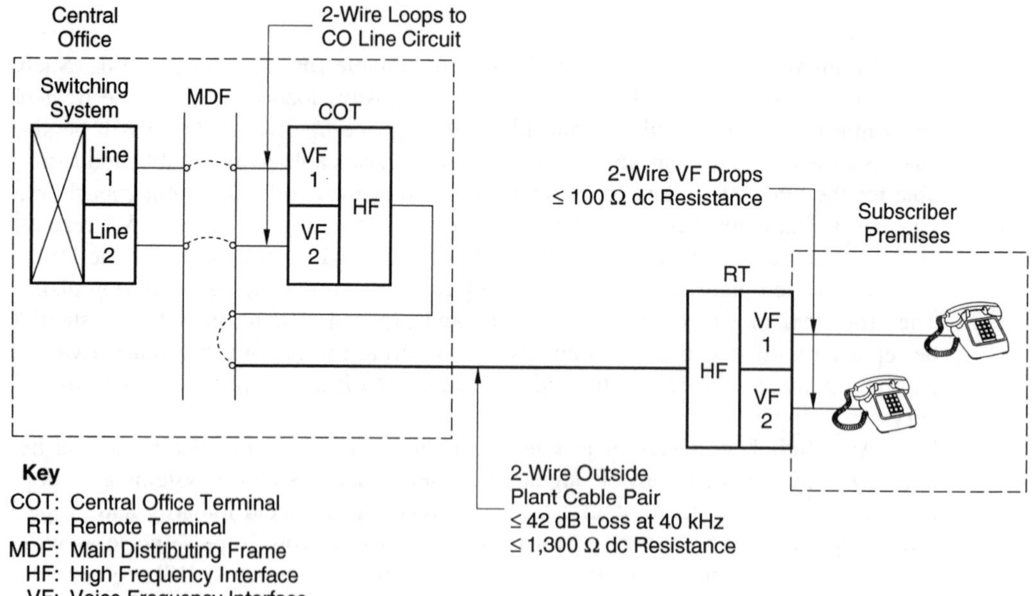

Figure 10-10 Digital pair gain system application using ISDN DSL technology

10.5 HDSL

10.5.1 Interference Considerations

As a universal digital transport system, the HDSL must coexist with other types of services in the loop plant, including other HDSLs and DSLs. The compatibility requirements are the same as the DSL, but the HDSL is specifically *not* required to be compatible with wideband analog systems, such as program audio channels and analog subscriber carrier.* Studies have shown that NEXT and impulse noise are important sources of interference to an HDSL. Self-NEXT (that is, NEXT caused by other HDSLs in the same cable) has been found to be the worst-case interference [12]. HDSLs are required to coexist with SRDLs, DSLs, and repeatered T1-carrier span lines, as well as conventional VF analog loops (POTS). The basic interference considerations for the ISDN DSL apply to HDSL.

10.5.2 HDSL Design

The HDSL is designed for service on twisted pair loops that fall within the carrier serving area (CSA) engineering concept, which is summarized in the form of design rules in Table 10-25 [13].

Although the HDSL is designed to work with any loop meeting CSA requirements, it is helpful to be able to calculate cable pair loss for specific situations. Table

Table 10-25 CSA Engineering Rules

RULE	DESCRIPTION
1	Nonloaded cable only
2	No more than two cable gauges (excluding short cable sections used for stubs or fuses)
3	Total bridged tap ≤ 2.5 kilofeet
4	No single bridged tap ≥ 2.0 kilofeet
5	The total amount of 26 AWG cable (used alone or in combination with another cable gauge) not to exceed nine kilofeet, including bridged tap
6	Total cable length, including bridged taps, of a multigauge cable that contains 26 AWG cable not to exceed the following: $$\text{Total cable length} \leq 12 - 3\frac{L26}{(9 - BT)}\ \text{kft}$$ where L26 = total length of the 26 AWG cable (excluding 26 AWG bridged taps) in kilofeet; BT = total length of the bridged taps in kilofeet
7	For 19, 22, and 24 AWG, in any combination, the maximum total cable length, including bridged tap, is 12 kilofeet

*DSL specifications mention compatibility only in general terms. Compatibility with specific services or systems, other than other DSLs, are not mentioned.

10-26 shows the 196 kHz loss for various cable gauges at 21 °C (68 °F), 38 °C (100 °F), and 60 °C (140 °F). The allowable loss is approximately 42 dB at 196 kHz and 38 °C.

Table 10-26 196 kHz Loss at Design Temperatures, Nonloaded Twisted Pair Exchange Cable

196 kHz Loss	19 AWG (dB/kft)	22 AWG (dB/kft)	24 AWG (dB/kft)	26 AWG dB/kft)
		APIC, FPIC, DEPIC		
21 °C	1.51	2.15	2.82	3.87
38 °C	1.55	2.23	2.95	4.08
60 °C	1.62	2.33	3.12	4.34
TC (%/10 °C)	1.9%	2.2%	2.7%	3.1%
		PULP INSULATED CABLE		
21 °C	1.76	2.42	3.10	4.00
38 °C	1.81	2.51	3.24	4.20
60 °C	1.88	2.62	3.42	4.46
TC (%/10 °C)	1.8%	2.1%	2.6%	3.0%

A rule of thumb is to increase the loss by 3% for every 10 °C increase in temperature. Usually, a 3 to 6 dB margin is used, which leaves an allowable EML of 36 to 39 dB. These loss values are not to be used as loop design constraints because HDSL terminal equipment is designed to meet error performance requirements under a variety of loop configurations, some of which may exceed 42 dB total loss. With the heavier pair gauges, the loss constraint of 36 dB (which includes approximately a 6 dB margin) will be met on loops greater than 12 kilofeet. For example, a 16-kilofoot loop consisting of only 22 AWG cable has a loss of 36 dB at 38 °C. Commerical products function normally on such a loop.

Figure 10-11 shows the loop loss at 196 kHz versus loop length for the four pair gauges at 38 °C. This illustration assumes no bridged taps. Table 10-22 and Figure 10-7, shown previously for the ISDN DSL, may be used to determine dc loop resistance for powering purposes on loops outside the CSA.

The engineering provisioning rules for the HDSL are identical to those required for other services within the CSA. It is reiterated here that, whereas it is not necessary to undertake wholesale removal of bridged taps in existing plant being used for HDSL service, the design of new plant will strive to eliminate bridged taps altogether. Figure 10-12 shows a plot of bridged tap loss at 196 kHz. Table 10-27 gives the loss for short bridged taps in various cable gauges. Although 19 AWG bridged taps are not specifically excluded from the CSA, they are to be avoided, as with the DSL. Again, loss is not necessarily a specific design constraint with the HDSL, so this table is provided only for reference.

EXAMPLE 10-5

The length of a buried feeder route using 26 AWG cable is 6.0 kilofeet. There are no bridged taps. Find the maximum allowable length of additional buried 24 AWG distribution cable that meets CSA criteria. Find the loss at 196 kHz of an HDSL loop using these cable combinations.

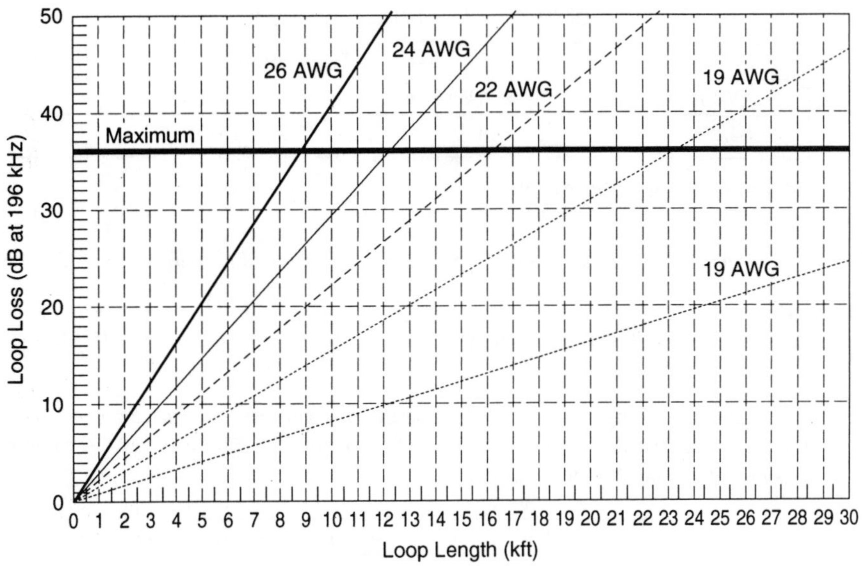

Figure 10-11 Loop loss at 196 kHz vs. loop length at 38 °C

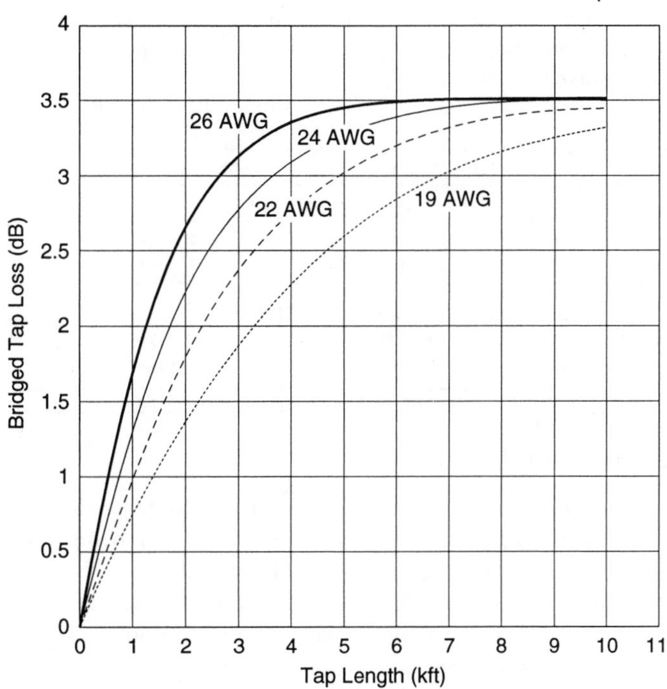

Figure 10-12 Bridged tap loss at 196 kHz vs. length

Table 10-27 Bridged Tap Loss at 196 kHz[a]

Bridged Tap Loss (dB/kft)			
19 AWG	22 AWG	24 AWG	26 AWG
0.642	0.829	1.004	1.176

[a]Valid for Bridged Taps ≤2.5 kft

The 26 AWG cable reduces the total allowable cable length from 12 kilofeet to

$$\text{Total cable length} = 12 - 3\frac{\text{L26}}{(9 - \text{BT})} = 12 - 3\frac{6}{9} = 10\ \text{kft}$$

Since the length of 26 AWG cable is 6.0 kilofeet, the maximum allowable length of the 24 AWG distribution cable is 10 − 6 = 4 kilofeet. The loss for the cable combinations is summarized in the table below.

AWG	Unit Loss	Segment Length	Segment Loss
26	4.08 dB/kft	6.0 kft	24.5 dB
24	2.95 dB/kft	4.0 kft	11.8 dB
		Total	36.3 dB

EXAMPLE 10-6

An HDSL loop is made up of 5.3 kilofeet of 22 AWG and 4.8 kilofeet of 24 AWG underground cable. The loop has a single 24 AWG bridged tap that is 0.6 kilofoot long. Find the loss at 196 kHz.

The loss for this loop is summarized in the table below.

AWG	Unit Loss	Segment Length	Segment Loss
		Main Cable	
24	2.95 dB/kft	4.8 kft	14.2 dB
22	2.23 dB/kft	5.3 kft	11.8 dB
		Bridged Taps	
24	1.004 dB/kft	0.6 kft	0.6 dB
		Total	26.6 dB

10.5.3 HDSL Loop Qualification

The HDSL was specifically designed to eliminate the requirements for significant loop qualification procedures; however, an area being converted to a CSA may require rehabilitation, including bridged tap audits and field rearrangements, to reduce bridged tap lengths. The dedicated plant and sealed plant concepts apply just as well to the CSA as to other plant administration concepts, including the DSL previously described.

Loops exceeding the CSA specified lengths are not suitable for use with the HDSL except in isolated cases. This would include dedicated, high-quality cables that extend beyond the CSA and meet the HDSL NEXT and loss requirements. In situations such as a CSA conversion in which existing loops are treated with load coils, build-out capacitors, VF repeaters, loop extenders, and induction neutralizing transformers, these devices must be removed before the pair can be used with an HDSL.

Analog carrier is not compatible with the HDSL, and it must be removed from the cables. The HDSL shares the deployment advantage with the ISDN DSL in that it, too, operates independently of tip-ring polarity.

The HDSL is sensitive to impulse noise, and cables with potential impulse noise sources must be identified and worked over as for the DSL. The loop qualification limits for the HDSL are similar to the ISDN DSL. The loop IL is adjusted to the appropriate frequency of 196 kHz, and the noise requirements are tightened due to the wider bandwidth of the HDSL signal detection circuits. To make this section stand by itself, the analog qualification table given previously for the DSL is repeated in Table 10-28. Table 10-29 gives the HDSL qualification test limits.

Table 10-28 Analog Loop Transmission Objectives for Reference in HDSL Qualification

PARAMETER	ACCEPTABLE LIMITS
Circuit noise, C-message	≤20 dBrnC
Circuit noise, 3 kHz flat	≤40 dBrn3kHz
Power influence	≤80 dBrnC
Circuit balance	≥60 dB

Table 10-29 HDSL Qualification Test Limits

TEST	LIMITS
Foreign voltage	±1 Vdc maximum
Insulation resistance	≥300 kΩ
Loop dc resistance	≤1,300 Ω, initial test value ± 30% Ω (operating)
Insertion loss (AML) at 196 kHz	≤42 dB (EML ≤ 36 dB)
Background noise, 50 kb (135 Ω impedance)	≤31 dBrn50kb
Impulse noise, 50 kb (135 Ω impedance)	0 counts in 30 minutes at 50 dBrn50kb threshold

10.6 Optical Fiber Loop Transmission Design

The design of optical fiber systems used in loop applications requires a number of important choices. As with virtually all other transmission media, the ultimate choices are based on tradeoffs between technology, economics, and service requirements. Table 10-30 shows some of the more important design variables.

Table 10-30 Design Variables for Loop Applications of Optical Fibers[a]

VARIABLE	ALTERNATIVES	CONSIDERATIONS
Optical fiber type	Singlemode Dispersion-shifted singlemode[b,c]	Attenuation, dispersion, mode-field diameter, cutoff wavelength, length, toughness, fatigue resistance
Cable design	Tight jacket Loose tube Ribbon	Fiber accessibility, environmental and mechanical performance, installation method, splicing method, fiber count, physical layout
Strength member	Dielectric Nondielectric	Grounding requirements, electrical protection, structural fastening method
Construction	Aerial Buried Underground Premises	Accessibility, armor or steam resistance requirements, route availability, hazards, electrical code requirements
System topology or architecture	Star Ring Bus	Network control, physical layout, logical layout
Splicing	Fusion Active mechanical Passive mechanical	Loss, strength, reflection requirements, temperature performance
Connecting	Push-pull Threaded Bayonet	Loss, construction, reconfiguration, reflection requirements, repeatability
Operating wavelength	1,310 nm[d] 1,550 nm	Distance, attenuation rate, dispersion rate, wavelength division multiplexing
Transmitter	Light emitting diode Laser diode Decision feedback (DFB)	Data rate, loss budget or output power, aging performance, linewidth, coupling loss, response time, coding scheme, bit error rate
Receiver	Positive-intrinsic-negative (PIN) diode Avalanche photodiode (APD) PIN/field effect transistor (FET) detector	Data rate, sensitivity, coupling loss, response time, coding scheme, bit error rate, dynamic range

[a]Adapted from [14]

[b]Multimode fiber is a common fiber type, but it is not used in modern loop applications.

[c]Dispersion-shifted fibers normally are not used in present loop applications of optical fibers but will be used in future broadband (video) applications.

[d]850 nm is a common optical wavelength, but it is not used with singlemode fibers in loop applications.

Typical FITL systems do not use intermediate regenerators. Therefore, for the purposes of this discussion, a *regenerator section* provides a passive end-to-end link between two fiber optic terminals. These terminals can be located at a central office and a remote terminal in a point-to-point configuration, as indicated in Figure 10-13. Systems using optical couplers and splitters are covered at the end of this section. An *optical path* includes the regenerator section plus optical transmitter and receiver.

The design of a fiber optic loop is based on two constraints:

- Dispersion tolerance (dispersion, or bandwidth, limit)
- System optical loss budget (power limit)

The tolerance of optical transmitter and receiver combinations to dispersion depends on the bit rate and transmitter spectral width. The bit rate determines the time between the transmitted pulses; a higher bit rate will have a shorter time between pulses, or shorter pulse period, than a lower bit rate. Pulses that are closer together in time are more subject to overlap as the pulse spreads out with distance along the fiber. Systems used in fiber loop transmission systems have a first-order *dispersion bit-rate product* (DBRP) of around 0.19 bits or, equivalently, 190,000 ps-Mbps. A corresponding first-order calculation of maximum system length can be made from

$$L_{max} = DBRP/(R\, t_{dm}\, \Delta\lambda)$$

where L_{max} = maximum fiber length, km
DBRP = dispersion bit-rate product, ps-Mbps
t_{dm} = material dispersion, ps/nm-km
$\Delta\lambda$ = transmitter full width, half maximum (FWHM) spectral width, nm
R = transmission line rate (including overhead), Mbps

Lower speed (DS-1 and DS-2 rate) fiber loop systems typically use light-emitting diode (LED) or laser diode (LD) transmitters with nominal FWHM spectral widths of approximately 100 and 10 nm, respectively. Higher speed (DS-3) systems typically use LD transmitters exclusively. In many lower speed systems, the actual line rate is twice the payload bit rate. Therefore, a system with a DS-2 source bit rate (6.3 Mbps) actually operates at a line rate of 12.6 Mbps.

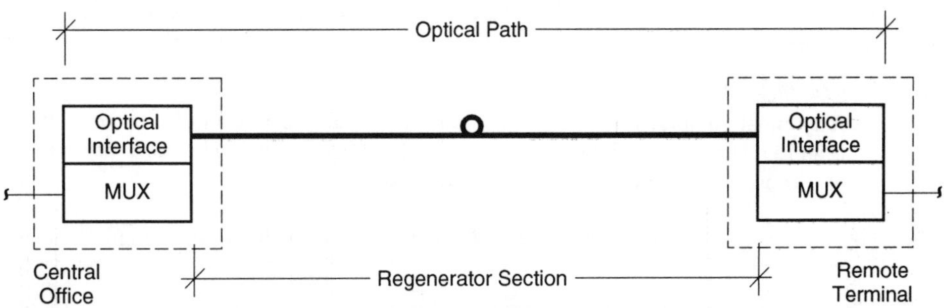

Figure 10-13 Fiber-in-the-loop path and section

Fiber loop terminal equipment data sheets will give the dispersion tolerance in ps/nm or just ps. When given in ps/nm, the pulse broadening is time units of picoseconds per nanometer of spectral width. In this case, the maximum dispersion limited length is given by

$$L_{max} = D_{max}/t_{dm}$$

where L_{max} and t_{dm} are defined above
D_{max} = maximum allowable dispersion of the terminal equipment, ps/nm

Typical values for D_{max} are 100 to 200 ps/nm. For a given set of terminal equipment, different values of D_{max} give different dispersion penalties (dispersion penalty is explained later). When terminal equipment tolerance to dispersion is given in ps, the bit rate already has been taken into account. For this situation, the maximum dispersion limited length is given by

$$L_{max} = d_{max}/(t_{dm}\,\Delta\lambda)$$

where L_{max}, t_{dm}, and $\Delta\lambda$ are defined above
d_{max} = maximum allowable dispersion of the terminal equipment, ps

EXAMPLE 10-7

Two systems are being evaluated. The first system uses an LD transmitter operating at a 44.7 Mbps line rate. The second system uses an LED transmitter operating at 12.6 Mbps line rate. The fiber optic cable has a dispersion of 2.8 ps/nm-km. Find the dispersion limited length of these systems.

$$L_{max} = 190{,}000 \text{ ps-Mbps}/(44.7 \text{ Mbps} \times 2.8 \text{ ps/nm-km} \times 10 \text{ nm}) = 150 \text{ km (DS-3 rate system)}$$

$$L_{max} = 190{,}000 \text{ ps-Mbps}/(12.6 \text{ Mbps} \times 2.8 \text{ ps/nm-km} \times 100 \text{ nm}) = 50 \text{ km (DS-2 rate system)}$$

EXAMPLE 10-8

A terminal equipment set has a maximum allowable dispersion of 150 ps/nm, and the fiber used with it has a maximum dispersion of 3.5 ps/nm-km. Find the dispersion limited length.

$$L_{max} = (150 \text{ ps/nm})/(3.5 \text{ ps/nm-km}) = 43 \text{ km}$$

EXAMPLE 10-9

A terminal equipment set has a maximum allowable dispersion of 4,000 ps, the transmitter is an LD with a spectral width of 10 nm, and the fiber used with it has a maximum dispersion of 3.0 ps/nm-km. Find the dispersion limited length.

$$L_{max} = (4{,}000 \text{ ps})/(3.0 \text{ ps/nm-km}) \times 10 \text{ nm} = 133 \text{ km}$$

The foregoing examples give the approximate limits based on dispersion for typical field conditions. Dispersion, conservatively, is not a factor in loop systems with bit rates less than approximately 50 Mbps and route lengths less than approximately 50 km. Calculations of the dispersion limit normally are not necessary but can be made easily, as shown above.

The loop system is broken into the specific design components, as shown in Figure 10-14. Associated with each component are specific design parameters and other considerations. These are explained in the following parameter tables and paragraphs. Below is a list of parameter tables.

Table 10-31 Optical Transmitter
Table 10-32 Optical Receiver
Table 10-33 Transmitter and Receiver Connectors
Table 10-34 Connector Pigtails
Table 10-35 Station Cable
Table 10-36 Fiber Optic Patch Panel Connector
Table 10-37 Outside Plant Cable
Table 10-38 Outside Plant Cable Splice

The electrical and optical characteristics of all components are subject to manufacturing variations, temperature drifts, aging drifts, and operational power penalties. How these effects are treated depends on the specific design procedure used. It is necessary to define where these variations apply so as to avoid duplication or omission during the design process. Some design procedures include these effects in the operating margin as separate considerations. The procedures given in this book follow the convention given in the Electronic Industries Association (EIA) fiber optic system standard [15].

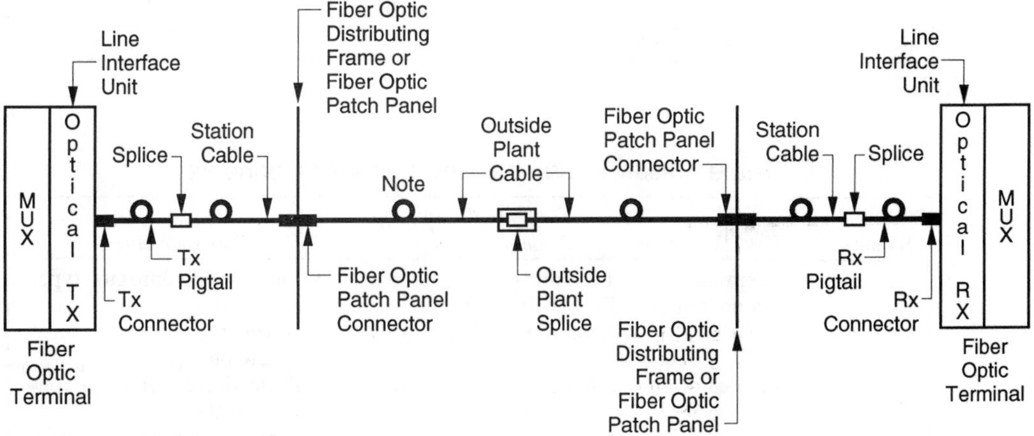

Figure 10-14 Fiber loop design components

Table 10-31 Optical Transmitter

Symbol	Parameter	Typical Values or Ranges	Considerations
λ_{tnom}	Nominal operating wavelength	1,310 or 1,550 nm	• Source type: LED LD
λ_{tmin} to λ_{tmax}	Wavelength range	±50 nm	
P_t	Output power	LED: −32.5 to −15.0 dBm LD: −12.0 to 0.0 dBm	• Temperature range: 0 to +50 °C indoor −40 to + 66 °C outdoor

Table 10-32 Optical Receiver

Symbol	Parameter	Typical Values or Ranges	Considerations
P_R	Sensitivity	−43.0 to −27.1 dBm (1E-9 BER) −41.5 to −29.6 dBm (1E-12 BER)	• Detector type: PIN APD
P_D	Dispersion power penalty	0 dB (unless given otherwise)	• Temperature range: 0 to + 50 °C indoor −40 to + 66 °C outdoor
R_P	Reflection power penalty	0 dB (unless given otherwise)	• Optical receiver wavelength must match transmitter
R_{max}	Maximum input power (saturation)	−10 to −13 dBm −23 to −26 dBm	

Table 10-33 Transmitter and Receiver Connectors

Symbol	Parameter	Typical Values or Ranges	Considerations
U_{con}	Maximum connector loss	Typical: 0.5 dB End-of-life: 0.8 dB Total (Tx + Rx): 1.5 dB	• Fiber optic connector type: Push-pull Bayonet Screw-on
N_{con}	Number of connectors	2 per system (Tx + Rx)	• Reflection requirements • Tx and Rx power parameters frequently are given at the end of the connector pigtail • Connector type determined by the Tx and Rx manufacturer

Table 10-34 Station Cable and Pigtail Splice

Symbol	Parameter	Typical Values or Ranges	Considerations
U_s	Maximum splice loss	Typical: 0.05–0.3 dB Average: 0.1–0.2 dB	• Splice type: Fusion Active mechanical Passive mechanical • Reflection requirements
N_s	Number of splices	System dependent	• Includes all installed splices between the transmitter or receiver and the fiber optic patch panel

Table 10-35 Station Cable

Symbol	Parameter	Typical Values or Ranges	Considerations
U_{sm}	Station cable per unit loss	Typical: 0.4 dB/km Range: 0.3–0.6 dB/km	• Loss usually is given at λ_{tnom}
l_{sm}	Station cable length	System dependent	• Station cable includes pigtails at transmitter and receiver
λ_{cc}	Fiber cutoff wavelength	$\lambda_{ccmax} < \lambda_{tmin}$	
MFD	Mode field diameter	Should be same as outside plant cable	• Fiber type (should be the same type as the outside plant fiber)

Table 10-36 Fiber Optic Patch Panel Connector

Symbol	Parameter	Typical Values or Ranges	Considerations
U_{con}	Maximum connector loss	Typical: 0.5 dB End-of-life: 0.8 dB	• Fiber optic connector type: Push-pull Bayonet Screw-on
N_{con}	Number of connectors	1 per patch panel (typically 2 per system)	• Reflection requirements • Connector type

The nominal transmitter wavelength, λ_{tnom}, and the wavelength range, λ_{tmin} to λ_{tmax}, determine the basic optical characteristics of the transmission system. Transmitter and receiver parameters used in the design are to be valid over the wavelength range specified.

Transmitter output power, P_T, and receiver sensitivity, P_R, include worst-case performance degradations due to manufacturing variations, temperature drifts, and

Table 10-37 Outside Plant Cable

SYMBOL	PARAMETER	TYPICAL VALUES OR RANGES	CONSIDERATIONS
MIFL or U_c	Maximum individual fiber loss, or cable per unit loss	Typical: 0.4 dB/km Range: 0.3–0.6 dB/km	• Loss usually is given at λ_{tnom}
l_t	Outside plant (OSP) cable length	System dependent	• Length includes installed cable plus slack at splice points plus allowances for additional repair cable and future rerouting
λ_{cc}	Fiber cutoff wavelength	$\lambda_{ccmax} < \lambda_{tmin}$	• Cable type and service requirements
MFD	Mode field diameter	8–10 μm	
ΔU_λ	Increase in U_c over transmitter wavelength range	0 dB	
ΔU_T	Increase in U_c over temperature range	0 dB	

aging drifts. These parameters are specified at the line side of the transmitter and receiver connectors (at the end of the connector pigtail, if equipped).

The receiver sensitivity normally does not include power penalties associated with dispersion or reflection, which are separately specified. The dispersion power penalty, P_D, is the additional power required by the receiver to account for total

Table 10-38 Outside Plant Cable Splice

SYMBOL	PARAMETER	TYPICAL VALUES OR RANGES	CONSIDERATIONS
U_s	Splice loss	Typical: 0.05 to 0.3 dB Average: 0.1–0.2 dB Maximum individual: 0.5 dB	• Reflection requirements
N_s	Number of splices	System dependent	• Includes all installed splices between the fiber optic patch panel at each end plus allowance for repair splices
ΔU_{sT}	Splice loss variation with temperature	0 dB	• Splice type: Fusion Active mechanical Passive mechanical

pulse distortion from intersymbol interference and mode partitioning noise. P_D is given at a specific bit error rate (BER) and at the maximum transceiver dispersion, D_{TR}. Where P_D is not explicit in manufacturer's data sheets, a value of 0.5 to 1.0 dB can be used. D_{TR} is the worst-case dispersion in picoseconds per nanometer of spectral width due to fiber length that the transmitter-receiver pair can tolerate to achieve the desired bit rate and BER. D_{TR} normally is not a direct consideration in loop fiber systems less than approximately 50 km.

The reflection power penalty, R_P, is the additional power required by the receiver due to reflection at the transmitter connector. R_P is specified at the maximum optical reflection, OR_{max}, that would give the same BER as without the reflection. Many manufacturers do not separately specify R_P in their data sheets. When not given separately, it already may be included in the transmitter and receiver power parameters.

The cutoff wavelength is λ_{cc}, above which the optical fiber supports only one propagation mode and below which multiple modes are supported. Operation below λ_{cc} may result in modal noise, modal distortion (increased pulse broadening), and improper operation of connectors, splices, and wavelength division multiplexer (WDM) couplers. Therefore, the minimum transmitter wavelength must be greater than the maximum allowed λ_{cc} to guarantee that the system operates entirely within the optical fiber's singlemode region; that is, $\lambda_{ccmax} < \lambda_{tmin}$ as shown in Figure 10-15. Pigtails, station cables, and outside plant cables may have different cutoff wavelengths, so this parameter must be checked for each component. However, if all components are supplied by the same manufacturer, the constraints will be met.

An important consideration in fiber loop design is splice loss, U_s. All modern splice types—fusion, active mechanical, and passive mechanical—have almost identical loss performance. The ultimate choice among these types may require other considerations, such as return loss, operational experience in similar environments, and cost.

Where a small number of splices is to be made on a route (say, less than 100), it is sufficient to specify a mean and maximum splice loss value. A typical specification would require a mean splice loss, U_s, of 0.1 to 0.15 dB for all splices, and a maximum individual splice loss of 0.5 dB.

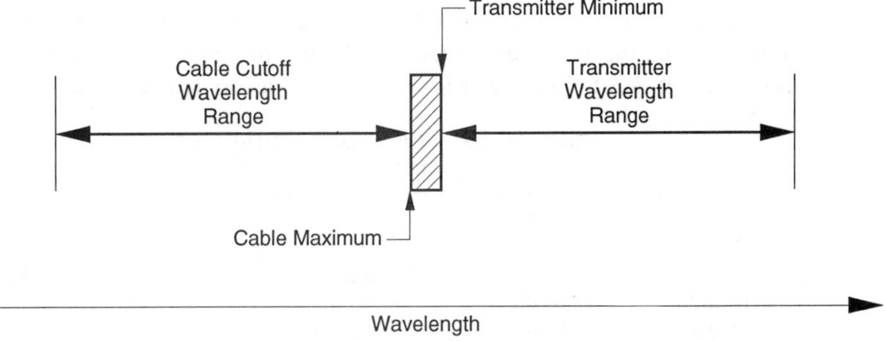

Figure 10-15 Cutoff wavelength requirements for singlemode operation

Where a larger number of splices is to be made, a modified statistical approach is used. With this method a mean splice loss, U_s, of 0.1 to 0.15 dB with a margin of 2σ (97.7%) or 3σ (99.9%) variation at 0.5 dB is specified. Any splice greater than $U_s + 1\sigma$ is required to be remade. If a splice loss less than this amount cannot be achieved after three or four tries, the splice is left alone unless it exceeds 0.5 dB. If it exceeds 0.5 dB, splice tooling or splicing techniques are checked and work continues on this splice until it is made less than 0.5 dB. Table 10-39 shows the values of σ when 2σ and 3σ is 0.5 dB. Similar tables can be easily built for other parameter values.

The number of splices, N_s, used in the system calculations must include:

- Entrance, transmitter, and receiver connector pigtail, and station cable splices
- Outside plant cable reel-end splices
- Outside plant cable splices at each cut required for construction
- Allowance for future damage and repair splices

In simplified or preliminary calculations, it is convenient to assign a value to N_s based on the number of splices per unit length of cable. Table 10-40 can be used when no other information is available. For example, if an underground route being planned is 10 km long, the estimated number of splices N_s = 10 km/(1.5 km/splice) = 7 (note rounding up to next highest integer). The values should be adjusted for reel lengths or other known factors.

For design convenience, the transmission link is divided at the fiber optic patch panel into three sections: transmitter and receiver station terminal equipment, and spliced fiber optic cable, as shown in Figure 10-16.

The system gain, G (also called the link budget or loss budget), is determined from:

$$G = NTP - NRP - M$$

where NTP = net Tx station power
NRP = net Rx station power
M = operating margin

The net Tx station power is the net of all losses between the transmitter output connector and the fiber patch panel, including the transmitter output power. Similarly, the net Rx station power is the net of all losses between the fiber patch panel

Table 10-39 Splice Loss Parameters for 2σ and 3σ = 0.5 dB

U_s (MEAN)	σ FOR 2σ = 0.5 dB	$U_s + 1\sigma$	σ FOR 3σ = 0.5 dB	$U_s + 1\sigma$
		Remake if >		Remake if >
0.10 dB	0.20 dB	0.30 dB	0.13 dB	0.23 dB
0.15 dB	0.18 dB	0.33 dB	0.12 dB	0.27 dB
0.20 dB	0.15 dB	0.35 dB	0.10 dB	0.30 dB

Table 10-40 Distance between Splices for Different Facility Classes

FACILITY CLASS	km/splice
Underground	1.5
Aerial	3
Buried	5

and the receiver input connector, including the receiver sensitivity and power penalties. In equation form, the relationships are:

$$NTP = P_T - U_{sm}l_{sm} - U_{con}N_{con}$$

$$NRP = P_R + P_D + R_P + U_{con}N_{con} + U_{sm}l_{sm}$$

In the above equations, NTP and NRP include the connector at the transmitter or receiver and at the patch panel; do not include these connector losses in the cable loss calculations. The operating margin, M, takes into account any losses not included in the other parameters, such as repair margin for future maintenance and unquantified component drift with age and temperature. It also can be used to compensate for uncertainty in the other parameters. For example, manufacturer's data sheets frequently are optimistic or may not include all the effects of aging or environmental factors. To compensate for this, the operating margin frequently is taken

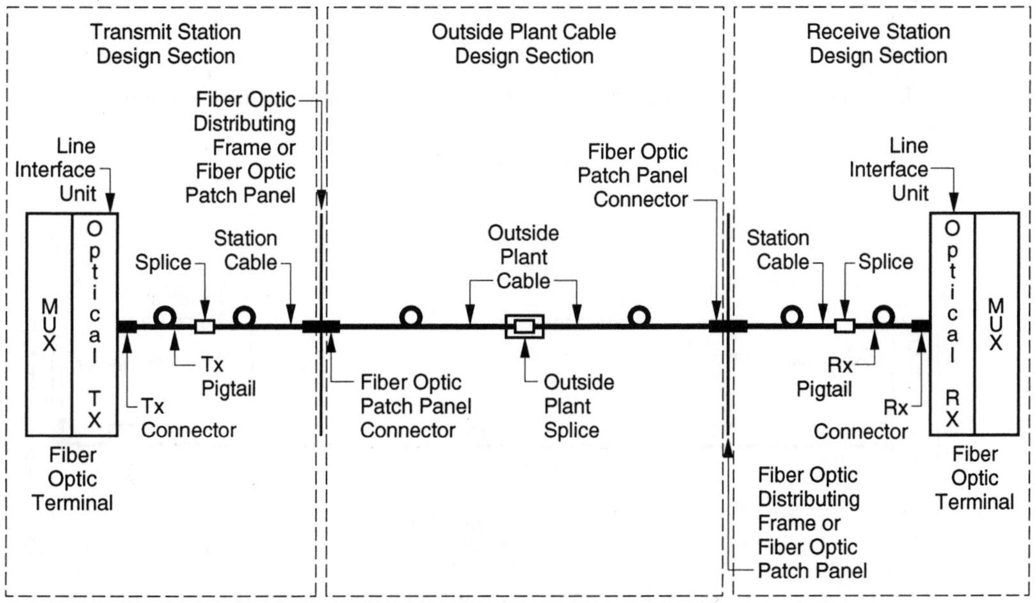

Figure 10-16 Fiber loop design sections

to be 2 to 3 dB for low capacity systems (DS-1 rate and below) and 3 to 6 dB for high capacity systems (DS-3 rate and above).

The system loss, L, which is due to the cable and intervening splices, is determined from

$$L = (U_c + U_{cT} + U_\lambda)l_t + (U_s + U_{sT})N_s$$

If $0 \text{ dB} \leq (G - L)$, then the fiber optic loop transmission objectives are attained. That is, the system gain (including margin) less the system loss must be a positive number. If $(G - L) < 0$, the system losses exceed the system gain, and not enough power is available at the receiver input to give the specified BER with margin M. If $(G - L + P_r) > \text{Rmax}$, an optical attenuator is needed to reduce the input power to the receiver. Attenuators are placed at the Rx end between the patch panel and receiver input connector, as shown in Figure 10-17. The effects of parameter variations are illustrated in Figure 10-18.

There are many other specific procedures for calculating link budgets, but all give the same final result. The detail differences are in the grouping of power penalties, and margin and component gains and losses. To illustrate another link budget, a typical design problem will be analyzed below.

When new routes are being planned or designed, one of the requirements is to determine the maximum allowable fiber unit loss (MIFL) in dB/km. The first step is to determine the system gain. In this case, the system gain is called G′ to distinguish it from the previous link budget method:

$$G' = P_T - P_R$$

where P_T and P_R are the transmitter power and receiver sensitivity previously defined. Next, the power penalties and desired operating margin are added together:

$$PP = P_D + R_P + M$$

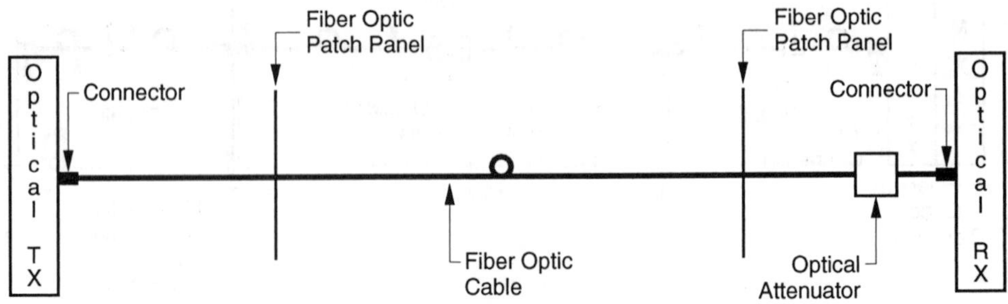

Figure 10-17 Using a fiber optic attenuator to reduce received power

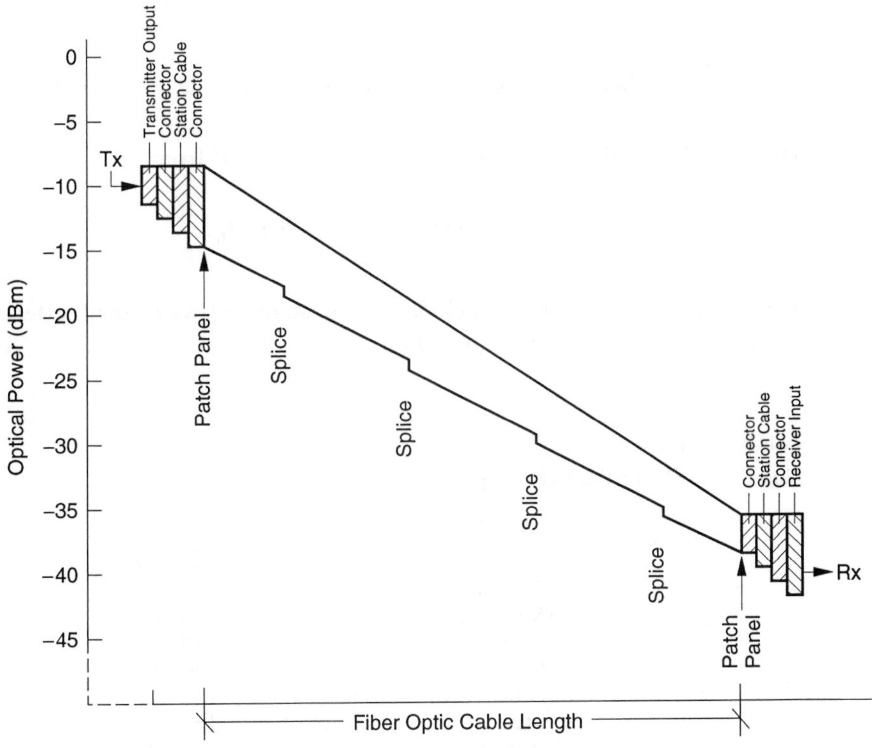

Figure 10-18 Effects of parameter variations on system loss budget

where these terms have been previously defined. The link budget in this case is found by subtracting the power penalties from the system gain:

$$\text{Link budget: } LB = G' - PP$$

Next, the component losses are added together to find the loss in the splices, connectors, and station cables:

$$CL = TCL + RCL + SL$$

$$\begin{aligned} \text{where Tx component loss: } TCL &= U_{sm}l_{sm} + U_{con}N_{con} \\ \text{Rx component loss: } RCL &= U_{con}N_{con} + U_{sm}l_{sm} \\ \text{Splice loss: } SL &= (U_s + U_{sT})N_s \end{aligned}$$

Now, the maximum allowable cable loss (MACAL) must be less than or equal to the link budget LB less the component losses, or:

$$MACAL = [(U_c + U_{cT} + U_\lambda)l_t] \leq LB - CL$$

Let

$$\text{Maximum individual fiber loss: MIFL} = U_c + U_{cT} + U_\lambda$$

Then, MIFL for the cable used on the route is found from:

$$\text{MIFL} = (\text{LB} - \text{CL})/l_t$$

The MIFL calculated above includes temperature and wavelength effects and is the value used in a purchase specification.

EXAMPLE 10-10

Find the MIFL for the fiber loop with the following characteristics:

Parameter	Value
Tx power	−15 dBm
Minimum Rx power	−35 dBm
Power penalties	6 dB
Connector loss (each)	0.5 dB
Number of connectors	4
Station cable loss (each)	0.6
Average splice loss	0.15 dB
Number of splices	20
Route length	20 km

$$\text{System gain: } G' = P_T - P_R = -15 - (-35) = +20 \text{ dB}$$
$$\text{Link budget: LB} = G' - \text{PP} = 20 - 6 = +14 \text{ dB}$$
$$\text{Component losses: CL} = \text{TCL} + \text{RCL} + \text{SL} = 2 \times 0.6 + 4 \times 0.5 + 20 \times 0.15 = 6.2 \text{ dB}$$

$$\text{Therefore, MIFL} = (\text{LB} - \text{CL})/l_t = (14 - 6.2)/20 = 0.39 \text{ dB/km}$$

The above design procedures covered a unidirectional point-to-point system with an optical transmitter at one end and an optical receiver at the other. A typical two-way system requires two fibers, one for each direction, as shown in Figure 10-19(a). A bidirectional point-to-point system using couplers can be designed for a single fiber, as shown in Figure 10-19(b). The design procedure is modified slightly to account for the IL of the coupler.

The link budget is the same as before (with the subscript "cpl" to indicate the coupled case):

$$\text{Coupled link budget: } \text{LB}_{cpl} = G' - \text{PP}$$

(a)

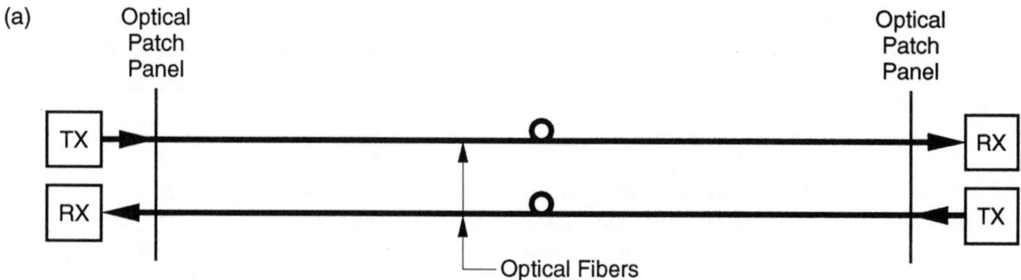

(b)

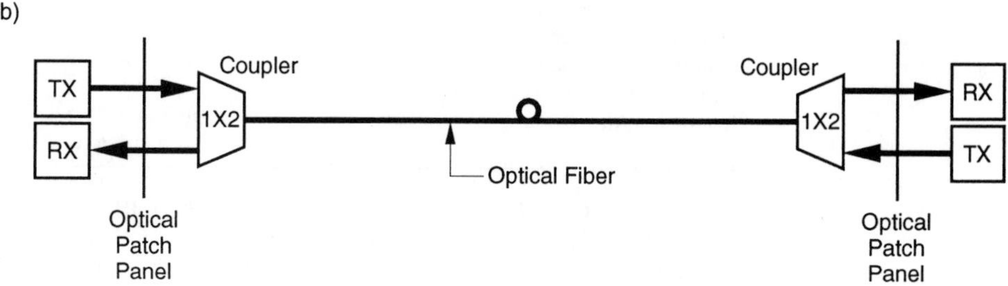

Figure 10-19 Uncoupled and coupled bidirectional fiber optic loops: (a) Uncoupled, 2-fiber system; (b) Coupled, 1-fiber system

However, the component losses now include the coupler insertion loss:

$$CL_{cpl} = TCL + RCL + CCL + SL$$

where Tx component loss: $TCL = U_{sm}l_{sm} - U_{con}N_{con}$
Rx component loss: $RCL = U_{con}N_{con} + U_{sm}l_{sm}$
Coupler component loss: $CCL = U_{cop}N_{cop}$
Splice loss: $SL = (U_s + U_{sT})N_s$

The coupler parameters above are U_{cop}, which is the coupling loss, and N_{cop}, which is the number of couplers in the link. For a simple system with one coupler at each end, $N_{cop} = 2$. The value of U_{cop} depends on the coupling ratio. For a 2:1 coupler, $U_{cop} = 4.0$ dB maximum (see Chapter 8 for other coupling ratios). Couplers may be connected to a system through connectorized patch panels, in which case the additional connector loss must be included in the link budget calculations.

The MACAL must be less than or equal to the link budget LB_{cpl} less the component losses, as before, or

$$MACAL_{cpl} = [(U_c + U_{cT} + U_\lambda)l_t] \leq LB_{cpl} - CL_{cpl}$$

The $MIFL_{cpl} = U_c + U_{cT} + U_\lambda$ and the MIFL for the cable used on the route is found from

$$MIFL_{cpl} = (LB_{cpl} + CL_{cpl})/l_t$$

EXAMPLE 10-11

Find the margin M for the coupled fiber loop with the characteristics listed below.

PARAMETER	VALUE
Tx power	−10 dBm
Minimum Rx power	−40 dBm
Reflection and dispersion power penalties	3 dB
Connector loss (each)	0.5 dB
Number of connectors	6
Station cable loss (each)	0.6
Average splice loss	0.15 dB
Number of splices	20
Coupler loss (each)	4.0 dB
Number of couplers	2
Route length	20 km
MIFL	0.4 dB/km

System gain: $G' = P_T - P_R = -10 - (-40) = +30$ dB

Power penalties: $PP = P_D + R_P + M = 3 + M$

Link budget $LB_{cpl} = G' - PP = 30 - 3 - M = +27 - M$ dB

Component losses: $CL_{cpl} = TCL + RCL + SL + CCL = 2 \times 0.6 + 6 \times 0.5 + 20 \times 0.15 + 2 \times 4.0 = 15.2$ dB

MIFL $l_t = (LB_{cpl} - CL_{cpl}) = 27 - M - 15.2$

Therefore, the margin $M = 27 - 15.2 - 0.4 \times 20 = 3.8$ dB

REFERENCES

[1] Reeve, W. *Subscriber Loop Signaling and Transmission Handbook: Analog*. IEEE Press, 1992.

[2] *Digital Data Service Dataport Loop Engineering Considerations*. Northern Telecom Ltd., Practice 368-5191-170.

[3] *Digital Data System Channel Interface Specifications*. AT&T PUB 62310, Nov. 1987, Addendum No. 1, Jan. 1988.

[4] *Secondary Channel in the Digital Data System: Channel Interface Requirements*. BELLCORE Technical Reference TR-NPL-000157, April 1986.

[5] Riley, E., Acuna, V. *Understanding Transmission*. Lee's ABC of the Telephone. Engineering Seminar I, 1976.

[6] *Transmission Design and Objectives, Resistance Engineering to Measured Limits, Customer Loops,* Section 832-100-072. Engineering Plant Series, GTE Practices. GTE Testmark Laboratories, Nov. 1982.

[7] Ahamed, S., et al. *A Tutorial on Two-Wire Digital Transmission in the Loop Plant*. IEEE Transactions on Communications, Vol. COM-29, No. 11, Nov. 1981.

[8] *IEEE Standard Methods and Equipment for Measuring the Transmission Characteristics of Analog Voice Frequency Circuits*. IEEE Standard 743-1984.

[9] American National Standards Institute. *Integrated Services Digital Network (ISDN): Basic Access Interface for Use on Metallic Loops for Application on the Network Side of the NT (Layer 1 Specification)*. ANSI T1.601-1988.

[10] *Engineering and Operations in the Bell System*. AT&T Bell Laboratories, NJ, 1983. Available from AT&T Customer Information Center.

[11] SNC Manufacturing Company, Inc., Oshkosh, WI.

[12] Generic Requirements for High-Bit-Rate Digital Subscriber Lines. BELLCORE Technical Advisory TA-NWT-001210, Oct. 1991.

[13] Functional Criteria for Digital Loop Carrier Systems. BELLCORE Technical Reference TR-TSY-00057. Available from BELLCORE Customer Service.

[14] *Just the Facts—A Basic Overview of Fiber Optics for Telephony Applications*. Corning, Inc., March 1992.

[15] *Single-Mode Fiber Optic System Transmission Design*. ANSI/EIA/TIA-559-1989. Available from Global Engineering Documents.

11 Premises Cabling Systems

Chapter 11 Acronyms

ASI	alternate space inversion	ISDN	integrated services digital network
BNC	bayonet-type coaxial connector	ISO	International Standards Organization
CPE	customer premises equipment	LAN	local area network
CSMA/CD	carrier sense multiple access with collision detection	MAU	media attachment unit
DEC	Digital Equipment Corporation	NEC®	National Electrical Code®
DSL	digital subscriber line	NEMA	National Electrical Manufacturers Association
ECTFE	ethylene chlorotrifluoroethylene (Halar®)	NEXT	near-end crosstalk
EIA	Electronic Industries Association	NT	network termination
EMI	electromagnetic interference	PBX	private branch exchange
FDDI	fiber distributed data interface	PTN	public telephone network
FEP	fluorinated ethylene propylene (Teflon®)	PVC	polyvinyl chloride
GRIN	graded index	SNR	signal-to-noise ratio
ICEA	Insulated Cable Engineers Association	ST	bayonet connector
IDC	insulation displacement connector	STP	shielded twisted pair
		TA	terminal adapter
		TE	terminal equipment
		TIA	Telecommunication Industries Association
		TP-PMD	twisted pair—physical media dependent
		UL®	Underwriters Laboratories, Inc.®
		UTP	unshielded twisted pair

11.1 General Considerations

In an analog environment, premises cabling systems have relatively low performance and serve as simple extensions of the subscriber loop. In a digital environment, however, the premises cabling systems play a much more important role in end-to-end performance. To round out the extensive discussions on digital loops in the previous chapters, this chapter provides information on high performance, structured premises cabling systems.

Premises cabling is any cabling that interconnects telecommunications equipment. Premises can be a single building or a group of buildings; for example, on a campus, as in Figure 11-1. This interconnection can be between pieces of terminal equipment or between terminal equipment and the telecommunication entrance facilities (network interface) of a local exchange carrier. In campus or building complex applications, premises cabling includes the outside plant cables between buildings when the cables are owned by the campus or building complex owner.

Other names for premises cabling are station wiring and inside wiring, and terminal equipment frequently is called customer premises equipment, or CPE. In the context of local area networks (LANs), terminal equipment locations frequently are called workstations. In the United States, premises telecommunication systems, including cables, are governed by the National Electrical Code (NEC)® and the Federal Communications Commission [1,2].

Premises cabling traditionally has been designed and installed for a particular application, such as voice telephony or vendor-specific LANs (for example, IBM token ring or Digital Equipment Corporation's (DEC) Ethernet). During a building's life, this approach leads to many layers of building wiring—one for each application or system—with each layer consisting of twisted pair cables and coaxial cables. The layers result in congested raceways and plenum systems and inadequate cabling system performance, as data transmission speeds and bandwidth requirements increase.

To meet the telecommunication requirements in new and retrofitted commercial buildings, a structured wiring system was standardized by the Electronic Industries Association (EIA) and Telecommunication Industries Association (TIA) in 1991 [3]. This standard, which is simply called EIA/TIA-568 in this chapter, defines a generic wiring system based on a hierarchical structure independent of vendors or network products. For simple applications consisting of one or two access lines, EIA/TIA issued a wiring standard for residential and light commercial buildings [4]. EIA/TIA also has issued other relevant standards for building telecommunication spaces and cabling system administration [5,6].

There are many different types of voice and data transmission systems that can be connected to premises cabling. For the most part, the detailed design of these systems is beyond the scope of this book. However, because the integrated services digital network (ISDN) is a good example of network, loop, and premises telecommunication integration, the premises application of ISDN warrants detailed explanation and is given in a later section.

Premises cabling normally is not considered a part of the subscriber's loop; however, loop design should consider premises cabling because it can affect the overall transmission characteristics of an end-to-end service. The actual effect will depend on the demarcation point location, type of interconnection between public telephone network (PTN) loops and network termination equipment, and the extent of the requirements for end-to-end service.

Traditionally, twisted pair cables have been used in building premises for voice applications, and coaxial cables have been used for data network applications. The trend in premises wiring systems, however, is toward using unshielded twisted pair (UTP) cables in a structured wiring system for both data and voice.

Shielded twisted pair (STP) cables are used with systems requiring 150 Ω

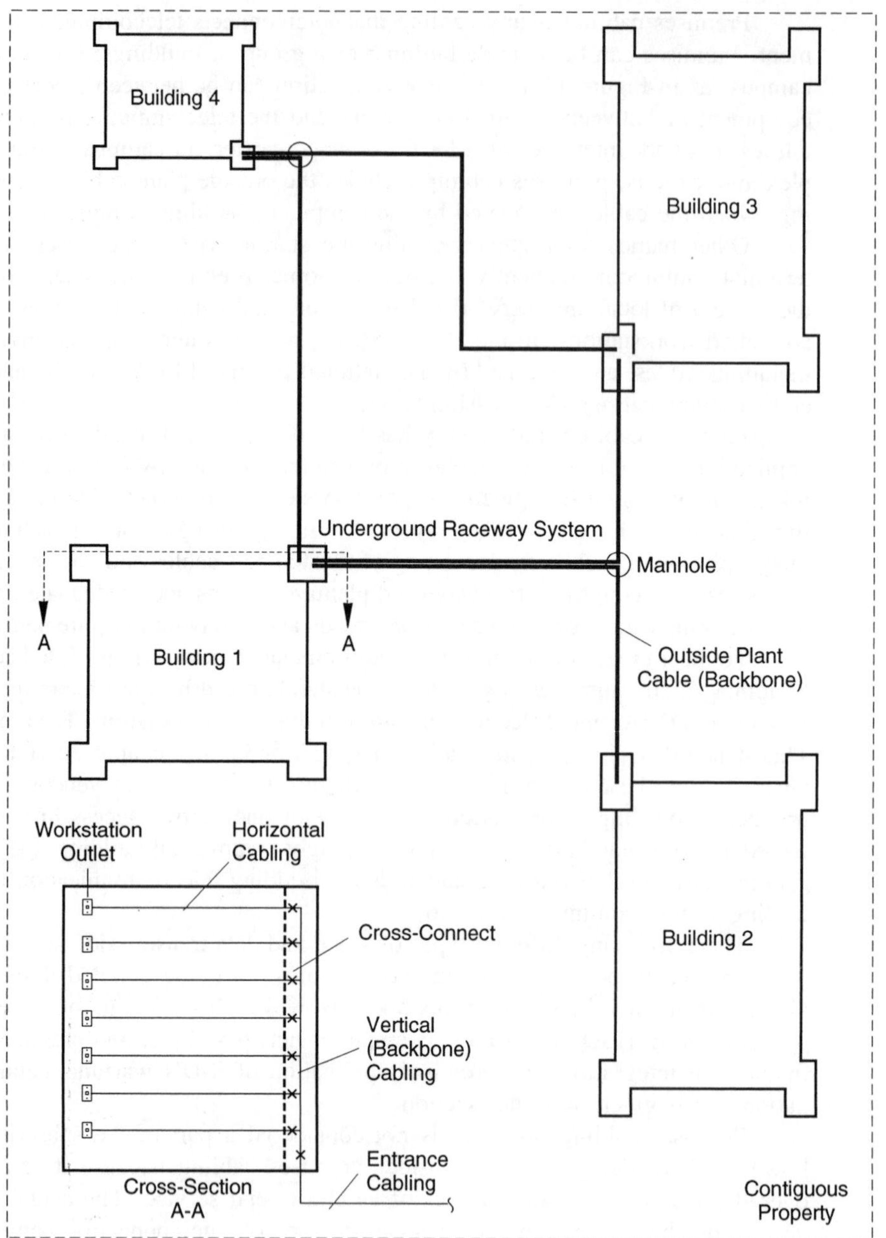

Figure 11-1 Premises—building or group of buildings

impedance or having particularly difficult interference problems. Fiber optics are used in installations covering a large area and in installations where large bandwidth is needed. All known future bandwidth requirements can be met with fiber optics, which can eliminate the need for coaxial cables in both horizontal and vertical wiring systems.

With a properly designed, structured wiring system, it may not be necessary to install anything other than UTP to the workstation outlet; however, many designers specify cables with UTP and STP under one sheath or jacket or, in some cables, UTP, STP, and optical fibers under one jacket.

The performance of UTP and STP notwithstanding, coaxial cables are used in LANs because they provide high immunity to the reception and radiation of electromagnetic interference (EMI). Coaxial cables frequently are applied in logical and physical ring and bus network topologies. They can be cabled in physical star topologies as well, but as would be expected, more cable is required. In systems covering a large area, coaxial cables are frequently used as backbone cabling.

The IEEE 802.3 LAN is an example of a computer LAN that can use several media types, including coaxial cables. This LAN is the standardized version of the Ethernet LAN developed by Xerox and Intel and popularized by DEC. The IEEE 802.3 LAN is based on the carrier sense multiple access with collision detection (CSMA/CD) protocol described in [7,8]. These standards describe several possible media, as follows:*

- 10Base2 (10 Mbps, baseband, 185 m Thicknet coaxial cable segment length)
- 10Base5 (10 Mbps, baseband, 500 m Thinnet coaxial cable segment length)
- 10BaseT (10 Mbps, baseband, twisted pair media)
- 10BaseF (10 Mbps, baseband, fiber optic media)
- 1Base5 (1 Mbps, baseband, 500 m coaxial cable segment length)

In some of the above designations—for example, 10Base5—the segment length is defined by the last character. The segment length is the maximum length of cable between two media attachment units (MAUs) or between an MAU and a termination.† MAUs also are called transceivers in this application. A typical network uses both the 10Base5 (Thicknet) and 10Base2 (Thinnet) cables. Thicknet cable is approximately the same physical size as the common RG-213/U coaxial cable (formerly RG-8/U) but has different performance requirements. Thinnet is similar to the well-known RG-58/U. Typically, Thicknet cable is used in backbone wiring and Thinnet is used in horizontal wiring.

Other types of coaxial cables, particularly RG-59/U (75 Ω), are used in LANs and broadband applications. These presently are not covered by EIA/TIA standards but are recognized by IEEE 802.3 and other IEEE standards [8]. RG-62/U and twinaxial cables are used in some other networks. However, as previously noted, twisted

*A 100 Mbps CSMA/CD LAN on twisted pair cable is being considered by the standards committee.

†The MAU nomenclature used here is taken from the previously cited standards. The term MAU also is used to indicate Multistation Access Unit used with the IBM token ring LAN. The IBM MAU is physically and electrically different than the media attachment unit described in the CSMA/CD standards.

pair cables are largely supplanting coaxial cables in new LANs, especially in horizontal wiring. Fiber optic cables are supplanting coaxial cables in backbone applications.

Fiber optic cables are available in several sizes and types. EIA specifies the performance of multimode fibers at nominal 850 and 1300 nm wavelengths [3]. The Insulated Cable Engineers Association (ICEA) also specifies multimode fiber performance at the same wavelengths [9]. Singlemode fibers can be used in backbone and horizontal wiring applications as well, but systems using multimode fibers and components are more economical in most LAN applications (the major savings with multimode systems are in the optical transmitters and receivers). The most significant advantage of singlemode fiber is its very high bandwidth over long distances.

Multimode fiber, in 90 m horizontal wiring, also has very high bandwidth: greater than 500 MHz. EIA/TIA-568 specifies the bandwidth of multimode fiber at 850 nm as 160 MHz-km, and 500 MHz-km at 1,300 nm. The bandwidth is length dependent. Bandwidth decreases with increasing length due to the propagation characteristics of multimode fibers. This dependence is nonlinear, and the bandwidth-length product given above applies to multimode fiber in one km lengths. At shorter lengths, the bandwidth is higher. The typical multimode fiber will have a bandwidth of more than several GHz in 90 m lengths. As a rule of thumb, for lengths longer than one km, the bandwidth decreases by approximately 50% for every additional one km.

11.1.1 Environmental Considerations

Premises cabling usually is not subjected to the same electrical or physical environment as outside plant cables; however, the cabling inside buildings presents another set of problems to be considered by the designer, builder, and users of such systems; these are noise (twisted pair cables), transmission performance, and fire hazard problems.

Although cables installed in buildings are not subjected to the low temperatures, harsh weather, and ultraviolet effects of outside cables, inside cables frequently are installed in exterior wall and air handling spaces. Temperatures can be high enough to degrade transmission, leading to intermittent performance problems. This often overlooked aspect of premises cabling design and other environmental considerations are discussed in the following sections.

11.1.2 Electromagnetic Compatibility

In twisted pair premises cabling systems, shielded cables are technically not always needed, and the current trend is to avoid them for economic reasons. This can lead to problems with EMI, especially in industrial environments. Premises wiring can be a noise source or noise receptor. With short runs and signal frequencies of less than approximately 20 MHz, signals from one system usually have negligible effect on other systems. On longer runs or higher frequencies, however, the lack of shielding can allow transmission degradation through noise pickup and crosstalk. Where twisted pair cables are used with LANs or other digital transmission systems, transmission performance, including crosstalk, impedance matching, and attenuation, is a very important issue.

Premises cabling systems frequently transport signals that are much different than those on outdoor cables. The signal transport systems in buildings are not always as robust as those designed for an outside environment. Line coding methods, in general, are different, which leads to different spectrum and crosstalk characteristics. Modern premises cabling systems carry a high percentage of digital transmissions (as opposed to analog or voice transmissions), and the transmission speeds are steadily increasing. The use of UTP for 16 megabits per second (Mbps) data transmission in office buildings is becoming commonplace, and fiber optic cabling routinely carries 125 Mbps data transmissions in buildings.*

As indicated above in twisted pair cable applications, digital transmissions can cause and are subject to crosstalk interference because of their wide frequency spectra. Such interference problems can be expected to increase unless the spectra of all transmission systems are coordinated. It is for these reasons that design and maintenance of premises cabling sometimes can be more difficult than outdoor transmission media. Fiber optic cables do not suffer from crosstalk (except potentially in coupled applications) or other EMI, and they are preferred in noisy environments.

11.1.3 Safety

In addition to the transmission problems facing users and designers of premises cabling systems, the uncontrolled deployment of polyolefin insulated and jacketed cables inside buildings can lead to safety problems even in minor fires. Flame spread and combustion byproducts (toxicity) are of concern in any building installation.

To reduce the safety hazard, the regulations covering premises cabling systems have become progressively tighter with each new edition of the NEC®.† The application or location of a particular part of the premises cabling determines the type of jacket and insulation required by the NEC®. The coverings are coded in Table 11-1 for metallic cables and in Table 11-2 for optical fiber cables, in descending order of fire resistance rating. More complete discussions of the various cable types used in communications and communications-related applications in buildings can be found in [10] and [11].

Generally, any listed cable may be used anywhere it is enclosed in a raceway; however, if the cable is not enclosed in an approved raceway, the cable must be listed for the application. Cables with a "P" suffix (for example, CMP) may be used in plenums and all other applications. Cables with an "R" suffix (for example, CMR) may be used in risers and all other applications except plenums. Cables with a "G" or no suffix (for example, CM) may be used in all applications except risers and plenums. Cables with an "X" suffix (for example, CMX), if less than 0.25 inch in diameter (this applies to most 2- to 4-pair cables), may be used only in one- or two-family dwellings and in multi-family dwellings if the cable is not concealed.

---

*The latter refers to fiber distributed data interface, or FDDI. The actual payload is 100 Mbps; the higher speed is needed for overhead associated with the transmission.

†See Article 800 of [1].

Table 11-1 Premises Wiring Coverings–Metallic Conductors, in Descending Order of Fire Rating

TYPE	COVERING
Multipurpose plenum cable	MPP
Communications plenum cable	CMP
Multipurpose riser cable	MPR
Communications riser cable	CMR
Multipurpose cable	MPG, MP
General purpose communications cable	CMG, CM
Limited use communications cable	CMX
Undercarpet communications cable	CMUC

Table 11-2 Premises Wiring Coverings–Optical Fibers, in Descending Order of Fire Rating

TYPE	COVERING
Nonconductive plenum cable	OFNP
Conductive plenum cable	OFCP
Nonconductive riser cable	OFNR
Conductive riser cable	OFCR
Nonconductive general purpose cable	OFNG
Conductive general purpose cable	OFCG
Nonconductive general purpose cable	OFN

Multipurpose cables can be used for communication applications as well as other applications (for example, signal, control, and alarm). Communications cables and wiring systems covered by the NEC® are required to have a minimum voltage rating of 300 V. Fiber optic cables with all-dielectric (nonconductive) construction greatly exceed this rating because there is nothing in the cable that conducts.

11.2 Structured Cabling Systems

A highly detailed manual is available for the design of structured cabling systems, so only a brief overview is provided here [11]. The complexity of premises cabling system installations varies with the application. In simple analog installations (one, two, or as many as four access lines and minimal customer premises equipment, all served on twisted pair cables), the wiring is almost always arranged in a star pattern, as shown in Figure 11-2. Here the terminal equipment is wired directly to the network interface device through simple premises cabling. Other arrangements commonly used in simple installations are the "daisy chain," where extensions are multipled (paralleled) with terminal equipment in the chain preceding it [5].

More complex systems will use a structured system. Structured systems use fully coordinated components for mechanical and electrical assembly of the cabling

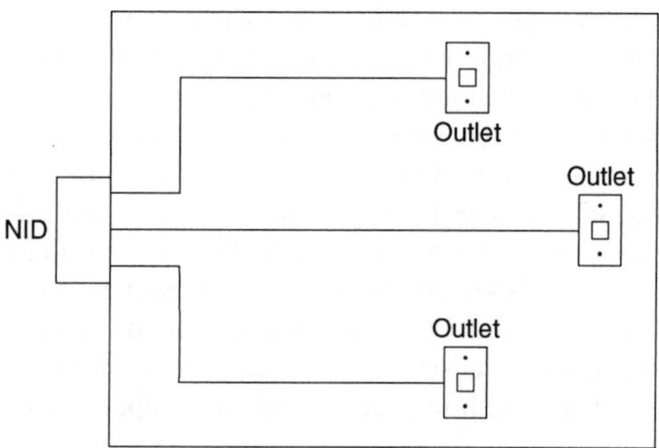

Figure 11-2 Simple premises wiring

between the network interface device, LAN electronics and workstations, optical fiber transmission equipment, and the various terminals or user equipment. A typical structured system is illustrated in Figure 11-3.

Structured cabling systems are commonplace with private branch exchanges, larger key systems (with more than 25 stations), and extensive LANs. In large build-

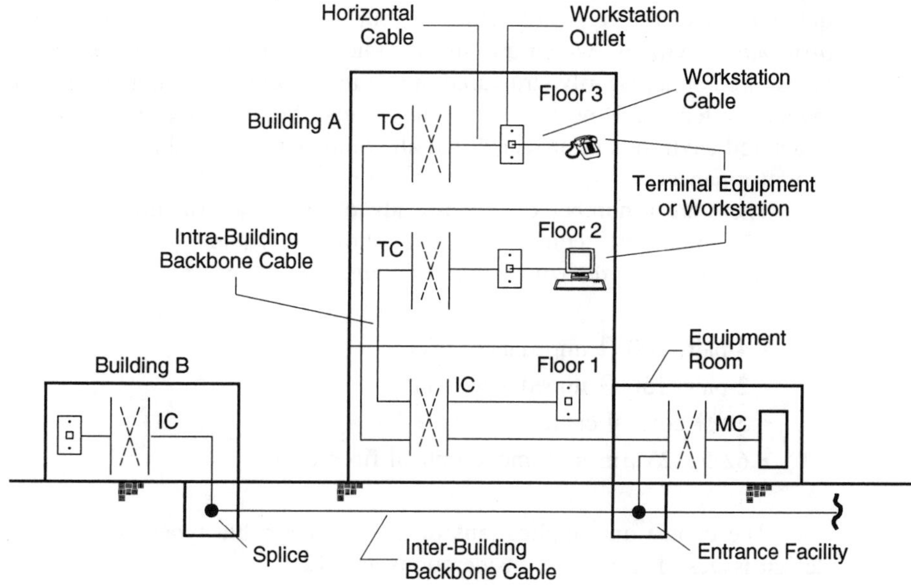

Key
IC: Intermediate Cross-Connect
MC: Main Cross-Connect
TC: Telecommunications Closet Cross-Connect

Figure 11-3 Structured cabling system

ing applications, the structured system interconnects not only regular telecommunication terminal equipment (telephone instruments, modems, facsimile machines, and the like) but also computer and other digital and analog equipment (for example, closed circuit television, security, and video conference systems).

With the integration of digital and analog signal transport in premises cabling, a large cost savings can be achieved not only in the initial installation but in daily administration of the premises network. Because installation of premises cabling systems can be very labor intensive, it is imperative, from an economical standpoint, that the designer consider total telecommunications requirements. This normally would include video, data transmission, security, energy management, and telecommunications. The bandwidth required could easily approach 500 MHz, or more, in more complex situations.

Structured cabling systems can consist of cables with:

- Twisted metallic pairs only (STP or UTP)
- Optical fibers only
- Both twisted metallic pairs and optical fibers
- Coaxial conductors
- Any combination of the above

A structured cabling system conforms to specific physical or topological requirements, such as types of cables, layout, length, and connectors. Some *premises distribution systems,* which are an integrated collection of interconnection components, are not necessarily structured in that they normally do not specify cable types, layout, or length. However, all premises distribution systems will work within a structured environment provided that all components and cables have the same level of performance.

There are a number of obvious advantages to a structured cabling system as compared to an unstructured system, as shown in Table 11-3.

EIA/TIA-568 recognizes the following cable types:*

- 4-pair, 100 Ω impedance, UTP
- 2-pair, 150 Ω impedance, STP
- 50 Ω coaxial cable
- 62.5/125 μm multimode optical fiber cable

The choice and applications of the various cable types constitutes important design issues. The major considerations are [12]:

---

*Other cabling types, such as 25-pair, singlemode optical fiber and compartmental core cables, can be used, but the standard does not recognize them as of this writing (1994). Singlemode fiber is under consideration for future issues of the standard.

Table 11-3 Comparison of Cabling Systems[a]

STRUCTURED CABLING SYSTEM	UNSTRUCTURED CABLING SYSTEM
Cross-connect flexibility	No (or minimum) cross-connect flexibility
Easy to administer	Difficult to administer
Easy to maintain	Difficult to maintain
Reliable	Relatively unreliable
Durable	Not durable
Permanent	Temporary
All work areas prewired	Wire as needed
Defined layout and distances	Free form
Requires planning	No planning required
Labeled, documented, and color coded	Not labeled, documented, or color coded
Wall outlets	No wall outlets
Concealed and in-the-wall cabling	Surface and on-the-wall cabling
Higher initial cost	Lower initial cost
Lower life cycle costs	Higher life cycle costs
Long life	Short life

[a]Data from [14]

- Network type:
  Analog voice
  Digital voice
  ISDN
  IEEE 802.3 CSMA/CD (Ethernet or 10BaseT)
  IEEE 802.5 (token ring)
  Fiber distributed data interface (FDDI)
- Bandwidth requirements
- Shielding requirements
- LAN node length requirements
- Number of network nodes
- Cabling system type (structured)
- Performance level of cabling system
- Types of connecting devices
- Space:
  Network and transmission equipment floorspace requirements
  Raceway requirements and availability
- Cable mechanical support requirements
- Cable jacket requirements to meet codes
- Future needs
- Industry standards and guidelines
- System life

Structured cabling system layout is specified in terms of cabling functions, including:

- Backbone (vertical) cabling
- Horizontal cabling
- Workstation cabling

Backbone cabling connects the cross-connect equipment in the main communications equipment room to the cross-connect equipment in the telecommunications closets on each floor. Backbone cabling generally is placed in a star topology between closets and the main communications room. Backbone cabling also is used to interconnect closets to accommodate bus or ring topologies. The maximum cable route lengths between the closets and main communications room depend on the system and backbone cable performance, as will be shown later. Backbone cables are one of four types:

- 100 Ω characteristic impedance UTP cable
- 150 Ω characteristic impedance STP cable
- 50 Ω characteristic impedance coaxial cable
- 62.5/125 μm multimode fiber optic cable
- Singlemode fiber optic cable

EIA/TIA-568 is based on a star topology in which each workstation is connected to a telecommunications closet via *horizontal wiring* that has a route length of less than 90 m. *Workstation* is a generic term used to describe computer workstations or PCs, telephones, integrated data, and voice telephones or other devices at the end-user's desk or work location.

A minimum of two communication jacks are installed at each workstation location. Normally, these jacks use a common faceplate over a single outlet box, as shown in the duplex outlet assembly drawing in Figure 11-4. One jack is connected to a 4-pair, 100 Ω UTP cable. The other jack is connected to another 4-pair, 100 Ω UTP cable; 2-pair, 150 Ω STP cable; 50 Ω coaxial cable; or 2-strand, 62.5/125 μm multimode fiber optic cable. The outlet shown has two 8-position/8-contact jacks for UTP cable. Coaxial or fiber optic connectors may be substituted for the blank filler plate below the twisted pair connectors.

Backbone, horizontal, and workstation wiring are illustrated in Figure 11-3. As indicated in this figure, the backbone cabling connects the main communications equipment rooms in different buildings (for example, in an office complex or campus) and connects main and intermediate cross-connects with telecommunications (or satellite) closets within a building. The horizontal cabling interconnects telecommunications closets with the workstation areas, and workstation (or terminal) cabling connects individual workstations with the horizontal cabling at an outlet.

In a structured cabling system, maximum distances are specified between cross-connection devices, such as are found in telecommunications closets and equipment rooms (backbone cabling), between telecommunications closets and outlets (horizontal cabling), and between workstations and outlets (workstation cabling). For EIA/TIA-568, these distances are illustrated in Figure 11-5.

In many office buildings, the horizontal cabling is constantly being rearranged.

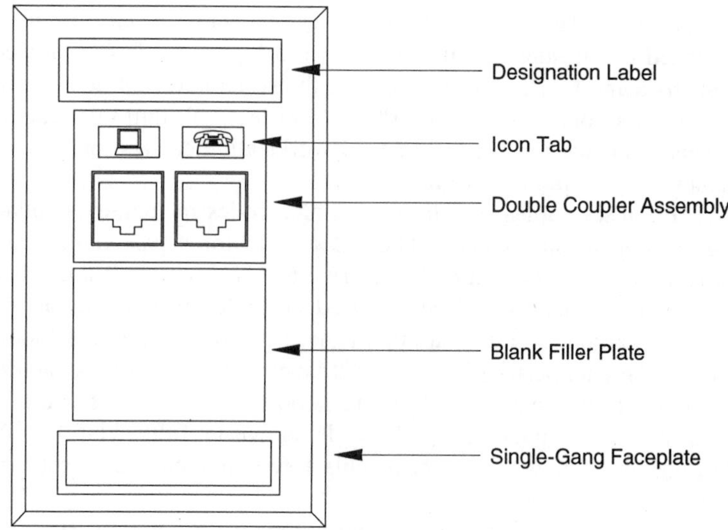

Figure 11-4 Typical duplex outlet assembly

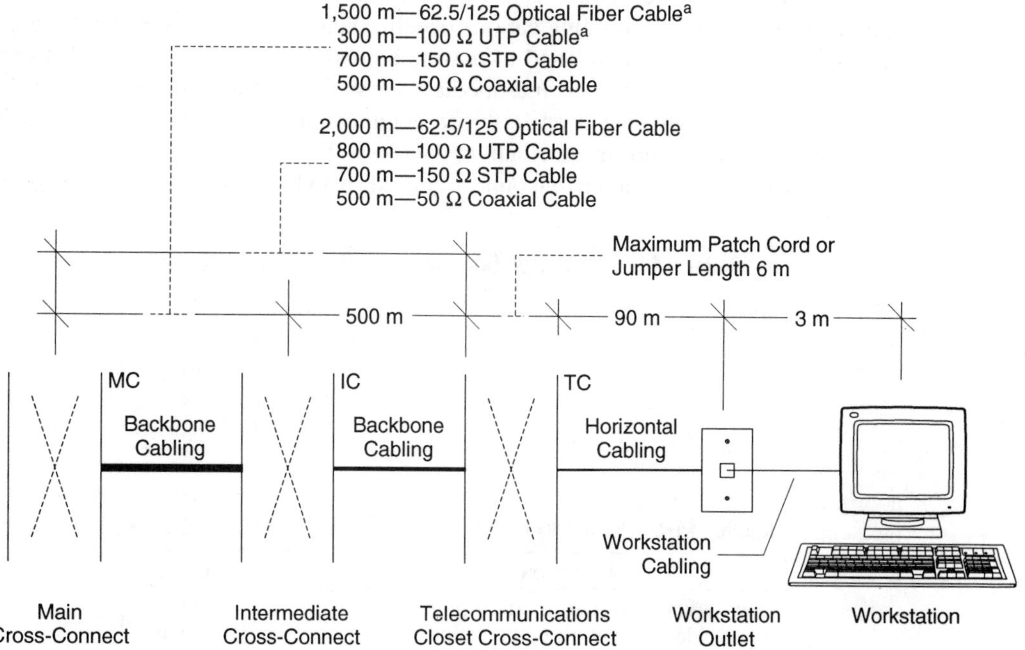

[a]Note: If the distance from IC to TC is less than maximum (500 m), the distance from MC to IC may be increased on a length-for-length basis, provided that the total distance from MC to TC does not exceed the maximum given.

Figure 11-5 Distance limitations for EIA/TIA-568 structured cabling system

Inexpensive UTP is widely used and is acceptable in all present LAN applications provided the distance limitations shown in Figure 11-5 are followed. To ensure that systems using UTP perform properly, the structured cabling system limits the length of UTP horizontal cabling to 100 m (330 feet). Within this length requirement, allowance is made for 3 m (10 feet) of workstation cabling and 6 m (20 feet) for patch cables in telecommunications closets.

Premises cables use the same color codes to identify conductors or fibers as are used in outside cables. The color code for 4-pair cables, as specified in the structured wiring system of EIA/TIA-568, is shown in Table 11-4.

A functional color code is used at cable terminations and cross-connects to simplify cabling administration. In these locations, terminal blocks used for interconnection (jumpering) have colored labels, and the various cables are marked with color-coded tags at each end. There is no universal acceptance of the color scheme used, but one standard gives the code shown in Table 11-5 [6]. An explanation of the functions follow. Different manufacturers may have different functions and colors in their product line.

The network side of the network demarcation point (network interface device) is color coded orange. The user-side of the network interface is color coded green. The cross-connects are used for network access connections.

Backbone cable terminations are vertical and horizontal cables between the equipment room and telecommunications (satellite) closets or service cabinets within the building. Three colors are used to mark the cross-connects. The cables between the main cross-connect and the intermediate cross-connect are connected to white fields. White fields also are used in telecommunications closets to terminate backbone cables if there is no intermediate cross-connect.

When an intermediate cross-connect is used, backbone cables are connected to

Table 11-4 EIA/TIA-568 Twisted Pair Color Code

Pair	Color Code (T/R)
1	White-Blue/Blue
2	White-Orange/Orange
3	White-Green/Green
4	White-Brown/Brown

Table 11-5 Premises Distribution System Color Scheme

Function	Color
Network side of demarcation point	Orange
User-side of network demarcation point	Green
Backbone cabling, first level	White
Backbone cabling, second level	Gray
Backbone cabling, inter-building	Brown
Horizontal cabling to outlets	Blue
Equipment (PBX, LAN, multiplex)	Purple
Key telephone systems	Red
Auxiliary connections	Yellow

gray fields at both the intermediate cross-connect and the telecommunications closets. Finally, inter-building backbone cables are connected to brown fields.

Auxiliary connections are made to yellow fields. These connections include signaling and conditioning equipment, external paging and voice mail systems, and call accounting systems. Horizontal cables from workstation (or information) outlets are connected to blue fields in the associated telecommunications closet or cabinet. Finally, common equipment such as private branch exchange (PBX) stations and trunks, LAN equipment, and multiplexers are connected to purple fields. Key systems are separately terminated on red fields. Figure 11-6 shows a typical installation using most of the colors.

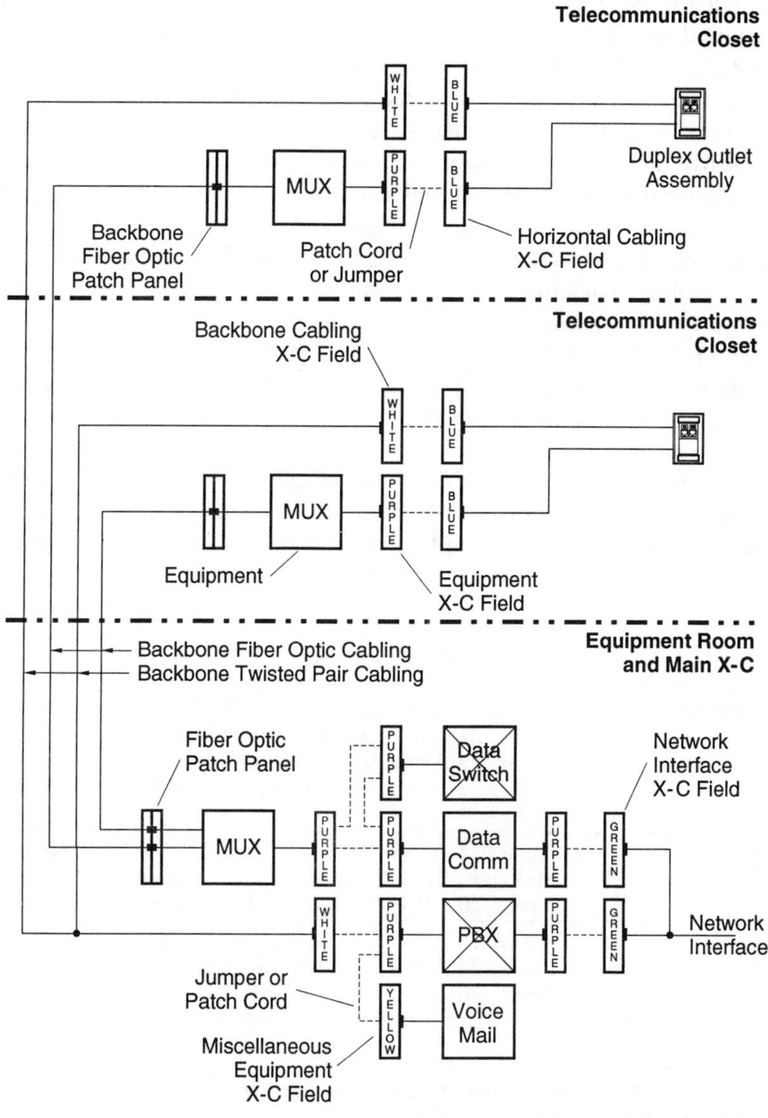

Figure 11-6 Premises distribution system application

11.3 Transmission Media Considerations

11.3.1 Twisted Pair

Twisted pair cables are ubiquitous in most office buildings. Until the 1980s, most premises cabling had limited length and was used in voice frequency applications. Because the length was relatively short, it had little effect on the transmission characteristics of the loop connected to it. LANs were traditionally based on coaxial cable transmission media and were entirely isolated from the loop.

In the early 1980s, IBM introduced performance specifications for twisted pair cabling to be used with their token ring LAN.* The IBM LAN is now commonly referred to as IEEE 802.5 Token Ring after the IEEE standard, although the two are not identical [13]. The IBM specifications list a number of mechanical and electrical tests that are required to verify the performance of various cable types. These tests include characteristic impedance, attenuation, resistance, crosstalk, capacitive and resistive unbalance, insulation resistance, temperature and humidity effects, and mechanical stress tests. The various cabling types are listed in Table 11-6.

As the use of personal computers and inexpensive LANs proliferated, the telecommunications industry, through the EIA and TIA, recognized the need for premises cabling standardization. This was necessary to allow continued use of new and embedded twisted pair cable plant on digital systems and LANs. Industry standards were publicly distributed in the early 1990s and continue to evolve as experience is gained with them [3–5]. Of these, EIA/TIA-568, the Commercial Building Telecommunications Wiring Standard, as previously discussed, has particularly far-reaching significance.

To aid in the selection of a suitable cable for a particular premises application, Underwriters Laboratories, Inc. (UL®), a nationally recognized third-party electrical testing laboratory, developed five performance *levels,* described below, for STP and

Table 11-6 IBM Cable Types

IBM Type	Description
1	2-pair, 22 AWG data, non-plenum cable
1P	2-pair, 22 AWG data, plenum cable
2	2-pair, 22 AWG data and 4-pair, 22 AWG voicegrade, non-plenum cable
2P	2-pair, 22 AWG data and 4-pair, 22 AWG voicegrade, plenum cable
3	Multipair, 22 or 24 AWG communications cable
6	2-pair, 26 AWG patch panel data, non-plenum cable
9	2-pair, 26 AWG data, non-plenum cable
9P	2-pair, 26 AWG data, plenum cable
9R	2-pair, 26 AWG data, riser cable

*The IBM Cabling System Technical Interface Specification, GA27-3773-1, covers cables with a nominal 150 Ω characteristic impedance.

UTP cables [15]. The jackets of cables tested under the UL® program are marked with the particular level for simple field identification.

At about the same time as UL®, EIA published Technical Service Bulletin TSB-36, which provides specifications for five performance *categories,* an approach similar to the UL® performance *level* program [16]. In 1992, National Electrical Manufacturers Association (NEMA) issued a proposed "Performance Standard for Premises Telecommunication Cables" that lists three performance *designations*—standard, low loss, and low loss extended frequency [17]. Also in 1992, BELLCORE issued a technical advisory for inside wiring cable that generally mirrors EIA/TIA-568. Finally, ICEA published a communication wire and cable standard for premises applications [18]. Both the UL® levels and EIA categories are described below, and equivalence is indicated where applicable. Because standards continually evolve, the reader is cautioned to refer to the most current standard and compare it to the following discussions.

The performance requirements for cabling components are specified in a number of standards [15–17,19,20]. These documents presently describe the performance requirements for the cable types listed in Table 11-7. The following paragraphs and tables describe the transmission aspects of the different standards. A later table compares each in spreadsheet form.

Table 11-7 Premises Cabling Performance Standards

Standard Body	Cable Types Covered
UL® [15]	5 performance levels for 22 and 24 AWG UTP and STP cable 2 types of 50 Ω coaxial cable
EIA/TIA [16]	5 performance categories of 22 and 24 AWG UTP 2 types of 62.5/125 μm multimode fiber optic cable
EIA/TIA [19]	3 performance categories of connection devices
NEMA [17]	3 performance levels of UTP and STP 2 types of 50 Ω coaxial cable
BELLCORE [20]	Same as EIA/TIA-568
ICEA [18]	Primarily station wire and riser cable; transmission performance not specified.

UL® Level I

Basic telecommunications and power-limited applications. Level I cable has no performance specifications and is generally used with voicegrade applications only.

UL® Level II

For low-speed data and LAN applications up to approximately 1 Mbps; similar to IBM Type 3 in 22 and 24 AWG conductor sizes; STP and UTP cable constructions and pair counts from two through twenty-five. Level II performance covers impedance, attenuation, and resistance, as shown in Table 11-8.

Table 11-8 UL® Level II Electrical Performance

Frequency	Impedance	Maximum Attenuation
256 kHz	90.0 to 120.0 Ω	4.0 dB/kft
512 kHz	87.0 to 117.5 Ω	5.7 dB/kft
772 kHz	85.0 to 114.0 Ω	6.8 dB/kft
1 MHz	84.0 to 113.0 Ω	8.0 dB/kft
AWG	Maximum dc Resistance at 20 °C	
24	28.6 Ω/kft	
22	18.0 Ω/kft	

UL® Level III

For moderate-speed data and LAN applications up to approximately 16 Mbps, such as 10BaseT; complies with EIA/TIA-568 for horizontal UTP and with EIA/TIA TSB-36 for Category 3 UTP cable. Level III performance covers a wider range of parameters than Level II, as shown in Table 11-9. NEMA designation "standard" for 24 AWG is approximately the same as UL® Level III, except that NEMA standard cable characteristics are specified to 20.0 MHz. Attenuation is not specified

Table 11-9 UL® Level III STP and UTP Electrical Performance

Frequency	Impedance (Ω)[a]	Minimum NEXT Loss (dB)[b]	Maximum Attenuation (dB/kft) 24 AWG	Maximum Attenuation (dB/kft) 22 AWG
64 kHz	125 ± 15%	Not specified	2.8	2.8
128 kHz	115 ± 15%	Not specified	Not specified	Not specified
150 kHz	Not specified	54	Not specified	Not specified
256 kHz	110 ± 15%	Not specified	4.0	4.0
512 kHz	Not specified	Not specified	5.6	5.6
772 kHz	102 ± 15%	43	6.8	6.8
1.0 MHz	100 ± 15%	41	7.8	7.8
4.0 MHz	100 ± 15%	32	17.0	17.0
8.0 MHz	100 ± 15%	28	26.0	26.0
10.0 MHz	100 ± 15%	26	30.0	30.0
16.0 MHz	100 ± 15%	23	40.0	40.0
dc Resistance (Ω/1,000 kft at 20 °C)			28.6	18.0
Maximum dc resistance unbalance (%)			5%	5%
Maximum pair mutual capacitance (pF/ft)			20	20
Maximum pair-to-ground capacitance unbalance (pF/kft)			1,000	1,000

NEXT = near-end crosstalk

[a]Although characteristic impedance limits are shown for discrete frequencies between 1.0 to 16.0 MHz, they also apply to any frequency in between.

[b]NEXT values shown are for the worst pair. The minimum NEXT coupling loss for any pair combination at 20 °C shall be determined from the following formula for all frequencies between 150 kHz and 16.0 MHz:

$$\text{NEXT}(F_{MHz}) > \text{NEXT}(772\text{ kHz}) - 15\,\text{Log}_{10}(F_{MHz}/0.772)$$

below 256 kHz, and crosstalk is not specified below 772 kHz. Also, the NEMA requirements for 22 AWG are tailored to that conductor size.

UL® Level IV

For high-speed data and LAN applications up to approximately 20 Mbps. Level IV performance requirements are similar to EIA/TIA TSB-36 for Category 4 UTP cable constructions. Level IV performance covers a higher frequency range than Level III, as shown in Table 11-10. NEMA performance designation "low loss" is equivalent to Level IV, except that the NEMA standard specifies attenuation performance down to 256 kHz.

UL® Level V

For high-speed data and LAN applications, particularly the twisted pair equivalent of FDDI. Level V performance requirements are similar to EIA/TIA TSB-36 for Category 5 UTP cable constructions. Level V performance covers a much higher frequency range than Level IV, as shown in Table 11-11. NEMA designation "low loss extended frequency" cables with nominal 100 Ω impedance are equivalent to Level V.

The performance requirements described in EIA TSB-36 parallel the UL® requirements, but only in the three higher levels and only for UTP as described below.

EIA Category 1

No specified requirements. Typical uses are for voice and low-speed data (less than 64 kbps).

EIA Category 2

No specified requirements. Typical uses are the same as EIA Category 1.

EIA Category 3

Applies to cables specified in EIA/TIA-568 and gives performance for frequencies up to 16 MHz. Typical uses include data rates up to 10 Mbps such as the 4 Mbps token ring (IEEE 802.3) and the 10 Mbps 10BaseT (IEEE 802.5). Most requirements are the same as UL® Level III. See Table 11-12.

EIA Category 4

Specifies performance for frequencies up to 20 MHz. Typical uses include data rates up to 16 Mbps such as the 16 Mbps token ring. Very similar to UL® Level IV except at the lower frequencies. See Table 11-13.

EIA Category 5

Specifies performance for frequencies up to 100 MHz. Typical uses include data rates up to 100 Mbps such as the metallic twisted pair equivalent of the FDDI, called twisted pair—physical media dependent (TP-PMD). It is very similar to UL® Level V; see Table 11-14.

Table 11-10 UL® Level IV STP and UTP Electrical Performance

Frequency	Impedance (Ω)	Minimum NEXT Loss (dB)[a]	Maximum Attenuation (dB/kft) 24 AWG	Maximum Attenuation (dB/kft) 22 AWG
772 kHz	102 ± 15%	58	5.7	4.5
1.0 MHz	100 ± 15%	56	6.5	5.5
4.0 MHz	100 ± 15%	47	13.0	11.0
8.0 MHz	100 ± 15%	42	19.0	15.0
10.0 MHz	100 ± 15%	41	22.0	17.0
16.0 MHz	100 ± 15%	38	27.0	22.0
20.0 MHz	100 ± 15%	36	31.0	24.0
dc Resistance (Ω/kft at 20 °C)			28.6	18.0
Maximum dc resistance unbalance (%)			5%	5%
Maximum pair mutual capacitance (pF/ft)			17	17
Maximum pair-to-ground capacitance unbalance (pF/kft)			1,000	1,000

[a]NEXT values shown are for the worst pair. The minimum NEXT coupling loss for any pair combination at 20 °C shall be determined from the following formula for all frequencies between 772 kHz and 20.0 MHz:

$$\text{NEXT}(F_{MHz}) > \text{NEXT}(772\text{ kHz}) - 15\,\text{Log}_{10}(F_{MHz}/0.772)$$

Table 11-11 UL® Level V STP and UTP Electrical Performance

Frequency	Impedance (Ω)	Minimum NEXT Loss (dB)[a]	Maximum Attenuation (dB/kft) 24 AWG	Maximum Attenuation (dB/kft) 22 AWG
772 kHz	102 ± 15%	64	5.5	Not specified
1.0 MHz	100 ± 15%	62	6.3	Not specified
4.0 MHz	100 ± 15%	53	13.0	Not specified
8.0 MHz	100 ± 15%	48	18.0	Not specified
10.0 MHz	100 ± 15%	47	20.0	Not specified
16.0 MHz	100 ± 15%	44	25.0	Not specified
20.0 MHz	100 ± 15%	42	28.0	Not specified
25.0 MHz	100 ± 15%	41	32.0	Not specified
31.25 MHz	100 ± 15%	40	36.0	Not specified
62.5 MHz	100 ± 15%	35	52.0	Not specified
100.0 MHz	100 ± 15%	32	67.0	Not specified
dc Resistance (Ω/kft at 20 °C)			28.6	18.0
Maximum dc resistance unbalance (%)			5%	5%
Maximum pair mutual capacitance (pF/ft)			17	17
Maximum pair-to-ground capacitance unbalance (pF/kft)			1,000	1,000

[a]NEXT values shown are for the worst pair. The minimum NEXT coupling loss for any pair combination at 20 °C shall be determined from the following formula for all frequencies between 772 kHz and 100 MHz:

$$\text{NEXT}(F_{MHz}) > \text{NEXT}(772\text{ kHz}) - 15\,\text{Log}_{10}(F_{MHz}/0.772)$$

Table 11-12 EIA Category 3 UTP Electrical Performance at 20 °C

Frequency	Impedance (Ω)[a]	Minimum NEXT Loss (dB)[b]	Maximum Attenuation (dB/kft) 24 AWG	Maximum Attenuation (dB/kft) 22 AWG
64 kHz	125 ± 15%	Not specified	2.8	2.8
128 kHz	115 ± 15%	Not specified	Not specified	Not specified
150 kHz	Not specified	54	Not specified	Not specified
256 kHz	110 ± 15%	Not specified	4.0	4.0
512 kHz	Not specified	Not specified	5.6	5.6
772 kHz	102 ± 15%	43	6.8	6.8
1.0 MHz	100 ± 15%	41	7.8	7.8
4.0 MHz	100 ± 15%	32	17	17
8.0 MHz	100 ± 15%	28	26	26
10.0 MHz	100 ± 15%	26	30	30
16.0 MHz	100 ± 15%	23	40	40
dc Resistance (Ω/kft at 20 °C)[c]			28.6	
Maximum dc resistance unbalance (%)[c]			5%	
Maximum pair mutual capacitance (pF/ft)[c]			20	
Maximum pair-to-ground capacitance unbalance (pF/kft)[c]			1,000	

Table 11-13 EIA Category 4 UTP Electrical Performance at 20 °C

Frequency	Impedance (Ω)	Minimum NEXT Loss (dB)[b]	Maximum Attenuation (dB/kft) 24 AWG	Maximum Attenuation (dB/kft) 22 AWG
64 kHz	Not specified	Not specified	2.3	2.3
150 kHz	Not specified	68	Not specified	Not specified
256 kHz	Not specified	Not specified	3.4	3.4
512 kHz	Not specified	Not specified	4.6	4.6
772 kHz	Not specified	58	5.7	4.5
1.0 MHz	100 ± 15%	56	6.5	5.5
4.0 MHz	100 ± 15%	47	13	11
8.0 MHz	100 ± 15%	42	19	15
10.0 MHz	100 ± 15%	41	22	17
16.0 MHz	100 ± 15%	38	27	22
20.0 MHz	100 ± 15%	36	31	24
dc Resistance (Ω/kft at 20 °C)[c]			28.6	
Maximum dc resistance unbalance (%)[c]			5%	
Maximum pair mutual capacitance (pF/ft)[c]			17	
Maximum pair-to-ground capacitance unbalance (pF/kft)[c]			1,000	

[a]Although characteristic impedance limits are shown for discrete frequencies between 1.0 to 16.0 MHz, they also apply to any frequency in between.

[b]NEXT values shown are for the worst pair. The minimum NEXT coupling loss for any pair combination at 20 °C shall be determined from the following formula for all frequencies between 150 kHz and 16.0 MHz:

$$\text{NEXT}(F_{MHz}) > \text{NEXT}(772\text{ kHz}) - 15\,\text{Log}_{10}(F_{MHz}/0.772)$$

[c]EIA/TIA-568 does not specify separate requirements for 22 AWG; however, 22 AWG is classified as Category 3 if it meets the performance requirements for 24 AWG.

Table 11-14 EIA Category 5 UTP Electrical Performance at 20 °C

FREQUENCY	IMPEDANCE (Ω)	MINIMUM NEXT LOSS (dB)[a]	MAXIMUM ATTENUATION (dB/kft) 24 AWG	MAXIMUM ATTENUATION (dB/kft) 22 AWG
64 kHz	Not specified	Not specified	2.2	2.2
150 kHz	Not specified	74	Not specified	Not specified
256 kHz	Not specified	Not specified	3.2	3.2
512 kHz	Not specified	Not specified	4.5	4.5
772 kHz	102 ± 15%	64	5.5	5.5
1.0 MHz	100 ± 15%	62	6.3	6.3
4.0 MHz	100 ± 15%	53	13	13
8.0 MHz	100 ± 15%	48	18	18
10.0 MHz	100 ± 15%	47	20	20
16.0 MHz	100 ± 15%	44	25	25
20.0 MHz	100 ± 15%	42	28	28
25 MHz	100 ± 15%	41	32	32
31.25 MHz	100 ± 15%	40	36	36
62.5 MHz	100 ± 15%	35	52	52
100 MHz	100 ± 15%	32	67	67
dc Resistance (Ω/kft at 20 °C)[b]			28.6	
Maximum dc resistance unbalance (%)[b]			5%	
Maximum pair mutual capacitance (pF/ft)[b]			17	
Maximum pair-to-ground capacitance unbalance (pF/kft)[b]			1,000	

[a]NEXT values shown are for the worst pair. The minimum NEXT coupling loss for any pair combination at 20 °C shall be determined from the following formula for all frequencies between 772 kHz and 100 MHz:

$$\text{NEXT}(F_{MHz}) > \text{NEXT}(772\text{ kHz}) - 15\,\text{Log}_{10}(F_{MHz}/0.772)$$

[b]EIA/TIA-568 does not specify separate requirements for 22 AWG; however, 22 AWG is classified Category 5 if it meets the performance requirements for 24 AWG.

The NEMA performance requirements cover *standard, low loss,* and *low loss extended frequency* cables, six pairs or less [17]. Standard and low loss and 100 Ω low loss extended frequency, for practical purposes, are equivalent to UL® Levels III, IV, and V, respectively, and EIA/TIA Categories 3, 4, and 5, respectively. NEMA also specifies a 150 Ω shielded cable that provides higher performance than either UL® or EIA cables. The NEMA 150 Ω STP specification covers only 22 and 26 AWG. The characteristics of the 22 AWG STP are specified to 100 MHz, while the characteristics of the 26 AWG STP are specified only to 16 MHz (attenuation) and 20 MHz (crosstalk). The 26 AWG STP can be used with LAN applications operating at 16 Mbps or less. Tables 11-15 through 11-19 list the performance data for NEMA cables.

Table 11-15 NEMA Standard UTP/STP Electrical Performance at 20 °C

Frequency	Impedance (Ω)	Minimum NEXT Loss (dB)[a]	Maximum Attenuation (dB/kft) 24 AWG	Maximum Attenuation (dB/kft) 22 AWG
256 kHz	Not specified	Not specified	4.0	3.2
512 kHz	Not specified	Not specified	5.6	4.6
772 kHz	102 ± 15%	43	6.8	5.7
1.0 MHz	100 ± 15%	41	7.8	6.8
4.0 MHz	100 ± 15%	32	17	15
10.0 MHz	100 ± 15%	26	30	25
16.0 MHz	100 ± 15%	23	40	34
20.0 MHz	100 ± 15%	22	47	40
dc Resistance (Ω/kft at 20 °C)			28.6	18.0
Maximum dc resistance unbalance (%)			5%	5%
Maximum pair mutual capacitance (pF/ft)			20	20
Maximum pair-to-ground capacitance unbalance (pF/kft)			1,000	1,000

[a]NEXT values shown are for the worst pair. The minimum NEXT coupling loss for any pair combination at 20 °C shall be determined from the following formula for all frequencies between 772 kHz and 20 MHz:

$$\text{NEXT}(F_{MHz}) > \text{NEXT}(772\text{ kHz}) - 15\,\text{Log}_{10}(F_{MHz}/0.772)$$

Table 11-16 NEMA Low Loss UTP/STP Electrical Performance at 20 °C

Frequency	Impedance (Ω)	Minimum NEXT Loss (dB)[a]	Maximum Attenuation (dB/kft) 24 AWG	Maximum Attenuation (dB/kft) 22 AWG
256 kHz	Not specified	Not specified	3.4	2.6
512 kHz	Not specified	Not specified	4.6	3.6
772 kHz	102 ± 15%	58	5.7	4.5
1.0 MHz	100 ± 15%	56	6.5	5.5
4.0 MHz	100 ± 15%	47	13	11
10.0 MHz	100 ± 15%	41	22	17
16.0 MHz	100 ± 15%	38	27	22
20.0 MHz	100 ± 15%	36	31	24
dc Resistance (Ω/kft at 20 °C)			28.6	18.0
Maximum dc resistance unbalance (%)			5%	5%
Maximum pair mutual capacitance (pF/ft)			17	17
Maximum pair-to-ground capacitance unbalance (pF/kft)			1,000	1,000

[a]NEXT values shown are for the worst pair. The minimum NEXT coupling loss for any pair combination at 20 °C shall be determined from the following formula for all frequencies between 772 kHz and 20 MHz:

$$\text{NEXT}(F_{MHz}) > \text{NEXT}(772\text{ kHz}) - 15\,\text{Log}_{10}(F_{MHz}/0.772)$$

Table 11-17 NEMA Low Loss Extended Frequency UTP/STP Electrical Performance at 20 °C

Frequency	Impedance (Ω)	Minimum NEXT Loss (dB)[a]	Maximum Attenuation (dB/kft) 24 AWG	Maximum Attenuation (dB/kft) 22 AWG
256 kHz	Not specified	Not specified	3.2	Not specified
512 kHz	Not specified	Not specified	4.5	Not specified
772 kHz	102 ± 15%	64	5.5	Not specified
1.0 MHz	100 ± 15%	62	6.3	Not specified
4.0 MHz	100 ± 15%	53	13	Not specified
10.0 MHz	100 ± 15%	47	20	Not specified
16.0 MHz	100 ± 15%	44	25	Not specified
20.0 MHz	100 ± 15%	42	28	Not specified
31.25 MHz	100 ± 15%	40	36	Not specified
62.5 MHz	100 ± 15%	35	52	Not specified
100 MHz	100 ± 15%	32	67	Not specified
dc Resistance (Ω/kft at 20 °C)			28.6	18.0
Maximum dc resistance unbalance (%)			5%	5%
Maximum pair mutual capacitance (pF/ft)			17	17
Maximum pair-to-ground capacitance unbalance (pF/kft)			1,000	1,000

[a]NEXT values shown are for the worst pair. The minimum NEXT coupling loss for any pair combination at 20 °C shall be determined from the following formula for all frequencies between 772 kHz and 100 MHz:

$$\text{NEXT}(F_{MHz}) > \text{NEXT}(772\text{ kHz}) - 15\log_{10}(F_{MHz}/0.772)$$

Table 11-18 NEMA 150 Ω, 26 AWG STP Electrical Performance at 20 °C

Frequency	Impedance (Ω)	Minimum NEXT Loss (dB)	Maximum Attenuation (dB/kft)
9.6 kHz	390 ± 15%	80	1.83
38.4 kHz	235 ± 15%	75	2.25
3.0 MHz	150 ± 10%	52	Not specified
4.0 MHz	150 ± 10%	52	10.0
5.0 MHz	150 ± 10%	52	Not specified
10.0 MHz	150 ± 10%	Not specified	13.1
12.0 MHz	150 ± 10%	34	Not specified
16.0 MHz	150 ± 10%	34	20.1
20.0 MHz	150 ± 10%	34	Not specified
dc Resistance (Ω/kft at 20 °C)			46.0
Maximum pair-to-ground capacitance unbalance (pF/kft)			457.2

Table 11-19 NEMA 150 Ω, 22 AWG STP Electrical Performance at 20 °C

Frequency	Impedance (Ω)	Minimum NEXT Loss (dB)	Maximum Attenuation (dB/kft)
9.6 kHz	270 ± 10%	80	0.9
38.4 kHz	185 ± 10%	75	1.5
3.0 MHz	150 ± 10%	58	Not specified
4.0 MHz	150 ± 10%	58	6.7
5.0 MHz	150 ± 10%	58	Not specified
10.0 MHz	150 ± 10%	Not specified	10.7
12.0 MHz	150 ± 10%	40	Not specified
16.0 MHz	150 ± 10%	40	13.7
20.0 MHz	150 ± 10%	40	Not specified
31.25 MHz	150 ± 10%	40	20.0
62.5 MHz	150 ± 10%	40	29.0
100 MHz	150 ± 10%	40	39.0
dc Resistance (Ω/kft at 20 °C)			17.4
Maximum pair-to-ground capacitance unbalance (pF/kft)			457.2

The BELLCORE cables are based on EIA categories and are equivalent in all respects except in pair counts. The EIA categories apply to small cables up to six pairs. The BELLCORE cables, on the other hand, apply to cables of any size used in building applications, generally 600 pairs or fewer. All transmission characteristics are the same except for NEXT, which is relaxed in the BELLCORE cables by 1 to 2 dB in cables larger than 25 pairs.

The final specification of interest is published by ICEA [18]. This specification details the mechanical and basic electrical requirements for cables up to 3,600 pairs and is patterned after the traditional telephony cable types, including station wire and riser cables. The mechanical characteristics are given in detail, but transmission performance is not specified. These cables are equivalent to EIA category 2.

Table 11-20 compares the cable specifications discussed above. Of particular importance are the applications in which these cables are used. Figure 11-7 shows some of the common applications for each performance category. Any application suitable for a given performance category cable can be used with a higher category cable. For example, Category 3 cable is suitable for LANs operating at 4 Mbps. This same LAN can be used on Category 4 or 5 cable. Also shown in this figure are the frequency ranges for each performance category. Cables should not be used with applications that use frequency components or spectra beyond the specified frequency range.

Table 11-20 Twisted Pair Cable Comparison[a]

Parameter	EIA	IBM	UL®	NEMA	BELLCORE	ICEA
Conductor Sizes	22, 24 AWG	22, 24, 26 AWG	22, 24 AWG	22, 24, 26 AWG	24 AWG	22, 24, 26 AWG
Impedance	100 Ω	150 Ω	100 Ω	100, 150 Ω	100 Ω	Not specified
Cable Sizes	<6 pairs	2 to 6 pairs	≤25 pairs	≤6 pairs	Any	≤3,600 pairs
Shielding	UTP	STP	STP/UTP	STP/UTP	UTP[b]	STP/UTP
Published Specification	EIA/TIA-568	GA27-3773-1	200-120	WC63	TA-NWT-000133	S-80-576
Performance	Category: 1–5	Type: 1–9	Level: I–V	Designation: Standard; Low loss; Low loss, extended frequency	Category: 1–5	Not specified
Equivalence to EIA/TIA-568	1	(none)	I	(none)	1	(none)
	2	Type 3	II	(none)	2	(none)
	3	(none)	III	Standard	3	(none)
	4	(none)	IV	Low loss	4	(none)
	5	(none)	V	Low loss, extended frequency	5	(none)

[a]Source: [21]

[b]The technical advisory does not preclude shielded twisted pair, or STP.

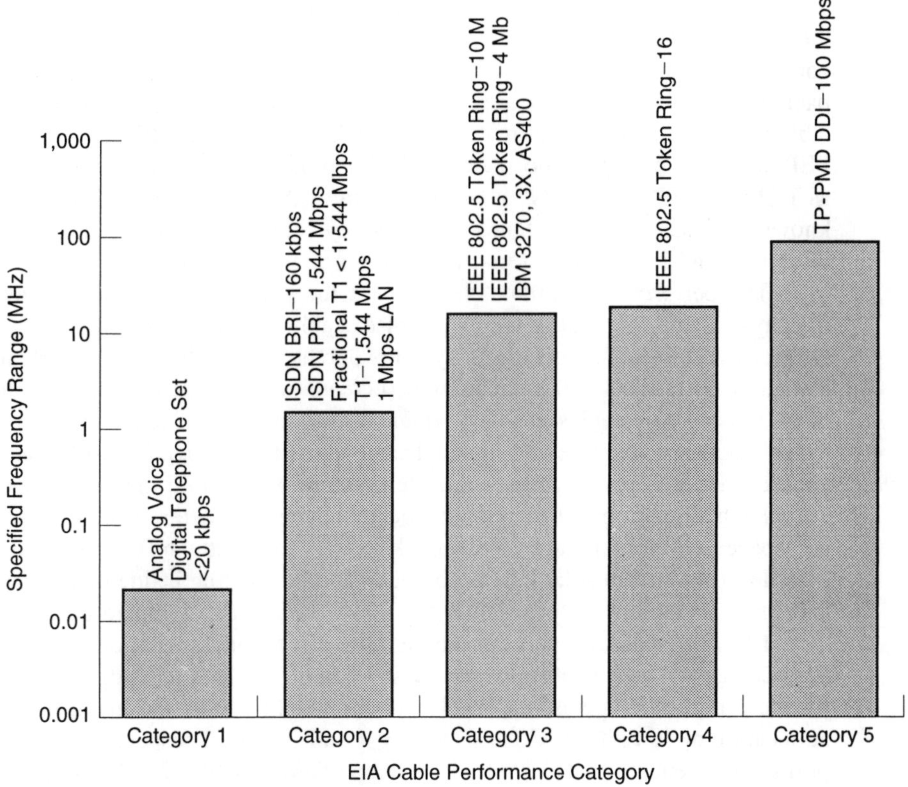

Figure 11-7 Twisted pair premises cable applications and frequency ranges, [21]

Cable performance requirements take into account the need for increasing data speeds and longer distances on premises environments. To achieve adequate performance, the following characteristics are optimized in premises cabling designs and constructions:

- Crosstalk
- Insulation performance
- Attenuation
- Impedance

Crosstalk is controlled by careful attention to capacitance unbalances between pairs and shields and by using random twist lengths and pair lays when the cable is manufactured. The higher the data speeds and frequencies, the tighter are the tolerances for these parameters. Crosstalk is independent of the insulation type for all practical purposes. The effects of crosstalk on signal-to-noise ratio (SNR) can be reduced by using a cable with low attenuation. This gives a higher received signal level for a given crosstalk level.

The electrical performance of twisted pair cables is inversely proportional to the insulation dielectric constant and dissipation factor. Cables with lower dielectric constant and dissipation factor have better attenuation characteristics and lower capacitance. Table 11-21 compares the characteristics for various insulation types. This table refers to various jacket materials known generically as FEP, ECTFE, and PVC. FEP is an acronym for fluorinated ethylene propylene, which is commonly known as Teflon®.* ECTFE stands for ethylene chlorotrifluoroethylene, which is commonly known as Halar®.† PVC stands for polyvinyl chloride, which is a quite common insulating and jacketing material available in many variations.

The performance parameters listed in the previous tables do not account for the physical environment in which premises cabling systems are expected to operate. Of greatest concern, especially in systems operating at bit rates greater than 64 kbps over long distances, is the attenuation increase with temperatures. The attenuation of all twisted pair cables at low frequencies increases with temperature mainly due to the resistance increase. At higher frequencies, the change in insulation dielectric constant and dissipation factor with temperature has a more marked effect.

High temperatures routinely can be encountered in exterior building walls, ceiling spaces (plenums), and mechanical rooms. Intermittent failures have been reported in LANs from solar heating of walls and the cabling inside them [22]. ECTFE and FEP insulations perform better than PVC under high temperature conditions.

EIA/TIA-568 accounts for temperature effects on attenuation by requiring that EIA Category 4 and 5 cables be tested at 40 and 60 °C and that adjustments be made for these higher temperatures. Tables 11-22 and 11-23 list the maximum allowable attenuation at 40 and 60 °C for Category 4 and 5 cables, respectively. For test purposes, the attenuation at the reference temperature (usually 20 °C) is increased by a factor of 0.3% per degree centigrade for all frequencies.

Although the standards do not address Category 1, 2, or 3 cables in terms of temperature performance, the attenuation for these cables is increased by approximately the same percentage as for Category 4 or 5. At frequencies below 1 MHz, all cables of a given gauge have approximately the same attenuation at the same frequency (within a few tenths of a decibel-per-kilofoot).

11.3.2 Twisted Pair Connection Devices

At frequencies below approximately 1 MHz, the transmission degradation introduced by twisted pair connection devices [jacks, plugs, patch cords, splices, insulation displacement connectors (punch-down blocks)] is negligible. At higher frequencies, crosstalk and insertion loss must be considered during the premises design and installation processes. Table 11-24 gives the maximum insertion loss requirements for connecting devices, as specified by EIA.

The minimum crosstalk performance requirements for Category 3, 4, and 5 components are given in Table 11-25 for various frequencies. The NEXT coupling loss for other frequencies can be found from

---

*Teflon® is a trademark of E.I. du Pont de Nemours & Company, Inc.

†Halar® is a trademark of Ausimont Corp.

Table 11-21 Comparison of Typical Premises Cabling Insulation Types[a]

Insulation Type[b]	Dielectric Constant	Dissipation Factor	Mutual Capacitance (pF/ft)	Attenuation (dB/100 ft)			Signal Delay at 20 MHz (ns/100 ft)
				1.0 MHz	10.0 MHz	20.0 MHz	
FEP	2.1	0.0005	14	0.6	1.8	2.7	2.5
Polyethylene	2.3	—	17.5	—	—	—	—
ECTFE	2.5	0.01	15	0.6	2.4	3.7	—
PVC-Non-Plenum	3.4	—	21	0.6	2.7	9.1	—
PVC-Plenum	3.6	0.04	18	0.9	3.7	5.5	8
XL Polyolefin	3.8	—	19	0.9	3.4	5.2	—

[a]Information courtesy of DuPont Company, Wire and Cable Group.

[b]Values are based on room-temperature tests with 6–7 mils insulation wall on commercial 24 AWG, 4-pair, UTP cable 100 m long.

Table 11-22 EIA Category 4 UTP Maximum Attenuation at Elevated Temperatures

Frequency	40 °C (dB/kft)	60 °C (dB/kft)
64 kHz	2.4	2.6
256 kHz	3.6	3.8
512 kHz	4.9	5.2
772 kHz	6.0	6.4
1.0 MHz	6.9	7.3
4.0 MHz	13.8	14.6
8.0 MHz	20.1	21.3
10.0 MHz	23.3	24.6
16.0 MHz	28.6	30.2
20.0 MHz	32.9	34.7

Table 11-23 EIA Category 5 UTP Maximum Attenuation at Elevated Temperatures

Frequency	40 °C (dB/kft)	60 °C (dB/kft)
64 kHz	2.3	2.5
256 kHz	3.4	3.6
512 kHz	4.8	5.0
772 kHz	5.8	6.2
1.0 MHz	6.7	7.1
4.0 MHz	13.8	14.6
8.0 MHz	19.1	20.2
10.0 MHz	21.2	22.4
16.0 MHz	26.5	28.0
20.0 MHz	29.7	31.4
25.0 MHz	33.9	35.8
31.25 MHz	38.2	40.3
62.5 MHz	55.1	58.2
100 MHz	71.0	75.0

Table 11-24 Maximum Insertion Loss for Structured Wiring System Components[a]

	Insertion Loss (dB)		
Frequency (MHz)	Category 3	Category 4	Category 5
1	0.4	0.1	0.1
4	0.4	0.1	0.1
8	0.4	0.1	0.1
10	0.4	0.1	0.1
16	0.4	0.2	0.2
20	Not specified	0.2	0.2
25	Not specified	Not specified	0.2
31.25	Not specified	Not specified	0.2
62.50	Not specified	Not specified	0.3
100.00	Not specified	Not specified	0.4

[a]Data from [19]

$$NEXT(f) = NEXT(f_0) - 20 \log (f/f_0)$$

where $NEXT(f_0)$ = NEXT coupling loss at f_0
f_0 = 16 MHz
f = frequency in question, MHz

All structured wiring systems are composed of patch panels, patch cords, and outlet jacks. The crosstalk performance for a complete set of commercially available components, including patch panels, patch cords, and outlet connectors (jacks), is listed in Table 11-26. These components meet EIA Category 5 requirements. Components should have at least 10 to 12 dB higher NEXT coupling loss than the cables connected to them.

Table 11-25 Minimum NEXT Coupling Loss of Structured Wiring System Components[a]

	NEXT Coupling Loss (dB)		
Frequency (MHz)	Category 3	Category 4	Category 5
1	58	>65	>65
4	46	58	>65
8	40	52	62
10	38	50	60
16	34	46	56
20	Not specified	44	54
25	Not specified	Not specified	52
31.25	Not specified	Not specified	50
62.50	Not specified	Not specified	44
100.00	Not specified	Not specified	40

[a]Data from [19]

Table 11-26 NEXT Coupling Loss of Typical Structured Wiring System Components

	NEXT Coupling Loss (dB)			
Frequency (MHz)	Connecting Block (IDC)[a]	Patch Cord	Connecting Block and Patch Cord Together	Outlet Connector
1	86	73	88	90
4	74	64	76	78
8	68	60	70	72
10	66	59	68	70
16	62	55	64	65
20	60	53	62	63
25	58	52	60	61
31.25	56	50	58	58
62.50	50	47	52	50
100.00	46	40	48	42

[a]IDC = insulation displacement connector

The crosstalk performance of a connection device is very sensitive to pair twist integrity at the connection. That is, a given connection device itself may exhibit high crosstalk coupling loss, but if the pair twist is not maintained right up to the device, the overall connection, including the cable, may have greatly degraded crosstalk coupling loss. This phenomenon is highly dependent on frequency, becoming worse with increasing frequency.

High performance premises cables have a shorter twist length to minimize crosstalk; this twist must be maintained up to within one-half inch of any connection (for example, jack or connector).

11.3.3 Optical Fiber and Coaxial Cable

EIA/TIA-568 specifies a graded index (GRIN) multimode optical fiber with nominal core/cladding dimensions of 62.5/125 μm and having the characteristics shown in Table 11-27. Two types of coaxial cables are specified by NEMA for use with the IEEE 802.3 10Base5 and 10Base2 LANs. The important electrical characteristics of these cables are summarized in Table 11-28.

Table 11-27 EIA/TIA-568 62.5/125 μm GRIN Multimode Optical Fiber Specifications[a]

Wavelength (nm)	Maximum Attenuation (dB/km)	Bandwidth (MHz-km)
850	3.75	160
1,300	1.5	500

[a]Data from [3]

Table 11-28 Coaxial Cable Characteristics at 20 °C per NEMA[a]

Parameter	System: 10Base5	System: 10Base2
Velocity factor	0.77	0.65
Maximum attenuation at 5 MHz	3.66 dB/kft	9.88 dB/kft
Maximum attenuation at 10 MHz	5.18 dB/kft	14.0 dB/kft
Nominal capacitance	26 pf/ft	26 pf/ft
Characteristic impedance	50 ± 2 Ω	50 ± 2 Ω
Center conductor dc resistance	3.05 Ω/kft	15.24 Ω/kft
Nominal outside diameter	0.375–0.405 inch	0.168–0.195 inch

[a]Data from [17]

11.3.4 Media Applications and Compatibility

The installation of premises cabling systems is labor intensive and therefore costly. To make maximum use of the cabling resource, it is desirable that as many applications as possible share the common cable sheath. For example, PBX or key system stations may be interconnected with the same cable used to interconnect computer workstations. Due to crosstalk, this is not recommended with some systems.

Generally, data transmission interfaces that are unbalanced with respect to ground cannot be mixed with other systems. The RS-232 interface, when extended using twisted pair cables, is an interface that is incompatible with just about everything else. Of course, the RS-232 interface can be extended using limited distance modems, voiceband modems, or optical fibers, which ease the compatibility constraints. Although the RS-232 standard limits the transmission distance to 50 feet on metallic cable, frequent attempts are made to extend this distance. Not only is this not recommended for performance reasons, it also is not recommended for EMI reasons.

Backbone cabling systems may be called upon to carry both analog and digital signals from more than one type of LAN, a PBX, key systems, and alarm systems. All baseband digital data transmission systems operating at speeds of 64 kbps or less generally are compatible with analog and digital PBX and key system station circuits as long as they use balanced transmission schemes. Transmission media choice depends to some extent on the applications shown in Table 11-29.

11.4 Jacks and Plugs

Three different connectors are specified at the workstation outlet by EIA/TIA-568, depending on the transmission medium: 100 Ω UTP, 150 Ω STP, and 50 Ω coaxial cable. The connector for IEEE 802.5 token ring LANs using 150 Ω STP is a self-mating connector. It connects to another connector like itself. This connector has shorting contacts to maintain continuity for automatic looping capability when the connector is not mated.

For thin coaxial cables used in IEEE 802.3 LANs, the bayonet-type coaxial connector (BNC) connector is specified. Thick coaxial cables usually use the N connector. EIA/TIA-568 does not specify the connector type for optical fiber cables, but the 2.5 mm ST (bayonet) and SC (push-pull) connectors are widely used in this application. It is conceivable that a workstation outlet will have one of each connector for maximum flexibility.

For cabling that is made up of 100 Ω impedance UTP, the 8-position/8-contact modular jack and plug (per ISO 8877) is used [14]. Some systems use a keyed version of the ISO 8877 jack and plug. The relationship between the number of positions and contacts in a modular connector is as follows: the number of positions specify the maximum number of spring contacts that may be installed in the connector, while the number of contacts indicates the actual number used in a particular configuration. An 8-position connector conceivably could utilize just one contact (although this is very unlikely), but has eight contacts available for installation. All commercial 8-position connectors do, in fact, have eight contacts.

Two wiring schemes are given in EIA/TIA-568 to accommodate the large base of existing systems that use the 8-contact connector. The first is the T568A connector and the second is the T568B connector. These commonly are called "TIA" and

Table 11-29 Media Applications[a]

Application or Interface	Optical Fiber	Coaxial Cable	UTP Level I or II	UTP/STP Level III or UTP Category 3	UTP/STP Level IV or UTP Category 4	UTP/STP Level V or UTP Category 5
Analog telephone set			√	√	√	√
Digital telephone set			√	√	√	√
EIA-RS232			√	√	√	√
EIA-422/449			√	√	√	√
EIA-423/449			√	√	√	√
ISDN			√ (Level II only)	√	√	√
IEEE 802.3 10Base2		√				
IEEE 802.3 10Base5		√				
IEEE 802.3 10BaseT				√	√	√
IEEE 802.3 10BaseF	√					
IEEE 802.5 token ring 4 Mbps				√	√	√
IEEE 802.5 token ring 16 Mbps				[b]	√	√
FDDI	√					
TP-PMD						√

[a]Data from [14] with additions

[b]Not normally recommended but may be used if installed cable meets qualification guidelines.

Key

√ indicates proper application

"AT&T," respectively, in commercial literature. The 8-contact modular connector specified in EIA/TIA-568 is physically identical to the 8-position connector used for PTN interfaces, as described in Chapter 2.

Although EIA/TIA-568 specifies only the 8-contact modular connector in two specific configurations, other connector types are used in premises applications: 6-position/6-contact modular connector, 50-position/50-contact ribbon connector, and the 8-position modular connector in various other configurations. While the 8-position modular connector is available in keyed and unkeyed versions, only the unkeyed version is used in the EIA/TIA-568. The connector is frequently, but incorrectly, called an RJ-45. This designation is incorrect because RJ-45 refers to a wiring pattern specified in the FCC rules and regulations that is not consistent with LAN connectors [2].

The 8-position connector is specified for twisted pair applications by IEEE 802.3 standard for the Ethernet or 10BaseT LANs [8]. The same connector is used with the twisted pair applications of the token ring LAN specified in the IEEE 802.5 standard (this standard also specifies a 6-position connector) defined in [13] and the ISDN S/T bus interface defined in [24]. The ISDN S/T bus is described later in section 11.5.

All 8-position connectors in the forementioned references are physically the same but are wired differently. Table 11-30 can be used to sort out the different wiring configurations. Where circuits have separate transmit and receive circuits, the directions normally are indicated by the nomenclature TX+/TX− or TD+/TD− for transmit and RX+/RX− or RD+/RD− for receive. The positive and negative symbols indicate the relative polarity of the framing pulses on each path. The EIA/TIA-568 connectors, designated T-568A and T-568B, are wired to specific pairs (by color code) rather than by electrical functions. The pair assignments for the EIA/TIA-568 connectors are shown in Figure 11-8, and the color codes associated with each conductor are shown in Table 11-31.

11.5 Premises Applications of ISDN

The ISDN uses standardized interface reference points, as shown in Figure 11-9. The ISDN digital subscriber line (DSL) terminates, for operational purposes, at the user premises in a device called a layer 1 network termination (NT1). The jurisdictional interface, or network demarcation point, is at the protector, but the operational interface is the NT1. The network side of the NT1 is called the U-reference (or U-loop), whereas the premises side of the NT1 is called the T (terminal) reference. The NT1 may be connected to a layer 2 network termination, or NT2, such as a PBX. The user side of the NT2 is designated the S (switched) reference. From a transmission engineering standpoint, the S and T references are the same. Therefore, no distinction will be made between the S and T references in this section. All references will be to the S/T bus.

The terminal equipment (TE) shown in Figure 11-9 are ISDN S/T bus compliant. TEs typically are ISDN telephone sets. The terminal adapters (TAs) are used

Table 11-30 8-Position Connector

Type/Contact No.	1	2	3	4	5	6	7	8
T568A (TIA) [3]	T-3	R-3	T-2	R-1	T-1	R-2	T-4	R-4
T568B (AT&T) [3]	T-2	R-2	T-3	R-1	T-1	R-3	T-4	R-4
10BaseT (IEEE 802.3) [8]	TD+	TD−	RD+			RD−		
Token Ring (IEEE 802.5) [13]			TX−	RX+	RX−	TX+		
ISDN TE (ANSI T1.605) [24]	PS3+	PS3−	TX+	RX+	RX−	TX−	PK2−	PK2+
ISDN NT (ANSI T1.605) [24]	PK3+	PK3−	RX+	TX+	TX−	RX−	PS2−	PS2+

Nomenclature for Table 11-30

T-1	Tip, pair 1
R-1	Ring, pair 1
T-2	Tip, pair 2
R-2	Ring, pair 2
TX+	Transmit, positive framing pulse
TX−	Transmit, negative framing pulse
RX+	Receive, positive framing pulse
RX−	Receive, negative framing pulse
TD+	Transmit data, positive framing pulse
TD−	Transmit data, negative framing pulse
RD+	Receive data, positive framing pulse
RD−	Receive data, negative framing pulse
PS+	Power source, positive polarity
PS−	Power source, negative polarity
PK+	Power sink, positive polarity
PK−	Power sink, negative polarity

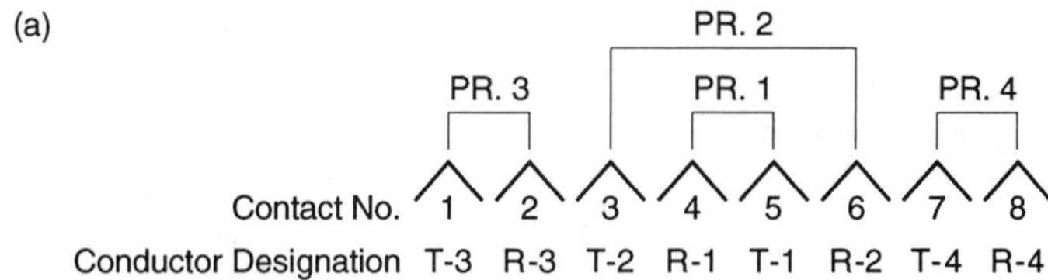

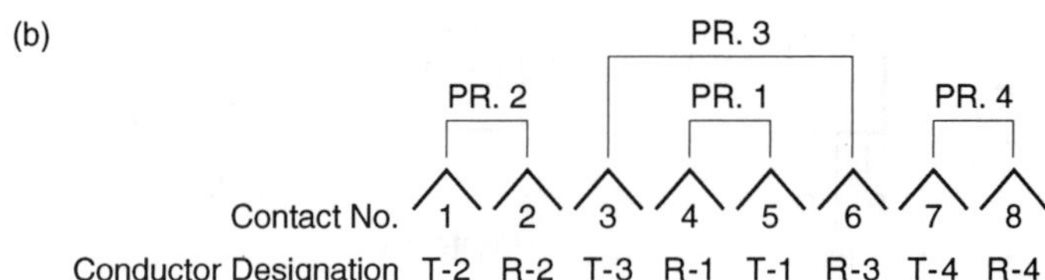

Figure 11-8 EIA/TIA structured wiring connectors: (a) T-568A connector wiring; (b) T-568B connector wiring

Table 11-31 EIA/TIA Structured Wiring Connector Color Code

Lead Designation	Base Color	Tracer Color	Shorthand
T-1	White	Blue	W-BL
R-1	Blue	White	BL-W
T-2	White	Orange	W-O
R-2	Orange	White	O-W
T-3	White	Green	W-G
R-3	Green	White	G-W
T-4	White	Brown	W-BR
R-4	Brown	White	BR-W

to connect non-ISDN terminal equipment, such as regular telephone sets, to the S/T bus.

The two aspects to be considered in ISDN premises design are the *operation mode* and the *wiring configuration* [24]. The operation mode defines the transmitter and receiver activity, whereas the wiring configurations define the interconnectivity. There are two operation modes: point-to-point and point-to-multipoint.

In the point-to-point operation mode, only one transmitter and one receiver are active on the S/T bus at a time in each direction. In the point-to-multipoint mode, more than one transmitter and receiver can be simultaneously active. Multipoint terminals *share* the 2B+D channels in the basic rate interface. When one terminal is active on a circuit-switched B-channel, no other terminal may use that B-channel. When both B-channels are being used, no other multipoint terminals may use the B-channels. The D-channel is used for signaling and packet data transmission and is shared by all multipoint terminals on a given S/T bus. The S/T bus operates at a line rate of 192 kbps, which includes the 2B+D, 144 kbps payload and 48 kbps of overhead. The S/T bus parameters are given in Table 11-32.

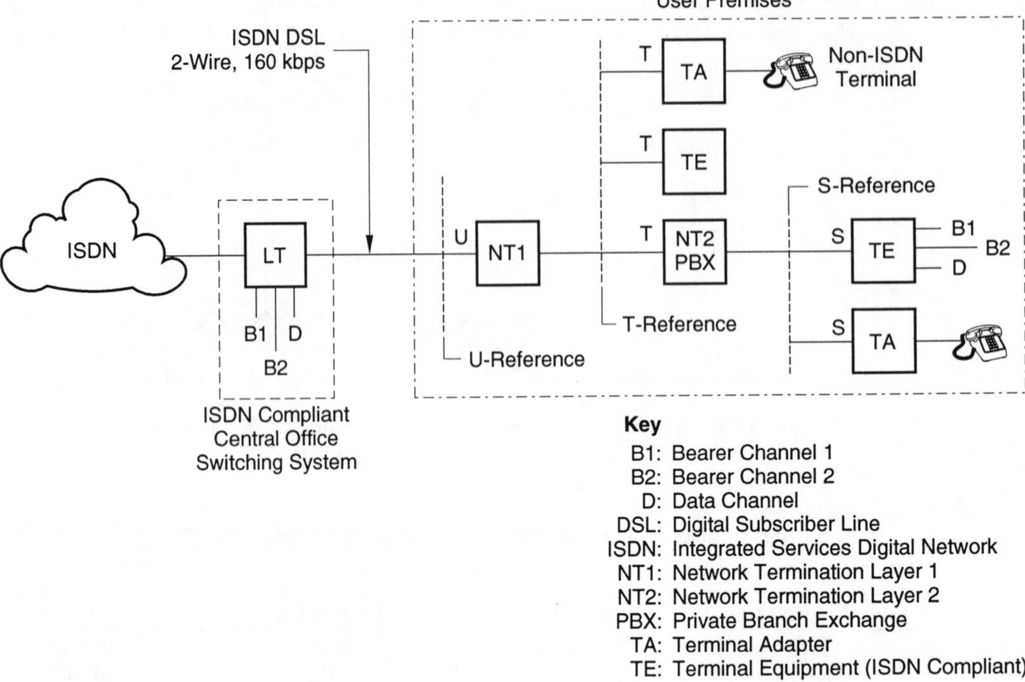

Figure 11-9 ISDN reference points

Table 11-32 ISDN S/T Bus Parameters

PARAMETER	REQUIREMENT
Line transmit rate Payload rate	192 kbps ± 100 ppm 144 kbps
Overhead rate Frame length	48 kbps 48 bits
Line code	Alternate space inversion (pseudoternary), 100% duty cycle
Nominal pulse height, base-to-peak Termination resistance	750 mV ± 10% 100 Ω ± 5%

Unlike the U-loop, the S/T bus is polarity sensitive, and proper conductor orientation must be maintained throughout the premises wiring system. Bridged taps are not allowed on the S/T bus under any circumstances, although one m stubs may be used to tap the bus. Each S/T bus must be properly terminated with 100 Ω resistors at each end to prevent signal reflections. The location of the terminating resistor depends on the wiring configuration. The S/T bus should not be placed on inter-building cable facilities because the TE and user-side of the NT1 are not designed to cope with the electrical environment encountered by outside cables.

The *wiring configurations* are defined as point-to-point and point-to-multipoint (the same as the operation modes). The multipoint configurations include the short passive bus and extended passive bus. The short passive bus and extended passive bus also include branched configurations, called short branched passive bus and extended branched passive bus, respectively.

These wiring configurations are shown schematically in the subparts of Figure 11-10, with applications guidelines summarized in each configuration's corresponding table. In all wiring configurations, four pairs connect the NT1 to the TE via the workstation outlet. Two pairs are used for transmission (one for each direction). The transmission pairs also can be used to provide simplex power to the TE in a phantom powering scheme shown in Figure 11-11.

Various other powering arrangements can be used. As shown in Figure 11-11, Auxiliary 1 Powering uses a third pair to power from the NT1 (power source 2) to TE (power sink 2). Auxiliary 2 Powering uses a fourth pair to power from a TE to an NT1 or another TE in TE-to-TE bus connections. Local powering also can be used as shown. In general, local powering should be used when more than two TEs share horizontal wiring. Different powering arrangements are handled at cross-connects located in the main communications room or telecommunications closets.

The S/T bus in the point-to-point wiring configuration uses terminating resistors at each end and has only one TE, as shown in Figure 11-10(a). This is the simplest wiring configuration to implement, and the full 2B+D data rate is available to the single workstation. See Table 11-33 for the basic point-to-point guidelines.

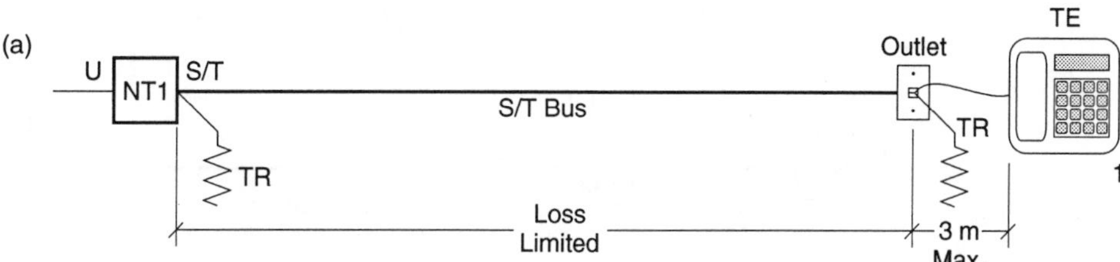

Figure 11-10(a) ISDN point-to-point wiring configuration

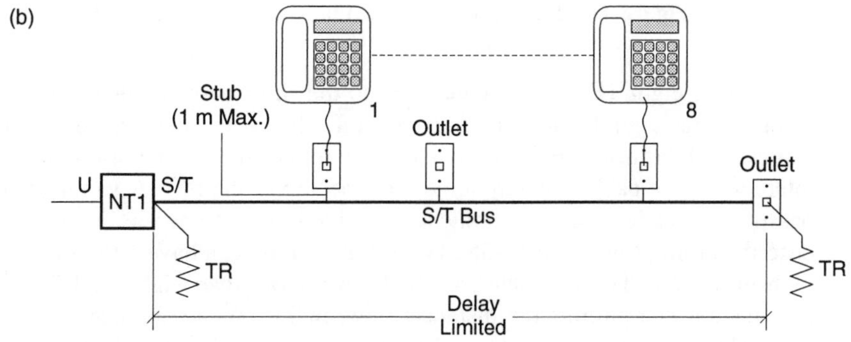

Figure 11-10(b) ISDN short passive bus wiring configuration

(c)

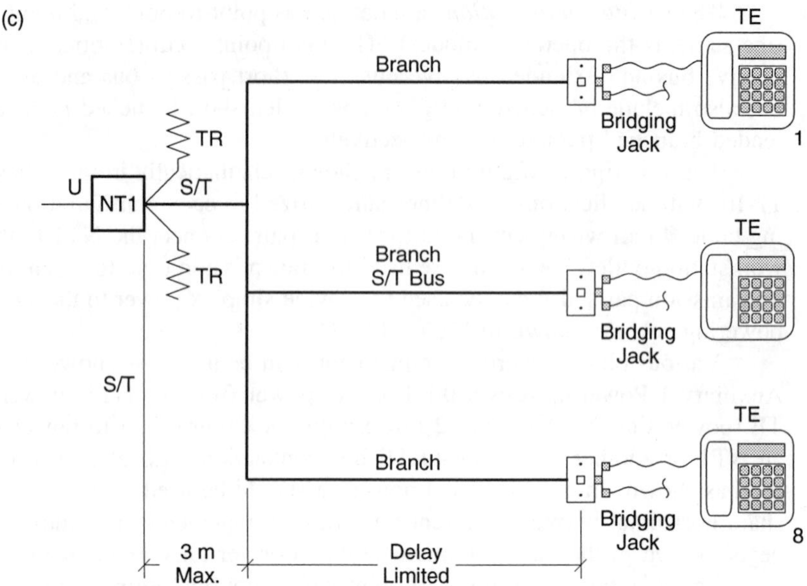

Figure 11-10(c) ISDN short branched passive bus wiring configuration

(d)

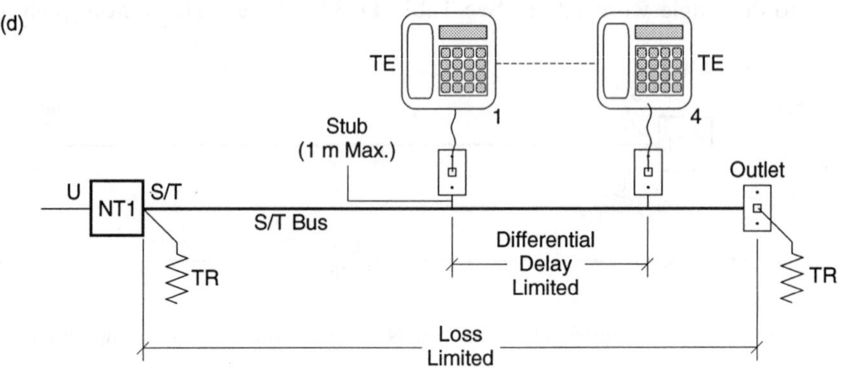

Figure 11-10(d) ISDN extended passive bus wiring configuration

Up to eight TEs can be connected to the short passive bus in Figure 11-10(b), using the basic guidelines summarized in Table 11-34. The separation between the nearest and farthest TE is limited by the roundtrip propagation delay due to the interconnecting cables and equipment connected to the bus. A variation of the short passive bus, called the short branched passive bus, is shown in Figure 11-10(c), and described further in Table 11-35. This configuration can have, at most, four branches with up to four TEs on each branch, but no more than eight total TEs. Roundtrip propagation delay limits the distance between TE on each branch.

The extended passive bus, shown in Figure 11-10(d) and described further in Table 11-36, allows the bus taps to be placed at a distance from the NT1. The length

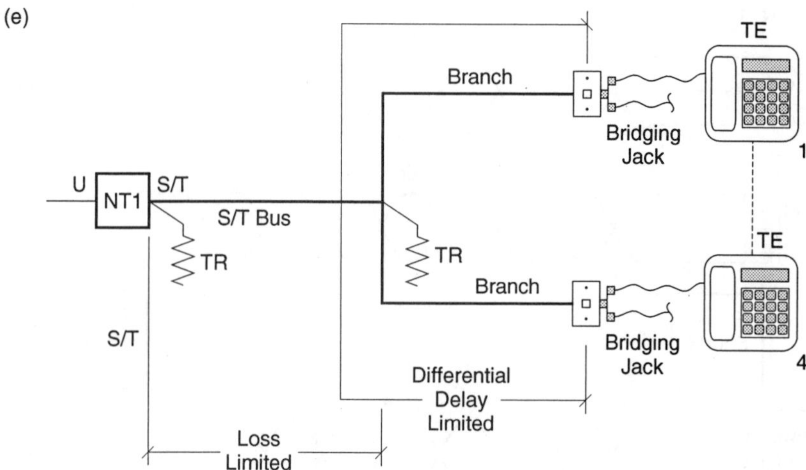

Figure 11-10(e) ISDN extended branched passive bus wiring configuration

of the bus is limited by the cable loss, and the distance between TEs is limited by the allowable differential delay between the TEs. The extended branched passive bus in Figure 11-10(e) is a variation with similar limitations. See Table 11-37 for a summary of corresponding guidelines.

The S/T bus is designed at 96 kHz, which is the Nyquist frequency for a system using a 192 kbps alternate space inversion (ASI) line code. The loss and delay characteristics of twisted pair cables used in premises applications are not specified at 96 kHz in the various structured cabling system performance tables given previously. To fill the gap, Table 11-38 gives the needed information for cables that would be classified as EIA/TIA-568, Category 2. Cables classified as Category 3, 4, and 5 have similar performance at 96 kHz. This table can be used to find the maximum bus and branch lengths for each wiring configuration.

The propagation delay is an important parameter in S/T bus design. Many wiring configurations are delay limited rather than loss limited. The roundtrip delay through each TE is 0.0875 μsecond and includes the delay associated with a 10 m connecting cord between the TE and workstation outlet. A TE can be equipped with a shorter cord.

Included in the delay calculations is the phase tolerance between the input and output of a TE (+15%, −7%). The S/T bus frame consists of 48 bits, with a total frame length of 250 μseconds. The bit interval is 250 μseconds/48 bits = 5.21 μseconds. Therefore, the phase tolerance accounts for +0.78 μsecond (+15%) and −0.36 μsecond (−7%), or 1.15 μseconds (22%) total, depending on the circumstances.

As shown in the application tables, the point-to-point wiring configuration allows a roundtrip propagation delay of 10 to 42 μseconds. The lower value assumes the TE is installed adjacent to the NT with a near-zero-length interconnecting cable. The delay in this case will be two bits due to frame offset (2 × 5.21 μseconds =

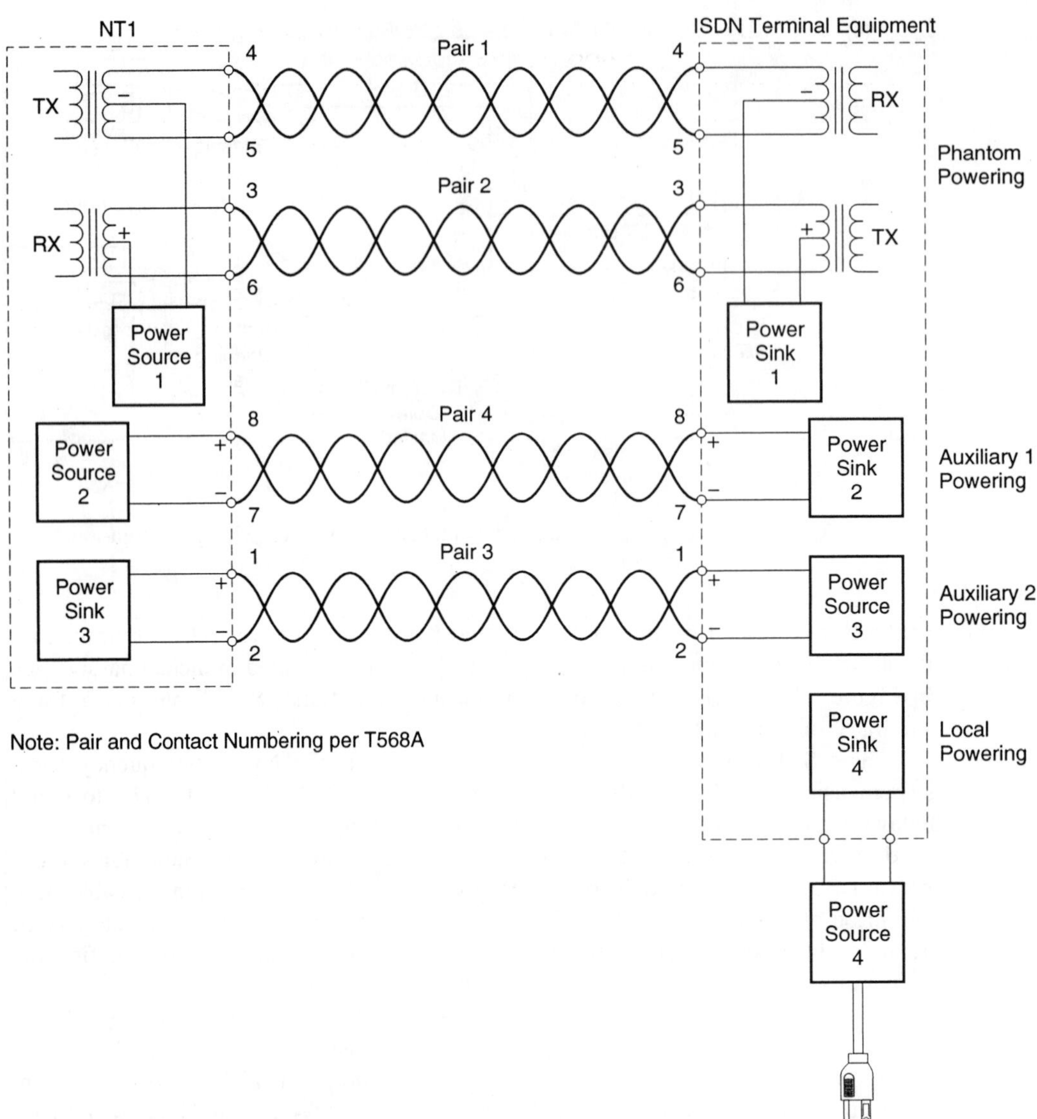

Figure 11-11 ISDN NT1 and terminal equipment powering arrangements

10.42 μseconds) less the TE negative phase tolerance of 0.36 μseconds. The 42 μseconds upper value assumes the TE is located at the very far end of the bus. The delay consists of 10.42 μseconds from the 2-bit frame offset, plus maximum delay in the cable, which is equal to the allowed processing time of 6 bits (6 × 5.21 μseconds =) 31.3 μseconds, plus the positive phase tolerance of 0.78 μsecond, for a total of 42.5 μseconds, rounded down to 42 μseconds. The point-to-point configuration normally is loss limited rather than delay limited. The operational length

Table 11-33 ISDN Point-to-Point Wiring Configuration Application Guidelines

RULE	DESCRIPTION
1	6 dB maximum S/T bus cable loss at 96 kHz between NT1 and TE outlet
2	10 to 42 μseconds total roundtrip delay, with fixed or adaptive timing (includes delay due to TE)
3	Maximum of one workstation outlet per S/T bus
4	Maximum of two S/T buses in star wiring configuration
5	One 100 Ω terminating resistor at NT1 for each S/T bus, and one 100 Ω terminating resistor at far end of each S/T bus
6	3 m maximum cable length between the NT1 and S/T bus
7	No bridged taps allowed on S/T bus

Table 11-34 ISDN Short Passive Bus Wiring Configuration Application Guidelines

RULE	DESCRIPTION
1	2 μseconds maximum roundtrip delay due to S/T bus cable (does not include delay due to TE)
2	10 to 14 μseconds total roundtrip delay due to TE with NT1 set for fixed timing; 10 to 13 μseconds for adaptive timing
3	1 m maximum stub length between bus tap and outlet
4	Maximum of eight TEs on S/T bus
5	One 100 Ω terminating resistor at NT1 and one 100 Ω terminating resistor at far end of S/T bus
6	3 m maximum cable length between the NT1 and S/T bus
7	No bridged taps allowed on S/T bus

Table 11-35 ISDN Short Branched Passive Bus Wiring Configuration Application Guidelines

RULE	DESCRIPTION
1	0.9 μsecond maximum roundtrip delay on each branch cable (does not include delay due to TE)
2	1 m maximum stub length between branch tap and outlet
3	Maximum of four branches
4	Maximum of four TEs per branch
5	Maximum of eight TEs, total
6	Two 100 Ω terminating resistors at NT1
7	3 m maximum cable length between the NT1 and S/T bus
8	No bridged taps allowed on S/T bus

Table 11-36 ISDN Extended Passive Bus Wiring Configuration Application Guidelines

Rule	Description
1	3.8 dB maximum S/T bus cable loss at 96 kHz between NT1 and terminating resistor at far end of S/T bus
2	10 to 42 μseconds total roundtrip delay, with 0 to 2 μseconds differential delay due to TE
3	0.5 μsecond maximum roundtrip delay between TE due to cable
4	1 m maximum stub length between bus tap and outlet
5	Maximum of four TEs
6	One 100 Ω terminating resistor at NT1, and one 100 Ω terminating resistor at end of S/T bus
7	3 m maximum cable length between the NT1 and S/T bus
8	No bridged taps allowed on S/T bus

Table 11-37 ISDN Extended Branched Passive Bus Wiring Configuration Application Guidelines

Rule	Description
1	3.8 dB maximum S/T bus cable loss at 96 kHz between NT1 and terminating resistor at S/T bus branching point
2	10 to 42 μseconds total roundtrip delay, with 0 to 2 μseconds differential delay due to TE
3	0.25 μsecond maximum roundtrip delay on each branch cable
4	1 m maximum stub length between branch tap and workstation outlet
5	Maximum of two branches
6	Maximum of two TEs per branch
7	One 100 Ω terminating resistor at NT1 and one 100 Ω terminating resistor at S/T bus branching point
8	3 m maximum cable length between the NT1 and S/T bus
9	No bridged taps allowed on S/T bus

Table 11-38 Premises Cable Characteristics at 96 kHz and 20 °C

Cable Type	AWG	Loss (dB/ft)	Nominal Capacitance (pf/ft)	Roundtrip Propagation Delay (ns/ft)
Inside riser	22	0.0016	15.7	3.2
Inside riser	24	0.0023	15.7	3.4
Inside riser	26	0.0032	15.7	3.6
3- or 4-pair station wire	24	0.0026	17.7	3.8
25-pair station wire	24	0.0029	21.6	4.0

objective of the point-to-point configuration is 3,280 feet (1 km), but this depends somewhat on the cable used.

The short passive bus can accommodate 10 to 14 μseconds of roundtrip delay with fixed timing, or 10 to 13 μseconds with adaptive timing. The lower value is based on the same assumptions as for the point-to-point wiring configuration. The upper value with fixed timing assumes the TE is located at the far end of the passive bus and consists of 10.42 μseconds from the 2-bit frame offset, plus the roundtrip delay of the unloaded bus cable, which is 2 μseconds, plus the delay due to eight TEs (8 × 0.0875 μsecond =) 0.7 μsecond, plus the TE positive phase tolerance of 0.78 μsecond = 13.9 μseconds, rounded to 14 μseconds. The maximum operational length of the short passive bus is in the range of 660 feet (200 m), but typical cables limit this to around 500 feet (150 m).

When the short passive bus uses adaptive timing, the roundtrip delay of the bus cable is cut in half to 1 μsecond due to the extra tolerance required for the adaptive timing (all other delays are the same). This results in a shorter bus length of around 250 to 300 feet (75 to 90 m). A similar evaluation can be made for the short *branched* passive bus by applying the maximum propagation delay requirements to the sum of the individual branches.

The cable delay requirements of the short passive bus are based on eight TEs. If fewer than eight TEs are installed, the bus cable can be made correspondingly longer. Each TE introduces a delay of 0.0875 μsecond, which is equivalent to approximately 20 feet (6 m) of bus cable. Therefore, the bus can be extended by approximately 20 feet for each TE not installed. Each installation should be evaluated based on the type of cable used.

The extended passive bus is intended for intermediate distances of 300 to 2,400 feet (100 to 7,500 m). The main difference between the extended passive and short passive buses is the grouping of the TE at the far end of the cable from the NT for the extended passive bus. This configuration restricts the distance between TE based on a differential roundtrip delay of 2 μseconds. This delay is based on a total TE phase tolerance of 22%, or 1.15 μseconds, plus the roundtrip delay due to the unloaded bus cable of 0.5 μsecond, plus the delay due to four TEs of 0.35 μsecond, for a total of 2 μseconds. The 0.5 μsecond delay from the cable corresponds to approximately 100 to 150 feet (30 to 50 m) of separation between TE attachments. The loss from the NT to the far end of the bus is restricted to 3.8 dB, which corresponds to 1,500 to 2,000 feet (500 to 600 m) for typical cables.

The extended passive bus can be adjusted if less than the maximum allowed four TEs are installed. The total length of the extended passive bus is always restricted to a loss of 3.8 dB, but the separation between TEs can be extended by an amount equal to the delay of each TE not installed. As with the short branched passive bus, the extended branched passive bus is restricted to the total delay requirements as applied to the sum of the branches.

The major components of the structured cabling system, including backbone cabling, horizontal cabling, and workstation cabling, adapt directly to the ISDN premises application. However, certain wiring configurations are more suitable for the star topology described in EIA/TIA-568. These are, in particular, the point-to-point, short branched passive bus, and extended branched passive bus. The short passive bus, and extended passive bus wiring configurations can be used in the structured wiring system, too, with proper attention to distance, loss, and delay limitations. The location of the NT somewhat affects the choice of wiring configuration. The NT may be located in the main communications equipment room or the telecommunications closets. Typical configurations are shown in Figure 11-12.

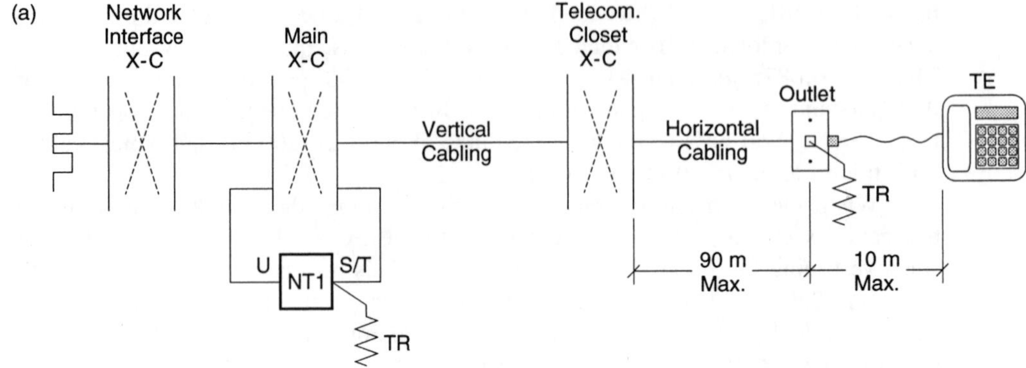

Figure 11-12(a) Application of point-to-point configuration in structured wiring system

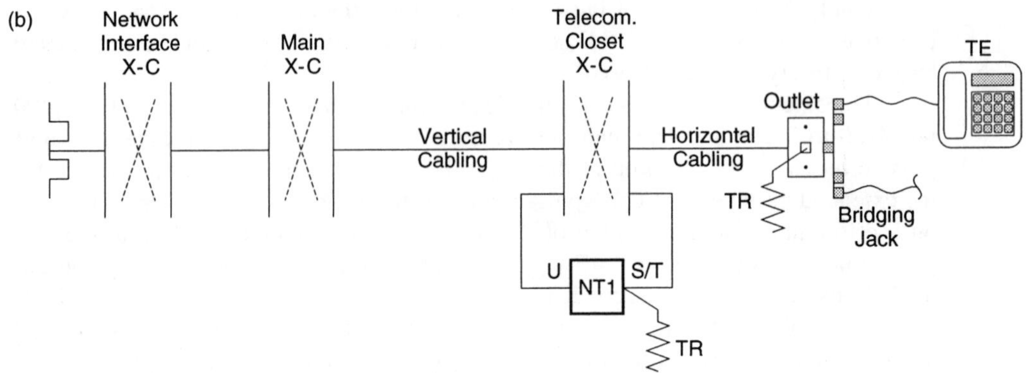

Figure 11-12(b) Application of short passive bus configuration in structured wiring system

For a typical ISDN premises wiring system, the network terminations, NT1, are connected to purple cross-connect fields located in the main equipment room, as shown in Figure 11-13(a). U-loop network access connections are jumpered to the green cross-connect field, and connections to the backbone cabling are through the white cross-connect field. If the NT1 equipment is located in a telecommunications closet, the connections are as shown in Figure 11-13(b). At each telecommunications closet, the white backbone cabling terminations are cross-connected to the blue horizontal cabling terminations. The terminating resistors are installed at the workstation outlet or at the cross-connects, depending on the configuration.

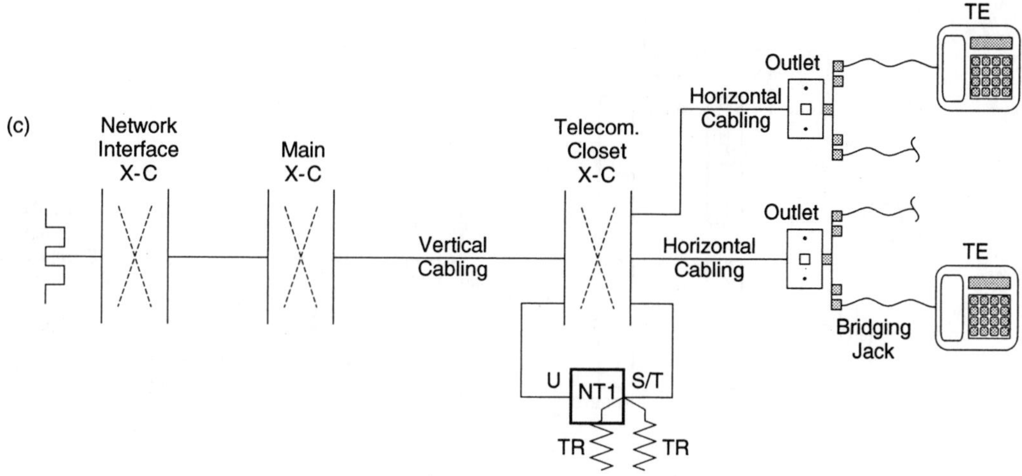

Figure 11-12(c) Application of short branched passive bus configuration in structured wiring system

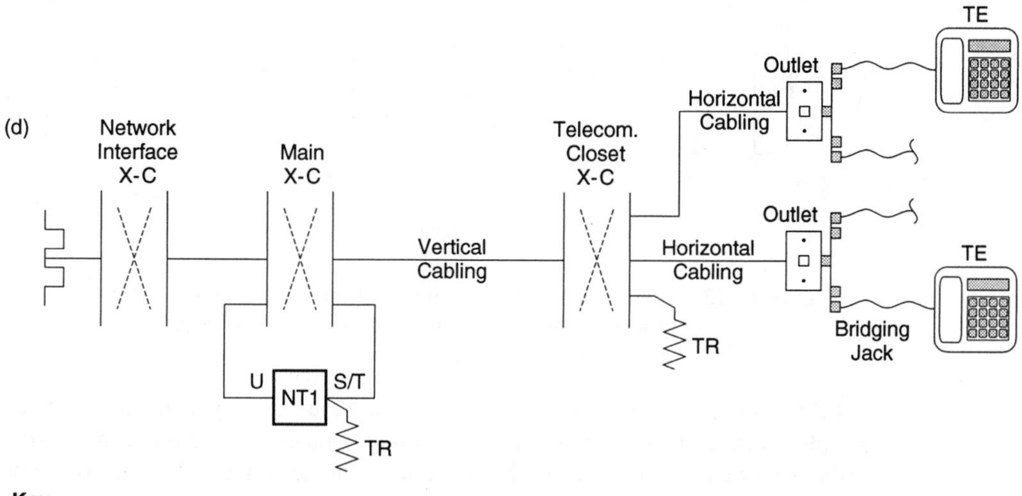

Key
NT1: Network Termination Layer 1
S/T: Switch/Terminal Reference
TE: ISDN Terminal Equipment
TR: Terminating Resistor
U: U Reference
X-C: Cross-Connect

Figure 11-12(d) Application of extended branched passive bus configuration in structured wiring system

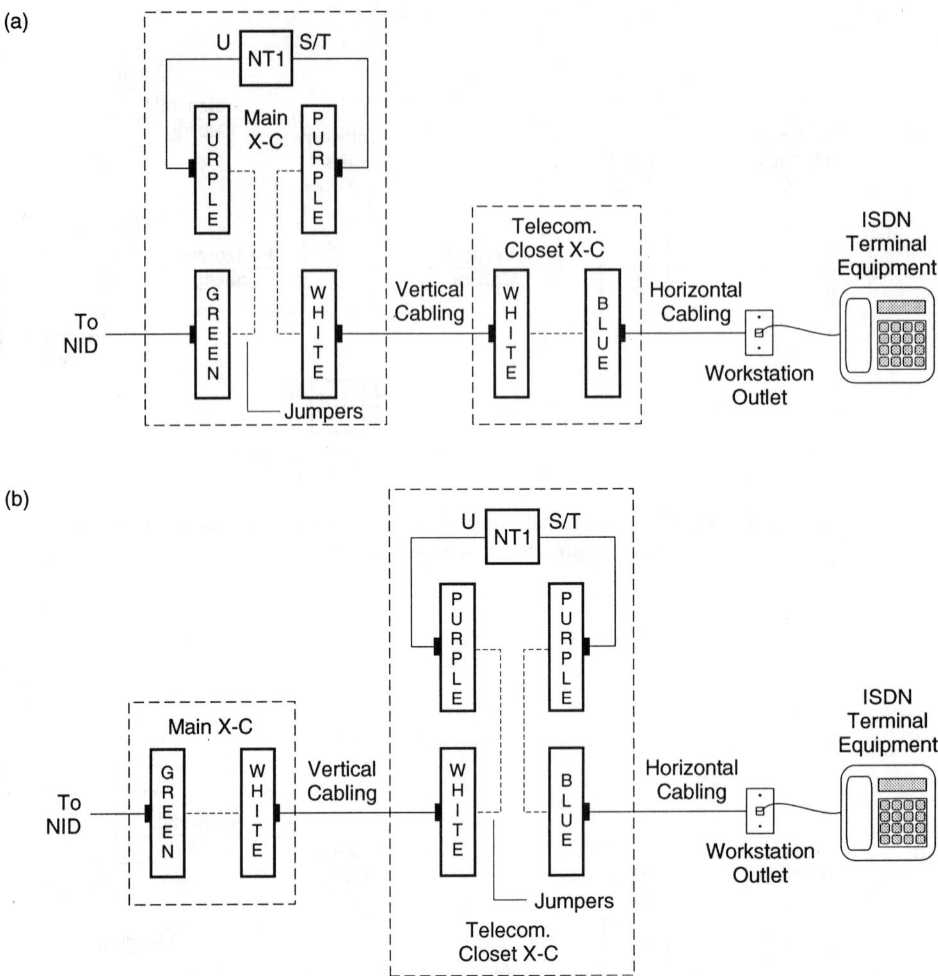

Figure 11-13 Examples of ISDN premises wiring connections: (a) NT1 located in the main equipment room; (b) NT1 located in a telecommunications closet

All ISDN premises connections, including cords, are through standard modular jack-and-plug arrangements. These are the familiar 8-position modular components previously described. Depending on the powering scheme, either four, six, or eight contacts are used (two, three, or four pairs). Each contact is marked either + or −. For the transmit and receive pairs, this indicates the relative polarity of the framing pulse. For the powering pairs, this indicates the relative polarity of the powering voltage.

The sense of transmit and receive direction is maintained throughout the premises wiring. The direction, but not the wiring, is reversed inside the TE, as shown in Table 11-39, to connect the NT transmit circuit to the TE receive circuit. The individual contacts are wired straight through at all connectors; that is, contact No.

Table 11-39 8-Contact Modular ISDN Premises Connector Assignments

Contact No.	Network Termination (NT)	Terminal Equipment (TE)	T568A Color Code	T568B Color Code
1	Power sink 3+	Power source 3+	P3, W-G	P2, W-O
2	Power sink 3−	Power source 3−	P3, G-W	P2, O-W
3	Receive +	Transmit +	P2, W-O	P3, W-G
4	Transmit +	Receive +	P1, BL-W	P1, BL-W
5	Transmit −	Receive −	P1, W-BL	P1, W-BL
6	Receive −	Transmit −	P2, O-W	P3, G-W
7	Power source 2−	Power sink 2−	P4, W-BR	P4, W-BR
8	Power source 2+	Power sink 2+	P4, BR-W	P4, BR-W

Key

P1 pair 1
P2 pair 2
P3 pair 3
P4 pair 4
W white
BL blue
O orange
G green
BR brown

8 is wired to contact No. 8 at all connectors, and so on. None of the leads is rolled or reversed except on the equipment side of the connector at the TE.

References

[1] National Fire Protection Association (NFPA). *National Electrical Code*. Quincy, MA: 1993.

[2] Federal Communications Commission Rules and Regulations. *Title 47 Code of Federal Regulations, Part 68*.

[3] Electronic Industries Association. *Commercial Building Telecommunications Wiring Standard*. EIA/TIA-568, July 1991. Available from EIA.

[4] Electronic Industries Association. *Commercial Building Standard for Telecommunications Pathways and Spaces*. EIA/TIA-569, Oct. 1990. Available from EIA.

[5] Electronic Industries Association. *Residential and Light Commercial Telecommunications Wiring Standard*. EIA/TIA-570, June 1991. Available from EIA.

[6] *Administration Standard for the Telecommunications Infrastructure of Commercial Buildings*. TIA/EIA-606, Feb. 1993.

[7] *IEEE Supplement to Carrier Sense Multiple Access with Collision Detection (CSMA/CD) and Physical Layer Specifications: Layer Management*. IEEE Std 802.3h-1990.

[8] *IEEE Supplement to Carrier Sense Multiple Access with Collision Detection (CSMA/CD) and Physical Layer Specifications: Systems Considerations for Multisegment 10 Mb/s Baseband Networks and Twisted-Pair Medium Attachment Unit and Baseband Medium, Type 10BASE-T*. IEEE Std 802.3i-1990.

[9] American National Standard for Telecommunications. *Insulated Cable Engineers Association Standard for Fiber Optic Premises Distribution Cable*. ANSI/ICEA-83-596-1988.

[10] Kaufman, S. "The 1990 National Electrical Code—Its Impact on the Communications Industry." *IAEI News*, Jan./Feb. 1990, pp. 34–39.

[11] *Telecommunications Distribution Methods Manual*. Building Industry Consulting Service International (BICSI), 1993.

[12] McDermott, P. *Twisted Pair Cable: Network Case Studies*. Paper given at Building Industry Consulting Service International Western Conference in Phoenix, AZ, Sept. 1992.

[13] *IEEE Standard for Local Area Networks: Token Ring Access Method and Physical Layer Specifications*. IEEE Std 802.5-1989.

[14] *OPEN DECconnect Building Wiring Components and Applications Catalog*. Digital Equipment Corporation, Nov. 1991.

[15] *UL's LAN Cable Certification Program*. Publication 200-120. Underwriters Laboratories, Inc., Nov. 1991. Available from UL.

[16] *Additional Cable Specifications for Unshielded Twisted Pair Cables*. Technical Systems Bulletin. EIA/TIA TSB-36, Nov. 1991.

[17] *Performance Standard for Premises Telecommunications Cables,* Draft NEMA WC 63-19XX. National Electrical Manufacturers Association, July 1992.

[18] Insulated Cable Engineers Association. *Standard for Communications Wire and Cable for Wiring of Premises*. ANSI/ICEA-80-576-1988.

[19] Electronic Industries Association. *Additional Transmission Specifications for Unshielded Twisted-Pair Connecting Hardware*. EIA/TIA TSB-40, Aug. 1992.

[20] *Generic Requirements for Network Inside Wiring Cable*. BELLCORE Technical Reference TR-NWT-000133, March 1993.

[21] Reeve, W. "Telecommunications in Rural Alaska, Part III—Terminal Equipment." *The Northern Engineer,* Vol. 23, No. 4 and Vol. 24, No. 1, Winter 1991/Spring 1992.

[22] "Sunshine, Attenuation and TEFLON® GE Solves Net-Failure Puzzle." *Communication News,* Jan. 1991.

[23] International Standards Organization. *Information Processing Systems: Interface Connector and Contact Assignments for ISDN Basic Access Interface Located at Reference Points 'S' and 'T.'* ISO-8877.

[24] American National Standard for Telecommunications. *Integrated Services Digital Network (ISDN)—Basic Access Interface for S and T Reference Points (Layer 1 Specification)*. ANSI T1.605-1991.

A Twisted Pair Cable Designation Cross Reference*

Each twisted pair cable manufacturer has a unique designation for each cable type, even though the cables may be identical, physically and electrically, to another manufacturer's products. These codes are used for ordering or in manufacturing and record keeping. Perhaps the most well known manufacturer is AT&T. Because it is used almost in a generic sense in telecommunication technical publications, the nomenclature used by AT&T will be described in this appendix. The AT&T coding scheme identifies:

- Cable design
- Conductor insulation
- Conductor gauge and material
- Sheath (jacket) type
- Pair count
- Outer protection

Coding is as follows:

a b c d—eeee—f

Cable Design (**a**—first letter of AT&T coding):

A—PIC filled, PIC riser, or pulp air core, 83 nF/mile
B—PIC air core, 83 nF/mile
C—Pulp MUP, pseudo-MUP, or high potential water resistant, 83 nF/mile
D—PIC—Steampeth or Ductpic®, 83 nF/mile or high-density (MAXPAC®), 87 nF/mile
[MAXPAC® is a high-density, 26 AWG cable for congested duct applications.]
G—High potential water resistant, 83 nF/mile
K—Compartmental (screened) core, 83 nF/mile
L—Low capacitance
M—Low capacitance, compartmental core (ICOT®)
N—Non-color coded

Conductor Insulation Type (**b**—second letter of AT&T coding):

B—Solid polyethylene-polyvinyl chloride (PE-PVC)

*Much of the information in this appendix was taken from [1].

C—Expanded polypropylene or dual expanded high-density polyethylene (both generically called dual expanded polyolefin or dual expanded foam)
D—Pulp
F—Filled dual expanded polyolefin with proprietary AT&T filling compound
G—Filled solid polyolefin with proprietary AT&T filling compound
H—Air core solid polyolefin, dielectric strength, per Table A-1 below
J—Filled solid polyolefin
K—Air core solid polyolefin, dielectric strength, per Table A-1 below
L—Filled dual expanded polyolefin
R—Expanded polyethylene-polyvinyl chloride (EPE-PVC)
S—350 V rms dielectric strength (designation not used with current cables)
N—500 V rms dielectric strength (designation not used with current cables)

Polyolefin material is generally high-density polyethylene, but insulation type J may be polypropylene. Insulation types H and K are stabilized polypropylene when used with Cable Design D (Steampeth).

Conductor Gauge and Material (**c**—third letter of AT&T coding):

A—22 AWG copper
B—19 AWG copper
C—17 AWG aluminum
D—20 AWG aluminum
F—22 AWG aluminum
K—24 AWG aluminum
M—24 AWG copper
R—25 AWG copper
T—26 AWG copper
W—28 AWG copper

Jacket (Sheath) Designation (**d**—fourth letter of AT&T coding):

A—Aluminum shield, polyethylene jacket (Alpeth)
C—Aluminum shield, steel armor, polyethylene jacket (Stalpeth)
D—Lead covered polyethylene jacket (Lepeth)
E—Polyethylene jacketed lead
F—Polyethylene jacketed Lepeth
G—Polyethylene, aluminum shield, polyethylene jacket (PAP)

Table A-1 Conductor Insulation Dielectric Strength for Insulation H & K

Designation	Gauge	Dielectric Strength
H	19	10,000 Vdc
H	22	8,000 Vdc
K	24	5,000 Vdc
K	26	3,000 Vdc

H—Polyethylene, bonded or nonbonded aluminum shield, steel armor, polyethylene jacket (PASP and bonded PASP)
J—Toll grade, polyethylene jacket (Tolpeth)
K—Toll grade, polyethylene jacket (Tolpeth)
L—Lead jacket
M—Aluminum shield, polyvinyl chloride (PVC) jacket (riser)
N—Bonded aluminum shield, Steampeth (see V)
P—Reinforced self-supporting Alpeth (see A)
S—Self-supporting Alpeth (see A)
T—Resin coated aluminum shield, polyethylene, aluminum, polyethylene (ARPAP)
U—Resin coated aluminum shield, polyethylene, aluminum, steel armor, polyethylene (ARPASP)
V—Aluminum shield, steel armor, polyethylene, polybutylene jacket (Steampeth)
W—Aluminum shield, steel armor, polyethylene jacket (ASP)
Y—Bonded aluminum shield, steel armor, polyethylene jacket (bonded ASP)
Z—Bonded Stalpeth (see C)

["Bonded" refers to bonding the underlying metal layer to the outer polyethyle jacket with a copolymer coating.]

The typical applications for each jacket type are shown in Table A-2.

Table A-2 Jacket Applications

JACKET DESIGNATION	TYPE	INDOOR RISER	AERIAL	DIRECT BURIED	UNDERGROUND
A	Alpeth		√		
C	Stalpeth				√
D	Lepeth		√		√
E	Jacketed lead			√	√
F	Jacketed lepeth			√	√
G	PAP			√	
H	PASP			√	
J	Tolpeth				
K	Tolpeth				
L	Lead				√
M	Riser	√			
N	Steampeth				√
P	Self-supporting		√		
S	Self-supporting		√		
T	ARPAP			√	√
U	ARPASP			√	
V	Steampeth				√
W	ASP			√	√
Y	ASP			√	√
Z	Stalpeth				√

Pair Count (**eeee**—fifth through eighth letters of AT&T coding):

Standard pair counts available:
6, 12, 18, 25, 37, 50, 75, 100, 150, 200, 300, 400, 600, 900, 1,200, 1,500, 1,800, 2,100, 2,400, 2,700, 3,000, 3,600, 4,200

Not all pair counts are available in all cable types and designations. For example, filled PIC is normally not available in pair counts larger than 1,200 except on special order.

Outer Protection (**f**—last letter of AT&T coding):

AT—Aerial tape armor [jute—a tough fiber obtained from the inner bark of the jute plant, two steel tapes, polyethylene]
BT—Buried tape armor
GT—Gopher-resistant tape armor, jute
MG—Modified gopher-resistant tape armor, polyethylene
MP—Soldered steel mechanical protection, polyethylene
UM—Unsoldered steel mechanical protection, polyethylene
LA—Light wire armor (jute, galvanized wire, jute)
SA—Submarine, single armor (same as LA but with heavier wire)
DA—Submarine, double armor (jute, galvanized steel wire, jute, galvanized steel wire, jute)

The application of outer protection types is shown in Table A-3.

Cables manufactured by AT&T to meet REA specification requirements use a different identification coding scheme, as shown below.

a b c d e—ffff

Cable Design (**a**—first letter of AT&T coding for REA):

K—Compartmental (screened) core
W—Standard exchange cable

Table A-3 Outer Protection Applications

Protection Type	Application
AT	Aerial cables requiring mechanical protection
BT	Buried cables requiring mechanical protection
GT	For lead cables requiring gopher protection
MG	For buried lead or Lepeth toll cables
P	For cables pulled through bored holes in roadways
UM	Replacement for MP on some cable types
LA	For cables installed in small streams and across gullies
SA	For cables installed in navigable lakes, streams, and coastal waterways
DA	For cables requiring more protection than type SA, such as rocky bottom or strong current

Core Type and Conductor Insulation (**b**—second letter of AT&T coding for REA):

A—Filled dual expanded polyolefin with proprietary AT&T filling compound, REA specification PE-89
B—Solid polyolefin, air core, REA specification PE-22
G—Filled solid polyolefin with proprietary AT&T filling compound, REA specification PE-39

Conductor Gauge and Material (**c**—third letter of AT&T coding for REA):

2—22 AWG copper
4—24 AWG copper
6—26 AWG copper
9—19 AWG copper

Jacket Material (**d**—fourth letter of AT&T coding for REA):

A—Coated aluminum shield, polyethylene jacket (CAP)
C—5 mil copper shield, polyethylene jacket
J—Copper alloy 194 armor and shield, polyethylene jacket
W—Coated aluminum shield, coated steel armor, polyethylene jacket (CACSP)

Listing (**e**—fifth letter of AT&T coding for REA):

R—REA listed

Pair Count (**ffff**—last four letters of AT&T coding for REA):

Standard pair sizes as shown above for fifth through eighth letters of AT&T coding.

Other industry designations are used with telecommunication cables, as shown below. These are very similar to those already discussed, but are more generalized. See General Telephone and Electronics Practice GTEP 920-000-070.

- Stalpeth
- Stalpeth double sheath (Stalpeth DS)
- Alvyn
- Alpeth-FPA, air core
- Alpeth-FPA, filled
- Alpeth-FPA-Figure 8, air core
- Alpeth-FPA, D-shield, filled
- Alpeth-FPA, D-shield, air core
- Alpeth-FPA, T-screen, filled
- Alpeth-FPA, T-screen, air core
- FPA-ASP, filled
- Alpeth-FPA-DS

The PIC cable core layouts differ between AT&T cables made to BELLCORE specifications and those made to REA and other common specifications in the larger pair sizes. BELLCORE specifications use what is called a "mirror-image" layout. Other specifications use a consistent 25-pair binder group and super-unit binder groups.

Pulp insulated cable cores follow either a layer or unit construction. Layer constructions use a variable pair count in each layer to achieve the desired overall pair count. Unit constructions use 25-pair units (19 AWG only), 50-pair units (22, 24, and 26 AWG), or 100-pair units (24 and 26 AWG), depending on the total pair count.

Reference

[1] *Outside Plant Engineering Handbook*. AT&T Document Development Organization, Winston-Salem, NC, Jan. 1990. Available from AT&T Customer Information Center.

B Exchange Cable Transmission Characteristics

The data for the following tables were taken from Appendix A of BELLCORE Technical Reference TR-NPL-000157. Typographical errors were corrected and the unit lengths adjusted to 1,000 ft (kft).* The data in the technical reference also appear in various forms in a number of other industry standards (all with unit lengths of one mile).

The data tables include, for each of the four common cable gauges (26, 24, 22, and 19 AWG) at temperatures of 120 °F (+49 °C), 70 °F (+21 °C), and 0 °F (−18 °C) and for frequencies from 1 Hz to 5 MHz:

- Primary constants (resistance R, inductance L, conductance G, and capacitance C)
- Secondary parameters characteristic impedance (real and imaginary parts, denoted Re Z and Im Z, respectively, in the tables) and propagation constant $\alpha + j\beta$ (attenuation α and phase delay β)

This information can be considered to be representative of polyolefin insulated cables (PIC) and pulp insulated cables in the field. However, some differences will be noted between the data in these tables and specific cable types, especially at higher frequencies. The following provides a description of each table title to locate particular combinations of gauges, constructions, and temperatures.

Table Title	Description	Table Title	Description
B 26-PIC-120	26 AWG PIC at 120 °F	**B** 26-PULP-120	26 AWG Pulp at 120 °F
B 26-PIC-70	26 AWG PIC at 70 °F	**B** 26-PULP-70	26 AWG Pulp at 70 °F
B 26-PIC-0	26 AWG PIC at 0 °F	**B** 26-PULP-0	26 AWG Pulp at 0 °F
B 24-PIC-120	24 AWG PIC at 120 °F	**B** 24-PULP-120	24 AWG Pulp at 120 °F
B 24-PIC-70	24 AWG PIC at 70 °F	**B** 24-PULP-70	24 AWG Pulp at 70 °F
B 24-PIC-0	24 AWG PIC at 0 °F	**B** 24-PULP-0	24 AWG Pulp at 0 °F
B 22-PIC-120	22 AWG PIC at 120 °F	**B** 22-PULP-120	22 AWG Pulp at 120 °F
B 22-PIC-70	22 AWG PIC at 70 °F	**B** 22-PULP-70	22 AWG Pulp at 70 °F
B 22-PIC-0	22 AWG PIC at 0 °F	**B** 22-PULP-0	22 AWG Pulp at 0 °F
B 19-PIC-120	19 AWG PIC at 120 °F	**B** 19-PULP-120	19 AWG Pulp at 120 °F
B 19-PIC-70	19 AWG PIC at 70 °F	**B** 19-PULP-70	19 AWG Pulp at 70 °F
B 19-PIC-0	19 AWG PIC at 0 °F	**B** 19-PULP-0	19 AWG Pulp at 0 °F

**Secondary Channel in the Digital Data System: Channel Interface Requirements*, BELLCORE Technical Reference TR-NPL-000157, Apr. 1986.

Table B 26-PIC-120

FREQ. (Hz)	R (Ω/kft)	L (mH/kft)	G (μS/kft)	C (nF/kft)	Re Z (Ω)	Im Z (Ω)	α (dB/kft)	β (Deg/kft)
1	92.58	0.1882	0.000	15.72	21,654.80	−21,642.81	0.019	0.12
5	92.58	0.1882	0.000	15.72	9,684.07	−9,679.21	0.041	0.27
10	92.58	0.1882	0.000	15.72	6,847.76	−6,844.14	0.059	0.39
15	92.58	0.1882	0.001	15.72	5,591.30	−5,588.10	0.072	0.48
20	92.58	0.1882	0.001	15.72	4,842.33	−4,839.32	0.083	0.55
30	92.58	0.1882	0.001	15.72	3,953.96	−3,951.07	0.102	0.67
50	92.58	0.1882	0.002	15.72	3,063.08	−3,060.13	0.131	0.87
70	92.58	0.1882	0.002	15.72	2,589.09	−2,585.97	0.155	1.02
100	92.58	0.1882	0.003	15.72	2,166.59	−2,163.18	0.186	1.23
150	92.58	0.1882	0.004	15.72	1,769.57	−1,765.68	0.227	1.50
200	92.58	0.1881	0.005	15.72	1,532.97	−1,528.65	0.262	1.73
300	92.58	0.1881	0.008	15.72	1,252.46	−1,247.35	0.321	2.13
500	92.58	0.1881	0.012	15.72	971.39	−964.98	0.414	2.75
700	92.59	0.1881	0.016	15.72	822.03	−814.52	0.489	3.26
1,000	92.59	0.1881	0.022	15.72	689.08	−680.19	0.584	3.90
1,500	92.59	0.1881	0.031	15.72	564.44	−553.62	0.713	4.79
2,000	92.60	0.1880	0.040	15.72	490.39	−477.94	0.820	5.55
3,000	92.61	0.1880	0.057	15.72	402.98	−387.78	0.998	6.84
5,000	92.63	0.1878	0.088	15.72	316.18	−296.63	1.273	8.95
7,000	92.66	0.1877	0.118	15.72	270.66	−247.58	1.487	10.72
10,000	92.71	0.1875	0.162	15.72	230.83	−203.32	1.745	13.06
15,000	92.82	0.1871	0.230	15.72	194.53	−161.02	2.073	16.51
20,000	92.94	0.1868	0.295	15.72	173.80	−135.34	2.323	19.67
30,000	93.24	0.1861	0.420	15.72	150.77	−104.35	2.686	25.59
50,000	94.06	0.1843	0.655	15.72	130.57	−72.93	3.129	36.94
70,000	95.17	0.1821	0.878	15.72	121.62	−56.59	3.399	48.17
100,000	97.34	0.1800	1.197	15.72	115.23	−42.76	3.669	65.21
150,000	101.56	0.1776	1.703	15.72	110.70	−30.96	3.985	93.97
200,000	106.40	0.1758	2.188	15.72	108.62	−24.79	4.255	122.93
300,000	117.92	0.1731	3.113	15.72	106.58	−18.66	4.806	180.95
500,000	141.35	0.1688	4.855	15.72	104.51	−13.69	5.876	295.71
700,000	163.30	0.1651	6.506	15.72	103.12	−11.45	6.880	408.50
1,000,000	192.04	0.1609	8.873	15.72	101.62	−9.56	8.212	575.06
1,500,000	231.57	0.1566	12.626	15.72	100.13	−7.80	10.050	849.94
2,000,000	264.88	0.1540	16.217	15.72	99.22	−6.75	11.601	1,122.96
3,000,000	320.71	0.1509	23.076	15.72	98.12	−5.51	14.205	1,665.79
5,000,000	409.18	0.1476	35.989	15.72	97.00	−4.27	18.336	2,744.56

Table B 26-PIC-70

FREQ. (Hz)	R (Ω/kft)	L (mH/kft)	G (μS/kft)	C (nF/kft)	Re Z (Ω)	Im Z (Ω)	α (dB/kft)	β (Deg/kft)
1	83.48	0.1868	0.000	15.72	20,562.46	−20,551.04	0.018	0.12
5	83.48	0.1868	0.000	15.72	9,195.60	−9,190.92	0.039	0.26
10	83.48	0.1868	0.000	15.72	6,502.38	−6,498.86	0.056	0.37
15	83.48	0.1868	0.001	15.72	5,309.30	−5,306.16	0.068	0.45
20	83.48	0.1868	0.001	15.72	4,598.12	−4,595.14	0.079	0.52
30	83.48	0.1868	0.001	15.72	3,754.58	−3,751.69	0.097	0.64
50	83.48	0.1868	0.002	15.72	2,908.66	−2,905.67	0.125	0.82
70	83.48	0.1868	0.002	15.72	2,458.60	−2,455.41	0.148	0.97
100	83.48	0.1868	0.003	15.72	2,057.43	−2,053.93	0.176	1.16
150	83.48	0.1868	0.004	15.72	1,680.47	−1,676.45	0.216	1.43
200	83.48	0.1867	0.005	15.72	1,455.83	−1,451.35	0.249	1.65
300	83.48	0.1867	0.008	15.72	1,189.51	−1,184.20	0.305	2.02
500	83.48	0.1867	0.012	15.72	922.69	−916.00	0.393	2.61
700	83.48	0.1867	0.016	15.72	780.91	−773.09	0.464	3.09
1,000	83.48	0.1867	0.022	15.72	654.74	−645.46	0.554	3.70
1,500	83.49	0.1867	0.031	15.72	536.48	−525.19	0.676	4.55
2,000	83.49	0.1866	0.040	15.72	466.25	−453.25	0.778	5.28
3,000	83.50	0.1866	0.057	15.72	383.39	−367.51	0.946	6.51
5,000	83.52	0.1864	0.088	15.72	301.19	−280.76	1.205	8.52
7,000	83.55	0.1863	0.118	15.72	258.16	−234.04	1.406	10.23
10,000	83.60	0.1861	0.162	15.72	220.58	−191.85	1.646	12.48
15,000	83.69	0.1857	0.230	15.72	186.46	−151.47	1.950	15.83
20,000	83.80	0.1854	0.295	15.72	167.08	−126.94	2.179	18.91
30,000	84.07	0.1847	0.420	15.72	145.71	−97.35	2.506	24.73
50,000	84.81	0.1830	0.655	15.72	127.25	−67.47	2.895	36.00
70,000	85.81	0.1808	0.878	15.72	119.21	−52.05	3.127	47.22
100,000	87.76	0.1786	1.197	15.72	113.56	−39.12	3.357	64.26
150,000	92.01	0.1763	1.703	15.72	109.61	−28.32	3.646	93.04
200,000	97.17	0.1745	2.188	15.72	107.80	−22.81	3.916	122.00
300,000	108.93	0.1716	3.113	15.72	105.92	−17.35	4.468	179.83
500,000	132.50	0.1670	4.855	15.72	103.87	−12.91	5.542	293.90
700,000	153.97	0.1631	6.506	15.72	102.45	−10.86	6.530	405.84
1,000,000	181.18	0.1587	8.873	15.72	100.90	−9.09	7.803	570.98
1,500,000	218.63	0.1543	12.626	15.72	99.35	−7.42	9.563	843.32
2,000,000	250.20	0.1515	16.217	15.72	98.39	−6.43	11.051	1,113.61
3,000,000	303.16	0.1482	23.076	15.72	97.23	−5.26	13.551	1,650.63
5,000,000	387.13	0.1447	35.989	15.72	96.02	−4.08	17.525	2,716.85

Table B 26-PIC-0

FREQ. (Hz)	R (Ω/kft)	L (mH/kft)	G (μS/kft)	C (nF/kft)	Re Z (Ω)	Im Z (Ω)	α (dB/kft)	β (Deg/kft)
1	70.73	0.1848	0.000	15.72	18,927.57	−18,917.02	0.016	0.11
5	70.73	0.1848	0.000	15.72	8,464.51	−8,460.10	0.036	0.24
10	70.73	0.1848	0.000	15.72	5,985.45	−5,982.07	0.051	0.34
15	70.73	0.1848	0.001	15.72	4,887.25	−4,884.19	0.063	0.41
20	70.73	0.1848	0.001	15.72	4,232.63	−4,229.69	0.073	0.48
30	70.73	0.1848	0.001	15.72	3,456.18	−3,453.27	0.089	0.59
50	70.73	0.1848	0.002	15.72	2,677.55	−2,674.49	0.115	0.76
70	70.73	0.1848	0.002	15.72	2,263.30	−2,260.00	0.136	0.90
100	70.73	0.1848	0.003	15.72	1,894.07	−1,890.40	0.162	1.07
150	70.73	0.1848	0.004	15.72	1,547.13	−1,542.89	0.199	1.31
200	70.73	0.1848	0.005	15.72	1,340.40	−1,335.64	0.229	1.52
300	70.73	0.1848	0.008	15.72	1,095.32	−1,089.66	0.280	1.86
500	70.73	0.1848	0.012	15.72	849.83	−842.68	0.362	2.40
700	70.73	0.1848	0.016	15.72	719.41	−711.03	0.427	2.85
1,000	70.73	0.1847	0.022	15.72	603.39	−593.44	0.509	3.41
1,500	70.74	0.1847	0.031	15.72	494.70	−482.58	0.621	4.20
2,000	70.74	0.1847	0.040	15.72	430.19	−416.23	0.714	4.87
3,000	70.75	0.1846	0.057	15.72	354.15	−337.10	0.868	6.01
5,000	70.77	0.1845	0.088	15.72	278.88	−256.93	1.102	7.89
7,000	70.79	0.1843	0.118	15.72	239.59	−213.68	1.284	9.49
10,000	70.83	0.1841	0.162	15.72	205.41	−174.55	1.498	11.62
15,000	70.91	0.1838	0.230	15.72	174.60	−137.06	1.764	14.82
20,000	71.00	0.1835	0.295	15.72	157.27	−114.27	1.961	17.80
30,000	71.23	0.1828	0.420	15.72	138.45	−86.81	2.235	23.50
50,000	71.86	0.1810	0.655	15.72	122.63	−59.32	2.545	34.70
70,000	72.71	0.1789	0.878	15.72	115.92	−45.35	2.724	45.92
100,000	74.36	0.1768	1.197	15.72	111.30	−33.81	2.902	62.99
150,000	78.63	0.1744	1.703	15.72	108.15	−24.53	3.158	91.80
200,000	84.24	0.1726	2.188	15.72	106.69	−19.98	3.430	120.75
300,000	96.35	0.1696	3.113	15.72	105.02	−15.48	3.986	178.29
500,000	120.12	0.1645	4.855	15.72	102.97	−11.81	5.068	291.36
700,000	140.91	0.1604	6.506	15.72	101.51	−10.03	6.031	402.10
1,000,000	165.98	0.1557	8.873	15.72	99.88	−8.41	7.221	565.24
1,500,000	200.52	0.1510	12.626	15.72	98.25	−6.88	8.869	833.97
2,000,000	229.66	0.1480	16.217	15.72	97.22	−5.97	10.266	1,100.40
3,000,000	278.59	0.1444	23.076	15.72	95.96	−4.89	12.617	1,629.20
5,000,000	356.27	0.1405	35.989	15.72	94.63	−3.81	16.366	2,677.58

Table B 24-PIC-120

Freq. (Hz)	R (Ω/kft)	L (mH/kft)	G (μS/kft)	C (nF/kft)	Re Z (Ω)	Im Z (Ω)	α (dB/kft)	β (Deg/kft)
1	58.23	0.1882	0.000	15.72	17,173.07	−17,163.43	0.015	0.10
5	58.23	0.1882	0.000	15.72	7,679.95	−7,675.80	0.033	0.22
10	58.23	0.1882	0.000	15.72	5,430.72	−5,427.44	0.047	0.31
15	58.23	0.1882	0.001	15.72	4,434.34	−4,431.31	0.057	0.38
20	58.23	0.1882	0.001	15.72	3,840.42	−3,837.46	0.066	0.43
30	58.23	0.1882	0.001	15.72	3,135.98	−3,132.98	0.081	0.53
50	58.23	0.1882	0.002	15.72	2,429.59	−2,426.33	0.104	0.69
70	58.23	0.1882	0.002	15.72	2,053.78	−2,050.23	0.123	0.81
100	58.23	0.1882	0.003	15.72	1,718.83	−1,714.83	0.147	0.97
150	58.23	0.1881	0.004	15.72	1,404.12	−1,399.46	0.180	1.19
200	58.23	0.1881	0.005	15.72	1,216.62	−1,211.36	0.208	1.38
300	58.23	0.1881	0.008	15.72	994.38	−988.08	0.254	1.69
500	58.23	0.1881	0.012	15.72	771.81	−763.83	0.328	2.18
700	58.23	0.1881	0.016	15.72	653.63	−644.26	0.387	2.59
1,000	58.23	0.1881	0.022	15.72	548.54	−537.40	0.461	3.10
1,500	58.24	0.1880	0.031	15.72	450.17	−436.59	0.562	3.82
2,000	58.24	0.1880	0.040	15.72	391.86	−376.21	0.646	4.43
3,000	58.26	0.1879	0.057	15.72	323.23	−304.12	0.783	5.49
5,000	58.29	0.1877	0.088	15.72	255.53	−230.94	0.991	7.23
7,000	58.32	0.1876	0.118	15.72	220.39	−191.37	1.149	8.73
10,000	58.38	0.1873	0.162	15.72	190.03	−155.53	1.334	10.75
15,000	58.52	0.1869	0.230	15.72	163.00	−121.15	1.559	13.84
20,000	58.68	0.1864	0.295	15.72	148.07	−100.30	1.721	16.76
30,000	59.09	0.1855	0.420	15.72	132.24	−75.39	1.941	22.45
50,000	60.25	0.1827	0.655	15.72	119.33	−51.11	2.193	33.76
70,000	61.82	0.1806	0.878	15.72	114.12	−39.17	2.353	45.20
100,000	64.38	0.1784	1.197	15.72	110.52	−29.48	2.530	62.54
150,000	69.59	0.1756	1.703	15.72	107.92	−21.76	2.801	91.60
200,000	75.53	0.1736	2.188	15.72	106.61	−17.93	3.078	120.66
300,000	87.31	0.1700	3.113	15.72	104.95	−14.03	3.614	178.17
500,000	108.79	0.1644	4.855	15.72	102.81	−10.71	4.598	290.91
700,000	126.86	0.1604	6.506	15.72	101.40	−9.04	5.436	401.69
1,000,000	149.64	0.1567	8.873	15.72	100.12	−7.56	6.495	566.59
1,500,000	180.97	0.1531	12.626	15.72	98.89	−6.17	7.953	839.40
2,000,000	207.36	0.1509	16.217	15.72	98.14	−5.34	9.183	1,110.71
3,000,000	251.60	0.1483	23.076	15.72	97.23	−4.36	11.248	1,650.73
5,000,000	321.70	0.1464	35.989	15.72	96.56	−3.37	14.484	2,732.16

Table B 24-PIC-70

FREQ. (Hz)	R (Ω/kft)	L (mH/kft)	G (μS/kft)	C (nF/kft)	Re Z (Ω)	Im Z (Ω)	α (dB/kft)	β (Deg/kft)
1	52.50	0.1868	0.000	15.72	16,306.82	−16,297.63	0.014	0.09
5	52.50	0.1868	0.000	15.72	7,292.58	−7,288.57	0.031	0.21
10	52.50	0.1868	0.000	15.72	5,156.83	−5,153.61	0.044	0.29
15	52.50	0.1868	0.001	15.72	4,210.72	−4,207.71	0.054	0.36
20	52.50	0.1868	0.001	15.72	3,646.77	−3,643.81	0.063	0.41
30	52.50	0.1868	0.001	15.72	2,977.88	−2,974.85	0.077	0.51
50	52.50	0.1868	0.002	15.72	2,307.15	−2,303.82	0.099	0.65
70	52.50	0.1868	0.002	15.72	1,950.32	−1,946.67	0.117	0.77
100	52.50	0.1868	0.003	15.72	1,632.29	−1,628.16	0.140	0.92
150	52.50	0.1867	0.004	15.72	1,333.50	−1,328.66	0.171	1.13
200	52.50	0.1867	0.005	15.72	1,155.49	−1,150.02	0.197	1.31
300	52.50	0.1867	0.008	15.72	944.50	−937.95	0.241	1.60
500	52.50	0.1867	0.012	15.72	733.25	−724.93	0.311	2.07
700	52.50	0.1867	0.016	15.72	621.10	−611.32	0.367	2.46
1,000	52.51	0.1867	0.022	15.72	521.40	−509.77	0.438	2.95
1,500	52.51	0.1866	0.031	15.72	428.12	−413.93	0.533	3.63
2,000	52.52	0.1866	0.040	15.72	372.85	−356.50	0.612	4.22
3,000	52.53	0.1865	0.057	15.72	307.87	−287.90	0.741	5.23
5,000	52.55	0.1863	0.088	15.72	243.87	−218.18	0.936	6.90
7,000	52.59	0.1862	0.118	15.72	210.75	−180.44	1.084	8.35
10,000	52.64	0.1859	0.162	15.72	182.25	−146.22	1.255	10.31
15,000	52.76	0.1855	0.230	15.72	157.04	−113.38	1.459	13.33
20,000	52.91	0.1850	0.295	15.72	143.23	−93.49	1.605	16.21
30,000	53.28	0.1841	0.420	15.72	128.79	−69.79	1.797	21.86
50,000	54.32	0.1814	0.655	15.72	117.22	−46.91	2.013	33.17
70,000	55.74	0.1792	0.878	15.72	112.62	−35.78	2.150	44.61
100,000	58.41	0.1770	1.197	15.72	109.50	−27.00	2.317	61.97
150,000	63.87	0.1743	1.703	15.72	107.21	−20.10	2.588	91.00
200,000	69.89	0.1721	2.188	15.72	105.96	−16.69	2.866	119.92
300,000	81.73	0.1683	3.113	15.72	104.31	−13.22	3.404	177.09
500,000	102.59	0.1623	4.855	15.72	102.12	−10.17	4.365	288.96
700,000	119.71	0.1581	6.506	15.72	100.67	−8.60	5.167	398.77
1,000,000	141.30	0.1543	8.873	15.72	99.33	−7.20	6.182	562.12
1,500,000	170.99	0.1505	12.626	15.72	98.03	−5.88	7.581	832.14
2,000,000	196.03	0.1482	16.217	15.72	97.23	−5.10	8.763	1,100.48
3,000,000	238.02	0.1454	23.076	15.72	96.26	−4.17	10.749	1,634.22
5,000,000	304.62	0.1425	35.989	15.72	95.26	−3.23	13.902	2,695.42

Table B 24-PIC-0

Freq. (Hz)	R (Ω/kft)	L (mH/kft)	G (μS/kft)	C (nF/kft)	Re Z (Ω)	Im Z (Ω)	α (dB/kft)	β (Deg/kft)
1	44.48	0.1848	0.000	15.72	15,010.30	−15,001.78	0.013	0.09
5	44.48	0.1848	0.000	15.72	6,712.82	−6,709.00	0.029	0.19
10	44.48	0.1848	0.000	15.72	4,746.90	−4,743.76	0.041	0.27
15	44.48	0.1848	0.001	15.72	3,876.04	−3,873.05	0.050	0.33
20	44.48	0.1848	0.001	15.72	3,356.95	−3,353.96	0.058	0.38
30	44.48	0.1848	0.001	15.72	2,741.27	−2,738.17	0.070	0.47
50	44.48	0.1848	0.002	15.72	2,123.91	−2,120.45	0.091	0.60
70	44.48	0.1848	0.002	15.72	1,795.49	−1,791.65	0.108	0.71
100	44.48	0.1848	0.003	15.72	1,502.79	−1,498.43	0.129	0.85
150	44.48	0.1848	0.004	15.72	1,227.82	−1,222.68	0.157	1.04
200	44.48	0.1848	0.005	15.72	1,064.01	−1,058.19	0.182	1.20
300	44.48	0.1848	0.008	15.72	869.90	−862.89	0.222	1.48
500	44.48	0.1848	0.012	15.72	675.58	−666.66	0.286	1.91
700	44.49	0.1848	0.016	15.72	572.47	−561.98	0.338	2.27
1,000	44.49	0.1847	0.022	15.72	480.85	−468.36	0.402	2.72
1,500	44.49	0.1847	0.031	15.72	395.19	−379.95	0.489	3.35
2,000	44.50	0.1846	0.040	15.72	344.49	−326.93	0.561	3.90
3,000	44.51	0.1845	0.057	15.72	284.98	−263.52	0.678	4.84
5,000	44.53	0.1844	0.088	15.72	226.58	−198.97	0.854	6.41
7,000	44.56	0.1842	0.118	15.72	196.51	−163.97	0.985	7.78
10,000	44.60	0.1840	0.162	15.72	170.82	−132.18	1.134	9.67
15,000	44.70	0.1835	0.230	15.72	148.38	−101.67	1.309	12.59
20,000	44.83	0.1831	0.295	15.72	136.31	−83.24	1.429	15.43
30,000	45.14	0.1822	0.420	15.72	123.97	−61.44	1.582	21.05
50,000	46.03	0.1795	0.655	15.72	114.36	−40.74	1.748	32.36
70,000	47.23	0.1774	0.878	15.72	110.62	−30.87	1.855	43.82
100,000	50.06	0.1752	1.197	15.72	108.13	−23.43	2.011	61.19
150,000	55.86	0.1725	1.703	15.72	106.24	−17.74	2.284	90.18
200,000	61.99	0.1700	2.188	15.72	105.06	−14.93	2.564	118.91
300,000	73.93	0.1658	3.113	15.72	103.42	−12.06	3.106	175.57
500,000	93.92	0.1595	4.855	15.72	101.15	−9.40	4.035	286.21
700,000	109.70	0.1550	6.506	15.72	99.63	−7.96	4.785	394.66
1,000,000	129.61	0.1509	8.873	15.72	98.21	−6.68	5.735	555.80
1,500,000	157.03	0.1469	12.626	15.72	96.82	−5.47	7.049	821.86
2,000,000	180.17	0.1444	16.217	15.72	95.95	−4.75	8.162	1,086.01
3,000,000	219.02	0.1413	23.076	15.72	94.88	−3.89	10.034	1,610.82
5,000,000	280.70	0.1370	35.989	15.72	93.41	−3.04	13.065	2,643.13

Table B 22-PIC-120

Freq. (Hz)	R (Ω/kft)	L (mH/kft)	G (μS/kft)	C (nF/kft)	Re Z (Ω)	Im Z (Ω)	α (dB/kft)	β (Deg/kft)
1	36.61	0.1882	0.000	15.72	13,614.46	−13,610.95	0.012	0.08
5	36.61	0.1882	0.000	15.72	6,088.90	−6,086.67	0.026	0.17
10	36.61	0.1882	0.000	15.72	4,305.83	−4,303.60	0.037	0.24
15	36.61	0.1882	0.000	15.72	3,515.97	−3,513.60	0.045	0.30
20	36.61	0.1882	0.000	15.72	3,045.16	−3,042.63	0.052	0.34
30	36.61	0.1882	0.001	15.72	2,486.76	−2,483.90	0.064	0.42
50	36.61	0.1882	0.001	15.72	1,926.86	−1,923.41	0.083	0.55
70	36.61	0.1882	0.001	15.72	1,629.02	−1,625.06	0.098	0.65
100	36.61	0.1882	0.002	15.72	1,363.59	−1,358.97	0.117	0.77
150	36.61	0.1881	0.002	15.72	1,114.27	−1,108.70	0.143	0.95
200	36.61	0.1881	0.003	15.72	965.77	−959.40	0.165	1.09
300	36.61	0.1881	0.005	15.72	789.83	−782.09	0.201	1.34
500	36.61	0.1881	0.008	15.72	613.79	−603.87	0.259	1.74
700	36.61	0.1881	0.010	15.72	520.43	−508.73	0.305	2.06
1,000	36.61	0.1881	0.014	15.72	437.56	−423.60	0.363	2.48
1,500	36.62	0.1880	0.021	15.72	360.18	−343.12	0.442	3.06
2,000	36.63	0.1880	0.027	15.72	314.47	−294.80	0.506	3.56
3,000	36.64	0.1878	0.040	15.72	260.96	−236.93	0.610	4.43
5,000	36.68	0.1876	0.065	15.72	208.77	−177.87	0.763	5.91
7,000	36.72	0.1874	0.088	15.72	182.16	−145.79	0.876	7.22
10,000	36.80	0.1871	0.123	15.72	159.72	−116.65	1.001	9.04
15,000	36.98	0.1865	0.181	15.72	140.54	−88.80	1.143	11.93
20,000	37.20	0.1859	0.236	15.72	130.52	−72.14	1.238	14.77
30,000	37.78	0.1845	0.345	15.72	120.57	−52.87	1.361	20.47
50,000	39.41	0.1811	0.557	15.72	112.99	−35.31	1.515	31.97
70,000	41.14	0.1788	0.764	15.72	110.04	−27.03	1.624	43.59
100,000	44.41	0.1763	1.066	15.72	107.93	−20.82	1.787	61.08
150,000	50.42	0.1731	1.559	15.72	106.16	−16.02	2.063	90.12
200,000	56.14	0.1703	2.040	15.72	104.97	−13.53	2.323	118.80
300,000	66.96	0.1657	2.982	15.72	103.25	−10.94	2.818	175.29
500,000	84.59	0.1597	4.810	15.72	101.14	−8.46	3.635	286.16
700,000	98.91	0.1563	6.590	15.72	99.97	−7.15	4.300	396.01
1,000,000	116.96	0.1532	9.202	15.72	98.91	−5.98	5.140	559.71
1,500,000	141.78	0.1502	13.450	15.72	97.88	−4.88	6.296	830.90
2,000,000	162.69	0.1484	17.606	15.72	97.27	−4.23	7.271	1,100.93
3,000,000	197.74	0.1470	25.732	15.72	96.75	−3.44	8.887	1,642.48
5,000,000	253.27	0.1455	41.507	15.72	96.26	−2.66	11.445	2,723.69

Table B 22-PIC-70

FREQ. (Hz)	R (Ω/kft)	L (mH/kft)	G (μS/kft)	C (nF/kft)	Re Z (Ω)	Im Z (Ω)	α (dB/kft)	β (Deg/kft)
1	33.01	0.1868	0.000	15.72	12,927.71	−12,924.34	0.011	0.07
5	33.01	0.1868	0.000	15.72	5,781.80	−5,779.59	0.025	0.16
10	33.01	0.1868	0.000	15.72	4,088.69	−4,086.44	0.035	0.23
15	33.01	0.1868	0.000	15.72	3,338.69	−3,336.28	0.043	0.28
20	33.01	0.1868	0.000	15.72	2,891.65	−2,889.05	0.050	0.33
30	33.01	0.1868	0.001	15.72	2,361.43	−2,358.49	0.061	0.40
50	33.01	0.1868	0.001	15.72	1,829.81	−1,826.24	0.078	0.52
70	33.01	0.1868	0.001	15.72	1,547.02	−1,542.91	0.093	0.61
100	33.01	0.1868	0.002	15.72	1,295.02	−1,290.21	0.111	0.73
150	33.01	0.1867	0.002	15.72	1,058.32	−1,052.52	0.135	0.90
200	33.01	0.1867	0.003	15.72	917.35	−910.71	0.156	1.04
300	33.01	0.1867	0.005	15.72	750.35	−742.28	0.191	1.27
500	33.01	0.1867	0.008	15.72	583.30	−572.94	0.246	1.65
700	33.01	0.1867	0.010	15.72	494.74	−482.52	0.290	1.96
1,000	33.01	0.1867	0.014	15.72	416.16	−401.58	0.345	2.35
1,500	33.02	0.1866	0.021	15.72	342.84	−325.02	0.418	2.91
2,000	33.02	0.1866	0.027	15.72	299.57	−279.02	0.479	3.39
3,000	33.04	0.1864	0.040	15.72	249.00	−223.89	0.576	4.23
5,000	33.07	0.1862	0.065	15.72	199.83	−167.56	0.719	5.65
7,000	33.11	0.1860	0.088	15.72	174.89	−136.92	0.822	6.93
10,000	33.19	0.1857	0.123	15.72	154.00	−109.09	0.936	8.71
15,000	33.34	0.1851	0.181	15.72	136.34	−82.53	1.062	11.57
20,000	33.54	0.1845	0.236	15.72	127.25	−66.72	1.145	14.40
30,000	34.06	0.1832	0.345	15.72	118.37	−48.55	1.250	20.09
50,000	35.54	0.1798	0.557	15.72	111.68	−32.21	1.382	31.60
70,000	37.45	0.1775	0.764	15.72	109.12	−24.81	1.491	43.23
100,000	40.82	0.1749	1.066	15.72	107.24	−19.27	1.654	60.69
150,000	46.89	0.1715	1.559	15.72	105.52	−14.99	1.930	89.57
200,000	52.64	0.1685	2.040	15.72	104.33	−12.77	2.192	118.08
300,000	63.14	0.1637	2.982	15.72	102.56	−10.38	2.675	174.13
500,000	79.84	0.1574	4.810	15.72	100.38	−8.05	3.456	284.02
700,000	93.42	0.1538	6.590	15.72	99.16	−6.81	4.094	392.81
1,000,000	110.53	0.1506	9.202	15.72	98.04	−5.70	4.900	554.80
1,500,000	134.07	0.1474	13.450	15.72	96.95	−4.66	6.012	822.97
2,000,000	153.92	0.1455	17.606	15.72	96.28	−4.04	6.950	1,089.74
3,000,000	187.22	0.1431	25.732	15.72	95.48	−3.30	8.527	1,620.95
5,000,000	240.02	0.1407	41.507	15.72	94.64	−2.56	11.031	2,677.96

Table B 22-PIC-0

Freq. (Hz)	R (Ω/kft)	L (mH/kft)	G (μS/kft)	C (nF/kft)	Re Z (Ω)	Im Z (Ω)	α (dB/kft)	β (Deg/kft)
1	27.97	0.1848	0.000	15.72	11,899.87	−11,896.69	0.010	0.07
5	27.97	0.1848	0.000	15.72	5,322.17	−5,319.98	0.023	0.15
10	27.97	0.1848	0.000	15.72	3,763.72	−3,761.42	0.032	0.21
15	27.97	0.1848	0.000	15.72	3,073.37	−3,070.88	0.040	0.26
20	27.97	0.1848	0.000	15.72	2,661.89	−2,659.19	0.046	0.30
30	27.97	0.1848	0.001	15.72	2,173.87	−2,170.77	0.056	0.37
50	27.97	0.1848	0.001	15.72	1,684.57	−1,680.78	0.072	0.48
70	27.97	0.1848	0.001	15.72	1,424.31	−1,419.94	0.085	0.56
100	27.97	0.1848	0.002	15.72	1,192.41	−1,187.27	0.102	0.67
150	27.97	0.1848	0.002	15.72	974.61	−968.40	0.125	0.83
200	27.97	0.1848	0.003	15.72	844.92	−837.80	0.144	0.96
300	27.97	0.1848	0.005	15.72	691.31	−682.65	0.176	1.17
500	27.97	0.1848	0.008	15.72	537.72	−526.60	0.226	1.52
700	27.97	0.1847	0.010	15.72	456.36	−443.23	0.266	1.81
1,000	27.97	0.1847	0.014	15.72	384.21	−368.55	0.316	2.17
1,500	27.98	0.1846	0.021	15.72	317.00	−297.85	0.383	2.69
2,000	27.98	0.1846	0.027	15.72	277.40	−255.32	0.438	3.14
3,000	27.99	0.1845	0.040	15.72	231.24	−204.27	0.526	3.93
5,000	28.02	0.1843	0.065	15.72	186.64	−152.00	0.652	5.28
7,000	28.05	0.1841	0.088	15.72	164.24	−123.53	0.742	6.51
10,000	28.12	0.1837	0.123	15.72	145.71	−97.68	0.838	8.25
15,000	28.25	0.1832	0.181	15.72	130.39	−73.12	0.941	11.07
20,000	28.42	0.1826	0.236	15.72	122.70	−58.62	1.006	13.89
30,000	28.86	0.1813	0.345	15.72	115.38	−42.20	1.087	19.59
50,000	30.11	0.1779	0.557	15.72	109.93	−27.73	1.190	31.10
70,000	32.27	0.1756	0.764	15.72	107.90	−21.62	1.299	42.74
100,000	35.81	0.1730	1.066	15.72	106.30	−17.05	1.463	60.16
150,000	41.95	0.1692	1.559	15.72	104.64	−13.52	1.742	88.82
200,000	47.75	0.1660	2.040	15.72	103.43	−11.68	2.006	117.06
300,000	57.80	0.1608	2.982	15.72	101.60	−9.59	2.472	172.49
500,000	73.19	0.1542	4.810	15.72	99.31	−7.46	3.203	281.00
700,000	85.72	0.1504	6.590	15.72	98.02	−6.32	3.801	388.28
1,000,000	101.52	0.1469	9.202	15.72	96.81	−5.30	4.558	547.87
1,500,000	123.29	0.1434	13.450	15.72	95.63	−4.35	5.605	811.73
2,000,000	141.66	0.1413	17.606	15.72	94.88	−3.77	6.491	1,073.88
3,000,000	172.50	0.1378	25.732	15.72	93.67	−3.10	8.008	1,590.32
5,000,000	221.46	0.1339	41.507	15.72	92.33	−2.42	10.433	2,612.59

Table B 19-PIC-120

Freq. (Hz)	R (Ω/kft)	L (mH/kft)	G (μS/kft)	C (nF/kft)	Re Z (Ω)	Im Z (Ω)	α (dB/kft)	β (Deg/kft)
1	18.26	0.1882	0.000	15.72	9,617.43	−9,612.55	0.008	0.05
5	18.26	0.1882	0.000	15.72	4,301.53	−4,298.38	0.018	0.12
10	18.26	0.1882	0.000	15.72	3,042.12	−3,038.94	0.026	0.17
15	18.26	0.1882	0.001	15.72	2,484.27	−2,480.89	0.032	0.21
20	18.26	0.1882	0.001	15.72	2,151.79	−2,148.18	0.037	0.24
30	18.26	0.1882	0.001	15.72	1,757.49	−1,753.42	0.045	0.30
50	18.26	0.1882	0.002	15.72	1,362.23	−1,357.32	0.058	0.38
70	18.26	0.1882	0.002	15.72	1,152.04	−1,146.41	0.069	0.46
100	18.26	0.1881	0.004	15.72	964.80	−958.23	0.082	0.55
150	18.26	0.1881	0.005	15.72	789.04	−781.14	0.101	0.67
200	18.26	0.1881	0.007	15.72	684.44	−675.40	0.116	0.77
300	18.27	0.1881	0.010	15.72	560.66	−549.70	0.141	0.95
500	18.27	0.1881	0.016	15.72	437.13	−423.08	0.182	1.24
700	18.27	0.1881	0.023	15.72	371.86	−355.28	0.213	1.47
1,000	18.27	0.1880	0.032	15.72	314.17	−294.42	0.253	1.78
1,500	18.28	0.1879	0.047	15.72	260.73	−236.61	0.305	2.21
2,000	18.29	0.1879	0.062	15.72	229.49	−201.71	0.346	2.60
3,000	18.31	0.1877	0.091	15.72	193.51	−159.65	0.411	3.28
5,000	18.36	0.1874	0.148	15.72	159.65	−116.44	0.500	4.52
7,000	18.43	0.1871	0.204	15.72	143.35	−92.97	0.559	5.68
10,000	18.56	0.1866	0.286	15.72	130.57	−71.95	0.618	7.39
15,000	18.85	0.1857	0.421	15.72	120.79	−52.65	0.678	10.25
20,000	19.22	0.1848	0.554	15.72	116.24	−41.84	0.719	13.16
30,000	20.07	0.1819	0.816	15.72	111.75	−30.29	0.780	18.97
50,000	22.17	0.1780	1.329	15.72	108.42	−20.69	0.889	30.68
70,000	24.57	0.1753	1.832	15.72	106.90	−16.61	0.999	42.34
100,000	28.03	0.1717	2.574	15.72	105.39	−13.45	1.156	59.64
150,000	33.44	0.1670	3.790	15.72	103.63	−10.88	1.403	87.96
200,000	38.12	0.1633	4.987	15.72	102.36	−9.41	1.620	115.85
300,000	45.97	0.1588	7.343	15.72	100.81	−7.68	1.984	171.14
500,000	58.41	0.1541	11.953	15.72	99.19	−5.95	2.563	280.66
700,000	68.52	0.1515	16.478	15.72	98.31	−5.03	3.034	389.43
1,000,000	81.25	0.1492	23.157	15.72	97.51	−4.21	3.628	551.83
1,500,000	98.76	0.1476	34.093	15.72	96.96	−3.43	4.438	823.08
2,000,000	113.51	0.1467	44.860	15.72	96.65	−2.96	5.119	1,093.92
3,000,000	138.23	0.1457	66.046	15.72	96.31	−2.41	6.261	1,635.04
5,000,000	177.58	0.1448	107.521	15.72	96.01	−1.86	8.078	2,716.55

Table B 19-PIC-70

Freq. (Hz)	R (Ω/kft)	L (mH/kft)	G (μS/kft)	C (nF/kft)	Re Z (Ω)	Im Z (Ω)	α (dB/kft)	β (Deg/kft)
1	16.47	0.1868	0.000	15.72	9,132.32	−9,127.63	0.008	0.05
5	16.47	0.1868	0.000	15.72	4,084.62	−4,081.48	0.017	0.12
10	16.47	0.1868	0.000	15.72	2,888.76	−2,885.55	0.025	0.16
15	16.47	0.1868	0.001	15.72	2,359.07	−2,355.63	0.030	0.20
20	16.47	0.1868	0.001	15.72	2,043.38	−2,039.68	0.035	0.23
30	16.47	0.1868	0.001	15.72	1,669.00	−1,664.81	0.043	0.28
50	16.47	0.1868	0.002	15.72	1,293.72	−1,288.64	0.055	0.37
70	16.47	0.1868	0.002	15.72	1,094.17	−1,088.33	0.065	0.43
100	16.47	0.1867	0.004	15.72	916.43	−909.60	0.078	0.52
150	16.47	0.1867	0.005	15.72	749.60	−741.37	0.095	0.64
200	16.47	0.1867	0.007	15.72	650.34	−640.91	0.110	0.74
300	16.47	0.1867	0.010	15.72	532.90	−521.46	0.134	0.91
500	16.47	0.1867	0.016	15.72	415.75	−401.08	0.172	1.18
700	16.47	0.1867	0.023	15.72	353.90	−336.59	0.202	1.40
1,000	16.48	0.1866	0.032	15.72	299.30	−278.66	0.239	1.69
1,500	16.48	0.1865	0.047	15.72	248.78	−223.59	0.288	2.11
2,000	16.49	0.1865	0.062	15.72	219.32	−190.30	0.327	2.48
3,000	16.51	0.1863	0.091	15.72	185.51	−150.15	0.387	3.15
5,000	16.56	0.1860	0.148	15.72	153.94	−108.88	0.467	4.35
7,000	16.62	0.1857	0.204	15.72	138.92	−86.50	0.520	5.50
10,000	16.73	0.1852	0.286	15.72	127.32	−66.53	0.571	7.20
15,000	17.00	0.1843	0.421	15.72	118.60	−48.35	0.623	10.07
20,000	17.33	0.1834	0.554	15.72	114.61	−38.26	0.657	12.97
30,000	18.19	0.1805	0.816	15.72	110.69	−27.71	0.714	18.79
50,000	20.38	0.1766	1.329	15.72	107.73	−19.14	0.822	30.48
70,000	22.81	0.1737	1.832	15.72	106.26	−15.51	0.933	42.09
100,000	26.29	0.1699	2.574	15.72	104.74	−12.69	1.091	59.27
150,000	31.53	0.1649	3.790	15.72	102.94	−10.32	1.332	87.38
200,000	35.97	0.1611	4.987	15.72	101.63	−8.94	1.539	115.03
300,000	43.41	0.1564	7.343	15.72	100.02	−7.31	1.888	169.80
500,000	55.20	0.1515	11.953	15.72	98.32	−5.67	2.443	278.20
700,000	64.79	0.1487	16.478	15.72	97.38	−4.80	2.896	385.76
1,000,000	76.87	0.1462	23.157	15.72	96.52	−4.02	3.469	546.21
1,500,000	93.50	0.1438	34.093	15.72	95.69	−3.29	4.258	812.28
2,000,000	107.52	0.1423	44.860	15.72	95.18	−2.85	4.925	1,077.29
3,000,000	131.04	0.1405	66.046	15.72	94.56	−2.33	6.045	1,605.43
5,000,000	168.34	0.1386	107.521	15.72	93.92	−1.80	7.828	2,657.46

Table B 19-PIC-0

FREQ. (Hz)	R (Ω/kft)	L (mH/kft)	G (μS/kft)	C (nf/kft)	Re Z (Ω)	Im Z (Ω)	α (dB/kft)	β (Deg/kft)
1	13.95	0.1848	0.000	15.72	8,406.26	−8,401.84	0.007	0.05
5	13.95	0.1848	0.000	15.72	3,759.96	−3,756.85	0.016	0.11
10	13.95	0.1848	0.000	15.72	2,659.23	−2,655.96	0.023	0.13
15	13.95	0.1848	0.001	15.72	2,171.70	−2,168.14	0.028	0.18
20	13.95	0.1848	0.001	15.72	1,881.13	−1,877.28	0.032	0.21
30	13.95	0.1848	0.001	15.72	1,536.58	−1,532.16	0.039	0.26
50	13.95	0.1848	0.002	15.72	1,191.22	−1,185.83	0.051	0.34
70	13.95	0.1848	0.002	15.72	1,007.60	−1,001.38	0.060	0.40
100	13.95	0.1848	0.004	15.72	844.07	−836.78	0.072	0.48
150	13.95	0.1848	0.005	15.72	690.62	−681.82	0.088	0.59
200	13.95	0.1848	0.007	15.72	599.35	−589.25	0.101	0.68
300	13.95	0.1848	0.010	15.72	491.41	−479.14	0.123	0.83
500	13.95	0.1847	0.016	15.72	383.84	368.09	0.158	1.09
700	13.96	0.1847	0.023	15.72	327.13	−308.54	0.185	1.30
1,000	13.96	0.1847	0.032	15.72	277.14	−254.98	0.219	1.57
1,500	13.97	0.1846	0.047	15.72	231.05	−203.99	0.263	1.96
2,000	13.97	0.1845	0.062	15.72	204.28	−173.12	0.297	2.31
3,000	13.99	0.1844	0.091	15.72	173.76	−135.83	0.350	2.95
5,000	14.03	0.1840	0.148	15.72	145.67	−97.49	0.418	4.12
7,000	14.08	0.1837	0.204	15.72	132.62	−76.77	0.461	5.25
10,000	14.18	0.1832	0.286	15.72	122.79	−58.45	0.502	6.95
15,000	14.40	0.1824	0.421	15.72	115.63	−42.02	0.541	9.81
20,000	14.69	0.1815	0.554	15.72	112.43	−33.05	0.568	12.72
30,000	15.55	0.1786	0.816	15.72	109.27	−24.00	0.619	18.55
50,000	17.88	0.1747	1.329	15.72	106.79	−16.94	0.728	30.21
70,000	20.34	0.1715	1.832	15.72	105.37	−13.95	0.839	41.74
100,000	23.85	0.1674	2.574	15.72	103.84	−11.61	0.998	58.76
150,000	28.87	0.1620	3.790	15.72	101.98	−9.54	1.231	86.56
200,000	32.95	0.1580	4.987	15.72	100.61	−8.28	1.425	113.87
300,000	39.81	0.1530	7.343	15.72	98.91	−6.78	1.751	167.91
500,000	50.70	0.1477	11.953	15.72	97.09	−5.28	2.273	274.72
700,000	59.56	0.1448	16.478	15.72	96.07	−4.47	2.700	380.57
1,000,000	70.75	0.1420	23.157	15.72	95.11	−3.75	3.240	538.24
1,500,000	86.15	0.1384	34.093	15.72	93.88	−3.09	3.999	796.91
2,000,000	99.15	0.1361	44.860	15.72	93.09	−2.69	4.644	1,053.57
3,000,000	120.97	0.1332	66.046	15.72	92.07	−2.21	5.733	1,563.02
5,000,000	155.41	0.1299	107.521	15.72	90.91	−1.72	7.466	2,572.47

Table B 26-PULP-120

FREQ. (Hz)	R (Ω/kft)	L (mH/kft)	G (μS/kft)	C (nF/kft)	Re Z (Ω)	Im Z (Ω)	α (dB/kft)	β (Deg/kft)
1	92.58	0.1835	0.106	15.51	23,625.25	−9,211.70	0.030	0.08
5	92.58	0.1835	0.107	15.50	10,619.93	−8,536.13	0.046	0.24
10	92.58	0.1835	0.109	15.50	7,245.15	−6,480.47	0.062	0.36
15	92.58	0.1835	0.110	15.49	5,380.42	−5,405.87	0.074	0.45
20	92.58	0.1835	0.112	15.49	5,010.50	−4,729.92	0.085	0.53
30	92.58	0.1835	0.115	15.49	4,058.58	−3,900.62	0.103	0.65
50	92.58	0.1835	0.121	15.48	3,123.51	−3,044.86	0.132	0.85
70	92.58	0.1835	0.127	15.48	2,632.71	−2,581.69	0.155	1.01
100	92.58	0.1835	0.137	15.48	2,198.49	−2,165.07	0.185	1.21
150	92.58	0.1835	0.153	15.47	1,792.81	−1,770.78	0.227	1.48
200	92.58	0.1835	0.170	15.47	1,551.97	−1,534.66	0.261	1.71
300	92.58	0.1835	0.204	15.46	1,267.15	−1,253.63	0.320	2.10
500	92.59	0.1835	0.276	15.45	982.41	−970.78	0.412	2.72
700	92.59	0.1835	0.352	15.44	831.30	−819.82	0.486	3.22
1,000	92.59	0.1835	0.474	15.44	696.89	−684.91	0.580	3.86
1,500	92.59	0.1834	0.692	15.43	570.93	−557.71	0.708	4.73
2,000	92.60	0.1834	0.927	15.42	496.11	−481.61	0.815	5.48
3,000	92.61	0.1833	1.435	15.41	407.79	−390.92	0.991	6.75
5,000	92.64	0.1831	2.572	15.39	320.05	−299.20	1.264	8.82
7,000	92.68	0.1830	3.833	15.38	274.03	−249.84	1.477	10.57
10,000	92.73	0.1827	5.902	15.37	233.72	−205.30	1.734	12.86
15,000	92.85	0.1823	9.720	15.35	196.93	−162.72	2.062	16.23
20,000	92.98	0.1818	13.899	15.34	175.89	−136.88	2.313	19.32
30,000	93.31	0.1810	23.095	15.32	152.44	−105.67	2.681	25.08
50,000	94.19	0.1790	43.973	15.29	131.74	−73.99	3.138	36.08
70,000	95.36	0.1768	67.320	15.27	122.54	−57.44	3.423	46.94
100,000	97.62	0.1746	105.837	15.25	115.95	−43.38	3.717	63.40
150,000	101.99	0.1722	177.169	15.22	111.26	−31.31	4.074	91.15
200,000	106.95	0.1704	255.453	15.20	109.10	−24.96	4.385	119.06
300,000	118.68	0.1676	428.026	15.17	107.03	−18.61	5.021	174.90
500,000	142.45	0.1632	820.474	15.13	104.95	−13.39	6.274	285.10
700,000	164.63	0.1594	1,259.741	15.13	103.46	−10.99	7.483	393.55
1,000,000	193.68	0.1551	1,984.848	15.13	101.85	−8.95	9.143	553.55
1,500,000	233.62	0.1507	3,064.651	15.13	100.24	−7.10	11.463	817.44
2,000,000	267.27	0.1480	4,168.863	15.13	99.25	−6.00	13.499	1,079.37
3,000,000	323.68	0.1448	6,428.052	15.13	98.05	−4.69	17.081	1,599.84
5,000,000	413.08	0.1413	11,080.367	15.13	96.81	−3.36	23.196	2,633.37

Table B 26-PULP-70

FREQ. (Hz)	R (Ω/kft)	L (mH/kft)	G (μS/kft)	C (nF/kft)	Re Z (Ω)	Im Z (Ω)	α (dB/kft)	β (Deg/kft)
1	83.48	0.1822	0.106	15.51	22,433.50	−8,746.51	0.028	0.07
5	83.48	0.1822	0.107	15.50	10,084.25	−8,105.51	0.044	0.23
10	83.48	0.1822	0.109	15.50	6,879.72	−6,153.53	0.059	0.34
15	83.48	0.1822	0.110	15.49	5,536.36	−5,133.12	0.070	0.43
20	83.48	0.1822	0.112	15.49	4,757.81	−4,491.26	0.080	0.50
30	83.48	0.1822	0.115	15.49	3,853.92	−3,703.78	0.098	0.62
50	83.48	0.1822	0.121	15.48	2,966.04	−2,891.17	0.125	0.81
70	83.48	0.1822	0.127	15.48	2,500.01	−2,451.35	0.148	0.96
100	83.48	0.1822	0.137	15.48	2,087.72	−2,055.73	0.176	1.15
150	83.48	0.1822	0.153	15.47	1,702.53	−1,681.30	0.215	1.41
200	83.48	0.1822	0.170	15.47	1,473.86	−1,457.06	0.248	1.63
300	83.48	0.1822	0.204	15.46	1,203.46	−1,190.17	0.303	2.00
500	83.48	0.1821	0.276	15.45	933.15	−921.52	0.391	2.58
700	83.48	0.1821	0.352	15.44	789.71	−778.13	0.462	3.06
1,000	83.48	0.1821	0.474	15.44	662.15	−649.95	0.550	3.66
1,500	83.49	0.1820	0.692	15.43	542.64	−529.07	0.671	4.50
2,000	83.49	0.1820	0.927	15.42	471.67	−456.74	0.773	5.21
3,000	83.51	0.1819	1.435	15.41	387.94	−370.50	0.939	6.42
5,000	83.53	0.1818	2.572	15.39	304.86	−283.21	1.196	8.41
7,000	83.56	0.1816	3.833	15.38	261.34	−236.20	1.207	10.08
10,000	83.61	0.1813	5.902	15.37	223.30	−193.73	1.636	12.29
15,000	83.72	0.1809	9.720	15.35	188.71	−153.09	1.940	15.56
20,000	83.84	0.1805	13.899	15.34	169.02	−128.41	2.170	18.56
30,000	84.13	0.1796	23.095	15.32	147.24	−98.61	2.503	24.23
50,000	84.93	0.1777	43.973	15.29	128.31	−68.46	2.906	35.15
70,000	85.98	0.1755	67.320	15.27	120.03	−52.83	3.153	46.00
100,000	88.02	0.1734	105.837	15.25	114.19	−39.66	3.406	62.46
150,000	92.40	0.1709	177.169	15.22	110.11	−28.61	3.735	90.23
200,000	97.67	0.1691	255.453	15.20	108.24	−22.93	4.044	118.15
300,000	109.64	0.1662	428.026	15.17	106.35	−17.25	4.680	173.81
500,000	133.53	0.1614	820.474	15.13	104.30	−12.59	5.938	283.35
700,000	155.23	0.1575	1,259.741	15.13	102.78	−10.39	7.127	390.99
1,000,000	182.73	0.1530	1,984.848	15.13	101.12	−8.46	8.726	549.63
1,500,000	220.56	0.1485	3,064.651	15.13	99.45	−6.72	10.962	811.08
2,000,000	252.46	0.1456	4,168.863	15.13	98.42	−5.67	12.928	1,070.40
3,000,000	305.97	0.1422	6,428.052	15.13	97.15	−4.43	16.396	1,585.30
5,000,000	390.82	0.1385	11,080.367	15.13	95.83	−3.18	22.329	2,606.81

Table B 26-PULP-0

FREQ. (Hz)	R (Ω/kft)	L (mH/kft)	G (μS/kft)	C (nF/kft)	Re Z (Ω)	Im Z (Ω)	α (dB/kft)	β (Deg/kft)
1	70.73	0.1803	0.106	15.51	20,649.83	−8,051.05	0.026	0.07
5	70.73	0.1803	0.107	15.50	9,282.50	−7,460.99	0.040	0.21
10	70.73	0.1803	0.109	15.50	6,332.78	−5,664.19	0.054	0.32
15	70.73	0.1803	0.110	15.49	5,096.25	−4,724.90	0.065	0.40
20	70.73	0.1803	0.112	15.49	4,379.61	−4,134.06	0.074	0.46
30	70.73	0.1803	0.115	15.49	3,547.61	−3,409.17	0.090	0.57
50	70.73	0.1803	0.121	15.48	2,730.36	−2,661.14	0.115	0.74
70	70.73	0.1803	0.127	15.48	2,301.42	−2,256.26	0.136	0.88
100	70.73	0.1803	0.137	15.48	1,921.94	−1,892.06	0.162	1.05
150	70.73	0.1802	0.153	15.47	1,567.43	−1,547.35	0.198	1.30
200	70.73	0.1802	0.170	15.47	1,356.99	−1,340.90	0.228	1.50
300	70.73	0.1802	0.204	15.46	1,108.15	−1,095.16	0.279	1.84
500	70.73	0.1802	0.276	15.45	859.44	−847.76	0.359	2.38
700	70.73	0.1802	0.352	15.44	727.51	−715.68	0.424	2.82
1,000	70.74	0.1802	0.474	15.44	610.20	−597.58	0.506	3.38
1,500	70.74	0.1801	0.692	15.43	500.35	−486.16	0.617	4.15
2,000	70.74	0.1801	0.927	15.42	435.16	−419.45	0.709	4.81
3,000	70.75	0.1800	1.435	15.41	358.32	−339.86	0.862	5.93
5,000	70.78	0.1798	2.572	15.39	282.22	−259.20	1.095	7.78
7,000	70.80	0.1797	3.833	15.38	242.47	−215.68	1.275	9.35
10,000	70.85	0.1794	5.902	15.37	207.87	−176.31	1.489	11.44
15,000	70.93	0.1790	9.720	15.35	176.61	−138.57	1.756	14.56
20,000	71.04	0.1786	13.899	15.34	158.99	−115.63	1.955	17.47
30,000	71.29	0.1777	23.095	15.32	139.78	−87.97	2.234	23.01
50,000	71.96	0.1758	61.208	15.29	123.51	−60.20	2.559	33.85
70,000	72.86	0.1737	67.320	15.27	116.60	−46.01	2.753	44.70
100,000	74.58	0.1715	105.837	15.25	111.82	−34.24	2.953	61.19
150,000	78.96	0.1691	177.169	15.22	108.56	−24.71	3.246	89.00
200,000	84.68	0.1673	255.453	15.20	107.07	−20.01	3.558	116.91
300,000	96.97	0.1642	428.026	15.17	105.40	−15.31	4.196	172.32
500,000	121.05	0.1590	820.474	15.13	103.38	−11.44	5.459	280.90
700,000	142.06	0.1549	1,259.741	15.13	101.82	−9.53	6.621	387.39
1,000,000	167.39	0.1501	1,984.848	15.13	100.09	−7.76	8.131	544.11
1,500,000	202.29	0.1453	3,064.651	15.13	98.34	−6.16	10.248	802.10
2,000,000	231.73	0.1423	4,168.863	15.13	97.24	−5.21	12.115	1,057.73
3,000,000	281.17	0.1385	6,428.052	15.13	95.88	−4.06	15.417	1,564.74
5,000,000	359.66	0.1346	11,080.367	15.13	94.44	−2.91	21.089	2,569.18

Table B 24-PULP-120

FREQ. (Hz)	R (Ω/kft)	L (mH/kft)	G (μS/kft)	C (nF/kft)	Re Z (Ω)	Im Z (Ω)	α (dB/kft)	β (Deg/kft)
1	58.23	0.1761	0.106	16.17	18,476.34	−7,426.75	0.023	0.06
5	58.23	0.1761	0.107	16.16	8,226.58	−6,670.77	0.037	0.20
10	58.23	0.1761	0.109	16.16	5,616.74	−5,046.52	0.050	0.30
15	58.23	0.1761	0.110	16.16	4,522.14	−4,205.41	0.060	0.37
20	58.23	0.1761	0.112	16.16	3,887.31	−3,677.78	0.069	0.43
30	58.23	0.1761	0.115	16.15	3,149.81	−3,031.51	0.083	0.53
50	58.23	0.1761	0.121	16.15	2,424.89	−2,365.47	0.107	0.69
70	58.23	0.1761	0.127	16.14	2,044.22	−2,005.23	0.126	0.82
100	58.23	0.1761	0.137	16.14	1,707.37	−1,681.29	0.150	0.98
150	58.23	0.1761	0.153	16.13	1,392.63	−1,374.77	0.183	1.20
200	58.23	0.1761	0.170	16.13	1,205.79	−1,191.21	0.212	1.39
300	58.23	0.1761	0.204	16.12	984.85	−972.71	0.259	1.70
500	58.23	0.1761	0.276	16.11	764.07	−752.73	0.333	2.20
700	58.23	0.1760	0.352	16.11	646.97	−635.26	0.393	2.61
1,000	58.24	0.1760	0.474	16.10	542.90	−530.20	0.468	3.13
1,500	58.24	0.1759	0.692	16.09	445.51	−431.03	0.570	3.85
2,000	58.25	0.1758	0.927	16.08	387.76	−371.62	0.655	4.47
3,000	58.27	0.1757	1.435	16.07	319.78	−300.69	0.795	5.52
5,000	58.31	0.1754	2.572	16.05	252.62	−228.70	1.008	7.27
7,000	58.36	0.1751	3.833	16.04	217.70	−189.81	1.171	8.76
10,000	58.44	0.1746	5.902	16.03	187.44	−154.60	1.362	10.77
15,000	58.61	0.1739	9.720	16.02	160.35	−120.85	1.598	13.80
20,000	58.80	0.1731	13.899	16.00	145.26	−100.40	1.771	16.66
30,000	59.29	0.1716	23.095	15.98	129.04	−75.93	2.013	22.17
50,000	60.61	0.1686	43.973	15.96	115.61	−51.89	2.303	33.07
70,000	62.34	0.1666	67.320	15.94	110.13	−39.92	2.495	44.08
100,000	65.11	0.1644	105.837	15.91	106.31	−30.11	2.713	60.72
150,000	70.62	0.1616	177.169	15.89	103.53	−22.19	3.046	88.59
200,000	76.80	0.1595	255.453	15.87	102.13	−18.23	3.383	116.41
300,000	89.00	0.1557	428.026	15.83	100.39	−14.15	4.041	171.31
500,000	111.06	0.1498	820.474	15.79	98.15	−10.60	5.268	278.43
700,000	129.62	0.1456	1,259.741	15.79	96.58	−8.79	6.361	383.62
1,000,000	153.00	0.1416	1,984.848	15.79	95.13	−7.16	7.809	539.88
1,500,000	185.14	0.1378	3,064.651	15.79	93.71	−5.68	9.832	797.88
2,000,000	212.23	0.1354	4,168.863	15.79	92.83	−4.79	11.614	1,054.01
3,000,000	257.63	0.1324	6,428.052	15.79	91.74	−3.73	14.761	1,562.90
5,000,000	329.58	0.1300	11,080.367	15.79	90.84	−2.64	20.132	2,579.72

Table B 24-PULP-70

FREQ. (Hz)	R (Ω/kft)	L (mH/kft)	G (μS/kft)	C (nF/kft)	Re Z (Ω)	Im Z (Ω)	α (dB/kft)	β (Deg/kft)
1	52.50	0.1748	0.106	16.17	17,544.33	−7,052.10	0.022	0.06
5	52.50	0.1748	0.107	16.16	7,811.63	−6,334.24	0.035	0.19
10	52.50	0.1748	0.109	16.16	5,333.46	−4,791.90	0.047	0.28
15	52.50	0.1748	0.110	16.16	4,294.08	−3,993.21	0.057	0.35
20	52.50	0.1748	0.112	16.16	3,691.29	−3,492.19	0.065	0.41
30	52.50	0.1748	0.115	16.15	2,991.00	−2,878.50	0.079	0.50
50	52.50	0.1748	0.121	16.15	2,302.67	−2,246.03	0.101	0.65
70	52.50	0.1748	0.127	16.14	1,941.23	−1,903.95	0.120	0.78
100	52.50	0.1748	0.137	16.14	1,621.40	−1,596.33	0.142	0.93
150	52.50	0.1748	0.153	16.13	1,322.57	−1,305.23	0.174	1.14
200	52.50	0.1748	0.170	16.13	1,145.18	−1,130.90	0.201	1.32
300	52.50	0.1748	0.204	16.12	935.44	−923.38	0.245	1.62
500	52.50	0.1748	0.276	16.11	725.87	−714.42	0.316	2.09
700	52.51	0.1747	0.352	16.11	614.75	−602.81	0.373	2.48
1,000	52.51	0.1747	0.474	16.10	516.01	−502.97	0.444	2.98
1,500	52.52	0.1746	0.692	16.09	423.64	−408.70	0.541	3.66
2,000	52.52	0.1745	0.927	16.08	368.90	−352.20	0.621	4.25
3,000	52.54	0.1744	1.435	16.07	304.51	−284.70	0.753	5.26
5,000	52.58	0.1741	2.572	16.05	241.02	−216.13	0.952	6.93
7,000	52.62	0.1738	3.833	16.04	208.07	−179.05	1.104	8.37
10,000	52.70	0.1733	5.902	16.03	179.64	−145.43	1.282	10.32
15,000	52.84	0.1726	9.720	16.02	154.33	−113.20	1.497	13.28
20,000	53.02	0.1718	13.899	16.00	140.34	−93.67	1.653	16.09
30,000	53.46	0.1703	23.095	15.98	125.48	−70.38	1.867	21.57
50,000	54.65	0.1674	43.973	15.96	113.39	−47.66	2.119	32.45
70,000	56.21	0.1653	67.320	15.94	108.54	−36.48	2.284	43.45
100,000	59.08	0.1631	105.837	15.91	105.22	−27.56	2.490	60.12
150,000	64.81	0.1604	177.169	15.89	102.78	−20.47	2.821	87.97
200,000	71.07	0.1581	255.453	15.87	101.47	−16.93	3.158	115.66
300,000	83.31	0.1541	428.026	15.83	99.75	−13.29	3.816	170.24
500,000	104.74	0.1479	820.474	15.79	97.48	−10.03	5.018	276.54
700,000	122.31	0.1435	1,259.741	15.79	95.87	−8.32	6.069	380.81
1,000,000	144.46	0.1395	1,984.848	15.79	94.37	−6.78	7.466	535.60
1,500,000	174.94	0.1354	3,064.651	15.79	92.89	−5.38	9.420	790.96
2,000,000	200.63	0.1329	4,168.863	15.79	91.97	−4.53	11.144	1,044.29
3,000,000	243.73	0.1298	6,428.052	15.79	90.82	−3.53	14.194	1,547.26
5,000,000	312.08	0.1265	11,080.367	15.79	89.61	−2.51	19.440	2,545.03

Table B 24-PULP-0

FREQ. (Hz)	R (Ω/kft)	L (mH/kft)	G (μS/kft)	C (nF/kft)	Re Z (Ω)	Im Z (Ω)	α (dB/kft)	β (Deg/kft)
1	44.48	0.1730	0.106	16.17	16,149.40	−6,491.36	0.021	0.05
5	44.48	0.1730	0.107	16.16	7,190.59	−5,830.54	0.032	0.17
10	44.48	0.1730	0.109	16.16	4,909.47	−4,410.82	0.044	0.26
15	44.48	0.1730	0.110	16.16	3,952.76	−3,675.61	0.052	0.32
20	44.48	0.1730	0.112	16.16	3,397.91	−3,214.41	0.060	0.38
30	44.48	0.1730	0.115	16.15	2,753.33	−2,649.49	0.073	0.46
50	44.48	0.1730	0.121	16.15	2,119.77	−2,067.27	0.093	0.60
70	44.48	0.1730	0.127	16.14	1,787.10	−1,752.35	0.110	0.71
100	44.48	0.1730	0.137	16.14	1,492.74	−1,469.14	0.131	0.86
150	44.48	0.1730	0.153	16.13	1,217.73	−1,201.13	0.160	1.05
200	44.48	0.1730	0.170	16.13	1,054.49	−1,040.61	0.185	1.21
300	44.48	0.1729	0.204	16.12	861.52	−849.51	0.226	1.49
500	44.49	0.1729	0.276	16.11	668.74	−657.03	0.291	1.93
700	44.49	0.1729	0.352	16.11	566.56	−554.20	0.343	2.29
1,000	44.49	0.1728	0.474	16.10	475.81	−462.17	0.408	2.74
1,500	44.50	0.1728	0.692	16.09	390.98	−375.21	0.497	3.38
2,000	44.50	0.1727	0.927	16.08	340.76	−323.06	0.570	3.93
3,000	44.52	0.1726	1.435	16.07	281.77	−260.69	0.689	4.87
5,000	44.55	0.1723	2.572	16.05	223.78	−197.22	0.869	6.44
7,000	44.59	0.1720	3.833	16.04	193.84	−162.83	1.004	7.80
10,000	44.65	0.1715	5.902	16.03	168.17	−131.61	1.160	9.66
15,000	44.77	0.1708	9.720	16.02	145.56	−101.66	1.345	12.53
20,000	44.94	0.1700	13.899	16.00	133.26	−83.54	1.475	15.29
30,000	45.30	0.1685	23.095	15.98	120.47	−62.07	1.648	20.71
50,000	46.30	0.1656	43.973	15.96	110.35	−41.44	1.846	31.59
70,000	47.62	0.1636	67.320	15.94	106.39	−31.46	1.978	42.61
100,000	50.63	0.1614	105.837	15.91	103.76	−23.88	2.170	59.30
150,000	56.68	0.1587	177.169	15.89	101.77	−18.02	2.500	87.13
200,000	63.04	0.1562	255.453	15.87	100.55	−15.09	2.837	114.64
300,000	75.36	0.1519	428.026	15.83	98.86	−12.07	3.497	168.75
500,000	95.89	0.1453	820.474	15.79	96.53	−9.22	4.661	273.89
700,000	112.08	0.1407	1,259.741	15.79	94.86	−7.65	5.654	376.86
1,000,000	132.52	0.1364	1,984.848	15.79	93.30	−6.23	6.976	529.56
1,500,000	160.66	0.1322	3,064.651	15.79	91.73	−4.94	8.830	781.18
2,000,000	184.40	0.1295	4,168.863	15.79	90.75	−4.17	10.471	1,030.55
3,000,000	224.27	0.1261	6,428.052	15.79	89.52	−3.24	13.383	1,525.10
5,000,000	287.57	0.1217	11,080.367	15.79	87.87	−2.32	18.444	2,495.66

Table B 22-PULP-120

FREQ. (Hz)	R (Ω/kft)	L (mH/kft)	G (μS/kft)	C (nF/kft)	Re Z (Ω)	Im Z (Ω)	α (dB/kft)	β (Deg/kft)
1	36.61	0.1743	0.106	16.17	14,649.91	−5,888.58	0.019	0.05
5	36.61	0.1743	0.107	16.16	6,523.00	−5,289.08	0.029	0.16
10	36.61	0.1743	0.109	16.16	4,453.73	−4,001.14	0.040	0.23
15	36.61	0.1743	0.110	16.16	3,585.88	−3,334.17	0.048	0.29
20	36.61	0.1743	0.112	16.16	3,082.57	−2,915.78	0.054	0.34
30	36.61	0.1743	0.115	16.15	2,497.88	−2,403.28	0.066	0.42
50	36.61	0.1743	0.121	16.15	1,923.21	−1,875.06	0.085	0.55
70	36.61	0.1742	0.127	16.14	1,621.48	−1,589.33	0.100	0.65
100	36.61	0.1742	0.137	16.14	1,354.51	−1,332.37	0.119	0.78
150	36.61	0.1742	0.153	16.13	1,105.12	−1,089.16	0.145	0.95
200	36.61	0.1742	0.170	16.13	957.11	−943.48	0.167	1.10
300	36.61	0.1742	0.204	16.12	782.18	−770.01	0.205	1.35
500	36.61	0.1742	0.276	16.11	607.49	−595.23	0.263	1.75
700	36.61	0.1741	0.352	16.11	514.96	−501.80	0.310	2.08
1,000	36.62	0.1741	0.474	16.10	432.84	−418.14	0.369	2.50
1,500	36.63	0.1740	0.692	16.09	356.17	−339.03	0.449	3.08
2,000	36.64	0.1739	0.927	16.08	310.85	−291.53	0.514	3.58
3,000	36.66	0.1737	1.435	16.07	257.76	−234.65	0.621	4.45
5,000	36.71	0.1732	2.572	16.05	205.84	−176.66	0.779	5.92
7,000	36.77	0.1728	3.833	16.04	179.26	−145.18	0.896	7.22
10,000	36.88	0.1722	5.902	16.03	156.70	−116.61	1.028	9.00
15,000	37.09	0.1712	9.720	16.02	137.20	−89.31	1.182	11.81
20,000	37.36	0.1701	13.899	16.00	126.82	−72.96	1.290	14.55
30,000	38.03	0.1680	23.095	15.98	116.26	−53.94	1.435	20.00
50,000	39.84	0.1648	43.973	15.96	108.24	−36.29	1.622	31.00
70,000	41.73	0.1626	67.320	15.94	105.06	−27.86	1.758	42.09
100,000	45.18	0.1601	105.837	15.91	102.80	−21.46	1.958	58.77
150,000	51.45	0.1568	177.169	15.89	100.90	−16.45	2.294	86.40
200,000	57.40	0.1539	255.453	15.87	99.64	−13.82	2.615	113.62
300,000	68.56	0.1491	428.026	15.83	97.82	−11.05	3.228	167.00
500,000	86.72	0.1427	820.474	15.79	95.58	−8.36	4.284	271.23
700,000	101.48	0.1391	1,259.741	15.79	94.24	−6.90	5.195	374.45
1,000,000	120.07	0.1357	1,984.848	15.79	93.00	−5.58	6.412	527.94
1,500,000	145.63	0.1324	3,064.651	15.79	91.78	−4.39	8.116	781.64
2,000,000	167.17	0.1304	4,168.863	15.79	91.02	−3.67	9.627	1,033.76
3,000,000	203.27	0.1284	6,428.052	15.79	90.29	−2.81	12.301	1,538.39
5,000,000	260.48	0.1264	11,080.367	15.79	89.54	−1.93	16.945	2,543.33

Table B 22-PULP-70

FREQ. (Hz)	R (Ω/kft)	L (mH/kft)	G (μS/kft)	C (nF/kft)	Re Z (Ω)	Im Z (Ω)	α (dB/kft)	β (Deg/kft)
1	33.01	0.1730	0.106	16.17	13,910.92	−5,591.52	0.018	0.05
5	33.01	0.1730	0.107	16.16	6,193.99	−5,022.23	0.028	0.15
10	33.01	0.1730	0.109	16.16	4,229.12	−3,799.25	0.038	0.22
15	33.01	0.1730	0.110	16.16	3,405.06	−3,165.91	0.045	0.28
20	33.01	0.1730	0.112	16.16	2,927.15	−2,768.61	0.052	0.32
30	33.01	0.1730	0.115	16.15	2,371.98	−2,281.94	0.063	0.40
50	33.01	0.1730	0.121	16.15	1,826.33	−1,780.34	0.080	0.52
70	33.01	0.1730	0.127	16.14	1,539.84	−1,509.00	0.095	0.62
100	33.01	0.1730	0.137	16.14	1,286.38	−1,264.96	0.113	0.74
150	33.01	0.1729	0.153	16.13	1,049.61	−1,033.98	0.138	0.91
200	33.01	0.1729	0.170	16.13	909.11	−895.62	0.159	1.05
300	33.01	0.1729	0.204	16.12	743.05	−730.84	0.194	1.29
500	33.01	0.1729	0.276	16.11	577.28	−564.78	0.250	1.66
700	33.01	0.1728	0.352	16.11	489.50	−475.98	0.294	1.98
1,000	33.02	0.1728	0.474	16.10	411.62	−396.45	0.350	2.38
1,500	33.03	0.1727	0.692	16.09	338.96	−321.20	0.425	2.93
2,000	33.03	0.1726	0.927	16.08	296.06	−275.99	0.487	3.41
3,000	33.05	0.1724	1.435	16.07	245.85	−221.81	0.587	4.25
5,000	33.10	0.1720	2.572	16.05	196.90	−166.51	0.734	5.66
7,000	33.15	0.1716	3.833	16.04	171.95	−136.45	0.842	6.92
10,000	33.25	0.1709	5.902	16.03	150.91	−109.16	0.963	8.67
15,000	33.45	0.1699	9.720	16.02	132.89	−83.12	1.101	11.45
20,000	33.69	0.1689	13.899	16.00	123.42	−67.57	1.195	14.17
30,000	34.29	0.1668	23.095	15.98	113.93	−49.60	1.321	19.60
50,000	35.92	0.1636	43.973	15.96	106.82	−33.12	1.483	30.60
70,000	37.98	0.1614	67.320	15.94	104.08	−25.56	1.617	41.70
100,000	41.54	0.1589	105.837	15.91	102.07	−19.83	1.816	58.36
150,000	47.84	0.1554	177.169	15.89	100.26	−15.36	2.152	85.85
200,000	53.83	0.1523	255.453	15.87	99.00	−13.01	2.473	112.90
300,000	64.65	0.1473	428.026	15.83	97.15	−10.46	3.073	165.88
500,000	81.85	0.1407	820.474	15.79	94.86	−7.92	4.088	269.18
700,000	95.84	0.1369	1,259.741	15.79	93.47	−6.54	4.967	371.41
1,000,000	113.46	0.1334	1,984.848	15.79	92.18	−5.29	6.143	523.30
1,500,000	137.72	0.1299	3,064.651	15.79	90.89	−4.16	7.792	774.17
2,000,000	158.16	0.1277	4,168.863	15.79	90.09	−3.48	9.258	1,023.25
3,000,000	192.46	0.1251	6,428.052	15.79	89.10	−2.67	11.871	1,518.22
5,000,000	246.85	0.1222	11,080.367	15.79	88.04	−1.84	16.416	2,500.63

Table B 22-PULP-0

Freq. (Hz)	R (Ω/kft)	L (mH/kft)	G (μS/kft)	C (nF/kft)	Re Z (Ω)	Im Z (Ω)	α (dB/kft)	β (Deg/kft)
1	27.97	0.1712	0.106	16.17	12,804.89	−5,146.91	0.016	0.04
5	27.97	0.1712	0.107	16.16	5,701.58	−4,622.83	0.026	0.14
10	27.97	0.1712	0.109	16.16	3,892.97	−3,497.06	0.034	0.20
15	27.97	0.1712	0.110	16.16	3,134.45	−2,914.06	0.041	0.26
20	27.97	0.1712	0.112	16.16	2,694.56	−2,548.33	0.048	0.30
30	27.97	0.1712	0.115	16.15	2,183.56	−2,100.32	0.058	0.37
50	27.97	0.1711	0.121	16.15	1,681.34	−1,638.55	0.074	0.48
70	27.97	0.1711	0.127	16.14	1,417.68	−1,388.75	0.087	0.57
100	27.97	0.1711	0.137	16.14	1,184.42	−1,164.06	0.104	0.68
150	27.97	0.1711	0.153	16.13	966.55	−951.37	0.127	0.83
200	27.97	0.1711	0.170	16.13	837.28	−823.95	0.146	0.96
300	27.97	0.1711	0.204	16.12	684.54	−672.17	0.179	1.18
500	27.97	0.1711	0.276	16.11	532.11	−519.15	0.230	1.54
700	27.97	0.1710	0.352	16.11	451.45	−437.29	0.270	1.82
1,000	27.98	0.1709	0.474	16.10	379.94	−363.92	0.321	2.19
1,500	27.98	0.1709	0.692	16.09	313.30	−294.44	0.390	2.71
2,000	27.99	0.1708	0.927	16.08	274.01	−252.65	0.446	3.16
3,000	28.01	0.1705	1.435	16.07	228.16	−202.50	0.536	3.94
5,000	28.05	0.1702	2.572	16.05	183.69	−151.20	0.666	5.29
7,000	28.09	0.1698	3.833	16.04	161.23	−123.28	0.761	6.49
10,000	28.17	0.1691	5.902	16.03	142.49	−97.93	0.864	8.19
15,000	28.34	0.1681	9.720	16.02	126.75	−73.80	0.978	10.92
20,000	28.54	0.1671	13.899	16.00	118.66	−59.50	1.054	13.62
30,000	29.05	0.1650	23.095	15.98	110.73	−43.18	1.152	19.06
50,000	30.44	0.1619	43.973	15.96	104.93	−28.51	1.281	30.06
70,000	32.72	0.1597	67.320	15.94	102.77	−22.24	1.414	41.19
100,000	36.43	0.1572	105.837	15.91	101.09	−17.50	1.613	57.81
150,000	42.81	0.1533	177.169	15.89	99.36	−13.81	1.949	85.10
200,000	48.83	0.1500	255.453	15.87	98.11	−11.86	2.272	111.90
300,000	59.18	0.1447	428.026	15.83	96.22	−9.62	2.852	164.30
500,000	75.03	0.1378	820.474	15.79	93.83	−7.29	3.809	266.30
700,000	87.94	0.1338	1,259.741	15.79	92.38	−6.02	4.641	367.11
1,000,000	104.22	0.1301	1,984.848	15.79	91.02	−4.86	5.760	516.74
1,500,000	126.64	0.1264	3,064.651	15.79	89.65	−3.83	7.330	763.59
2,000,000	145.56	0.1241	4,168.863	15.79	88.77	−3.20	8.730	1,008.34
3,000,000	177.33	0.1204	6,428.052	15.79	87.41	−2.47	11.252	1,489.53
5,000,000	227.77	0.1163	11,080.367	15.79	85.89	−1.71	15.652	2,439.59

Table B 19-PULP-120

Freq. (Hz)	R (Ω/kft)	L (mH/kft)	G (μS/kft)	C (nF/kft)	Re Z (Ω)	Im Z (Ω)	α (db/kft)	β (Deg/kft)
1	18.26	0.1705	0.106	15.98	10,388.93	−4,140.24	0.013	0.03
5	18.26	0.1705	0.107	15.98	4,638.18	−3,751.05	0.021	0.11
10	18.26	0.1705	0.109	15.97	3,166.36	−2,840.18	0.028	0.16
15	18.26	0.1705	0.110	15.97	2,549.22	−2,367.23	0.033	0.20
20	18.26	0.1705	0.112	15.97	2,191.41	−2,070.29	0.038	0.24
30	18.26	0.1705	0.115	15.96	1,775.86	−1,706.37	0.046	0.30
50	18.26	0.1705	0.121	15.96	1,367.60	−1,331.07	0.059	0.38
70	18.26	0.1705	0.127	15.95	1,153.34	−1,127.96	0.070	0.45
100	18.26	0.1705	0.137	15.95	963.85	−945.22	0.083	0.55
150	18.26	0.1704	0.153	15.94	786.95	−772.15	0.102	0.67
200	18.27	0.1704	0.170	15.94	682.04	−668.41	0.117	0.78
300	18.27	0.1704	0.204	15.93	558.19	−544.75	0.143	0.95
500	18.27	0.1703	0.276	15.92	434.79	−419.92	0.184	1.24
700	18.27	0.1702	0.352	15.92	369.64	−353.02	0.216	1.48
1,000	18.28	0.1702	0.474	15.91	312.05	−292.94	0.256	1.78
1,500	18.29	0.1700	0.692	15.90	258.64	−235.88	0.309	2.21
2,000	18.30	0.1698	0.927	15.89	227.36	−201.46	0.351	2.59
3,000	18.34	0.1695	1.435	15.88	191.23	−160.01	0.418	3.27
5,000	18.41	0.1688	2.572	15.87	156.94	−117.48	0.512	4.46
7,000	18.50	0.1681	3.833	15.85	140.21	−94.39	0.576	5.58
10,000	18.66	0.1670	5.902	15.84	126.82	−73.67	0.643	7.21
15,000	19.01	0.1653	9.720	15.83	116.20	−54.55	0.716	9.90
20,000	19.44	0.1635	13.899	15.81	110.99	−43.74	0.768	12.60
30,000	20.39	0.1609	23.095	15.79	106.11	−31.90	0.846	18.06
50,000	22.65	0.1573	43.973	15.77	102.46	−21.88	0.981	29.02
70,000	25.19	0.1547	67.320	15.75	100.81	−17.56	1.116	39.94
100,000	28.82	0.1511	105.837	15.73	99.19	−14.18	1.309	56.06
150,000	34.44	0.1461	177.169	15.70	97.29	−11.39	1.613	82.36
200,000	39.30	0.1424	255.453	15.68	95.92	−9.78	1.887	108.13
300,000	47.44	0.1376	428.026	15.64	94.23	−7.86	2.363	159.02
500,000	60.35	0.1325	820.474	15.60	92.45	−5.89	3.166	259.30
700,000	70.83	0.1296	1,259.741	15.60	91.37	−4.81	3.868	358.82
1,000,000	84.04	0.1269	1,984.848	15.60	90.37	−3.83	4.819	507.02
1,500,000	102.21	0.1248	3,064.650	15.60	89.57	−2.95	6.149	753.90
2,000,000	117.51	0.1235	4,168.863	15.60	89.08	−2.42	7.343	999.81
3,000,000	143.16	0.1220	6,428.052	15.60	88.48	−1.78	9.498	1,489.91
5,000,000	183.99	0.1203	11,080.367	15.60	87.87	−1.14	13.323	2,466.35

Table B 19-PULP-70

FREQ. (Hz)	R (Ω/kft)	L (mH/kft)	G (μS/kft)	C (nF/kft)	Re Z (Ω)	Im Z (Ω)	α (dB/kft)	β (Deg/kft)
1	16.47	0.1692	0.106	15.98	9,864.89	−3,931.37	0.013	0.03
5	16.47	0.1692	0.107	15.98	4,404.26	−3,561.77	0.020	0.10
10	16.47	0.1692	0.109	15.97	3,006.72	−2,696.83	0.026	0.16
15	16.47	0.1692	0.110	15.97	2,420.73	−2,247.71	0.032	0.20
20	16.47	0.1692	0.112	15.97	2,080.98	−1,965.74	0.036	0.23
30	16.47	0.1692	0.115	15.96	1,686.43	−1,620.14	0.044	0.28
50	16.47	0.1692	0.121	15.96	1,298.80	−1,263.74	0.056	0.36
70	16.47	0.1692	0.127	15.95	1,095.38	−1,070.84	0.066	0.43
100	16.47	0.1692	0.137	15.95	915.49	−897.27	0.079	0.52
150	16.47	0.1692	0.153	15.94	747.58	−732.87	0.097	0.64
200	16.47	0.1691	0.170	15.94	648.02	−634.31	0.111	0.74
300	16.47	0.1691	0.204	15.93	530.50	−516.81	0.136	0.91
500	16.47	0.1690	0.276	15.92	413.47	−398.15	0.174	1.18
700	16.48	0.1690	0.352	15.92	351.71	−334.52	0.205	1.40
1,000	16.48	0.1689	0.474	15.91	297.18	−277.34	0.242	1.70
1,500	16.49	0.1687	0.692	15.90	246.67	−223.00	0.292	2.11
2,000	16.51	0.1685	0.927	15.89	217.15	−190.18	0.332	2.48
3,000	16.53	0.1682	1.435	15.88	183.15	−150.62	0.394	3.13
5,000	16.60	0.1675	2.572	15.87	151.10	−110.01	0.480	4.30
7,000	16.68	0.1668	3.833	15.85	135.62	−87.97	0.537	5.40
10,000	16.83	0.1658	5.902	15.84	123.39	−68.25	0.596	7.01
15,000	17.14	0.1641	9.720	15.83	113.83	−50.18	0.660	9.70
20,000	17.53	0.1623	13.899	15.81	109.21	−40.05	0.705	12.40
30,000	18.48	0.1597	23.095	15.79	104.94	−29.20	0.776	17.86
50,000	20.83	0.1561	43.973	15.77	101.73	−20.23	0.909	28.82
70,000	23.38	0.1532	67.320	15.75	100.16	−16.38	1.044	39.68
100,000	27.03	0.1495	105.837	15.73	98.55	−13.36	1.237	55.71
150,000	32.47	0.1444	177.169	15.70	96.63	−10.79	1.535	81.80
200,000	37.07	0.1405	255.453	15.68	95.23	−9.27	1.797	107.35
300,000	44.79	0.1355	428.026	15.64	93.49	−7.45	2.256	157.76
500,000	57.03	0.1302	820.474	15.60	91.64	−5.59	3.030	257.02
700,000	66.98	0.1272	1,259.741	15.60	90.51	−4.56	3.710	355.43
1,000,000	79.52	0.1244	1,984.848	15.60	89.44	−3.63	4.633	501.85
1,500,000	96.77	0.1216	3,064.650	15.60	88.39	−2.80	5.932	744.00
2,000,000	111.32	0.1198	4,168.863	15.60	87.72	−2.30	7.101	984.61
3,000,000	135.72	0.1176	6,428.052	15.60	86.88	−1.71	9.211	1,462.92
5,000,000	174.41	0.1152	11,080.367	15.60	85.96	−1.10	12.949	2,412.71

Table B 19-PULP-0

FREQ. (Hz)	R (Ω/kft)	L (mH/kft)	G (μS/kft)	C (nF/kft)	Re Z (Ω)	Im Z (Ω)	α (dB/kft)	β (Deg/kft)
1	13.95	0.1674	0.106	15.98	9,080.56	−3,618.74	0.012	0.03
5	13.95	0.1674	0.107	15.98	4,054.17	−3,278.46	0.018	0.10
10	13.95	0.1674	0.109	15.97	2,767.79	−2,482.25	0.024	0.14
15	13.95	0.1674	0.110	15.97	2,228.42	−2,068.81	0.029	0.18
20	13.95	0.1674	0.112	15.97	1,915.72	−1,809.24	0.033	0.21
30	13.95	0.1674	0.115	15.96	1,552.58	−1,491.07	0.041	0.26
50	13.95	0.1674	0.121	15.96	1,195.85	−1,162.93	0.052	0.34
70	13.95	0.1674	0.127	15.95	1,008.67	−985.32	0.061	0.40
100	13.95	0.1674	0.137	15.95	843.15	−825.47	0.073	0.48
150	13.95	0.1674	0.153	15.94	688.69	−674.05	0.089	0.59
200	13.95	0.1674	0.170	15.94	597.13	−583.24	0.102	0.68
300	13.95	0.1673	0.204	15.93	489.11	−474.94	0.125	0.84
500	13.96	0.1673	0.276	15.92	381.62	−365.50	0.160	1.09
700	13.96	0.1672	0.352	15.92	324.97	−306.75	0.188	1.30
1,000	13.97	0.1671	0.474	15.91	275.02	−253.91	0.222	1.57
1,500	13.98	0.1669	0.692	15.90	228.89	−203.61	0.266	1.96
2,000	13.98	0.1668	0.927	15.89	202.02	−173.19	0.302	2.30
3,000	14.01	0.1664	1.435	15.88	171.25	−136.47	0.357	2.93
5,000	14.06	0.1658	2.572	15.87	142.62	−98.73	0.431	4.06
7,000	14.13	0.1651	3.833	15.85	129.05	−78.30	0.479	5.14
10,000	14.26	0.1641	5.902	15.84	118.57	−60.14	0.526	6.74
15,000	14.52	0.1623	9.720	15.83	110.60	−43.72	0.576	9.43
20,000	14.85	0.1605	13.899	15.81	106.80	−34.65	0.611	12.13
30,000	15.80	0.1580	23.095	15.79	103.38	−25.30	0.675	17.60
50,000	18.27	0.1545	43.973	15.77	100.73	−17.87	0.808	28.54
70,000	20.85	0.1513	67.320	15.75	99.25	−14.69	0.942	39.33
100,000	24.52	0.1472	105.837	15.73	97.66	−12.19	1.136	55.21
150,000	29.73	0.1419	177.169	15.70	95.70	−9.93	1.423	81.02
200,000	33.96	0.1378	255.453	15.68	94.25	−8.54	1.670	106.25
300,000	41.08	0.1326	428.026	15.64	92.43	−6.87	2.103	156.00
500,000	52.38	0.1270	820.474	15.60	90.48	−5.15	2.838	253.79
700,000	61.58	0.1238	1,259.741	15.60	89.28	−4.21	3.485	350.63
1,000,000	73.18	0.1208	1,984.848	15.60	88.13	−3.34	4.367	494.52
1,500,000	89.16	0.1170	3,064.650	15.60	86.71	−2.59	5.621	729.92
2,000,000	102.65	0.1146	4,168.863	15.60	85.79	−2.14	6.751	962.92
3,000,000	125.29	0.1115	6,428.052	15.60	84.58	−1.59	8.795	1,424.28
5,000,000	161.02	0.1079	11,080.367	15.60	83.21	−1.03	12.409	2,335.55

C Cable Core Assembly Drawings

The following core assembly drawings and tables are provided courtesy of Essex Group, Inc. and apply only to twisted pair polyolefin insulated cables (PIC).* These are representative of the telecommunications cable industry, although some minor differences exist between manufacturers. This is particularly true in larger cable sizes, with the designation of S-units and SD-units described below. However, all manufacturers provide adequate means of identifying specific pairs in any cable size. Cable assemblies sometimes are called lay-ups, and the terms *unit* and *group* usually are used interchangeably.

All twisted pair cables are assembled from groups of conductor pairs. The basic group size is 25 pairs. However, cables with fewer than 25 total pairs are assembled in 12- and 13-pair or smaller subgroups. A 25-pair cable may be assembled from a 12-pair and a 13-pair subgroup or from 25 pairs, as shown in Figure C-1(a). The 25-pair group (and cable) and 12- and 13-pair subgroups are shown in Figure C-1(a), (b), and (c), respectively. Each circle denotes a twisted pair.

A 50-pair cable may be assembled from two 25-pair groups or from two 12- and two 13-pair subgroups shown in Figure C-2. The color code for a 25-pair group is given in Table C-1. No separate identification is used for subgroups. In these, only the pairs from 1 through 12 or 1 through 13 in Table C-1 apply.

Cables with 75, 100, 150, 200, 300, and 400 pairs are assembled from 25-pair groups, as shown in Figures C-3 through C-8. Each 25-pair group is identified by a colored ribbon, according to Table C-2. A 600-pair cable may be assembled from twenty-four 25-pair groups or from twelve 50-pair super-units, or S-units, as shown in Figure C-9. The S-unit is made from two 12- and two 13-pair subgroups, as shown in Figure C-1(d). Each 50-pair S-unit, consisting of four subgroups, is wrapped with a ribbon, colored according to Table C-3. The 900-pair cable is assembled from S-units, as shown in Figure C-10.

Cables 1,200 pair and larger are assembled from super-units of 100 pairs each, called SD-units. Each SD-unit, shown in Figure C-1(e), consists of four 25-pair groups and is wrapped with a ribbon, colored according to Table C-4. Standard cable sizes of 1,200, 1,500, 1,800, 2,100, 2,400, 2,700, 3,000, 3,300, and 3,600 pairs are shown in Figures C-11 through C-19. Some manufacturers do not make 2,100- or 3,300-pair cables.

Compartmental core, or internally screened, cables usually are made with service pairs (sometimes, but incorrectly, called spare pairs) in addition to transmission pairs. Services pairs are used for fault locating systems and order wire. Typical core

---

*Drawings and tables from Bulletin 3030-0888 © 1988 by Essex Group, Inc. Used with permission.

assemblies are shown in Figures C-20 through C-26. Other service pair/transmission pair combinations may be available from some manufacturers. For example, a 28-pair cable may have 25 transmission pairs and three service pairs, for a total of 28 pairs, rather than 24 transmission pairs, plus four service pairs, as shown in Figure C-20. Service pairs use a separate color code, as shown in Table C-5, to distinguish them from transmission pairs.

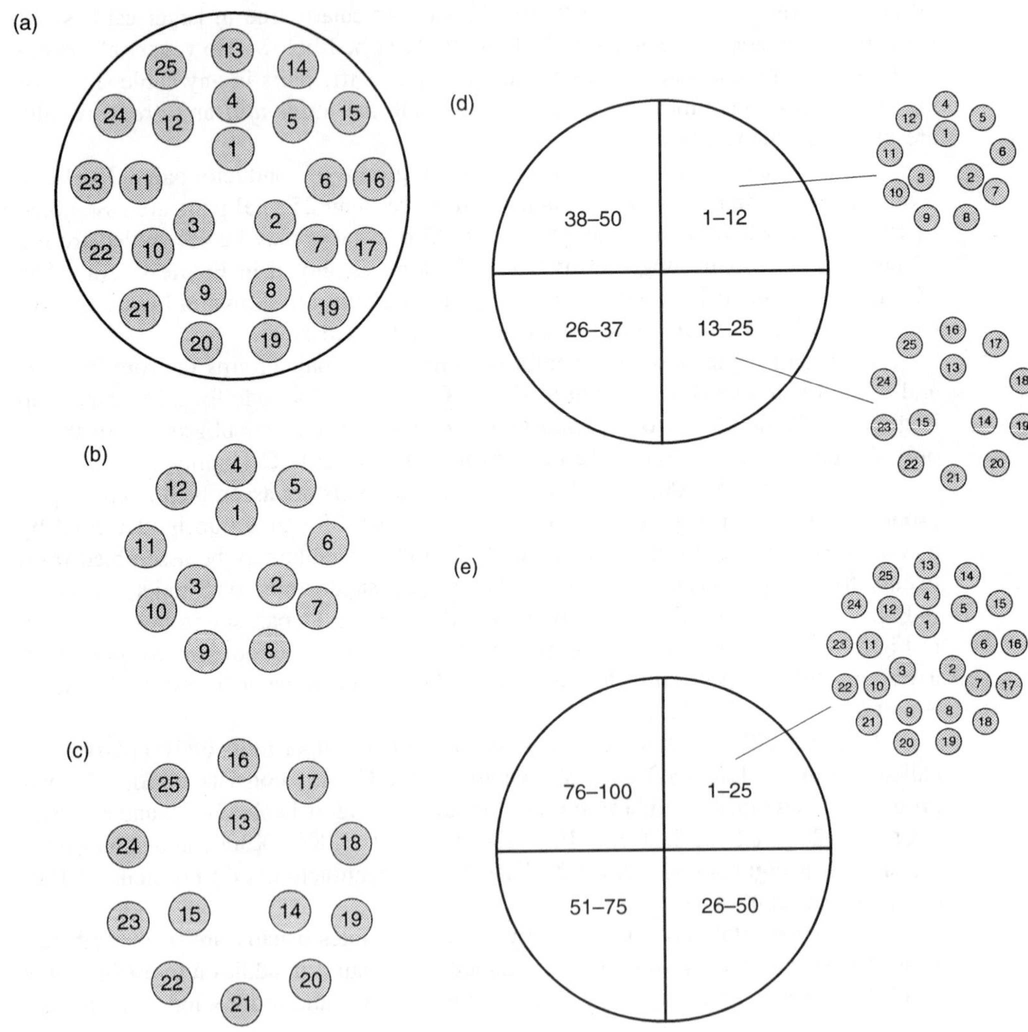

Figure C-1 Twisted pair cable assemblies: (a) 25-pair group and 25-pair cable; (b) 12-pair subgroup; (c) 13-pair subgroup; (d) 50-pair S-unit; (e) 100-pair SD-unit.

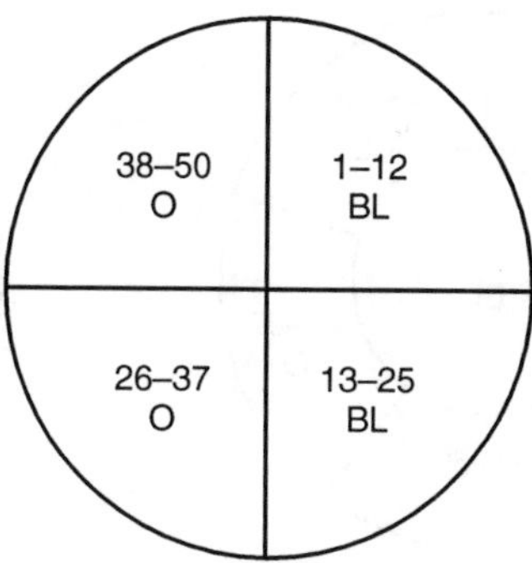

Figure C-2 50-pair

Table C-1 Pair No. Color Code

PAIR NO.	TIP	RING
1	White	Blue
2	White	Orange
3	White	Green
4	White	Brown
5	White	Slate
6	Red	Blue
7	Red	Orange
8	Red	Green
9	Red	Brown
10	Red	Slate
11	Black	Blue
12	Black	Orange
13	Black	Green
14	Black	Brown
15	Black	Slate
16	Yellow	Blue
17	Yellow	Orange
18	Yellow	Green
19	Yellow	Brown
20	Yellow	Slate
21	Violet	Blue
22	Violet	Orange
23	Violet	Green
24	Violet	Brown
25	Violet	Slate

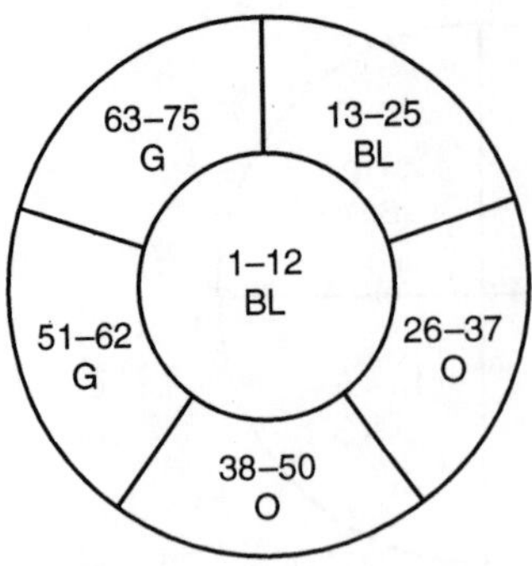

Figure C-3 75-pair

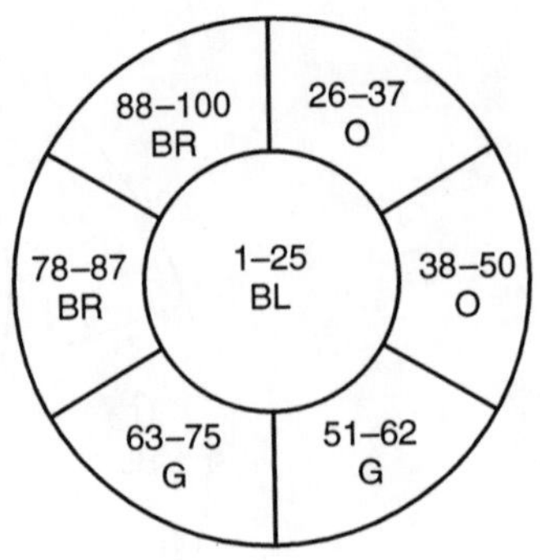

Figure C-4 100-pair

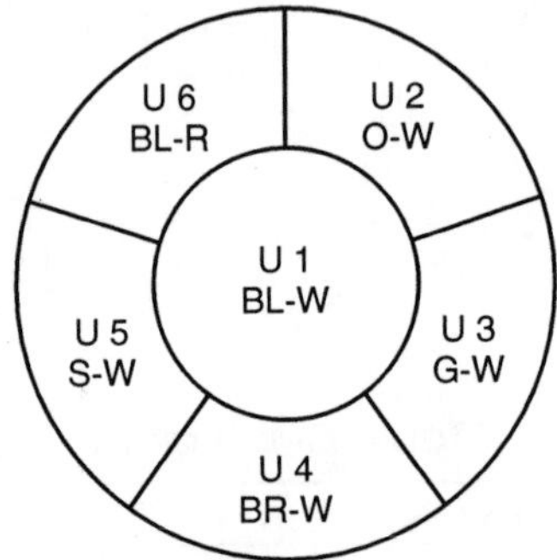

Figure C-5 150-pair

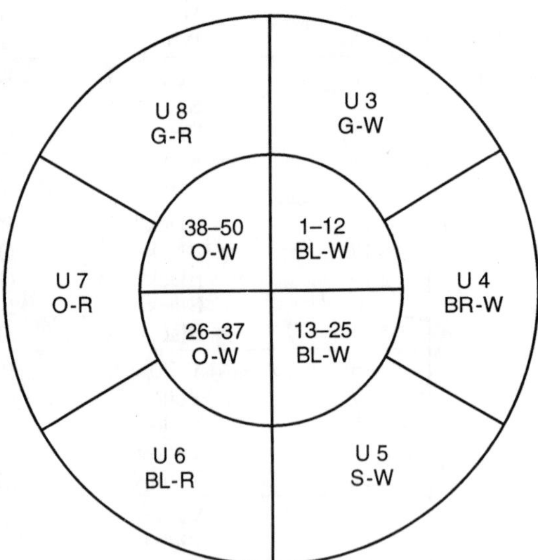

Figure C-6 200-pair

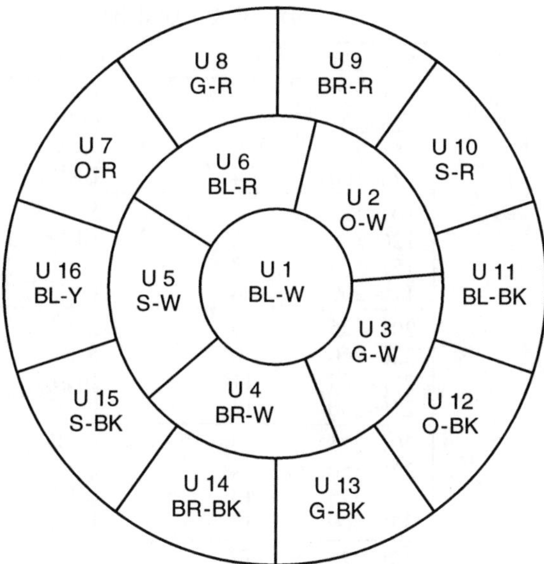

Figure C-7 300-pair

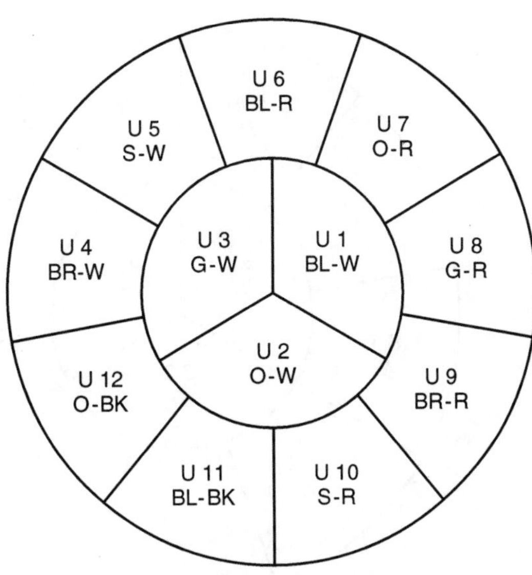

Figure C-8 400-pair

Table C-2 Unit Binder

PAIR NO.	UNIT NO.	UNIT BINDER COLOR
1–25	1	BL-W
26–50	2	O-W
51–75	3	G-W
76–100	4	BR-W
101–125	5	S-W
126–150	6	BL-R
151–175	7	O-R
176–200	8	G-R
201–225	9	BR-R
226–260	10	S-R
251–275	11	BL-BK
276–300	12	O-BK
301–325	13	G-BK
326–350	14	BR-BK
351–375	15	S-BK
376–400	16	BL-Y
401–425	17	O-Y
426–450	18	G-Y
451–475	19	BR-Y
476–500	20	S-Y
501–525	21	BL-V
526–550	22	O-V
551–575	23	G-V
576–600	24	BR-V

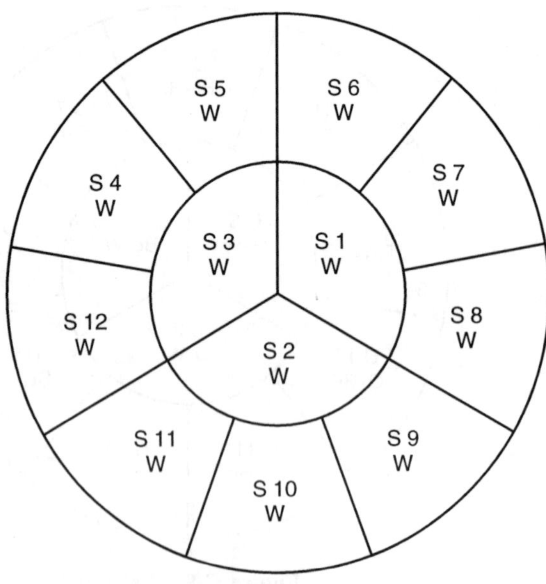

Figure C-9 600-pair

Table C-3 Super Unit Binder Color Code and Pair Numbers

UNIT BINDER COLOR	WHITE	RED	BLACK
BL-W	S-1	S-13	S-25
O-W	1–50	601–650	1,201–1,250
G-W	S-2	S-14	S-26
BR-W	51–100	651–700	1,251–1,300
S-W	S-3	S-15	S-27
BL-R	101–150	701–750	1,301–1,350
O-R	S-4	S-16	S-28
G-R	151–200	751–800	1,351–1,400
BR-R	S-5	S-17	S-29
S-R	201–250	801–850	1,401–1,450
BL-BK	S-6	S-18	S-30
O-BK	251–300	851–900	1,451–1,500
G-BK	S-7	S-19	S-31
BR-BK	301–350	901–950	1,501–1,550
S-BK	S-8	S-20	S-32
BL-Y	351–400	951–1,000	1,551–1,600
O-Y	S-9	S-21	S-33
G-Y	401–450	1,001–1,050	1,601–1,650
BR-Y	S-10	S-22	S-34
S-Y	451–500	1,051–1,100	1,651–1,700
BL-V	S-11	S-23	S-35
O-V	501–550	1,101–1,150	1,701–1,750
G-V	S-12	S-24	S-36
BR-V	551–600	1,151–1,200	1,751–1,800

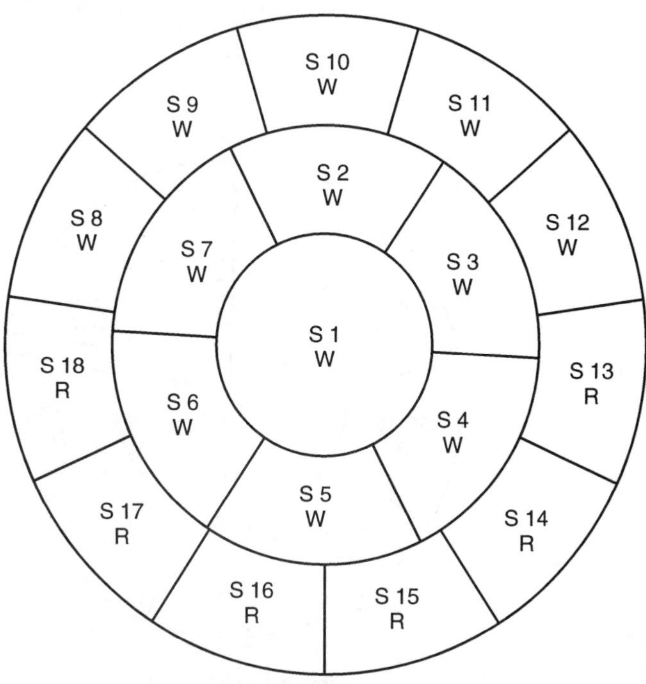

Figure C-10 900-pair

Table C-4 SD Unit Binder Color and Pair Numbers

Unit Binder Color	White	Red	Black	Yellow	Violet	Blue
BL-W O-W G-W BR-W S-W	SD-1 1–100	SD-7 601–700	SD-13 1,201–1,300	SD-19 1,801–1,900	SD-25 2,401–2,500	SD-31 3,001–3,100
BL-R O-R G-R BR-R S-R	SD-2 101–200	SD-8 701–800	SD-14 1,301–1,400	SD-20 1,901–2,000	SD-26 2,501–2,600	SD-32 3,101–3,200
	SD-3 201–300	SD-9 801–900	SD-15 1,401–1,500	SD-21 2,001–2,100	SD-27 2,601–2,700	SD-33 3,201–3,300
BL-BK O-BK G-BK BR-BK S-BK	SD-4 301–400	SD-10 901–1,000	SD-16 1,501–1,600	SD-22 2,101–2,200	SD-28 2,701–2,800	SD-34 3,301–3,400
BL-Y O-Y G-Y BR-Y S-Y	SD-5 401–500	SD-11 1,001–1,100	SD-17 1,601–1,700	SD-23 2,201–2,300	SD-29 2,801–2,900	SD-35 3,401–3,500
BL-V O-V G-V BR-V	SD-6 501–600	SD-12 1,101–1,200	SD-18 1,701–1,800	SD-24 2,301–2,400	SD-30 2,901–3,000	SD-36 3,501–3,600

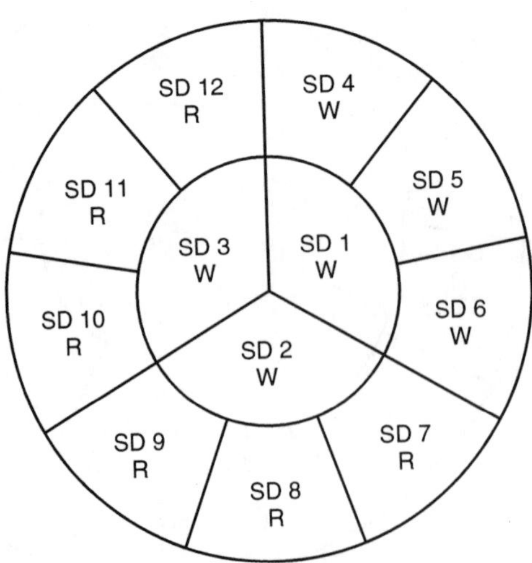

Figure C-11 1,200-pair

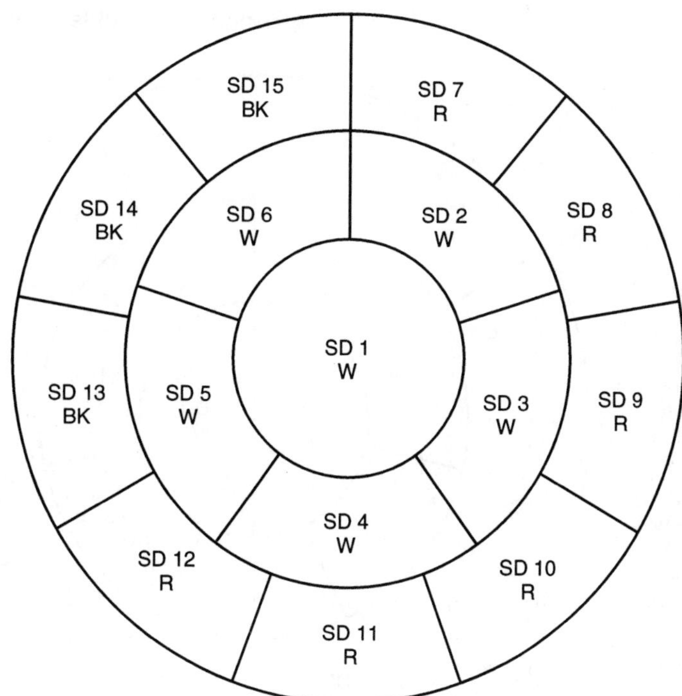

Figure C-12 1,500-pair

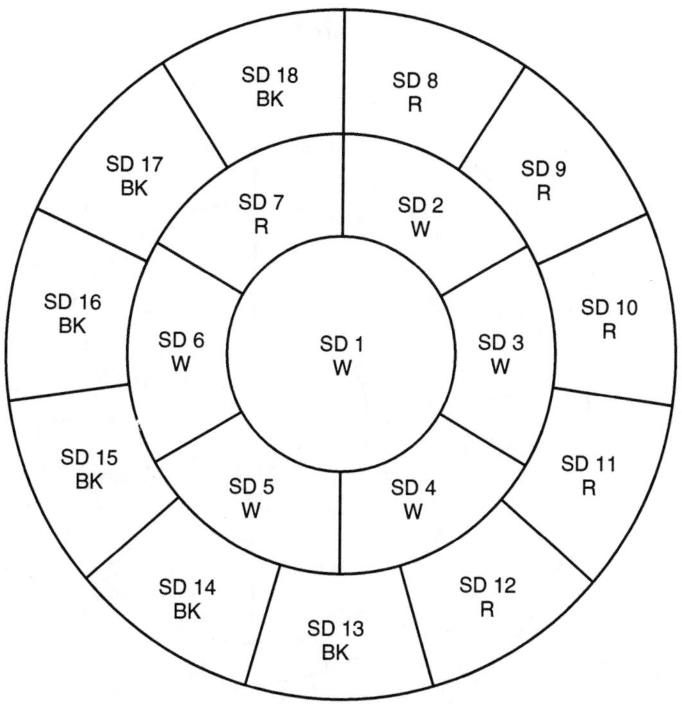

Figure C-13 1,800-pair

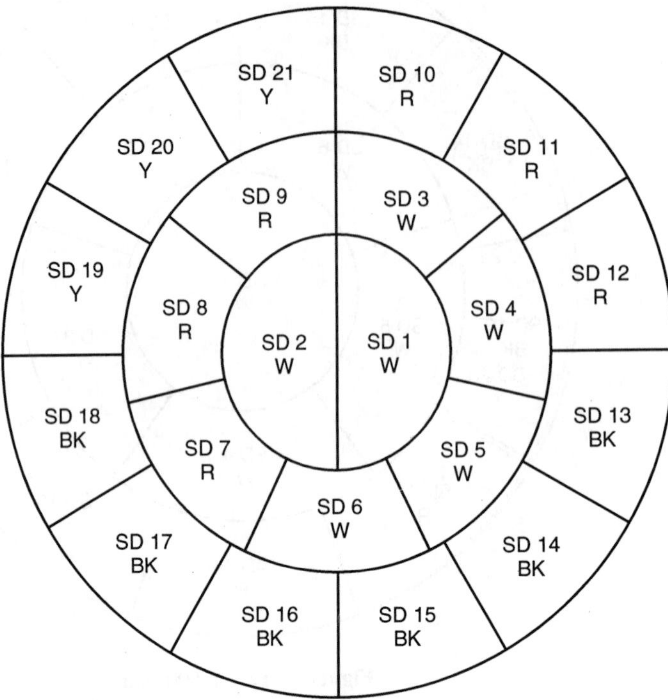

Figure C-14 2,100-pair

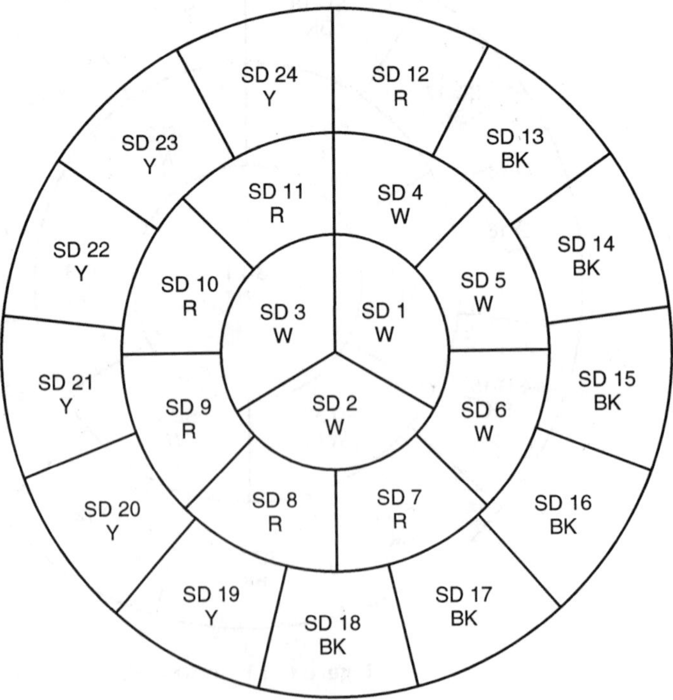

Figure C-15 2,400-pair

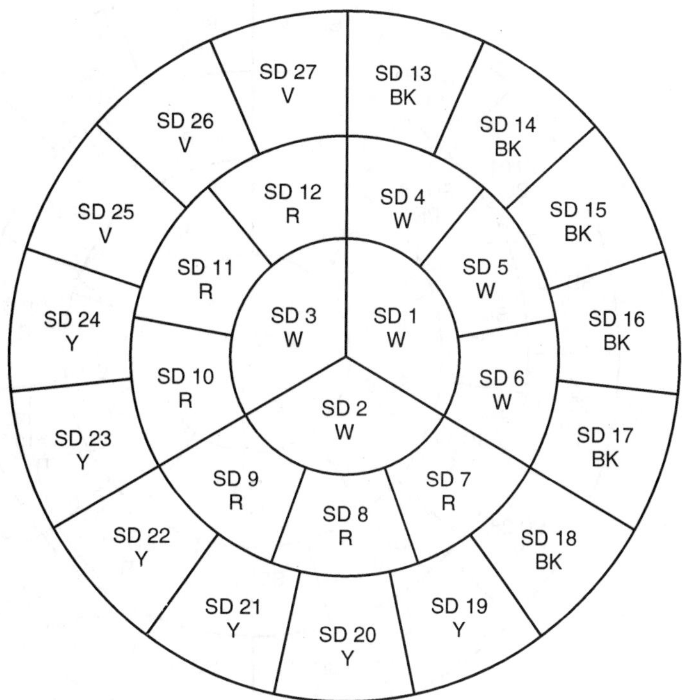

Figure C-16 2,700-pair

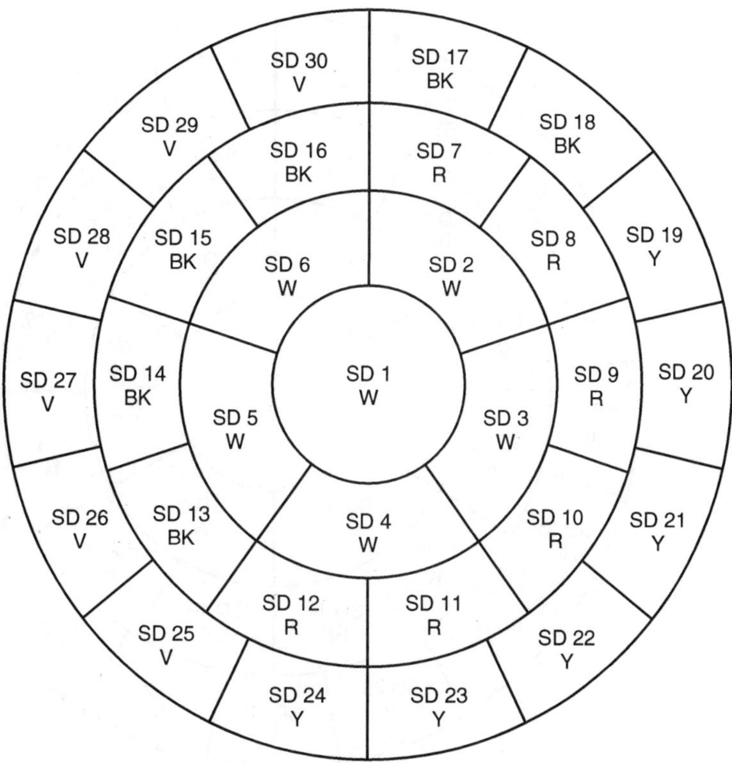

Figure C-17 3,000-pair

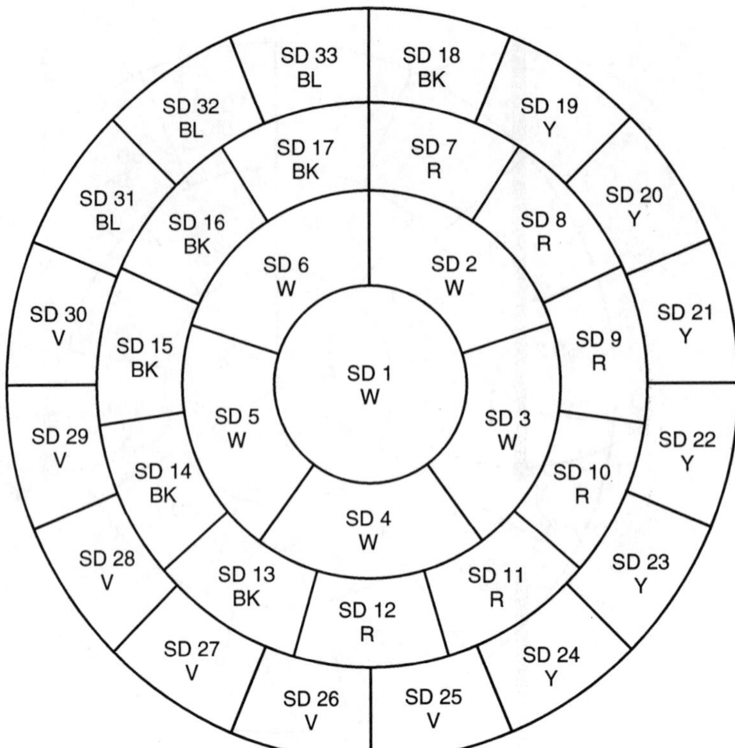

Figure C-18 3,300-pair

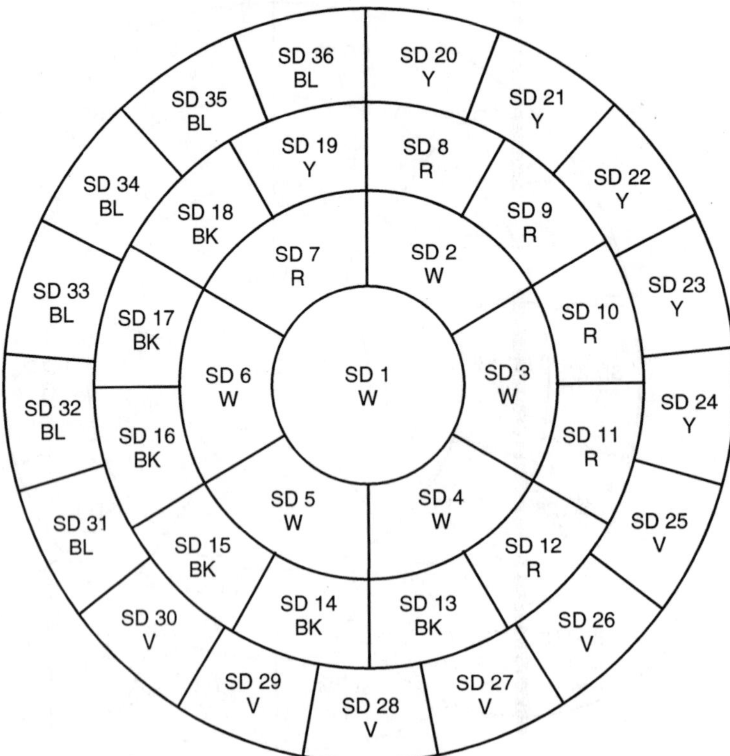

Figure C-19 3,600-pair

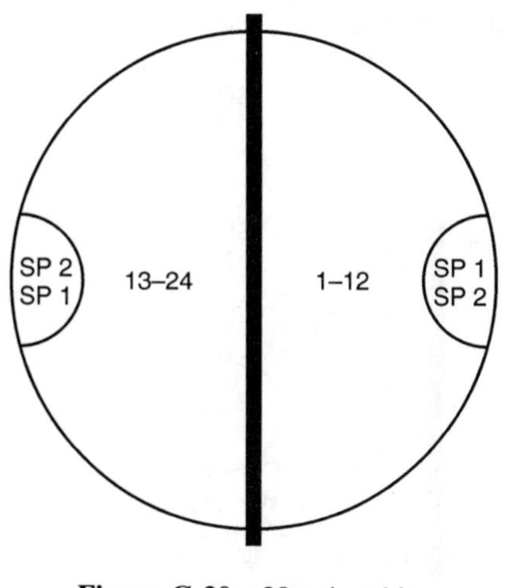

Figure C-20 28-pair cables

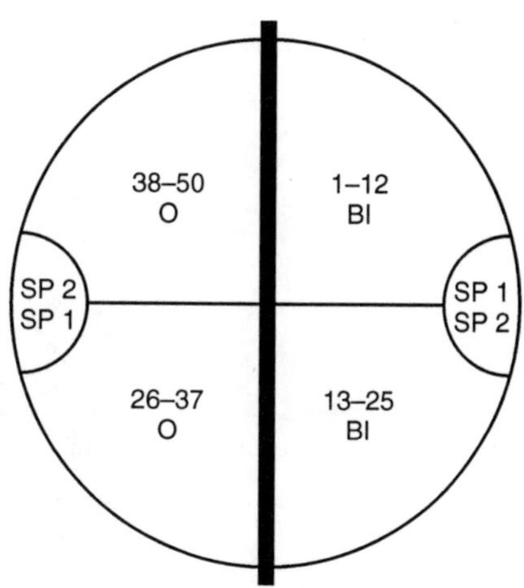

Figure C-21 54-pair

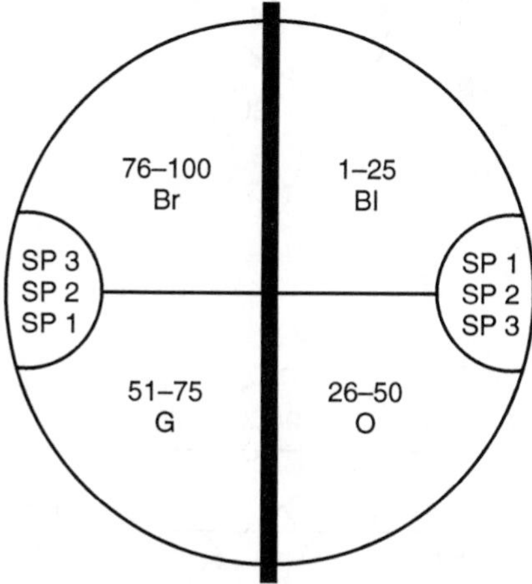

Figure C-22 106-pair

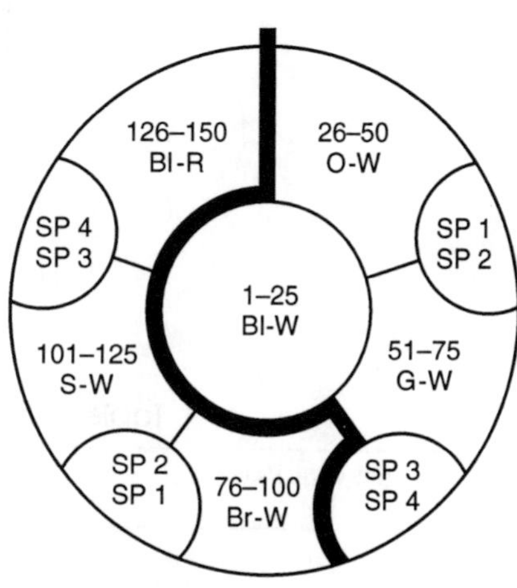

Figure C-23 158-pair

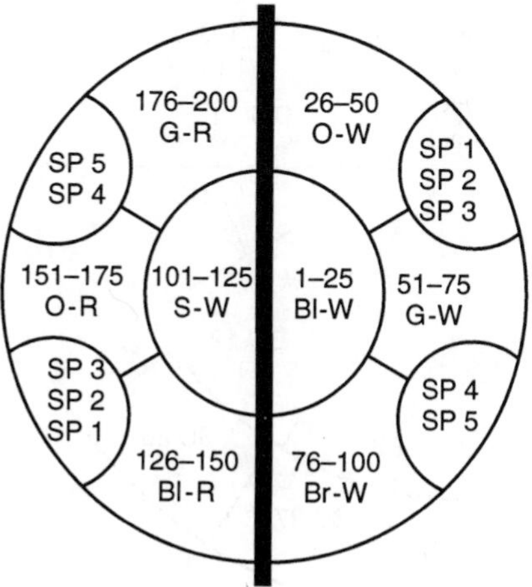

Figure C-24 210-pair

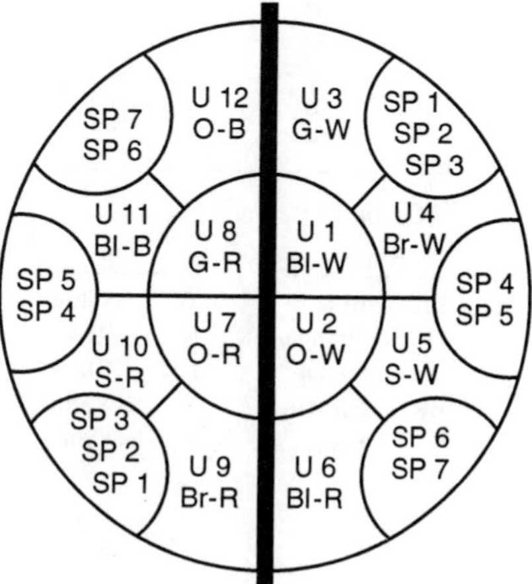

Figure C-25 314-pair

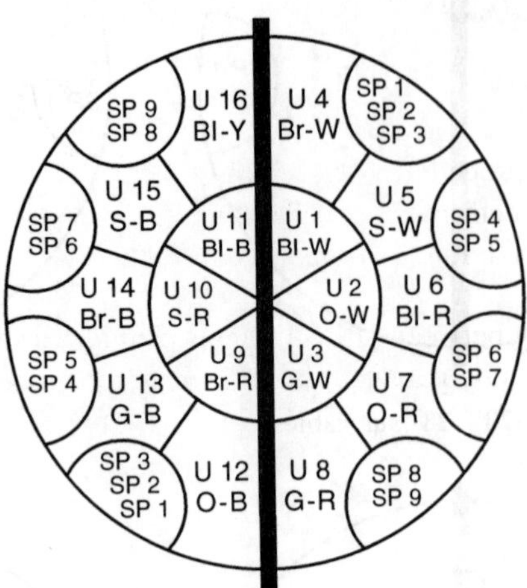

Figure C-26 418-pair

Table C-5 Color Code for Service Pairs

Service Pair No.	Tip Conductor	Ring Conductor
1	White	Red
2	White	Black
3	White	Yellow
4	White	Violet
5	Red	Black
6	Red	Yellow
7	Red	Violet
8	Black	Yellow
9	Black	Violet

Color Key

Ring		Tip	
BL:	Blue	W:	White
O:	Orange	R:	Red
G:	Green	BK:	Black
BR:	Brown	Y:	Yellow
S:	Slate	V:	Violet

D Operational Information for DSX-1 Patch Panels

The following procedures are provided courtesy of ADC Telecommunications, Inc. and apply to electromechanical digital signal cross-connect systems (DSX-1) operating at the DS-1 rate.* The procedures also apply to other similar cross-connect systems.

Jumpering between DSX-1 panels provides a semipermanent connection, which may be bypassed by patch cords. A typical connection by jumpers is shown in Figure D-1.

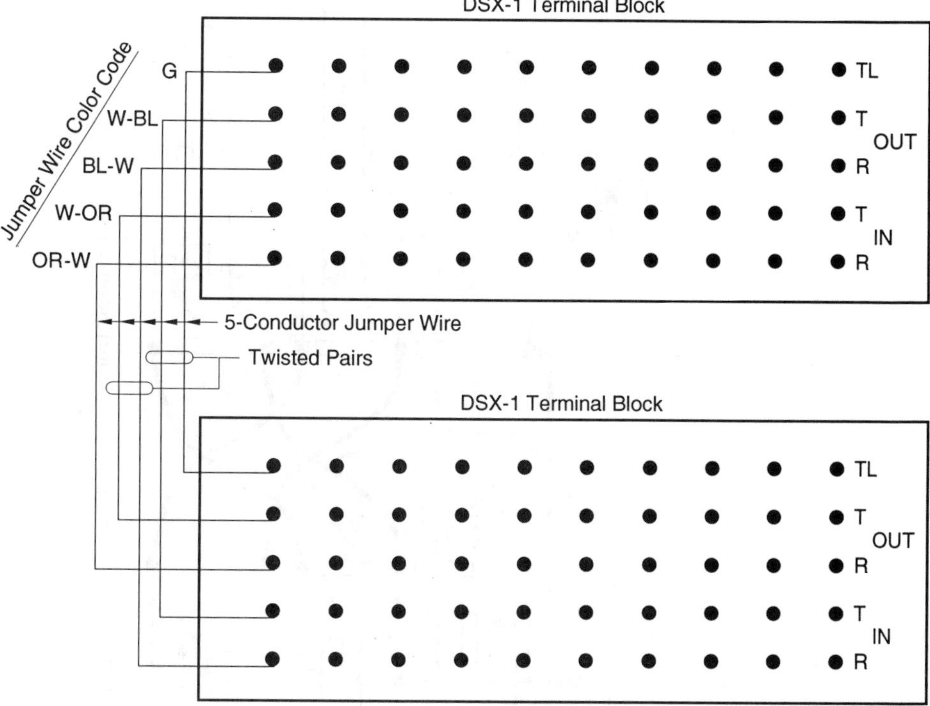

Figure D-1 Typical DSX-1 jumper connections

*Figures D-2 through D-9 and the corresponding procedures are from practice ADCP-80-328 © 1992 by ADC Telecommunications, Inc. Used with permission.

Procedure 1. Cross-Connect Circuit Identification

This procedure is used to identify the equipment terminated at the opposite end of a cross-connect jumper.

Step	Procedure
1	Insert a patch plug (either dummy or functional) into the monitor jack of the known equipment termination.
2	Observe all tracer lamps. The tracer lamp above the monitor jack with the plug inserted and the tracer lamp above the monitor jack of the unknown cross-connected equipment termination will both flash for about 30 seconds and then remain lit.

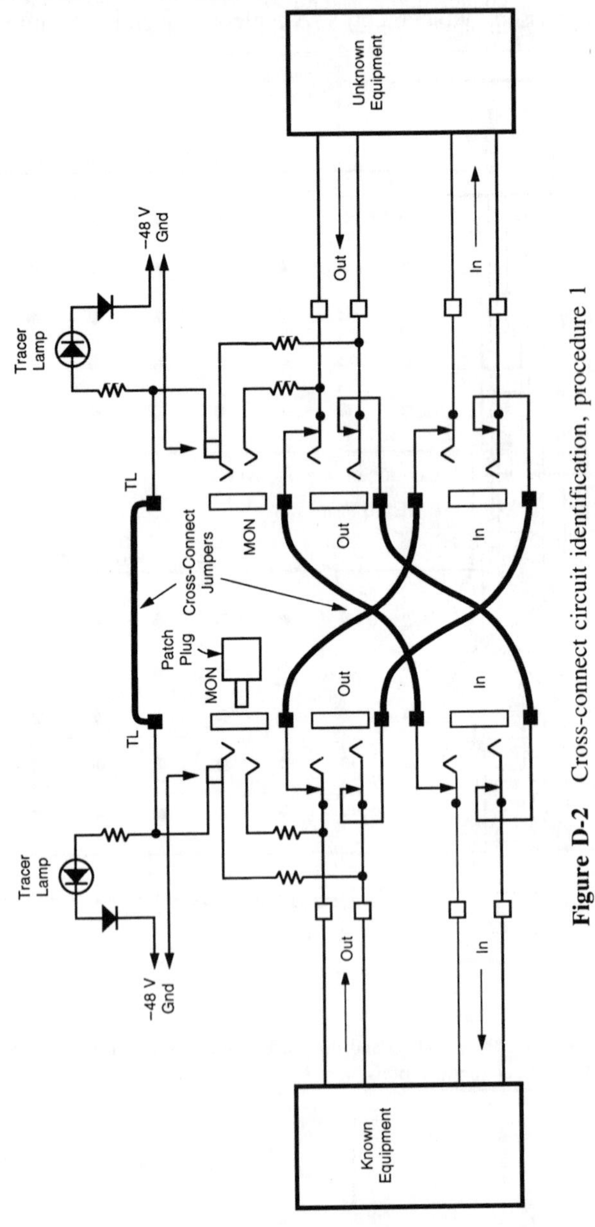

Figure D-2 Cross-connect circuit identification, procedure 1

Procedure 2. Office Equipment Patch-Around

This procedure is used to rearrange office equipment cross-connections temporarily, using patch cords or cables at the DSX.

Step	Procedure
1	Observe the DSX designation cards and strips to identify equipment units to be temporarily cross-connected, keeping in mind that the corresponding permanent cross-connections will be overridden.
2	Interconnect equipment as desired by inserting patch cords into the IN and OUT jacks on the DSX. *Note:* The permanent cross-connections to the corresponding cross-connect terminals are automatically disabled until the patch cords are unplugged from the DSX. This is considered an out-of-service patch.

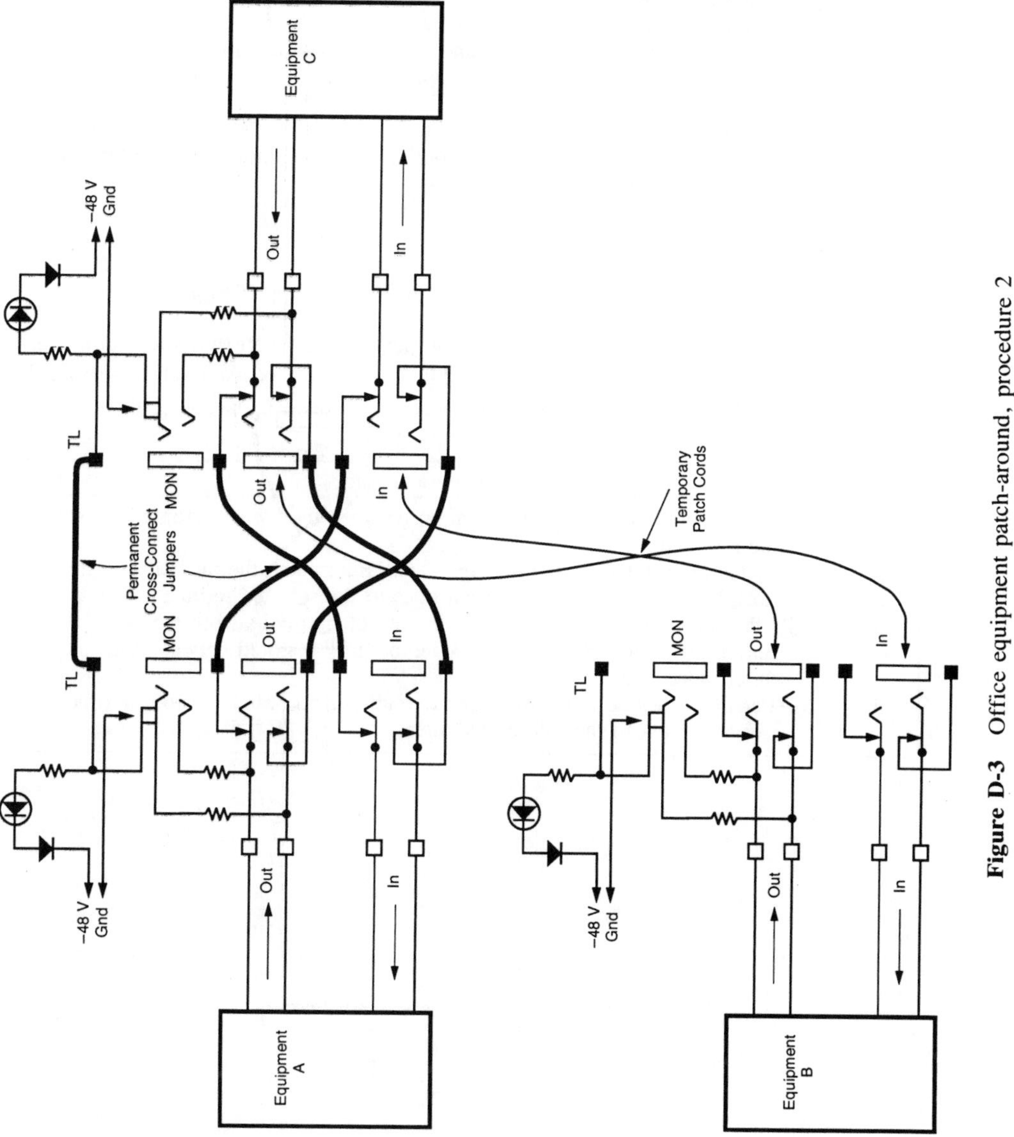

Figure D-3 Office equipment patch-around, procedure 2

Procedure 3. Digital Office Equipment Test

This procedure is used to test two digital office equipment units wired together at the DSX system. The units can be tested either with or without interruption of the circuit cross-connect.

Many of the office test equipment units and signal sources are accessible at the DSX Miscellaneous Jackfield. These can be patched directly from the Miscellaneous Jackfield to the DSX jacks.

Step	Procedure
1	To test a circuit at the cross-connect point of the two equipment units without interrupting the cross-connect circuit, plug the appropriate test unit into the desired monitor jack and perform the tests. *Note:* When a plug is inserted into the monitor jack, the tracer lamps of the corresponding cross-connected circuits will flash for about 30 seconds and then remain lit.
2	To open a cross-connect circuit and test an equipment unit whose output is wired to the DSX, plug the appropriate test unit into the OUT jack of the circuit to be tested.
3	To open a cross-connect circuit and test an equipment unit whose input is wired to the DSX, plug the appropriate test unit into the IN jack of the circuit to be tested.

Procedure 4. Restoration of Interoffice Service

This procedure is used to restore partial service between two sites temporarily by rerouting circuits through a third site. This procedure is used when service between the two sites is damaged (such as would be caused by a cut cable).

Step	Procedure
1	Select a site at which communication between the two interrupted sites will be rerouted.
2	At sites interrupted by the damaged span line, patch the interrupted circuits to DSX jacks connected to span lines to the selected reroute site.
3	At the selected reroute site, patch the rerouted circuits from the two interrupted sites. Service is now temporarily restored between the interrupted sites.
4	After the span line is repaired, remove all patch cords to restore normal service between the interrupted sites.

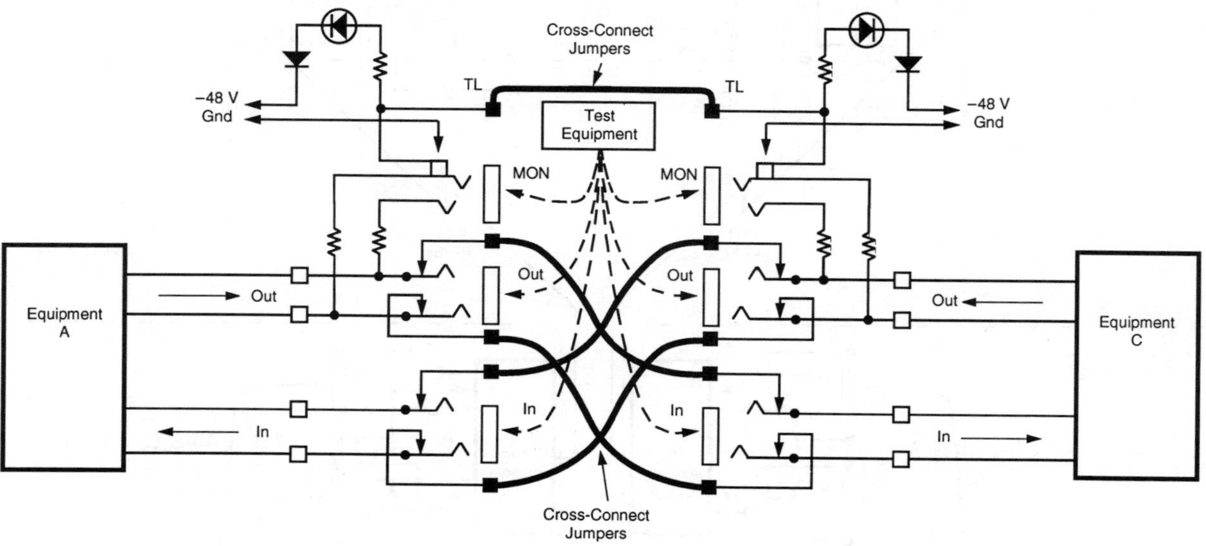

Figure D-4 Digital office equipment test, procedure 3

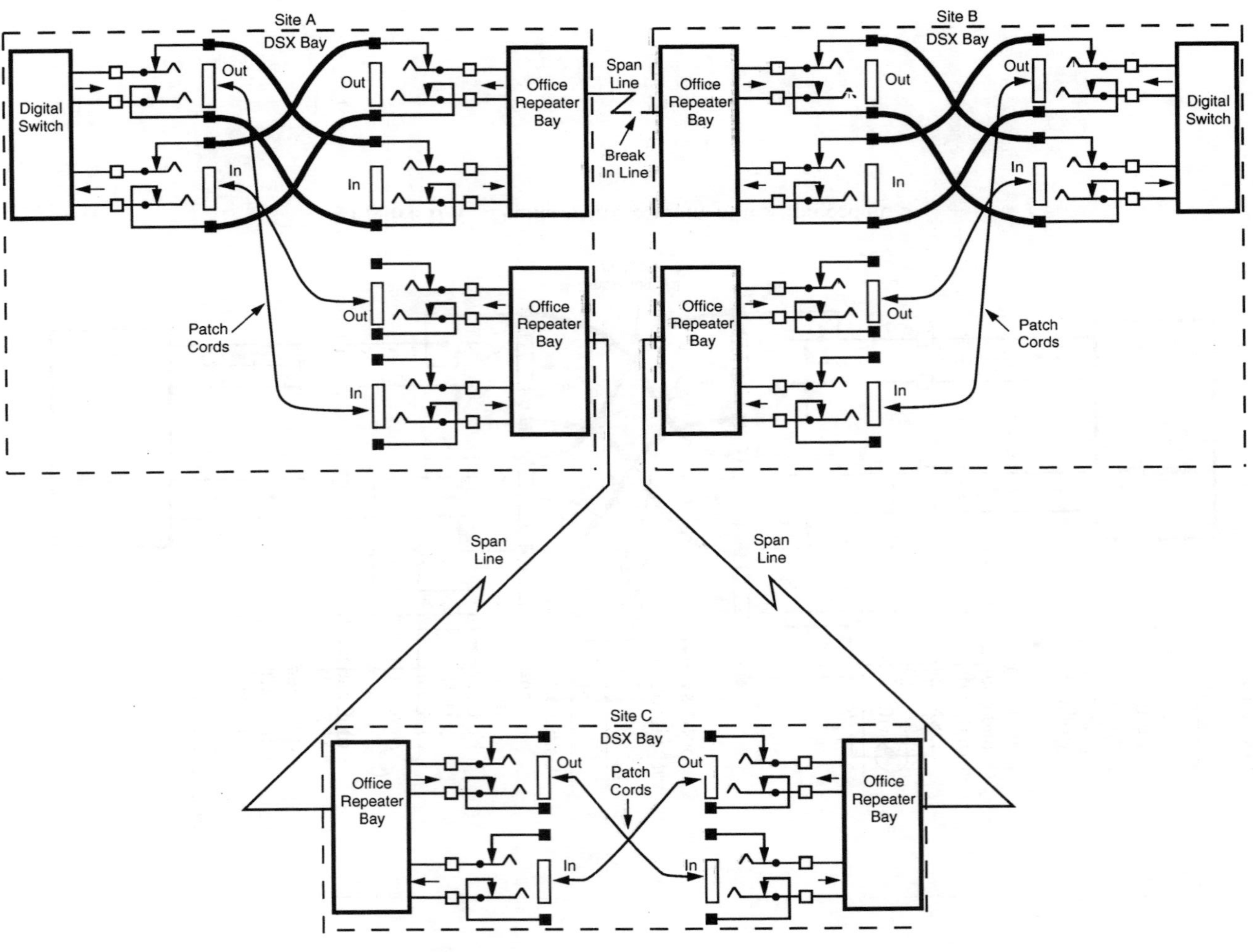

Figure D-5 Restoration of interoffice service, procedure 4

PROCEDURE 5. DIGITAL SWITCH INSTALLATION AND CUTOVER

This procedure is used to replace an analog switch with a digital switch.

STEP	PROCEDURE
1	Wire the new digital switch to equipment IN/OUT terminals on the DSX and record the connections on DSX designation strips/cards.
2	Using patch cords, connect the existing office repeater bay DSX appearance to the existing channel bank DSX appearance. These patch cords will be in parallel with the existing cross-connect jumpers.
3	Remove the cross-connect jumpers between the existing office repeater bay and channel bank appearances.
4	Install new cross-connect jumpers between the existing office repeater bay DSX appearance and the new digital switch DSX appearance.
5	Remove the patch cords installed in step 2 to make the cutover to the digital switch.

PROCEDURE 6. IN-SERVICE PATCHING

This procedure is used to rearrange working circuits or to restore service on a failed circuit without interrupting service when patches are installed and removed.

STEP	PROCEDURE
1	Using patch cords: a) Connect the channel bank MON (monitor) jack to a bridging repeater IN jack. b) Connect the bridging repeater OUT jack to the standby facility IN jack. c) Connect the standby facility MON jack to a bridging repeater IN jack. *Note:* The other office must complete steps 1a, 1b, and 1c before step d is performed; otherwise service will be interrupted. d) Connect the bridging repeater OUT jack to the channel bank IN jack. *Note:* The other office must complete step 1d defore step 2 is performed.
2	Insert a terminating plug into the standby facility OUT jack.
3	Insert a terminating plug into the channel bank OUT jack.
4	To remove the patch cord, after the failure is repaired, perform the above steps in *exactly* the reverse order; otherwise service will be interrupted.

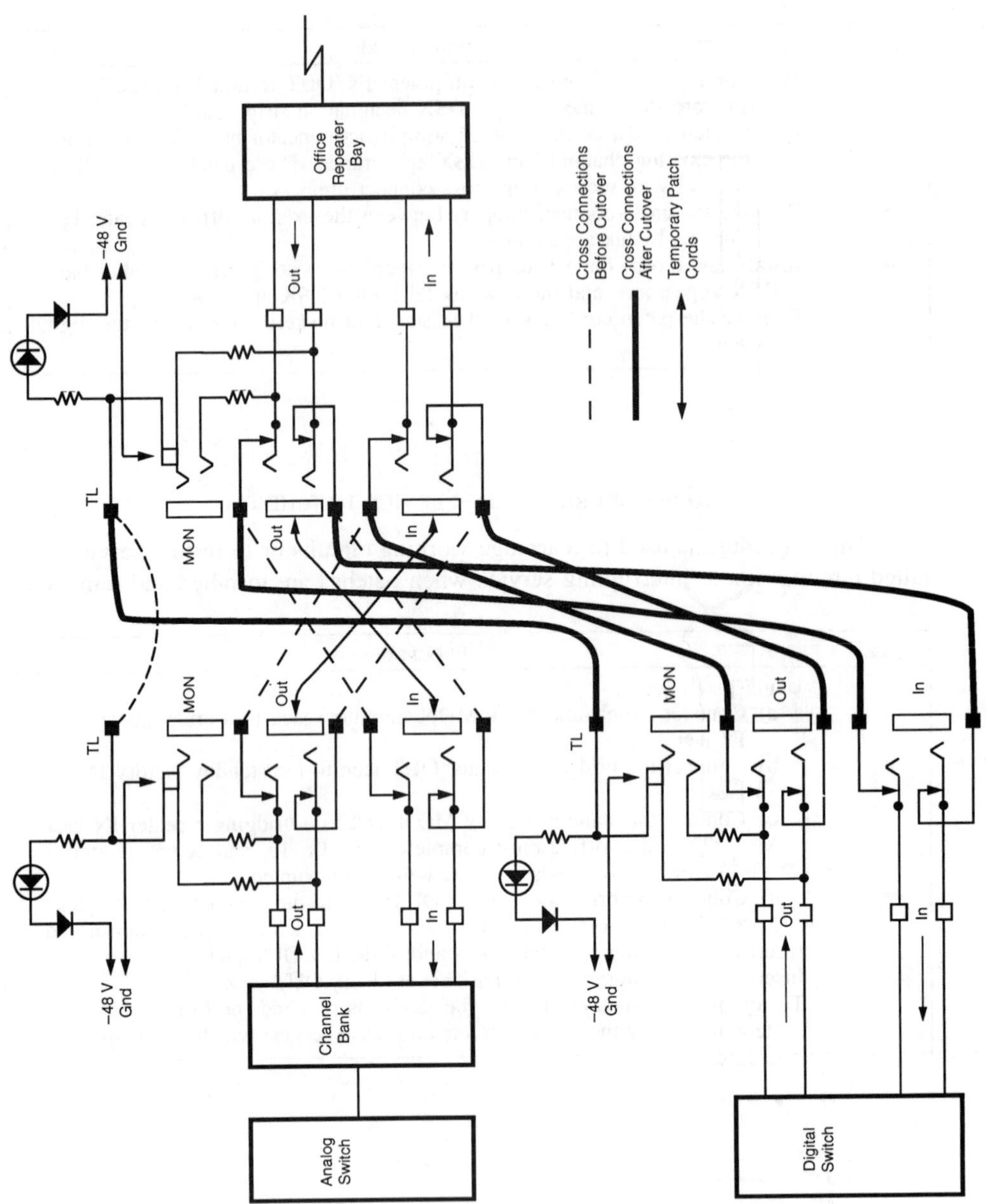

Figure D-6 Digital switch installation and cutover, procedure 5

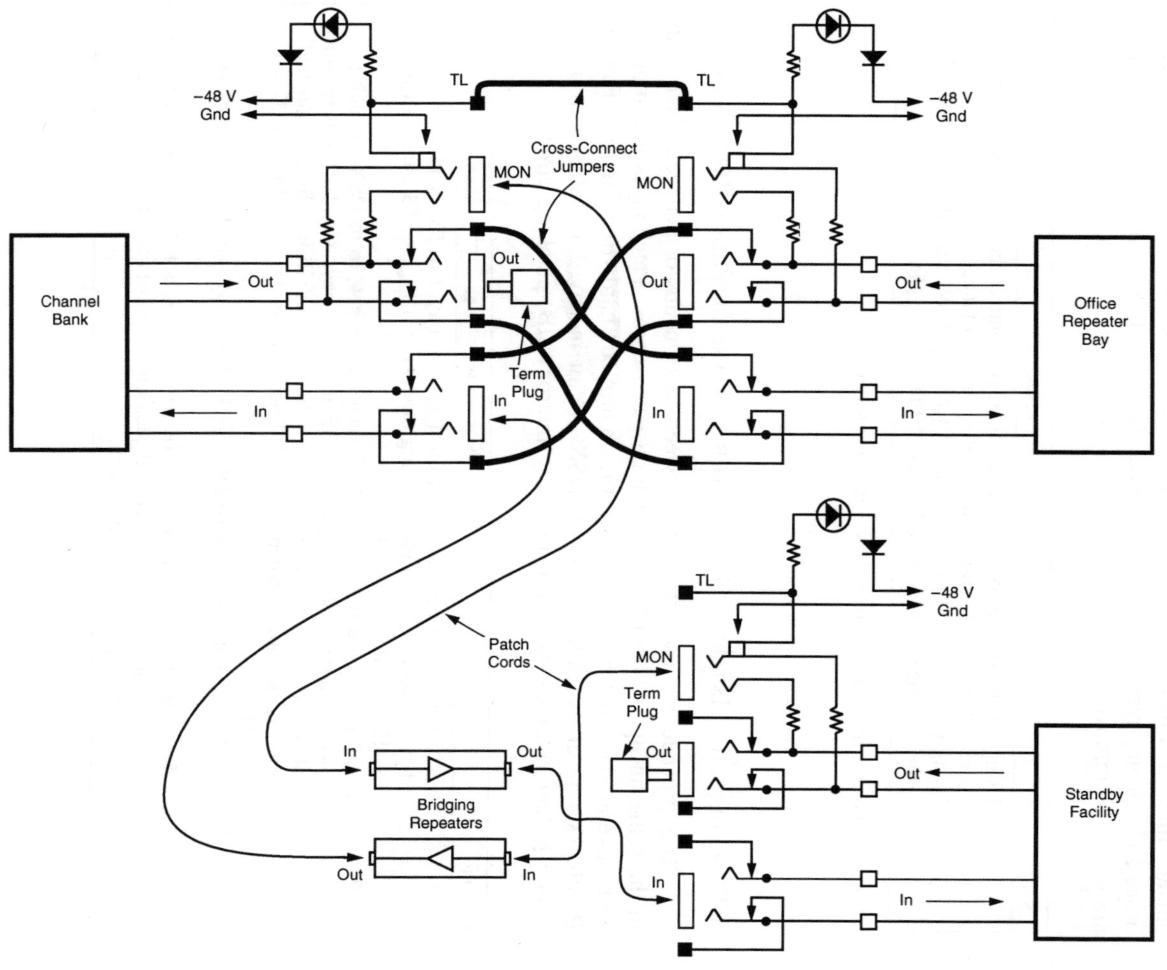

Figure D-7 In-service patching, procedure 6

Procedure 7. System Restoration Using a Maintenance Line When No Bridging Repeaters Are Available

This procedure is used to temporarily patch between two offices using a maintenance line. This method is similar to Procedure 4 except there is no intermediate office and bridging repeaters are not required. Service will be interrupted when patches are placed or removed.

Step	Procedure
1	Using patch cords at office A and office B, connect the OUT jacks of the digital switches to the IN jacks of the maintenance line.
2	Using patch cords at office A and office B, connect the OUT jacks of the maintenance line to the IN jacks of the digital switches.

Note: Placing and removing patch cords will interrupt service.

Procedure 8. Interbay Patching by Means of Short Patch Cords

This procedure is used to patch between bay lineups or between distant bays in the same lineup, without the use of long patch cords. Patching is accomplished with several short patch cords, using the multipled, hardwired Interbay Patching Panels. A typical example is when a DSX circuit in bay 2 is patched to a DSX circuit in bay 18 of the same lineup, using Interbay (IB) patching circuit 3.

Step	Procedure
1	Install patch cord between DSX OUT jack in bay 2 to IB IN jack 3R on bay 4. Busy LEDs (light-emitting diodes) on jack 3 of IB panels in bays 8, 12, 16, and 20 will light.
2	Install patch cord between IB OUT jack 3L on bay 16 to DSX IN jack in bay 18. Busy LEDs on jack 3 of IB panels in bays 8 and 12 will remain lit, LEDs on jack 3 of IB panels in bays 16 and 20 will be extinguished.
3	Install patch cord between DSX OUT jack in bay 18 to IB IN jack 3L on bay 16.
4	Install patch cord between IB OUT jack 3R on bay 4 to DSX IN jack on bay 2. *Note:* Busy LEDs on jack 3 of IB panels on bays 8 and 12 will remain lit until patches are removed. IB jack circuit 3 between bays 16 and 20 may be used for another patch.

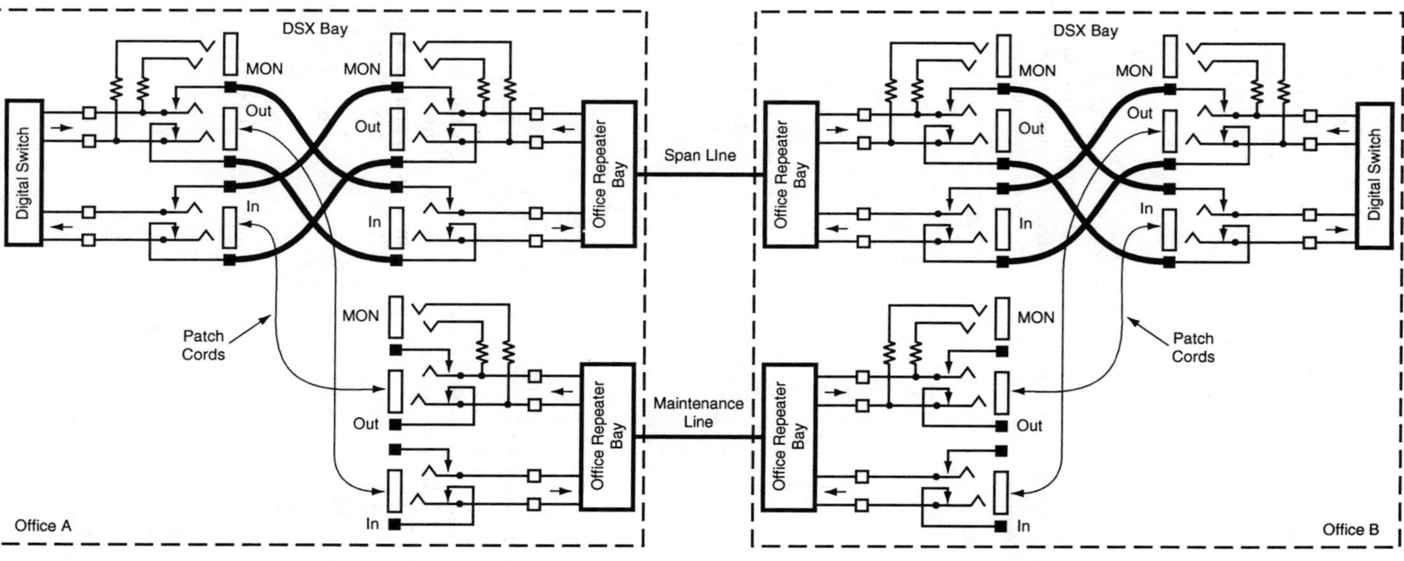

Figure D-8 System restoration using a maintenance line, procedure 8

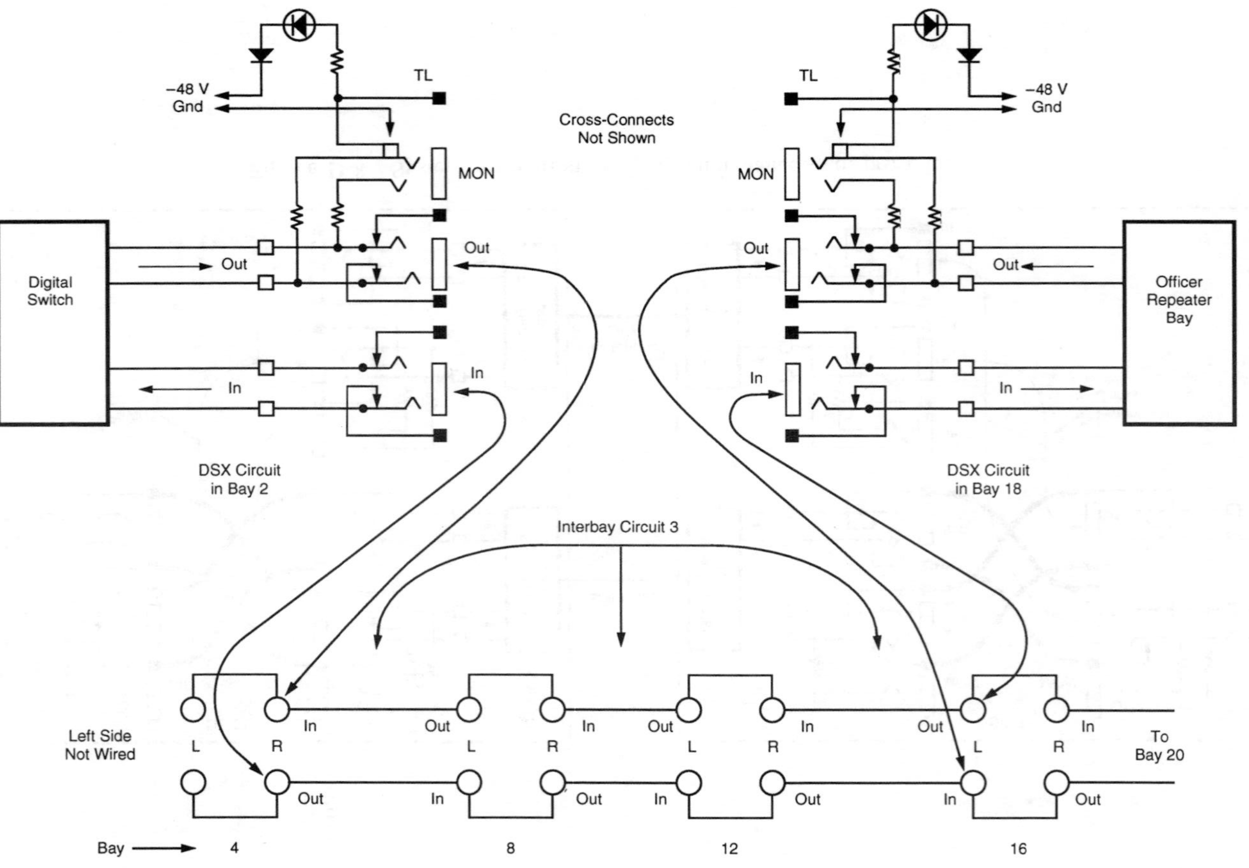

Figure D-9 Interbay patching by means of short patch cords, procedure 8

Index

A

B

C

D

E

M

P

Q

T

U

V

W

X

Y

Z